John,

Thanks for your friendship over the years. I appreciate all your help in the HAC and your many useful comments during our runs.

My best always,

Steve

4-1-03

WEED SCIENCE

Principles and Practices

WEED SCIENCE

Principles and Practices

FOURTH EDITION

Thomas J. Monaco

North Carolina State University
Raleigh, North Carolina

Stephen C. Weller

Purdue University
West Lafayette, Indiana

Floyd M. Ashton

University of California
Davis, California

JOHN WILEY & SONS, INC.

This book is printed on acid-free paper. ♾

Published simultaneously in Canada.

Library of Congress Cataloging-in-Publication Data:

Monaco, Thomas J.
Weed science : principles & practices / Thomas J. Monaco, Stephen C. Weller, Floyd M. Ashton.--4th ed.
p. cm.
Rev. ed. of: Weed science / Floyd M. Ashton. 3rd ed. c1991.
Includes bibliographical references (p.).
ISBN 0-471-37051-7 (cloth: alk. paper)
1. Weeds--Control. 2. Herbicides. I. Weller, Stephen C. II. Ashton, Floyd M. III. Ashton, Floyd M. Weed science. IV. Title.

SB611.M58 2001
632'5—dc21

2001026926

Printed in the United States of America

10 9 8 7 6 5 4 3 2 1

This textbook is dedicated to the memory of each of these authors:
Glenn Klingman, Fred Warren, and Dan Hess for their tremendous contributions
to the advancement of weed science.

CONTENTS

PREFACE

Weeds affect everyone and should be of concern to all. The nature of weeds and how they interact with human activities form the basis of the discipline of weed science. Weeds may poison livestock or seriously slow their weight gain. They can cause allergic reactions in humans, such as hay fever and dermatitis. Weeds create problems in recreational areas such as golf courses, parks, and fishing and boating sites. They are troublesome in industrial areas, irrigation and drainage systems and along highways and railroads. In addition, crop yield and quality can be reduced by the competition and interference of weeds. Although weed science is an ever-expanding discipline, it has always involved an understanding of weed biology and ecology. Such knowledge is basic to designing effective and efficient methods of managing weeds and reducing their impact on human activities. Weed management is expensive and directly increases the price of food and fiber. It can also affect the environment, both beneficially and detrimentally. For example, various methods designed to manage weeds reduce crop loss and increase labor efficiency, but others, such as tillage, can increase soil erosion, and the use of herbicides must be constantly monitored to reduce or eliminate their effects on the environment and human health.

Since 1940, there have been greater advances in the science of weed control than in all of its previous history. In this book, the old and reliable methods of weed control are integrated with newer techniques to design truly integrated multiple-factor weed management systems. The use of multitactic, fully integrated production practices results in more effective and less expensive weed control programs. These programs help reduce the cost of food and fiber production. Although this textbook places great emphasis on the chemical management of weeds, we realize that the discipline of weed science is rapidly moving toward reduced reliance on herbicides. This is a positive change; however, total elimination of herbicides for weed control in cropping systems will not happen in the near future, and herbicides will continue to play an important role in most weed management programs. This text has been written to provide the practitioner with important background information on all aspects of weed management. Informed practitioners know how best to use the available tools. They can determine the most effective, economical, and environmentally sound practices to maintain a sustainable agriculture in the twenty-first century and provide a reliable supply of food and fiber for the world.

Designed mainly for college classroom instruction in the principles and practices of weed science, this fourth edition follows the format of previous editions but has been completely revised to provide more in-depth coverage of all topics. The detailed discussion of these topics also makes it useful to county agents, farm advisors, extension specialists, consultants, herbicide development personnel, research scientists, and farmers. The first third of the book deals with principles of weed science, the second third with herbicides, and the last third with weed control practices in specific crops or situations. Thus, this book brings together the modern philosophy of weed science and the techniques of weed control.

In recent years weed scientists and the general public have become increasingly interested in the environmental impact and safety of the various methods of weed control, especially in regard to the use of herbicides. Such concern has led chemical companies to stress the development of new herbicides of low mammalian toxicity that can be used at very low rates. These products minimize environmental impact and maximize safe use. There has also been an increase in studies on the biology and ecology of weeds directed toward more effective control methods. An increased emphasis on such new directions is a feature of this fourth edition. Useful Web sites relating to and expanding on these topics are provided in each chapter.

This fourth edition also provides more complete details on the use of herbicides than given in earlier editions. The sections on herbicide mechanism of action contain information useful to advanced students as well as undergraduates. It can provide a better understanding of how herbicides work and how they can be used more efficiently to control weeds. We hope that this will increase the value of this book to those using these chemicals in the field. However, nothing in this text is to be construed as recommending or authorizing the use of any weed control practice or chemical. A current manufacturer's or supplier's label is the final word for the use of any herbicides, including method and time of application, rates at which a product is to be used, permissible crops or situations, weeds that can be controlled, and special precautions. Label recommendations must be followed—regardless of any statements in this book. All long-term weed management systems will, by necessity, be part of an integrated approach as described herein.

Herbicide usage is continually being revised. The Environmental Protection Agency's policy and the recent passage of the Food Quality Protection Act require the re-registering of pesticides and increased testing related to health issues. Both this policy and the law are having a major impact on the registration of pesticides and their uses. There will be many changes in the availability and allowable uses of herbicides in the future; therefore, it is critical to check all current registrations before use and to employ a more holistic approach in weed management, with the use of multiple tactics.

The common names of herbicides are generally used throughout the text. However, trade names and chemical names are also supplied. Trade names have been given as a convenience to the reader, not as an endorsement of any particular product. In addition to the text, the appendixes provide an assortment of information useful to students and weed control practitioners, including listings for herbicidal effectiveness for the control of many common weeds; commercially available herbicides cross-referenced by common name, trade name, and chemical manufacturer; conversion factors; herbicide concentration calculations; weight of dry soil, length of row required for one acre; mixing instructions for available commercial liquid and dry materials; and common cropland weeds in the United States.

ACKNOWLEDGMENTS

With great appreciation, we acknowledge the contributions of our weed science colleagues at our respective universities, the individuals and organizations that provided illustrations, and the information provided by the many instructors in the Purdue University Herbicide Action Short Course. Special thanks are given to the late Dan Hess and to David Bridges, Donn Thill, Ron Turco, Rex Liebl, John Jacetta, Peter Goldsbrouch, Clark Throssell, Gail Ruhl, and Harvey Holt, who provided excellent and invaluable information used in the preparation of the various chapters and to Nancy Petretic, Lacretia Rothenberger, Lonni Kucik for their technical assistance. Stephen Weller acknowledges the members of the Elite Runners Association for the most helpful interludes they provided during the writing of this text, especially the thought-provoking discussions on agriculture and philosophy with Howard Zelaznik.

We are also grateful for the valuable contributions to this text provided by the love, support, patience, and assistance of our wives, Jenny S. Monaco, Kathleen M. Weller, and Theo E. Ashton, during the preparation of our manuscript. Finally, we acknowledge the students of weed science for whom this text was prepared. We hope that the information provided will help these students to maintain and improve the food production systems of the world for the betterment of all humankind.

THOMAS J. MONACO
STEPHEN C. WELLER
FLOYD M. ASHTON

Raleigh, North Carolina
West Lafayette, Indiana
Davis, California
January 2002

PART I
Principles

1 Introduction to Weed Science

Weed science is the scientific discipline that studies plants that interfere with human activity. Areas of study range from basic biological and ecological investigations to the design of practical methods of managing weeds in the environment. The overall goal of weed management is to design the most appropriate methods in a variety of situations that ensure a sustainable ecosystem and a minimum influence of nuisance weeds.

The first question is "What is a weed?" Before a plant can be considered a weed, humans must provide a definition. Many varying definitions have been developed for weeds, depending on each particular situation where they occur and the plants involved. For the purpose of this book, we define a *weed* as *a plant growing where it is not desired, or a plant out of place*—some plant that, according to human criteria, is *undesirable*. We decide for each particular situation which plants are or are not desired in terms of how they affect our health, our crops, our domesticated animals, or aesthetics. For example, some people consider a dandelion in a lawn a weed and want to control it, whereas others feel the dandelion is desirable and do not control it. The same thinking is involved for any weed situation, whether in a crop field, a pasture, a body of water, or in a noncropland or natural site.

Weeds are also classed as pests and included with insects, plant diseases, nematodes, and rodent pests. A chemical used to control a pest is called a *pesticide,* and a chemical used specifically for weed control is known as a *herbicide*.

Weed control is the segment of weed science that most people are familiar with and where the greater part of education and training is focused. The methods employed to manage weeds vary, depending on the situation, available research information, tools, economics, and experience. Improved agricultural technology over the centuries has contributed greatly to increased food production (Warren, 1998) and a related increase in our standard of living. Advances in weed control practices have been an important part of these gains.

Weed control in human endeavors is as old as the growing of food crops and has progressed from intense human inputs to methods involving less human energy and increasing inputs from other sources (Figure 1-1). For thousands of years humans have achieved amazing advances in weed control. Before 10,000 B.C., weeds were removed from crops by hand. The efforts of one person could hardly feed that person, and starvation was common. Later, farmers substituted a sharp stick or other wooden tools for fingers. By 1000 B.C., crude hoes dragged by an animal through a field helped reduce human labor in seedbed preparation, and, later, metal hoes dragged by a horse or ox through a field became common, although subsistence farming was still the norm. In 1731, in his book, *Horse-Hoeing Husbandry*, Jethro Tull proposed planting

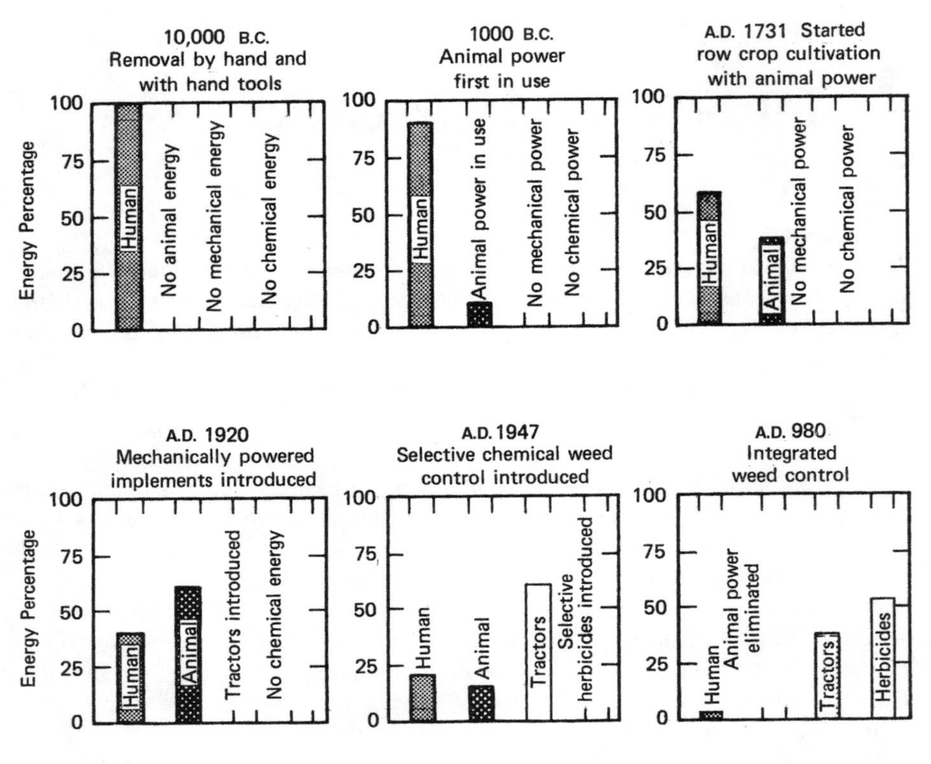

Figure 1-1. Energy sources providing weed control at different times. Data shown for 1920, 1947, and 1990 are for the United States.

crops in rows to permit "horse-hoeing" and was among the first to use the word *weed* with its present spelling and meaning. With this advancement, one farmer could now provide food for 4 people. Less than 200 years later (by 1920), tractors started to replace horses in most agricultural situations and one farmer could now produce enough food for 8 people. Progressively, and with increasing momentum, humans learned to use their bare hands, hand tools, horsepower, and tractor power to manage weeds. All these methods still used brute force to control weeds. However, with the introduction of herbicides in 1947, one farmer could now feed 16 people. During the intervening years, many new herbicides have been developed and extensively used, resulting in chemical energy becoming the major tool of weed control in the United States and other countries (Figure 1-2). In 1990, one farmer could feed 75 people. This means that multitudes of people who previously worked on farms mainly hoeing weeds have been able to pursue other jobs and provide inputs into a wide variety of goods and services that have helped to increase our standard of living. As we continue to investigate new approaches to weed management, additional chemical, cultural,

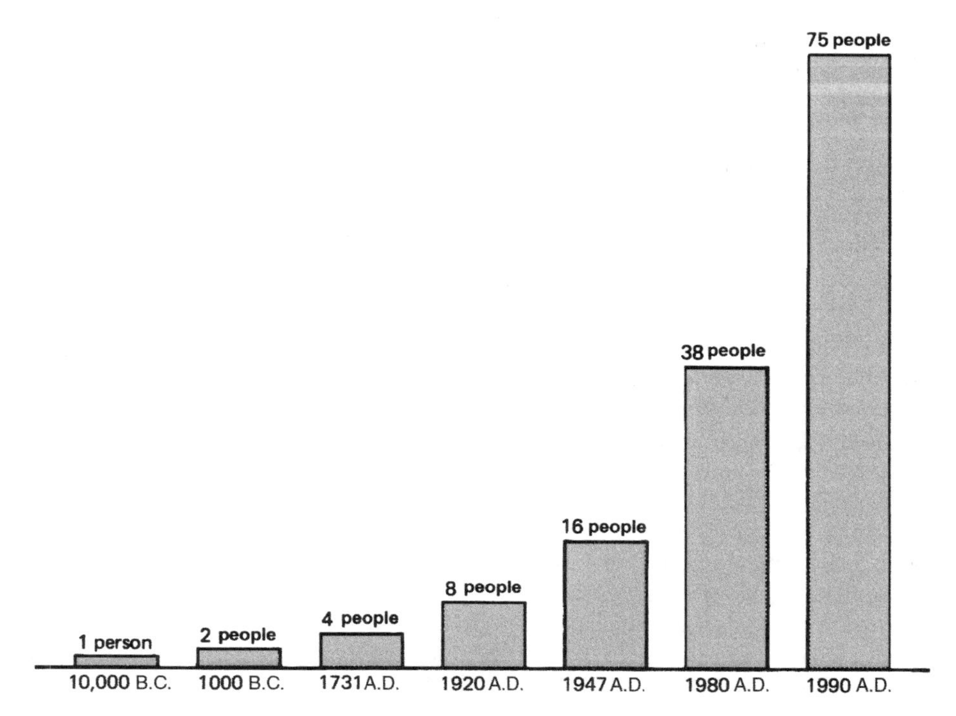

Figure 1-2. Crop energy output per farmer or the number of people fed by one farmer. Data for 1920, 1947, 1980, and 1990 are for the United States.

biological, and biotechnology-based practices will provide ever improved tools to permit a sustainable agriculture.

THE FUTURE

In the future, weed control methods presently being intensively researched will allow expanded weed control options beyond herbicides and mechanical methods in both agricultural and nonagricultural weed management. Biological control by insects and plant disease organisms, predictive modeling of weed/crop interactions, and the use of herbicide antidotes, more competitive crops, allelopathy, and genetic engineering/genomics will become more common as their reliability is improved.

The overall objective of additional approaches is to discover new, more environmentally acceptable weed management tools that not only control weeds effectively, but improve our understanding of weed ecology/biology and allow more sustainable management of the agroecosystem. Biological control of weeds by insects and plant disease organisms has had considerable success in several weed management situations, and ongoing research will lead to additional uses of biologicals. Considerable biological research involves the potential introduction of natural control species from an invasive weed species site of origin (Watson, 1993). The use of herbicide antidotes or safeners (Hatzios and Wu, 1996) to protect crop plants has been successful for some herbicides in certain crops—for example, chloroacteamide herbicides in corn and sorghum. One of the greatest recent changes in weed control has occurred through the genetic transformation of crops with herbicide-resistant genes and the incorporation of herbicide resistance through conventional breeding. In 1999 and 2000, more than 50% of the U.S. soybean acreage and more than 30% of the corn acreage was planted to cultivars resistant to one of several herbicides. Genetic engineering offers tremendous potential in all areas of weed science for improved understanding of plants and of weed control. Genetic engineering, along with recent advances in sequencing the genome of *Arabidopsis* (and in the future, other plants), will allow a clear determination of specific gene function. Such knowledge will permit gene manipulation and modification in our agricultural endeavors, such as the discovery of genes that contribute to weediness, competitiveness, allelopathy, dormancy, or a plant's being a perennial, among functions (Weller et al., 2001; Gressel, 2000). Genes of interest in weed science, once discovered, may then be engineered into crops or used to manipulate weeds to achieve a desirable effect in crop productivity and reduced weed influences. One area in particular where genetic engineering may play a role is allelopathy. Allelopathy (Rizvi and Rizvi, 1992) results from any direct or indirect inhibitory or stimulatory effect by one plant (including microorganisms) on another through the production and release into the environment of a chemical compound. Although no commercial breakthroughs have yet occurred in engineering plants to produce higher levels of allelochemicals, several such genes have been identified in *Arabidopsis*. Genetic engineering of crop plants or cover crops with genes for allelochemicals could allow major strides in developing plants useful in weed management. The future for weed science is exciting, as there are many opportunities for challenging basic and applied approaches for weed management in our environment, as reviewed by Hall et al. (2000).

WEED IMPACTS

Weeds should be everybody's business, as they affect everyone in one way or another. They not only reduce crop production and increase the cost of agricultural products, but they also cause problems for the general public in many other ways—for example, in regard to health and maintaining home landscaping recreational areas and other noncrop areas. Specific problems include lower crop and animal yields, less efficient land use, higher costs of insect and plant disease control, poorer-quality products, more water management problems, and lower human efficiency.

Lower Plant and Animal Yields

Weed control is an expensive but necessary part of agricultural production, directly affecting the price of food and other agricultural products. However, such products would be less abundant and more expensive without modern agricultural and weed science technology. Weeds reduce yields of plant and animal crops. Plant yields are primarily reduced by competition between the weed and the crop for soil water, soil nutrients, light, and carbon dioxide. Certain weeds may also reduce plant yields by releasing allelopathic compounds into the environment (Figure 1-3).

Livestock yields may be reduced by weeds, whose growth allows less pasture or range forage, or by poisonous or toxic plants that cause slower growth or death in animals (Figure 1-4).

Less Efficient Land Use

The presence of weeds on a given piece of land can reduce the maximum efficiency of the use of that land in a number of ways. These include increased costs of production and harvest, reforestation, and noncropland maintenance, as well as reduced plant growth, root damage resulting from cultivation, limitation of the crops that can be grown, and reduced land values.

Higher Costs of Insect and Plant Disease Control

Weeds harbor insect and disease organisms that attack crop plants. For example, the carrot weevil and carrot rust-fly may be harbored by the wild carrot, only later to attack the cultivated carrot. Aphids and cabbage root maggots live in mustard and later attack cabbage, cauliflower, radish, and turnips. Onion thrips live in ragweed and mustards

Figure 1-3. Weeds often reduce the yield of crops through allelopathic effects by exudates from roots or leaves. This figure shows the effect of leaf leachates from various grasses on the growth of peach trees. From left to right: Control (fertilizer water); leachate from fescue + fertilizer; leachate from common bermudagrass + fertilizer; leachate from coastal bermudagrass + fertilizer; leachate from hybrid turf bermudagrass + fertilizer. Trees were watered daily for 6 weeks. Note trees watered with the various grass leachates all had reduced growth compared to the control trees even though they received similar amounts of fertilizer.

Figure 1-4. Photo of a Friesan-Holstein cow with forage-induced photosensitization. Ingestion of St. Johnswort can cause this problem which can lead to reduced growth, reduced milk production and skin infection and overall loss of productivity. The nonpigmented skin becomes red and blistered with accompanying hair loss. Black skin is not affected. (S.B. Hooser, Purdue University).

and may later prey on an onion crop. The disease of curly top on sugar beets is carried by insect vectors that live on weeds in wastelands. Many insects overwinter in weedy fields and field borders.

Disease organisms such as black stem rust may use the European barberry, quackgrass, or wild oat as a host prior to attacking wheat, oats, or barley. Some virus diseases are propagated on members of the weedy nightshades. For example, the virus causing "leaf roll" of potatoes lives on black nightshade. It is thought that aphids carry the virus to potatoes. A three-way harboring and transmission of a mycoplasma disease from weeds to citrus has been discovered in California. Leaf hoppers transmit the disease organism, citrus stubborn disease (*Spiroplasma citri*), to and from diseased periwinkle and to and from London rocket (*Sisymbrium irio*). These weedy plants act as a source of the disease organism to infect citrus trees.

Poorer-Quality Products

All types of crop products may be reduced in quality. Weed seeds and onion bulblets in grain and seed, weedy trash in hay and cotton, spindly "leaf crops," and scrawny vegetables are a few examples.

Livestock products may be lower priced or unmarketable because of weeds; for example, onion, garlic, or bitterweed flavor in milk, and cocklebur in wool, reduce the quality of the products. Poisonous plants may kill animals, slow their rates of growth, or cause many kinds of abnormalities (Figure 1-4).

More Water Management Problems

Aquatic weeds can be a major problem in irrigation and drainage systems, lakes, ponds, reservoirs, and harbors. They restrict the flow of water (Figure 1-5), interfere with commercial and recreational activities, and may give off undesirable flavors and odors in domestic water supplies. Their control is often difficult and expensive. Terrestrial weeds growing at the edges of aquatic sites can also be a problem. Chapter 27 is devoted to the weed problems and control methods on these sites.

Lower Human Efficiency

Weeds have been a plague to humans ever since they gave up the hunter's life. Traveling in developing nations, one may feel that half the world's population work in the fields, stooped, moving slowly, and silently weeding. These people are a part of the great mass of humanity that spends a lifetime simply weeding. Many young people doing such work in Africa, Asia, and Latin America can never attend school; women do not have time to prepare nutritious meals or otherwise care for their families. Modern weed-control methods integrated into the economies and cultures of developing nations provide relief from this arduous chore and give nations the opportunity to improve their standards of living through more productive work.

Figure 1-5. In St. Cloud, Florida, the State Mosquito Control Board sprayed ditches with diuron to eliminate weeds and improve drainage. The weed-control program reduced the expenses of mosquito control enough to permit a one-third savings in the total budget. *Right*: One year after hand weeding. *Left*: One year after chemical treatment with diuron. (E.I. du Pont de Nemours and Company.)

Weed control constitutes a large share of a farmer's work required to produce a crop. This effort directly affects the cost of crop production and thus the cost of food. It affects all of us, whether we farm or not.

Weeds reduce human efficiency through allergies and poisoning. Hay fever, caused principally by pollen from weeds, alone accounts for tremendous losses in human efficiency every summer and fall. Poison ivy, poison oak, and poison sumac cause losses in terms of time and human suffering; children occasionally die from eating poisonous plants or fruits.

COST OF WEEDS

The cost of weeds to humans is much higher than generally recognized. Because weeds are so common and widespread, people do not fully appreciate their significance in terms of losses and control costs. Although relatively accurate estimates have been made of losses and control costs on farms in the United States, other areas of economic impact are much more difficult to estimate. The latter include noncropland, recreational areas, homesite maintenance, aquatics, livestock, and human efficiency, as well as many others.

Weeds are common on all 485 million acres of U.S. cropland and almost one billion acres of range and pasture. In U.S. agriculture, weeds are estimated to reduce yields by 12% annually, or approximately $36 billion in lost revenue (USBC, 1998). In addition, another $4 billion is spent each year on herbicides to control these weeds (Pimentel et al., 1999), and more than $3 billion for cultural and other methods of control.

Bridges (1992) surveyed weed losses in 46 crops in the United States. The annual monetary loss caused by weeds in crops using current Best Management Practices (BMP) with herbicides was estimated at $4.1 billion, and this cost increased to $19.6 billion, or a 4.9-fold increase if herbicides were not available. For both categories of weed control, approximately 82% of the monetary loss occurred in field crops, 5% in noncitrus fruit crops, 3% in citrus crops, 1% in tree nuts, and 9% in vegetables (Bridges, 1992).

Other areas where weeds are costly include pastures and rangelands, lawns, gardens and golf courses, and aquatic sites. Estimated costs of controlling weeds were $5 billion in pastures and rangelands, $1.5 billion in lawns, gardens, and golf courses, and $100 million for aquatic situations (Pimentel et al., 1999). A high percentage of these costs is related to control of nonindigenous (alien) plant species that have been introduced into the United States. There are more than 1000 alien plant introductions that have become major weed pests in cropping systems.

Another example of high costs related to the nonherbicide approaches to weed control relates to a new policy enacted in 2000 by the Los Angeles County School Board (*Wall Street Journal*, 2000). A nonherbicide weed control policy for school district land was begun with an estimation that manual weed control would take only one-sixth of groundskeeping time and cost $650,000 for equipment and 15 full-time weeders. The result was that in less than a year, more than 50% of the groundskeeping time was spent torching, digging, or pulling weeds, at a cost of $1.5 million and requiring the work of 37 full-time employees. These examples demonstrate that a significant increase in efficiency of food production and land management can result from effective, efficient, and integrated weed-control strategies on lands used for human activities.

PREVENTION, MANAGEMENT, AND ERADICATION

Prevention

Prevention, which means stopping a given weed species from contaminating an area, is often the most practical means of controlling weeds. Prevention in agriculture is best accomplished by (1) making sure that new weed seeds are not carried onto a farm in contaminated crop seeds, feed, or on machinery, (2) preventing weeds on the farm from going to seed, and (3) preventing the spread of perennial weeds that reproduce vegetatively. Prevention is a method of weed management that, if properly employed, could greatly reduce weed problems worldwide.

Control

Control is the process of limiting weed infestations. In crops, the weeds are limited so that they have minimal effect on crop growth and yield. The degree of control is usually a matter of economics, a balance between the costs involved and the increase in profits due to the control of the weeds and the types of production systems and tools being used. Each farmer has the ability to decide what level of weed control is suitable to reach the objectives of the cropping system. On noncropland, it is often desirable to remove essentially all vegetation for a specific period of time. Weeds are thus limited to a level that does not allow them to interfere with human activities.

Most biological control programs using a highly specific insect or plant disease organism as a control agent are based on obtaining adequate economic management of a weed, but not eradication.

Eradication

Eradication is the complete elimination of all living plants, including their vegetative propagules and seeds. Eradication is much more difficult than prevention or control. In general, it is justified only for the elimination of a serious weed in a limited area—for example, a perennial weed in a small area of a field, around fuel storage tanks, or in railroad yards.

The most difficult part of eradication is the elimination of the vegetative propagules and seeds in the soil. Seeds of many weeds may remain dormant for a number of years, and in this dormant state they are not usually killed by standard weed control practices. Vegetative propagules and many weed seeds can be killed by soil fumigation, such as by methyl bromide or other chemicals (Chapter 7). This practice is expensive and methyl bromide will not be available after 2005 because of environmental concerns, which has necessitated research to find suitable alternatives. Persistent soil-applied herbicides are also used; however, because of their long soil persistence, they prevent the growth of desirable species for substantial periods of time.

LITERATURE CITED AND SUGGESTED READING

Bridges, D. C., ed. 1992. *Crop Losses Due to Weeds in the United States, 1992*. WSSA, Lawrence, KS.

Gressel, J. 2000. Molecular biology of weed control. *Transgenic Res.* **9**:355–382.

Hall, J. C., L. L. van Eerd, S. D. Miller, M. D. K. Owen, T. S. Prather, D. L. Shaner, M. Singh, K. C. Vaughn, and S. C. Weller. 2000. Future research directions for weed science. *Weed Technol.* **14**:647–658.

Hatzios, K. K., and J. Wu. 1996. Herbicide safeners: Tools for improving the efficacy and selectivity of herbicides. *J. Envir. Sci. Health.* **B31**:545–553.

Herbicide Handbook, 7th ed. 1994, and 1998 *Supplement*. WSSA, Lawrence, KS.

Herbicide tolerant crops: 1994. Their value to world agriculture. *Proc. Brighton Crop Protection Conf.—Weeds* **2**:637–660.

Pimentel, D., L. Lach, R. Zuniga, and D. Morrison. 1999. Environmental and economic costs associated with non-indigenous species in the United States. *Bioscience* **50**:53–65.

Rizvi, S. J. H., and V. Rizvi. 1992. *Allelopathy: Basic and Applied Aspects*. Chapman and Hall, London.

USBC. 1998. U.S. Bureau of the Census, Statistical Abstract of the United States, 1996. 200th ed. Washington, DC, U.S. Government Printing Office.

Warren, Susan. Tangled Up in Green: Weeds Run Amok on School Playgrounds. *Wall Street Journal*, October, 05, 2000. Vol 236, Issue 67, p A1.

Warren, G. F. 1998. Spectacular increases in crop yields in the United States in the twentieth century. *Weed Technol.* **12**:752–760.

Watson, A. K., ed. 1993. *Biological Control of Weeds Handbook*. WSSA Monograph Ser. #7. WSSA, Lawrence, KS.

Weller, S. C., R. A. Bressan, P. B. Goldsbrough, T. B. Fredenburg, and P. M. Hasegawa. 2001. The impact of genomics on weed management in the 21st century. *Weed Sci.* **49**:282–289.

WEB SITES

Weed Science Society of America

http://www.wssa.net/

National Center for Food and Agriculture Policy (NCFAP)

Pesticide use data for U.S. agriculture

http://www.ncfap.org

Go to "Pesticide Use" icon.

For chemical use, see the manufacturer's or supplier's label and follow these directions. Also see the Preface.

2 Weed Biology and Ecology

The biology of weeds is concerned with their taxonomy, genetics, establishment, growth, and reproduction. The ecology of weeds is concerned with the development of a single species within a population of plants and the development of all populations within a community on a given site. The numerous factors of the environment have a pronounced influence on all of these processes and systems. The environment and the living community are considered to be an ecosystem, and in an agricultural situation are considered an agroecosystem.

Genetic background and environment are the master factors governing life. The genes of a plant determine what it becomes by controlling life form, growth potential, method of reproduction, length of life, and so on. The environment largely determines the extent to which these life processes proceed by influencing the expression of the genes within the plant.

Knowledge of weed biology and environmental management practices makes it possible to shift plant populations and communities in desired directions. This is the principle behind crop production that theoretically optimizes the growth environment of the crop but minimizes the potential of unacceptable pest levels. For example, cultivation in a crop field makes the environment favorable to the crop plants by removing competing weeds. The use of proper grazing and/or fertilization management in pastures and range areas maximizes the growth environment for desirable species by minimizing the growth of yield-reducing weeds (see Chapter 25). Other examples of environmental management include mechanical removal of undesirable species from forestlands (Chapter 28) and the use of herbicides.

Understanding the basic biology of a weedy plant, how it responds to its environment (ecosystem), its place of origin and similarity (crop mimics with parallel evolution) or dissimilarity (independent evolution) with crop plants can provide needed insight to weed managers on specific practices to reduce weed influences in given situations. A good discussion of weed evolution is provided by Harlan and de Wet (1965). At present, the weakest link in our weed management programs is the lack of basic biological and ecological information. This lack of information has necessitated that most effective weed management programs are designed to remove problem weeds by brute physical or chemical means. The recent emphasis in research on obtaining a better understanding of weed biology/ecology and the interactions within the agroecosystem will allow for the design of more balanced ecologically and environmentally based weed management systems. The purpose of such systems will be to provide consistent and acceptable weed control and ensure the sustainability of our agricultural systems. A further consideration addressed in more detail at the end of the chapter is the issue of invasive plants not only in our agriculture lands but also in our general environment and the consequences of such plants in regard to ecosystem balance, biodiversity, and loss of native species.

WEED CHARACTERISTICS

Weeds have been defined as plants growing where they are not wanted or as undesirables. In most instances, weeds are plants that take advantage of disturbed sites, having characteristics that allow them to efficiently capture available resources and grow prolifically. Weeds have been described by different authors as colonizers or pioneer species in a disturbed field (Bridges, 1995), as ruderals, which are plants growing in waste places, along roadsides, or in rubbish, or as plants found in highly disturbed but potentially productive environments. Weeds are usually herbs with a characteristic short life span and high seed production. Such plants occupy the earliest stages of succession. For students to understand more clearly what constitutes a weed, we must clarify what *undesirable* means. In this regard, Navas (1991) defined a weed as "a plant that forms populations that are able to enter habitats cultivated, markedly disturbed or occupied by man, and potentially depress or displace the resident plant populations which are deliberately cultivated or are of ecological and/or aesthetic interest." Bridges (1995) suggests that Navas's definition provides a useful description of a weed that recognizes the ecology and biology of the plant as well as the impact on humans. Therefore, weeds are plants that are equally well adapted to environmental disturbances as our crops. Weeds can thrive under the conditions generated by the agriculture field practices of tillage, irrigation, fertilization, and row spacing that minimize normal growth-limiting stresses of drought, low fertility, limited light, and high pest levels.

TABLE 2-1. Ideal Characteristics of Weeds

1.	Germination requirements fulfilled in many environments
2.	Discontinuous germination (internally controlled) and great longevity of seed
3.	Rapid growth through vegetative phase to flowering
4.	Continuous seed production for as long as growing conditions permit
5.	Self-compatibility but not complete autogamy or apomixy
6.	Cross-pollination, when it occurs, by unspecialized visitors or wind
7.	Very high seed output in favorable environmental circumstances
8.	Production of some seed in wide range of environmental conditions; tolerance and plasticity
9.	Adaptations for short-distance and long-distance dispersal
10.	If a perennial, vigorous vegetative reproduction or regeneration from fragments
11.	If a perennial, brittleness, so as not to be drawn from ground easily
12.	Ability to compete interspecifically by special means (rosette, choking growth, and allelochemicals)

From Baker (1974).

For a plant to consistently be considered a weed, it must have certain characteristics that set it apart from other plants and allow success in invading, becoming established, and persisting in an agricultural setting. These ideal characteristics of weeds were described by Baker (1974) and are listed in Table 2-1. As seen in Table 2-1, these characteristics give weedy plants an enormous ability to survive in disturbed environments (genetic plasticity).

WEED CLASSIFICATION

How many plants have these weedy characteristics? Relatively few, in fact. There are approximately 250,000 species of plants in the world, but only about 200 species are considered to be major weed problems (Holm, et al., 1977). In addition to this small number of species there are relatively few plant families that contain major weeds. Of the 300 plant families, 75 families comprise 75% of all flowering plants, and of these only 12 families comprise 68% of the world's worst weeds (Holm, et al., 1977). Within these 12 families, just 3 families comprise 43% of the world's worst weeds, with 37% being in the Poaceae (grass family) and Asteraceae (composite family). Most of the major families of weeds also contain members that are major crops, such as grains in the Poaceae, beans/peas in the Leguminosae, and vegetables in the Solanaceae and Brassicaceae, to name a few. Other families have few crop representatives but many weeds, such as the Asteraceae.

The definition of a weed and the plant characteristics that contribute to its weediness are good to know. However, other factors such as habitat, growth form or seed type, and life cycle are important in identifying the most appropriate management practices for weeds in humans' various plant-related activities and useful in determining specifically what weed is present in any given situation.

Habitat refers to whether the weed grows in a terrestrial or an aquatic environment. Weeds can be a problem in both habitats and can include epiphytic and parasitic types.

Growth form or seed type can be used to classify plants in 3 categories. *Gymnosperms*, such as pines, have seeds not enclosed in an ovary. Examples include larch, fir, spruce, hemlock, Douglas fir, cedar, and redwood. Most gymnosperms are not considered to be weeds. *Monocots*, or flowering plants with one seed or cotyledon, generally have narrow leaves with parallel veins, but some monocots have large leaves with palmate-type veins, such as water hyacinth. Examples include lilies, irises, sedges, grasses, palms, orchids, cattails, sugar cane, and banana. Many of our most serious weed problems are monocots. An important distinction is that all grasses are monocots, but not all monocots are grasses. *Dicots*, or flowering plants with two seed leaves or cotyledons, include maple, oak, pigweed, common lambsquarters, and sunflower. Many of our most serious weed problems are dicots.

Life cycle refers to a plant's life span, season of growth, and method of reproduction and determines the methods needed for management or eradication. Plants have been divided into three life cycle categories: *annual, biennial,* and *perennial.*

Annual plants complete their life cycle (from seed to seed) in 1 year or less. Normally, they are considered easy to control. This is true for any one crop of weeds. However, because of an abundance of dormant seed and fast growth, annuals are very persistent and they actually cost more to control than perennial weeds. Most common weeds are annuals, and there are two types—winter and summer.

Winter annuals germinate in autumn or winter and overwinter as a rosette, resume growth in early spring, and produce fruit and seed and die by midsummer. The seeds often lie dormant in the soil during the summer months. In this group, high soil temperature (125°F or above) has a tendency to cause seed dormancy—to inhibit seed germination. Examples include chickweed, downy brome, hairy cress, cheat, sheperds purse, field pennycress, corn cockle, cornflower, and henbit. These weeds are most troublesome in winter-growing crops such as winter wheat, winter oats, and winter barley.

Summer annuals germinate in the spring, grow through the summer, and mature, form seed, and die by autumn. The seeds lie dormant in the soil until the next spring. Summer annuals include cockleburs, morningglories, pigweeds, common lambsquarters, common ragweed, crabgrasses, foxtails, and goosegrass. These weeds are troublesome in summer crops like corn, sorghum, soybeans, cotton, peanuts, tobacco, and many vegetables.

A *biennial* plant lives more than 1 but less than 2 years. During the first phase of growth, the seedling usually develops vegetatively into a rosette. Following a cold period, vegetative growth resumes followed by floral initiation, fruit set, and death. There is confusion between the biennials and winter annuals because winter annuals normally live during 2 calendar years and during 2 seasons. Biennials generally grow later into the second season and tend to be larger plants. Examples include wild carrot, common mullein, bull thistle, wild lettuce, and common burdock. Several biennials are weed problems in minimum- or no-tillage systems and perennial crops.

A *perennial* plant lives for more than 2 years and is characterized by renewed growth year after year from the same root system. Most perennials reproduce by seed, and many are able to spread vegetatively. They are classified as simple, creeping, or woody.

Simple herbaceous perennials reproduce by seed and have no natural means of spreading vegetatively unless injured or cut; the cut pieces may produce new plants. For example, a dandelion or dock root cut in half longitudinally may produce two plants. The roots are usually fleshy and may grow very large. Examples include common dandelion, dock, buckhorn plantain, broadleaf plantain, and pokeweed.

Creeping herbaceous perennials reproduce by seed and by vegetative means, including creeping aboveground stems (stolons), creeping underground stems (rhizomes), or a spreading root system that contain buds. Examples include red sorrel, perennial sowthistle, quackgrass, bermudagrass, johnsongrass, and field bindweed. Some weeds maintain themselves and propagate by means of tubers, which are modified rhizomes adapted for food storage. Examples include purple and yellow nutsedge and Jerusalem artichoke.

In all cases, creeping perennials have tremendous vegetative reproductive capacity and are the most difficult weed problems to manage regardless of the tools used. Cultivators and plows often drag pieces about a field. Herbicides applied and mixed into the soil may reduce the chances of establishment of such pieces. Continuous and repeated cultivation, or mowing for 1 or 2 years, and use of persistent herbicides is often necessary for control. An eradication program requires the killing of seedlings as well as the dormant seeds in the soil.

Woody perennials are plants whose stems have secondary thickening and an annual growth increment. These plants can be weed problems in pastures and many perennial-cropping systems. Examples include poison ivy, wild brambles, and multiflora rose.

FACTORS RELATING TO WEED ESTABLISHMENT AND SURVIVAL

Environmental Factors

The environmental factors to be considered in relation to weed biology, ecology, and management include climatic, physiographic, and biotic aspects.

Climatic factors include the following:

Light (intensity, quality, and duration including photoperiod)
Temperature (extremes, average, frost-free period)
Water (amount, percolation, runoff, and evaporation)
Wind (velocity, duration)
Atmosphere (CO_2, O_2, humidity, toxic substances)

Physiographic factors include the following:

Edaphic (soil factors including pH, fertility, texture, structure, organic matter, CO_2, O_2, water drainage)
Topographic (altitude, slope, exposure to the sun)

Biotic factors include the following:

Plants (competition, released toxins or stimulants, diseases, parasitism, soil flora)
Animals (insects, grazing animals, soil fauna, humans)

Many of the most common weeds have a broad tolerance to environmental conditions. In fact, that is a major reason that they are so common and troublesome. For example, common lambsquarters, common chickweed, and shepherdspurse grow on almost all types of soils. Rarer species such as saltgrass, halogeton, and alkali heath are usually found only on alkali soils.

Similar environmental requirements of certain weeds and selected crop species produce some rather common crop–weed associations. Examples include mustard in small grain, barnyardgrass in tomatoes, burning nettle in lettuce, common chickweed in celery, pigweeds in sugar beets, and red rice in rice.

Competitive Factors

Competition between weeds and crops generally implies an inhibition of crop growth by weeds. However, more technically, competition is one of several types of interference among species or populations. Interference refers to all types of positive and negative interaction between species. Such interference can involve physical factors such as space, light, moisture, nutrients, and atmosphere or some type of chemical interaction. Competition between weeds and crops is generally associated with a negative interference involving physical factors that induces decreased growth in both types of plants because of an insufficient supply of a necessary growth factor (water, nutrients, etc.). Competition can be both within a species (intra) when two or more plants of the same species coexist in time and space and between species (inter) when two or more species coexist as described. Allelopathy (discussed in the next section) is interference between plants based on a chemical influence. Amensalism, another type of negative interference, can be defined as the inhibition of one species by another. However, in contrast to competition, which involves the removal of a resource, amensalism involves the addition of something to the environment. These concepts are more thoroughly discussed in books cited at the end of the chapter by Radosevich et al., 1997; Harper, 1977; and Rizvi and Rizvi, 1992.

Weeds are considered to compete with crops primarily for soil nutrients, soil moisture, light, and carbon dioxide. The degree of direct competition can be reduced to some extent by certain crop cultural practices based on our knowledge of weed biology and ecology. These methods include planting times, spacing, and herbicide placement. Weeds, as mentioned earlier, are able to compete quite well with crops in the less stressful field environment encountered in agriculture because of their characteristic (Baker, 1974) high seed production, leading to high population numbers, rapid germination, very rapid early growth, and long duration (life cycle).

Much research in weed science has focused on competition between crops and weeds, with the objective of reducing weed interference in the cropping cycle (Figure 2-1). This research has shown that the time or period of weed competition is important and that competition early in the season usually reduces crop yields far more than late-season weed growth (Bridges, 1995). Although late-season weed growth may not seriously reduce yields, it often makes harvesting difficult, reduces crop quality, reinfests the land with weed seeds, and may favor the overwintering of insect and disease pests. This time is known as the period of competition, or the weed-free period required during the cropping cycle to obtain optimum yields, and defines for each crop the time period when weeds must be controlled. Zimdahl (1980) reviewed numerous experiments on the effects of both density and duration of weeds on yields of many crops.

Another focus of competition research has related to determining the density of weeds and their effect on crop yields, with the basic goal to establish economic/weed and action thresholds in order to determine when or if weed control must be employed.

Figure 2-1. Competition between weeds and crops. Corn growing with heavy weed competition (*left*) as compared with corn growing with no weed competition (*right*). (R. Liebl, BASF.)

The economic threshold is the weed density at which the value of loss due to weed competition exceeds the cost of control. Action thresholds may include other factors such as the effect of weed seed production and its effects on subsequent weed management. The concept of weed thresholds has generated much interest, as well as disagreement, among weed scientists in regard to what level of weed survival and seed production is acceptable for long-term weed control (Cousins, 1987). In terms of the influence on reducing the number of weed seeds in the soil, some argue that the only acceptable threshold is zero and others maintain that a zero threshold is unrealistic. A thorough discussion of these concepts for those interested is provided in the works listed at the end of this chapter by Swanton et al., 1999; Norris, 1999; Dekker, 1999; and in WSSA Symposium, Weed Technol., 1992.

In addition to the generation of weed thresholds, a considerable amount of recent research on weed–crop competition involves the development of computer models to assist growers and advisers in making weed management decisions. Several models have been developed and are being used in weed management decision making for several crops. These include models for weed management in soybeans (Wilkerson et al., 1991; Rankins et al., 1998), in corn and sugar beets (Shribbs et al., 1990; Lybecker et al., 1991), and in cereals, sugar beets, corn, and sorghum (Striliani and Resina, 1993). Those interested in a more in-depth discussion of weed–crop competition from an ecological perspective are referred to Chapter 5 of Radosevich et al., (1997), and for discussion of the theory and practice of modeling crop–weed interactions to Kropff and van Laar (1993).

ALLELOPATHY

Allelopathy refers to chemical interactions between plants (microbes and higher plants), including stimulatory as well as inhibitory influences (Molisch, 1937; Rice, 1984; Putnam and Tang, 1986). In weed science, however, the inhibitory effects of weeds on crop yields are the main interest in regard to this phenomenon. Such effects may originate from a direct release of the toxin(s) from the living plant or from a leaching of decaying plant litter, residues, or root tissues. Microorganisms have also been implicated in the release of the toxin or the modification of nontoxic compounds to toxic compounds from the nonliving plant residue.

Many weed species, perhaps as many as 90, may interfere with plant growth through allelopathic mechanisms (Putnam and Tang, 1986). Perennial weeds, including quackgrass, johnsongrass, bermudagrass (Figure 2-2), and nutsedges, have often been implicated.

Rice (1984) classified allelopathic agents into 14 chemical categories (plus a miscellaneous group) that are either secondary compounds produced by plants (not involved in basic metabolism) or associated with the shikimic acid and acetate pathways. Most of the allelopathic compounds that have been isolated and identified have one or more rings and many have quite complicated chemical structures.

The unequivocal proof that a postulated allelopathic phenomenon is not some other type of interference is quite difficult. It has been suggested that a specific protocol similar to the established procedure for proof of disease (Koch's postulates) be followed (Fuerst and Putnam, 1983; Putnam and Tang, 1986).

Putnam and Tang (1986) suggest that "allelopathy is now a maturing scientific discipline" and that the future of allelopathy will probably be involved in ecosystem management, pest management, and the development of novel agricultural chemicals. Duke and Lydon (1987) discuss further information on the potential of natural plant products for pest management.

Figure 2-2. Allelopathic effect of bermudagrass on growth of young peach trees. Tree growing in bermudagrass for 2 years (*left*) as compared with tree growing in bare ground for 2 years (*right*). Both trees received fertilizer and irrigation.

REPRODUCTION OF WEEDS

Weeds multiply and reproduce by both sexual and asexual (vegetative) means. Sexual reproduction requires fertilization of an egg by sperm. This usually proceeds via the pollination of a flower, which subsequently produces seed. The viable seed then has the potential of producing a new plant. Asexual reproduction involves the development of a new plant from a vegetative organ such as a stem, root, leaf, or modifications of these basic organs. These include underground stems (rhizomes), aboveground stems (stolons), tubers, corns, bulbs, and bulblets.

Although several environmental factors influence both vegetative and seed reproduction in a general way, the influence of day length or photoperiod can be quite specific. These terms refer to the relative lengths of day and night. In many plants, flowering and/or the development of certain vegetative reproduction organs are controlled by the photoperiod. This may control the development of propagules of a given species on a particular site and limit its geographic distribution.

Following a discussion of several aspects of the reproduction of weeds from seeds, a later section of this chapter briefly considers the reproduction of weeds by vegetative means.

Seed Dissemination

Seeds in general have no method of movement; therefore, they must depend on other forces for dissemination. Regardless of this fact, they are excellent travelers. The spread of seeds, plus their ability to remain viable in the soil for many years (dormant), poses one of the most complex problems of weed control. This fact makes "eradication" nearly impossible for many seed-producing weeds.

Weed seeds are scattered by (1) crop seed, grain feed, hay, and straw, (2) wind, (3) water, (4) animals, including humans, (5) machinery, and (6) weed screenings.

Crop Seed, Grain Feed, Hay, and Straw

Weeds are probably more widely spread through crop seeds, grain feed, hay, and straw than by other means.

Studies of wheat, oats, and barley seed sown by farmers in six North Central states reveal the seriousness of the problem. Researchers took seed directly from drill boxes and analyzed for weed seeds. About 8% of the samples contained primary noxious weeds, and about 45% contained secondary noxious weed seeds. About 80% of the seed had gone through a "recleaning" operation. Much of the recleaning was of limited benefit, however, because only part of the weed seeds was removed.

In the aforementioned study, *certified, registered,* and *foundation* seeds were free of primary noxious weeds and contained only a small percentage of the common weeds (Furrer, 1954).

Farmers often believe that a low percentage of weed seeds on the seed label means that the few weeds present are of little importance. This may be a serious mistake (see Table 2-2). One dodder plant may easily spread to occupy 1 square rod during a single

TABLE 2-2. Field Dodder[a] and Its Rate of Planting in Contaminated Legume Seed Sown at the Rate of 20 lb/acre

Dodder Seed by Weight (%)	No. of Dodder Seeds (per lb of legume seed)	No. of Dodder Seeds Sown	
		Per acre	Per square rod
0.001	8	160	1
0.010	80	1,600	10
0.025	200	4,000	25
0.050	400	8,000	50
0.100	800	16,000	100
0.250	2,000	40,000	250

[a]There are 550,000 to 800,000 dodder seeds per pound.

season; thus, only 0.001% dodder seed is enough to completely infest a legume crop the first year!

The prevalence of weed seed in legume seed is clearly shown by data collected by official state seed analysts, where in 3643 samples, weed seeds averaged from 0.10 to 0.38% by weight. These figures indicate that certain weed species can completely infest a field the first year, despite an extremely low percentage of weed seeds present in the crop seed. These percentages are often below legal tolerances, which makes it necessary to state their presence on the seed label.

Weeds are commonly spread through grain feed, hay, and straw. Where straw is used for mulching, it is important that the straw be free of viable weed seeds as well as grain seeds. The grain seed in the straw may prove to be a weed under some circumstances. Most of the grain seed will germinate and die if the straw is kept moist for 30 days with temperatures favorable to germination.

As shown later in this chapter, an appreciable number of weed seeds in grain and hay are viable after passing through the alimentary canal of an animal. If the manure is allowed to compost, the weed seeds will be killed.

Wind

Weed seeds have many special adaptations that help them spread. Some are equipped with parachute-like structures (pappus) or cottonlike coverings that make the seed float in the wind. Common dandelion, sow thistle, Canada thistle, wild lettuce, some asters, and milkweeds are examples (Figure 2-3).

Water

Weed seed may move with surface water runoff, in natural streams and rivers, in irrigation and drainage canals, and in irrigation water from ponds. Some seeds have

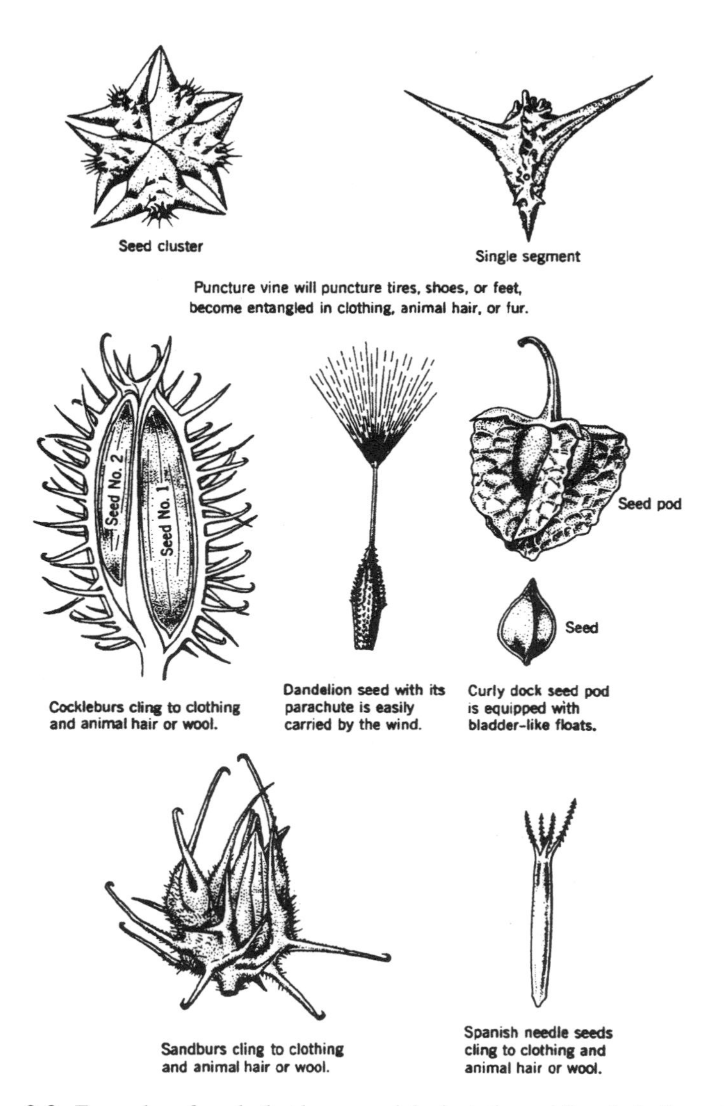

Figure 2-3. Examples of seeds that have special adaptations aiding their dispersion.

special structures to help them float in water. For example, curly dock has small pontoon arrangements on the winged seed covering. Other seeds are carried in moving water or along the river bottom. Flooded areas from river overflows are nearly always heavily infested with weeds.

Irrigation water is a particularly important means of scattering seed. In 156 weed seed catches in Colorado, in three irrigation ditches, 81 different weed species were found. In a 24-hour period, several million seeds passed in a 12-foot ditch.

Scientists have found great variation in the length of time that seeds remain viable in fresh water. Results shown in Table 2-3 are based on weed seed suspended in bags (luminite screen sewn with nylon thread) at 12- and 48-inch depths in a freshwater canal at Prosser, Washington. Water could circulate freely within the bags. The bags were removed periodically over a 5-year period and germination counts were made.

There was little or no variation in germination between the 12- and 48-inch depths, but there was considerable difference in viability among species. Clearly, some seeds can be stored in fresh water for 3 to 5 years and still germinate. In some cases, storage in water tended to break dormancy and increased the percent germination; this was especially true after 2 to 4 months in the water (Bruns and Rasmussen, 1958).

Seed production from plants along irrigation ditches, drainage ditches, and reservoirs is probably the major source of weed seed contamination of irrigation waters. Every effort should be made to keep the banks of these areas free of seed-producing plants.

Animals

Animals, including humans, are responsible for scattering many seeds. They may carry the seed on their feet, clinging to their fur or clothes, or internally (ingested seed). Many seeds have specially adapted barbs, hooks, spines, or twisted awns that cling to the fur or fleece of animals or to people's clothing. Sandburs, cockleburs, sticktights, and beggar-ticks are examples. Others may imbed themselves in an animal's mouth, causing sores. Examples are wild barley, downy brome, and various needlegrasses. Other seeds have cottonlike lint or similar structures that help them cling to fur or clothes. Annual bluegrass and bermudagrass seed will stick to the fur of rabbits and dogs. Mistletoe seeds are sticky and become attached to the feet of birds. Birds often carry away fleshy fruits containing seeds for food.

TABLE 2-3. Germination of Weed Seed After Storage in Fresh Water

Field bindweed	After 54 months, 55% germinated.
Canada thistle	After 36 months, about 50% germinated. After 54 months, none germinated.
Russian knapweed	After 30 months, 14% of seed still sound; none germinated after 5 years.
Redroot pigweed	After 33 months, 9% still sprouted.
Quackgrass	None sprouted after 27 months.
Barnyardgrass	After 3 months less than 1% germinated; none germinated after 12 months.
Halogeton	After 3 months less than 1% germinated; none germinated after 12 months.
Hoary cress	After 2 months germination dropped to 5% or less, and dropped to zero after 19 months.

TABLE 2-4. Percentage of Viable Seeds Passed by Animals Based on Total Number of Seeds Fed

	Percentage of Viable Seeds Passed by					
Kind of Seeds	Calves	Horses	Sheep	Hogs	Chickens	Average
Field bindweed	22.3	6.2	9.0	21.0	0.0	11.7
Sweetclover	13.7	14.9	5.4	16.1	0.0	10.0
Virginia pepperweed	5.4	19.8	8.4	3.1	0.0	7.3
Velvetleaf	11.3	4.6	5.7	10.3	1.2	6.6
Smooth dock	4.5	6.5	7.4	2.2	0.0	4.1
Pennsylvania smartweed	0.3	0.4	2.3	0.0	0.0	0.6
Average	9.6	8.7	6.4	8.8	0.2	6.7

From Harmon and Keim (1934).

Many weed seeds pass through the digestive tracts of animals and remain viable. Weed seedlings are often found germinating in animal droppings. Table 2-4 shows the results of feeding weed seeds to different kinds of livestock and washing the seed free from the feces. The figures are given as a percentage of viable seeds recovered.

Machinery

Machinery can easily carry weed seeds, rhizomes, and stolons. Harvesting equipment, especially combines, often spreads weed seeds. Cultivation equipment, tractors, and tractor tires frequently carry soil that may include weed seeds. Moreover, cultivation equipment may drag rhizomes and stolons, dropping them later to start new infestations.

Weed Screenings

Most weed seeds have a reasonably high feed value. Common ragweed seed, when chemically analyzed, had 20.0% crude protein, 15.7% crude fat, 18% nitrogen-free extract, and 4.83% ash. Because of their relative cheapness, weed seed screenings are often included in livestock feeds (see Tables 3-5 and 25-1).

Over the years, considerable attention has been given to the problem of destroying the viability of weed seeds in screenings. Seeds are usually finely ground, or soaked in water and then cooked. Fine grinding with a hammer mill has reduced to a minimum the hazard of scattering live weed seeds.

Pelleting destroyed the seed's viability when high temperatures were used. Fish solubles with high protein content were added to screenings after heating to about 200°F. The mixture was pressed into pellets. Germination tests indicated that the seeds were no longer viable with that amount of heat.

TABLE 2-5. Number of Seeds Produced per Plant, Number of Seeds per Pound, and Weight of 1000 Seeds

Common Name	Number of Seeds (per plant)	Number of Seeds (per lb[a])	Weight of 1000 Seeds (g)
Barnyardgrass	7,160[b,c]	324,286	1.40
Buckwheat, wild	11,900	64,857	7.0
Charlock	2,700	238,947	1.9
Dock, curly	29,500	324,286	1.4
Dodder, field	16,000[c]	585,806	0.77
Kochia	14,600	534,118	0.85
Lambsquarters	72,450	648,570	0.70
Medic, black	2,350	378,333	1.2
Mullein	223,200	5,044,444	0.09
Mustard, black	13,400[d]	267,059	1.7
Nutsedge, yellow	2,420[b]	2,389,474	0.19
Oat, wild	250[b]	25,913	17.52
Pigweed, redroot	117,400[b]	1,194,737	0.38
Plantain, broadleaf	36,150	2,270,000	0.20
Primrose, evening	118,500	1,375,757	0.33
Purslane	52,300	3,492,308	0.13
Ragweed, common	3,380[b]	114,937	3.95
Sandbur	1,110[b]	67,259	6.75
Shepherdspurse	38,500[b,c]	4,729,166	0.10
Smartweed, Pennsylvania	3,140	126,111	3.6
Spurge, leafy	82,100[b,c]	6,053,333	0.07
Sunflower, common	7,200[b,c]	69,050	6.57
Thistle, Canada	680[b,c]	288,254	1.57

[a]Calculated from the weight of 1000 seeds.
[b]Many immature seeds also present.
[c]Many seeds shattered.
[d]Yield of one main stem.
From Stevens (1932).

NUMBER AND PERSISTENCY OF WEED SEEDS

Number in the Soil

Some species produce a remarkable number of living seeds per acre. Wild poppies seriously infested the Rothamsted (England) Experiment Station in 1930, with an estimated 113 million poppy seeds per acre.

In Minnesota, weed seed counts at four different locations on 24 different plots showed 98 to 3068 viable weed seeds per square foot of soil, 6 inches deep. Converted to a per acre basis, this is between 4.3 million and 133.0 million seeds per acre in the upper 6 inches.

Soil samples were taken on ten plantations in Louisiana heavily infested with johnsongrass. The average number of viable johnsongrass seeds per acre was

1,657,195. A sugar cane crop was grown for 3 years without permitting the addition of new johnsongrass seed; after 3 years the johnsongrass seed population in the upper 2.5 inches of soil dropped to 1.3% of the original number.

Number Produced by Plants

The persistence of annual and biennial weeds depends mainly on their ability to reinfest the soil. The first infestation of most perennial species depends on seed. Obviously, if we could control the production of seed, we could eventually eliminate many species.

One scientist, reporting on 245 species, found that the number of weed seeds produced by one plant ranges from 140 for leafy spurge to nearly a quarter of a million seeds for common mullein. Of these, 23 species were selected, as listed in Table 2-5. Only plump, well-developed seeds, from well-developed plants growing with comparatively little competition, were counted. Those 23 species averaged 25,688 seeds per plant. A second report listed 263 species. One witchweed plant was found to produce as many as one-half million seeds.

Age of Seed and Viability

The length of time a seed is capable of producing a seedling varies widely with different kinds of seeds and with different conditions. Certain seeds keep their viability for many years, whereas others die within a few weeks after maturing if they do not find a suitable environment for germination. Many seeds die before they are able to germinate, as suggested by studies of Forcella (1992), Forcella et al. (1992), and Gross and Renner (1989), who reported that 80% or more of seeds in the soil seedbank are dead or die between fall and spring. Fenner (1995) suggests that the vast majority of weed seeds fall into the later categories of limited soil life. Yet the great number of seeds produced by weeds still result in a vast reservoir for future infestation, and knowledge of seed longevity and germination requirements and dormancy is important.

A short-lived species, silver maple, has seeds with about 58% moisture when shed, and they will germinate immediately. However, when moisture content drops to 30 to 34%, they die. In nature this occurs within a few weeks.

Lotus *(Nelumbo nucifera)* seeds found in a lakebed in Manchuria were still viable after approximately 1000 years. The seeds were 1040 ± 210 years old, as measured by a radioactive carbon dating technique. The seeds had been covered in deep mud and very cold water.

To determine the longevity of seed, Duvel placed 107 species in porous clay pots and buried them 8, 22, and 42 inches deep in the soil. He removed samples at various intervals and left others undisturbed. His findings, reported by Toole and Brown (1946), were as follows:

After 1 year, seed of 71 species germinated.

After 6 years, seed of 68 species germinated.

After 10 years, seed of 68 species germinated.
After 20 years, seed of 57 species germinated.
After 30 years, seed of 44 species germinated.
After 38 years, seed of 36 species germinated.

After 38 years,
- 91% of jimsonweed seed germinated.
- 48% of mullein seed germinated.
- 38% of velvetleaf seed germinated.
- 17% of evening primrose seed germinated.
- 7% of lamb's-quarter seed germinated.
- 1% of green foxtail seed germinated.
- 1% of curly dock seed germinated.

The actual percentage of germination is not as important as the fact of survival. Programs to eradicate plants with long seed dormancy are doomed to failure until techniques are developed to break (100%) seed dormancy in soil.

The three depths did not greatly influence the data; however, seed at the 42-inch depth lived slightly longer. The 8-inch depth is far below the ideal depth for most seeds to germinate. If one sample had been buried 1 inch deep or less, greater differences caused by depth would have been likely.

In another test started in 1879, Beal mixed seeds of 20 different weed species with sand and buried the mixtures in uncorked bottles, with the opening tipped downward. After 20 years, 11 of the species were still alive. After 40 years, 9 species were still alive. These 9 were redroot pigweed, prostrate pigweed, common ragweed, black mustard, Virginia pepperweed, evening primrose, broadleaf plantain, purslane, and curly dock.

After 50 years, moth mullein germinated for the first time. Therefore, the seeds were still alive at 40 years but did not germinate because they were dormant. After 70 years, moth mullein had a germination rate of 72%, evening primrose 17%, and curly dock a few percent. After 80 years, moth mullein still had a germination rate of 70 to 80%, but with the other two species only a few seeds germinated. It appeared that these two species were nearing the end of their survival period.

Thus, many weed seeds retain their viability for 40 years and longer when buried deep in the soil. Scientists believe that their longevity depends, to a great extent, on the seeds being buried deep, with a reduced oxygen supply available. If brought to the surface with other conditions favorable, many of the seeds will germinate. It is evident that repeated cultivation for several years without opportunity of reinfestation effectively reduces the weed seed population in a soil.

Depending on climatic conditions, abandoned cultivated cropland will usually return to forest or grassland vegetation. If cultivated after 30 to 100 years, the original field weeds immediately appear. The weed seeds will have remained dormant while buried deep in the soil.

GERMINATION AND DORMANCY OF SEEDS

Germination includes several steps that result in the quiescent embryo changing to a metabolically active embryo as it increases in size and emerges from the seed. It is associated with an uptake of water and oxygen, use of stored food, and, normally, release of carbon dioxide. For a seed to germinate, it must have an environment favorable for this process. This includes an adequate, but not excessive, supply of water, a suitable temperature and composition of gases (O_2/CO_2 ratio) in the atmosphere, and light for certain seeds. Specific requirements for seed germination differ for various species (see Buhler and Hoffman, 2000). Although these factors are optimal, a seed may not germinate because of some kind of dormancy.

Dormancy is a type of resting stage for the seed. Dormancy may determine the time of year when a seed germinates, or it may delay germination for years and thus guarantee the viability of the seed in later years. Five environmental factors affect seed dormancy: *temperature, moisture, oxygen, light,* and *the presence of inhibitors* including allelopathic effects. Other factors directly related to the seed and its dormancy include impermeable (to water, oxygen, or both) seed coats, mechanically resistant seed coats, immature embryos, and afterripening.

Temperature

The temperature that favors seed germination varies with each species. There is a *minimum* temperature below which germination will not occur, a *maximum* temperature above which germination will not occur, and an *optimum,* or ideal, temperature when seeds germinate most quickly. Thus, some seeds germinate only in rather cool soils, whereas others do so only in warm soils.

The temperature requirements for most crop seeds are well established, and farmers recognize these requirements and plant accordingly. Cotton, for example, requires relatively high temperatures for germination, whereas the small grains will germinate at relatively cool temperatures.

Russian pigweed seed has germinated in ice and on frozen soil. Wild oat may germinate at temperatures of 35°F. Comparatively low temperatures (between 40 and 60°F) are necessary before certain winter annuals will germinate. High temperatures may cause a secondary type of seed dormancy, especially with some winter annual weeds. Wormseed mustard was introduced to secondary dormancy by temperatures of 86°F, and many summer annuals require temperatures of 65 to 95°F to germinate. Alternating temperatures are often better than a constant temperature for seed germination.

When redroot pigweed seed (a summer annual) was placed in germinators at 68°F, some seed remained dormant for more than 6 years. The seeds could be induced to germinate at any time in three ways: (1) by raising the temperature to 95°F, rubbing with the hand, and replacing at 68°F, (2) by partial desiccation, or (3) by alternating the temperatures.

Temperature alone does not completely explain the periodicity of seed germination. Often the seeds have another form of dormancy that temporarily stops germination. This may be a survival mechanism to keep the plant from germinating immediately upon maturity in a season not suited to the plant.

Seeds may lie dormant for as little as several weeks or as much as several years. For example, cheat and hairy chess have a primary dormancy of 4 to 5 weeks after maturity. During this period they will germinate only if subjected to low temperatures (59°F or below). However, if the seeds are stored for 4 to 5 weeks, germination will then readily occur at temperatures of 68 to 77°F.

Moisture

Germination is normally a period of rapid expansion and high rates of metabolism or cell activity. Much of the expansion is simply an increase in water, expanding cell walls. If water content of the seed is reduced, the activity of enzymes—and consequently metabolism—slows down. The amount of moisture contained in seeds may determine their respiratory rate. During germination the seed respires at a very rapid rate. Many seeds cannot maintain this high rate of respiration until they reach a moisture content of 14% or more. Thus, in dry soils the seed remains dormant.

Dry seeds can tolerate severe conditions; some have been kept in boiling water for short intervals without injury and others in liquid nitrogen (–310°F). When moist enough for germination, the same seeds may be killed by cold temperatures of 30°F or warm temperatures of 105°F.

Oxygen

In addition to the right temperature and sufficient moisture, germination depends on oxygen. Aerobic respiration requires more free oxygen than anaerobic respiration; thus, some seeds start germination under anaerobic conditions and then shift to aerobic respiration when the seed coat ruptures.

The percentage of oxygen found in the soil varies widely, depending on soil porosity, depth in the soil, and organisms in the soil that use oxygen and release CO_2 (microorganisms, roots, etc.). The percentage of oxygen in the soil is usually inversely proportional to the percentage of CO_2. In swampy rice land there may be less than 1% oxygen in the soil atmosphere; in freshly green-manured land, 6 to 8%; and where corn is growing rapidly, 8 to 9%, as compared with about 21% in a normal atmosphere. The percentage of carbon dioxide may range from 5 to 15% under such conditions, as compared with 0.03% for normal air.

Different species vary considerably in the amount of oxygen needed for seeds to germinate. Wheat seed germinated well when the replenished oxygen supply of the soil was 3.0 mg/m^2 /hr or more. It failed to germinate when the rate was below 1.5 mg. Rice seed germinated at 0.5 mg. Broadleaf cattail and some other aquatic plants germinate better at low oxygen concentrations than with normal air.

The effect of different oxygen concentrations on the seeds of field bindweed, leafy spurge, hoary cress, and horse nettle was determined. Oxygen concentrations of 5% produced little to no germination of hoary cress, horse nettle, and leafy spurge. At 10% oxygen and below, these three weeds germinated at a rate far below normal, whereas bindweed was reduced somewhat. The highest percentage germination was found at about 21% oxygen (normal air) for leafy spurge and hoary cress, but the best germination for horse nettle was at 36% oxygen and for field bindweed at 53%.

Wild oat and charlock germination can be greatly suppressed by reducing the oxygen supply by soil compaction. Cultivation increased sixfold the number of wild oat seeds that germinated and the number of charlock twofold, as compared with compacted plots. Cultivation increased soil aeration and thus increased the content of oxygen in the soil atmosphere.

Excess water in the soil cuts down seed germination of most plant species. Researchers believe that lower germination is related to a smaller supply of oxygen in waterlogged soils, rather than merely excess water.

Many small-seeded weed species germinate only in the upper 1 to 2 inches of soil, mostly in the upper 1 inch. A limited number, however, germinate below 2 inches. Seeds germinate deeper in sandy soils than in clay soils as a result of better aeration or better oxygen supply in the sands. Some seeds buried deep in the soil do not germinate but lie dormant for many years. When brought to the surface, they germinate promptly. Aeration, involving increased oxygen supply, is probably responsible.

Through the use of herbicides, successive crops of weed seedlings may be killed without disturbing the soil. Few to no viable weed seeds may remain in the upper soil layer. The soil surface may then remain relatively free of weeds. Repeated treatments to kill annual weeds before they produce seed is especially useful in areas where the soil is not disturbed by cultivation, such as in some perennial crops, permanent sod, or turf areas.

Light

Some kinds of seed germinate best in light, others in darkness, and others germinate readily in either light or darkness. Among several hundred species in which the role of light has been investigated, about half require light for maximum germination. The length of day and quality (color) of light are also influential. Here is a brief review of the electromagnetic spectrum (color of wavelengths).

Name	Wavelength (nm)
X ray	10–150
Ultraviolet	Below 400
Visible spectrum	
Violet	400–424
Blue	424–491
Green	491–575
Yellow	575–585
Orange	585–647
Red	647–700
Infrared	Greater than 700

Germination of lettuce seed was promoted by radiation at 660 nm (red light) and inhibited at 730 nm (infrared light). A later study with lettuce showed that the inhibitory effect of infrared light could be reversed by red light. Regardless of the number of alternating periods of red and infrared light to which the seeds were exposed, the final type of light determined the percentage germination. For example, when the final light was infrared, germination was about 50%, but when the final light was red, germination reached almost 100%. Without light, germination fell to about 8%.

Here are a few examples of species and the light requirement of their seeds for germination.

Germination Favored by Light		Germination Favored by Darkness	Germination in Either Light or Darkness	
Bluegrass	Dock	Onion	Salsify	Wheat
Tobacco	Primrose	Lily	Bean	Rush
Mullein	Buttercup	Jimsonweed	Clover	Toadflax

These effects vary from species to species. In some seeds, the light requirement can be replaced by afterripening in dry storage, alternating the daily temperature, higher temperatures, and treatment in potassium nitrate or gibberellic acid solutions.

Seed Coat Impermeability to Water, Oxygen

Seed coats may be waterproof, which can prevent the seed from absorbing water. Such seed will not germinate even if soil moisture is plentiful. Seed that fails to germinate because of a waterproof seed coat is called *hard seed.* Hard seed is common in annual morningglory, lespedeza, clovers, alfalfa, and vetch. Researchers believe that many weed species have hard seeds.

Some seed coats are impermeable to oxygen but not water. Cocklebur has two seeds per fruit, one set slightly below the other. The lower seed usually germinates during the first spring, and the upper seed remains dormant until the next year. Both can be made to germinate immediately by breaking the seed coats or by simply increasing the oxygen supply. Seeds of ragweed, several grasses, and lettuce also show this type of dormancy.

As with waterproof seed coats (hard seeds), anything that breaks the seed coat—scarification, acids, soil microorganisms—will break this type of dormancy. Under laboratory conditions, oxygen dormancy can usually be broken by increasing the oxygen supply. Cultivation of the soil often has a similar effect by increasing the oxygen level in the upper layer of soil.

Mechanically Resistant Seed Coats

A tough seed coat may forcibly enclose the embryo and prevent germination. While the seed absorbs oxygen and water, it builds pressures in excess of 1000 psi. The seed will quickly germinate if the seed coat is removed. Pigweed, wild mustard, shepherd's purse, and pepperweed have this type of dormancy. As long as the seed coat remains moist, it remains tough and leathery—for as long as 50 years. Any factor that weakens the seed coat will help break dormancy. Drying at temperatures of 110°F or mechanical or chemical injury to the seed coat may break this type of dormancy.

Immature Embryos

The outside of the seed may appear fully developed, but it may have an immature embryo that needs more growth before the seed can germinate. Therefore, the seed appears dormant, even though the embryo is slowly growing and developing. Seeds of orchids, holly, smartweed, and bulrush show this type of dormancy. Dormancy in seeds and its importance to weed science are discussed in articles, cited at the end of this chapter, by Li and Foley, (1997), Forcella, (1998), Khan, (1997), Hilhorst, (1995), and Foley and Fennimore, (1998).

Afterripening

In some species the embryos appear completely developed, but the seed will not germinate even though the seed coat has been carefully removed to permit easy absorption of water and oxygen. Light and darkness have no effect. In this case germination occurs normally after a period of *afterripening*. Occasionally, cool temperatures for several months will end this type of dormancy. Afterripening is especially common in the grass (Poaceae), mustard (Brassicaceae), smartweed (Polygonaceae), rose (Rosaceae), and pink (Caryophyllaceae) families.

Afterripening is a physiological change of a complex physicochemical nature. Although the exact processes are not completely understood, they may be associated with changes in the storage materials present, substances promoting germination may appear, or substances inhibiting germination may disappear.

GERMINATION AS AFFECTED BY CERTAIN CONDITIONS

Burning

Burning fields after weed seeds have matured gives only partial and erratic destruction of seeds. Seeds that lie on the ground may readily escape, whereas those held on the plants often burn completely. The degree of weed seed destruction depends largely on the intensity of the heat, and this in turn on dryness and the amount of litter and debris to be burned.

The aftereffects of burning are usually pronounced, especially after forest fires. Weed species absent, for the most part, from the area for many years may suddenly appear after a fire and dominate other vegetation.

Data and experience clearly show that moderate heat may end seed dormancy. Five other factors that normally follow a fire may also terminate dormancy: (1) greater alternation of temperature in the upper soil layers between day and night, (2) more light reaching the surface soil, (3) removal of litter, (4) removal of competition by other plants, and (5) removal of plants previously living in the area that had soil-inhibiting substances that prevented seed germination. With removal of such plants, those substances are no longer present. This is probably an allelopathic effect.

Cutting (Stage of Maturity)

Cutting weeds to prevent seed production is a common recommendation (Table 2-6). The practice is important in agricultural croplands, in turf, and in hay crops.

In South Dakota, Canada thistle and perennial sow thistle heads were removed from the plants and dried at daily intervals after the flowers opened. The experiment was continued for 3 years. Perennial sow thistle heads harvested 3 days after blooming had 0.0% viable seed; 6 days after blooming they had an average of 6% viable seed; and 8 days after blooming, 65% viable seed.

Canada thistle harvested 6 days after blooming had an average of 0.03% viable seed; 8 days after blooming, 6.7% viable seed; and 11 days after blooming, 73% viable seed.

Another study compared removal of the heads as previously described with the effect of cutting the entire plant and leaving the heads on the plant during drying. Results of the two methods were similar; thus, little seed development takes place after either type of cutting.

In summary, either mowing in or before the bud stage prevents viable seed production. If seeds reach medium ripeness, probably a large percentage of viable seeds will be produced. In the case of many species, cutting after that time does little or no good in preventing viable seed production. With other species, mowing is not effective because some of the heads are short and missed by the mower, and these produce seed.

Storage in Silage

Many seeds lose their viability in silage. Many weed seeds lose their germinating power 10 to 20 days after being placed in silage. Some reports indicate, however, that a number of seeds will germinate after being in a silo for periods up to 4 years. These

TABLE 2-6. Germination of Weed Seeds Cut at Various Stages of Maturity

	Percent Germinated		
Weed Seeds	Cut When Dead Ripe	Cut When in Flower	Cut in Bud Stage
Groundsel, ragwort	72	80	0
Sowthistle, common	100	100	0
Groundsel, common	90	35	0
Sea aster	90	86	0
Dandelion	91	0	0
Catsear, spotted	90	0	0
Canada thistle	38	0	0

TABLE 2-7. Effect of Length of Time in Cow Manure on the Viability of Various Weed Seeds

Kind of Seeds	Percent Viability Before Burial	Percent Viability after Storage			
		1 Month	2 Months	3 Months	4 Months
Velvetleaf	52.5	2.0	0.0	0.0	0.0
Field bindweed	84.0	4.0	22.0	1.0	0.0
Sweetclover	68.0	22.0	4.0	0.0	0.0
Pepperweed	34.5	0.0	0.0	0.0	0.0
Smooth dock	86.0	0.0	0.0	0.0	0.0
Smartweed	0.5	0.0	0.0	0.0	0.0
Cocklebur	60.0	0.0	0.0	0.0	0.0
Puncturevine	52.0	0.0	0.0	0.0	0.0

From Harmon and Keim (1934).

variations are possibly a result of differences in the silage as to moisture content, temperature, and amount of organic acids produced (Tildesley, 1937).

Storage in Manure

When manure is spread fresh from the stable, viable weed seeds are usually spread with it. But if manure is stored, heating and decomposition will begin and, in time, the weed seeds are destroyed (Table 2-7).

Of those listed in Table 2-7, only three weeds showed any viability after 1 month storage. All seeds were destroyed at the end of 4 months.

Stoker et al. (1934) found complete destruction of hoary cress and Russian knapweed seeds in moist, compacted chicken manure at the end of 1 month. However, seed of field bindweed was still viable at the end of 4 months.

These tests were conducted during the summer in cow or chicken manure. Horse and mule manure tend to heat, whereas cow and chicken manure do not. Decomposition of the weed seed would be more rapid at the higher temperature. If the manure were frozen or cold, the seed would live longer.

If the edges or the outside of the manure pile dries out, decomposition slows down and viable weed seeds would likely persist. Therefore, manure should be turned occasionally to kill all seeds. Well-rotted manure is free of viable seeds.

DISSEMINATION BY RHIZOMES, STOLONS, TUBERS, ROOTS, BULBS, AND BULBLETS

Although weeds reproduce and spread most widely by means of seeds, they also multiply by vegetative methods. Rhizomes, stolons, tubers, roots, bulbs, and bulblets are all vegetative or asexual means of reproduction. Many perennial weeds classified as "serious" reproduce vegetatively, and most of these also reproduce by means of seeds. Vegetative organs occasionally have a short period of dormancy.

Most plants spread slowly by vegetative means alone. Without help from humans and their cultivation equipment, weeds such as quackgrass, field bindweed, johnsongrass, bermudagrass (Figure 2-4), nutsedge, and Canada thistle would spread vegetatively less than 10ft/year. However, the rhizomes, stolons, roots, and tubers are dragged about the field with soil-tillage equipment. Wherever these plant pieces drop, a new infestation is likely. Disc-type cultivation equipment is less likely to drag the plant parts than are shovels, sweeps, and plows.

Repeated tillage will kill most plants possessing rhizomes, stolons, roots, or tubers. If cut off and in dry soils, the vegetative parts may quickly dry and die, thus preventing new growth.

In most soils the cut vegetative parts quickly take root and establish new plants. Under such conditions, repeated tillage may exhaust the underground food reserves. Most such weeds are killed through *carbohydrate starvation* by repeated tillage for 1 to 2 years.

Some chemicals mixed into the soil will retard the development of new roots, especially after the plant has been cut off. With a combination treatment of an effective herbicide plus repeated cultivation, many serious perennial weeds may be controlled in a short time. For example, johnsongrass can be controlled by trifluralin plus repeated cultivation.

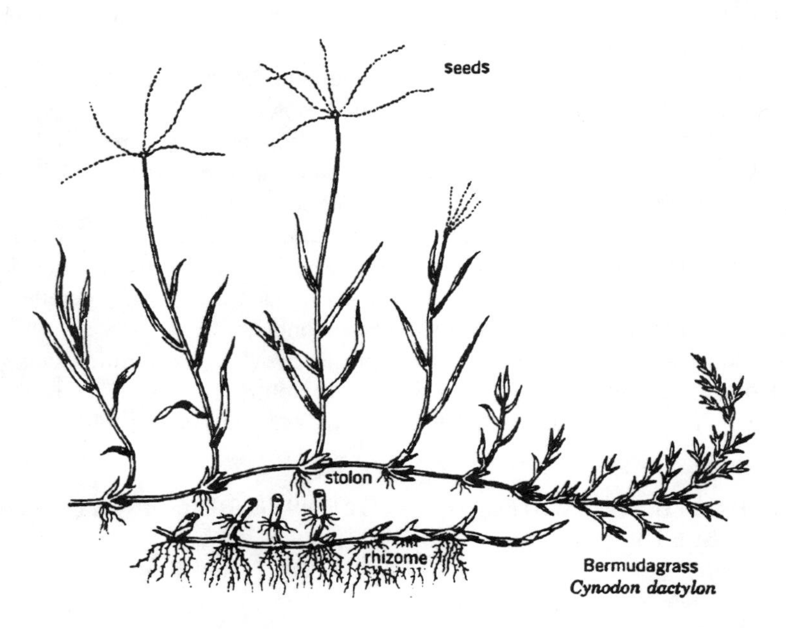

Figuure 2-4. Bermudagrass reproduces by seeds, stolens, and rhizomes.

INVASIVE PLANT SPECIES

Definitions and Characteristics

In Chapter 1 we defined a weed as *a plant growing where it is not desired, or a plant out of place*—some plant that, according to a human criterion, is *undesirable*. However, there is increasing concern about the potential damage that can occur to our environment because of invasive species. Invasive species include all types of biological organisms, of which plants are a part. So, are weeds and invasive plant species synonymous? The answer is not always. President Clinton in a February 3, 1999 Executive Order relating to invasive species defined alien, invasive and native species. "*Alien species* means with respect to a particular ecosystem, any species, including its seeds, eggs, spores or other biological material capable of propagating that species, that is not native to that ecosystem; *Invasive species* means an alien species whose introduction does or is likely to cause economic or environmental harm or harm to human health; and *Native species* means with respect to a particular ecosystem, a species that, other than as a result of an introduction, historically occurred or currently occurs in that ecosystem" (Executive Order on Invasive Species by President Clinton, February 3, 2000).

Many alien species were introduced into the United States for food, fiber, or ornamental purposes (Pimentel et al., 1999). There are an estimated 5000 alien plant species that have escaped and now exist in U.S. ecosystems as compared with 17,000 native U.S. plants. As discussed earlier in this chapter, all weeds have certain characters that make them successful in disturbed environments (Table 2-1), and invasive species tend to have these same characters. Not all alien species become problems. The potential for problems to develop exists for any introduced plant, because within the new ecosystem there are generally no natural enemies that would control their spread and reproduction. In terms of our major weeds, approximately 65% of all weeds in cropland in the United States are non-natives, (Westbrooks, 1998; other good sources include: Mooney and Hobbs 2000; and Cox, 1999), so although, a high percentage of U.S. cropland weeds are alien, not all are.

To illustrate the potential for future problems from alien plant species, it is useful to refer to Holm (1978), who stated that there are 6741 plant species recognized as weeds somewhere in the world, while in the United States there are approximately 1365 weedy species considered to be of foreign origin (WSSA, 1989). These numbers suggest that there are many weedy species found throughout the world that in theory could be introduced into the United States as aliens and potentially become invasive weeds.

Costs of Invasive Species

Awareness by all citizens of the importance of containing the proliferation of alien species in any particular country is important if alien weed introductions are to be controlled in the current global economy. Alien species can arrive in a country in many ways. Some are introduced through ignorance as a solution to a man-made problem such as erosion. For example, kudzu was introduced to reduce soil erosion and has now invaded large areas of the United States. Another example is water hyacinth, which was introduced as an ornamental and escaped to waterways. Many species arrive by accident as contaminants brought into a country by travelers or in cargo shipments. Once an invasive species becomes established, it affects the ecosystem and people in numerous ways. The cost both to the environment and to the economy is enormous (estimated to be in the billions of dollars) when invasive weeds are found on the land or water, whether in agriculture, ranching, forestry, industry, or in parks, natural areas, refuges, or grasslands. Most people agree that invasive species are costly in terms of economics in relation to control, health-related problems, loss of land utility, lost crop value and, probably most important, in the effect they have on the balance and diversity within our native ecosystems. Loss of native species and the resultant lowered biodiversity upset ecosystem balance, endanger the integrity of our land, and have a long-term negative effect on our ability to fully utilize and enjoy the land. There is a great need to implement programs to minimize the potential threats of invasive species in the United States; various government and private organizations have programs to study and address these problems.

Invasive Weed Management Programs

The most important recent action to address invasive species is the Executive Order of February 3, 2000. In this Executive Order, alien, native, and invasive species were defined, it described a federal plan to address questions concerning invasive species from a research and action perspective, and the Invasive Species Council was established and directed to facilitate programs and develop an Invasive Species Management Plan. This order, in effect, brought the problem of invasive species to the attention of the public and began a visible, active government process to deal aggressively with the problem.

A *National Strategy for Invasive Plant Management*, which is part of the Federal Interagency Committee for the Management of Noxious and Exotic Weeds (FICMNEW) (1997), has defined "goals, objectives and opportunities that will result in the preservation of cropland, parks, preserves, forests, waterways, wetlands, rangelands, urban green spaces and their associated uses and industries." The FICMNEW is composed of 17 federal agencies. This plan has been supported by most other federal, state, private commodity and public interest groups, and professional scientific societies that either deal with or are concerned about invasive species. The program has three main goals: (1) effective prevention, (2) effective control, and (3) effective restoration of previously infested areas. Within each goal there are further details regarding how it can be accomplished. Basically, the plan involves minimizing potential problems and, when problems are discovered, to implement plans for reducing and removing the invasive species with eventual restoration of infested sites to their natural balance, and to increase public awareness. More information regarding this program and those of other groups is available at the Web sites listed at the end of this chapter (FICMNEW, The Nature Conservancy, and *The Invasive Weed Fact Book*). For example, The Nature Conservancy (Web site) lists the "dirty dozen" invasive species, six of which are plants: purple loosestrife, tamansk, leafy spurge

(Figure 2-5), hydrilla, miconia, and Chinese tallow. Yet these are but a few of the species that cause concern throughout the United States (Figure 2-6). The information does point out that many other invasive species are "despoiling our ecosystems and imperiling our native plants and animals," and that these examples are just the "worst of a bad lot." The six other species in the dirty dozen are the zebra mussel, flathead catfish, rosy wolfsnail, green crab, balsam wooly adelgid, and the brown tree snake.

Many government agencies have responsibilities for weeds, including management, regulation, and research, among which are the Animal and Plant Health Inspection Service (APHIS); U.S. Forest Service; U.S. Fish and Wildlife Service; National Park Service; Bureau of Land Management, Bureau of Reclamation; U.S. Geological Service; Bureau of Indian Affairs; and the executive branch Departments of Agriculture, Defense, and Transportation. APHIS, the frontline agency responsible for preventing the introduction of foreign weeds into the United States, works with state and local government and private organizations to eradicate alien weeds and to regulate the introduction of biocontrol agents. Many U.S. government agencies and granting organizations support research into invasive weeds. In addition, individual states and local governments are involved in weed management. Such involvement includes enactment and enforcement of local seed and weed laws (noxious weed

Figure 2-5. Leafy spurge infesting rangelands in North Dakota. (Courtesy of Rodney Lym, North Dakota State University.)

Figure 2-6. Lustra infesting ditch banks and wetlands. (Courtesy of Rodney Lym, North Dakota State University.)

laws), which prohibit or restrict movement or sale of certain types of plants within their jurisdiction, and various funding programs for activities and research to address weed problems. The current situation regarding invasive species is a concern of all citizens. The problem is not yet out of control, and with the current activities in both the public and private sectors along with improved education and research programs, an increasing public awareness will help minimize future problems resulting from invasive species.

LITERATURE CITED AND SUGGESTED READING

Aldrich, R. J. 1984. *Weed-Crop Ecology: Principles in Weed Management.* Brenton Publications, North Scituate, MA.

Altieri, M. A., and M. Liebman. 1988. *Weed Management in Agroecosystems: Ecological Approaches.* CRC Press, Boca Raton, FL.

Baker, H. G. 1974. The evolution of weeds. *Ann. Rev. Ecol. Syst.* **5**:1–24.

Beal. W. J. 1911. The vitality of seeds buried in the soil. *Proc. Sec. Promot. Agric. Sel.* **31**:21–23.

Bridges, D. C. 1995. Ecology of weeds. In *Handbook of Weed Management Systems*, ed. by A. E. Smith, Chapter 2. Marcel Dekker, New York.

Bruns V. F., and L. W. Rasmussen. 1958. The effects of fresh water storage on the germination of certain weeds. III. Quackgrass, green bristlegrass, yellow bristlegrass, watergrass, pigweed and halogeton. *Weeds* **6**:42–48.

Buhler, D. D., and M. L. Hoffman. 2000. *Andersen's Guide to Practical Methods of Propagating Weeds and Other Plants*. Weed Science Society of America, Lawrence, KS.

Cousins, R. 1987. Theory and reality of weed control thresholds. *Plant Prot. Quart.* **2**:13–20.

Cox, G. W. 1999. *Alien Species in North America and Hawaii—Impacts on Natural Ecosystems*. Covelo, CA.

Dekker, J. 1999. Soil seed banks and weed management. In *Expanding the Context of Weed Management*, ed. by D. D. Buhler, pp. 139–166. Food Products Press, New York.

Duke, S. P., and J. Lydon. 1987. Herbicides from natural products. *Weed Technol.* **1**:122–128.

Executive Order on Invasive Weeds, February 3, 2000.

Fenner, M. 1995. Ecology of seed banks. In *Seed Development and Germination*, ed. by J. Kigel and G. Galili. Marcel Dekker, New York.

Foley, M. E., and S. A. Fennimore. 1998. Genetic basis for seed dormancy. *Seed Sci Res.* **8**:173–182.

Forcella, F. 1992. Prediction of weed seedling densities from buried seed reserves. *Weed Res.* **32**:29–38.

Forcella, F. 1998. Real-time assessment of seed dormancy and seedling growth for weed management. *Seed Sci Res.* **8**:201–209.

Forcella, F., R. G. Wilson, K. A. Renner, J. Dekker, R. G. Harvey, D. A. Alm, D. D. Buhler, and J. Cardina. 1992. Weed seedbanks of the U.S. cornbelt: Magnitude, variation, emergence and application. *Weed Sci.* **40**:636–644.

Fuerst, E. P., and A.R. Putnam. 1983. Separating the competitive and allelopathic components of interference: Theoretical principles. *J. Chem. Ecol.* **9**:937–944.

Furrer, J. D. 1954. The farmer and his seed. *North Central Weed Cont. Conf. Proc.* **11**:26–27.

Gould, F. 1991. The evolutionary potential of crop pests. *Amer. Sci.* **79**:496–507.

Gross, K. L., and K. A. Renner. 1989. A new method for estimating seed numbers in the soil. *Weed Sci.* **37**:836–839.

Harlan, J. R., and J. M. J. de Wet. 1965. Some thoughts about weeds. *Econ. Bot.* **19**:16–24.

Harmon G. W., and F. D., Keim. 1934. The percentage and viability of weed seeds recovered in the feces of farm animals and their longevity when buried in manure. *J. Am. Soc. Agron.* **26**:762–767.

Harper, J. L. 1977. *Population Biology of Plants*. Academic Press, New York.

Hilhorst, H. W. M. 1995. A critical update on seed dormancy. I. Primary dormancy. *Seed Sci. Res.* **5**:61–73.

Holm, L. 1978. Some characteristics of weed problems in two worlds. *Proc. WSWS* **31**:3–12.

Holm, L. G., D. L. Pluknett, J. V. Pancho, and J. P. Herberger. 1977. *The World's Worst Weeds: Distribution and Biology*. University of Hawaii Press.

Khan, A. A. 1997. Quantification of seed dormancy: Physiological and molecular considerations. *Hortscience* **32**:609–614.

Kropff, M. J., and H. H. van Laar, eds. 1993. *Modeling Crop-Weed Interactions*. CAB International, Manila.

Li, B., and M. Foley. 1997. Genetic and molecular control of seed dormancy. *Trends in Plant Sci.* **2**:384–389.

Lybecker, D. W., E. E. Schweizer, and R. P. King. 1991. Weed management decisions in corn based on bioeconomic modeling. *Weed Sci.* **39**:124–129.

Molisch, H. 1937. *Der Einfluss einer Planze auf die andere- Allelopathie*. Fisher, Jena.

Mooney, H. A., and R. J. Hobbs, eds. 2000. *Invasive Species in a Changing World*. 457 pp. Island Press. Covelo, CA.

Navas, M. L. 1991. Using plant populations biology in weed research: A strategy to improve weed management. *Weed Res.* **31**:171–179.

Norris, R. F. 1999. Ecological impacts of using weed thresholds. In *Expanding the Context of Weed Management*, ed. by D. D. Buhler, pp. 31–58. Food Products Press, New York.

Pimentel, D., L. Lach, R. Zuniga, and D. Morrison. 1999. Environmental and economic costs associated with non-indigenous species in the United States. *Bioscience* **50**:53–65.

Putnam, A. R., and C. S. Tang, eds. 1986. *The Science of Allelopathy*. John Wiley & Sons, Inc., New York.

Radosevich, S., J. Holt, and C. Ghersa. 1997. *Weed Ecology: Implications for Management*. 2d ed. John Wiley & Sons, Inc., New York.

Rankins, A., Jr., D. R. Shaw, and J. D. Byrd Jr. 1998. HERB and MSU-HERB field validation for soybean (*Glycine* max) weed control in Misssissippi. *Weed Technol.* **12**:88–96.

Rice, E. L. 1984. *Allelopathy*. Academic Press, Orlando, FL.

Rizvi, S. J. H., and V. Rizvi. 1992. *Allelopathy: Basic and Applied Aspects*. Chapman and Hall, London.

Shribbs, J. M., D. W. Lybecker, and E. E. Schweizer. 1990. Bioeconomic weed management models for sugarbeet (*Beta vulgaris*) production. *Weed Sci.* **38**:436–444.

Shribbs, J. M., E. E. Schweizer, L. Hergert, and D. W. Lybecker. 1990. Validation of four bioeconomic weed management models for sugarbeet (*Beta vulgaris*) production. *Weed Sci.* **38**:445–451.

Stevens, O. A. 1932. The number and weight of seeds produced by weeds. *Am. J. Bot.* **19**:784–794.

Stoker, G. L., D. C. Tildesley, and R. J. Evans. 1934. The effect of different methods of storing manure on the viability of certain weed seeds. *J. Am Soc. Agron.* **26**:600–609.

Striliani, L., and C. Resina. 1993. SELOMA: Expert system for weed management in herbicide-insensitive crops. *Weed Technol.* **7**:550–559.

Swanton, C. J., S. Weaver, P. Cowan, R. Van Acker, W. Deen, and A. Shrestha. 1999. Weed thresholds: Theory and applicability. In *Expanding the Context of Weed Management*, ed. by D. D. Buhler, pp. 9–30. Food Products Press, New York.

Tildesley, W. T. 1937. A study of some ingredients found in ensilage juice and its effect on the vitality of certain weed seeds. *Sci Agric.* **17**:492–501.

Toole, E. H., and E. Brown. 1946. Final results of the Duvel buried seed experiment. *J. Agric. Res.* **72**:201–210.

Westbrooks, R. G., ed. 1998. *Invasive Plants: Changing the Landscape of America—Fact Book*. Federal Interagency Committee for the Management of Noxious and Exotic Weeds. U.S. Department of Agriculture, Natural Resources Conservation Service. Brooksville, FL.

Wilkerson, G. G., S. A. Modena, and H. D. Coble. 1991. HERB: Decision model for postemergence weed control in soybean. *Agron. J.* **83**:413–417.

WSSA. 1989. *Composite List of Weeds*. WSSA, Lawrence, KS.

WSSA symposium on ecological perspectives on utility of thresholds for weed management. 1992. *Weed Technol.* **6**: 182–235.

Zimdahl, R. L. 1980. *Weed-Crop Competition*. International Plant Protection Center, University of Oregon, Corvallis.

WEB SITES FOR INVASIVE WEEDS

Federal Interagency Committee for the Management of Noxious and Exotic Weeds (FICMNEW)

http://refuges.fws.gov/FICMNEWFiles/eo.html

The Nature Conservancy

http://www.tnc.weeds.ucdavis.edu/

For chemical use, see the manufacturer's or supplier's label and follow these directions. Also see the Preface.

3 Integrated Weed Management

INTRODUCTION

Weed control is essential for successful crop production, as weeds are ever present in the soil and can potentially reduce crop yields every year. Weed populations in a field are relatively constant from year to year, whereas insect and disease outbreaks, although they can have dramatic effects, can be sporadic. Farmers and vegetation managers can plan a weed management program based on prior knowledge of the weeds to expect. This plan should include a well-reasoned approach to weed management that stresses integration of control tactics with all other practices that affect the agroecosystem and links weed control to the overall pest management approach. Integrated weed management (IWM) therefore includes the application of many types of technology and supportive knowledge in the deliberate selection, integration, and implementation of effective weed control strategies, with consideration of the economic, ecological, and sociological consequences. IWM is a component of integrated pest management (IPM). Most descriptions of IPM mention three elements: (1) multiple tactics of pest management used in a compatible manner, (2) pest populations maintained below levels that cause economic damage, and (3) conservation of environmental quality (Thill et al., 1991). An IWM system for a single crop in a single year is relatively simple; however, for long-term IWM to be successful, it must link the farmer's attitude, knowledge, preferences, and abilities with available tools that best fit each situation. Planning demands knowledge of the weed and cropping history on the site, knowledge of weed biology and ecology, and knowledge of weed control methods. The farmer must then use this knowledge to manage the system to obtain good high-quality crop yields while minimizing and, over time reducing, the harmful effects of weeds. A successful IWM system is effective, economically and ecologically sound, stresses integration of control tactics with all other practices that influence the ecosystem, and links weed control to the larger picture of ecosystem management (Thill et al., 1991).

The basic principles related to weeds and IWM must be understood and considered in designing and implementing an effective IWM system. These principles include factors discussed in detail in Chapter 2: What is a weed, the basic resources that weeds and crops compete for, factors affecting weed seed emergence, weed growth and reproduction, length of interaction, and the general ecology and population biology of weedy plants. In field crops, weed problems are seldom the result of a single species. Commonly, one to four species of annual weeds will dominate a cropland population, with often ten or more other weed species occupying the site in low numbers.

Environmental conditions that favor germination and emergence of crop plants usually favor the germination and emergence of weeds. Consequently, weeds emerge at or near the time when the crop emerges and, if not managed, will interfere with crop growth and reduce yields and harvest efficiency. The presence of weed material can also reduce crop quality and storage life.

This chapter covers the many factors and practices to consider in developing an IWM plan. For each situation, different challenges must be addressed, but, in general, the level of weed management will help to determine the tools, inputs, and knowledge necessary to achieve the weed control objectives on the site. Commonly (as discussed in Chapter 1), weed management practices are grouped into three levels: prevention, control, and eradication. Weed management practices, including prevention and various approaches to control, will be discussed in some detail in this chapter. Eradication, or the complete elimination of a weed species from an area, including both live plants and reproductive parts (seeds and vegetative reproduction structures), is often very difficult to achieve in large-scale infestations. For eradication to have any chance for success, thus avoiding the ever increasing costs associated with weed management, the principles of prevention and control discussed in the following sections must be used in an integrated and forceful manner by good land managers. Eradication is never a cheap or easy method to manage weed infestations, as some type of fumigation or elimination by mechanical brute force is almost always required.

WEED MANAGEMENT PRACTICES

Achievement of the desired level of weed suppression requires the use of specific weed management practices. Although these are discussed as separate strategies, it is important to remember that the most effective and economical weed control plan will always employ several approaches. Each component contributes to the overall level of weed control. Omitting or reducing the control achieved from one or more components increases the level of control needed from the remaining weed control practices. Integration of weed control practices is discussed at the end of this chapter. The six main areas of weed control tactics are (1) scouting, (2) prevention, (3) mechanical practices, (4) cultural practices, (5) biological control, and (6) chemical control. An interesting book by Bender (1994) discusses weed management without the use of chemicals, and students are encouraged to look at this book for additional information on weed management.

SCOUTING

Scouting involves knowing specifically what weeds are present in a given field, an estimation of their number (density), their location, and, over time, whether shifts in location or weed types are occurring. The techniques used for scouting include walking fields, drive-bys, and, as discussed in Chapter 7, field mapping using on-site sampling coupled with global positioning systems (GPS) and geographic information systems (GIS). Scouting begins before the cropping season but must continue throughout the entire season. Information gathered prior to and during a cropping cycle can allow a farmer to plan an appropriate management scheme to minimize weed interference and then use the best tools available for weed management. In addition to scouting, it is important to maintain good records of the management tools used and their effectiveness both in managing weeds and in reducing weed seed return to the soil. Many new weed management tools such as HERB (Rankins et al., 1998) use scouting information to design the most appropriate IWM program for a crop. WeedCast 2.0 is a program that forecasts three aspects of weed phenology, weed emergence potential, emergence timing, and seedling height. This information can then be used by the weed manager to make informed decisions for the cropping system.

PREVENTION

Prevention means stopping a new weed from invading an area or limiting weed buildup in a field. Prevention is practiced by (1) not planting crop seed contaminated with weed seed, (2) not carrying weed seeds or vegetative propagules into an area with machinery, contaminated manure, irrigation water, transplants or nursery stock, or growth media or soil, (3) not allowing weeds to go to seed and recharge the soil seedbank, (4) eliminating weeds from fencerows and other areas adjacent to fields, and (5) stopping the spread of vegetatively reproducing perennial weeds. Many of these issues are discussed in Chapter 2 regarding the biology and ecology of weeds. Good scouting aids in all these measures, especially in early detection of localized infestations of perennial weeds and of weeds that escape herbicide control and may be the beginning of a herbicide-resistant population. Prevention, when faithfully employed, can be a cost-effective and practical way to control weeds. This is particularly true for discouraging outbreaks of new problem weeds. Unfortunately, perfect (100%) weed control is needed to prevent seed production by a general weed population in a field. This can be very difficult and uneconomical to achieve, even in extremely high-value crops. Limiting weed seed production is a desirable goal, but totally preventing it is usually practiced only for isolated occurrences of new weeds. As discussed in regard to invasive weeds, prevention is the important first tool to stop the import and establishment of alien plant species, which can become serious weed problems in our ecosystems. There are numerous federal (Federal Seed Act of 1939, Federal Noxious Weed Act of 1974), state, and local noxious weed and seed laws governing control, movement, and distribution of contaminated crop seed and importation and movement of noxious and alien species. These laws were written and are enforced to prevent weed problems.

MECHANICAL PRACTICES

Tillage, hand weeding, mowing, mulching burning, and flooding are considered mechanical weed control methods. Tillage is the mechanical disturbance of the soil involving soil preparation, followed by planting, cropping, harvest, and post-harvest soil management. Primary tillage is the initial ground breaking in preparation for crop production, and secondary tillage is additional soil movement to smooth and level the ground prior to planting. Many specialized tillage implements are used during the crop sequence as a weed management tool. There are a number of types of tillage used in agriculture today, ranging from conventional tillage to conservation tillage. Categories can be defined by the type of primary tillage performed, the amount of plant residue left on the soil surface, and the ultimate objective for the system, which involves minimizing soil erosion.

Conventional tillage varies according to geographical region; however, a general rule is that it leaves much less than 30% residue on the soil surface after planting. Clean (primary) tillage leaves essentially no residue on the soil surface and uses a moldboard plow, disk, chisel plow, or subsoiler, which leave increasing surface roughness and residue, respectively. The moldboard plow is not used to any great extent anymore in most regions because of its negative effects on soil conservation. This tillage is followed by additional secondary tillage with a disk, field cultivator, or harrow to level the ground prior to planting.

Conservation tillage has been defined as any tillage system that leaves at least 30% residue cover on the soil surface after planting and has been redefined by the Soil Conservation Service (SCS) as crop residue management (CRM) (MidWest Plan Service (MWPS), 1992). The main types of conservation tillage include no-till, ridge-till, and mulch-till (MWPS, 1992), and the terms used to describe the primary tillage systems are *conservation, reduced tillage*, and *minimum tillage*. No-till involves no soil disturbance during the cropping sequence other than for planting, which uses specialized no-till planters. Ridge-till does not disturb the soil from harvest to planting, and crops are grown on ridges that are formed annually. A planter equipped with sweeps, disk row cleaners, coulters, or horizontal disks, is used. The planter removes 1 to 2 inches of soil, surface residue, and weeds from the row and leaves a residue-free strip of moist soil on top of the ridge (~3 inches higher than furrows) where seed is planted. Mulch-till refers to conservation tillage other than no-till or ridge-till and is usually performed with a chisel plow, blade plow, rod weeder, disk, or field cultivator, which leaves at least 30% residue on the soil surface for erosion control (MWPS, 1992).

Conventional and conservation tillage systems can also be characterized by the amount of soil pulverization, inversion, and plant residue cover associated with each type of tillage (Table 3-1). The objective of tillage is to prepare a crop seed bed, and weed control effects are secondary. However, each tillage type both directly and indirectly impacts weed management. Tillage kills perennial weeds by physically damaging the vegetative reproductive parts that can accelerate microbial attacks on the plants. Tillage operations can also leave reproductive organs on the soil surface, exposing them to freezing and/or drying conditions, and repeated cultivation can weaken the plant through carbohydrate depletion. Reductions in tillage can increase perennial weed levels (Table 3-2). Both simple and creeping perennial weed populations can be greater with reduced tillage.

TABLE 3-1. Effect of Tillage Systems on Soil Pulverization, Inversion, and Plant Residue Cover

		Soil		Residue Cover (%)		
Tillage System	Tillage Operation	Pulverization Rank[a]	Inversion Rank[a]	After Corn	After Soybean	After Small Grain
Conventional	Moldboard Plow, disk twice	1	1	5	2	5
Conservation	chisel plow,[b] field cultivate	2	2	30	10	30
No-tillage	—	3	3	80	60	80

[a]1 = highest; 3 = lowest
[b]Chisel points
Adapted from D. R. Griffith, J. V. Mannering, and J. C. Box. 1986. Soil and moisture management with reduced tillage. In *No-Tillage and Surface Tillage Agriculture*, ed. by M. A. Srauge and G. B. Triplett, pp. 19–57. John Wiley & Sons, Inc., New York.

Tillage, especially soil inversion, buries weed seed and places seed in a less favorable germination environment. Seed burial can reduce the weed population the year after heavy seed production by uncontrolled weeds. However, the reservoir of dormant buried weed seed serves as a source of continuing future weed problems, which argues against the need for heavy tillage with moldboard plows as a primary weed management tool. Reductions in tillage levels have reduced the utility of many herbicides that require physical soil incorporation for activity. Plant residue levels (Table 3-1) present in various tillage systems also affect weed control, as the residue can intercept soil-applied herbicide and reduce the amount of herbicide reaching the soil (Figure 3-1). Herbicides on plant residue may be washed to the soil with

TABLE 3-2. Effect of Tillage System on Perennial Weed Populations[a]

	Weeds per Acre		
Tillage Treatment	Hemp Dogbane	Common Milkweed	Canada Thistle
Moldboard plow	880	0	0
Chisel plow	925	3	3
No-till	1850	34	5

[a]Population counts taken after 4 years of treatment with the tillages. The counts are the average of those from continuous corn and corn-soybean rotation patterns.
From R. S. Fawcett, 1982. Can you control weeds with reduced tillage? *Proc. 34th Iowa Fertilizer and Agricultural Chemical Dealers Conference*. Iowa State University Coop. Ext. Bul. CE-1720. Ames.

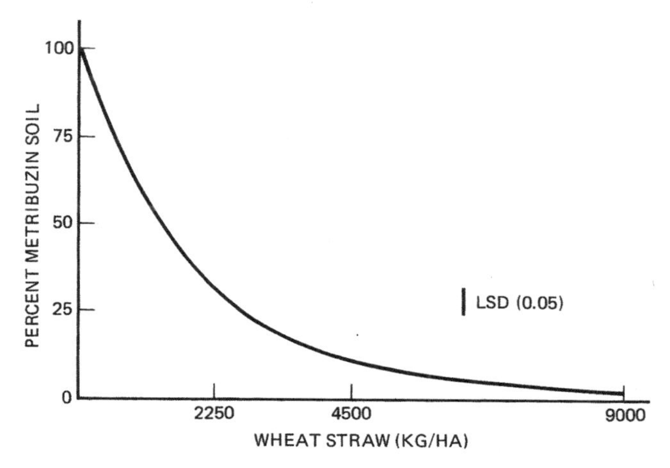

Figure 3-1. Influence of wheat straw on the amount of metribuzin reaching the soil surface. (P. A. Banks and E. L. Robinson, 1982, *Weed Sci.* **30**:164, published with permission.)

subsequent rain; however, herbicides that are volatile or subject to photodegradation may be lost from plant residue before rain occurs. Even with rain, herbicides can remain trapped on the plant residue (Figure 3-2), and weed control can suffer as a result of reduced herbicide availability.

Secondary tillage can also contribute to weed suppression. A large portion of the potential weed population can germinate and emerge before crop planting if planting is delayed. Reworking the ground with harrows just before planting will kill these weeds. Such weeds can also be killed by herbicides. In stale seedbed culture and various reduced and no-till systems, the initial flush of weeds is often allowed to

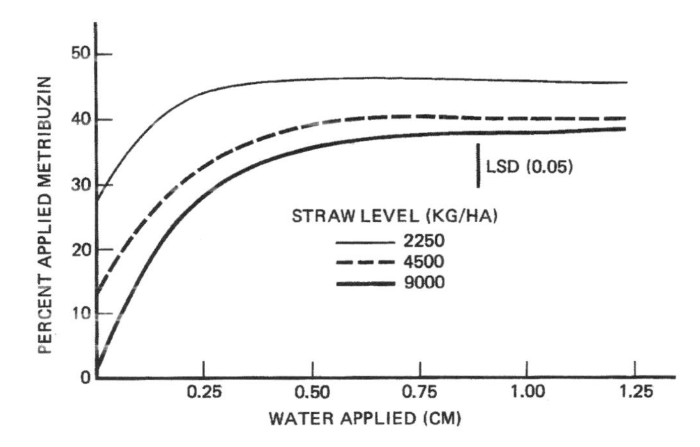

Figure 3-2. Metribuzin reaching the soil surface as influenced by wheat straw and irrigation. (P. A. Banks and E. L. Robinson, 1982, *Weed Sci.* **30**:164, published with permission.)

germinate and is then killed with herbicide without further soil disturbance other than minimal disturbance at crop planting. If the need for tillage immediately before planting is eliminated buried weed seeds are not brought near the soil surface, where another flush of germination can occur, and the weed population immediately after planting is reduced. Unfortunately, delaying planting may not be practical for farmers with large acreages to plant, and delayed planting can also decrease the potential crop yield. Stoller and Wax (1973) and Buhler et al. (1997) provide useful data on the periodicity of weed seed germination that can be useful to growers interested in timing planting and cultivation based on weed growth biology.

Row Crop Cultivation

The primary objective of cultivation is to control weeds, and primary and secondary tillage are aimed at preparing a suitable crop seedbed. The main purpose of growing crops such as corn and soybeans in rows is to allow mechanical weed control between the rows, and the original row widths were designed to allow the passage of draft animals without damaging the crop. Herbicides and machine cultivation has allowed a decrease in row widths. Annual weeds are buried or uprooted, and a great variety of mechanical devices kill weeds by these methods. Tillage with properly arranged and adjusted tools when weed seedlings are small (< 2.5 inches) can consistently control 100% of the weeds in the tilled area. In a crop seeded in 30-inch rows and cultivated to leave a 3-inch band untilled in the rows, tillage controls 90% of the weeds. In a field with a weed population of 1,000,000 per acre, even a level of weed control as high as 90% is of little value by itself, because the 100,000 weeds remaining per acre are more than enough to destroy the crop. However, cultivation is an extremely important part of an IWM system that includes other practices to deal with the weeds remaining in the rows. Weeds must be small for consistent control by tillage; therefore, in some cases tillage must be repeated several times per season.

Seedlings of perennial weeds are easily controlled by cultivation, but older plants can often escape. Larger weeds can also clog cultivation equipment. Effective cultivation needs dry soil both at the surface and below the depth of cultivation, as dry soil promotes desiccation of uprooted weeds and proper soil moisture avoids damage to the soil structure. Cultivation of wet soil can lead to soil compaction, and weeds (especially perennials) simply reroot after being moved. The same problem can occur if rainfall happens immediately after cultivation.

The criteria of optimal weed size and soil moisture are limitations to the use of cultivation for weed control. These can be especially critical if cultivation is used as the sole means of weed control, as untimely rains can delay the use of cultivation and result in large, uncontrollable weeds that greatly interfere with crop growth. Another concern is the cost of fuel. When fuel and farm labor were less expensive than they are today, mechanical weed control was more cost-effective.

Cultivation is often used as a complement to herbicides. Cultivation can control the weeds escaping the herbicide and extend the longevity of the weed suppression; with herbicides like the dinitroanilines, the inhibition of root growth can be increased by cultivation.

Using herbicides with cultivation helps to overcome another disadvantages of relying only on cultivation for a weed control program. A herbicide can be used to control weeds directly in the crop row, as these weeds are generally inaccessible to the mechanical action of cultivation. This technique has also been used for between-row cultivation after herbicide banding in the crop row. Exceptions to the practice are the use of a rotary hoe to control small weeds in corn and soybeans and the use of a retracting tree weeder in orchards and nurseries. These rapidly moving implements uproot small weed seedlings in the rows of small corn and soybean seedlings or in a tree row. A rotary hoe operates at a shallow depth, which uproots the small weed seedlings but does not kill the deeper-rooted crop plants. Rotary hoes work best when the crop is slightly wilted, as fully turgid crops are more susceptible to breakage.

Some types of cultivation are not considered highly effective for weed control within the crop row. However, with the use of specially designed cultivators and careful operation, some measure of weed suppression in the crop row can be obtained by burying small seedlings with soil thrown into the row by cultivation (as described for ridge-till).

Reduced tillage systems can limit the use of cultivation for weed control. The soil surface in such systems is often hard, which restricts the penetration of cultivation tools, and plant residues on the soil surface can clog the cultivator. In no-tillage systems there is no use of cultivation, so there is, by necessity, a higher reliance on herbicides or other means of weed control.

In summary, cultivation is an excellent component of a weed control program. It can provide effective weed control, especially when combined with other available tools.

Hand Weeding

Pulling out unwanted weeds by hand or by hoeing is the oldest method of selective weed control; it remains a very safe and effective method against most weeds in most crops. The major disadvantages are the expense and increased potential for crop injury if such methods are performed carelessly. Arguably, a major social benefit of modern weed control methods is the release of workers from the drudgery of manual weed control. Hand labor is used as part of an IWM system in high-value crops to bring the level of weed control to 100% after most weeds have been killed by less expensive methods. However, the high cost of labor makes hand weeding economically unattractive and, in many cases, impossible for most farmers.

In any IWM system, the main control components are naturally directed against major species of the weed flora. Quite commonly, minor and insignificant species are not controlled and, in the absence of competition from major species, can proliferate and in time become dominant in the weedy flora. Hand labor is usually the most effective method for destroying these few surviving weeds and preventing a future problem (often called roguing). Hand removal of weeds missed by initial control methods is an underutilized tool in situations where weeds are developing resistance to herbicides.

Mowing

Mowing can effectively prevent seed formation on tall annual and perennial weeds, deplete food reserves of the vegetative reproductive organs of perennial weeds, and favor competitive crops adapted to mowing. Unfortunately, mowing can also favor weeds that grow and reproduce below the cutting height. Repeated mowing can cause a shift in the dominant biotype of a weed species, from an upright growing form to a more prostrate form.

Effective prevention of seed formation by mowing requires cutting weeds before flower formation. Pollination, fertilization, and production of viable weed seed occur so soon after flower appearance in a number of weed species that mowing flowering weeds is often a cosmetic solution.

Mowing kills existing shoot growth, but mowed plants can produce additional flushes of shoot material. The previously dormant lateral buds may start to grow, with more new stems developing. Thus, the stand may appear to thicken; however, this is desirable if the existing stand is mowed repeatedly. The new stems grow at the expense of the below-ground stored food, and repeated cutting hastens food depletion and the death of the plant.

Some annual weeds sprout new stems below the mower cut. This growth can often be controlled by cutting rather high at the first mowing and sufficiently lower at the second mowing to cut off sprouted stems. By the second mowing, the stem is often hard and woody and cannot develop new sprouts below the cut. This procedure is effective on bitter sneezeweed, horseweed, and many other weeds.

A single mowing will often not prevent seed production. The new stems produced below the initial cut will flower and form seeds. Two or three mowings will be needed to ensure prevention of seed formation.

Repeated mowing not only prevents the seed production of perennial weeds, but may also starve the underground parts. Cutting the leaves removes the food (photosynthate)-producing organs. Moreover, the regrowth stimulated by the cutting (as described earlier) draws on the stored food supply. Even following these guidelines, it may take 2 or more years of such treatment to completely kill a perennial weed stand. A small amount of reproductive organs left after 1 year of mowing can easily reestablish the weed problem.

The best time to begin mowing is usually when the underground root reserves are at a low level, between full leaf development and flower appearance (Figure 3-3).

These principles of timing for mowing can also be applied to the use of herbicides for perennial weed control.

Mowing for harvest and maintenance of hay, pasture, turf, and cover crops helps eliminate tall-growing weeds. Crop plants grown in these situations are adapted to mowing, and cutting favors them over nonadapted weeds. As is generally the case, the combination of mowing and its use with a competitive crop is more effective for weed suppression than either element alone. Weeds that are also adapted to mowing, forming rosettes or mats, or growing close to the sod, are not controlled by mowing. This is why mowing does not kill established plants of weeds such as common dandelion, buckhorn plantain, violets, bermudagrass, crabgrass, and goosegrass.

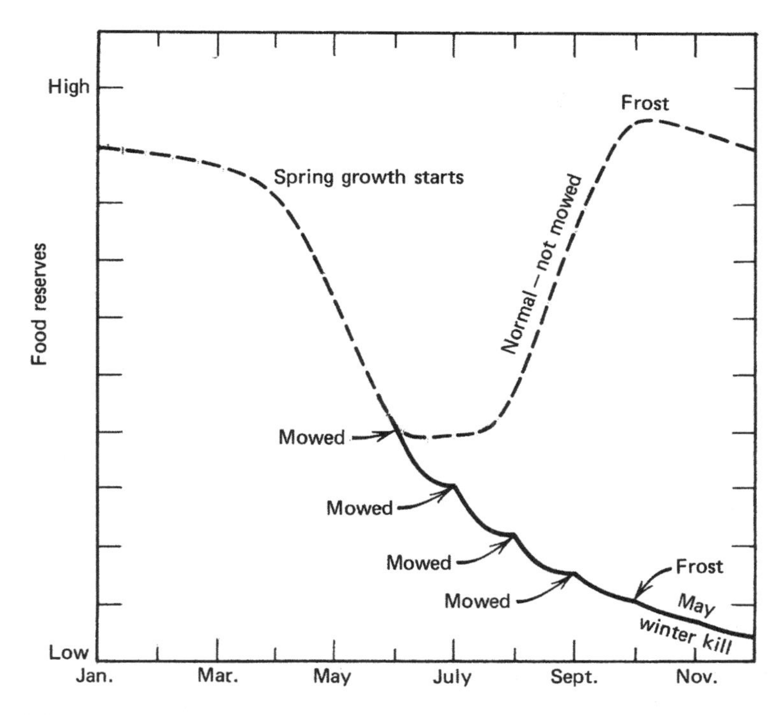

Figure 3-3. Food reserves of a perennial unmowed plant as compared with reserves of a repeatedly mowed plant.

Mulches

Mulches stop weed growth by restricting the penetration of sunlight to the soil surface, and in the case of surface mulches of cover crops have the potential to release inhibitory (allelopathic) chemicals into the soil environment that inhibit weed seedling growth. Many weed seeds require light to stimulate germination, so mulches reduce the germination of such seeds. Seedlings that do not require light can germinate, but if light is restricted the seedlings emerging from the soil are killed through starvation by lack of photosynthesis or, if allelopathic mulches are present, may die because of chemical inhibition of growth. Perennial weeds are not well controlled by most types of mulches as they have sufficient plant reserves to begin growth and to emerge through the mulch in the absence of light and in the presence of most allelochemicals.

Mulches can be either nonliving or living material of enough density to restrict light penetration. Nonliving materials used for mulches include wood chips and bark chips, shredded bark, pine needles, sawdust, straw, leaves, grass clippings, paper, cardboard, compost, polyethylene in various colors, and stones. The cost of these materials and the labor intensity of their use restricts widespread applications of these mulches to high-value and low-acreage crops and in the urban landscape. Applications for polyethylene mulches are mainly in field production of vegetables, flowers, and small fruit such as strawberries, and the other materials are mainly used in landscaping applications for urban and industrial sites with ornamentals and in home gardens. Polyethylene can influence the microclimate of the soil dramatically (temperature and moisture) and weeds can be a problem along the edges of the mulch and in the holes where plants are growing.

There has been considerable research investigating the use of living plants (living mulch or cover crops) that are suppressed or killed with herbicides and then used as mulch prior to crop emergence in large acreage crops. An example is the use of *crownvetch (Coronilla varia* L.) as living mulch in corn. Herbicides are used to suppress, but not kill, the crownvetch to avoid competition with the corn. Crownvetch recovers later in the season, and the living mulch is maintained. The dense cover provided by crownvetch helps reduce weed growth. Another example is the use of subterranean clover (*Trifolium subterraneum*) that senesces after the main crop has become established (Enache and Ilnicki, 1990). The use of legumes such as crownvetch as living mulches also has potential for supplying some of the nitrogen requirement of the companion crop. Moreover, mulch protects the soil from erosion. A major problem with living mulches is potential competition with the companion crop, which may occur when insufficient mulch suppression is obtained with the herbicide. A fine line is usually drawn between the quantity of herbicide needed to manage the living mulch and the amount that will not kill it.

Another mulch system for larger acreage crops that eliminates the competition problem is the use of a killed (by herbicides, sometimes by mowing and/or rolling) cover crop of cereal grains, other grasses, and legume and brassica species, as mulch or green manure (Creamer, et al., 1995). The cover crop is planted in the fall and then killed in the spring prior to planting the crop, and the aboveground biomass is left on the soil surface. Wheat, winter rye, oats, rye grass, and some legumes such as red clover and hairy vetch have shown promise for use as cover crop mulches (Figure 3-4). The cover crop biomass acts to suppress weed growth as a physical barrier and through exudation of allelopathic chemicals. Cover crops can greatly reduce weed growth and density (Masiunas et al., 1995; Barnes and Putnam, 1983). The amount of biomass present is important, as are the complete coverage of the soil surface and the timing of cover crop kill prior to crop planting (Smeda and Weller, 1996). Other functions of the cover crop include nutrient capture and recycling, breaking down plow pans, water retention, and serving as a source of organic matter when the cover crop is plowed into the soil prior to planting.

Research with a rapeseed (*Brassica napus*) green manure in potatoes (Boydston and Hang, 1995) and crimson clover (*Trifolium incarnatum*) in sweet corn (Dyck et al., 1995) resulted in improved crop growth. The beneficial effects were attributed to improved nutrient availability, reduced soil pathogens, and weed control resulting from allelopathic influences. Problems associated with cover and green manure crops include inconsistent biomass production on a year-to-year basis, and hence inconsistent weed control. In years with dry winters or in dry climatic areas, the cover crop can deplete soil moisture. The growth of some crops planted into the allelopathic mulches can be inhibited because of the presence of allelochemicals and because the

Figure 3-4. Comparison of a rye cover crop mulch versus no cover on weed control in tomatoes. *Left*: Numerous weeds and low-vigor tomatoes with no rye mulch; *Right*: No weeds present with a rye mulch and vigorous tomatoes.

soil temperature and moisture levels may not be optimal for rapid growth. The cover crop can contribute to increased soil pathogen and insect populations, and some immobilization of nutrients can occur during the initial phases of cover crop decomposition, all of which can result in reduced crop growth. The exact contribution of the physical and allelopathic effects of such mulches is hard to determine, but regardless of these problems, in many experiments, season-long weed suppression with good crop growth and yields was accomplished without the need for in-season herbicide applications.

Another consideration in using any type of mulch for weed control is that the soil covering can restrict the use of other weed control methods. It is unlikely that all weeds can be controlled by the mulch, and the mulch can interfere with cultivation, hoeing, mowing, and herbicide applications reaching the soil (Figure 3-1). Thick layers of organic mulches around plants can encourage rot at the base of the plant stems, and certain soil insects can become a problem. Potential benefits of mulches, beyond weed suppression, include soil moisture conservation, lower soil temperatures with organic mulches (although higher temperatures can occur under polyethylene mulches), protection of the soil from erosion, and added organic matter to the soil. The relative advantages and disadvantages of a particular mulch must be evaluated for each situation. Wetter and cooler soil can delay spring planting and slow crop development, but moisture conservation may help to avoid later plant drought stress. Warmer soil and a more uniform moisture supply with polyethylene mulch can improve the growth and yield of vegetables such as tomatoes and peppers. The key consideration is how well a particular mulch fits into the overall weed control objectives in terms of efficiency and cost.

Burning

Fire can be used to remove undesirable plants from ditch banks, roadsides, and other waste areas, to remove undesirable underbrush and broadleaf species in conifer forests, and for annual weed control in some row crops. Burning must be repeated at frequent intervals if it is to control most perennial weeds. In alfalfa and western mint, burning can control weeds, diseases, and some insects. Environmental air quality laws may restrict burning as a weed control tactic in the future.

In waste areas, if vegetation is green, a preliminary searing will usually dry the plants enough so that they will burn by their own heat 10 to 14 days later.

Proper burning techniques can favor conifer trees over hardwood species in forestry. This *controlled burning* can also remove undesirable underbrush if done at regular intervals. Controlled burning reduces the hazard of uncontrolled forest fires.

Flaming has been used most successfully for selective weed control in cotton. Special propane burners are used to direct a flame at the base of the cotton plants. The hard woody cotton stem escapes injury, but young weed seedlings are killed. Two passes are normally done a few days apart for best results. Proper adjustment and speed of operation are essential to avoid crop injury. Flaming is done in other crops as well, and specialized row crop, backpack, and shielded flamers are available. The technique is similar to row cultivation or the use of a directed spray herbicide application (see the later section "Chemical Weed Control") in that it requires a size difference between the weed and the crop for effective weed suppression and crop safety. Technique is important, and flaming the weeds early, while they are small, is most efficient. Efficient flaming conserves fuel and does not toast the weeds but results in a drooping and wilting of the weeds within a few hours. Many systems now incorporate a water shield, which sprays a thin layer of water over the crop plants with a flat fan nozzle to protect them from the flame heat. However, as fuel costs continue to increase, flaming is less economical as a technique for weed control.

Flooding

Flooding is used to control weeds in rice fields, as water-saturated soil limits oxygen availability, which prevents many seeds from germinating but does not inhibit rice seed germination. Aquatic plants can tolerate the flooded conditions and are not controlled by this technique. The common rice weeds—red rice, barnyardgrass and arrowhead—also tolerate flooding. Flooding has limitations for further use in weed control in rice production. Further reliance on flooding will increase rice production costs and reduce yields, and over time certain weeds will begin to adapt for survival against flooding. Flooding has little effect on weed seeds in the soil.

Perennial weeds can also be controlled by prolonged (3 to 8 weeks) flooding at a depth of 6 to 10 inches and has been used to control established perennials such as silverleaf nightshade, camelthorn, and Russian knapweed in the western United States. Flooding for perennial control requires a good water source and is expensive because it requires creating dikes and maintaining the water level for prolonged periods. These requirements limit the wider use of flooding for perennial weed control. Moreover, some perennial vegetative reproductive buds will enter dormancy as a result of the flooding and are not killed.

CULTURAL PRACTICES

Crop selection, rotation, variety selection, planting date, plant population and spacing, plus fertility and irrigation are all cultural practices that affect weed management. Farmers should keep in mind that cultural practices will impact weed interference and should always consider how effectively the methods employed can minimize weeds.

Crop Selection

Selection of a crop determines strategies for the subsequent battle with weeds. Crop selection will determine the level of weed control needed for efficient crop production and, in many cases, which weeds will be most competitive. Some crops by nature are not competitive, such as many vegetables, whereas others tend to be more competitive, such as small grains. Weeds are opportunistic and occupy ecological niches not utilized by the crop (Cardina et al., 1999); the farmer's practices must reduce unused niches. The farmer must realize that the crop grown will determine the level of weed control needed for efficient production and, in many cases, which weeds will be most troublesome. The crop grown will also determine available weed control options (cultural, mechanical, biological, and herbicide) and the degree of integration necessary to effectively manage weeds. The potential monetary return from a particular crop will also determine the economics of weed control practices. Unfortunately, crop selection can rarely be made solely from a weed management perspective. Climate, soil adaptability, history, market availability, and the potential economic return are all factors that must be considered by a farmer in deciding on a crop. If a crop is chosen that is inherently noncompetitive with weeds, the farmer has to realize that extra effort must be employed to combat weeds if acceptable yields are to be obtained.

Crop characteristics that have been shown to be most important in helping crops compete with weeds include rapid germination and root development, early aboveground growth and vigor, rapid establishment of leaf area and canopy, development and duration of a large leaf area, and greater plant height (Calloway, 1992; Challaiah et al., 1986). All these characteristics allow the crop to establish dominance in the field and minimize the ability of weeds to compete for essential growth resources. Rapid closure of the crop canopy over weeds decreases sunlight and directly limits weed growth, limited light on the soil surface can reduce subsequent weed seed germination, and limited weed growth will reduce weed seed flowering and seed production. The effect of light levels on growth of giant foxtail is illustrated in Figure 3-5; depending on the companion crop, interception of light will vary.

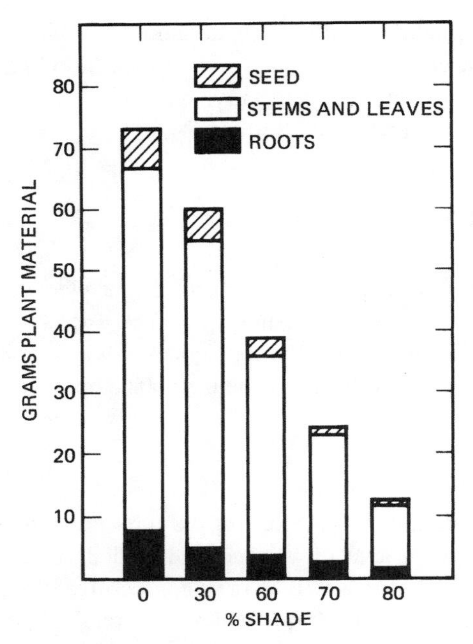

Figure 3-5. Effect of shade on weight of roots, stems, leaves, and seeds of giant foxtail. (E. L. Knake, 1972, *Weed Sci.* **20**:588, published with permission.)

Practices such as creating narrow rows and increased crop density promote early crop canopy closure and help maximize the effect of crop competition. These practices can be used in less competitive dry bean crops to reduce weed influences (Malik et al., 1993). Another example is soybean cultivars that have been genetically engineered to be resistant to glyphosate herbicide. The common practice is to drill the soybeans in narrow rows, allow crop and weed emergence and growth for 2 to 3 weeks, and then apply glyphosate to the field to kill the weeds. As the weeds die, the soybeans continue to grow and crop canopy closure reduces or eliminates subsequent weed competition. Conversely, conditions such as poor crop growth, planter skips, and wide row spacing, which delay crop canopy closure, will make weed control, especially late in the season, more difficult and costly. A good example is the increased difficulty in controlling grass weeds in fields of corn grown for seed as compared with corn grown for feed. The inbreds grown for seed production grow slowly, and their final height is much less than that of hybrid corn. They are planted in wide rows to allow passage of crews for detassling, and other hand labor operations.

The greatest amount of weed suppression due to crop competition occurs when a dense perennial sod, turf, or hay crop is grown. The heavy perennial growth, combined with cutting for harvest, can greatly reduce weed seedling establishment, seed production, and perennial growth. Because of the survival of dormant weed seed in the soil, a perennial crop will rarely eliminate a weed from an area. However, raising a perennial crop or other dense cover crop, especially for several years in succession, can reduce a problem weed population to manageable levels. Examples of crops useful for weed suppression include alfalfa, buckwheat, sudangrass, and densely planted small grains.

Crop Rotation

Crop rotations help prevent the buildup of weeds adapted to a particular cropping system. Certain weeds are more common in some crops than others. Pigweed, lamb'squarter, common ragweed, velvetleaf, cocklebur, foxtail species, and crabgrass are found in summer-cultivated crops such as corn. Mustards, wild oat, wild garlic, chickweed, and henbit are associated with fall-sown small grains. Pastures often contain perennial weeds such as ironweed and thistles. Changing crops changes the cultural conditions (planting date, crop competition, fertility, etc.) that a weed must tolerate. Rotating crops also often means that a different set of management tools (especially herbicides) will be used. The overall success of crop rotation in managing weeds depends on the ability to control the weeds in each crop grown in the rotation. Rotation will prevent a weed species from becoming dominant in a field but will also maintain a diversity of weed species in the same area.

Crop rotation historically was very important for managing weed problems. Today, rotation is used more for managing diseases and insects than weeds. Rotation requires the farmer to have additional knowledge and to use additional equipment to manage the various rotational crops. Even with an abundant supply of fertilizers and diverse herbicides that make it possible to minimize the need of crop rotation for weed control, there are still sound reasons to rotate crops for environmental and pest management reasons. For example, corn rotated with soybeans consistently yields more than corn grown continuously in the same field. Rotation of vegetable crops is important to avoid buildup of soil diseases that reduce crop yields. However, rotation is not an option with long-term perennials such as orchards, forest trees, nurseries, and perennial forages. Some of the benefits of rotation can be retained in monoculture cropping systems by the selection of a variety of herbicides, especially those differing in mode of action, and the use of various cultural practices, especially cultivation. Herbicide diversity and cultivation help prevent the development of resistant weed populations that are adapted to an unchanging herbicide program and crop.

Problems tend to arise when farmers do not rotate their crops and pest management strategies in an integrated manner. For example, in the past the corn–soybean rotation avoided the buildup of corn rootworm in the corn cycle, as rotation for 1 year to soybean broke the insect life cycle. However, the insect has adapted to these cropping strategies to be able to survive on soybean and has once again become a major corn problem. Similar examples are available in weed control. With the availability of a variety of glyphosate-resistant crops, there will be a tendency to continually use glyphosate for weed control even as we rotate crops. This is poor management, and it will become necessary to rotate herbicide-resistant crops with nonresistant crops to avoid a buildup of weeds not well controlled by glyphosate. The same holds true for herbicides that inhibit branch chain amino acids and can be used in many of our major acreage crops. There is a law of nature that holds true for agriculture that one should always remember: "Mother Nature deplores a vacuum." Repeated use of any successful pest management practice without appropriate integration with a variety of other tactics and rotation over time will result in that tactic's selecting for its own extinction. There are many good examples of this phenomenon in weed science, and they are called *herbicide-resistant weeds* (Chapter 18).

Crop Varieties

Development of new higher-yielding crop varieties is generally done under conditions of minimal weed, insect, and disease interference. Normal variety development schemes yield little information on the differential competitive ability of cultivars. However, more vigorous, faster-growing, and taller crop varieties are likely to be better competitors. Differential competitive ability among soybean varieties has been amply demonstrated, as shown in Table 3-3. Ennis (1976) estimated that selection of competitive soybean cultivars could provide up to 80% control of selected weeds in the crop. Suppression of less than 50% would not be sufficient to eliminate competitive yield losses but would contribute to overall control when other weed control practices were also used. Generally, growers select cultivars with the highest yield potential, and weed control is implemented to allow expression of the true yield potential.

In the future, through breeding and genetic engineering, there will be greater emphasis on creating crop varieties that are more competitive with weeds by emerging earlier, growing faster, and possibly being allelopathic (Weller, 2001; Gressel, 2000). These issues are addressed in more detail in Chapter 30; however, as our knowledge of weed biology and ecology expands and general knowledge of genes associated with competitiveness are identified, there will be substantial progress in this area (Pester et al., 1999). The importance of creating more competitive crops involves the addition of new tools, beside herbicides and cultivation, to reduce the negative effects of weeds

TABLE 3-3. Yield Reductions in Selected Soybean Varieties Due to Johnsongrass or Cocklebur Competition

	Competing Weed	
	Johnsongrass	Cocklebur
Soybean Variety	(% soybean yield reduction)	
Davis	34	56
Lee	41	67
Semmes	23	53
Bragg	24	57
Jackson	30	67
Hardee	23	26

From C. G. McWhorter and E. E. Hartwig. 1972. Competition of johnsongrass and cocklebur with six soybean varieties, *Weed Sci.* **20**:56–59, published with permission.

on our ability to grow food and to safeguard our environment from excessive use of chemical crop protectants.

Planting Date

The trend in crop production is for earlier planting to increase yields. The resulting longer exposure to sunlight is primarily responsible for the higher yields associated with this practice. Early planting can establish adapted crops before weeds emerge and provide the crop with a competitive edge.

There are some disadvantages to early planting for weed control. Early planting means soil-applied herbicides may have to persist longer in the environment for the most effective weed control. It also eliminates the cultivation done just before later planting, which often destroys the first flush of germinating weed seedlings.

Unfortunately, any advantages gained for weed control by delayed planting are often outweighed by decreased crop yield potential, especially for agronomic crops. Certain short-season vegetable crops can be delayed to avoid early weed flushes; however, even this may not be possible when early planting is dictated by market demands and price premiums. Therefore, the cost of using delayed planting as a weed control strategy must be weighed for each crop-weed-environment situation.

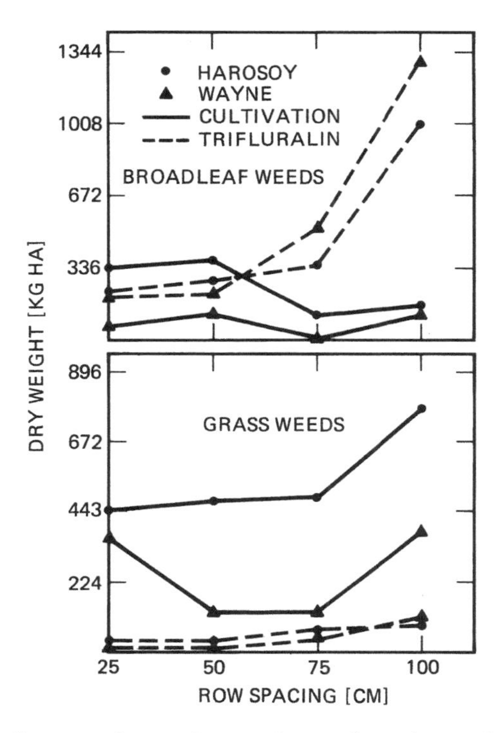

Figure 3-6. Effects of row spacing, soybean variety, and weed control method on weight of weeds. (L. M. Wax and J. W. Pendleton, 1968, *Weed Sci.* **6**:462, published with permission.)

TABLE 3-4. Chemical Composition of Corn and Corn Weeds in September

Plant	Growth Stage	Mean Percentage Composition, Air-Dry Basis				
		Nitrogen	Phosphorus	Potassium	Calcium	Magnesium
Corn	Total plant, early milk stage	1.20	0.21	1.19	0.18	0.15
Pigweed	Green plants, 25–50% seeds ripe	2.61	0.40	3.86	1.63	0.44
Lambsquarters	Green plants, 25–50% seeds ripe	2.59	0.37	4.34	1.46	0.54
Smartweed	Green plants, 25–50% seeds ripe	1.81	0.31	2.77	0.88	0.56
Purslane	Lush plants, seeds partly ripe	2.39	0.30	7.31	1.51	0.64
Galinsoga	Lush plants, seeds partly ripe	2.70	0.34	4.81	2.41	0.50
Ragweed	Green plants, seeds partly ripe	2.43	0.32	3.06	1.38	0.29
Crabgrass	After bloom, seeds not ripe	2.00	0.36	3.48	0.27	0.54
Average for weeds		2.36	0.34	4.23	1.36	0.50

From J. Vengris et al. 1953. *Agron. J.* **45**:213.

Plant Population and Spacing

Historically, crops were planted in rows spaced widely enough to allow passage of draft animals pulling cultivation equipment to control weeds. Development of herbicides removed this constraint and allowed adoption of narrow-row production systems that can produce higher crop yields and permit higher plant densities to aid in weed suppression. In most close-row systems, herbicides play an important role in early-season weed control so the crop can gain a competitive advantage (Figure 3-6). Cultivation becomes problematic in close-row culture and is difficult to achieve without some crop damage, even with the use of specialized cultivation equipment. The major advantage of close-row spacing and high plant density is the more rapid establishment of a closed-crop canopy and the resulting reduction in light reaching the soil surface to support weed growth.

Fertility and Irrigation

Crops and weeds generally require and will compete for the same nutrients. Changes in soil fertility levels have a great influence on the competitive interactions between weeds and crops. Weeds respond in a positive manner to increasing nutrient levels, which allow them to better compete with the crop for other necessary growth factors. Nitrogen is generally the nutrient of greatest concern in weed competition. Increasing the nitrogen supply can increase crop yields, have no effect, or reduce crop yields when weeds are present. Final crop yield with increased nitrogen and weed competition is rarely as great as yield without competition. Nitrogen has also been shown to stimulate the germination of dormant weed seeds (Cavers and Benoit, 1989), which can increase weed density and the level of competition in the field. Using extra fertilizer is not generally an efficient way to avoid crop losses resulting from weed competition. Some weeds use more fertilizer than needed for growth. These "luxury consumers" may actually benefit from the fertilizer more than the crop (Table 3-4). One method to reduce some of these interactions is to band the fertilizer near the crop row to preferentially place the crop at a competitive advantage over weeds in accessing the nutrients.

TABLE 3-5. Effect of Quackgrass Interference on Yield of Irrigated and Nonirrigated Soybeans

Soybean Treatment		
Irrigation	Quackgrass	Soybean yield (ton/acre)
No	No	1.3
Yes	No	1.2
No	Yes	0.7
Yes	Yes	1.0

From F. L. Young, D. L. Wyse, and R. L. Jones. 1983. Effect of irrigation on quackgrass (*Agropyron repens*) interference in soybeans. (*Weed Sci.* **31**:720–727, published with permission.)

Like fertilizer, added moisture, through irrigation, can overcome a portion of crop yield loss due to weed competition, but full crop yield potential will unlikely be reached (Table 3-5). Also like fertilizer, additional water can favor weeds because many weeds have high water requirements—that is, higher amounts of water used per plant weight.

BIOLOGICAL CONTROL

Biological control of weeds involves the use of any organism, or management practice using an organism, to reduce or eliminate the potential detrimental effects of weed populations. Classical biological control is associated with the use of insects, pathogens, herbivores, or parasites that naturally attack weeds; however, it can be expanded to include the previously mentioned uses of competitive crops, cover crops, living mulches, green manures, and any organisms associated with these practices that can reduce weed growth. There are two common approaches used in the introduction of classical biological control agents into a system. The first approach is the *inoculative* or *classic* method whereby an organism is released, reproduces, and disperses on its own in habitats with the target weed. When this approach is successful, there are no recurring weed control costs. The second approach is called *inundative* or *augmentative*, using an agent sometimes referred to as a bioherbicide or a mycoherbicide. In this method, the weed is controlled in the area where an abundant supply of the agent (usually a fungal pathogen) is applied. This method is unlike the inoculative approach, in that the biocontrol agent is applied repeatedly and does not remain in an active form in the environment over time (Wapshere et al., 1989). Organisms useful as mycoherbicides must be easily cultured in the laboratory, produce high amounts of inoculum, be highly virulent yet specific, nontoxic to nontarget plants and animals, and formulated to be easily applied, effective, and consistent in activity.

Although there is much public awareness about the concept of biological control, there are currently no examples in which biocontrol agents have been successfully used to control a spectrum of weeds in an IWM cropping system. The major successes for biological weed control have occurred in relatively undisturbed areas such as rangeland and aquatic weed infestations. Two major characteristics of classical biological control are that they affect one weed species only and that the effect progresses slowly. Because of these characteristics, biological control is not well suited for weed control in crops because cropland almost always contains a complex of weeds. A particular crop may have two dozen weeds commonly associated with its culture, and these weeds must be controlled rapidly to protect the crop from permanent injury. Other tactics must be used in association with the bioagent to control the weeds it does not attack. A biological control program may be best suited for management of a single weed species that is poorly controlled by other available management techniques. Unfortunately, the profitability of a biological control agent that attacks only a single species, perhaps only in one crop, is generally low. Commercial companies are generally unwilling to undertake the costs of discovery, research, development, production, and marketing for such a product. In most cases in which commercial products have been developed, they are of the inundative type. Most of the work and expense required for discovery and research in biological weed control is done by the federal government and universities.

The action of biological agents on weeds can be slow. Reductions in weed growth may be too slow to avoid crop yield loss resulting from weed competition if the biological approach is not complemented by other tactics in an IWM approach. This is primarily a problem when the objective of the control strategy is to quickly reduce the weed population below an economic threshold. A longer-term objective can be the establishment of an ecological balance between the biological control agent and the weed. This approach is not constrained by the time factor inasmuch as the goal is to eventually prevent the weed population from exceeding a threshold. Long-term objectives are easier to adopt in rangeland and aquatic situations.

When an ecological balance is achieved with biological control, in effect, a permanent solution to the weed problem is an obvious benefit. However, a long-term balance between the biological control agent and the weed population means that some portion of the weed population will continue to survive. Survival of some weeds is necessary to ensure survival of the biological control agent. As discussed in Chapter 2, the presence of even a relatively few weeds in an annual crop may be enough to cause significant yield loss.

Perennial crops, rangeland, and aquatic areas represent relatively stable ecosystems if properly managed. Such stability can allow a buildup of biocontrol agent populations introduced to control problem weeds and improve the effectiveness of the biocontrol strategy. Annual crop production is the opposite of stability. Populations of biocontrol agents must be plastic in their ability to survive annual tillage operations, cultivations, crop rotations, and pesticide applications.

Economics favors biocontrol of weeds in rangelands and aquatic areas. These are often large acreages with difficult accessibility, on which little can be spent for weed control technology. Crop production, in contrast, represents a comparatively high economic investment for which the potential economic return from weed control promotes greater use of active weed management strategies.

Criteria for Biological Control Agents

The most important criterion for any potential biological control agent is that it attacks only the target weeds and no other plants. Fear of accidentally introducing new pests has diminished as our knowledge of biological control agents has grown. Plant-feeding insects are often very specific in their feeding preferences, and the danger of insects switching feeding preferences seems remote. However, as discussed in Chapter 2, the introduction of an alien species must be done carefully to avoid potential calamity resulting from its becoming an invasive species.

Although a successful biological control agent will attack only one type of plant, it should be remembered that just as definitions of a weed can differ, there can be different opinions on the worth of a plant. Some weeds were imported as ornamentals and are still used this way. A plant can be an important source of forage in one situation and a weed in others. The total economic impact of biological control agents must be evaluated before their introduction. It is difficult to reverse or contain biological control programs following their release.

An introduced inoculative biological control agent should reproduce quickly and build its population fast enough to effect weed control. The biological control agent should also be adapted to the new environment and be free of its own parasites, predators, and diseases to ensure a high survival rate of the agent. Inundative agents must be applied at high initial amounts but must also be able to quickly cause their effect on the weed under a variety of environmental conditions. The major concern in all biological approaches, from a practitioner's view, is their consistency in effecting weed control.

Types of Biological Weed Control Agents

Insects Plant-attacking insects are currently the most widely used biological control agents for weed control. They have a specific host range, can be mobile (which promotes their dispersion), and can destroy both vegetative and reproductive portions of weeds. Insect attacks can also predispose weeds to attack by other factors such as disease; in fact, research is investigating a combination approach of insects as disease vectors for biological control (Kennedy, 1999). The action of several biocontrol agents on a weed is often more effective than attack by only one agent.

The most outstanding example of biological weed control concerns the prickly pear or cactus (*Opuntia* spp.) in Australia. The prickly pear was originally planted for ornamental purposes, but it spread rapidly from 1839 to 1925, covering 60 million acres and threatening much of the cultivated land. In seeking control measures, research scientists found insects that attacked the prickly pear and no other plants. A moth borer (*Cactoblastis cactorum*) from Argentina was the most effective. It tunneled through stems, underground bulbs, and roots. The damage was even more effective after several bacterial rot organisms were accidentally introduced into the wounds caused by the insect.

The moth borer was released in 1926 and by 1931 had multiplied to such numbers that nearly all the prickly pears were destroyed by midseason. With little or no food supply, many of the moth borers starved. Several years followed when prickly pear increased, with a later increase in the moth borer. Several "waves" of each type of growth occurred before equilibrium or a "balance of nature" was reached. Thus, biological methods control rather than eradicate. This is especially true of weed species reproducing from seeds that may lie dormant in the soil for years.

There are many other examples of weeds, especially perennials, found in noncropland for which effective biological control agents have been employed: tansy ragwort (*Senecio jacobaea*), a poisonous biennial plant found in California and Oregon controlled by a complex of two insects, the cinnabar moth (*Tyria jacobaeae*) that feeds on foliage and the ragwort flea beetle larvae (*Longitarsus jacobaeae*) that feeds on the root crowns and stems of the plant; Saint-John's-wort (*Hypericum perforatum*), a poisonous plant controlled by leaf-eating beetles (*Chrysolina* spp.) (see

Figure 3-7); pamakani (*Eupatorium adenophorum*) in Hawaii controlled by a stem gallfly (*Procecidochares utilis*); rush skeletonweed (*Chondrilla juncea*) in Australia controlled by the gall mite (*Aceria chondrillae*); curse (*Clidemia hirta*) in Fiji controlled by a shoot-feeding thrip (*Liothrips urichi*); leafy spurge (*Euphorbia esula*) controlled by six different species of the root-boring flea beetle (*Aphthona* spp) and by the leafy spurge hawkmoth (*Hyles euphorbia*), which eats leaves and flowers, and a stem-boring beetle (*Obera erythrocephala*), which consumes the roots; Russian thistle (*Salsola iberica*) controlled by a moth (*Coleophora* spp); diffuse and spotted

Figure 3-7. Saint-John's-wort or Klamath weed, control by *Chrysolina quadrigemina* (beetles) at Blocksburg, California. *Top:* Photograph taken in 1946. The foreground shows weeds in heavy flower, whereas the rest of the field has just been killed by beetles. *Middle*: Portion of the same location taken in 1949 when heavy cover of grass had developed. *Bottom*: Photograph taken in 1966 showing degree of control that has persisted since 1949. Similar results were reported throughout the state. (C. B. Huffaker, University of California, Berkeley.)

knapweed (*Centaurea* spp.) controlled by a seedhead gall fly (*Urophora quadrifasciata* and *U. affinis*) and a root-boring beetle (*Sphenoptera jugoslavica*) along with nine other genera; Dyers woad (*Isatis tinctoria*) controlled by the Eurasian rust fungus (*Puccinia thlaspeos*) that causes root deterioration; and musk thistle (*Senecio jacobaea*) controlled by a weevil (*Rhinobyllus conicus*) that was released in 1997. Although this weevil does manage musk thistle, it also has begun to feed on five native thistles, which makes it an example of a biological agent that probably should not have been introduced inasmuch as it is not species specific.

For aquatic weeds, several biological control agents have been released. These include the South American flea beetle (*Agasicles hygrophilia*) for alligatorweed (*Alternanthera philoxeroides*); two weevils, *Neochetina eichhorniae* and *N. bruchi*, and a moth, *Sameodes albiguttalis*, for waterhyacinth (*Eichhornia crassipes*); two leaf-mining beetles (*Gallerucella calmariensis* and *G. pusilla*) that defoliate the plant; a root-mining weevil (*Hylobius transversovittatus*) for purple loosestrife (*Lythrum salicaria*); and the Australian weevil (*Oxyops vitiosa*) for melaleuca (*Melaleuca quinquenervia*) control in Florida. A milfoil-feeding weevil (*Euhrychiopsis lecontei*) shows promise for control of Eurasian watermilfoil (*Myriophyllum spicatum*), and a leaf-mining fly (*Hydrellia pakistanae*) is being tested for hydrilla (*Hydrilla verticillata*) control. Several useful Web sites for biological control of weeds are listed at the end of this chapter.

Pathogens The inoculum (classic) and inundative (mycoherbicide) methods are both used for employing plant pathogens, primarily fungi, for biological control of weeds. The mycoherbicide approach offers the best potential for extension of biological weed control into nontraditional disturbed areas and is being used commonly in citrus groves and rice fields.

An example of the classic approach is the use of the rust fungus (*Puccina chondrilla*) to control rush skeletonweed in areas of the western United States. The fungus was previously successful in reducing rush skeletonweed populations in Australia. The specificity of this biocontrol agent is illustrated by the difficulty in introducing it to the United States. Initial attempts to infect skeletonweed in the United States with the rust from Australia failed, but rust from Italy infected and spread through California and Oregon. The strain from Italy did not establish in Idaho; however, a second strain from Italy established on southern, but not northern, biotypes of skeletonweed in Idaho. Additional screening on infected U.S. plants was needed to find a rust that would infect rush skeleton weed in Washington State. Spores of the rust need six hours of dew during periods of darkness for infection to be successful. This requires aerial spreading of the spores in the arid western United States for quick dispersal during the brief environmental conditions favorable for infection.

Most of the biological control fungal agents being studied or commercially developed attack the developing plant and are generally applied to the plant foliage. Three commercial products have been registered at various times under the names DeVine, BioMal, and Collego. DeVine (*Phytophthora palmivora*) acts against the stem and roots of stranglevine (*Morrenia odorata*). BioMal (*Colleotrichum gloesporiodes* f. sp. *Malvae*) is registered for control of round leaf mallow (*Malva pusilla*) in small grains and lentils. Collego (*Colleotrichum gloesporiodes*) controls northern jointvetch (*Aeschynomene virginica*) in rice and soybean. It promotes anthracnose and results in plant lesions and necrosis. All of these biological control agents are used like herbicides and applied to the leaves of the target weed. Collego is sold as a two-component formulation of dry fungal spores plus a liquid spore-rehydrating agent that is mixed before use. Care must be taken to avoid fungicide use 1 week before and 3 weeks after Collego application. In addition, prior to Collego application, aerial application equipment is cleaned with a suspended charcoal solution to ensure removal of any contaminating pesticides from the sprayer. These precautions illustrate the difficulty of integrating a mycoherbicide into crops that require extensive fungicide use.

Extensive research is being conducted on many fungal pathogens and bacterial biocontrol agents, as summarized by Kennedy (1999). Many have shown good potential and in the future may become useful as additional biorational approaches to weed management in crops. Among the organisms being studied are *Colletrichum coccodes* for the control of eastern black nightshade (*Solanum ptycanthum*), *Phomopsis convolvulus* for control of field bindweed (*Convolvulus arvensis*), *Bipolaris sorghicola* for johnsongrass (*Sorghum halepense*), and *Sclerotinia sclerotiorum* for Canada thistle (*Cirsium canadensis*), spotted knapweed (*Centauria maculosas*), and dandelion (*Taraxacum officinale*).

Herbivores Grazing animals such as geese, goats, sheep, and cattle have been used selectively to control weeds in crops, pastures, and noncropland. Geese of the white Chinese breed have found specialized use for control of small seedling grass weeds in a number of broadleaf row crops, especially cotton. They have also been used in orchards, nurseries, and for some perennial crops. Additional control measures are needed to manage broadleaf weeds where the geese are used. Sheep and goats can be used to improve grazing conditions for cattle. They can remove weedy annual broadleaf species and brush, which allows improved range grass growth for cattle feed. There is little food competition between goats and cattle if sufficient shrubs and broadleaf plants are available for the goats. Sheep and goats can be used to help avoid *Senecio* poisoning in cattle. About 20 times as much *Senecio* on the basis of body weight is needed to poison sheep and goats as needed for a cow. Sheep and goats will readily graze *Senecio*, and stocking sheep and goats with cattle can help prevent cattle poisoning.

A number of possible animal management costs and efforts can be associated with the use of grazing animals for weed control. These include feeding for a balanced diet, fencing, herding, sheltering, protection from predators, and general care.

Aquatic weed eating animals and fish are promising for weed management in aquatic sites. The manatee or sea cow (*Trichecchus manatus*), a mammal, consumes large amounts of vegetation, but its use is limited to tropical areas and in many places it is endangered.

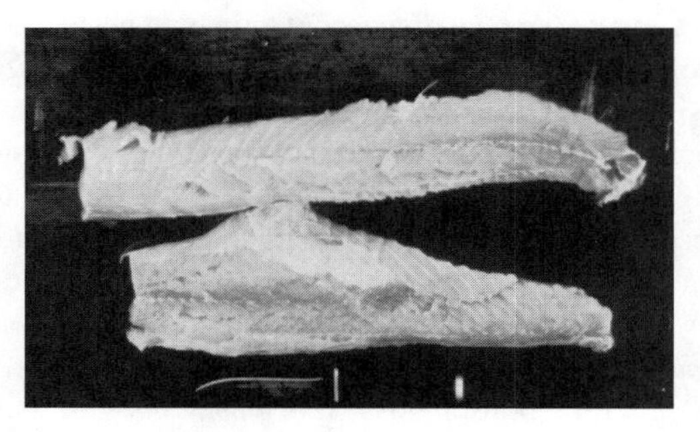

Figure 3-8. *Top*: The Chinese grass carp, or white amur, grows rapidly by grazing on water weeds. *Bottom*: A valuable by-product of this activity is a highly nutritious fish flesh. (Richard R. Yeo, USDA, SEA, ARS, and Leroy Holms, Madison, WI.)

Waterfowl such as geese, ducks, and swans can be used for removing floating and submerged plants from small ponds. Unfortunately, waterfowl can interfere with use of the pond because of their aggressiveness and their deterioration of the water quality and bank stability. Waterfowl must also be protected from predators.

Plant-eating fish offer one of the more promising biological approaches to aquatic weed control. These are imported fish that feed exclusively on vegetation. The Congo tilapia (*Tilapia melanopleura*) can feed on phytoplankton and unicellular algae when young, and older Tilapia also feed on larger plants. Tilapia are sensitive to cold and die at temperatures below 46 to 48°F. Annual stocking of Tilapia has been used for aquatic weed management in irrigation ditches in southern California.

The grass carp or white amur (*Ctenophryngodon idella*) (Figure 3-8) can survive in water below an ice cover. The grass carp has a life span of several years and feeds exclusively on filamentous algae, chara, weeds below the water surface, and duckweed. The fish consumes its weight in vegetation each day and grows to 20 to 50 lb. As with any introduced species, there is some concern about competition of grass carp with desirable fish species and other environmental impacts. The grass carp does not appear to reproduce naturally in the United States. Thus, its population will not crowd out desirable game fish. Commercial hatcheries use hormonal treatments and special environmental conditions to raise fish for release. The grass carp also does not roil sediments and cloud water, as does the common carp. Although a number of states have outlawed the importation of the grass carp, this fish is gaining wider acceptance and use.

CHEMICAL WEED CONTROL

The use of chemicals that selectively kill weeds in crops is an integral part of many modern weed management systems. The specific *pesticides* for controlling weeds are called *herbicides*. Selectivity is the key to the widespread utility of herbicides and is discussed in detail in Chapter 5. Chemicals were used in ancient times to control unwanted vegetation. The Roman army would salt fields of their enemies, preventing the growth of all plants, both crops and weeds. Weeds were controlled selectively in the early twentieth century with the use of inorganic salts such as sulfuric acid, but selectivity was limited to a few crops and special care was required in use of the chemical.

Modern selective herbicides were born during the Second World War with the discovery of the herbicidal properties of synthetic plant growth regulators. The Weed Science Society of America (WSSA, *Herbicide Handbook* (1994) and *Supplement*, 1998) lists more than 250 chemicals that are herbicidal, and these chemicals are marketed throughout the world under thousands of trade-named products for use in weed management. Discovery, production, and sale of herbicides are a multibillion-dollar worldwide industry. Because of the importance of herbicides to modern agriculture, a large portion of this book is devoted to their characteristics, use, environmental behavior, and toxicology (health considerations). This information is vital to all practitioners and researchers so that these chemicals can be used properly and in the safest manner possible.

Herbicide Classification

Herbicides can be grouped in numerous ways, including chemical similarity, mechanism of action (how they kill the plant), herbicide movement within the plant (mobile versus immobile), selectivity (selective versus nonselective), and application and use patterns. This book, as explained more completely in Chapter 8, uses mechanism of action groups for simplicity and clarity of discussion.

Adoption of Herbicides

The widespread adoption of herbicides by farmers is linked to advantages for weed management in several areas. Herbicide use increased largely at the expense of cultivation for weed control. However, it can result in higher crop yields than reliance on cultivation alone. It is worth reemphasizing that the best and most efficient weed management scheme will utilize all available weed control tactics as appropriate.

Herbicides can control weeds beyond the reach of the cultivator. Weeds directly within the crop row, in closely seeded (drilled) crops such as small grains, and in no-tillage can be managed with herbicides much more easily than with cultivation. Control of weeds that compete directly with the crop in the row is especially important to minimize crop yield losses.

Herbicides, as compared with cultivation, help reduce the labor and time needed for effective weed management. These reductions can directly lead to increased economic return for a farmer. Reduced time and labor requirements for weed control also allow farming of greater acreage despite the shrinkage of available farm labor.

Crop yield is increased for some crops, such as corn, by early planting. Herbicides allow planting into soils too wet for a final tillage to kill any emerged weeds. Crops can also be planted before the soil warms enough to promote weed seed germination and seedling growth. Herbicides applied at planting or later will control weeds as they germinate without having to wait for a tillage operation to kill them as they emerge.

Reduced tillage systems, particularly no-tillage, are now more feasible because of herbicides. Without herbicides, primary tillage, secondary tillage, and cultivation would still be needed to suppress weed populations. Herbicides are available for many crops to kill existing vegetation before planting, to keep weeds from becoming a problem later, and to control any escaping weeds.

Hand weeding in many cropping systems is no longer necessary because of herbicides. This has freed labor from the drudgery and hard work of hand pulling or hoeing weeds. Freedom from manual weed control has allowed workers to seek better employment elsewhere or to improve the efficiency of the total farm enterprise. This is an advantage where there is work available for workers released from weed control but has been criticized by some when employment is not available for displaced workers. As with the introduction of any new technology, herbicide use can bring sociological as well as agricultural change. This is more of a problem in less developed countries where the labor supply exceeds available nonagricultural work opportunities.

Effective herbicide programs have increased the available cropping system choices available to farmers as these choices are not as much affected by existing weed problems. Effective herbicide programs aid growing crops in closely spaced rows for higher crop yields and ease of mechanical harvest.

Herbicide use has resulted in energy-efficient and economical weed control. Farmers have adopted herbicides for weed control because the chemicals increased profit, weed control efficacy, production flexibility, and reduced time and labor requirements for weed management.

Herbicide use is not without potential problems. Farmers who use herbicides need to be concerned about possible crop injury. Selectivity can be reduced under adverse environmental conditions (see Chapter 5) or can be marginal for some herbicides even with good growing conditions. There is also the danger of injury to nontarget plants in adjacent fields and, in some cases, areas far removed from herbicide applications. Herbicides can potentially move off target as volatile gases, in water running off a field, and attached to dust particles or sediments in runoff water. There have been cases in which an aircraft flying to treat a crop field leaked herbicide over a wide area of its flight path. There must also be concern for potential environmental damage associated with herbicide use. Herbicide residues in soil can restrict or prevent rotational crop growth. Groundwater (water in the water table) and surface water contamination by herbicides must be addressed. The effects of herbicides on other life forms in the environment, such as fish, animals, birds, invertebrates, microorganisms, and humans, must be evaluated, and there is a need to make sure that chemical residues in our food supply are nonexistent or at minimal allowable levels.

The danger of human toxicity resulting from herbicide exposure should be of paramount concern, including not only the danger of immediate (acute) effects but also the complications from long-term (chronic) exposure. Such effects occur through direct exposure of herbicide production workers and applicators to high amounts of chemicals and the possible indirect exposure of others to very low levels of herbicides in food and water. These can all be controversial and emotional subjects but must be considered in terms of the benefits versus the risks associated with herbicide use, as discussed further in Chapter 4.

Many of the worst problems that have occurred with herbicides could have been avoided through their proper use. Herbicide users have a responsibility to select the right herbicide for the weed problem and the crop, to handle and store the herbicide with respect, and to apply the chemical correctly according to the label. There are many sources of information, including chemical manufacturers and their representatives as well as literature from the Cooperative Extension Service and others, to help with proper herbicide use. Above all, it is imperative to read and follow the herbicide product label directions before even using the chemical.

Time of Herbicide Application

One of the major distinguishing characteristics between different herbicide programs is the time the chemical is applied. These timings are defined with respect to the stages of both weed and crop growth. In the broadest sense, herbicides can be applied either directly to the soil (soil active) or directly to the foliage of the weeds (foliar applied). Some herbicides are effective with only one of these applications, whereas others can be applied either way. More specific application timings are *preplanting*, *preemergence*, and *postemergence*. Examples are given in Table 3-6.

Preplanting Preplanting treatments are made anytime before crop planting. Soil fumigation and preplow, early preplant, and preplant incorporation treatments are

TABLE 3-6. Examples of Preplant, Preemergence, and Postemergence Herbicide Use Defined by Crop, Weed, or Both

Application Type	Crop Stage	Weed Stage	Example
Preplant	Preplant	Pre	Early preplant application of atrazine in corn to control annual weeds
Preplant	Preplant	Post	Application of 2,4-D before soybean planting in no-till to control perennial weeds
Preplant incorporated	Preplant	Pre	Trifluralin application in soybeans to control annual weeds
Preemergence	Preemergence	Preemergence	Acetachlor in corn or soybeans to control annual grass weeds
Postemergence	Preemergence	Postemergence	Paraquat or glyphosate after planting in no-tillage, but before corn emergence, to control existing weeds
Postemergence	Postemergence	Postemergence	Sethoxydim use in established alfalfa to control grass weeds
Postemergence	Postemergence	Preemergence	Simazine in an apple orchard or a lay-by application in a row crop

examples of preplanting applications. Soil fumigation places a nonselective herbicide in the soil to eliminate many existing weed seeds and reproductive structures. The soil fumigant, such as methyl bromide, must dissipate from the soil before planting for crop safety. Preplow treatments are applied to the soil prior to primary tillage for seedbed preparation.

Early preplant applications generally use herbicides that persist in soils and that are applied to no-till fields 2 or more weeks prior to planting. These early applications may be done before any weeds emerge prior to planting. The residual herbicide prevents early weed growth and can reduce the need for herbicide control of existing weeds at planting. Firmer soil in no-till fields than in tilled fields allows sprayer passage early in the season without the threat of getting stuck in muddy tilled fields. Early treatment of the soil surface also increases the likelihood of any subsequent rainfall to move the herbicide into the soil before weed emergence. Soil-applied herbicides must be moved into the soil to be active on weeds. Early preplant treatments can increase the reliability of herbicide treatments. Good weed control relying on herbicides is essential for success in no-tillage. Early preplant treatments also let farmers and other herbicide applicators begin their work early in the season. This can be a big advantage when many acres must be treated.

There are some disadvantages to early preplant programs. Early herbicide application can reduce the period of effective weed control after planting. Herbicides in the soil generally last only a few weeks for weed control (see Chapter 6 for a discussion of the fate of herbicides in soils). Additional herbicide treatments at planting or later may be necessary following early preplant applications to ensure adequate long-term weed management.

Preplant incorporation is still fairly common with soil-applied herbicides. Equipment for this purpose is discussed in Chapter 7. Incorporation (mixing) of the herbicide into the soil before planting can offer several advantages. The advantages include less reliance, as opposed to surface application, on rainfall to move the herbicide into the soil, more uniform distribution of the herbicide in the soil and thus more consistent weed control, and improved control of some weeds that germinate deep in the soil and some perennial weeds. Applying a herbicide before planting helps ensure good weed control during seedling growth of the crop.

Disadvantages of preplant incorporation include the monetary cost of the tillage operation, the need for more equipment and time, possible soil drying and erosion losses due to the tillage, and the potential for improper incorporation (too deep or streaked in the soil), causing reduced weed control (Thompson et al., 1981). Higher herbicide rates can be needed to offset the herbicide dilution in the soil. However, some herbicides must be incorporated to stop losses resulting from herbicide volatility (gaseous loss) or ultraviolet degradation that would otherwise happen if the chemicals remain on the soil surface. Incorporation can also be difficult to coordinate with aerial or contract (custom sprayed) ground application of herbicides. It may not be possible to use herbicide incorporation in a no-tillage system, and it may be contrary to the objectives of reduced tillage programs.

Preemergence Preemergence treatments made shortly after crop planting but before weeds emerge are a very common way to use soil-applied herbicides. There are several benefits of this type of application. Herbicides are often more concentrated after preemergence application in the upper soil layers than when the chemical is mechanically mixed into the soil. The higher herbicide concentration can produce better weed control of shallow-germinating weed seedlings. Longer residual control is also possible, as preemergence-applied herbicides are not as subject to leaching (downward movement of the herbicide in the soil with water) below the weed seed germination depth as some incorporated herbicides. Preemergence herbicide applications can be made with planting equipment, thus avoiding an extra trip across the field. Weeds are controlled early in the crop growth, minimizing competitive effects, and preemergence applications are suitable for a variety of tillage practices. Aerial and contract applications are easy to arrange with preemergence treatments. Greater crop safety can be effected with preemergence-applied herbicides because of the spatial separation of the herbicide-treated soil layer from the crop seed. Preemergence herbicide applications can be made to soil that would be too wet for effective incorporation.

The most severe limitation for preemergence herbicide treatments is the requirement for rainfall or (irrigation water) to move the herbicide into the soil to achieve weed control. Delay in rainfall can result in loss of weed control. Premergence applications may not be feasible in arid regions with limited access to irrigation water. The majority of preemergence herbicides are most effective against recently germinated weeds or small seedlings. Application soon after crop planting is necessary for effective weed control. Unfortunately, this may slow the planting operation; however, many farmers apply the herbicide immediately after the crop seed is placed in the soil and covered. Finally, although preemergence applications are often safer on crop plants than incorporated treatments, high rainfall can move a concentrated band of herbicide from the soil surface into the crop root zone and result in crop injury.

Postemergence Postemergence applications are made after emergence of the specified crop and/or weed and have recently become a primary method with many agricultural crops. Postemergence application is the only herbicide application strategy that is not strongly influenced by the soil environment. Both preplant and preemergence application rates must be adjusted for soil texture (relative proportions of sand, silt, and clay) and soil organic matter. Although some postemergence herbicides can have soil activity, the primary foliar activity allows postemergence applications to be made in areas with high organic matter soils, where soil applications would be totally ineffective. The high rates of soil-applied herbicides needed on organic soils can favor the use of postemergence herbicides. A second advantage of postemergence applications is that they are made after the weed problem appears. This can eliminate unneeded preventative applications or allow only infested parts of the field to be treated. Postemergence treatments do not take time during planting and make aerial and contract ground applications very feasible. Many, but not all, postemergence herbicides have little soil activity, which eliminates the threat of injury to rotational crops. However, farmers must be familiar with the replant restriction portion of the product label for any postemergence herbicide so as to avoid injury to rotational crops.

The major disadvantage of postemergence herbicides is the often limited time over which they can be effectively and safely applied. There can be restrictions on both the size of the weeds effectively controlled and the crop size for selectivity. The critical period for weed control can be lengthened in some cases by increasing the herbicide rate. Of course, this incurs extra costs. For maximum herbicide effectiveness, the optimum conditions for weed treatment and crop safety must coincide. Environmental conditions (weather too hot, cold, dry, wet, or wet soil) can impact or delay postemergence applications and prevent full control. The amount of area that some farmers must spray can prevent optimum application timing for postemergence control on all fields. Finally, any delay in controlling weeds increases the potential for yield losses due to weed competition.

Another limitation to use of postemergence herbicides is the relatively limited spectrum of weeds controlled by many of these herbicides. Often, postemergence herbicides are effective in controlling only broadleaf or grass weeds. More than one herbicide is frequently required for control of the total weed population in a field. Recently, with the release of glyphosate-resistant crops, some of these problems have been eliminated because of the ability of glyphosate to kill many types of both grass and broadleaf weeds. Beyond the general group of weeds controlled (broadleaf or grass) by any one postemergence herbicide, the response of specific weed species to a particular herbicide will vary widely. This may require that more than one herbicide be used for control of the entire weed population. Postemergence herbicides can also have less selectivity than soil-applied herbicides. Moreover, with foliar-applied herbicides there is a greater danger of spray drift harming nontarget plants removed from the treatment area.

Specialized postemergence applications include *directed applications* and *lay-by applications*. Directed applications achieve selectivity by specialized application equipment that allows minimal contact of the spray solution with sensitive crop parts. Directed spray operations use contact-type herbicides (such as paraquat) to control weeds in an established crop. The spray is directed to the base of the crop plant to avoid contact with the crop foliage and prevents crop injury (Figure 3-9). Systemic herbicides, for example 2,4-D or dicamba, are used as directed postemergence applications in corn. This treatment avoids actual contact of the herbicide with sensitive crop parts by directing the spray. In the case of 2,4-D or dicamba in corn, protection of the growing point (apical meristem) of the corn is the important factor.

Selectivity can also be gained by treating only weeds growing above the crop without contacting the crop below. Systemic herbicides, such as glyphosate, are used for this treatment. Specialized equipment (see Chapter 7) is employed to accomplish this. Unfortunately, allowing the weeds to remain in a field until they overtop the crop means that significant competition has occurred prior to the treatment. However, treatment of perennials growing above the crop can reduce the weed population in the

Figure 3-9. A directed spray of a herbicide in which the nozzles direct the spray across the row, killing small weeds. The plant stem is tolerant, but the leaves would be killed. (Spraying Systems Co.)

field. These applications can also help prevent harvest problems from weed infestations. In addition, over-the-top herbicide treatments cost little to apply.

A lay-by application is made with or following the last cultivation before it is impossible to move equipment through the field because of crop size. Lay-by applications are often soil treatments intended to extend the period of residual weed control.

Area of Application

Herbicides are applied broadcast, as a band, or as a spot treatment (Figure 3-10). Broadcast treatment, or blanket application, is uniform application to an entire area.

Band treatment usually means treating a narrow strip directly over the seeded row. The space between the rows is not chemically treated but is usually cultivated for weed control. This method reduces chemical cost because the treated band is often one-third of the total area, with comparable savings in chemical costs. In addition, when the chemical has a long period of residual soil toxicity (remains toxic in the soil for a long

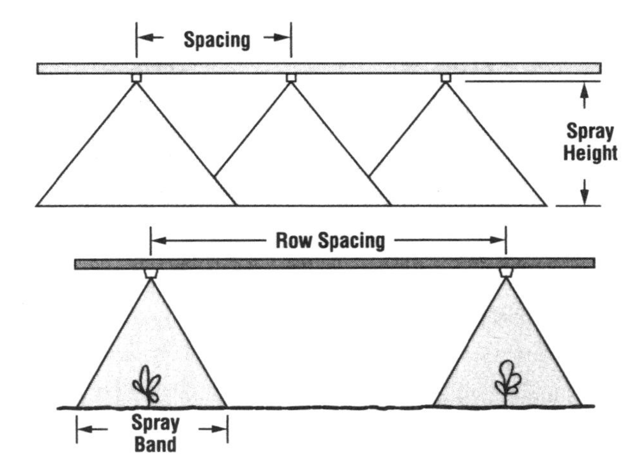

Figure 3-10. Broadcast and banded spray applications of herbicides. (Spraying Systems Co.)

time), the smaller total quantity of the chemical reduces the residual danger to the succeeding crop.

Spot treatment is a treatment made to a restricted area, usually to control an infestation of weed species requiring special treatment. Soil sterilant treatments or nonselective herbicides (sacrificing any crop present) are often used on small areas of seriously threatening perennial weeds to prevent their spread.

Herbicide Selection

Once the decision to use herbicides as part of a weed management program is made, several points should be considered before determining which herbicide(s) to use.

1. Will the herbicide(s) adequately control the weed species present? As discussed in a later section, knowledge of the weeds present is vital to the success of a herbicide application. Extension service information, herbicide labels, sales literature, personal experience, and retailers can help to identify the best herbicide choice(s). A number of states are developing computer programs to aid in the herbicide selection process. It is also good to consider whether two or more herbicides applied separately or as a tank mixture are needed to adequately control the weeds.

2. Is the crop sufficiently tolerant of the herbicide? Herbicides are not generally marketed unless the crop will tolerate a herbicide application rate twice that needed to control susceptible weeds. This is called a 2x safety factor. However, the crop tolerance of various herbicides does differ. This is especially true under favorable environmental conditions for crop growth or if a high herbicide rate is used to control all the weeds. Tolerance can also vary among crop varieties. It is necessary to rely on both past experience with herbicide use and published information to assess the threat of yield-reducing crop injury versus the benefit of herbicide control of weeds.

3. What are the crop rotation plans? As discussed earlier in considering crop rotation, it is important to avoid using a herbicide that will not allow the desired crop rotation. This limitation is primarily due to herbicide residues in the soil. Only when the weed problem cannot be managed in another way should the use of herbicides that will limit cropping sequence flexibility be considered.

4. What is the danger to nontarget plants? Movement of herbicides off the application site can potentially injure adjacent and far removed crops and other plants. Volatile (as a gas) herbicide movement is the prime danger, but transfer with dust, soil, and water to untreated areas can also occur. This danger should be assessed in selecting particular herbicides and herbicide formulations. Formulation can have a large impact on volatility, as discussed in Chapter 7. The use of volatile herbicides is discouraged or outlawed to prevent damage to sensitive nontarget plants. For example, the use of volatile 2,4-D formulations is outlawed in some states during the period of time from tobacco planting until harvest. Tobacco is very sensitive to damage from 2,4-D vapors.

5. Is the soil suited for the herbicide choice? Soil organic matter, clay content, and pH can all affect the toxicity and persistence of herbicides. Herbicides applied to soils low in organic matter and/or clay may be too toxic and may damage crops.

Alternatively, herbicides in these soils may leach (wash downward in the soil) too quickly for an effective length of weed control. The opposite extreme can occur on soils with high organic matter and/or clay, where the herbicide is not available for weed control because it is bound to these soil constituents. These soils can require impractically high rates of soil-applied herbicides or reliance on foliar-applied (postemergence) herbicides for weed control.

Soil pH can also limit herbicide choices. Some herbicides such as the triazines are very quickly degraded under low soil pH, whereas both triazines and sulfonylureas can be too persistent with high soil pH.

6. Are there other environmental dangers from the herbicide use? There are both high-risk herbicides and high-risk areas of the United States for potential groundwater contamination (Chapter 4). Farmers in these areas must practice caution in their herbicide choices. Care should also be exercised if contamination of surface water with a herbicide is likely. Selection of herbicides should be restricted to those that are neither prohibited from application near water nor extremely toxic to aquatic organisms.

7. Is the herbicide economical to use? It is assumed that herbicide users will always weigh whether the potential increased economic return warrants herbicide use. This can be difficult to predict precisely for a soil application, but models are being developed for decisions on the economics of postemergence herbicide use.

DEVELOPING AN INTEGRATED WEED MANAGEMENT PROGRAM

Although this chapter discusses weed management tactics separately, designing a weed control program involves more than simply selecting weed control techniques. A weed management program integrates the various tactics into a long-term strategy for dealing with weeds in a field and on the farm. Both the short-term and long-term impacts of the weed control system on the weed population should be weighed. In addition, the environmental, cultural, economic, and management factors discussed in the next section must be considered in planning a weed control strategy. Most components of an IWM system can be planned in advance of planting. Wherever possible, alternate components should be factored into an IWM system plan to provide flexibility in the event of different conditions than expected.

Knowledge of the System

All successful weed control programs begin with an open mind as to the various tactics available to control the existing weed problem. This is followed by a well-designed plan to integrate the various tools into an effective weed control system based on the crop, the environment, and the objectives of the farm. Knowledge of each field is a critical starting point: What are the present weeds that have caused problems in the past? How are they distributed within the field? What is the past cropping history? and What tools have worked in the past? It is impossible to judge the potential impact of various weed control tactics on a given weed population in a particular field without this knowledge. With no knowledge, a farmer is by necessity required to rely on high herbicide inputs in anticipation of a weed problem. A characteristic of U.S. agriculture that sometimes prevents a suitable knowledge base is that a high percentage of acres farmed are rented land. Rented land may have been farmed by a variety of people with a variety of crops and a patchwork of weed control practices pieced together to address the immediate weed problems in a particular year, with no long-range weed management objective in mind. Even though this situation is common, the important point for a farmer when managing weeds is to base the program on the best available information about the field and his or her past experience with weed management in the target crop. Such an approach will allow the farmer to obtain the best weed control results from an integrated approach.

INTEGRATED WEED MANAGEMENT—IMPLEMENTATION

Steps in the Plan

In order for a weed management system to work, the beginning point is correct identification of the existing weed problem. It is impossible to judge the potential impact of various weed control tactics on a weed population without knowing the weed species present. The number of different weeds infesting a crop is relatively limited, so proper species identification is not an impossible task. Some of this knowledge can be gained from past experience and records kept on the field in question. If the field is new, it is worthwhile to check the edges of the field and other areas of poor weed control to establish a list of weed species present. This can be done early in the growing season and at harvest. Keeping good field records will aid not only the weed control program but also the total crop production scheme. Records can include field maps that show areas of weed infestations and note the abundance of individual weeds. New weeds can be located and considered for eradication efforts. The field records help to prioritize the weed control needs with respect to the most economically damaging and troublesome weeds. It may be possible to treat only certain parts of the field, with obvious savings in weed control costs.

The second step is to implement the weed management plan within a particular crop. We have discussed in some detail the various tools available to the farmer for use in an IWM program. These main areas of weed management include scouting, prevention, and mechanical, cultural, biological, and chemical control. Within each of these areas are many options available for the farmer to develop a flexible and effective weed management program. The farmer has a great challenge in integrating these tactics to accomplish the objectives of the farm within each particular cropping system relating to economic stability and environmental stewardship in the production of quality crops. No one system will work for all crops, and not all tools can be used each year; these must be designed for each situation and crop. The farmer must plan the weed management scheme considering the overall objectives of the enterprise.

A final important step in any endeavor is to evaluate the yearly success of a weed management program to verify and identify tactics to retain in the future and areas where change is necessary. A good long-term objective of any weed management program is to decrease the weed problems. If weed problems stay the same or increase, the management plan needs to be altered.

The increased emphasis on integrated crop management, reduction in pesticide use, and better conservation practices in agroecosytems to ensure sustainability will affect our use of herbicides. IWM will necessitate improving our methods for understanding weed problems within a field and result in improved application technology, use of decision aids (Chapter 7), and better nonchemical alternatives. Herbicides will remain a main tool for most farmers for control of weeds. The combination of better decision aids and improved technology to determine when and if herbicide applications are necessary will allow farmers and others to obtain the best weed control with the minimum chemical use. These techniques will be based on a clear understanding of the field situation, weed biology, alternate control methods, and the limitations imposed by a particular cropping situation. There will be a reduction in the total amount of herbicide used, but the improvement in efficiency and safety of herbicide use will allow these valuable tools to continue to contribute to agriculture productivity.

LITERATURE CITED AND SUGGESTED READING

Altieri, M. A., and M. Liebman. 1988. *Weed Management in Agroecosystems*. CRC Press, Boca Raton, FL.

Barnes, J. P., and A. R. Putnam. 1983. Rye residues contribute weed suppression in no-tillage cropping systems. *J. Chem Ecol.* **9**:1045–1057.

Bender, J. 1994. *Future Harvest: Pesticide-Free Farming*. University of Nebraska Press, Lincoln. 159 pp.

Bowman, G. 1997. *Steel in the Field: A Farmer's Guide to Weed Management Tools*. Sustainable Agriculture Network, Beltsville, MD.

Boydston, R. A., and A. Hang. 1995. Rapeseed (*Brassica napus*) green manure crop suppresses weeds in potato (*Solanum tuberosum*). *Weed Technol.* **9**:669–675.

Buhler, D. D., ed. 1999. *Expanding the Context of Weed Management*. Food Products Press, New York.

Buhler, D. D., R. G. Hartzler, and F. Forcella. 1997. Implications of weed seedbank dynamics to weed management. *Weed Sci.* **45**:329–336.

Calloway, M. B. 1992. A compendium of crop varietal tolerance to weeds. *Am. J. Alt. Agric.* **7**:169–180.

Cardina, J., T. M. Webster, C. P. Herms, and E. E. Regnier. 1999. Development of weed IPM: Levels of integration for weed management. In *Expanding the Context of Weed Management*, ed. by D. D. Buhler. Food Products Press, New York.

Cavers, P. B., and D. L. Benoit. 1989. Seed banks in arable lands. In *Ecology of Soil Seed Banks*, ed. by M. A. Leck, V. T. Parker, and R. L. Simpson. Academic Press, San Diego.

Challaiah, O., C. Burnside, G. A. Wicks, and V. A. Johnson. 1986. Competition between winter wheat (*Triticum aestivum*) cultivars and downy brome (*Bromus tectorum*). *Weed Sci.* **34**:689–693.

Creamer, N. G., B. Plassman, M. A. Bennett, R. K. Wood, B. R. Stinner, and J. Cardina. 1995. A method for mechanically killing cover crops to optimize weed suppression. *Am. J. of Altern. Agric.* **10**:157–162.

Dyck, E., M. Liebman, and M. S. Erich. 1995. Crop-weed interference as influenced by a leguminous or synthetic nitrogen source: I. Double cropping experiments with crimson clover, sweet corn, and lambsquarters. *Agric. Ecosys. and Envir.* **56**:93–108.

Enache, A. J., and R. D. Ilnicki. 1990. Weed control by subterranean clover (*Trifolium subterraneum*) used as a living mulch. *Weed Technol.* **4**:534–538.

Ennis, W. B., Jr., 1976. Modern methods for controlling pests. *Proc. World Soybean Conf.* 1975, pp. 375–386.

Gressel, J. 2000. Molecular biology of weed control. *Transgenic Res.* **9**:355–382.

Herbicide Handbook, 7th ed. 1984 Weed Science Society of America, Lawrence, KS.

Herbicide Handbook Supplement to 7th ed. 1998. Weed Science Society of America, Lawrence, KS.

Hodges, L., and R. E. Talbert. 1990. Adsorption of the herbicides diuron, terbacil, and simazine to blueberry mulches. *Residue Rev.* **52**:1–26.

Kennedy, A. C. 1999. Soil microorganisms for weed management: *Alternaria cassiaeis* induces blight in sicklepod (*Cassia obtusifolia*) and can be used in soybean and peanut. In *Expanding the Context of Weed Management*, ed. by D. D. Buhler. Food Products Press, New York.

Malik, V. S., C. J. Swanton, and T. E. Michaels. 1993. Interaction of white bean (*Phaseolus vulgaris* L.) cultivars, row spacing, and seeding density with annual weeds. *Weed Sci.* **41**:62–68.

Masiunas, J. B., L. A. Weston, and S. C. Weller. 1995. The impact of rye cover crops on weed populations in a tomato cropping system. *Weed Sci.* **43**:318–323.

MidWest Plan Service (MWPS). 1992. *Conservation Tillage Systems and Management.* Midwest Plan Service, Ames, IA.

Miller, D. A. 1996. Allelopathy in forage crop systems. *Agron. J.* **88**:854–859.

Pester, T. A., O. C. Burnside, and J. H. Orf. 1999. Increasing crop competitiveness to weeds through crop breeding. In *Expanding the Context of Weed Management*, ed. by D. D. Buhler. Food Products Press, New York.

Rankins, A., Jr., D. R. Shaw, and J. D. Byrd Jr. 1998. HERB and MSU-HERB field validation for soybean (*Glysine Max*) weed control in Mississippi. *Weed Technol.* **12**:88–96.

Ross M. A., and C. A. Lembi. 1999. *Applied Weed Science*. 2d ed. Prentice-Hall, Upper Saddle River, NJ.

Smeda, R. J., and S. C. Weller. 1996. Use of rye cover crops for weed management in processing tomatoes. *Weed Sci.* **44**:596–602.

Stoller, E. W., and L. M. Wax. 1973. Periodicity of germination and emergence of some annual weeds. *Weed Sci.* **21**:574–580.

Swanton, C. J., and S. F. Weise. 1991. Integrated weed management: Rationale and approach. *Weed Technol.* **5**:657–663.

Thill, D. C., J. M. Lish, R. H. Callihan, and E. J. Bechinski. 1991. Integrated weed management—A component of integrated pest management: A critical review. *Weed Technol.* **5**:648–656.

Thompson, L., Jr., W. A. Skroch, and E. O. Beasley. 1981. *Pesticide Incorporation: Distribution of Dye by Tillage Implements.* AG-250. NC Cooperative Extension Service.

Wapshere, A. J., E. S. Deelfosse, and J. M. Cullen. 1989. Recent developments in biological control of weeds. *Crop Prot.* **8**:227–250.

Weller, S. C., R. A. Bressan, P. B. Goldsbrough, T. B. Fredenburg, and P. M. Hasegawa. 2001. The impact of genomics on weed management in the 21st century. *Weed Sci.* **49**:282–289.

WEB SITES

Biological Control

Biological Control: A Guide to Natural Enemies, Cornell University: http://www.nysaes.cornell.edu

Go to "entomology," then search with "Biocontrol of Weeds" and scroll to "Weed-Feeders Introduction."

The Cutting Edge in Weed Science—News and Advancements: http://utahweeds.tripod.com

Scroll to "Cutting Edge" and select, then go to Biological Weed Control Agents

"Biological Control Virtual Information Center:"

http://ipmwww.ncsu.edu/biocontrol

Decision Aid Web Sites

WeedCast 2.0. forecasts three aspects of weed phenology: weed emergence potential, emergence timing and seedling height. Provides vital information for making weed management decisions. http://www.morris.ars.usda.gov

Select "Software and Equipment," then select "Weed Ecology and Management" to obtain WeedCast.

University of Minnesota. *Cultural and Chemical Weed Control Guide in Field Crops*: http://www.extension.umn.edu

Select "Crops," then select "Weed Control" and scroll to "Cultural and Chemical Weed Control in Field Crops."

University of Nebraska. *Guide for Weed Management in Nebraska*: http://www.ianr.unl.edu

Select "Publications," then select "Find Information;" select "Weeds" and scroll to "Weed Management Guide."

University of Florida. *Weed Management in Field Crops and Pasture Grasses*: http://edis.ifas.ufl.edu/MENU_WG:Field_Crops_and_Pasture

North Carolina State University. *North Carolina Agricultural Chemicals Manual*: http://ipm.www.ncsu.edu/agchem/agchem.html

University of California. *Pest Management Guidelines*: http://www.ipm.ucdavis.edu

Select "How to Manage Pests."

For chemical use, see the manufacturer's or supplier's label and follow these directions. Also see the Preface.

4 Herbicide Registration and Environmental Impact

The term *pesticide* includes not only herbicides, insecticides, and fungicides, but also insect repellants, plant growth regulators, disinfectants, and even swimming pool chemicals. To ensure that no unreasonable adverse effects to human health or the environment occur, any pesticide product must be registered with the U.S. Environmental Protection Agency (EPA). Pesticides approved for use by the EPA are granted a license or "registration" permitting their distribution, sale, and use according to requirements set by the EPA. States also require registration. When state registrations differ from federal registrations, they usually involve special local needs or emergency pest problems (see 24(c) and Section 18 registrations in the following section of this chapter).

Pesticides are classified as general-use or restricted-use pesticides. General-use pesticides are relatively nontoxic to humans, whereas restricted-use pesticides are more toxic to humans and require a warning label, precautionary safety handling procedures, and a special permit for their use. Restricted-use pesticides require that they be applied by a certified applicator; and mixers, loaders, and applicators of restricted-use pesticides must wear protective clothing.

PESTICIDE REGISTRATION

The registration of a pesticide is a scientific, legal, and administrative process. EPA assesses a wide variety of potential human health and environmental effects associated with use of a product, considering the particular site or crop on which it is to be used, the amount, frequency, and timing of its use, and the recommended storage and container disposal practices. For evaluation of a pesticide registration application, the registrant must provide data from tests done according to specific EPA guidelines conducted under recognized good laboratory practices (GLP). Results of these tests determine whether a pesticide has the potential to cause adverse effects on humans, wildlife, fish, or plants, including endangered species and nontarget organisms, as well as possible contamination of surface water or groundwater from leaching, runoff, and spray drift. The potential human risks evaluated include short-term toxicity and long-term effects such as cancer and disorders of the reproductive system. A pesticide will be registered only if it is determined that it can be used to perform its intended function without unreasonable adverse effects on humans or the environment. EPA also must approve the specific language that appears on each pesticide label, and the product can be legally used only according to the label directions.

FEDERAL PESTICIDE LAWS

Prior to 1970, the United States Department of Agriculture (USDA) registered pesticides. In 1970, President Richard Nixon's Reorganization Plan created the Environmental Protection Agency (EPA). Among many other activities, EPA oversees the registration of pesticides.

The *Federal Insecticide, Fungicide, and Rodenticide Act (FIFRA)* requires all pesticides sold or distributed in the United States (including imported pesticides) to be registered by EPA. Unregistered pesticides, or pesticides registered for other uses, can be used when approved by EPA and the states to address emergencies (FIFRA Section 18) or a state's special local needs (FIFRA Section 24(c)). FIFRA requires that the use of each registered pesticide must be consistent with directions contained on the label.

The *Federal Food, Drug, and Cosmetic Act (FFDCA)* regulates the establishment of pesticide tolerances. A *tolerance* is the maximum permissible level of a pesticide residue allowed in or on commodities used for human food and animal feed. FIFRA governs the sale, distribution, and use of a pesticide through the registration process and enforcement of the requirements on the pesticide label. FFDCA provides the means of policing pesticide residue levels in food through pesticide residue tolerances. EPA will not register the use of a pesticide unless all needed tolerances, or exemptions from tolerance, have been established.

Under the *Food Quality Protection Act (FQPA)*, a 1996 law that amended both FIFRA and FFDCA, EPA must adhere to additional criteria for the registration of pesticides, including new considerations of exposure for infants and children and consideration of all risks posed by pesticides with similar modes of action. Under FQPA, EPA must find that a pesticide poses a "reasonable certainty of no harm" before it can be registered. FQPA replaced the "Delaney clause," which prohibited the registration of any compound that caused cancer at any test rate in test animals. FQPA also requires EPA to accelerate the registration of reduced-risk pesticides and complete the reregistration of older pesticides. Specific changes mandated by FQPA to pesticide registration are listed in Table 4-1.

Guidelines for Registration of Herbicides

Approximately 120 to 150 individual guideline studies must be completed and evaluated by EPA in order to grant a registration for a new active ingredient. The exact numbers of studies required vary. Registration of a new active ingredient is very expensive; costs range from $20 to 25 million and can take from 6 to 10 years to complete. All costs associated with the registration process are borne by the manufacturer. There is increasing emphasis on safety aspects in the registration process, especially for human and animal health, the herbicide's fate in the environment, and its environmental impact.

TABLE 4-1. Specific Changes to Pesticide Registration Mandated by FQPA

A pesticide chemical includes all active and inert ingredients of such pesticide.
A safety finding about a pesticide in establishing a tolerance must:

1. Determine that there is reasonable certainty no harm will result from aggregate exposure to the pesticide chemical residue, including all anticipated dietary exposures and all other exposures for which there is reliable information.
2. Assess exposures of major population subgroups to pesticide residues in the home, garden, school, and any other nonoccupational source; this is in addition to assessing exposure to pesticide residues in food and water.
3. Determine potential estrogenic effects once test methodology is developed.

The EPA is also required to specifically assess the risk of a pesticide to infants and children considering:

a. Available information on food consumption patterns among infants and children
b. Susceptibility of infants and children to the effects of pesticides, including neurological effects, from pre- or postnatal exposures
c. Cumulative effects on infants and children of such residues and other substances that have a common mechanism of toxicity in order to ensure that there is "a reasonable certainty of no harm" to infants and children

The statute further provides that an additional tenfold margin of safety shall be applied for infants and children to take into account potential pre- and postnatal exposures and the completeness of the submitted data with respect to exposure and toxicity unless EPA determines that no hazard exists.

Guideline study requirements fall into specific areas. Complete information on the requirements for the various studies can be obtained from the EPA Office of Pollution Prevention and Toxics. Each of these categories provides important information necessary to make informed, scientifically based decisions concerning pesticide approval and subsequent use and involve the following data: *plant protection or efficacy, the product's chemical and physical characteristics, environmental fate, residue chemistry, wildlife and aquatic acute toxicology, reentry time frame after application, effects on nontarget insects*, and *potential for spray drift*.

Information Necessary for Herbicide Registration

Efficacy data involve the demonstration that the herbicide is effective for the stated purpose in the field.

Product chemistry data include composition and analytical methods for the technical and formulated products and the environment and crop residues.

Wildlife and aquatic acute toxicology data involve determination of toxicity (inherent capacity of a known amount of a substance to produce injury or death) values of the parent compound and its formulations on experimental animals. These data, which help in assessing the potential toxic effects on humans and other animals, include acute (short-term effect), subacute, and chronic (long-term) toxicity. Toxicology information is required for laboratory animals, soil microorganisms, and wildlife, including birds, fish, other aquatic animals, and insects. The toxicology of degradation products is also often required, along with first-aid and diagnostic information that is found in the Materials Safety Data Sheets (MSDSs). The MSDSs are provided to all purchasers of pesticides and must include information on product ingredients, physical and chemical properties, health and physical hazards, primary routes of chemical entry, exposure limits, precautions for safety in use, emergency first aid, and responsible party contacts. "Safety in use" is determined by the toxicity of the herbicide and applicator exposure (exposure level and length of time of exposure), which is a total estimate of the hazards of use of a particular herbicide.

Toxicology terms are used by the EPA to indicate levels of toxicity of a particular herbicide. *LD* is a lethal dose, and LD_{50} is a dose that will kill 50% of a population of test animals.

LC is a lethal concentration, and LC_{50} is the concentration that will kill 50% of the animals tested. LC_{50} values are expressed in terms of milligrams of the substance, as a mist or dust, per liter (mg/l) of air.

Acute oral refers to a single dose taken by mouth or ingested.

Acute dermal and *skin effects* refer to a single dose applied directly to the skin. Acute oral LD_{50} and acute dermal LD_{50} are expressed in terms of milligrams of the substance per kilogram (mg/kg) of body weight of the test animal.

Inhalation refers to exposure through breathing or inhaling, and *eye effects* refers to a single dose applied directly to the eye. A toxicity category is assigned to every pesticide according to the criteria listed in Table 4-2.

Signal words are found on all pesticide labels. A signal word is a one-word summary of the product's toxicity to humans, and there are three signal words in decreasing order of toxicity: *DANGER* (highly toxic, Category I), *WARNING* (moderately toxic, Category II), and *CAUTION* (slightly toxic, Categories III and IV).

Most herbicides (>90%) have relatively low toxicity in higher animals and are in the Caution category. Those most toxic are in the Warning category (based on acute oral LD_{50}) and include bromoxynil, cyanazine, diallate, difezoquat, diquat, endothall (amine), and paraquat. No herbicides currently in use are in the Danger category.

All pesticide labels must state "Keep Out of Reach of Children."

Environmental fate and residue chemistry data include field stability, rate of degradation and degradation products formed, movement of the herbicide to ground and surface waters, and potential for crop residues (see the later section "Pesticide Tolerances").

The environment is infinitely complex and includes the totality of the land, air, and water that surround us, as well as their interactions. Many climatic, edaphic, biotic, and social factors influence an ecological community and its organisms. In most human manipulation of the environment, an environmental impact assessment must be provided to public agencies for approval prior to completion of the proposed activity. The EPA requires environmental fate and residue chemistry data for

TABLE 4-2. Toxicity Categories for Pesticides

	Hazard Indicators				
Category	Oral LD_{50} (mg/kg)	Inhalation LC_{50} (mg/liter)	Dermal LD_{50} (mg/kg)	Eye Effects	Skin Effects
I	50 or less	0.2 or less	200 or less	Corrosive, corneal opacity, not reversible within 7 days	Corrosive
II	51 to 500	0.21 to 2.0	201 to 2000	Corneal opacity, reversible within 7 days, irritation persisting for 7 days	Severe irritation at 72 hours
III	501 to 5000	2.1 to 20	2001 to 20,000	No corneal opacity, irritation reversible within 7 days	Moderate irritation at 72 hours
IV	>5000	>20	>20,000	No irritation	Mild or slight irritation at 72 hours

pesticides to ensure that chemicals placed in the environment have minimal effects on the ecosystem.

In the specific case of weed control, the practices employed to manage weeds can affect the environment and can have either a beneficial or a negative impact. Beneficial effects can include reduced soil erosion and silting of streams, increased wildlife habitat, and natural beauty. Detrimental effects can include interference with human activities and a reduction in the aesthetics of the environment or an unacceptable effect on a plant community's soils and water quality. Cultural practices for land preparation and weed control may contribute to environmental degradation by increasing wind and water erosion of soil.

Information on the physical and molecular fate of herbicides in the environment is essential in determining their environmental impact and suitability for use. The fate of herbicides in plants (Chapter 5) and soil (Chapter 6) is discussed in greater detail later. Obviously, a herbicide cannot have an environmental impact unless it enters the environment. Herbicides enter the environment on application and almost always have some impact, usually as a beneficial biotic response of controlling target weed species with no detrimental effects. However, the potential negative effects are of interest from the point of registration suitability and center on their toxicology in regard to human health and wildlife, as discussed earlier, and their potential for contamination of natural waters. The Council for Agricultural Science and

Technology (CAST, 1987) report entitled "Health Issues Related to Chemicals in the Environment: A Scientific Perspective," is an excellent article on pesticides and their potential health impacts in the environment. The EPA ensures that all pesticides used in the United States have undergone extensive testing prior to registration.

Effects of Herbicides on Human Health

Most herbicides are relatively nontoxic to humans, because their action at the molecular level is usually at a site that is specific to plants or microorganisms but not to higher animals. Furthermore, all chemicals developed for herbicide use in recent years have low mammalian toxicity. The few older herbicides with higher mammalian toxicity are being phased out or are classified as restricted-use pesticides and require special handling. However, all chemicals, synthetic and natural, are toxic and should be handled with due caution. Even aspirin and table salt (sodium chloride) have significant oral LD_{50} values, 1.2 and 3.3 g/kg, respectively, and are included in Category III for toxicity.

The greatest health hazards of herbicides are to people who handle or are otherwise exposed to large quantities—for example, industrial manufacturing, formulation, and distribution personnel and those involved in applying herbicides in the field: applicators, mixers, loaders, and aircraft flagmen. To the best of our knowledge, herbicide-induced injury to farmworkers has not occurred as a result of entering treated areas or handling a commodity from a treated area. Reentry restrictions after application (if any) are noted on the herbicide label.

Indirect exposure of the general public to low levels of a herbicide may occur through the ingestion of contaminated food or water. In general, these indirect exposures to herbicides present little hazard, because the levels of exposure and mammalian toxicity are low. Absence or safe levels of herbicides in food or animal feed are assured by residue analysis of these products as established by the registration procedures outlined earlier.

Herbicides in Natural Waters The presence of herbicides in natural waters must be prevented. An adequate water supply is one of the world's most precious resources, and it must be kept safe for all plants and animals (Messersmith, 1988). A great deal of attention has been given to pesticides in natural waters, and numerous publications deal with this topic; several references are provided at the end of the chapter. The CAST (1989) report includes a discussion of the Safe Drinking Water and Toxic Enforcement Act of 1986 and its subsequent renewals. This Act was designed to protect the public health against the harmful effects of chemicals in drinking water. It requires EPA to specify the contaminants that may have any adverse effect on public health and to control their concentrations within safe levels.

The great interest in this topic stems from the fact that trace amounts of pesticides have been detected in both surface water and groundwater. Surface water includes streams, rivers, and lakes, and groundwater is the water that occurs in the earth below the water table (see Figure 6-2). The pesticides most likely to become contaminants in natural waters are those used in large quantities. When detected, even these are usually found in concentrations below 1 part per billion (CAST, 1987). However, they have occasionally been found at concentrations as great as 20 to 50 parts per billion.

Surface Water The surveillance of surface water for pesticides has been conducted for many years (Guenzi, 1974). In 1957, the Public Health Service established surveillance stations on major rivers and the Great Lakes. During the 1960s, the Department of the Interior implemented a program for continuous monitoring of major streams. State programs have also been established, and now the U.S. EPA has major responsibilities in this area.

Water runoff from pesticide-treated land is a major source of pesticide contamination of surface waters. However, some surface water contamination may occur by lateral movement through shallow groundwater. Local contamination can occur as a result of pest control procedures (e.g., for mosquitoes and aquatic weeds) whereby the pesticide is applied directly to surface water and when water retention procedures are inadequate.

Groundwater Recently, traces of many types of agricultural chemicals such as fertilizers, insecticides, and herbicides have been detected in groundwater. This issue was addressed by Hunnicut (1995), and the National Research Council (1986). The depth of groundwater and the time required for surface water to reach the groundwater pool vary with climatic and geologic conditions. Depending on these conditions, the upper boundary of groundwater may range from a few feet to hundreds of feet below the soil surface. The time for surface water to reach these depths may range from a few days to centuries. Because pesticides are generally bound and/or degraded (see Meyer and Thurman, 1996, and Racke and Coats, 1990) as they pass through the soil profile with water, shallow groundwater has a greater potential for pesticide contamination than deep groundwater.

Groundwater contamination is of particular concern because water from this source is pumped for domestic, agricultural, and industrial use. Leaching of pesticides with water from pesticide-treated land is the major source of pesticide contamination of groundwater (see Figure 6-2). Factors affecting the leaching of herbicides through soil are discussed in Chapter 6. Despite media claims to the contrary, health effects of pesticides are documented carefully [see preceding sections], and in their calculations, EPA leaves a generous safety factor.

Movement Most herbicides also have chemical properties that restrict movement from the soil to contaminate water sources. However, some herbicides have been detected at low levels in surface waters. When movement occurs, the major loss of herbicides from fields is due to leaching and runoff. Leaching is related to the solubility of the herbicide, how strongly it attaches to the soil, how fast it degrades, and the timing and amount of water that moves through a soil profile. The rate of movement of a leachable herbicide is related to two processes: movement through small pores in the soil, which is slow; and preferential flow in large pores, which is rapid movement. Infiltrating water can be intercepted by drainage tiles in a field and moved rapidly to surface supplies. Any herbicide that moves with the water can enter the surface water.

Runoff, or the overland flow of water, can transport dissolved materials to rivers and streams. This is the major source of pesticide contamination of surface waters. However, some surface water contamination may occur by lateral movement through shallow groundwater.

There are several sources of information related to pesticide movement in the soil and the potential for contamination of water sources listed at the end of this chapter.

Off-Site Vegetation

The major sources of the adverse effects of herbicides to off-site vegetation are drift, spray, and volatility. When such an effect occurs, it is usually near the area treated and could have been prevented by more careful application or the use of a nonvolatile formulation (Chapter 7).

Data on the potential for spray drift determine the potential for aerial off-site movement of the pesticide during application and the potential for it to result in damage to adjacent vegetation, animals, or humans. Possible particle drift from herbicides can pose problems to neighbors, field workers, and the environment. Applicators must take measures to keep the product in the field where it is applied, as misapplication or drift can endanger the public and may be illegal. The applicator should choose products based on their reduced potential for off-site movement. However, under certain climatic and/or topographic conditions, drift injury to off-site vegetation may occasionally occur some distance away. Such conditions may result, for example, from atmospheric inversion layers in valleys or alternating land-sea airflows. The leaching of relatively persistent herbicides into the rooting zone of trees outside a treated area has caused tree injury.

Wildlife

The major adverse effect of herbicides on wildlife is indirect inasmuch as herbicides are relatively nontoxic to higher animals. These indirect adverse effects are primarily related to the removal of vegetation that provides food and habitat for wildlife. However, in some cases the vegetative shifts induced by herbicides can be beneficial to certain species. Deer populations often increase when native grasses and forbs replace heavy brush stands removed by herbicides.

Reentry Time Frame

The data on the reentry time frame include studies relating to foliar and soil dissipation of the pesticide and dermal and inhalation exposure potential after pesticide application. These studies allow a determination of the time required after pesticide application before reentry into a treated area by humans is safe. All labels have a reentry statement.

Pesticide Tolerances for Food and Feed

EPA is required to establish a tolerance (the maximum legal residue limit (MRL)) on each commodity with a labeled use. A tolerance represents the "worst case" expected concentration that could occur when the material is applied at the maximum-labeled rate, the maximum number of applications, and the minimum preharvest interval. Essentially, the tolerance is an enforcement tool used to indicate non-approved pesticide use. The number of residue studies performed to establish a tolerance varies by crop and represents all relevant cultural areas of the country. Additional studies are required to establish tolerances on any processed fractions that may be derived from the treated produce, as well as in animals that may consume treated forage.

The U.S. Food and Drug Administration and the U.S. Department of Agriculture periodically check samples of fresh produce, meat, and poultry for pesticide residues. Both federal and state authorities have the power to seize food that contains amounts of pesticide that exceed tolerance levels established by EPA.

EPA estimates of human dietary exposure for risk assessment also use the maximum allowable residue levels established by the tolerance. As a result, an EPA risk assessment, in most cases, will assume that 100% of the crop is treated with the herbicide at the maximum rate and frequency and at the minimum preharvest interval. Strictly speaking, a pesticide tolerance has nothing to do with any toxicology study and has no health significance; it is an enforcement tool only, <u>not</u> the "safe" limit of the pesticide in a particular food commodity.

In summary, data generated by the aforementioned studies include efficacy, general chemistry, environmental chemistry, crop residues, toxicology, fate in the environment, environmental impact, and determination of tolerances. These data are used by the EPA in determining the usefulness and safety of a particular pesticide and are based on the concept of risk-benefit. If the known benefits outweigh any known risks, the pesticide can be registered. If future data indicate that the risk is greater than originally shown, the registration of a particular pesticide can be rescinded.

THE PESTICIDE LABEL

EPA must approve all label language before a herbicide can be sold or distributed in the United States. The intent of the label is to provide clear directions for achieving effective product performance while minimizing risks to human health and the environment; it is a <u>legal document</u> that permits the applicant to distribute and sell the product. It is a violation of federal law for anyone to use a pesticide in a manner inconsistent with its labeling instructions. The label must show the trade name, registrant name, active ingredients, (names and amounts), inactive ingredients (amounts) use classification (general or restricted), net weight of or measure of contents, directions for use, a signal word, and a warning or precautionary statement. The signal word and warning and precautionary statements are mainly concerned with toxicological, environmental, physical, and/or chemical hazards.

TYPES OF REGISTRATIONS UNDER FIFRA

Federal Registration: Section 3 Registration. EPA is authorized to register pesticides for use throughout the United States under FIFRA Section 3, based on registration steps as outlined in Table 4-3. A Section 3 registration represents the main registration of a product label. EPA can also regionally restrict the registration of some pesticides to certain states. In addition, states, tribes, and territories can further restrict EPA-registered pesticide products.

Federal Registration: Section 5 Registration

The EPA allows manufacturers to field-test pesticides under development through FIFRA Section 5, the Experimental Use Permit (EUP). Manufacturers are required to obtain experimental use permits before testing new pesticides, or new uses of pesticides, if they conduct experimental field tests on 10 acres or more of land or 1 acre or more of water per pest. Biopesticides and genetically engineered crops expressing plant-protectant traits also require an EUP when used in field trials. The data required

TABLE 4-3. Steps Involved in Obtaining a Section 3 EPA Registration

1. Completion of primary reviews of approximately 120 to 150 studies submitted under FIFRA Section 3
2. Six EPA committee reviews:
   a. Hazard ID Assessment
   b. Cancer Assessment
   c. Reproduction and Developmental Toxicity
   d. Metabolism Assessment
   e. Mechanism of Toxicity
   f. Risk Assessment
3. Development of an Environmental Fate and Effects Summary
4. Development of a Health Effects Risk Characterization including Reference Dose (RfD), Dietary Risk and Exposure Scenario (DRES/DEEM), and Food Quality Protection Act (FQPA) Assessment
5. Benefits and Economic Assessment (BEAD)
6. Mitigation negotiations with registrant
7. Development of Health Effects Risk Summary
8. Completion of EPA documents:
   a. Pesticide Fact Sheet
   b. Final Pesticide Tolerance
   c. Registration Division Memorandum
9. Health Effects Division concurrence
10. Registration Division concurrence
11. Office of General Council legal review
12. Publication of a Federal Register Notice
13. Approval signature by Office of Pesticide Programs director
14. EPA Notice of Registration with stamped accepted label

for an EUP are similar to those required for a Section 3 registration and include product chemistry; proposed experimental label; toxicology data, including wildlife, residue, and environmental fate data; performance data summary; temporary tolerance proposal; and the proposed experimental program.

EUPs are time-limited and conditional registrations that are granted for 1 or 2 years. A renewal of an EUP is required for additional years. All application sites are subject to both state and federal inspections. A Federal Register Notice establishing a temporary tolerance is required if the treated crop is to be harvested and sold. If no tolerance is established, then all crops treated under the EUP must be destroyed. EUP conditions of registration always require an acreage limitation, a limit on the total quantity of pesticide available for use, a reporting of any adverse effects, and a final report summarizing the disposition of all pesticide shipped under the EUP. The EUP is also subject to additional special requirements that EPA may determine are necessary. In all cases, *state notification and registration are required prior to product use.*

State-Specific Registrations: Section 24(c)

Under FIFRA Section 24(c) Special Local Need Registration (SLN), states can register a new pesticide product for any use, or an additional use of a federally registered product, as long as there is both a demonstrated "special local need" for such a product and a tolerance, or an exemption from a tolerance, has been established. SLN registrations issued by a state are considered federal registrations under Section 3 of FIFRA but allow distribution and use only within the state of issuance for the special local need.

The state, not the EPA, determines whether there is a special local need for an SLN registration, and this label is useful on the day the state grants it. However, EPA can reject a state's special local need registration within 90 days of the state's action. Because the states normally discuss potential SLNs with the EPA before applying, disapproval is rare. Except in cases of environmental hazard, a product already sold under an SLN that is later disapproved may still be used according the SLN label.

In most situations, it is the federal pesticide registrant who applies to the state for an SLN registration. However, anyone can request an SLN registration, and grower groups can be particularly effective in obtaining a new use under an SLN label. Good communication between the registrant, grower groups, agricultural extension, state agencies, and other contributors is essential for a successful application to the state.

A special local need is defined as an existing or imminent pest problem within a state for which the state lead agency, based on satisfactory supporting information, has determined that an appropriate federally registered pesticide product is not sufficiently available. The actual criteria for "not sufficiently available" is set by individual state regulatory policy, and such criteria vary widely between states.

Traditionally, SLN registrations are granted for a new use rate; a new method or timing of application; a new crop or site; a new pest; a changed rate; an application in a particular soil type; a new product or different formulation; resistance management within an integrated pest management (IPM) program; a use that is less restrictive than required by the Section 3 label, if the state can demonstrate that the proposed change will not cause an unreasonable effect on humans or the environment; a use that is more restrictive than required by the Section 3 label, or restricts the product to a subset of the Section 3 label; and a less hazardous product to prevent pollution and reduce risk. All states may issue 24(c) registrations to more than one product for the same use in the same state if the additional registrations are necessary to provide full economic control of the pest problem and sufficient efficacy and economic data are provided to support that use. An example of such a situation is the requirement of different pesticides to provide control at early and late stages of the life cycle of the weed, insect, or disease.

State-Specific Registrations: Section 18

FIFRA Section 18—Emergency Exemption—allows state and federal agencies to permit the use of an unregistered pesticide in a specific geographic area for a limited time to address an emergency. Such situations usually arise when growers and others encounter a pest problem on a site for which there is no registered pesticide available. A Section 18 can also be issued when there is a registered pesticide that would be more effective than the currently available products but is not yet approved for use for the particular pest. Emergency exemption Section 18s can also be approved for reasons of public health and quarantine. EPA must find that the pesticide use proposed in an emergency exemption poses no unreasonable adverse effects and that there is a reasonable certainty of no harm to human health or the environment.

Section 18 emergency exemptions are not issued at the request of the product registrant. Only growers, grower groups, local agricultural departments, and other interested parties, such as extension specialists, may request emergency exemptions.

FIFRA allows four specific types of Section 18s: (1) *Specific Exemption Section 18*, may be authorized in emergency conditions to avert a significant economic loss to the grower or a significant risk to endangered or threatened species, the environment, or beneficial organisms; (2) *Crisis Exemption Section 18*, to be utilized in an emergency condition when the time from discovery of the emergency to the time when the pesticide is needed is insufficient to allow normal agency review.

A crisis exemption is issued by the head of a state or federal agency, the governor of a state, or an official designee, and usually requires that the pest exceed specific thresholds for pest populations and economic damage. A crisis Section 18 cannot be granted for a first time use of a new active ingredient on any crop that will be used for food; (3) *Quarantine Exemption Section 18*, to control the introduction or spread of any pest new to or not known to be widely prevalent or distributed throughout the United States; and (4) *Public Health Exemption Section 18,* to control a pest causing a significant risk to human health.

MINOR ACREAGE CROPS

In contrast to the major crops such as corn, soybeans, and wheat that are grown extensively worldwide on millions of hectares, minor crops are cultivated in limited areas. Minor crops include most horticulture and specialty species—vegetables, fruits, nuts, ornamentals, and some forages. Because the market for herbicide sales is small for minor crops and the cost of development and registration of herbicides for specific uses on these crops can exceed the potential financial return, chemical companies are reluctant to develop herbicides specifically for a minor crop. Thus, most herbicides marketed for minor crops were previously developed for a major crop and, as a result, the tolerance of many minor crops may be less than optimal or even unknown. Therefore, herbicides must be used with greater care in minor crops. Normally, at least a twofold safety factor is desired; that is, twice as much herbicide is required to induce crop injury as is needed for weed control. An additional factor that deters a company's interest in registration of herbicides for minor crops is the risk/profit ratio. Because minor crops are high in value, the liability resulting from crop injury can be high relative to the profit made on the limited amount of chemical sold.

There is little incentive for chemical companies to register pesticides for minor crops, thus the number of available pesticides is low and few new chemicals are becoming available. This problem is compounded by the fact that many of the pesticides now registered for use in minor crops are old chemistries that are undergoing reregistration because of FQPA requirements and, as a result, may be lost to minor crop growers. Chemical companies do obtain some minor crop registrations for their herbicides previously registered for use in major crops when the economics are favorable. When minor crop registrations are submitted, a chemical company can use the chemistry, toxicology, and environmental data used for registration of the herbicide for use on a major crop.

IR-4

Another approach to the registration of pesticides for minor crops involves the government agency IR-4 (Interregional Project 4). This project, housed at Rutgers University, coordinates and supports federal registration and state research directed toward the development of the information required for the registration of a pesticide for minor crops and other minor uses by EPA. The main purpose of IR-4 is to assist in gaining pesticide registrations for minor crops. This project started as a small USDA regional project but has increased greatly in size, funding, and use in recent years. The federal government supports IR-4, but state involvement is also an integral part of the total program. Priorities are established by IR-4 in collaboration with states and researchers regarding chemical pest control needs in minor crops. Once the need is established, a coordinated program between federal and state research personnel to develop the required efficacy and crop residue data is initiated. Chemical company information derived from existing registrations of the pesticide is also used. The information is assembled into a petition that IR-4 submits to EPA to establish a residue tolerance in the case of food crops. Upon EPA's approval of a residue tolerance, the manufacturer can petition for a registration. IR-4 has reduced chemical companies' costs associated with data gathering and tolerance petition preparation and, as a result, has been responsible for the labeling of thousands of pesticides for minor crops that otherwise would have never occurred.

LITERATURE CITED AND SUGGESTED READING

CAST. 1987. *Health Issues Related to Chemicals in the Environment: A Scientific Perspective.* Council for Agricultural Science and Technology. Ames, IA.

Dietary pesticide risk assessment. 1992. *Rev. Environ. Contam. Toxicol.* **127**:23–67.

Guenzi, W. D., ed. 1974. *Pesticides in Soil and Water.* Soil Science Society of America, Madison, WI.

Hunnicut, R. C. 1995. *Mechanisms of Pesticide Movement into Ground Water.* CRC Publications, Boca Raton, FL.

Messersmith, C. E. ed. 1988. Symposium on groundwater contamination by herbicides. *Weed Technol.* **2**:206–227.

Meyer, M. T., and E. M. Thurman, eds. 1996. Herbicide metabolites in surface and ground water. *ACS Symposium Series 630. Am. Chem. Soc.*, Washington, DC.

National Research Council. 1986. *Pesticides and Groundwater Quality: Issues and Problems in Four States.* National Academy Press, Washington, DC.

Racke, K. D., and J. R. Coats, eds. 1990. *Enhanced Biodegradation of Pesticides in the Environment.* ACS. Symp. 426. American Chemical Society, Washington, DC.

Somasundaram, L., and J. R. Coats, eds. 1991. *Pesticide Transformation Products: Fate and Significance in the Environment.* ACS Symposium Series 459. American Chemical Society, Washington, DC.

Whitford, F., D. T. Barber, D. Scott, C. R. Edwards, and J. Carvetta. 1994. *Pesticides and the Label.* Bulletin #PPP-24. Purdue Pesticide Programs, Purdue University Cooperative Extension Service, West Lafayette, IN.

Whitford, F., D. Gunther, J. Contino, R. Doucette, B. Amber, and J. Castleman. 1996. *Pesticides and Material Safety Data Sheets: An Introduction to the Hazard Communication Standards.* Bulletin #PPP-37. Purdue Pesticide Programs, Purdue University Cooperative Extension Service, West Lafayette, IN.

WEB SITES

U.S. EPA Office of Pesticide Programs

http://www.epa.gov/pesticides

IR-4 Program

http://aesop.rutgers.edu/~ir4/

California Department of Pesticide Regulation

http://www.cdpr.ca.gov/index.htm

For chemical use, see the manufacturer's or supplier's label and follow the directions. Also see the Preface.

5 Herbicides and the Plant

When a herbicide comes in contact with a plant, its action is influenced by the morphology and anatomy of the plant as well as numerous physiological and biochemical processes that occur within the plant and the environment. These processes include (1) absorption, (2) translocation, (3) molecular fate of the herbicide in the plant, and (4) effect of the herbicide on plant metabolism. The interaction of all these factors with the herbicide determines the effect of a specific herbicide on a given plant species. When one plant species is more tolerant to the chemical than another plant species, the chemical is considered to be selective. These topics and methods to influence selectivity will be discussed in this chapter.

The life processes of plants are many and varied; they are complex and delicately balanced. Disturb one of these processes, even only slightly, and a chain of events may be initiated that changes plant growth and development. Minor changes in the environment may also result in major changes in the life processes of a plant. For example, many perennial plants remain dormant below ground all winter. When the soil temperature increases a few degrees in the spring, a complex series of reactions is initiated that results in the beginning of another annual cycle of growth.

There are different concepts of the terms *mode of action* and *mechanism of action* of herbicides within the scientific community. However, the National Academy of Science book entitled *Weed Control* (Anon., 1968) stated, "The term 'mode of action' refers to the entire sequence of events from introduction of a herbicide into the environment to the death of plants. 'Mechanism of action' refers to the primary biochemical or biophysical lesion leading to death."

The following terms will also be used repeatedly in the subsequent discussion of how herbicides kill plants and are applied: (1) *herbicide*—a chemical that kills or inhibits growth of plants; (2) *contact herbicide*—a herbicide that causes injury only to tissue to which it is applied; (3) *mobile herbicide*—a herbicide that moves or translocates in a plant; (4) *symplast*—total living protoplasmic continuum of a plant; it is continuous throughout the plant, and there are no islands of living cells; the phloem is a component of the symplast, and long-distance symplastic transport is via the phloem; (5) *apoplast*—total nonliving cell-wall continuum of a plant; the xylem is a component of the apoplast, and long-distance apoplastic translocation is via the xylem; (6) *burndown*—refers to applying a foliage-active herbicide before planting to kill undesired vegetation; (7) *preplant incorporated*—refers to applying a herbicide to the soil before planting the crop and then mixing it with the soil; (8) *preplant*—refers to applying a herbicide to the soil surface before planting the crop; (9) *preemergence*—refers to applying a herbicide after planting but before the crops and weeds emerge; (10) *directed*—refers to applying a herbicide so contact with the crop is minimized; and (11) *postemergence*—refers to applying a herbicide after emergence of the weeds or the crop, including broadcast and spot treatments.

ABSORPTION

Absorption of applied herbicides occurs through shoot and root tissue. To be effective in killing weeds, herbicides applied postemergence must move through the leaf surface to the living parts (symplast) of plant cells. Although leaves are the most important absorption site, absorption can occur through any aboveground plant part where the herbicide is present. After preemergence application, absorption is primarily through root tissue but can also occur through shoot tissue in contact with the treated soil. As with postemergence applications, herbicides applied preemergence must absorb into the plant and move to the symplasm for activity. Root absorption can be important after a postemergence herbicide application when the herbicide reaches the soil and moves in the soil to the root zone.

Absorption by Leaves and Stems

There are several factors that affect absorption by influencing the amount and distribution of herbicide on a plant surface, among which are the following:

1. The surface tension of the spray solution
2. The inherent wettability of the leaf surface
   a. The amount of cuticular wax and physical structure of the wax
   b. The hairiness (number of trichomes) on the leaf surface
3. Leaf orientation with respect to incoming spray droplets
4. The total leaf area per plant (probability of intercepting a spray droplet)

Most herbicide applications are made with water as the carrier. Water has a very high surface tension due to hydrogen bonding between water molecules. Thus, a water-based spray solution has difficulty wetting the waxy surface of weeds. The surface tension of water can be easily reduced by the addition of a surfactant.

The type of surface wax (epicuticular wax) and the surface topography can have a great influence on the ease with which a spray solution is able to wet the leaf surface and penetrate into the plant. Figure 5-1 shows some of the differences in leaf surface characteristics between grass and broadleaf species. Generally smooth leaf surfaces, devoid of crystalline epicuticular wax but containing amorphous wax (e.g., many dicot species), are relatively easy to wet. Leaf surfaces covered with crystalline epicuticular wax (e.g., many grass species) are much more difficult to wet. Retention of spray solution on difficult-to-wet leaf surfaces increases with smaller droplet size and as the surface tension of the droplet at the moment of impact decreases.

Large differences in absorption between species in some cases can account for herbicide selectivity between weed and crop. The most commonly cited examples of

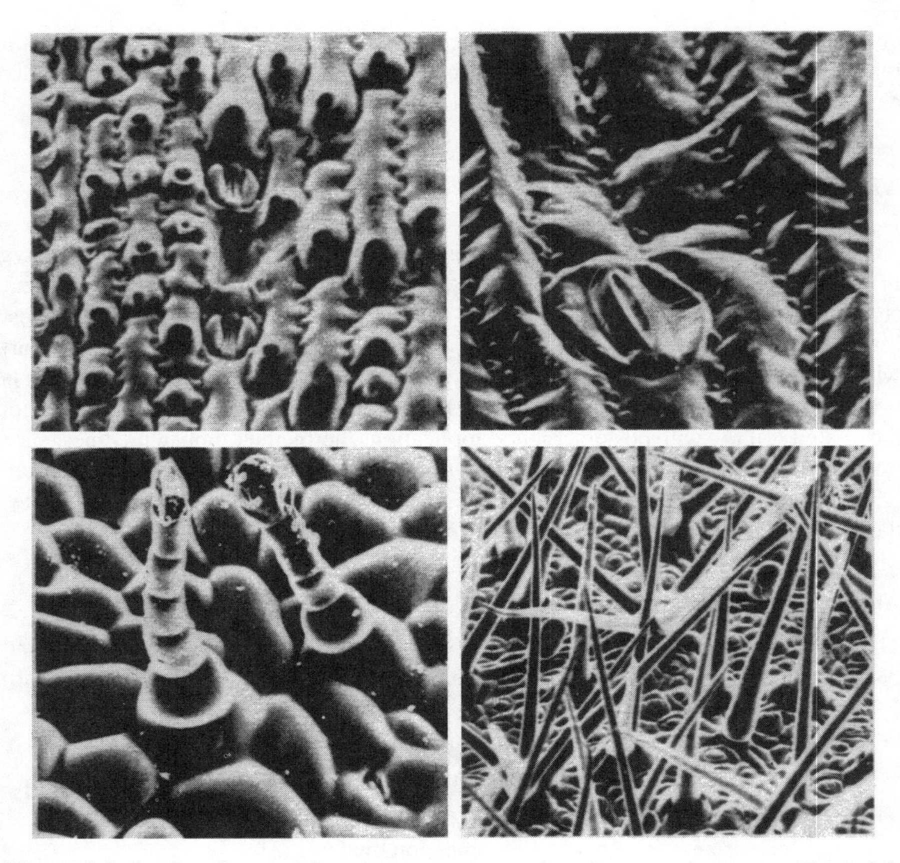

Figure 5-1. Leaf surfaces as they appear on a scanning electron microscope. *Upper left*: Bermudagrass (450x). *Upper right*: Nutsedge (1050x). *Lower left*: Redroot pigweed (350x). *Lower right*: Velvetleaf (170x). (D. E. Bayer and F. D.Hess, University of California, Davis.)

tolerant crop species are onion and *Brassica* species such as cabbage, because they absorb a minimal amount of a herbicide as a result of their thick leaf cuticles. Santier and Chamel (1992) showed that glyphosate absorption was 94% through thin tomato fruit cuticles, but only 1 to 6% for thicker cuticles of box-tree leaves, rubber plant leaves, and pepper fruit.

Total carrier volume, droplet size, and droplet number per area can have an effect on the absorption and performance of herbicides. Knoche (1994) published a very detailed analysis of the literature on droplet size and carrier volume. Carrier volume, droplet size, and droplet number per area are different variables and thus must be analyzed separately. A significant factor for carrier volume is how it influences concentration of herbicide within droplets. Movement of herbicides across plant cuticles is a diffusion process; thus, the higher herbicide concentration per unit area covered by spray in low-volume applications results in a larger concentration gradient across the cuticle, which increases the "driving force" for diffusion into the leaf (Figure 5-2).

Unfortunately, the influence of carrier volume on herbicide performance is not always consistent for different herbicides applied under different conditions. Generally, application volumes for most postemergence herbicides range from 10 to 40 gallons per acre. These volumes result in optimal herbicide activity. Further reductions in carrier volume below 10 gallons per acre or increases above 40 gallons per acre tend to decrease activity. The optimal carrier volume for every postemergence herbicide is specified on its product label and should be used in making applications. It is important to remember that the herbicide concentration per unit area of leaf can change dramatically if some droplets roll off the leaf or dry quickly. Spray droplets are influenced by the surface tension of the spray solution and waxiness of the leaf surface. Decreasing droplet size more frequently enhances herbicide performance on difficult-to-wet plant species, probably because of better adhesion of smaller droplets. The performance of systemic herbicides tends to increase as droplet size decreases, whereas contact herbicides require good leaf coverage with larger spray droplets for better activity. Spray droplet size also influences penetration of the plant canopy, as

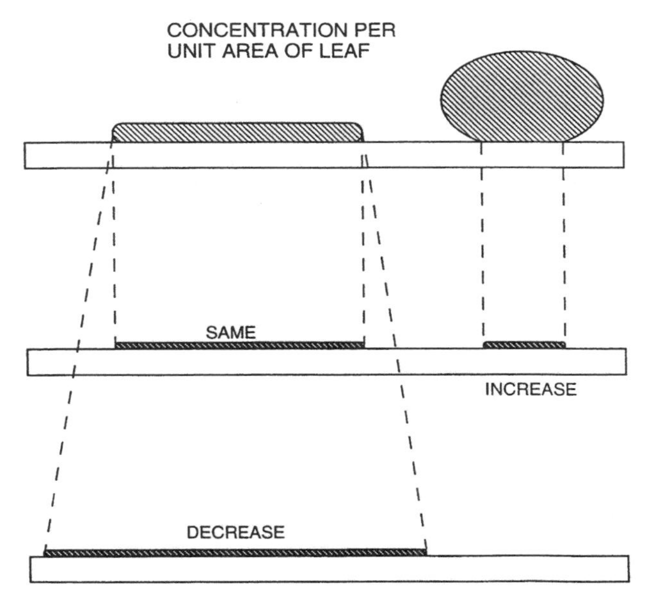

Figure 5-2. Effect of droplet spread and drying on herbicide concentration per unit area of leaf surface. *Left*: Droplets (*top*) and (*middle*) have same concentration per unit area versus (*bottom*), which has spread more and has decreased concentration per unit area. *Right*: Droplet (*top*) has lower concentration per unit area than middle droplet even though area covered is the same, because there is a higher concentration of herbicide in the lower droplet due to carrier evaporation.

larger droplets penetrate farther into the canopy and spray drift is more likely with small droplets.

Once the herbicide has contacted the plant surface, five things can happen to the active ingredient (Figure 5-3). It may

1. Volatilize and be lost to the atmosphere or be washed off by rain.
2. Remain on the outer surface in a viscous liquid or crystalline form.
3. Penetrate the cuticle but remain absorbed in the lipoid (wax) components of the cuticle.
4. Penetrate the cuticle, enter the cell walls, and then translocate prior to entering the symplasm. This is called apoplastic translocation, which includes movement in the xylem.
5. Penetrate the cuticle, enter the cell walls, and then move into the internal cellular system (through the plasmalemma) for symplastic translocation, which includes phloem movement.

Rainfall can wash a significant amount of a herbicide from the leaf surface. Anionic (negatively charged) salt herbicides (e.g., sodium salts) are water soluble and do not

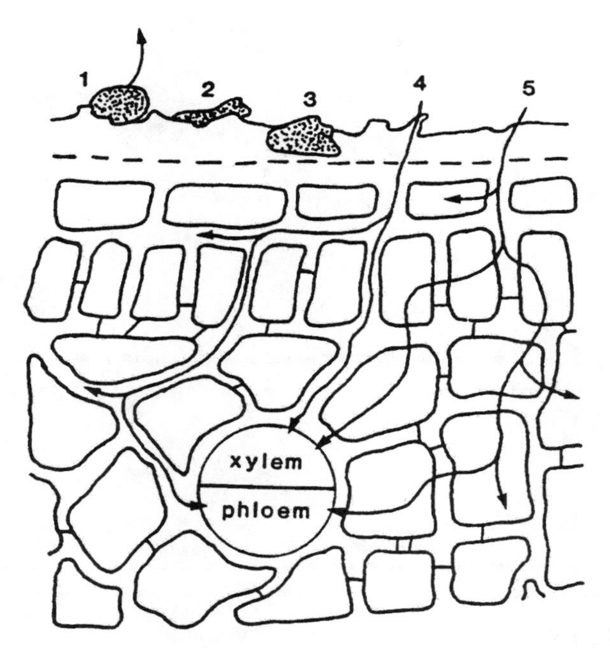

Figure 5-3. Five things that can happen to a herbicide once it contacts the leaf surface: (1) volatilize or be washed off by rain; (2) remain on the surface; (3) penetrate the cuticle and remain in the cuticle; (4) penetrate the cuticle, enter the cell walls, and translocate in xylem to the symplast; (5) penetrate the cuticle, enter the cell walls, and translocate in phloem to the symplast.

penetrate the cuticle rapidly or absorb to the cuticle surface. These characteristics result in greater amounts of these herbicides that can be washed from the leaf surface by rain within 24 hours after application. Cationic (positively charged) salt herbicides (e.g., paraquat) are water soluble but rapidly adsorb to the negatively charged cuticle and thus are less subject to removal from leaves by rain. Lipophilic herbicides (usually formulated as an EC or flowable) have low water solubility but are readily absorbed into the lipophilic cuticle. These characteristics make both cationic and lipophilic herbicides less subject to loss from the leaf surface by rainfall.

The physical form of the herbicide spray droplet on the leaf can have a significant impact on its activity. Crystallization of the herbicide active ingredient on the leaf surface reduces effectiveness (e.g., Hess and Falk, 1990; MacIsaac et al., 1991; Nalewaja et al., 1992). If the active ingredient is a solid in pure form at ambient temperatures, crystallization may occur on the leaf surface as the water and formulation components (solvents) evaporate from the spray droplet.

Dew does not result in herbicide wash-off from leaves, but it can rehydrate (redissolve) salt herbicides on leaf surfaces. Because herbicide absorption into leaves is most rapid when herbicides are in true solution on the leaf surface, there may be a burst of absorption in the morning as the dew redissolves the salt herbicides on the leaf surface.

Ions present in the spray water can also influence herbicide performance, and most reports indicate that ions in spray solutions decrease herbicide performance. For example, glyphosate activity was reduced when cations (e.g., calcium, sodium, and magnesium) and anions (e.g., bicarbonate, chloride, and sulphate) were present in the spray water (de Villiers and du Toit, 1993; Thelen et al., 1995). The decreased activity was thought to be due to formation of ionic complexes between glyphosate and the ions, decreasing glyphosate absorption. Tralkoxydim and sethoxydim are also known to be negatively influenced by ions (e.g., sodium bicarbonate) in the spray water (de Villiers, 1994; Nalewaja et al., 1994). In this instance, ions in the spray water reduced the speed of herbicide absorption, which was thought to allow for greater UV (ultraviolet) degradation of unabsorbed herbicide on the leaf surface.

Environmental factors 1 to 2 weeks before and immediately after herbicide application can influence absorption of postemergence herbicides. High light coupled with low relative humidity and, most important, low soil moisture tend to induce synthesis of leaf cuticles with increased lipophilic character; thus, when herbicide—particularly, water-soluble herbicide—application occurs, performance decreases. In one study, haloxyfop efficiency was reduced from 92% in nonstressed johnsongrass and crabgrass to 12% in water-stressed plants (Peregoy et al., 1990). This difference was due to decreased absorption and translocation of haloxyfop in stressed plants. Low relative humidity during and after treatment results in a dehydrated cuticle, which can reduce absorption of water-soluble herbicides that need a well-hydrated cuticle for optimum absorption. For example, glufosinate activity was reduced from complete death at 95% relative humidity, to 30% growth inhibition at 40% relative humidity (Anderson et al., 1993). Rain or irrigation can eliminate water stress and overcome reduced absorption due to a dehydrated cuticle, but it will not overcome reduced absorption due to the change in cuticle composition. In most instances, the best postemergence herbicide performance is obtained when the treated weed is growing in moderate temperatures (65 to 85°F) rather than in high (greater than 95°F) or low (less than 40°F) temperatures. Generally, as temperature increases, diffusion rates (permeance) of organic molecules within cuticles increase (Baur and Schönherr, 1992). Although the decrease in herbicide performance in high temperatures is not well understood, it is probably related to the overall influence of temperature stress on metabolic processes in the plant.

The overall chemical character of the cuticle is lipidlike (fatlike). The cuticle consists of cutin, epicuticular wax, embedded wax, and pectin (Figure 5-4). Its thickness ranges from 0.1 to 10.0 μm, and it has an overall negative charge at physiological pH. The cuticle has both hydrophilic (water-loving) and lipophilic (wax-loving) characteristics. The bulk of the cuticle volume consists of cutin, and it has a relatively even distribution of wax and water-loving components. Embedded and epicuticular wax are primarily wax loving but do have 10 to 20% water-loving components. Wax present on a leaf surface is either crystalline or amorphous. Pectin is present as strands at the base of the cuticle next to the cell wall; however, pectin strands also extend into the cuticle proper in many species. The pectin components tend to be more water loving.

Lipophilic/hydrophobic properties of herbicides are important in understanding how they absorb across plant cuticles. Lipophilic (or wax-loving) herbicides (oil-soluble, water-emulsifiable, or water-dispersible formulations) are able to move through the cuticular barrier by simple diffusion in association with the dominant

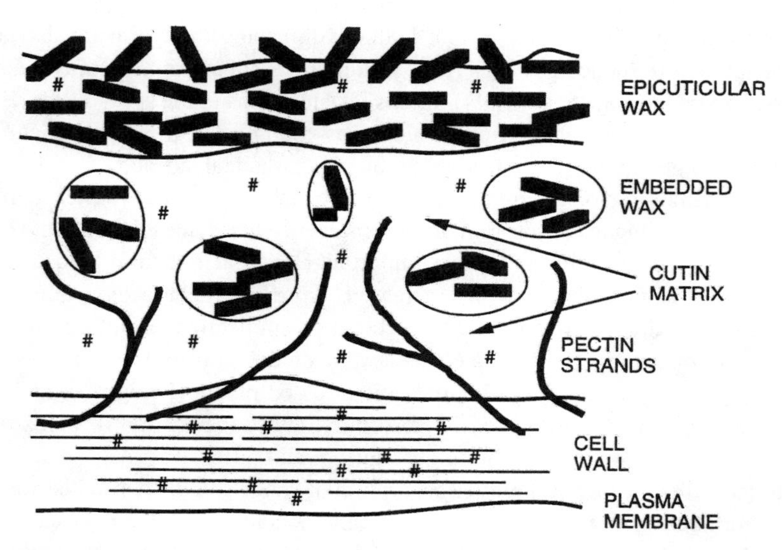

Figure 5-4. The chemical composition of the cuticle of a plant leaf. The cuticle consists of epicuticular wax, embedded wax, cutin and pectin located above the cell wall, and the plasma membrane. See text for more detail.

waxy components. The rate of herbicide movement across the cuticle is dependent on the transport properties (solubility or permeance within the cuticle) and the driving force (concentration gradient). Diffusion of lipophilic herbicides across the epicuticular layer is primarily through those waxes that are in the amorphous state. The cuticle is known to be thinner at some locations on the leaf surface (e.g., at the base of the epidermal hairs and over the guard cells), which can provide preferential penetration sites for oil-soluble herbicide formulations.

Water-soluble herbicide formulations (e.g., salts) are also able to enter the plant through the cuticle surface by simple diffusion. However, because of their low permeance within the cuticle, their rate of movement is significantly less than that of lipophilic herbicides. This reduced permeance often results in less total herbicide absorbing into the plant. The epicuticular waxes are the most significant barriers to absorption of water-soluble herbicides. For example, Tan and Crabtree (1992) found that removing the epicuticular wax from apple leaves increased 2,4-D absorption by about 2.5 times. Once diffusion has occurred across the epicuticular wax, there are hydrophilic/polar components in the cuticle to help in the absorption process. These hydrophilic cuticular components consist of the portion of the cutin that is hydrophilic, or water-loving, and the pectin strands. The overall cuticle contains water, which imparts a hydrophilic component to the cuticle that allows diffusion of water-soluble herbicides. In addition to the natural absorption via diffusion, there are often breaks in the cuticle layer caused by wind, rain, insects, and other agents that can increase the absorption of water-soluble herbicides.

Stomatal penetration of spray solutions is not a common occurrence (Schönherr and Bukovac, 1972). However, if the surface tension of the spray solution is reduced enough, stomatal penetration can occur. One thing to remember is that in most weeds, the stomates are on the underside of the leaf; therefore, for most plants, stomatal penetration is not a major means of spray entry for any foliar-applied compound.

Root Absorption of Herbicides from Soil

With the exception of gaseous fumigants (e.g., methyl bromide), herbicides affect plants only after germination begins. Thus, entry of herbicides into ungerminated seeds during water imbibition may occur, but is not considered important to the ultimate death of the plant. Of importance is absorption of herbicides into roots as they begin to grow out from the seed. Absorption of herbicides by roots is not as limited as absorption into leaves. The primary reason for this is that no significant wax layer or cuticle is present at the locations where most of the herbicide absorption occurs. The most important pathway of entry is comigration of the herbicide with water being taken into the plant in the root hair zone (zone of differentiation) of the root tips. Root hairs greatly increase the surface area of roots available for uptake of water and herbicides. There is no known specialized "route of entry" for herbicides into root tissue; they go along with the mass flow of water because they are dissolved in the water. Even though the water solubility of some herbicides absorbed into roots is low (0.3 ppm for trifluralin), the solubility is adequate to deliver the needed dose to the site of action. Herbicide in the vapor phase in the soil may be important, but it will still have to diffuse through the film of water surrounding the root tissue at the root tip.

Whereas there is no significant cuticular barrier on the surface of roots at the location where most herbicides enter (root hair zone), there is a lipophilic barrier located in the root endodermis. In the endodermal layer of cells, all radial walls contain a band (Casparian strip) heavily impregnated with a lipophilic substance called suberin. This barrier is known to be impermeable to water and thus forces the movement of water and dissolved substances (primarily ions) to cross the plasma membrane that separates the symplast from the apoplast. Most ions are not able to readily diffuse across the plasma membrane, thus some type of transport mechanism is required (see pumps, carriers, in the following sections). This process allows the plant to maintain a correct ion balance in the water that is to be translocated throughout the plant in the xylem. What happens to herbicides at the endodermis is not completely clear. Herbicides probably also enter the symplast by diffusion through the plasmalemma, as most herbicides can readily diffuse across the plasma membrane.

Shoot Absorption of Herbicides from Soil

Shoot absorption is an important means of entry for many soil-applied herbicides that are active on germinating seeds or small seedlings (e.g., carbamothioates and chloroacetamides). Before emergence, a shoot has a poorly developed cuticle and probably no wax layers, making it more easily penetrated by herbicides. In addition, the Casparian strip barrier is not present in shoot tissues. Shoot absorption is a particularly important route of herbicide entry in grass species, but less so for dicot weeds. Shoot zone entry is by diffusion, from herbicide dissolved in the soil solution in contact with the shoot tissue or, probably more important, diffusion from herbicide present in the vapor phase of the soil (e.g., EPTC). Herbicides known to have a major route of entry in the shoot zone are primarily growth inhibitors having their site of action in the shoot meristem as it emerges through the soil.

Absorption Across Plant Membranes

All biochemical target sites for herbicide action are located within the living cell (symplast). For herbicides to reach their target site, they must cross the membrane located at the cell wall (plasma membrane; also termed the plasmalemma) and often an additional organelle membrane (e.g., chloroplast envelope). Most herbicides cross membranes by simple diffusion, although a few have been shown to move with the help of specific carriers.

In the case of herbicides crossing the plant membranes by simple diffusion, movement is related to the concentration difference (gradient) across the membrane. The herbicide moves from a region of higher concentration to a region of lower concentration. When the concentration is equal on both sides of the membrane, accumulation stops. The herbicide concentration gradient across the membrane is the driving force that moves the herbicide across the membrane. The second important parameter for diffusion is the partition coefficient of the herbicide within the membrane, which represents its ability to "dissolve" in the membrane. Lipophilic (oil-soluble) herbicides are able to move more freely across the membrane (Figure 5-5, top right) than hydrophilic (water-soluble) herbicides (Figure 5-5, bottom right).

Some herbicides can move across the plasma membrane against a concentration gradient. These herbicides have an ionizable group [e.g., carboxylic acid group (COOH)] as part of the molecule. In this case the hydrogen ion (H^+) gradient across the membrane can serve as an additional driving force for herbicide movement. The aqueous environment outside the cell has a lower pH (more hydrogen ions) than the aqueous environment inside the cell. This pH difference is created by the action of the ATPase hydrogen ion pump. Depending on the pK_a of the ionizable group on the

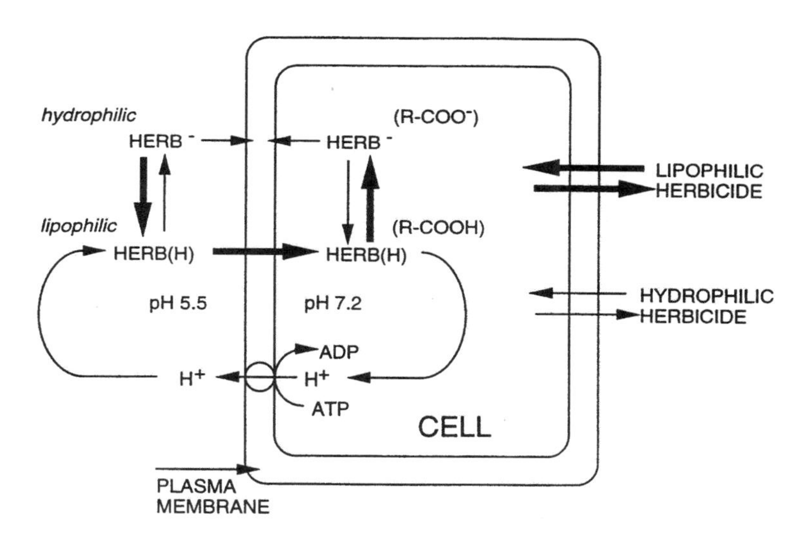

Figure 5-5. Movement of herbicides across plant membranes can occur by diffusion or by use of the H^+ gradient. Simple diffusion is the process by which many herbicides cross the membrane, and movement is related to a concentration difference (gradient) across the membrane. No metabolic energy is required. When the herbicide concentration is equal on both sides of the membrane, accumulation stops. Lipophilic herbicides more readily diffuse across the membrane than do hydrophilic herbicides. Some herbicides move across the plasma membrane against a concentration gradient. Most of these herbicides have an ionizable group (COOH) as part of the molecule, and the hydrogen ion (H^+) gradient across the membrane is the driving force for movement. The aqueous environment outside the cell has a lower pH (more hydrogen ions) than the aqueous environment inside the cell. This pH difference is created by the action of the ATPase hydrogen ion pump. Depending on the pKa of the ionizable group on the herbicide, an equilibrium will be established between the ionic form (–) (more water soluble) and the protonated form (+) (more lipid soluble) on each side of the membrane. Because of the pH difference between the inside and outside of the cell, the ratio of the protonated form, HERB(H), to the ionized form, $HERB^-$, will favor the protonated form, HERB(H), on the outside as compared with the inside. The protonated form, HERB(H) (more lipophilic), readily diffuses into the cell, whereas the ionized form, $HERB^-$ (water soluble), has difficulty diffusing out of the cell. Thus, the herbicide can build up inside the cell against a concentration gradient (often termed ion trapping).

herbicide, an equilibrium will be established between the ionic form (more water soluble) and the protonated form (more lipid soluble) on each side of the membrane. "Protonated" means that a hydrogen ion (also referred to as a proton) is associated with the ionic group. Because of the pH difference between the inside and the outside of the cell, the ratio of the protonated to ionized form will favor the protonated form on the outside as compared with the inside.

The protonated form (more lipophilic) readily diffuses into the cell and loses its proton, and this ionized form (water soluble) has a difficult time diffusing out of the cell. Thus, the herbicide can build up inside the cell against a concentration gradient (often termed ion trapping) (Figure 5-5). Examples of herbicides that have different types of ionizable groups are bentazon, chlorsulfuron, 2,4-D, and sethoxydim. A detailed discussion of herbicide movement by ion trapping can be found in Sterling (1994), as listed at the end of this chapter.

Only three herbicide types have been shown to move across the plasma membrane by use of a carrier-mediated process. These are paraquat, 2,4-D, and glyphosate. Paraquat moves across the membrane on the putrescine carrier. Auxinic herbicides (e.g., 2,4-D) cross the plasma membrane using the auxin carrier. Glyphosate crosses plant membranes using the phosphate carrier (Denis and Delrot, 1993). Characteristic of the involvement of carriers in herbicide absorption are (1) accumulation of the herbicide against a concentration gradient and (2) reduced absorption after treatment with metabolic inhibitors [see Sterling (1994) for additional information].

TRANSLOCATION

Once a herbicide has penetrated the leaf or stem cuticle or the root epidermis, there are still many barriers that can affect movement to its site of action. The herbicide can be moved into a portion of the cell not containing a site of action (e.g., vacuole), resulting in the herbicide's becoming compartmentalized and no longer available for transport to its site of action. A number of herbicides are conjugated (chemically bound) and adsorbed onto cellular components or in some way inactivated in the plant roots or leaves and do not move to other parts of the plant. A number of selectivity mechanisms that differ between weeds and crops are the result of differential compartmentation or inactivation by metabolism (Hess, 1985).

Short-distance herbicide movement across a few cell layers occurs by simple diffusion, ion trapping, or, for a few herbicides, carrier-mediated processes. Some herbicides (e.g., trifluralin applied preemergence and atrazine applied postemergence) require only diffusion to reach their target site of action.

Assuming the herbicide is not immobilized in some manner, it is available for long-distance movement in the plant through the xylem and phloem transport systems. In both systems herbicides dissolved in water move along with the mass flow of water. The general directions of flow are shown in Figure 5-6 and are described in detail in the following sections.

Xylem

In roots, the xylem tissue is located inside the endodermis (termed the *stele region*), often as a collection of cells in the very center of the root. As xylem cells become mature and functional in the root hair zone, the end walls are digested away. The symplasts then disintegrate, leaving hollow cells with no end walls between adjacent xylem cells. Thus, the xylem is composed of nonliving cells (apoplast) connected end to end, without end walls, which provide a series of long, very thin tubes that extend from the root to the shoot. Xylem tubes continue through leaf petioles and into all sections of the leaves. The main function of the xylem is to supply all living cells in the plant with a constant supply of water, as well as nutrients that are dissolved in the water absorbed by the roots.

Herbicides that enter the plant roots may move upward in the xylem with the flow of water. The majority of preemergence herbicides transported in the xylem system enter the root near the root hair zone. This is because herbicide molecules dissolved in water move with the mass flow of water and most of the water transported in the xylem enters roots at the root hair zone.

Water and substances dissolved in the water move in the xylem by two mechanisms. The mechanism that is most often operating in plants is termed *transpirational pull*. When the relative humidity is less than saturated, water is "pulled" up the xylem as a result of water evaporating from the leaf surface (transpiration). The water evaporating from the leaf can provide an adequate driving force (pull) for this mechanism to function. At 100% relative humidity the water potential of air equals zero. At 98% relative humidity the water potential of air decreases to –27.5 bars, which would be adequate to support a capillary column of water 920 feet high. As relative humidity decreases, the water potential continues to decrease; as the plant transpires, a large negative water potential is created in the leaves. A water potential gradient thus exists through the continuous water column in the xylem network from the shoots to the roots. The plant roots have more negative water potential than the soil solution, so water and substances dissolved in the water readily move into the roots. The overall mass movement of water and any dissolved substances (e.g., herbicides) into the roots will be toward the more negative water potential, created by the evaporating water at the surface of the leaf.

The second mechanism for xylem flow occurs with high soil moisture and high relative humidity. In this case, water moves in the xylem as a result of root pressure. However, this mechanism of xylem flow is of minor importance in herbicide movement within the plant.

Because of the mass flow of water and dissolved substances in the xylem system during xylem translocation, and the apparent diffusion of herbicides through the plasma membrane at the endodermis, most, but not all, herbicides are xylem mobile. There are several explanations for a lack of xylem mobility: (1) Herbicides may adsorb to apoplastic or symplastic cellular components, (2) herbicides may become compartmentalized in cellular components (e.g., vacuoles or plastids), and (3) herbicides may become conjugated to cellular substrates that are not xylem mobile or may be degraded to inactive forms. Because of the need for herbicides to enter the symplast to reach their sites of action, it is unlikely that any herbicides are not xylem mobile as a result of an inability to cross the plasmalemma at the endodermis.

Phloem

Movement of herbicides in the phloem follows the same path to the same locations as the products of photosynthesis (sugars). Translocation in the phloem system is from "source" (photosynthesizing mature leaves) to "sink" [plant parts using these sugars for growth (e.g., cellulose), maintaining metabolism (e.g., respiration) or storing the sugars for future use (e.g. starch)]. Examples of sinks are roots, underground storage organs (e.g. tubers and rhizomes), young developing leaves, all meristematic zones, flowers, and developing fruits.

The principal type of cells in the phloem of higher plants are sieve elements, which when joined end to end are termed *sieve tubes*. Sieve elements are living cells (symplast) that have no nuclei. The end walls (sieve plates) of sieve elements contain connective pores, which join sieve elements together. The loading and unloading of photosynthate (primarily sucrose) into and out of phloem at sources and sinks are important in understanding the movement of substances in the phloem. Sugar enters phloem companion cells or directly into sieve elements by being moved from the cell wall (apoplast), across the plasmalemma to the symplast, against a concentration gradient, by the use of metabolic energy (ATP). An ATPase is responsible for pumping hydrogen ions into the wall, which then reenters the phloem with sucrose transport (termed *symport*). The sugar concentration is from 1.5 to 2.0 times higher in the sieve elements than in the mesophyll cells. In sink regions sugars are unloaded from the phloem, most likely by an active process, and then utilized or stored.

An important feature of most growth-regulator herbicides (e.g., 2,4-D), as well as herbicide classes such as sulfonylureas and imidazolinones, and the herbicide glyphosate, is an ability to be transported in the phloem. Herbicides applied to the leaves are able to move to the roots of perennial plants. Because the phloem movement of herbicides is associated with sugar production and transport, environmental conditions favoring optimum photosynthesis (high light, adequate soil moisture, and moderately warm temperatures) will maximize herbicide movement. In addition, it is very important not to kill the leaf and stem tissues rapidly, because transport is via living phloem tissue. Rapid foliage kill will result in poor transport and poor root kill. Sometimes two or three small doses of a herbicide of this type will give better results than a single large dose that kills too rapidly. A useful feature of glyphosate is the lack of rapid leaf injury even at high concentrations.

A question arises as to why some herbicides are able to translocate in the phloem and others are not. According to collected data, herbicides are readily able to move across membranes, so most herbicides are apparently able to enter the phloem. Herbicides with rapid contact action, however, are not phloem mobile because they destroy the plasma membrane as they come in contact with sieve elements. Phloem function requires sieve elements to have intact live plasma membranes. Thus, contact herbicides destroy the system used to transport them. Other herbicides do not translocate any appreciable distance in the phloem because they diffuse across membranes too easily. Atrazine, for example, can diffuse into the phloem, but as the water (and dissolved atrazine) in the phloem begins to move, the atrazine diffuses back out. In leaves and petioles the xylem and phloem systems are in proximity (vascular bundles). Xylem and phloem flows are in opposite directions, and water flow in the xylem (transpiration stream) is more rapid than in the phloem. Thus, the net direction of movement in the vascular bundle for herbicides (e.g., the triazines), which can freely move between the symplast (phloem) and the apoplast (xylem), is in the direction of the transpiration stream.

Why then, can some herbicides remain in the phloem long enough to translocate? There are two mechanisms for explaining herbicide movement in the phloem. The first mechanism becomes apparent when the structures of phloem-mobile herbicide molecules are studied. Many phloem-mobile herbicides have ionizable groups (mostly carboxyl groups or other weak acids), which exist in a noncharged form (COOH) or a charged form (COO^-). These forms are in equilibrium, with the favored species being dependent on the pH of the surrounding medium. The pH outside the phloem is approximately 5 and the pH within the phloem is approximately 8. This pH gradient is established during the active loading of sucrose into sieve elements by the involvement of an ATPase that "pumps" hydrogen ions into the cell wall. In the apoplast (outside the phloem), ionizable groups on herbicides with appropriate pK_a values will favor the protonated form (COOH). The hydrophilic-lipophilic balance of the protonated molecule is more lipophilic than the anion (COO^-), and this form of the herbicide readily diffuses across the plasmalemma into the sieve element (into the phloem). Because of the higher pH in the phloem, the equilibrium is shifted toward the nonprotonated (anionic) form (COO^-) as the H^+ disassociates from the herbicide molecule. This form does not readily penetrate the plasmalemma, so the herbicide becomes trapped in the sieve element (as described in Figure 5-5) and moves to the sink, where it can be unloaded at its site of action. For this principle to work, herbicides must have significantly different degrees of dissociation between pH 5 and pH 8. Herbicide movement in the phloem can be inhibited with uncouplers of phosphorylation; however, the inhibition is the result of a loss of the pH gradient that drives movement into the phloem.

There are exceptions to all phloem-mobile compounds containing an ionizable group with an appropriate pK_a. A second mechanism for phloem mobility occurs for compounds with intermediate membrane permeability coefficients. Once diffused into the phloem, these compounds can remain long enough for phloem translocation to occur.

If a herbicide is xylem mobile and applied to roots, it will move more or less uniformly to aboveground leaf and stem tissues that are not in direct contact with the herbicide. Within leaves, preferential accumulation sometimes occurs at the leaf margins. If this same herbicide is applied postemergence to the base of a mature leaf, it will move throughout the leaf but often accumulates at the margins. As water (vapor pressure of 18 mm Hg) evaporates from the leaf through the stomata and cuticle, herbicides that have a much lower vapor pressure (e.g., $< 10^{-5}$ mm Hg) stay in the leaf. Water that flows toward the margins of the leaves will move the herbicide toward the leaf margins, where it concentrates. Under normal growing conditions, very little of a xylem-mobile herbicide that is applied will be exported from the treated leaf. In opposite fashion, a phloem-mobile herbicide applied to a mature leaf will move out of the leaf and translocate to the plant sinks. Figure 5-6 demonstrates the general

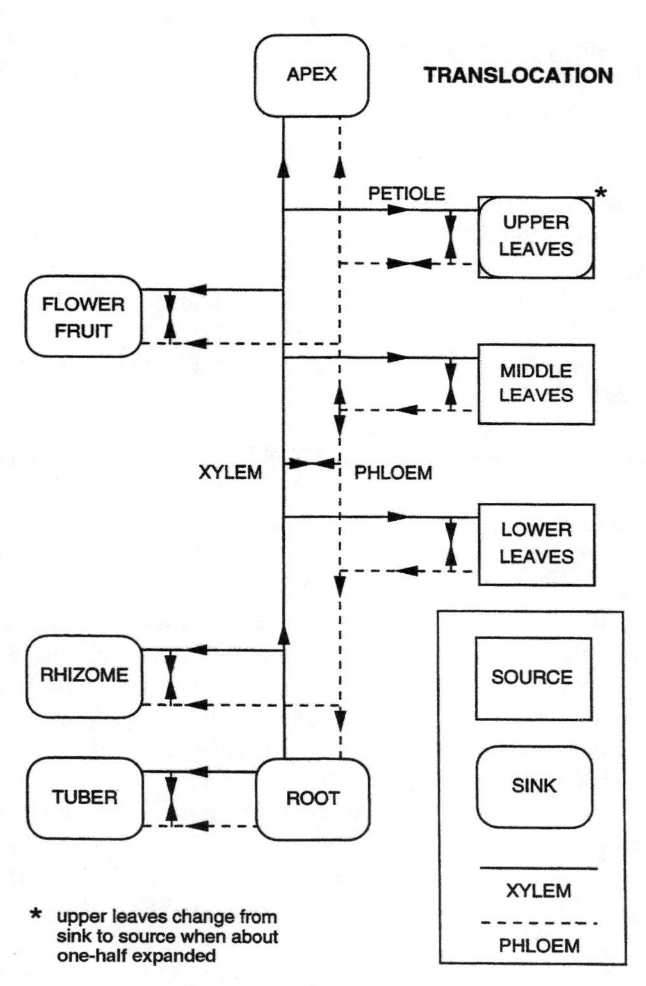

Figure 5-6. Translocation direction of solutes in the xylem and phloem of a plant. Solid lines (—) indicate movement direction in the xylem, and dashed lines (-----) indicate movement in the phloem. Movement of solutes (including herbicides) in xylem is unidirectional from the root upward to tubers, rhizomes, leaves, flowers, fruit, and apex. Movement in phloem is from source (□) to sinks (▢). Phloem movement can be upward to sinks such as the apex, flower, and fruits, and downward to rhizomes, tubers, and roots. More mature lower leaves and middle leaves are sources. Moreover, interchange of materials (➨) can occur between xylem and phloem constituents.

TABLE 5-1. Relative Mobility and Primary Translocation Pathway(s) of Herbicides[a,b]

Good Mobility			Limited Mobility			
Apoplast	Symplast	Both	Apoplast	Symplast	Both	Little or No Mobility
Bentazon	Glyphosate	AMA	Bromoxynil	Oxadiazon	AMS	Bensulide
Carbamothioates		Amitrole		Phenoxys	Aryloxyphenoxys	DCPA
Carbamates[c]		Arsenicals[d]	Difenzoquat			Dinitroanilines
Chloroacetamides		Asulam	Diquat		Endothall	Diphenylethers
Diclobenil		Clopyralid	Ethofumesate			
			Fluridone		Napthalam	
		Dicamba	Paraquat		Propanil	
Napropamide		Fosamine				
Norflurazon		Imidazolinones				
Pyrazon		Pyridines				
TCA		Sethoxydim				
Triazines		Sulfonylureas				
Uracils						
Urea						

[a]Translation rate may vary considerably in different species. Some herbicides may also move from the symplast to the apoplast and vice versa.
[b]When a class of herbicides is given, see class section in herbicide chapters for individual herbicides.
[c]Except for asulam, which has good mobility in both the symplast and apoplast.
[d]Organic arsenicals.

translocation patterns of constituents (herbicides) moving in the phloem and the xylem. Usually postemergence phloem-mobile herbicides applied to mature leaves on the lower portion of the plant will translocate preferentially to roots, and those applied to the upper mature leaves will move to developing leaves and shoot meristem. Once such herbicides arrive in the roots, some (e.g., 2,4-D) diffuse out into the surrounding soil solution (for review, see Coupland, 1989). Table 5-1 shows the relative mobility and primary translocation pathways of herbicides.

SELECTIVITY, MOLECULAR FATE, AND METABOLISM OF HERBICIDES

Selectivity is perhaps the most important concept of modern weed science. Because of this phenomenon weeds can be controlled in crops with herbicides. To be useful, the herbicide (or mixture of herbicides) must provide an acceptable level of weed control while not injuring the crop to the degree that yield loss occurs. A herbicide is selective to a particular crop only within certain limits. The limits are determined by a complex interaction between the plant, the herbicide, and the environment (Ashton and Harvey, 1987). There are numerous ways in which this is accomplished, some of which may be surprising. In practice, the overall crop tolerance is often the result of a combination of two or more of the selectivity mechanisms described in the following sections.

Molecular Fate and Metabolism

Clearly, the difference between a herbicide's metabolism in the weed and in the crop is one of the main mechanisms for selectivity. In addition, where selectivity is not adequate for a particular herbicide in a specific crop, genetic engineering has been used to introduce genes that can detoxify the herbicide through metabolism (e.g., bromoxynil in cotton, glufosinate in many crops). The mechanism of metabolism varies significantly between different crop-herbicide combinations; thus, a simple table showing how each herbicide class is metabolized is not possible. For example, chlorimuron-ethyl undergoes ring hydroxylation in corn, and glutathione conjugation in soybean. However, there are two enzyme-based reaction types that dominate with respect to metabolism leading to selectivity in crop plants: oxidation and conjugation.

The overall metabolism of herbicides in plants can be divided into four phases (Figure 5-7). Phase I is a direct change in the herbicide structure brought about by oxidation, reduction, or hydrolysis reactions. Phase II is conjugation to cell constituents such as glucose, glutathione, or amino acids. Many herbicides undergo Phase I prior to Phase II; however, some herbicides can be directly conjugated with no preoxidation. During Phase III conjugates are transported across cell membranes into the vacuole or cell wall, where further processing (Phase IV) can occur to yield

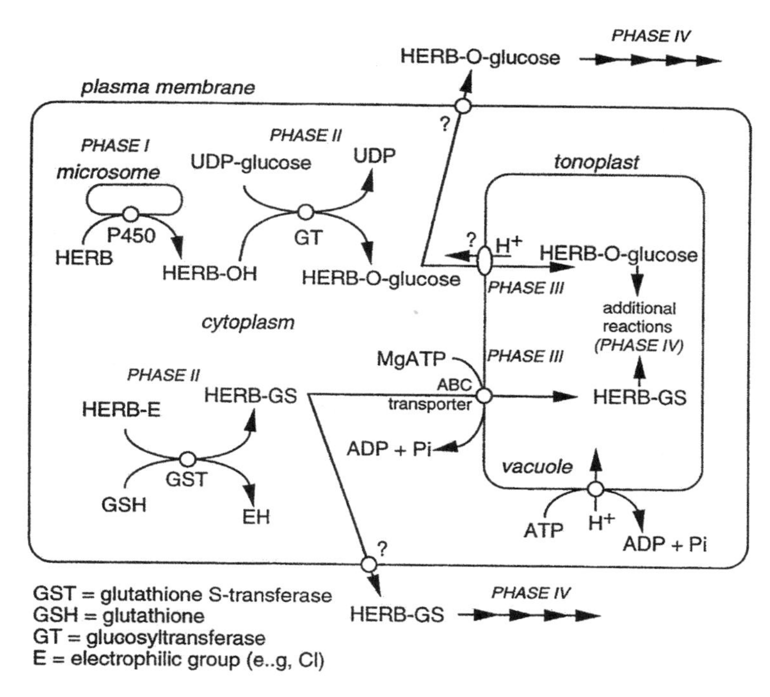

Figure 5-7. The overall metabolism of herbicides in plants can be divided into four phases. Phase I, which generally involves P450 enzymes, is a direct change in the herbicide structure brought about by oxidation, reduction, or hydrolysis reactions. Phase II reactions involve conjugation of the herbicide molecule to cell constituents such as glucose, glutathione, or amino acids. Many herbicides undergo Phase I prior to Phase II; however, some herbicides can be directly conjugated with no preoxidation. Phase III involves herbicide conjugates being transported across cell membranes into the vacuole or cell wall, where they are stored or further processed by Phase IV reactions, which result in complete inactivation of insoluble and/or bound herbicide residues.

insoluble and/or bound residues (Figure 5-7). All these processes result in loss of herbicide activity.

Oxidation reactions (Phase I) within plants that lead to herbicide detoxification are most often based on P_{450} enzymes (also called cytochrome P_{450} or MFO enzymes—mixed function oxidase enzymes). These enzymes bind molecular oxygen, catalyze its activation, and incorporate one of its atoms into the herbicide. The second oxygen atom is reduced to form water. NADPH (reduced form of nicotinamide adenine dinucleotide phosphate) provides the electrons needed for activation of the oxygen.

Examples of oxidation reactions for herbicides include hydroxylation of aromatic rings (2,4-D, dicamba, primisulfuron, bentazon), hydroxylation of alkyl groups (chlortoluron, prosulfuron, chlorsulfuron), and hydroxylation followed by loss of oxidized carbon [N-dealkylation (chlortoluron, diuron) and O-dealkylation (metoxuron)]. The diversity of P_{450} enzymes accounts for the large differences in herbicide metabolism by different plant species.

Following oxidation, many herbicides are rapidly conjugated to a sugar (glycosylated). Glycosylated herbicides are then processed further and become "bound" residues in the extracellular matrix or are stored as water-soluble metabolites in the vacuole. Examples of herbicides conjugated in this manner include metribuzin and 2,4-D.

Where conjugation is the primary mechanism of herbicide metabolism (preoxidation not involved), glutathione is the primary substrate and glutathione S-transferase (the primary enzyme) is involved in deactivating the herbicide. Herbicides directly conjugated by this means include the chloroacetamides, aryloxyphenoxypropionates, triazines, and carbamothioates.

Several GST (Glutathione S-tranferase) isozymes exist in plants, and they differ in their herbicide specificity. These isozymes account for the vast differences in the susceptibility of different plant species to different herbicides. Once GSH-conjugates of herbicides form, they are further processed, as discussed earlier for glycosylated herbicides.

Examples of other metabolic inactivation reactions of herbicides are reduction (N-deamination in metribuzin); hydrolysis in phenmedipham, propanil, and chloro-s-triazines; and amino acid amide conjugation in 2,4-D.

Activation of Herbicides Through Metabolism

The classic example of activation of a herbicide through metabolism is beta oxidation of 2,4-DB (which is not herbicidal) to 2,4-D. In legume crops, beta-oxidation of 2,4-DB to 2,4-D does not occur; thus, they are protected from any herbicidal effects. Other plant types, including many weeds, beta oxidize 2,4-DB to 2,4-D and are killed. Wild oat is killed by imazamethabenz as a result of rapid de-esterification of the inactive methyl ester to the phytotoxic acid, and wheat metabolizes imazamethabenz to an inactive form. A final example of plant metabolism activating herbicides is the carbamothioate herbicides that are activated by undergoing sulfoxidation to the more reactive sulfoxide form in susceptible weeds.

Placement of the Herbicide in Time or Space

In placement, the sensitivity of the crop to the herbicide must be considered, so the herbicide is kept away from the crop in time or space. This is accomplished by applying the herbicide weeks to months prior to when the crop is planted. In some instances application may be in the fall for a spring-planted crop. Although the crop may have some tolerance to the herbicide, soil degradation of the herbicide that occurs prior to crop planting further reduces the risk of injury.

Another method is to apply postemergence herbicides to control emerged weeds before the crop emerges. Timing is critical, because if the crop has started to emerge, damage can occur. This technique works best in cool-season crops inasmuch as more time is usually available for herbicide application prior to crop emergence. Contact nonresidual herbicides (e.g. paraquat, glyphosate, and glufosinate) are often used this way because the herbicide cannot damage the crop after moving into the soil. This technique is extensively used in no-till soybeans and corn to kill weeds present when the crop is planted.

Some herbicides not inherently selective may become selective when they are in specific vertical positions in the soil profile (Figure 5-8). Such selectivity depends on the different rooting habits of crop and weed. A herbicide that readily leaches below the rooting zone of a shallow-rooted crop can be used to control deep-rooted weeds without injuring the crop. Conversely, a herbicide that remains near the soil surface can control shallow-rooted weeds in a deep-rooted crop.

A residual preemergence herbicide absorbed by roots can be selective when it is placed on the soil surface (or incorporated shallow) and the roots of perennial crops or the seeds of annual crops remain below the herbicide zone. Examples include simazine, diuron, and terbacil on perennial crops such as sugar cane, asparagus, citrus, coffee, apples, peaches, and woody ornamentals.

Certain herbicides applied on the soil surface over large-seeded crops planted deep (e.g., peanuts, beans, corn and cotton) are also selective, because roots rapidly grow away from the herbicide if contained in the upper zones of the soil. Shallow incorporation of trifluralin coupled with deep sowing of wheat is used to control green foxtail in spring wheat. For this "placement" mechanism to work, the herbicide must not be highly mobile in the soil.

A herbicide can be applied so that most of it covers the weed but little of it contacts the crop. This can be done by using shielded or directed sprays or wick-wiper applicators. Shielded sprays prevent the herbicidal spray from touching the crop while the weeds are covered with the spray. The spray nozzles are simply placed under a hood, or the crop is covered with a shield (Figure 5-9). Directed sprays are less cumbersome than shielded sprays and can be used when the crop can tolerate a small amount of the herbicide.

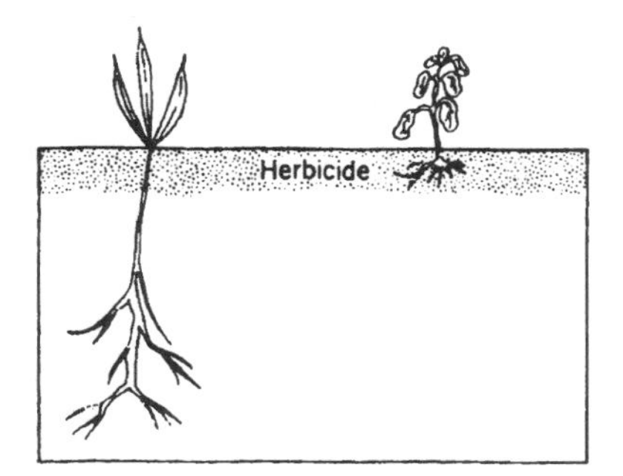

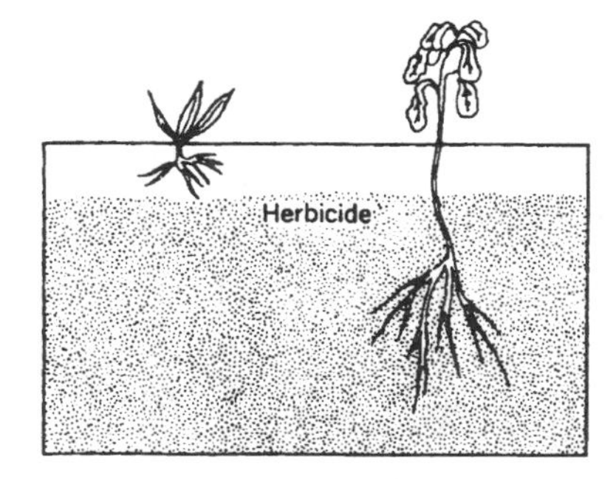

Figure 5-8. Differential leaching by herbicides alters selectivity. *Left*: A herbicide that remains near the soil surface can injure shallow-rooted weeds (*right*) and not injure the deep-rooted crop (*left*). *Right*: A herbicide that leaches from soil surface into the lower soil profile can injure deep-rooted weeds (*right*) and not injure a shallow-rooted crop (*left*).

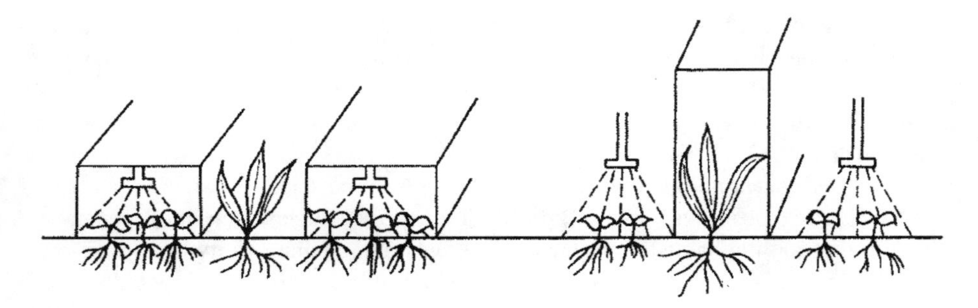

Figure 5-9. Shielded sprays protect crops from being sprayed with herbicides. *Left*: Spray confined within shields. *Right*: Crop covered with a shield.

Examples include contact herbicide sprayed under tree crops, vineyards, cotton, soybeans, and corn. Often the stems of crop plants have more tolerance to the herbicide than leaves (perhaps due to reduced uptake); thus, if a small amount of herbicide is sprayed on the stem tissue, injury is minimal. This is accomplished in row crops by using drop nozzles and nozzles that minimize spray drift, carefully controlling the nozzle height and direction of the nozzles, and when the crop is taller than the weeds, (for example, 2,4-D in corn) (Figure 5-10). Wick-wiper applicators are used where the weeds are higher than the crop. The herbicide solution is wiped on the weeds; very little, if any, herbicide contacts the crop (Figure 7-12).

Anatomical Differences Between Crop and Weed

The amount of spray retention by foliage after postemergence applications can affect selectivity. This selectivity is usually due to the crop plant's having a waxy cuticle that repels the spray solution. Examples include onions, peas, cereal grains, *Brassica* vegetable crops, and conifers. Medium to high spray volumes usually provide better

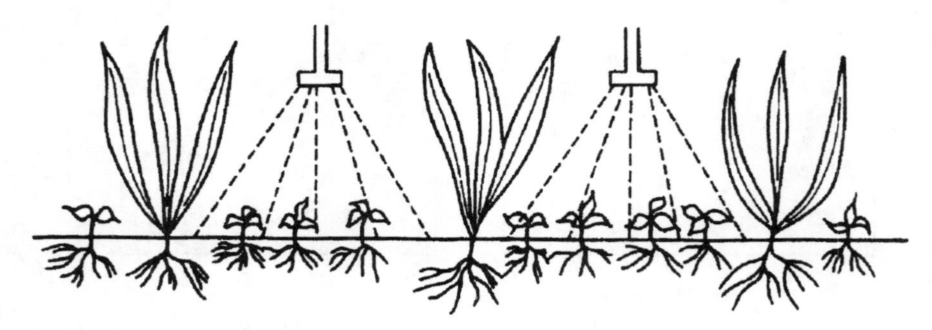

Figure 5-10. Directed sprays are aimed toward the base of the crop plant, favoring minimum coverage of the crop and maximum coverage of weeds.

selectivity, and adding an adjuvant can decrease selectivity as a result of enhanced adhesion of the spray droplets.

Differences in leaf shape, size, and orientation between weed and crop can provide some selectivity differences. This is most common for controlling dicot weeds in small grain crops (the grain leaves retain less herbicide because of shape, orientation, size, and granular epicuticular wax). Postemergence selectivity can be due to the growing point of the crop being protected from direct contact by the herbicide while the growing point of the weed is exposed. The best example is dicot weed control (growing point not well protected by emerging leaves) in small grains (growing point well protected by the whorls of emerging leaves). The herbicide must not have a high degree of phloem mobility for this selectivity mechanism to work. For more detail regarding the influence of plant morphology on herbicide absorption, see the review by Hess (1987).

Preemergence selectivity can be due to a difference in root morphology between the weed and the crop. Grass weeds usually have a fibrous root system, whereas dicot crops usually have a taproot system. Thus, growth inhibitor herbicides, such as trifluralin, applied to the soil come directly in contact with the growing root tips in grass weeds, but not with those of the deeper-rooted dicot crops. For this selectivity mechanism to be useful, the water solubility and soil binding characteristics of the herbicide must be such that movement is restricted to the upper soil profile. Morphology differences within stem tissue of grass plants can provide differences in selectivity. The growing point of many grass weeds (crabgrass and wild oat) are more exposed to herbicide-treated soil than wheat and barley where the growing point is protected inside the coleoptile.

Resistance at the Site of Action

There are surprisingly few instances of crop selectivity being due to resistance at the herbicide site of action in the cell. One example is the tolerance of carrots to dinitroanilines. However, a mechanism of introducing selectivity into crops through genetic engineering is to add an enzyme to the crop having an altered binding site so the herbicide is no longer active. Examples are glyphosate-resistant soybeans and cotton and imidazolinone-resistant corn.

Tolerance at the site of action may be due to the herbicide's being present as a proform where the metabolism needed for changing the molecule to its active form does not occur in the crop. Perhaps the best-known example is the application of 2,4-DB to a legume and the deesterification of inactive imazamethabenz-methyl ester to the active acid, which is much more rapid and complete in weeds than in wheat (as described earlier under Activation of Herbicides through Metabolism).

Internal Factors Other Than Metabolism

Differences in the kind and/or quantity of food reserves in seeds can be the basis for selectivity. Some crops with large food reserves in the seed are able to rapidly grow away from the location of the herbicide (cotton fields treated with trifluralin) or outlast the inhibition that occurs prior to the herbicide's being metabolized (corn fields treated with triazines).

There are several examples of selectivity being due to differences in translocation to the site of action between the weed and the crop (Hess, 1985). This is often the result of the herbicide's being removed from solution by metabolism or compartmentation prior to or after moving in the xylem or phloem. For example, norflurazon can be compartmentalized in the glands of glanded cotton. Linuron remains in the roots of parsnip and carrot, thus keeping it away from the chloroplast site of action in the leaves. The difference in soybean cultivar sensitivity to metribuzin is thought to be due to reduced translocation (due to glucose conjugation in the roots), coupled with more of the translocated metribuzin remaining in the leaf veins of the tolerant cultivars.

Differences in Crop and Weed Susceptibility at Different Stages of Growth

There are herbicide applications in which selectivity is due to the crop's being dormant (little growth and development is occurring) when small winter annual weeds are germinating and vigorously growing. An example is early application of herbicides to alfalfa during its dormant period in the early spring. Conifer species are more tolerant of 2,4-D prior to bud break in the spring and after rapid terminal growth has ceased in the fall. Application at these times can be used to control dicot plants competing with conifer growth.

Selectivity may simply be due to a difference in age (development state) between the weed and the crop. This is a selectivity mechanism for controlling seedling weeds in established perennial crops. Many annual crops become more tolerant to herbicides as their growth stage (age) increases. For example, tomatoes are much more tolerant to metribuzin after they reach the 5-to-6 leaf stage. In this case, the difference in selectivity between the older crop and the younger (more succulent weed) is usually due to a combination of the differential in uptake, translocation, and metabolism between the weed and the crop.

Localized Application of Adsorbents

Activated charcoal is used to increase crop tolerance to certain herbicides. The charcoal is usually applied as a slurry to the roots during the transplanting process (e.g., strawberries) or in a narrow band over the row at time of seeding (e.g., some grass seed and vegetable crops). Any herbicide applied cannot reach the absorption region of the roots because of its being adsorbed by the charcoal.

Use of Safeners to Protect the Crop

Otto Hoffman conceived the concept of using chemicals to achieve selectivity between weed and crop in the late 1940s. As a result of this pioneering research, the use of herbicide antidotes, or safeners, in weed control is a commercially accepted technology.

With respect to how crop safeners are used with herbicides, there are two types of application. One is application of the safener to the seed prior to planting, and the other involves mixing the safener directly with the herbicide (a premix). Which application method to use depends on the activity of the safener. If the safener has activity on weeds as well as crops, then it must be used as a crop seed treatment. If the safener reduces crop injury without reducing weed control, then the safener is usually mixed with the herbicide. The latter is the preferred application method. Safeners are effective in preemergence and postemergence applications for reducing herbicide phytotoxicity to grass (*Poaceae*) crops; however, most commercial applications are preemergence.

There are two primary mechanisms that define how most safeners reduce crop injury, both of which are related to enhancing the metabolic detoxication of the herbicide. Many safeners enhance the level of glutathione (GSH) and/or the glutathione S-transferase enzyme that conjugates the herbicide to GSH, thus inactivating the herbicide. The most consistent correlation related to safener action is an increase in GST activity.

The other mechanism of safening is an enhancement of activity or quantity of cytochrome P_{450} monooxygenase enzymes in plants. These enzymes can inactivate herbicides by various oxidation reactions as described earlier. Following oxidation, the herbicide is conjugated to glucose or other natural plant constituents, which usually occurs through a second enzyme system (for example, glucosyltransferases in the case of glucose conjugation). In some systems, glucosyltransferase enzymes are also enhanced by safeners.

There are further reports that safeners reduce the uptake and/or translocation of the herbicide in the crop plant. This may explain part, but not all, of the safening action of these compounds.

Unfortunately, some safeners are reported to cause mild phytotoxicity to crop plants when applied without the herbicide. For example, it is well known that naphthalic anhydride causes a mild phytotoxicity (chlorosis and growth inhibition) under some growing conditions. One problem in treating seeds with safeners prior to planting is that the mild phytotoxicity can increase as the time the safener is exposed to the seed increases. With naphthalic anhydride, the phytotoxicity to the crop increases with the increased time the safener is in contact with the seed during storage. This problem has thus far prevented naphthalic anhydride from being introduced to the commercial market.

Role of the Herbicide

The various aspects of herbicides relative to selectivity include molecular configuration, concentration, formulation, and chemical combinations.

Molecular Configuration Variations in molecular configuration of a herbicide change its properties, which in turn modify its effect on plants. This is illustrated in Figure 5-11, which shows the herbicides trifluralin and benefin. The only difference is that a methyl group (-CH_2-) is moved from one side of the molecule to the other. Chemical structures of herbicides are modified during discovery and development by chemical companies to alter phytotoxicity and selectivity.

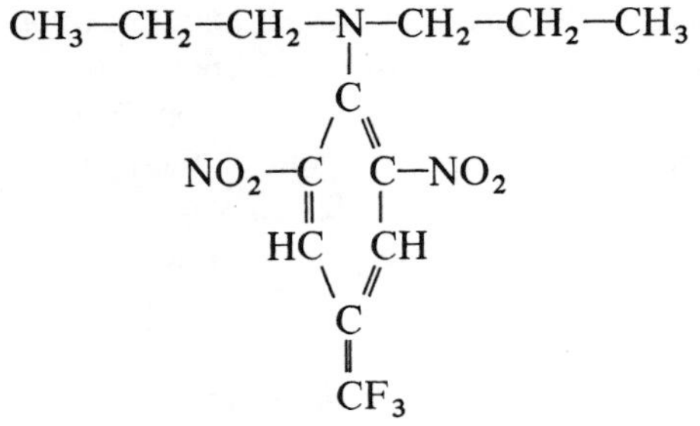

```
CH3—CH2—CH2—N—CH2—CH2—CH3
            |
            C
           / \\
    NO2—C     C—NO2
        ||    |
        HC    CH
          \  //
            C
            ||
            CF3
```

trifluralin

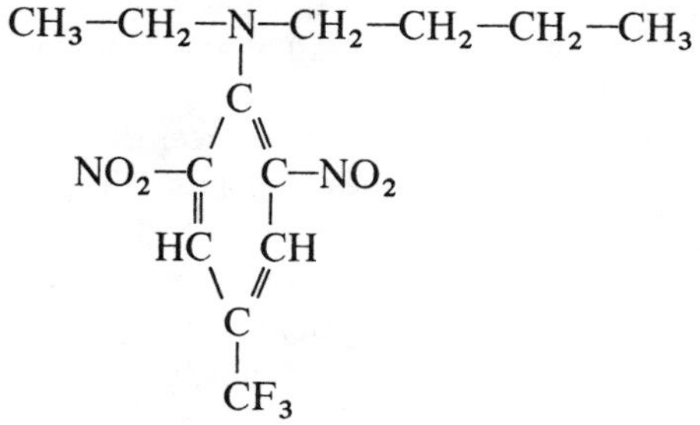

```
CH3—CH2—N—CH2—CH2—CH2—CH3
        |
        C
       / \\
NO2—C     C—NO2
    ||    |
    HC    CH
      \  //
        C
        |
        CF3
```

benefin

Figure 5-11. The chemical structures of trifluralin and benefin are quite similar; however, there are major differences in plant selectivity.

Concentration Concentration may determine whether a herbicide inhibits or stimulates metabolism and growth of a plant. The endogenous plant growth regulator indole-3-acetic acid (IAA) inhibits respiration and growth at high concentrations but stimulates them at low concentrations. In many ways, herbicides like 2,4-D act similarly to IAA. In fact, 2,4-D at low concentrations is a common component of tissue culture growth media that stimulate cell division.

Formulation The formulation of a herbicide is vital in determining whether it is selective or not with regard to a given species. A good example is the granular form that permits a herbicide to "bounce off" a crop and fall to the soil. In this instance, the crop may be susceptible to a spray application, yet tolerant to the herbicide if it can be kept off the foliage. Substances known as adjuvants and surfactants are often added to improve the application properties of a liquid formulation; these additives may increase or decrease phytotoxicity. The addition of nonphytotoxic oils or surfactants to liquid atrazine or diuron formulations induces foliar contact activity in these normally soil-active herbicides. The addition of herbicide antidotes, safeners, or protectants (as discussed earlier) to formulations or as seed coatings is used to increase the crop tolerance to certain herbicides.

Chemical Combinations Herbicides are often mixed with fertilizers, fungicides, insecticides, nematicides, or other herbicides to facilitate application. Occasionally, tank mixing of herbicides with other herbicides, pesticides, or fertilizers can alter selectivity. Therefore, it is advisable to use such combinations only when they are specifically recommended on the product label.

Role of the Environment

Dominant environmental factors that affect selectivity include soil type, rainfall or overhead irrigation, and soil-herbicide interactions. Details of soil-herbicide interactions are presented in Chapter 6.

In general, herbicide characteristics, soil type, and the amount of water received after herbicide application from rainfall or overhead irrigation determine the vertical position of a specific herbicide in the soil. Adsorption, the tenacity with which a herbicide molecule is bound to soil particles, will strongly affect its movement in the soil. Low adsorption, high water solubility, high amounts of overhead water, and coarse soil types favor leaching of the herbicide into the soil profile. Some herbicides

Figure 5-12. Grape leaves exposed to 2,4-D. *Left*: Low level exposure. *Right*: High level exposure.

are extremely resistant to leaching, whereas others readily move with water. The movement is normally downward, but the herbicide may move upward as water evaporates from the soil surface.

The temperature of the environment in which a plant is growing has considerable influence on the rate of its physiological and biochemical processes. For example, the optimum temperature for the germination of seeds of different species varies greatly (e.g., spinach 41°F and cantaloupe 77°F). The selectivity of various plants to herbicides also varies as the temperature differentially affects their physiological and biochemical processes.

The effect of temperature on the rate of these processes is often expressed as the temperature coefficient, or Q_{10}. A Q_{10} of 2 means that the rate of a chemical reaction is doubled for each increase in temperature of 10°C. Most reactions of herbicides that influence plant growth are chemical in nature. Therefore, a change from 15°C (59°F) to 25°C (77°F) may result in a doubling of the activity of the herbicide.

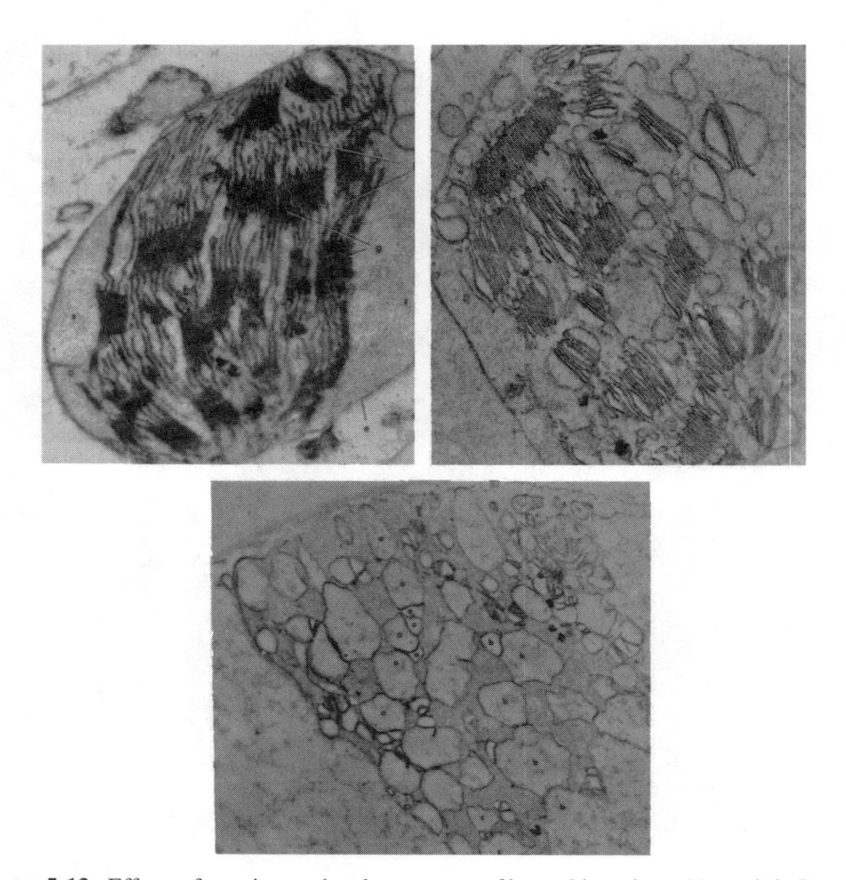

Figure 5-13. Effects of atrazine on the ultrastructure of bean chloroplasts. *Upper left*: Control. *Upper right*: Moderate injury. *Lower:* Severe injury.

GROWTH AND PLANT STRUCTURE

All the selectivity factors discussed in this chapter can play an important role in how herbicides are most effectively used to obtain the desired weed control while minimizing crop injury. Normal growth and plant structure are the result of previous normal biochemical or biophysical processes. Therefore, abnormal growth and plant structure caused by herbicides must be preceded by altered biochemical or biophysical processes induced by the herbicide.

Herbicides may induce abnormal plant growth through morphological, anatomical, and cytological effects. However, these effects are generally specific for a given herbicide on a particular plant species. Thus, a given abnormal growth symptom in the field often suggests which herbicide group induced the injury, as will be described in more detail in Chapters 8 through 17. Knowledge of plant and herbicide symptomology has become increasingly important with time, as it is related to herbicide litigation (Chapter 29).

Among the abnormal responses induced by herbicides are (1) seed emergence failure, shoot inhibition, and root swelling after germination, (2) leaf chlorosis, (3) abnormal leaf form (Figure 5-12), (4) stem swelling, (5) cell division inhibition, (7) chloroplast destruction (Figure 5-13), and (8) membrane disruption followed by necrosis.

LITERATURE CITED AND SUGGESTED READING

Anon. 1968. *Principles of Plant and Animal Pest Control*. Vol. 2, *Weed Control*, Pub. 1975. National Academy of Science, Washington, DC.

Anderson, D. M., C. J. Swanton, J. C. Hall, and B. G. Mersey. 1993. The influence of temperature and relative humidity on the efficacy of glufosinate-ammonium. *Weed Res.* **33**:139–147.

Ashton, F. M., and W. A. Harvey. 1987. *Selective Chemical Weed Control*. Bulletin 1919. University of California, Division of Agriculture and Natural Resources, Berkely.

Bauer, H., and J. Schönherr. 1992. Determination of mobilities of organic compounds in plant cuticles and correlation with molar volumes. *Pestic. Sci.* **35**:1–11.

Briskin, D. P. 1994. Membranes and transport systems in plants: An overview. *Weed Sci.* **42**:255–262.

Coupland, D. 1989. Factors affecting the phloem translocation of foliage-applied herbicides. In *Mechanism and Regulation of Transport Process.*, ed by R. K. Atkin and D. R. Clifford. pp. 85–112. British Plant Growth Regulator Group, Monograph 18. Bristol, UK.

Denis, M. H., and S. Delrot. 1993. Carrier-mediated uptake of glyphosate in broad bean (*Vicia faba*) via a phosphate transporter. *Physiol. Plant.* **87**:569–575.

de Villiers, B. L. 1994. Salts in carrier water affect tralkoxydim activity. *S. African J. Plant and Soil* **11**:186–188.

de Villiers, B. L., and D. du Toit. 1993. Chemical composition of carrier water influences glyphosate efficacy. *S. African J. Plant and Soil* **10**:178–182.

Hess, F. D. 1985. Herbicide absorption and translocation and their relationship to plant tolerance and susceptibility. In *Weed Physiology*. Vol. 2, *Herbicide Physiology*, ed. by S. O. Duke, pp. 191–214. CRC Press, Boca Raton, FL.

Hess, F. D. 1987. Relationship of plant morphology to herbicide application and absorption. In *Methods of Applying Herbicides*, ed. by C. G. McWhorter and M. R. Gebhardt, pp. 19–35. WSSA Monograph 4. Lawrence, KS.

Hess, F. D., and R. H. Falk. 1990. Herbicide deposition on leaf surfaces. *Weed Sci.* **38**:280–288.

Holloway, P. J. 1993. Structure and chemistry of plant cuticles. *Pestic. Sci.* **37**:203–206.

Knoche, M. 1994. Effect of droplet size and carrier volume on performance of foliage-applied herbicides. *Crop Protection* **13**:163–178.

MacIsaac, S. A., R. N. Paul, and M. D. Devine. 1991. A scanning electron microscope study of glyphosate deposits in relation to foliar uptake. *Pestic. Sci.* **31**:53–64.

Nalewaja, J. D., R. Matysiak, and T. P. Freeman. 1992. Spray droplet residual of glyphosate in various carriers. *Weed Sci.* **40**:576–589.

Nalewaja, J. D., R. Matysiak, and E. Szelezniak. 1994. Sethoxydim response to spray carrier chemical properties and environment. *Weed Technol.* **8**:591–597.

Peregoy, R. S., L. M. Kitchen, P. W. Jordan, and J. L. Griffin. 1990. Moisture stress effects on the absorption, translocation, and metabolism of haloxyfop in johnsongrass (*Sorghum halepense*) and large crabgrass (*Digitaria sanguinalis*). *Weed Sci.* **38**:331–337.

Santier, S., and A. Chamel. 1992. Penetration of glyphosate and diuron into and through isolated plant cuticles. *Weed Res.* **32**:337–347.

Schönherr, J., and M. J. Bukovac. 1972. Penetration of stomata by liquids. *Plant Physiol.* **49**:813–819.

Sterling, T. M. 1994. Mechanisms of herbicide absorption across plant membranes and accumulation in plant cells. *Weed Sci.* **42**:263–276.

Tan, S., and G. D. Crabtree. 1992. Effects of nonionic surfactants on cuticular sorption and penetration of 2,4-dichlorophenoxy acetic acid. *Pestic. Sci.* **35**:299–303.

Thelen, K. D., E. P. Jackson, and D. Penner. 1995. The basis for the hard-water antagonism of glyphosate activity. *Weed Sci.* **43**:541–548.

Zang, Z., G. E. Coats, and A. H. Boyd. 1994. Germination and seedling growth of sorghum hybrids after seed storage with safeners at varying humidities. *Weed Sci.* **42**:98–102.

For chemical use, see the manufacturer's or supplier's label and follow these directions. Also see the Preface.

6 Herbicides and the Soil

Numerous soil factors, herbicide characteristics, the diversity of plant species, and climatic variation make interactions of herbicides with soils complex. There are at least ten different soil variables of importance that can interact with the more than 150 available herbicides and hundreds of different plant species. The complexities of the herbicide–soil–weather–plant interactions are enormous but are important to understand when using herbicides. Understanding herbicide–soil interactions is important for predicting effectiveness and reducing the negative effects of a herbicide on plants and the environment. Comprehensive reviews of herbicide and pesticide behavior in soils can be found in Kearney and Kaufman (1975, 1976, and 1988).

Herbicides are applied directly to the soil as (1) *preplanting* treatments and (2) *preemergence* treatments. The time of application may refer to the crop or to the weed. Some preplanting treatments are mechanically mixed into the soil, whereas others are left on the surface. When mechanically incorporated into the soil, a herbicide is usually immediately effective on seeds germinating in the area—with no added moisture. When applied to the soil surface, the herbicide must be assisted by

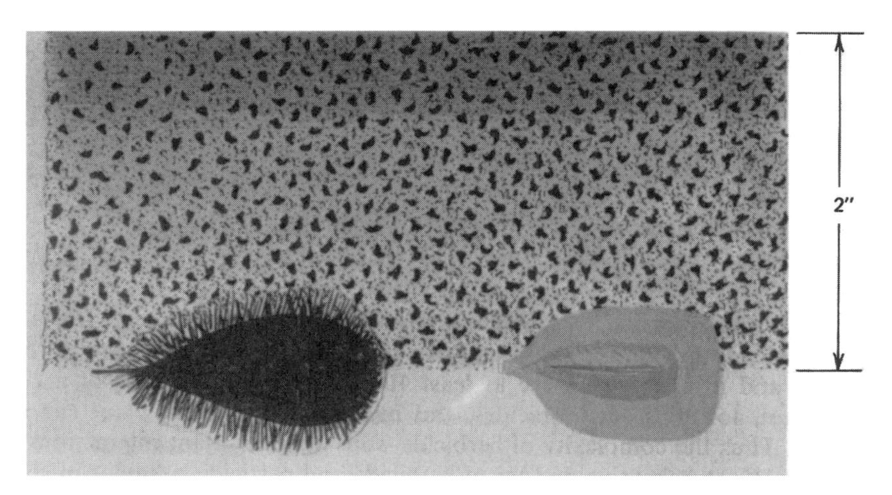

Figure 6-1. Many preemergence-type herbicides require soil incorporation to be effective in the absence of adequate rainfall or overhead irrigation. Soil incorporation places the herbicide in the area of the soil profile where most of the weed seeds germinate, the upper 2 inches. (North Carolina State University.)

water (either rain or irrigation) to move below the soil surface to effectively come in contact with germinating seeds so that it can act.

The success of an incorporated preplanting treatment or a preemergence treatment depends largely on the presence of an effective concentration of the herbicide in the upper 2 inches of soil. This is where most annual-weed seeds germinate. In addition, there must be a relatively low concentration of the herbicide in the zone where the crop seeds germinate, unless the crop seed is tolerant to the chemical (see Figure 6-1).

The herbicide-treated area may remain weed free long after the chemical has dissipated if all the initially germinating weed seedlings are killed, no further viable weed seeds sprout, and the soil is not disturbed by tillage. This happens because most weed seeds will not germinate if buried deeply in the soil; however, any disturbance can bring seeds closer to the soil surface, where germination can occur.

For effective soil sterilization, the chemical must remain active in the rooting zone to kill both germinating seeds and growing plants.

PERSISTENCE IN THE SOIL

The length of time that a herbicide remains active or persists in the soil is extremely important. The persistence of the herbicide should be long enough to achieve effective weed control to prevent weed competition with the crop, followed by quick dissipation in the soil to inactive components. However, residual toxicity or long-term presence in the soil in an active form does occur in some herbicides. Herbicide residues are important, as they relate to phytotoxic aftereffects *(carryover)* that may prove injurious to subsequent crops or plantings. In such cases, excessive chemical persistence may restrict crop rotation options available to the farmer and may cause environmental problems.

Factors that affect the persistence of a herbicide in the soil are classified as either degradation processes or transfer processes (Weber et al., 1973) and involve herbicide and soil characteristics, soil biota, and the environment. Degradation processes that break down herbicides and change their chemical composition are (1) biological decomposition, and (2) abiotic decomposition, which includes chemical decomposition and photodecomposition. Kearney and Kaufman (1975, 1976, 1988) provide reviews of the *degradation of herbicides*. Transfer processes important in determining what happens to herbicides in the soil are (1) adsorption by soil colloids, (2) leaching or movement through the soil, (3) volatility, (4) surface runoff, (5) removal by higher plants, and (6) absorption and exudation by plants and animals (see Figure 6-2). The first 5 factors are discussed later in this chapter.

Table 6-1 gives water solubility and sorption index characteristics of herbicides that determine their relative persistence in soil and potential for leaching. This table is based on experimental work and observations. In general, the persistence values are developed under conditions favorable for rapid herbicide decomposition. Herbicides persisting 30 days or less may be used to control weeds present at the time of treatment. Those persisting 30 to 90 days will protect the crop only during a short period early in the growing season. This is generally adequate for many annual row

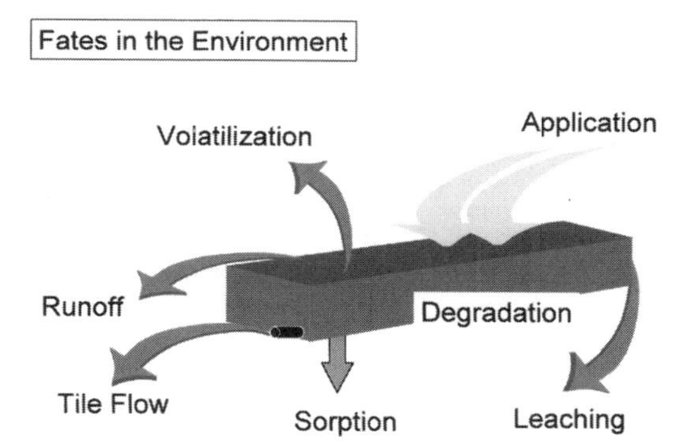

Figure 6-2. Processes influencing the behavior and fate of herbicides in the environment. Transfer processes are characterized by the herbicide molecules remaining intact on the soil (sorption), undergoing photo-, chemical, or biological degradation, volatilizing, being leached, and moving with surface runoff or through soil tiles. Herbicides are also removed by being absorbed into weeds and crops. (R. Turco, Purdue University.)

crops that produce a dense canopy and thereby suppress weed growth through shading. Those providing 90 to >144 days of control may protect the crop for the entire growing season and are useful in perennial crops such as orchards and vineyards. Those providing more than 12 months of control are used primarily for total vegetation control in noncrop situations where persistence is desirable (see Figures 6-3 and 6-4).

Both the physical nature of the herbicide and the structure of soil control the fate of a chemical. Most soil aggregates contain clay, organic matter, water, and microorganisms, with which the herbicide can interact. Overlaid on the complex soil structure are the three possible phases in which organic herbicides can exist. Chemicals can occur in one, two, or all three of the phases of vapor, solid, or liquid at any given time. The particular phase in which the chemical occurs will exert a significant impact on the fate of the herbicide.

How a herbicide behaves in the environment can be traced to its chemical structure and function, which will control how a chemical can be applied and used. Four major attributes of the herbicides are most important: water solubility; retention by organic matter or soil(K_{OD}, which is the sorption coefficient); vapor pressure (potential to volatilize); and soil half-life ($T_{1/2}$ or persistence). A key to determining the environmental fate of a herbicide is understanding the chemical's interaction with water. A herbicide that is highly water soluble will tend to remain in the soil water, whereas a herbicide that has lower solubility will try to escape from the soil water (Figure 6-5).

Water solubility is a reflection of the polarity of the chemical and is determined by the maximum amount of chemical that will dissolve in pure water at a specified temperature and pH. In general, the more polar a chemical, the higher its water solubility.

TABLE 6-1. Herbicide Characteristics That Determine Their Effectiveness, Persistence, and Potential for Leaching[a]

Herbicide	Major Site of Uptake	Water Solubility[b]	Sorption Index[b] (KOC)	Soil Half-Life (days)	Leaching Potential[c] (ppm)
Alachlor	Shoot, some root	240	170	15	Medium
Atrazine	Root, some shoot	33	100	60	High
Benefin	Germinating seed, shoot	<1	9000	40	Low
Butylate	Germinating seed, shoot	44	400	13	Small
Clomazone	Root, some shoot	1100	300	24	Medium
Chlorimuron	Root, some shoot	1200	110	40	High
Chlorsulfuron	Root	7000	40	160	High
Cyanazine	Root, some shoot	170	190	14	Medium
Cycloate	Germinating seed, shoot	95	430	30	Medium
Dicamba	Root	400,000	2	14	High
Ethofumesate	Shoot, some root	50	340	30	Medium
Ethalfluralin	Shoot	<1	4,000	60	Low
EPTC	Germinating seed, shoot	344	200	6	Small
Imazaquin	Root, some shoot	160,000[E]	20[E]	60	High
Imazethapyr	Root, some shoot	200,000[E]	10[E]	90	High
Linuron	Root, some shoot	75	400	60	Medium
Metolachlor	Shoot, some root	530	200	90	High
Metribuzin	Root, some shoot	1220	60[E]	40	High
Metsulfuron	Root	9500	35	120	High
Pendimethalin	Shoot	<1	5000	90	Low
Picloram	Foliage	200,000[E]	16	90	High
Pronamide	Root	15	200	60	High
Propachlor	Shoot, some root	613	80	6	Low
Simazine	Root	6	130	60	High
Sulfometuron	Root	70	78	20	Medium
Terbacil	Root	710	55	120	High
Trifluralin	Shoot	<1	8000	60	Low

[a]Water solubility, sorption index, soil half-life, and leaching potential values are from the *Soil Conservation Service Pesticide Properties Database Technical Guide for Nebraska*, Section II-D-5.
[b]Water solubility and sorption index for these herbicides were measured at pH 7. $K_{oc} = K_D$/organic carbon.
[c]Soil texture and structure will affect leaching potential.
[E] = an estimate—a wide range of values have been reported.
Adapted from *Nebguide G92-1081-A*, by Moomaw et al. (1996).

The vapor pressure of a chemical is defined as the pressure of the gas that is in equilibrium with the solid/liquid phase at a given temperature; it gives an indication of the tendency of the chemical to escape from a surface as a gas. All chemicals have a vapor pressure.

Figure 6-3. Total-vegetation-control herbicides may provide annual weed control for up to 2 years. (E.I. duPont de Nemours and Company.)

Herbicide *sorption* is defined as the retention of a chemical on or in a solid phase (in this case, the soil). The partition coefficient (K_D) is the ratio of herbicide bound to soil as compared with the amount left in the water surrounding the soil. K_D is calculated by determining, in a soil–water solution, the amount of herbicide adsorbed by the soil divided by the amount in the water phase. The smaller the K_D value, the greater the concentration of herbicide in solution.

Figure 6-4. Annual weed control for up to 2 years may be provided with 4 to 20 lb/acre (4.48 to 21.12 kg/ha) of simazine. (Syngenta Crop Protection, Inc.)

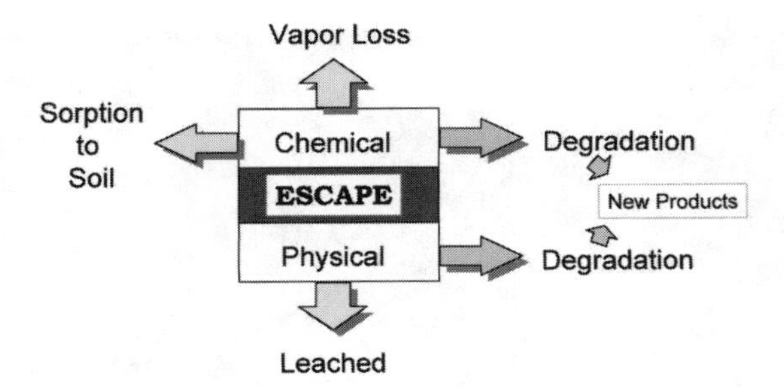

Figure 6-5. The key to determining the environmental fate of a herbicide is understanding its interaction with water. In general, herbicides able to reside in the soil water (highly soluble) will remain in the water and can leach. Less soluble herbicides try to escape from the water and can sorb to the soil, volatilize, or be degraded. (R. Turco, Purdue University.)

The soil sorption index is referred to as the K_{OC}. K_{OC} values are K_D values normalized for the organic carbon content of the soil, and, in essence, they express sorption as if the organic matter alone controlled adsorption. This is a fairly good assumption in most surface soils. $K_{OC} = K_D$ divided by the amount of organic carbon in the soil. K_{OC} is an expression of the tendency for herbicide sorption by soil organic carbon (organic matter). As in the K_D, the smaller the K_{OC}, the less likely it is that the herbicide will be adsorbed by the soil and thus the greater the potential for leaching. Values for each of these coefficients (water solubility, K_D, and K_{OC}) are provided for every herbicide in the *WSSA Herbicide Handbook* (1994, 1998) and are useful in predicting the behavior of the herbicides in different soil types.

Herbicide soil half-life ($T_{1/2}$) is the integrated result of all herbicide loss pathways that act upon the parent herbicide when it is in the soil environment. The $T_{1/2}$ is important because it affects the efficacy period, exposure to environmentally important transport processes, and potential carryover to the next crop.

All the various characteristics of herbicides and their interactions with soils are discussed in detail in the following sections, regarding their influence on herbicide fate once application is made into the soil environment.

Degradation Processes

Biological Decomposition Soil is a mixture of inert and living materials. Sand, silt, clay, and organic matter make up the nonliving fraction. The living fraction is composed of bacteria, fungi, algae, nematodes, protozoa, worms, and hundreds of other organisms. The living fraction should be thought of as the *machine* that drives the reactions that occur in the soil. Management of soils for optimum plant growth and quality is an indirect effort to control and, in some cases, to overcome the normal functioning of soil microorganisms. The living portions of the soil force the turnover of nitrogen, phosphorus, sulfur, and iron, and partially control the final fate of soil-applied herbicides. Biological decomposition of herbicides includes detoxification by soil microorganisms and higher plants.

Microbial Decomposition. The primary microorganisms important for herbicide decomposition are bacteria and fungi. They must have food for energy and growth. Organic compounds in the soil provide this food supply, except to a group of organisms that feed on inorganic sources.

Microorganisms use all types of organic matter, including organic herbicides. Some chemicals are easily decomposed (easily utilized by the microorganisms), whereas others resist decomposition. Other factors beside food supply may affect the growth and rate of multiplication of microorganisms: temperature, water, oxygen, mineral nutrient supply, the specific chemical structure, and the degree to which the chemical is adsorbed in the soil. Most soil microorganisms have slowed metabolism at 40°F, with a temperature of 75 to 90°F being the most favorable for microbial growth. Without water, most microorganisms become dormant or die. Aerobic organisms are very sensitive to an adequate oxygen supply, and a deficiency of nutrients, such as nitrogen, phosphorus, or potash, may reduce microorganism growth. Thus, a herbicide may remain active in the soil for a considerable time if the soil is cold, dry, or poorly aerated or if other conditions are unfavorable to the microorganisms. The adsorption of a herbicide into the soil (based on K_D and K_{OC}) can result in a reduced ability of microorganisms to break it down, as the herbicide is less readily available (or accessible) in the soil solution. The various atoms that make up the herbicide also affect microbial activity. Chemicals containing halogens (Cl, F, Br, or I) will decompose slowly. Another consideration in some fields where fumigation or sterilization has been used is that these processes kill most microorganisms. A lack of microorganisms results in no decomposition of any herbicide residue in the soil until a microbe population can reestablish.

Soil pH also influences the growth of microorganisms. In general, the bacteria and actinomyces are favored by soils having a medium to high pH, and their activity is reduced below pH 4.5. Fungi tolerate all normal soil pH values. In normal agricultural soils, fungi predominate at low pH.

Thus, a warm, moist, well-aerated, fertile soil with optimal pH is most favorable for higher populations and activity of microorganisms. Under these ideal conditions, microbes can quickly decompose most organic herbicides. Microbial decomposition of many herbicides follows typical growth curves for bacterial populations (Figure 6-6).

At the usual rate of herbicide application on farmlands, the total number of organisms is seldom changed to any great extent, because the herbicide may benefit one group of organisms and injure another group. When herbicides have been decomposed, the microorganism population returns to normal. The biological activity of most herbicides applied at rates recommended for cultivated crops disappears in less than 12 months (Table 6-2). Therefore, no long-term effect on the microorganism population of the soil is expected. The length of soil persistence for commonly used herbicides is shown in Table 6-2.

Higher Plant Decomposition. Herbicides absorbed from the soil by higher plants are generally changed or metabolized (see Chapter 5). A small amount can remain in

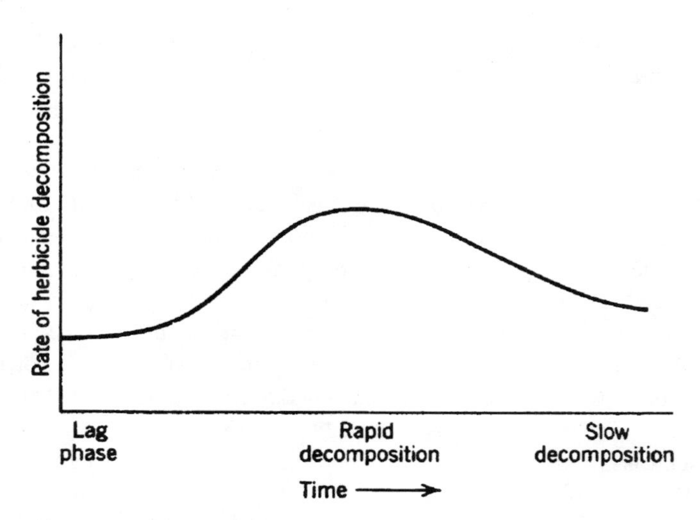

Figure 6-6. Rate of herbicide decomposition by microorganisms. (A. D. Worsham, North Carolina State University.)

its original form and can be stored or exuded. These topics are dealt with later in this chapter.

Chemical Decomposition Chemical decomposition is the breakdown of a herbicide by a chemical process or reaction in the absence of a living organism. This may involve reactions such as oxidation, reduction, and hydrolysis. For example, atrazine will slowly hydrolyze in the presence of water, rendering it ineffective as a herbicide. Another example is the hydrolysis of sulfonylurea herbicides in low pH or acidic soils (see Chapter 5 for an explanation of these reactions).

In soil saturated with water, oxygen will likely become a limiting factor. Under such conditions, anaerobic degradation of organic compounds can be expected. It has not been established whether this is a chemical or microbial process, but probably both are involved. Trifluralin, under standing water, was completely degraded in 7 days at 76°F in nonautoclaved soils, whereas only 20% had degraded at 38°F.

Photodecomposition Photodecomposition, the breakdown of a substance by light, has been reported for many herbicides. The process begins when the herbicide molecule absorbs light energy, which causes excitation of its electrons, and may result in breakage or formation of chemical bonds. Most herbicides are white, or nearly so, and have peak light absorption in the ultraviolet range (220–324 nm), whereas yellow compounds such as the dinitroanilines and dinoseb have absorption peaks at about 376 nm. Solar energy below 295 nm reaching the surface of the earth is considered to be negligible. Some of the breakdown products of photodecomposition are similar to those produced by chemical or biological means.

TABLE 6-2. Persistence of Biological Activity at the Usual Rate of Herbicide Application in a Temperate Climate with Moist-Fertile Soils and Summer Temperatures[a]

1 Month or Less	1–3 Months	3–12 Months[b]		Over 12 Months[c]
Acifluorfen	Azafenidin	Acetachlor	Hexazinone	Borates
Amitrole	Bromoxynil	Alachlor	Imazamethabenz	Bromacil
AMS	Butylate	Ametryn	Imazaquin	Chlorates
Bentazon	Chlorpropham	Atrazine	Imazethapyr	Chlorsulfuron
Cacodylic acid	Cycloate	Benefin	Isoxaben	Fluridone[e]
Carfentrazone	Desmedipham	Bensulide	Metribuzin	Imazapyr
2,4-D	Dithiopyr	Chloransulam	Napropamide	Picloram
2,4-DB	EPTC	Chlorimuron	Norflurazon	Prometon
Diclofop	Flumioxazin	Clomazone	Oryzalin	Tebuthiuron
Diquat[f]	Fluthiamide	Clopyralid	Oxadiazon	Terbacil
DSMA	Haloxyfop	Cyanazine	Oxyfluorfen	
Endothall	Imazamox	DCPA	Pendimethalin	
Glufosinate	Isoxaflutole	Dicamba	Pronamide	
Glyphosate	Linuron	Dichlobenil	Pyrithiobac	
Fluazifop	Mecoprop	Difenzoquat	Quinclorac	
Fenoxaprop	Metolachlor	Diuron	Simazine	
Lactofen	Metsulfuron	Ethalfluralin	Sulfometuron	
Metham	Naptalam	Ethofumesate	Sulfentrazone	
Methyl bromide	Nicosulfuron	Flumetsulam	Triasulfuron	
MCPA	Pebulate	Fluometuron	Trifluralin	
MCPB	Primisulfuron	Fluridone[d]		
MSMA	Prometryn	Fomesafen		
Paraquat[f]	Propachlor			
Phenmedipham	Pyrazon			
Propanil	Quizalofop			
Pyridate	Rimsulfuron			
Sethoxydim	Siduron			
	Sulcotrione			
	Thifensulfuron			
	Thiobencarb			
	Triallate			

[a]These are approximate values and will vary, as discussed in the text.
[b]At higher rates of application, some of these chemicals may persist at biologically active levels more than 12 months.
[c]At lower rates of application, some of these chemicals may persist at biologically active levels for less than 12 months.
[d]In water.
[e]In soil.
[f]Although diquat and paraquat molecules may remain unchanged in soils, they are adsorbed so tightly that they become biologically inactive.

Chemicals applied to soil surfaces are frequently lost, especially if an extended period without rain follows herbicide application. It is entirely possible that photodecomposition is responsible for the losses. However, other factors that may account for such loss should not be overlooked. Volatilization accentuated by high soil-surface temperatures, biological and chemical degradation, and adsorption are a few of the factors that should be considered in explaining the disappearance of herbicides from soils. Shallow incorporation with a rotary hoe or harrow is recommended to prevent photodecomposition and volatilization if rainfall does not occur within 5 to 7 days of herbicide application.

Transfer Processes

Soil colloids (from the Greek word *kolla* meaning "glue") have very high adsorptive capacities. In regard to soil, *colloid* refers to the microscopic (1 μm or less in diameter) inorganic and organic particles in the soil. These particles have an extremely large surface area in proportion to a given volume. It has been calculated that one cubic inch of colloidal clay may have 200 to 500 square feet of particle surface area.

Many clay particles react chemically like the negative radical of a weak acid, such as COO^- in acetic acid. Thus, the negative clay particle attracts to its surface positive ions (cations) such as hydrogen, calcium, magnesium, sodium, and ammonium. These cations are rather easily displaced, or exchanged (from the clay particle) for other cations. They are known as *exchangeable ions*. This replacement is called *ionic exchange* or *base exchange*.

The base-exchange capacity of a soil is expressed as milliequivalents (meq) of hydrogen per 100 g of dry soil. A soil with a base-exchange capacity of 1 meq can adsorb and hold 1 mg of hydrogen (or its equivalent) for every 100 g of soil. This is equivalent to 10 ppm or 0.001% hydrogen.

Adsorptive capacity, and thus exchange capacity, is closely associated with inorganic and organic colloids of the soil. Inorganic colloids are principally clay. There are two principal groups of clay: *kaolinite* and *montmorillonite* (see Table 6-3). Kaolinite is a *nonexpanding* clay; adsorption occurs on the external surface of the clay particle. Montmorillonite is an *expanding* clay; adsorption can occur on external and internal surfaces. Because of this difference, montmorillonite clays have an adsorptive capacity of three to seven times that of kaolinite clays.

Kaolinite clays have a tendency to predominate in areas of high rainfall and warm-to-hot temperatures. Thus, clay soils of the tropics and the southeastern United

TABLE 6-3. Characteristics of Some Common Clay Minerals

Characteristics	Montmorillonite	Vermiculite	Illite	Kaolinite
Type of layering	2:1	2:1	2:1	1:1
Type of swelling	Expanding	Limited expanding	Nonexpanding	Nonexpanding
CEC (meq/100g)	80–120	120–200	15–40	2–10
Specific surface (m^2/g)	700–750	500–700	75–125	25–50

From Weber (1972).

States are principally kaolinite. Aluminum and iron oxides are important constituents of kaolinic soils.

The montmorillonite clays predominate in areas of moderate rainfall and moderate temperatures. They are typical of corn belt soils.

Organic colloids involve the humus (organic matter) of the soil. Organic colloids have a very high adsorptive capacity, about 4 times the base-exchange capacity of a montmorillonite clay and perhaps 20 times that of kaolinite on a weight basis.

As mentioned earlier, the K_D and K_{OC} for each herbicide provide information relating to its potential for binding to the soil. Some pesticides will bind through ionic bonding; however, most herbicides tend to be nonionic, resulting in the soil organic carbon becoming an important factor for binding by partitioning.

How Do Herbicides Undergo Sorption to Soil?

The interaction of herbicides with negatively charged soil colloids (clays, organic matter) is dependent on the chemical nature of the herbicides (Table 6-4). Positively charged (cationic) herbicides such as diquat and paraquat (bipyrdilums) are held to soil by ionic bonds, much like potassium and calcium are held (cation exchange). However, unlike potassium and calcium, paraquat and diquat bound to clays are not readily displaced through ion exchange.

TABLE 6-4. Classification of Herbicides According to Their Ionic Properties

Weak Acids[a]	Cationic[b]	Weak Bases[c]	Nonionic[d]
Aliphatics	Bipyrdiliums	Pyridinones	Amides
Amino		Thiadiazoles	Anilides
Benzoic		Triazines	Diphenylethers
Benzonitriles		Triazoles	Dinitroanilines
Imidazolinones			Cyclohexanones
Organic arsenicals			N-phenylheterocycles
Phenylacetic			Isoxazoles
Phenolics			Phenylcarbamates
Phenoxy			Pyridazinones
Phenoxypropionates			Pyridines
Phthalamic			Carbamothiates
Phthalic			Ureas
Pyridine			Semicarbones
Pyrimidinyloxybenzoates			
Sulfonylureas			
Triazolopyrimidines			
Uracils			

[a]At neutral pH, anions or negatively charged forms predominate.
[b]Exists in a cationic or positively charged form.
[c]At acid pH, cations or positively charged forms exist.
[d]At all pH values, uncharged forms exist.

Basic herbicides, such as the triazines, can become cations in low-pH (acid) soils and adsorb to soil particles by changing from a net negative charge (nonionic sorption) to a net positive charge (ionic state). Thus the activity of s-triazine herbicides such as atrazine is greater in high pH (basic) soils than in acid soils (Best et al., 1975). This happens because a higher percentage of the herbicide molecules, which are cations in the acid soil, are rendered unavailable to plants through adsorption to soil colloids.

Weak acids such as 2,4-D (phenoxy), dicamba (benzoic), and picloram (pyridine) will lose a H^+ ion and go from a net zero (nonionic) to a net negative state that is not readily adsorbed, because they have the same negative charge as the soil particles. However, small amounts may be retained by organic matter and positively charged soil colloids such as iron and aluminum hydrous oxides.

Small amounts of neutral or nonionic (molecular form) herbicides have no charge and little tendency to gain or lose a H^+. These herbicides tend to interact with soil organic matter in a process called partitioning. Partitioning does not involve a major charge transfer, and adsorption by soil particulate matter is through relatively weak physical forces. Adsorption of nonionic herbicides generally increases as their water solubility decreases. For example, highly water-insoluble herbicides such as the dinitroanilines are adsorbed in large quantities by the organic matter fraction of soils.

Good crop growth requires a pH between 5 and 7 to be maintained. At these pH levels, it is important to know that most herbicides will be in the nonionic form. Problems arise with herbicides becoming ionized at pH extremes; either acid at pH 4.5 or below or basic at pH 7.5 or above. The information in Table 6-4 is useful for determining which herbicides would be most affected in their activity and residual soil life by extremes of soil pH.

Much evidence supports the fact that organic matter and clay (especially montmorillonite) play important roles in determining herbicide phytotoxicity and residual persistence through adsorption, leaching, volatilization, and biodegradation. Observations in research work as well as in the field have shown the following:

1. Soils high in organic matter require relatively large amounts of most soil-applied herbicides for weed control.
2. Soils high in clay content require more soil-applied herbicide than sandy soils for weed control.
3. Soils high in organic matter and clay content have a tendency to retain herbicides for a longer time than sand. The adsorbed herbicide may be released so slowly that the chemical is not effective as a herbicide.

The importance of soil organic matter on the activity of soil-applied herbicides was identified in research conducted by Weber et al. (1987). These authors reported that weed control attained with alachlor, metolachlor, metribuzin, and trifluralin was highly correlated with humic matter or organic matter (see Table 6-5). Scientists have attempted to develop equations for predicting safe, effective rates of herbicides for various soil types. Of the several chemical and physical properties of soils that were measured, organic matter gives the best prediction of performance (Weber et al.,

TABLE 6-5. Correlation Coefficients (*r*) of Herbicide Rates (kg ai/ha) Required for 80% Weed Control at 4 Weeks After Application and Soil Humic Matter or Organic Matter Content

	Correlation Coefficient (*r*)	
Herbicide	Humic Matter	Organic Matter
Alachlor	0.90	0.87
Metolachlor	0.89	0.91
Metribuzin	0.95	0.88
Trifluralin	0.93	0.88

From Weber et al. (1987).

1987). The K_{OC} of a herbicide is useful in allowing a prediction of the adsorption potential of a herbicide on the organic matter of a soil. For example, the K_{OC} of trifluralin is 7000 ml/g, which indicates strong sorption to soil, whereas atrazine has a K_{OC} in the range of 100 ml/g, which indicates moderate soil sorption. The use of other soil properties with organic matter in the equations did not greatly improve predictability. Such prediction equations are useful for many soils; however, the theoretical value of a given soil may be considerably different from that observed in practice. Therefore, such values should be used as guidelines but not as absolute recommendations.

These principles can be used when there are herbicide spills, misapplication, or a desire to safen a herbicide by the use of activated carbon. "Activated" carbon is one of the most effective adsorptive materials known. It has been chemically and thermally activated to be nonpolar in nature and is ground to an extremely fine particle size (92–98% of the particles pass through a 325 mesh sieve) to achieve very high surface area (1400–1800m/g). Activated carbon has been used to protect plants from herbicides, as its nonpolar surface is very adsorptive of nonpolar herbicides (likes attract likes), which prevents the herbicide from coming in contact with the plant root or seed. Roots of strawberry plants have been coated with activated carbon prior to setting the plants in herbicide-treated soil (Poling and Monaco, 1985). Moreover, bands of "activated"' carbon have been placed over previously seeded rows soon after planting and before a preemergence herbicide treatment. Well-decomposed organic matter presumably has properties similar to those of "activated" carbon.

These facts indicate that a certain amount of herbicide is required to saturate the adsorptive capacity of a soil. Above this "threshold level," higher rates will greatly increase the amount of herbicide in the *soil solution* and thus increase the herbicide's toxicity to the pest.

Therefore, the nature and strength of the "adsorption linkage" or "bonding" are of considerable importance for both cations and anions. Apparently the nature and characteristics of the colloidal organic matter, as well as the clay, may affect the tenacity of this bonding (Figure 6-7).

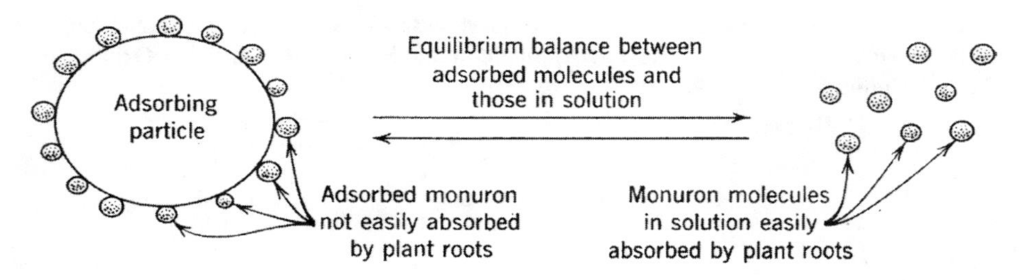

Figure 6-7. Plants absorb herbicides in the soil solution more easily than herbicides that are adsorbed on soil colloids.

In summary, various soils show large differences in their adsorptive capacities for herbicides. In practice, however, the range of herbicidal rates of application is much less than might be predicted from the very wide ranges in adsorptive capacity of the soils. As mentioned earlier, reference to the herbicide label is the best guide for determining application rates for various soil types.

Leaching is the downward movement of a substance dissolved in water through soil. Leaching may determine herbicide effectiveness, may explain selectivity or crop injury, or may account for a herbicide's removal from the soil. Preemergence herbicides are frequently applied to the soil surface. Rain or irrigation moves the chemical into the upper soil layers, and weed seeds germinating in the presence of the herbicide are killed. Large-seeded crops such as corn, cotton, and peanuts planted below the area of high herbicidal concentration may not be injured (see Figure 6-1). In addition to the protection offered by depth, crop tolerance to a herbicide through physiological processes is also desirable.

The extent to which a herbicide is leached is determined principally by

1. Adsorptive relationships between the herbicide and the soil
2. Solubility of the herbicide in water
3. Amount of water passing downward through a soil

Solubility is sometimes cited as the principal factor affecting the leaching of a herbicide. Simple calculation of the amount of water in a 4-inch rainfall disproves this assumption. A 4-inch rainfall weighs nearly 1,000,000 lb/acre. If you apply 1 pound of herbicide per acre, this equals 1 ppm of the herbicide in water; thus, if the herbicide is soluble to the extent of 1 ppm, you might expect a 4-inch rain to remove essentially all of the herbicide from the surface inch of soil.

The interrelationship between the binding of herbicides to the soil and water solubility can be demonstrated with 2,4-D. Salts of 2,4-D are water soluble and readily leach through porous, sandy soils. Soils with high organic-matter content adsorb 2,4-D, reducing the tendency to leach. The ester formulations of 2,4-D have low water solubility, and their tendency to leach is reduced by both the low solubility and adsorption by the soil. In general, solubility of a herbicide and its adsorption on soil are inversely related; that is, increased solubility results in less adsorption.

The immobility of paraquat in soil stresses the importance of recognizing the influence of adsorption and water solubility on the leaching process. Paraquat is completely water soluble but does not leach in soils because it is a cation and is held very tightly by soil colloids.

To restate the point, the strength of "adsorption bonds" is considered more important than water solubility in determining the leaching of herbicides. Organic-matter content in the soil is the most important single factor determining the adsorptive capacity of the soil. The second most important is the clay fraction.

A leaching chemical can move downward in one of two pathways. Matrix flow is a slow movement through the small pore spaces in the bulk soil. The water and the herbicide interact with the small pores on the way down the profile. In contrast, preferential flow is a rapid movement of water in the large channels and flow paths. Leaching loss of herbicides is most severe when a rainfall event follows soon after an application, primarily via preferential flow. Besides downward leaching with water, herbicides are known to move upward in the soil, driven by the capillary movement of soil water. As water evaporates from the soil surface, more water moves slowly upward and may carry with it soluble herbicides. As the water evaporates, the herbicide is deposited on the soil surface. Volatile herbicides such as EPTC also move upward and laterally in open soil pores in the vapor state. In areas where furrow irrigation is practiced, lateral movement of herbicides in soil with the irrigation water also occurs.

Volatilization. All chemicals, both liquids and solids, have a vapor pressure. Water is an example of a liquid that will vaporize, and naphthalene (mothballs) is an example of a solid that will vaporize. At a given pressure, the vaporization of both liquids and solids increases as the temperature rises. The vapor pressure and water solubility are the keys to a chemical's potential to volatilize, and these are expressed together in the Henry's law constant.

The Henry's constant is very similar to the sorption coefficient used to describe the magnitude of adsorption a chemical can undergo. The K_H value is a partition coefficient that describes the distribution of a chemical between air and water.

$$K_H = \frac{C_V}{C_l}$$

Where C_V is the concentration of the chemical at the water-phase interface and C_l is the concentration of the chemical in the liquid at the water–vapor interface. K_H is highly temperature dependent and increases with increasing temperature.

Herbicides may vaporize and be lost to the atmosphere as either phytotoxic or nonphytotoxic gases. The volatility of some soil-applied herbicides, along with their vapor pressures, is shown in Table 6-6. The toxic volatile gases may drift to susceptible plants. The *ester* forms of 2,4-D are volatile, and the vapors or fumes can cause injury to susceptible crops such as cotton, tomatoes, and grapes (see Figure 5-12).

TABLE 6-6. Volatility and Vapor Pressures of Some Soil-Applied Herbicides[a]

Herbicide	Volatility	Vapor Pressure[b]
	Must Be Mechanically Soil Incorporated	
Balan	Moderate to high	7.8×10^{-5}
Eptam	High on wet soil; moderate on dry soil	3.4×10^{-2}
Eradicane	High on wet soil; moderate on dry soil	3.4×10^{-2}
Ro-Neet	High on wet soil; low on dry soil	6.2×10^{-3}
Sonalan	Moderate	8.2×10^{-5}
Sutan+	High on wet soil; low on dry soil	1.3×10^{-2}
Treflan	Moderate to high	1.1×10^{-4}
	Do Not Require Mechanical Soil Incorporation	
Ally	Low	2.5×10^{-12}
Atrazine	Low to moderate	2.9×10^{-7}
Banvel	Moderate	9.24×10^{-6}
Bladex	Low	1.6×10^{-9}
Classic	Low	4×10^{-12}
Command	Moderate	1.44×10^{-4}
Dual	Low	1.3×10^{-5}
Kerb	Low	8.5×10^{-5}
Lasso	Low	1.6×10^{-5}
Lorox/Linex	Low	1.7×10^{-5}
Nortron	Low	6.45×10^{-7}
Oust	Low	5.5×10^{-16}
Princep	Low	2.2×10^{-8}
Prowl	Moderate	9.4×10^{-6}
Pursuit	Low	<10–7 (60°C)
Ramrod	Very low	7.9×10^{-5}
Sencor/Lexone	Low	1.2×10^{-7} (20°C)
Scepter	Low	$<2 \times 10^{-8}$ (45°C)
Sinbar	Low	3.1×10^{-7}

[a]Adopted from *Nebguide G92-1081-A*, by Moomaw et al., 1999.
[b]Data from *Herbicide Handbook* of the Weed Science Society of America, Lawrence, KN, 1984. At 25°C unless noted in ( ). Volatility ranges: Volatile herbicides—10^{-1} to 10^{-4} mm Hg; intermediate volatility herbicides—10^{-4} to 10^{-6} mm Hg; nonvolatile herbicides—$>10^{-6}$ mm Hg.

Certain herbicides may move in a porous soil as a gas. EPTC is thought to move in this way. This was clearly shown in experiments where injecting EPTC into the soil provided a much wider area of weed control than just at the injection point.

The importance of volatilization and the loss of a herbicide from the soil surface is often underestimated. In volatility studies, EPTC volatilized from a free-liquid surface at the rate of about 5 lb/acre per hour at 86°F (Ashton and Sheets, 1959). This high rate of vaporization could easily explain the loss of the herbicide. EPTC, trifluralin, and other volatile soil-applied herbicides are usually mechanically mixed into the soil soon after application to reduce loss.

Interactions with the soil and the soil's condition will impact the losses of herbicides due to volatilization. For example, the soil organic matter (adsorption), water content of the soil, and incorporation depth can affect the amount of herbicide that will volatilize. Greater volatilization can occur from a wet soil than a dry soil, as fewer binding sites exist in the wet soil for the herbicide and the loss of water draws the herbicide up to the surface. As organic matter increases, more herbicide is bound to the soil and less will volatilize. In addition to these factors, herbicide characteristics relating to vapor pressure and water solubility also play a role. A volatile herbicide incorporated into the soil will volatilize less than if it is left on the soil surface. Figure 6-8 illustrates the influence of soil moisture level, herbicide K_D and K_H, as it relates to higher volatility in a wet soil versus a dry soil.

Codistillation with water evaporating from the soil surface (steam distillation) is another means by which a volatile herbicide may be lost. This process has not been extensively studied but may be of considerable importance in view of the immense amount of water lost from the soil surface through evaporation.

Herbicides with very low vapor pressures, such as atrazine, may also volatize from a surface over an extended period of time, especially if exposed to high temperatures. Soil surface temperatures have been measured as high as 180°F. This can also occur in greenhouses, where generally nonvolatile herbicides can volatilize when extreme soil temperatures occur following their application to the greenhouse soil.

Rain or irrigation water applied to a dry or moderately dry soil will usually leach a surface-applied herbicide into the soil or aid in its adsorption by the soil. Once adsorbed by the soil, the loss by volatility is usually reduced.

Surface Runoff Herbicides applied to the soil surface may dissolve in rainwater and leach into the soil. However, heavy rains may carry the dissolved herbicide away from the treated area. Severe runoff, which causes erosion, can also carry adsorbed herbicides on the eroding soil particles. *Washoff* is the term used to describe such losses. Cultural practices that minimize erosion, such as conservation tillage (with

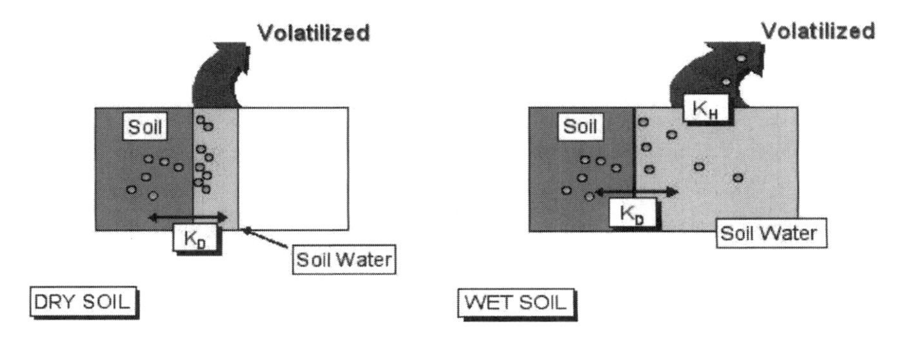

Figure 6-8. In a dry soil (*left*), there is greater soil surface available for herbicide sorption and less volatilization occurs. In a wet soil (*right*), there is less opportunity for sorption of a herbicide because the herbicide is competing for soil-binding sites with water. As water evaporates from a wet soil, it carries volatile herbicides with it. (R. Turco, Purdue University.)

large amounts of plant residues on the soil surface) or contour plowing, will help minimize washoff losses of herbicides.

Removal by Higher Plants Herbicides may be absorbed by the crop or surviving weeds and stored or given off in their original form. Usually, however, the herbicide molecule is altered in the plant by metabolism, and the herbicide breakdown products are either used by the plant or discharged back into soil solution. In some cases, herbicides are retained within the tissues of the plant, thereby delaying decomposition.

Herbicides may be removed from treated fields if the compounds are present in harvested plant parts, but the amounts removed are nearly always insignificant. For example, 1 ppm of a herbicide in a 10-ton hay crop amounts to only 0.02 lb/acre removed in the hay. However, plants can degrade sizable amounts of herbicide during a cropping season (Weber et al., 1973).

Removal of herbicides from the soil by plants may not be a major factor in the persistence of herbicides under most conditions; however, it has been used to help remove persistent herbicides from soils where they were applied as soil sterilants and the planting of ornamentals was desired (e.g., corn for simazine or atrazine removal).

Exudation Herbicides that are absorbed by plants and microorganisms can also be *exuded* or discharged from inside the organism to the surrounding environment. The herbicide can be in an altered form or the original form. Generally, this does not represent a significant percentage of the amount the herbicide absorbed.

Figure 6-2 shows the interrelations of the processes that lead to detoxification, degradation, and disappearance of herbicides in the environment.

LITERATURE CITED AND SUGGESTED READING

Ashton, F. M., and T. J. Sheets. 1959. The relationship of soil adsorption of EPTC to oats injury in various soil types. *Weeds* **7**:88–90.

Best, J. A., J. B. Weber, and T. J. Monaco. 1975. Influence of soil pH on *s*-triazine availability to plants. *Weed Sci.* **23**:378–382.

Hunnicut, R. C., and D. J. Schabacker, eds., 1994. *Mechanisms of Pesticide Movement into Ground Water.* CRC Press, Boca Raton, FL.

Herbicide Handbook. 7th ed. 1994. Weed Science Society of America. Lawrence, KS.

Herbicide Handbook Supplement to 7th ed. 1998. Weed Science Society of America. Lawrence, KS.

Kearney, P. C., and D. D. Kaufman, eds. 1975, 1976, 1988. *Herbicides*. Vols. 1, 2, and 3. Dekker, New York.

Poling, E. B., and T. J. Monaco. 1985. Activated charcoal root dips enhance herbicide selectivity in strawberries. *HortScience* **20**:251–252.

Weber, J. B. 1972. Interaction of organic pesticides with particulate matter in aquatic and soil systems. In *Fate of Organic Pesticides in the Aquatic Environment*, ed. by R. F. Gould, pp. 55–122. Adv. Chem. Ser. 111, American Chemical Society, Washington, DC.

Weber, J. B., T. J. Monaco, and A. D. Worsham. 1973. What happens to herbicides in the environment? *Weeds Today* **4**:16–17.

Weber, J. B., M. R. Tucker, and R. A. Isaac. 1987. Making herbicide rate recommendations based on soil test. *Weed Technol.* **1**:41–45.

Weber, J. B., L. R. Swain, H. J. Strek, and J. L. Sartori. 1986. Herbicide mobility in soil leaching columns. In *Research Methods Weed Science*. 3d ed., Chapter IX, pp. 189–200. Southern Weed Science Society, Champaign, IL.

Weber, J. B., G. G. Wilkerson, H. M. Linker, J. W. Wilcut, R. B. Liedy, S. Senseman, W. W. Witt, M. Barrett, W. K. Vencill, D. R. Shaw, T. C. Mueller, D. K. Miller, B. J. Brecke, R. E. Talbert, and T. E. Peeper. 2000. A proposal to standardize soil/solution herbicide distribution coefficients. *Weed Sci.* **48**:75–88.

WEB SITES FOR HERBICIDE AND SOIL INTERACTIONS

Factors That Affect Soil-Applied Herbicides, Nebguide G92-1081-A, by Moomaw et al., 1996.

http://ianrwww.unl.edu

Go to "Publications," then "Publication Search," and enter "g1081."

Reducing Herbicide Movement to Surface and Groundwater, Factsheet, by David R. Pike. University of Illinois.

http://ext.agn.uiuc.edu

Select "Crop Science Extension," then select "Weed Science" and scroll down to "Water Contamination" (surface.ground) and select

For chemical use, see the manufacturer's or supplier's label and follow these directions. Also see the Preface.

7 Formulations and Application Equipment

A newly synthesized herbicide is not suitable for use in the field. It must be formulated before it can be applied. There are several different types of formulations, depending on the characteristics of the herbicide and its uses. Various types of equipment have been developed for application of these several types of formulations. This chapter covers these formulations and the equipment required for their application to the targets.

FORMULATIONS

Herbicides are formulated to facilitate their handling, storage, and application and to improve their effectiveness under field conditions. A formulation chemist can change the formulation of a chemical to affect its solubility, volatility, toxicity to plants, and numerous other characteristics. This is accomplished by changing the chemical form (e.g., acid to ester) or using adjuvants, including surfactants. An adjuvant is any substance in a herbicide formulation or added to the spray tank to modify herbicidal activity or application characteristics. A surfactant (surface-active agent) is a material that improves the emulsifying, dispersing, spreading, wetting, or other properties of a liquid by modifying its surface characteristics. Adjuvants and surfactants are considered to be inactive ingredients even though they can have a pronounced effect on the performance of the product.

The active ingredient (ai) is the chemical in a herbicide formulation primarily responsible for its phytotoxicity and is identified on the product label. The concentration of the active ingredient on the label is commonly given as a percentage for solid formulations and pounds per gallon for liquid formulations. However, concentrations of certain herbicide derivatives are usually expressed as their acid equivalents (ae). In this case, the value refers to the concentration of the theoretical mass of the parent acid rather than that of the derivative. Phenoxy-type herbicides are universally handled in this manner (e.g., 2,4-D).

Additional information on pesticide formulations can be found in WSSA (1982), and Foy and Pritchard (1996).

Types of Formulations

Herbicides are usually formulated so they can be applied with a liquid (usually water) or solid carrier. The primary purpose of the carrier is to allow a uniform spray distribution of the herbicide over a large area. These formulations include (1) water or oil solubles, (2) emulsifiable concentrates, (3) wettable powders, (4) water-dispersible liquids and granules, (5) granules, and (6) pellets. The selection of the herbicide formulation to use for weed control depends on the species to be controlled, the crop involved, the equipment available, and environmental conditions.

Water or Oil Solubles These formulations are liquids or particulate solids that readily dissolve in the carrier, water or oil, to form a solution. A solution is a physically homogeneous mixture of two or more substances and is clear in appearance. The dissolved constituent is the solute, and the dissolving substance is the solvent.

Liquid water solubles (S or SL) require little tank agitation to dissolve in water, and soluble powders or granules (SP, WSP or SG) need more agitation. Once these materials are dissolved, further agitation is not required. These points also apply to oil solubles (OS), except that the solvent is oil rather than water.

The salts of most herbicides are soluble in water. Conversely, the esters of many herbicides are soluble in oil, and for these herbicides oil may be used as the carrier. For example, many inorganic and amine salts of 2,4-D are soluble in water, whereas their ester formulations are soluble in oil. However, there are also oil-soluble amine formulations of 2,4-D.

Emulsifiable Concentrates Emulsifiable concentrates (E or EC) are mixtures of the herbicide and emulsifying agents dissolved in an organic solvent. The emulsifying agents enable the emulsifiable concentrate to be dispersed in water, forming an emulsion that is used as the spray mixture. An emulsion is one liquid *dispersed* in another liquid, each maintaining its original identity. The droplets (EC) are referred to as the dispersed phase, and the liquid (water) they are suspended in is the continuous phase. Emulsions appear milky. Emulsifiable concentrates require tank agitation to form the emulsion and maintain it during spraying. Gels (GL) are thickened emulsifiable concentrates packed in water soluble bags in pre-measured amounts.

Invert emulsions are mixtures of an emulsifiable concentrate in which oil is the carrier and continuous phase, rather than water. These emulsions may be mayonnaise-like and too viscous to spray with conventional equipment. Special equipment has to be designed for application of invert emulsions. Although not widely used, they are primarily intended to reduce spray drift.

Wettable Powders Wettable powders (WP) are finely ground herbicide particles intended to be suspended in water for spraying. They may also contain clay as a diluent, synthetic silica as an anticaking agent, and various adjuvants. This wettable powder suspension gives a cloudy appearance to the liquid.

Wettable powders require tank agitation for initial suspension and during application to prevent settling of the solid particles. The inert materials in these formulations are often abrasive and can cause excessive wear to nozzles and certain pumps (especially roller pumps).

Water-Dispersible Liquids and Granules Water-dispersible liquids (WDL) and water-dispersible granules (WDG) are similar to wettable powders. However, a WDL is already suspended in a liquid and WDG is an aggregate of granule size made of finely ground particles. Water-dispersible liquids are also commonly referred to as liquids (L) or flowables (F), and water dispersible granules are also commonly referred to as dry flowables. Suspension concentrates (SO), aqueous concentrates (AO), and microencapsulated (ME) formulations are also in this group. These formulations overcome some of the mixing and dust exposure problems of wettable powders. However, like wettable powders, these formulations require tank agitation for initial suspension and during application to prevent settling of solid particles.

Granules and Pellets Herbicides can also be applied directly as dry formulations without spraying in a liquid carrier. Some chemicals, such as sodium borates and chlorates, are applied at high enough rates so that crystals of the chemical can be uniformly applied. However, most herbicides are so active that they must be formulated with an inert solid carrier to achieve a uniform application. Many carriers are used, including clays, vermiculite, starch, plant residues, and dry fertilizers.

Granules (G) are smaller than pellets (P). Granules are generally spread mechanically, and pellets are often spread by hand for spot treatment. The advantages of these formulations over a spray are that (1) water is not needed for application, (2) application equipment may be less costly to purchase and maintain, and (3) granules can pass through plant residues to the soil surface in conventional tillage systems. Disadvantages of these formulations relative to spray are that (1) they weigh more and are bulky, (2) they can be more expensive, (3) application equipment is more difficult to calibrate, and (4) uniform application is a problem.

Encapsulation Encapsulation is the incorporation of the herbicide into very small capsules, generally 10 μm or less. The capsules are suspended in a liquid system and sprayed. The primary advantage of encapsulation appears to be extending the period of weed control by a herbicide. Differential release times can be accomplished by altering the nature of the encapsulating material. These formulations are being researched as to their potential, although the principle of "controlled release," including encapsulation, has been used for some time in other fields (Mervosh et al., 1995). Figure 7-1 shows the effect of starch encapsulation on effectiveness of EPTC when applied for weed control in corn.

Herbicide Storage*

No matter whether a small inventory or a large inventory of herbicides is being stored, the key to proper storage is to limit the chance of accidental human or environmental

*From Whitford et al., 1997.

Figure 7-1. Effect of starch encapsulation of EPTC on weed control effectiveness. *Left*: Control plot, no herbicide application; *Center*: Surface application of starch-encapsulated EPTC resulted in effective weed control; *Right*: EPTC applied to soil surface, no incorporation, vapor loss, and no herbicidal activity. (M. V. Hickman, Purdue University.)

exposure. The storage facility (cabinet or building) must follow EPA (Environmental Protection Agency) regulations in the following areas:

General Information
Clean and neat pesticide storage site
Current, on-site pesticide inventory
Posted emergency phone numbers
Labels and MSDS (Materials Safety Data Sheets) on file
Accurate storage inspection log maintained

Pesticide Containers
Containers marked with purchase date (old pesticide inventory used first)
Insecticides, herbicides, and fungicides segregated
Pesticides stored in original containers
Labels legible and attached to containers
Container caps tightly closed
No reused pesticide containers present
Pesticides stored off floor and low to ground
Dry formulations stored on pallets
Feeds stored separately from pesticides
Used containers rinsed and punctured
Rinsed and unrinsed containers separated

Spills and Disposal
Storage area free of spills and leaks
Shovel and absorbent materials available
Floor drains sealed (if present)

Safety Information
No Smoking signs posted
Safety equipment separated from pesticides
Fire extinguisher in good working order
Storage room locked
Storage room posted: *Pesticides: Keep Out*
Storage site well lit and ventilated

Spray Additives

Materials (adjuvants) are often added to spray mixtures to modify or enhance herbicidal activity or aid in the spraying operation. Use of adjuvants, other than as formulation agents, is primarily restricted to postemergence herbicide applications. Adjuvants can be divided into two general categories: spray modifiers and activators (Kirkwood, 1994). The most common type of adjuvants are (1) activity enhancers, such as surfactants, oils, organosilicones, and fertilizers, (2) spray modifiers such as stickers and drift control agents, and (3) utility modifiers such as compatibility and antifoam agents (McWhorter, 1982). However, not all adjuvants are always beneficial; some may have no effect or even decrease the desired effect. Therefore, adjuvants should not be placed in the spray mixture unless suggested on the label or recommended by knowledgeable authorities.

The proceedings of a symposium on herbicide adjuvants containing many articles relating to their properties and uses is a good source of additional detailed information for weed science students (*Weed Science*, 2000).

Adjuvants According to Use

Activity Enhancers Surfactants derive their name from the term *surface-active agents*. Most surfactant molecules are composed of a lipophilic long-chain hydrocarbon (alkyl) group and a hydrophilic polar group. Surfactants are generally classified according to the nature of the polar segment of the molecule. Among the types of surfactants are cationic (positive charge), anionic (negative charge), zwitterionic (having both a positive and a negative charge, depending on the water pH), and nonionic (no charge). Nonionic surfactants dissociate little in water, whereas the others are charged when dissolved in water. Because adjuvants contain both lipophilic (oil-like) and hydrophilic (waterlike) properties, they can interact with the lipophilic plant surfaces, lipophilic herbicides, hydrophilic herbicides, and water. The most common surfactants for use with herbicides are nonionic, and most emulsifiers are blends of anionic and nonionic types. In general, a blend with a high proportion of the anionic types will improve performance in cold water and soft water, whereas a blend with a predominance of nonionic types will usually perform better in warm water and hard water.

Surfactants concentrate and act at the surface of the liquid in which they are dissolved because their molecules have both polar and nonpolar segments (Figure 7-2). The polar segment is attracted to water (hydrophilic), and the nonpolar segment is attracted to oil-like compounds (lipophilic). Most agricultural nonionic surfactants have chains of ethylene oxide (EO, $-CH_2-CH_2-O$), also called oxyethylene or ethyoxylate, as the polar (hydrophilic) groups. The number of EO units in the polar portion are referred to as the "moles of ethylene oxide." Common nonionic surfactants are alcohol, alkylamine, and alkylphenol ethoxylates. Propylene oxide [PO,$-(CH_2)_3-O-$] or butylene oxide [$-(CH_2)_4-O-$] can be built into the EO chain to reduce its hydrophilic nature, which makes the surfactant more compatible with lipophilic herbicides (Butselaar and Gonggrijp, 1993). The more EO or PO units on the surfactant, the more polar the surfactant. The constituent at the end of the EO or PO chain (called the end cap) can further modify the polarity of the surfactant. For example, a methoxy ($-O-CH_3$) is a less polar end cap than a hydroxy (-OH) end cap.

At very low concentrations surfactants are soluble in water; however, as the surfactant concentration is raised to that commonly used in weed control, the lipophilic groups associate with one another to form micelles (Figure 7-2). The surfactant concentration where micelle formation occurs is called the critical micelle concentration (CMC). These micelles can emulsify lipophilic substances, including herbicides, oils, and perhaps cuticular components. In a mixture of oil and water, the lipophilic portion orients itself into the oil droplets and the hydrophilic portion is within the water. This is how an emulsifier, a type of surfactant, facilitates the suspension of an oil-like herbicide in a water carrier. Another type of surfactant is a

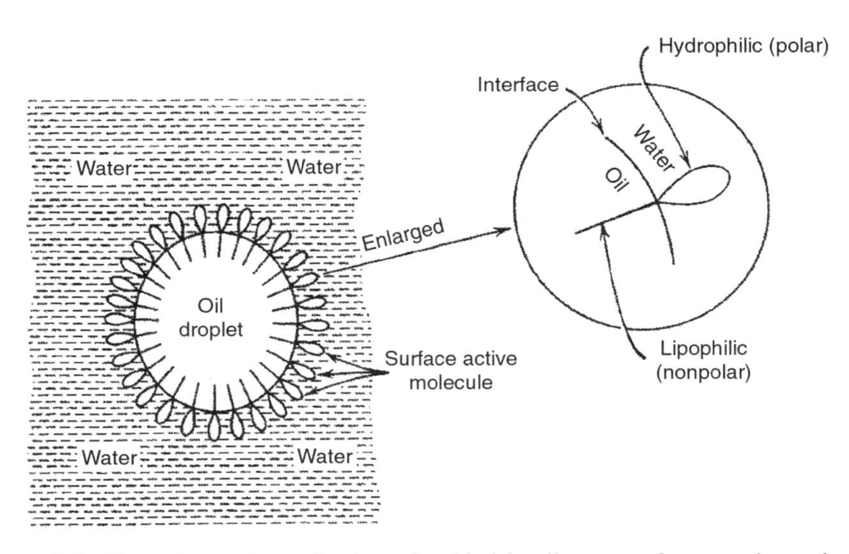

Figure 7-2. The surface-active molecule tends to bind the oil-water surfaces together, reducing interfacial tension.

wetting agent, which is oriented with the polar segment in the water droplets and the lipophilic segment protruding from the water droplets. This allows the herbicide spray to spread over a normally repellent leaf surface.

Surfactants are often assigned a "hydrophilic-lipophilic balance" (HLB) value. HLB is a quantitative measure of the polarity of surfactant molecules. HLB uses a scale of 0 to 20, with higher numbers being more hydrophilic than lower numbers. The HLB can be determined experimentally, it is also possible to estimate the value by calculation or by observing dispersibility in water (Table 7-1). Lipophilic surfactants are assigned HLB numbers of 8 and below. Surfactants with HLB numbers between 9 and 11 are intermediate, and those with HLB numbers above 11 are hydrophilic in nature. Surfactants used as wetting agents have HLB values of 7 to 9. A surfactant with optimal HLB for a particular herbicide can be predicted on the basis of the water solubility of the herbicide—low HLB surfactants for water-insoluble herbicides, and high HLB surfactants for water-soluble herbicides. HLB values are very useful in the selection of surfactants for formulating emulsions. They also serve to emphasize that all surfactants are not equally suitable for all uses. It is most important to select a surfactant appropriate for the intended use, and the herbicide label can assist in the proper selection; also see (Table 7-2).

An important function of surfactants used as adjuvants is to reduce the surface tension of a spray solution. This allows increased wetting of leaves and spreading of the spray to achieve more intimate contact between the spray droplet and plant surface (Figure 7-3). Surface tension is the tendency of surface molecules of a liquid to be attracted toward the center of the liquid body. Spray droplet spreading occurs when the surface tension of the droplet is less than the surface tension of the leaf surface. The degree of effectiveness of the surfactant on droplet spreading can be determined by measuring the contact angle between the droplet and the surface. Any substance that will bring the herbicide into more intimate contact with the leaf surface and keep it in a soluble form has potential of aiding absorption. Surfactants achieve this by

1. Causing a more uniform spreading of the spray solution and a uniform wetting of the plant
2. Helping spray droplets to stick to the plant, resulting in less runoff

TABLE 7-1. Approximation of HLB by Water Solubility

Behavior When Added to Water	HLB Range
No dispersibility in water	1–4
Poor dispersion	3–6
Milky dispersion after vigorous agitation	6–8
Stable milky dispersion (upper end almost translucent)	8–10
From translucent to clear	10–13
Clear solution	13+

From Becher (1973).

TABLE 7-2. Surfactant Class, Chemical Composition, and Herbicides Commonly Used with Each Class of Surfactant

Surfactant Class	Composition	Commonly Used With
Nonionic (NIS)	Linear or nonyl-phenyl alcohols and/or fatty acids	Quizalofop, fluazifop fenoxaprop, paraquat
Crop Oil Concentrates	Paraffinic oil (80–90%) + surfactants (15–20%)	Atrazine, quizalofop, fluazifop, fenoxaprop, bentazon sethoxydim, clethodim
Methylated Seed Oils	Fatty acids from seed oils esterified with methyl alcohol	Nicosulfuron, primisulfuron, imazethapyr
Organosilicones	Composed of silicone or blends of silicone with surfactants	Bentazon, acifluorfen, lactofen fomesafen, mixtures of these
Nitrogen–Surfactant	Surfactant + some nitrogen (AMS, 28% ammonium nitrate)	Glyphosate, chlorimuron, imazethapyr, thifensulfuron

Adapted from Miller and Westra, 1996.

3. Ensuring that droplets do not remain suspended on hairs, scales, or other surface projections
4. Partially solubilizing the lipoidal plant cuticle substances (controversial)
5. Preventing crystallization of the active ingredient on the leaf surface by acting as a solvent
6. Slowing the drying of, and increasing the water retention in, spray droplets once on the leaf surface

At low to normal spray volumes (10 to 40 gal/acre), adjuvants usually increase the effectiveness of herbicidal sprays by increasing the coverage of the plant surface; herbicide absorption is thereby also increased. However, at high spray volumes (>100 gal/acre) adjuvants may decrease effectiveness by causing excessive "runoff" from the plant onto the ground. Herbicidal selectivity may be lost by the addition of an adjuvant if the desired selectivity among species depends on differential wetting and thus absorption.

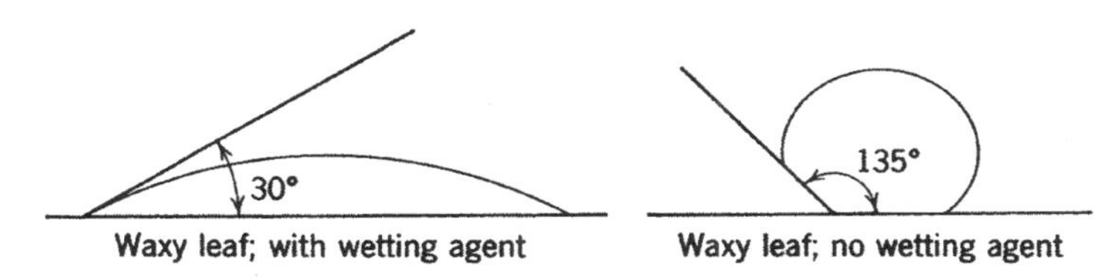

Figure 7-3. *Left*: Water droplets containing a wetting agent will spread in a thin layer over a waxed surface. *Right*: Pure water will stand as a droplet, with only a small area containing the surface wax.

Surfactants of the nonionic type are commonly used with contact herbicides such as paraquat and many postemergence grass and broadleaf specific herbicides to enhance activity at 0.1 to 0.5% by volume of the spray mixture. Surfactants are usually available at 50 to 100% active ingredient. Surfactants having a dominant lipophilic character (HLB of 0 to 8) increase the fluidity of cuticle components and result in easier diffusion of lipid soluble herbicides (ECs, WPs, flowables, and dispersibles) suspended in the spray water across the cuticle. Surfactants having a dominant hydrophilic character (HLB of 11 to 20) increase water retained in the cuticle, resulting in increased presence of hydrophilic routes for herbicide entrance and result in easier diffusion of water-soluble herbicides (salts and acids) dissolved in the spray water across the cuticle.

Oils used in agriculture are of two primary types: *refined oils* (petroleum-based) and *seed oils* (sometimes called vegetable oils). Petroleum (*refined*) oils are specific "cuts" from the distillation of petroleum, and crop oils and crop oil concentrates are highly refined and purified paraffinic nonphytotoxic oils plus surfactants that are used to increase foliar activity of certain herbicides. Originally the oil-surfactant mixture contained 2 to 5% surfactant (referred to as crop oil). However, crop oil use with herbicides has largely been replaced by crop oil concentrate (COC) that contains 15 to 20% surfactant in the oil-surfactant mixture. Crop oil is used at 1 to 4% (volume to volume) in water primarily as dormant oil sprays for control of certain insects in fruit and ornamentals, whereas COC is used at 1% by volume in herbicide spray mixtures. Although the ability of COC to enhance foliar activity of various herbicides such as atrazine, linuron, bentazon, and fluazifop appears to be associated with increased uptake by treated leaves, other factors may also be involved. For example, COC delays crystallization of atrazine on the leaf surface, which keeps the herbicide in an absorbable form for a longer time. COC can also reduce volatile and photodegradative loss of some herbicides.

Highly purified paraffin-based nonphytotoxic oils have been used as carriers for oil-soluble herbicides applied postemergence; however, the application volumes originally used are no longer practical. Research on delivery systems using air-assist spray nozzles (McWhorter et al., 1988) and positive displacement pumps (Hanks and McWhorter, 1991) has increased interest in using oils as carriers for oil-soluble herbicides at ultralow volumes ($\frac{1}{4}$ to 1 gallon per acre).

Seed oils are extracted from plants by pressing or solvent extraction. The hydrocarbon chain length of seed oils is primarily even numbered, with 16 to 18 carbons. Seed oils must be refined to remove gums, mucilages, phospholipids, short-chain ketones, acids, and other hydrocarbons to make them nonphytotoxic. There are two main types of seed oils used. The first are triglycerides and the second are methylated oils. Seed oils are primarily triglycerides when isolated and are generally not directly used in agriculture applications. For use in agriculture, the methylated form is obtained by hydrolyzing the triglyceride to yield free fatty acids. The free fatty acids are then esterified by reaction with methyl alcohol, and this form is combined with a surfactant for use as an adjuvant. The composition of triglyceride oils varies, depending on the seed source, and fatty acid composition can influence efficacy (Nalewaja, 1994). The performance of methylated seed oils can be comparable to and in some cases better than that of COC.

Organosilicone surfactants are often composed of a trisiloxane backbone (lipophilic or hydrophobic portion), with an ethylene oxide chain (hydrophilic portion) attached to one of the silicon atoms. Using a mix of EO and PO or modifying the EO end cap can reduce the lipophilicity. Organosilicone surfactants cause a tremendous reduction in the surface tension of water-based spray solutions and cause substantially greater spreading of the spray droplet than would be predicted by the reduction in surface tension. This increased spreading is thought to be due to the compact size of the lipophilic portion of the trisiloxane moiety, allowing it to transfer readily from the liquid/air interface to the leaf surface as the drop moves across the leaf (Ananthapadmanabhan et al., 1990). These surfactants provide improved rainfastness of the spray droplet and have humectant properties. Even though the increased spreading would in most cases be thought to increase the rate of droplet evaporation, these surfactants tend to slow the drying of the droplet. One problem with organosilicones is that they are unstable when the pH of the spray solution is not within the range of 6 to 8. Hydrolysis of the silicon-oxygen bonds occurs under acidic as well as basic conditions. Buffering the spray solution to a neutral pH can reduce this effect. Other disadvantages are that the degree of spreading is lost when such surfactants are mixed with other nonorganosilicone adjuvants, the extreme surface activity can cause excess spray tank foaming, eye and skin contact must be avoided, these surfactants are expensive, and they are not effective with all herbicides.

Salts of fertilizers are used as adjuvants in water-based spray solutions to increase the activity of foliar-applied herbicides. Common additives are ammonium sulfate, ammonium nitrate, and urea plus ammonium nitrate (e.g., 28% UAN) added at a concentration of 2 to 5%. The exact mechanism of action is not known, although increased herbicide absorption into the plant cells has been reported. For instance, thifensulfuron absorption into velvetleaf was increased from 4% to 45% when 28% UAN was added to the spray solution (Fielding and Stoller, 1990).

Ammonium sulfate has been shown to reduce the precipitation (crystallization) of glyphosate on the plant surface (MacIsaac et al., 1991). Ammonium sulfate has been used as a surfactant to overcome decreased herbicide activity due to antagonism caused by cations (Ca, Na, K, and Mg) in the water used as a spray carrier for certain herbicides, such as 2,4-D, bentazon, dicamba, acifluorfen, imazethapyr, glyphosate, nicosulfuron, and clethodim (Nalewaja and Matysiak, 1993a, 1993b; McMullan, 1994; Nalewaja et al., 1995). Adding ammonium sulfate to the spray solution has often been shown to reduce the antagonism between herbicides such as bentazon plus sethoxydim (Wanamarta et al., 1993), primsulfuron plus atrazine, and dicamba plus bentazon (Hart et al., 1992).

Spray Modifiers Spray modifiers, such as stickers, increase the adhesion of spray solutions to treated plant surfaces and are often used in conjunction with wetting agents (referred to as spreader-stickers). Film-forming vegetable gels, emulsifiable resins, emulsifiable mineral oils, waxes, and water-soluble polymers have been used as stickers.

Drift control agents are materials that thicken the spray solution and thereby increase droplet size and reduce the number of very small satellite droplets. These materials include swellable polymers and hydroxyethyl cellulose or polysaccharide gums and are used at concentrations of 0.1 to 1.0% of the volume. Invert emulsions are also used to reduce spray drift.

Drift control agents are of great value when herbicide applications are made near sensitive nontarget plants, even though they increase application costs. The appropriate spray equipment and operating conditions must be used with these thickened solutions.

Utility Modifiers Utility modifiers are adjuvants that are used to reduce or avoid application problems and/or increase the usefulness of a formulation. Emulsifiers and other surfactants of herbicide formulations can cause foaming with agitation of the spray mixture. Thus, *antifoam agents* are used to prevent or reduce excessive foaming in the spray tank. Antifoam agents are typically silicones and are used at 0.1% by volume. Kerosene or diesel fuel added to the spray tank at the same rate can often inhibit foaming.

Compatibility agents are used to help with mixing and/or application problems that may occur when a combination of pesticides is used. They can also be used when herbicides are applied in combination with a suspension, slurry, or true solution of fertilizers. Compatibility agents can counter separation problems that occur with hard or cold water.

HERBICIDE DRIFT

Herbicides may drift through the air from the target site and cause considerable damage if they contact susceptible plants. Movements through the air may result from *spray drift* or *volatility drift*.

Spray Drift

About one-third of all misapplications of agricultural pesticides are the result of drift. Drift is the movement of spray particles and vapors off-target, causing less effective control and possible injury to susceptible vegetation and wildlife. The amount of drift depends primarily on (1) droplet size, (2) wind velocity, and (3) height above the ground where the spray is released. In drift control, the emphasis should lie mainly in two areas: (1) timing of the spray application relative to weather phenomena and (2) reducing driftable particles.

The size of the droplet depends primarily on pressure, nozzle design, and surface tension of the spray solution. In general, low pressures produce large droplets and high pressures produce small droplets. Different nozzle designs produce different droplet sizes; small nozzle orifices produce small droplets. The lower the surface tension of a spray solution, the smaller the droplets. The importance of droplet size on spray drift is illustrated in Table 7-3. The smaller the size of the droplet, the longer it takes for the droplet to reach the ground and the greater distance it will travel. Droplets of 150 μm or smaller provide the greatest potential for drift.

Windy conditions are the most obvious weather problem. Ideally, herbicides should be sprayed when there is no wind. However, because this is not always possible, some guidelines should be established to minimize the spray drift hazard. Measure wind speed, and plan pesticide applications when wind speeds are somewhere between 0 and 5 mph for best results. In addition to droplet size and spray release height, these guidelines should consider herbicide characteristics and the distance to and type of surrounding vegetation. Herbicides should be sprayed only when winds are less than 3 mph and never when winds are greater than 5 mph. Excessive spray drift often occurs at greater wind velocities. Wind speeds in excess of 10 to 15 mph carry small spray particles locally downwind from the target, damaging susceptible plants in their path. Wind velocities are usually lowest just before sunrise and just after sunset and throughout the night.

The height above the ground that the spray is released is important for two reasons. First, the greater the height, the longer it takes the drop to reach the ground and the greater the distance of the drift. Second, wind velocities are usually lower close to the ground than at higher elevations. Spray uniformity depends on good overlap patterns between nozzles. For 20-inch nozzle spacing, the boom should generally be 17 to 19 inches above the target. Adjust the boom height above the ground for soil-applied herbicides and above the canopy for plant targets. Excess boom height reduces the chance that small spray droplets will reach the target before they decelerate or evaporate.

Therefore, spray applications from aircraft present a greater drift hazard than those from ground sprayers. Air currents produced by an aircraft have a major effect on the trajectory of particles released from it. Any aircraft, rotary (helicopter) or fixed wing, produces updrafts at the wing tips and downdrafts under the middle of the aircraft (Figure 7-4).

TABLE 7-3. Spray Droplet Size and Its Effect on Spray Drift

Droplet Diameter (μm)	Type of Droplet	Number of Droplets/in^2 from 1 gal of Spray/Acre	Time Required to Fall 10 ft in Still Air	Distance Droplet Will Travel in Falling 10 ft with a 3 mph Breeze
0.5	Brownian Max.	—	6750 min	388 miles
5	Fog	9,000,000	66 min	3 miles
100	Mist	1,164	10 sec	440 ft
200	Drizzle	195	3.8 sec	17 ft
400	Fine rain	28	2.0 sec	9 ft
500 ($\frac{1}{50}$ in.)	Rain	9	1.5 sec	7 ft
1000 ($\frac{1}{25}$ in.)	Heavy rain	1.1	1 sec	4.4 ft

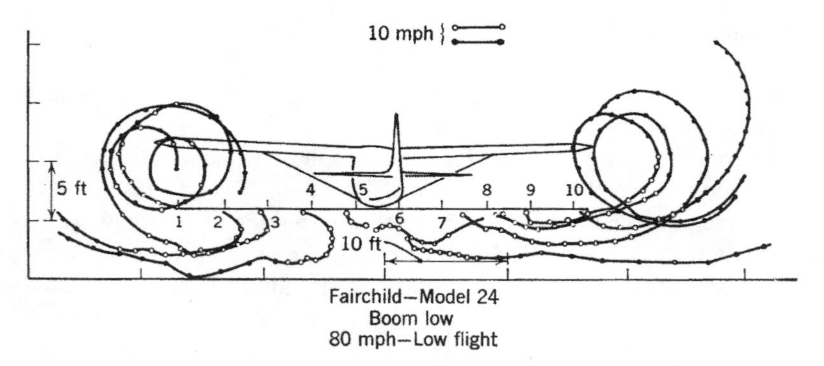

Figure 7-4. Air circulation from a high-wing monoplane.

Volatility Drift

Volatility refers to the tendency of a chemical to vaporize or give off fumes. The amount of fumes or vapors emitted is related to the vapor pressure of the chemical. Herbicides with vapor pressures of 1×10^{-4} mm Hg or greater at 25°C are more likely to be affected by volatility, and greater care must be employed in their application. Herbicides in this vapor pressure range include many of the carbamothioates, the dinitroanilines, clomazone, and the growth regulator herbicides such as 2,4-D. Calm air is often a sign of an atmospheric inversion—a condition that causes air near the earth's surface to stagnate. Atmospheric inversions are the cause of vapor drift. Small spray particles can be held in stable air and carried for miles before being deposited off-target, causing damage to crops sometimes quite remote from the application area. Vapor drift of soil-applied herbicides in this group can be reduced by soil incorporation and not applying them to wet soils. Vapor drift may damage susceptible plants or may simply reduce, through loss, the effectiveness of the herbicide treatment.

The volatility of 2,4-D has perhaps received more attention than that of any other chemical. Figures 7-5 and 7-6 indicate differences in the injury potential of plants when exposed to different salt and ester formulations of 2,4-D. The amine and sodium salts of 2,4-D have little or no volatility hazard, whereas the ester formulations vary from low to high volatility potential.

The length and structure of the alcohol portion of the 2,4-D ester molecule directly affects its volatility. In general, the longer the carbon chain in the part contributed by the alcohol, the lower the volatility. Those esters made from five carbon alcohols or fewer are usually considered volatile. Inclusion of oxygen as an ether linkage in the alcohol portion of the molecule will also reduce the volatility of an ester of 2,4-D.

Comparing the relative volatility of various formulations, Grover (1976) assigned 2,4-D amine a relative volatility of 1, low volatile esters of 2,4-D (propylene glycol butyl, butoxy-ethanol, and isooctyl) a relative volatility of 33, and a high volatile ester of 2,4-D (butyl) a relative volatility of 440. Thus, where 2,4-D-susceptible plants are grown, the 2,4-D amine form should be chosen. In addition, application methods that keep spray drift to a minimum should be used.

Figure 7-5. Tomato plants exposed to different 2,4-D formulations. *1*. Sodium salt—slight to no injury. *2*. Diethanolamine salt—no injury. *3*. Trietrhanolamine salt—no injury. *4*. Butyl ester—serious injury to death. *5*. Ethyl ester—serious injury to death. (Klingman, 1947.)

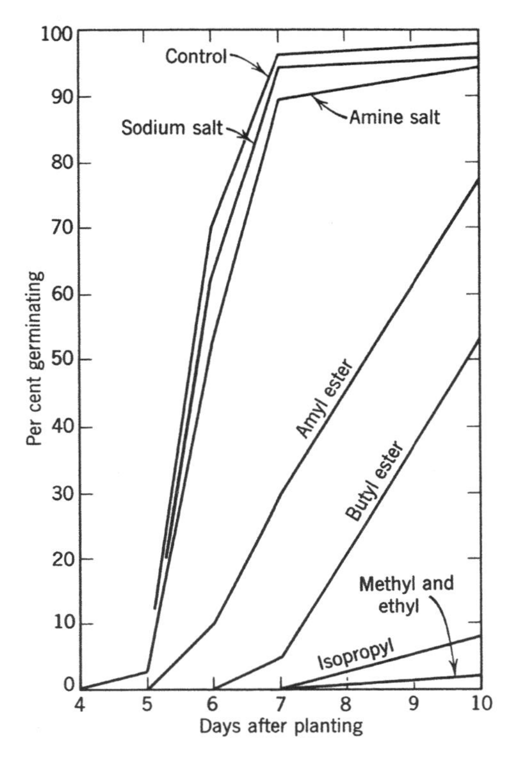

Figure 7-6. Germination of pea seeds after being exposed to different 2,4-D formulations. (Mullison and Hummer, 1949.)

Soil surfaces exposed to direct sunlight often reach 180°F. At this temperature, some chemicals quickly volatilize and may be carried away in the wind, and therefore are hazards to susceptible plants. Moreover, the herbicidal effect of the treatment may be lost.

APPLICATION EQUIPMENT

Safe and effective use of herbicides requires proper selection, calibration, and operation of the application equipment. A wide variety of equipment is used to apply herbicides. Selection of the specific type of equipment depends primarily on the weed, crop, herbicide, formulation, and the advanced technology available. Other considerations include whether the herbicide will be applied broadcast covering the entire area, in narrow bands, as individual spot treatments, or to a particular part of the plant. Superimposed on these factors are physiographic, edaphic, and climatic factors.

Most herbicide formulations, with the exception of granules, are usually applied to soil or plants as sprays with water as the diluent or carrier. Granular formulations are applied to soil by mechanical spreaders similar to those used for broadcasting seed or fertilizer. Herbicide application may also involve (1) mechanical incorporation into the top 1 to 6 inches of soil, (2) subsurface horizontal layering a few inches below the soil surface, or (3) injection into soil or water (lakes, reservoirs, irrigation or drainage systems).

Conventional Sprayer

Spraying is the most common method of application of a herbicide formulation. Sprays can be applied uniformly to the target, with the spray volume varying from 1 to 500 gal/acre. Lower volumes (1 to 5 gal/acre) are usually aircraft applications, and very high volumes (>100 gal/acre) may be required for thorough coverage in postemergence applications to dense vegetation in rights of ways and some noncropland applications. Spray volumes for both soil and foliar applications by ground rigs are usually in the range of 10 to 60 gal/acre.

The most frequently used equipment to apply herbicide sprays on farms is a low-pressure sprayer. Such sprayers can deliver from close to 0 to 200 psi, as contrasted with high-pressure air blast sprayers that can develop pressures of 600 psi and are used to deliver insecticides to tall trees.

Although sprays can be applied by hand-pump-type sprayers on limited areas (e.g., home lawns), most spray applications are made using tractor-, jeep-, truck-, trailer-, or aircraft-mounted equipment. This type of equipment has several essential components, including a tank, a pump, nozzles (generally on a boom), filters or strainers, pressure gauges, pressure regulators, shutoff valves, and connecting hoses (Figure 7-7).

Tanks A tank must have sufficient capacity; be easy to fill, drain, and clean; be corrosion resistant; and be equipped with appropriate openings, hoses, an agitation

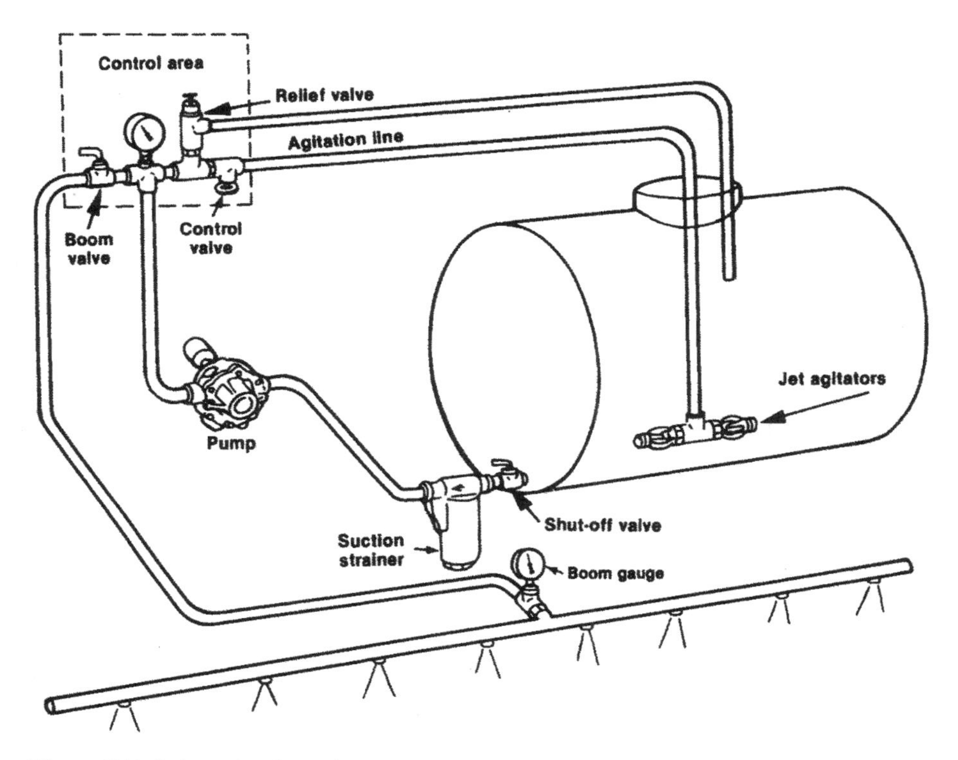

Figure 7-7. Schematic of a typical hydraulic sprayer used to apply herbicides. (North Dakota State University Extension Service.)

device, and connections. The tank must also have markings (on transparent tanks) or a gauge (on opaque tanks) to ensure accurate measurement of solution volumes.

A tank for a spray mixture can vary in size, composition, and design, depending on the nature of the job it is meant for. It may be a 55-gallon drum, a 1000-gallon trailer-mounted tank, or a railroad tank car (spraying railroad rights of ways). Materials used for tanks include galvanized steel, stainless steel, aluminum, fiberglass, polyethylene, and polypropylene. Stainless steel is the most durable and corrosion resistant, but also the most expensive. Galvanized steel is durable and inexpensive but prone to corrosion. Aluminum is durable, and although it is somewhat more expensive than galvanized steel, it is much more corrosion resistant. Fiberglass, polyethylene, and polypropylene are lightweight and corrosion resistant, but are less durable than metal tanks and require good support to avoid breakage. Fiberglass costs about as much as aluminum, and "poly" types are the least expensive; however, fiberglass tanks can be repaired whereas poly tanks cannot.

Although there is considerable flexibility in the possible tank design, it must have some provision for agitation of the spray mixture. Agitation is necessary because emulsions and wettable powders must have continuous movement to maintain a homogeneous mixture. Hydrolytic agitation of the spray mixture is most common; however, mechanical agitation with paddles is also used. Hydrolytic agitation is usually accomplished by recycling a portion of the spray mixture from the pump and expelling it into the spray mixture in the tank, using either a simple pipe with holes or a special agitator fitting. A separate agitator line should be used, not merely a bypass.

Pumps There are many types of pumps for liquids, each with certain advantages and disadvantages. The main pump types are centrifugal or turbine, roller, diaphragm, piston, gear, and flexible impeller types. Because only the first three of these pumps are commonly used in the United States for herbicide applications, they are the only ones discussed.

Power supplies to drive a pump include (1) tractor power take-off, (2) gasoline engine or electric motors as direct drive, (3) ground wheel tractor drive, and (4), on airplanes, a small propeller to drive the pump. The pump selected should have the necessary capacity in gallons per minute (GPM) to provide the desired volume and agitation requirements, plus a 10 to 20% excess to offset lost performance due to wear. Horsepower (Hp) needed to drive the pump, assuming an efficiency of 50 to 60%, can be estimated by the formula

$$\text{Hp} = \frac{\text{GPM} \times \text{psi}}{857}$$

To offset inefficiency in the power source, electric motors and gasoline engines should be rated at 33% and 33 to 67% higher horsepower, respectively.

Centrifugal pumps are the most commonly used pumps in low-pressure spray systems. They develop pressure using centrifugal force created by rapidly rotating blades in a chamber. Because they are nonpositive displacement pumps, they are not self-priming. However, such a pump is easily primed by placing it below the tank and using a small vent at the top of the pump to allow trapped air to escape. These pumps wear well even with wettable powders and deliver a high capacity, 70 to 130 GPM at 30 to 40 psi. However, because these pumps have to operate at speeds of 3000 to 4500 rpm, a step-up from the power take-off (PTO) is required. Turbine pumps are similar to centrifugal pumps but operate at speeds low enough to allow direct drive from the PTO.

Roller pumps are preferred by many operators for low-pressure sprayers because they are inexpensive, operate at PTO rpm, and are easily repaired. They are self-priming positive displacement pumps that move a constant volume of liquid each pump cycle. Therefore, they require a pressure relief valve or control device to divert unsprayed solution back to the tank. They are capable of delivering 5 to 40 GPM at 40 to 280 psi. Abrasive wear of rollers by wettable powders and other materials in spray solutions is the major disadvantage of roller pumps.

Diaphragm pumps use the movement of a flexible diaphragm to alternately pull liquid into a chamber through an intake valve and expel it through an outlet valve. They are positive displacement self-priming pumps that need less power to operate than other pumps. They develop moderate pressures and can deliver 15 to 50 GPM. They tend to be more expensive than pumps of other types, and the diaphragms can be affected by some herbicide solutions. However, they are excellent for the application of wettable powder formulations.

Nozzles Nozzles could be considered to be the most important part of the sprayer and other parts to merely facilitate their proper operation. A nozzle converts the spray mixture into spray droplets for delivery to the soil or plant. Several nozzles are often spaced along the length of a spray boom. The boom should be rigid during the spray operation for an accurate application. A typical fan-type nozzle has four parts: body, strainer, tip, and cap. The strainer (filter) is placed immediately ahead of the nozzle tip to filter the liquid to prevent nozzle clogging. Filters are classed by mesh size (number of openings per square inch), the most common types for herbicides being 50 and 100 mesh. One hundred mesh filters are used with nozzles having a flow rate below 0.2 gal/min, and 50 mesh screens used with flow rates between 0.2 and 1.0 gal/min.

Nozzle flow rate depends on design (orifice size) and the conditions of its operation, especially pressure, which largely determine the uniformity and rate of the application. They also influence the size and uniformity in size of the droplets. As pressure increases, flow rate increases, although this increase is not linear. The flow rate varies with the square root of the pressure increase, meaning that the pressure must increase four times to double the flow rate. At the ideal pressure, droplet formation occurs near the nozzle tip with the formation of uniform small droplets across the width of the spray pattern. At low pressures, liquid escapes from the nozzle tip as a liquid film. This film ligaments and then forms relatively large droplets at the outer edges as it expands. At high pressures, very small droplets are formed immediately at the nozzle tip. These small droplets may be of fog and mist size, subject to drift from the target site creating a potential hazard. Spray pressures for herbicide applications may range from as low as 5 psi to as high as 50 psi; however, the usual rate is 20 to 40 psi. Lower pressures and higher rates are used when penetration of dense vegetation is desired.

Faulty nozzles or faulty operation of nozzles can cause uneven spray patterns that may result in several-fold variations from the desired application rate. Such unevenness causes crop injury due to higher rates of application than recommended or can result in lack of weed control due to lower rates of application than recommended. These variations may appear as narrow strips of a few inches to a foot or so wide.

Nozzles are constructed of many types of materials, including hardened stainless steel, stainless steel, brass, aluminum, nylon, plastic, and ceramic. Hardened stainless steel, stainless steel, and ceramic are the most resistant to wear and corrosion but are the most expensive. Nylon and other plastics resist wear somewhat less well and may swell with the use of some formulations. Brass spray tips are relatively inexpensive but wear rapidly with abrasive spray solutions and can corrode with some fertilizers. Many nozzles are now constructed with plastic bodies and hardened stainless steel spray orifices to reduce overall cost and lengthen life. Spray applicators must maintain nozzles that are accurate and should frequently check nozzle output and replace worn nozzles to avoid uneven application.

Although there are many types of nozzles, herbicides are normally applied with regular flat-fan, even flat-fan, flooding, or whirl-chamber hollow cone nozzles (Figure 7-8). Other nozzles used in certain situations include rotary-disk nozzles, when it is essential to minimize spray drift, off-center, double outlet flat-fan, and boomless types.

Regular flat-fan nozzles are general-purpose nozzles suited for broadcast applications of preemergence and postemergence herbicides when penetration of foliage is not needed. These nozzles deliver more spray in the center of the spray pattern than at the edges (Figure 7-8). The spray patterns must overlap 40 to 50% for uniform applications. Several spray angles are available (65°, 73°, 80°, and 110°); selection is determined by nozzle spacing along the boom and height of the nozzle tip above the target. Pressure should be restricted to 15 to 30 psi to minimize drift. Variations in the flat-fan design are available. *Low-pressure* (LP) and *extended range* (XR) nozzles are designed to provide better spray distribution over a range of pressures, resulting in less drift at low pressures and better coverage at high pressures. *Reduced drift* nozzles are designed to produce larger droplets and a reduced number of small droplets, which minimizes off-target spray. *Air induction* nozzles are designed to produce larger droplets for less drift by producing air-filled drops from a venturi air aspirator. *Twin orifice* nozzles are designed with two orifices that direct one flat-fan spray 30° forward and a second 30° to the rear. These nozzles produce smaller droplets for improved coverage and better penetration of crop residue and plant foliage, and have better spray distribution along the boom than hollow cone nozzles.

Even-edge flat-fan nozzles are similar to regular flat-fan nozzles, except that they deliver an equal amount of spray across the spray pattern (Figure 7-8). These nozzles should be used only to apply a band of herbicide over a crop row. Spray angle and height of the nozzle tip above the target determine the width of the band. These nozzles are also available in LP, XR, reduced drift, and other types.

Flood nozzles deliver a spray pattern similar to that of flat-fan nozzles, but the distribution of droplets is less uniform (Figure 7-8). They are particularly useful for herbicide and herbicide-fertilizer applications to soil. Large, coarse droplets are produced at an operating pressure of 8 to 25 psi. Although this pressure reduces drift, 100% overlapping patterns should be used to offset the less uniform spray pattern. Flood nozzles can be operated in any orientation, from spraying straight down to straight back, provided that the overlap is maintained. The "Turbo Floodjet" nozzle is a variation on the original flood nozzle design. This nozzle design incorporates a pre-orifice, which produces larger droplets for less drift and more uniform distribution, and hence better spray coverage.

Whirl-chamber hollow cone nozzles are commonly used for spraying directly from a herbicide soil-incorporation implement. These nozzles have a whirl chamber above a cone-shaped outlet and produce a hollow cone pattern with a fan angle. Raindrop

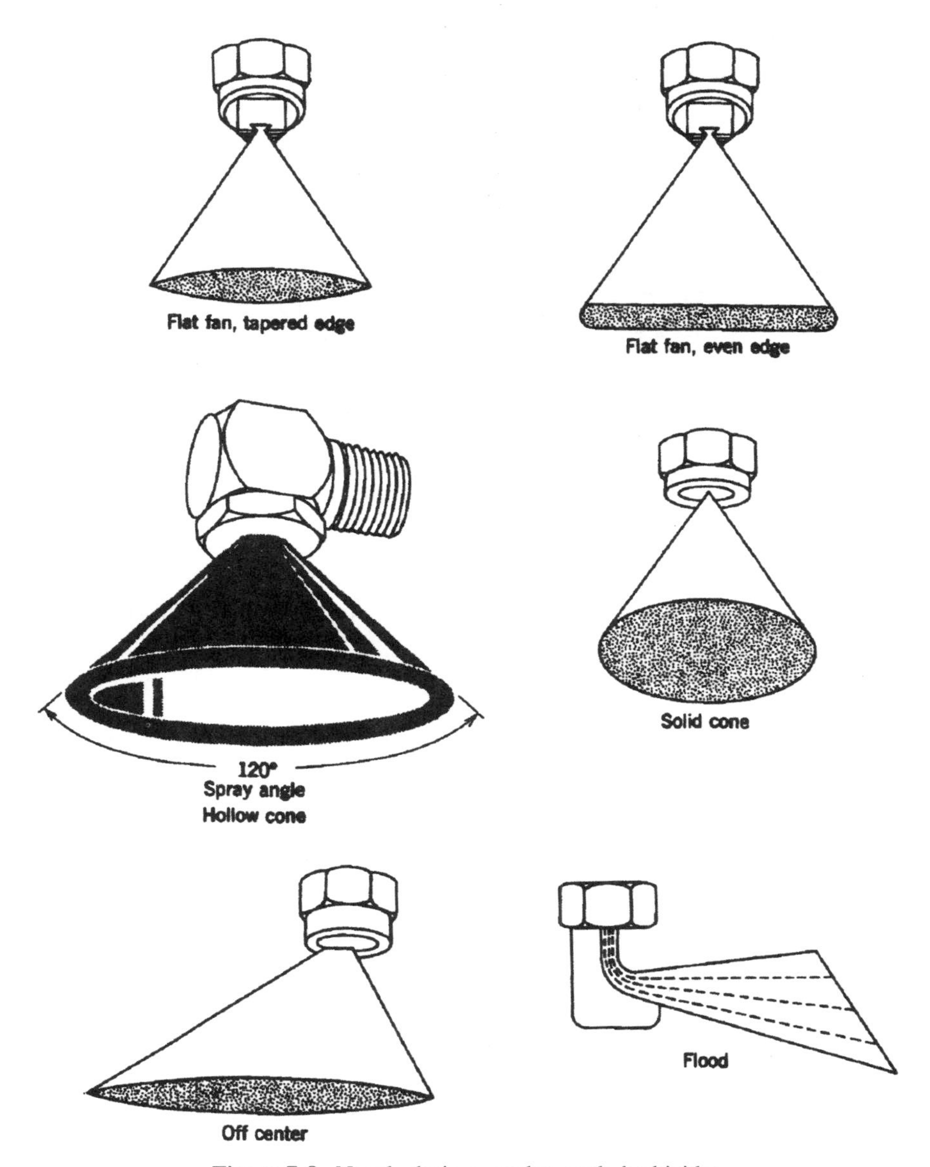

Figure 7-8. Nozzle designs used to apply herbicides.

nozzles produce a hollow cone pattern with large droplets and have been designed for both soil and foliar application of herbicides (Figure 7-9).

Off-center nozzles emit a pattern directed to one side of the nozzle and are commonly used at the end of a boom to increase the spray width. The spray pattern is very uneven, and these nozzles are primarily useful to deliver herbicides to relatively inaccessible areas.

Figure 7-9. Raindrop (brand) nozzle spraying at 40 psi. (Photo by Ann Hawthorne, Delavan Corporation.)

Double outlet flat-fan nozzles have two orifices to obtain a wide spray angle of up to 150°. These nozzles are used on drop pipes to deliver herbicide to the base of the plant or to the ground under tall crops.

Boomless nozzles are often used to spray areas with rough terrain and those not easily accessible with a boom sprayer. These nozzles cover a wide swath of 30 to 60 feet; however, the spray distribution is not as uniform as with a boom sprayer.

Spinning-disc nozzle sprayers use centrifugal force to form droplets as the spray solution exists from the nozzle (Figure 7-10). This is in contrast to all previously described nozzle types that use hydraulic energy to produce droplets. In the spinning-disc nozzle, the spray solution is directed to the center of a spinning disc and the droplets are formed at the disc edge as the solution is thrown off the surface. The disc edge may be either smooth or have teeth. Herbicides are applied at a total spray volume of only 1 to 3 gal/acre (10 to 30 l/hectare). A major advantage of a rotary-disc nozzle is controlled droplet application (CDA) that produces consistently large spray droplets of around 250 micron that minimize the potential of spray drift and fall in a precise 1.2 m wide circular pattern and the lower diluent volumes compared to other nozzles. Other spray width nozzles are available for narrower or wider banding. These sprayers are used worldwide for band spraying of herbicides in orchards, forestry and plantations for general clean-up, border, pathway or strip spraying of weeds.

Other Components *Filters or strainers* are usually used at the tank opening, in the spray lines, and at the nozzles. Mesh number, as mentioned earlier, indicates strainer size. Sixteen to 20 mesh screens are used at the tank opening to remove extraneous material and large lumps of herbicide. Strainers of 40 to 50 mesh are used in the suction lines to prevent foreign materials from the tank damaging the pump. However, because it is very important that the inlet of a centrifugal pump not be restricted, a strainer no smaller than 20 mesh with a diameter several times larger

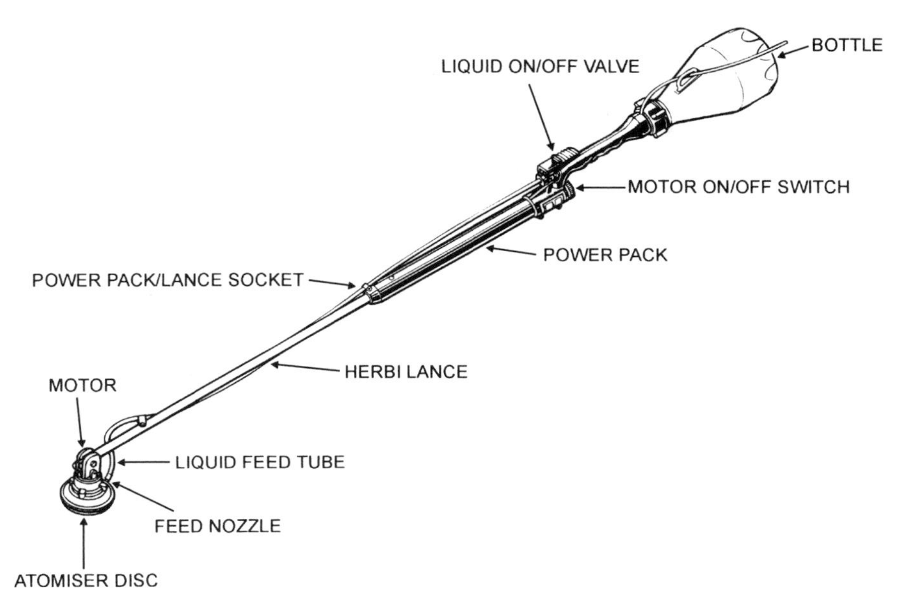

Figure 7-10. The Microfit Herbi uses a spinning disc to produce uniform size droplets at low gallonage per acre. Larger units are available for tractor, all terrain vehicles and aircraft application. (Micron Sprayer Limited, Herefordshire, UK.)

than the suction line should be used. Strainers are also needed to prevent clogging by particulate material. Nozzle strainers, as discussed earlier, are usually 50 or 100 mesh.

A good *pressure gauge* is essential for proper nozzle operation. Its range should be relatively narrow, only slightly exceeding the intended range of spraying pressures. A gauge with a pressure range that is too broad makes accurate pressure determinations difficult.

A *pressure regulator* or *relief valve* is required with positive displacement pumps (e.g., roller pumps) to prevent damage to the pump that can occur when spraying is interrupted. Although a centrifugal pump does not require a pressure relief valve, a special throttling valve should be used for accurate control of spray pressure. The throttling valve can be electrically controlled and operated from the tractor cab.

Spray lines or hoses must be of sufficient strength to withstand the pressures or vacuums expected and made of material that will tolerate the chemical to be used. If the sprayer is built so that a vacuum may develop between the spray tank and the pump, a heavy-walled hose (with a metal interior) or metal pipe should be used in place of a regular hose to prevent collapse. A vacuum may develop from clogged suction strainers in the hose, the inlet opening held to the wall of the tank, or from collapsed or twisted hose. Insufficient liquid flow will damage some pumps and will reduce the efficiency of all pumps.

Oil-resistant hoses made of neoprene, plastic, or other oil-resistant materials may be most satisfactory with oils and oil-like herbicides. In addition to having a longer life, oil-resistant hoses resist absorption of the chemical, making it less difficult to remove the herbicide.

Pressures in a hydraulic system are the same regardless of hose size, minus any friction loss involved in liquid movement through the hose. Therefore, hoses should be chosen that are large enough not to restrict liquid flow.

Calibrating the Sprayer

Several methods can be used to determine the number of gallons of spray to be applied per acre, which allows a determination of how much herbicide to put in the spray tank to obtain the proper herbicide dosage (rate). The three basic steps involved in proper calibration are (1) select the proper herbicide product and application equipment, (2) determine the size of the treatment area, and (3) determine the amount of herbicide product needed. Regardless of the method used, it is imperative that the sprayer is functioning properly and each nozzle delivers the same volume of spray in a uniform pattern. Four methods of calibrating a sprayer are briefly discussed in the following paragraphs. Most university cooperative extension services have information relating to proper sprayer calibration, as do nozzle companies, chemical companies, and private consultants.

Spraying an Area of a Known Size and Measuring the Amount of Spray Applied Perhaps the most accurate method is to actually spray an area of known size and measure the volume of spray mixture used. In practice, one starts with a full tank and measures the gallons required to refill the tank after spraying a specific area. The gallons per acre can be readily calculated from these measurements by dividing the gallons sprayed in the calibration run by the area sprayed, which equals gallons of spray per acre. Obviously, the size of the area sprayed must be large enough to give an accurate measurement of the area sprayed and the volume used. In general, aircraft sprayers require 5 to 10 acres, tractor sprayers require 1 to 2 acres, and a hand sprayer 1000 square feet.

Measurement of Gallons of Spray Delivered This method uses a measurement of the volume of spray mixture delivered per unit of time and the spray width. Gallons per acre are calculated by use of the following two formulas.

The acres sprayed per hour can be calculated from the tractor speed, spray width, and two constants, as follows:

$$\text{Acres sprayed per hour} = \frac{\text{mph} \times 5280\ (\text{ft/mile}) \times \text{spray width (ft)}}{43{,}560\ (\text{ft}^2/\text{acre})}$$

The gallons per acre can be calculated from the preceding value (acres sprayed/hr) and the measured spray volume per unit time as follows:

$$\text{Gallons/acre} = \frac{\text{Gallons applied/hr}}{\text{Acres sprayed/hr}}$$

For example, a sprayer traveling at 6 miles per hour with a swath width of $16\frac{1}{2}$ feet will spray 12 acres per hour. If the sprayer applies 60 gallons of spray per hour, the sprayer will apply 5 gallons per acre.

Prepared Tables Prepared tables that give nozzle spacing, pressures, speed, and various nozzle sizes giving various gallons of spray per acre are available from some nozzle manufacturers. The proper nozzle size can be selected from these tables. The disadvantage to this method is that there is no assurance that the speeds and pressures used in the spraying are correct.

Special Measuring Devices Special measuring devices and prepared charts or graphs can be used. The spray is usually collected for a prescribed period of time or distance. The amount of spray collected is then converted, through tables or charts, into gallons per acre. Glass or plastic containers with the table printed directly on them are available to collect the spray. This technique has the same disadvantages as those for using prepared tables. *The table determinations and measuring device methods should be used as a guide to set up the sprayer. An actual calibration run as described for this method should be used to confirm sprayer output in gallons per acre.*

In all cases, check to make sure that the nozzles are all functioning properly and that the spray width and height are correct. All nozzles should have an output within ±5% of each other.

Cleaning the Sprayer

After use, a sprayer and tank should always be cleaned (both inside and out) before the next use or before it is stored. Many pesticides, if left in the system, can cause corrosion and sprayer damage. The best approach is to finish spraying with as little pesticide solution left in the tank as possible. Immediately after use, rinse the sprayer with clean water. This is sufficient if the same herbicide will be applied the following day. If another chemical is to be used, special solutions may be necessary to completely rid the system of any herbicide residue.

Always check the pesticide label to determine the specific cleaning procedure for the sprayer. A solution of detergent (2.2 lbs/100 gal water) is usually acceptable for most herbicides. First flush the sprayer with water, then circulate the cleaning solution and follow with a rinsing with clean water two times. Always flush the boom and loosen and clean all nozzles with a soft brush.

Some oil-soluble herbicides (chemicals that form emulsions with water) may require special cleaning procedures. These chemicals can sometimes result in a gummy residue in the tank that may require flushing with water and a solvent like kerosene, diesel fuel, or fuel oil. After the tank is clean, rinse thoroughly with a detergent, followed by clean water as mentioned earlier. Certain oil-soluble herbicides, such as 2,4-D esters, are usually the most difficult to remove, and sometimes residues cannot be easily eliminated from hoses. After the aforementioned rinses, add 1 quart of household ammonia to 25 gallons of water; circulate this solution in the system and flush the nozzles. Allow the solution to sit in the system overnight, then remove and thoroughly flush the system two times with clean water. The 2,4-D salts are water soluble and removed by thorough washing with water. Check the spray on susceptible plants, such as tomatoes, to make certain that the 2,4-D or other herbicide has been removed from the sprayer.

For wettable powder herbicides, examine the tank to see that none of the wettable powder remains in the bottom; for any such residue, a thorough rinsing with water is usually sufficient.

Other Types of Applicators

Shielded or *hooded sprayers* use angled shields over the spray nozzle to protect the crop plant from the herbicide and increase herbicide selectivity and/or to reduce the influence of wind on herbicide placement. Shielded sprayers are used for directed sprays in row crops and in high-value fruit and nursery crops grown in rows.

Airfoil needle and *small-diameter orifice booms* are used to apply herbicide solutions from helicopters. The nozzles deliver a stream of solution through narrow-diameter tubes into the air with reduced turbulence, and the stream breaks into uniform droplets that resist drift.

Recirculating sprayers, *roller applicators* and *rope wicks* are specialized equipment to treat weeds that are taller than the crop. The recirculating sprayer directs the spray horizontally as a narrow stream above the crop canopy, hitting only the tops of tall weeds. Spray that is not intercepted by tall weeds is caught and recycled through the sprayer (Figure 7-11).

Roller applicators use a rotating carpeted roller to wipe herbicide onto tall weeds.

Rope-wick applicators use a straight section of plastic pipe with rows of wicks exposed to wipe tall weeds (Figure 7-12).

Although these applicators do an excellent job on tall weeds, their use must be delayed until the weeds are taller than the crop. Thus, crop yields are reduced before these methods are used; they were developed mainly for eliminating weeds that would hinder crop harvesting. These applicators were also developed to allow the use of relatively nonselective herbicides like glyphosate in many field crop situations. With the development of glyphosate-resistant crops (Chapter 18), there is less need for these types of applicators.

Variable rate herbicide application (VRT), a concept that is in its infancy, applies preemergence and postemergence herbicide to the areas of a field where they are needed. Variable rate fertilizer applications have been used for a number of years, and the equipment used can be adapted to apply soil herbicides. The objective is to apply the herbicide to the soil at the appropriate rate, based on soil texture, soil organic matter, soil pH, and the cation exchange capacity. These soil factors and how they vary within a field can have either a pronounced or a minimal effect on the herbicide activity. The goal is to reduce the rate in some areas of the field and increase the rate in other areas, as needed to achieve uniform weed control. Work is under way in many

Figure 7-11. Recirculating sprayer equipped with solid-stream nozzles to apply herbicides to weeds that are taller than the crop. Spray caught in the box is returned to the tank to be sprayed again. (Southern Weed Science Laboratory, USDA-ARS, Stoneville, MS.)

Figure 7-12. Rope-wick applicator. Herbicide-soaked ropes are attached to a horizontally positioned pipe reservoir. (Southern Weed Science Laboratory, USDA-ARS, Stoneville, MS.)

labs to develop systems using VRT coupled with field mapping of weed infestations and GIS (geographic information systems) technology for applying herbicides in a field precisely where the weed problem exists, not necessarily over the entire field.

The goal of postemergence variable rate herbicide application is to treat only those areas where weeds are present. Locating the weed and then differentiating it from the crop is the major challenge in VRT at present. The patented Patchen, WeedSeeker is an example of an applicator that uses advanced optics and computer circuitry to sense green plant tissue—that is, whether a weed is present. When a weed enters the sight of the sensor unit, a spray nozzle is signaled to deliver a precise amount of herbicide and sprays only weeds, not bare ground. The Patchen has been tested for use in rights of way (railroad and utilities), on noncropland, along roadsides, on irrigation ditch banks and roadbeds. Other potential uses include airport runways, golf courses, paved parking lots with medians, dirt and gravel parking lots, parks and hiking trails, and the like. This technology is exciting. In the future, as more sophisticated and effective automated equipment and improved sensors are developed, VRT equipment for spraying herbicides in cropland (which allows sensors to differentiate weeds from crops) will become more common and will result in less herbicide use but more effective weed management (Medlin et al., 2000). Wet blade application is accomplished by the patented Burch Wet Blade Mower, whereby herbicide is applied to vegetation through the action of mowing blades. This is a revolutionary nonspray closed application system capable of delivering 2.5 or less gpa. Herbicide solution is delivered to an area beneath the cutting blade and is applied directly to cut stems of vegetation. This applicator has shown the greatest utility for weed management in rights of way, turf, pastures, and noncropland areas.

Hand guns are used to apply herbicides and fertilizers on home sites and public areas, primarily to turf. The precision of application is low with handguns, and they should be used only by qualified applicators.

Hand sprayers and *backpacks* are designed for small areas and are most effective for spot treatments. Most common home sprayers do not use compressed gas or have pressure gauges or pressure controls. Pressure is obtained by air introduced by the operator and continuously varies during the spray operation. These sprayers are limited to small areas and are common in the horticulture and home landscaping industries. The applicator has to develop a feel for the amount of pressure and spray application technique that must be used to ensure uniform and complete coverage (see Whitford et al., 1999, for more details on types of handheld liquid sprayers and calibration). Backpack sprayers, pressurized and powered with compressed gas (CO_2 or N_2), can be quite accurate, as they are set up like hydraulic sprayers, with pressure gauges, precision nozzles, and so forth, except that they use gas to pressurize the system rather than a pump.

Granular applicators apply herbicide granules with equipment similar to dry fertilizer spreaders (Figure 7-13). These units consist of a hopper containing the herbicide, a flow rate controller, and a distribution system. The three main types are drop, rotary, and pneumatic spreaders. These applicators are very common in the home landscape for application of fertilizer and herbicide mixtures (see Whitford et al., 1999, for details on granular applicators for the home landscape).

Figure 7-13. Applicator for banding granular materials over plant rows at planting time. (Gandy Company.)

Drop types have a full-length agitator mounted on the spreader axle that rotates over a series of openings in the hopper through which the granules drop to the ground. The size of the openings can be adjusted to obtain the appropriate application rate with various products. Slight overlapping of the wheels is necessary during application to obtain uniform coverage. The advantage of drop spreaders is that the application rate is equal over the entire hopper length. Problems include skips due to improper overlap, excessive overlap, clogging of the openings, and improper application when turning corners.

Rotary spreaders drop the granules out of one or more adjustable openings at the bottom of the hopper onto a rotating plate, which spreads the granules in a semicircular arc. The width of application can be much greater with rotary spreaders than with drop spreaders. The uniformity of application can vary greatly with these applicators. As differing sizes and weights of granules from different products spread differently, it is hard to overlap properly, clogging is frequent, and turning changes the speed of the rotating plate.

The typical granule is formulated at 10% or less active ingredient. This low percentage of active ingredient is helpful when using drop and rotary applicators, as the normal variation in precision of application is accounted for with these products.

Pneumatic spreaders use a combination of air, tubes, and deflectors to distribute the granules. Granules are gravity fed and metered into a chamber and trapped into a high-speed airstream that moves the granules through delivery tubes to the exit point, where the deflector plates evenly distribute the granules to the soil. Properly calibrated, these applicators are quite accurate.

Soil Incorporation Equipment

Virtually all types of cultivation equipment have been used to mix herbicides into the soil. However, not all implements work equally well for every situation or herbicide. The most common equipment used includes a disk harrow, power-driven tillers, field cultivators, rolling cultivators, a rotary hoe, and a combination of these. For all broadcast applications, a disc harrow or ground-driven rolling cultivator can be used. For band applications, power-driven rotary tillers (rototiller, Roterra, and power-driven harrows) are usually used. Figure 7-14 shows the procedure for incorporation of a herbicide into a preformed bed.

The disc harrow is probably the most common incorporation tool. A second discing is common at right angles to the first. Speeds must be fast enough to effectively mix the soil. A small disc harrow (finishing disc) is better adapted to incorporation than a large, heavy disc. Disc blades should be spaced no more than 8 inches apart. Large discs recently available may incorporate the chemical too deeply in the soil, especially in sandy and sandy loam soils. Depth-gauge wheels set for the disc to cut 4 to 5 inches deep may help. Following the disc with a heavy spike-tooth harrow may further help to mix the herbicide evenly in the soil. The herbicide manufacturer's label should be checked for the depth of incorporation and equipment recommended for each herbicide. Recently, there has been more interest in one-pass shallow incorporation (sometimes called surface blending), and the field cultivator has become more commonly used. A secondary tillage implement then follows the field cultivator.

Regardless of the type of soil incorporation equipment to be used, three major factors must be considered in addition to the proper application rate: depth of incorporation, soil conditions, and correct ground speed.

Depth of incorporation, or mixing of the herbicide with the soil, is critical. If herbicides requiring shallow incorporation are placed too deeply, they may lose some of their effectiveness by dilution in a greater volume of soil. If the depth at which volatile herbicides are mixed is too shallow, some volatility loss may occur. The more volatile herbicides (e.g., EPTC) require deep mixing of 2 to 4 inches. Less volatile

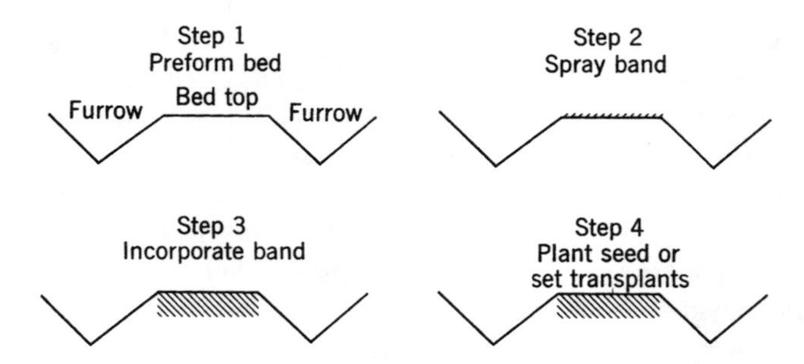

Figure 7-14. Steps in making a band application of a soil-incorporated herbicide to a preformed bed. All of these steps can be performed in a single pass through the field by placing all the equipment required on one tractor.

herbicides (e.g., trifluralin) may need to be mixed to only a moderate depth of 1 to 2 inches.

Soil conditions may result in erratic weed control when large clods are present at the time of incorporation. A fine seedbed with clods of less than 0.5 inch in diameter is recommended. A herbicide does not readily mix with excessively wet soil, which can also increase volatility losses by decreasing the adsorption on soil particles. The development of a plow hard pan, soil compaction, or loss of soil structure may occur by working soil with excessive moisture.

Ground speed of 5 to 6 mph is necessary to obtain adequate mixing of the herbicide into the soil with discs, harrows, and ground-driven rolling cultivators. Lower speeds (1 to 2 mph) can produce a "streaking" and poor weed control. With power-driven rotary tillers, ground speeds of 1.5 to 3.0 mph are recommended. Good results have been obtained at a tiller speed of about 500 rpm and 2 mph ground speed. The tillers must be close enough together to provide a "clean sweep" of the soil. This may require L-shaped knives.

For an overall flat preplant soil-incorporation treatment followed by bedding (listing and bed forming), the lister shovels should always be set to run 1 to 2 inches shallower than the depth of incorporation. This setting prevents placing untreated soil on top of the bed. For a band application of a preplant soil-incorporation treatment to preformed beds, care must be taken not to remove treated soil or place untreated soil on top of the beds by subsequent planting or cultivating operations (Figure 7-14).

Subsurface Layering Equipment

The spraying of a horizontal layer of herbicide a few inches below the soil surface (subsurface layering) has produced effective control of several hard-to-control perennial weeds. An example is field bindweed control with trifluralin or dichlobenil. This method is also called the spray-blade method or simply blading or layering.

The spray-blade method uses a blade with backward-facing nozzles attached under the leading edge. Modified sweeps and V-shaped blades have been used. The concentrated layer of herbicide acts as a "protective wall," preventing the weed shoots from growing through it; thus, the deeper storage roots starve and the plant dies. Any disturbance of this layer by cultivation or natural cracking of the soil permits the shoots to emerge, and control is therefore greatly reduced.

Injection

Herbicides are injected into soil or water to control terrestrial and aquatic weeds. The most common chemical injected into soil is methyl bromide. Methyl bromide is a broad spectrum fumigant that controls many soil pests, such as plant diseases and insects as well as weeds, and is scheduled to be phased out of the U.S. market by 2005 because it is claimed to be an ozone depleter. This material is usually applied by a commercial applicator because of its toxicity and because the treated area must be covered by a gas-tight tarpaulin for 24 to 48 hours. Methyl bromide is applied with chisel-type applicators that inject the chemical 6 to 8 inches below the soil surface. The chisel-injection units are spaced not more than 12 inches apart.

Herbigation

Herbicides are, in some cases, applied through irrigation systems, often through sprinkler and furrow systems but most commonly through center pivot systems. A specialized injector, in addition to the fertilizer injector, must be used. The herbicides are precisely metered into the water by the injector, thoroughly mixed, and then applied to the soil surface by the irrigation water. The injectors and equipment used are very precise; however, total accuracy is only as good as the water distribution system, and this system is usually less accurate than a spray application. Government regulation of *chemigation* (application of pesticides through an irrigation system) concerns the suitability of particular chemicals and is designed to protect ground and surface water from contamination and to ensure that suitable injectors are used, applicators are qualified, and farmworkers and the general public are protected from chemical exposure. The application of pesticides by chemigation is under the jurisdiction of the U.S. EPA and specific state agencies. The U.S. Environmental Protection Agency (EPA) requires labels of all registered pesticides to say whether or not they can be used in chemigation, and each chemical manufacturer specifies on the product label the application method and chemigation directions (see chemigation Web sites for further information).

To control aquatic weeds, herbicides are injected into flowing waters (irrigation and drainage canals) or injected from a boat into static waters (lakes and reservoirs). See Chapter 27 for details of aquatic weed control.

Aircraft

Aircraft sprayers are less commonly used to apply herbicides than other pesticides because of the drift hazards. In addition, the precision of application is somewhat less than that of ground sprayers. Despite these limitations, aircraft are especially adapted for spraying or applying granular formulations to areas not readily accessible to ground equipment, such as utility lines and firebreaks through remote woody areas, flooded rice fields, very large pasture or range areas, and large cereal grain fields.

Both fixed wing and rotary wing (helicopters) aircraft are used. In general, the components of sprayers for aircraft are similar to those of ground sprayers. Included are a tank, agitators, a pump, a boom, valves, screens, nozzles, a pressure regulator, and a pressure gauge. Because the design of aircraft sprayers requires special engineering knowledge, the details of design are not discussed here.

LITERATURE CITED AND SUGGESTED READING

Ananthapadmanabhan, K. P., E. D. Goddard, and P. Chandler. 1990. A study of solution interfacial and wetting properties of silicone surfactants. *Colloids and Surfaces* **44**:281–297.

Becher, P. 1973. The emulsifier. In *Pesticide Formulations*, ed. by W. van Valkenburg, p. 65–80. Dekker, New York.

Butselaar, R. J. and W. R. Gonggrijp. 1993. New fatty-amine based adjuvants: Mode of action and applications. *Pestic. Sci.* **37**:212–215.

Davis, D. E., H. H. Funderburk, Jr., and N. G. Sonsing. 1959. The absorption and translocation of C^{14}labeled simazine by corn, cotton and cucumber. *Weeds* **7**:300–309.

Fielding, R. J., and E. W. Stoller. 1990. Effects of additives on the efficacy, uptake, and translocation of the methyl ester of thifensulfuron. *Weed Sci.* **38**:172–178.

Foy, C. L., and D. W. Pritchard, eds. 1996. *Pesticide Formulation and Adjuvant Technology.* CRC Press, Boca Raton, FL.

Grover, R. 1976. Relative volatilities of ester and amine forms of 2,4-D. *Weed Sci.* **24**:26–28.

Hamilton, R. J. 1993. Structure and general properties of mineral and vegetable oils used as spray adjuvants. *Pestic. Sci.* **37**:141–146.

Hanks, J. E., and C. G. McWhorter. 1991. Variables affecting the use of positive displacement pumps to apply herbicides in ultra-low volume. *Weed Technol.* **5**:111–116.

Hart, S. E., J. J. Kells, and D. Penner. 1992. Influence of adjuvants on the efficacy, absorption, and spray retention of primsulfuron. *Weed Technol.* **6**:592–598.

Kirkwood, R. C. 1994. Surfactant–pesticide plant interactions. *Biochem. Soc. Transactions* **22**:611–616.

Klingman, G. C. 1947. New chemicals control weeds. *North Carolina Research and Farming* **6**:3,4,12,13.

Knoche, M. 1994. Organosilicone surfactant performance in agricultural spray application: A review. *Weed Res.* **34**:221–239.

MacIsaac, S. A., R. N. Paul, and M. D. Devine. 1991. A scanning electron microscope study of glyphosate deposits in relation to foliar uptake. *Pestic. Sci.* **31**:53–64.

McMullan, P. M. 1994. Effect of sodium bicarbonate on clethodim or quizalofop efficacy and the role of ultraviolet light. *Weed Technol.* **8**:572–575.

McWhorter, C. G. 1982. The use of adjuvants. In *Adjuvants for Herbicides.* WSSA, Lawrence, KS.

McWhorter, C. G., F. E. Fulgham, and W. L. Barrentine. 1988. An air-assist spray nozzle for applying herbicides in ultra-low volume. *Weed Sci.* **36**:118–121.

Medlin, C. R., D. R. Shaw, P. D. Gerard, and F. E. LaMastus. 2000. Using remote sensing to detect weed infestations in *Glycine max. Weed Sci.* **48**:393–398.

Mervosh, T. L., E. W. Stoller, and W. F. Simmons. 1995. Effects of starch encapsulation on clomazone and atrazine movement in soil and clomazone volatilization. *Weed Sci.* **43**:445–453.

Miller, P., and P. Westra. 1996. *Herbicide Surfactants and Adjuvants.* Bulletin #0.559, Crop Series. Colorado State University Cooperative Extension, Fort Collins, CO.

Mullison, W. R., and R. W. Hummer. 1949. Some effects of the vapor of 2,4-Dichlorophenoxyacetic acid derivatives on various field-crop and vegetable seeds. *Bot. Gaz.* **111**:77–85.

Nalewaja, J. D. 1994. Esterified seed oil adjuvants. *NCWSS Proc.* **49**:149–156.

Nalewaja, J. D., and R. Matysiak. 1993a. Optimizing adjuvants to overcome glyphosate antagonistic salts. *Weed Technol.* **7**:337–342.

Nalewaja, J. D., and R. Matysiak. 1993b. Spray carrier salts affect herbicide toxicity to Kochia (*Kochia scoparia*). *Weed Technol.* **7**:154–158.

Nalewaja, J. D., T. Praczyk, and R. Matysiak. 1995. Salts and surfactants influence nicosulfuron activity. *Weed Technol.* **9**:587–593.

Ozkan, H. E. 1995. Herbicide application equipment. In *Handbook of Weed Management Systems*, ed. by A. E. Smith, Chapter 6. Marcel Dekker, New York.

Wahlers, R. L., J. D. Burton, E. P. Maness, and W. A. Skroch. 1997. A stem cut and blade delivery method of herbicide application for weed control. *Weed Sci.* **45**:829–832.

Wahlers, R. L., J. D. Burton, E. P. Maness, and W. A. Skroch. 1997. Physiological characteristics of a stem cut and blade delivery method of application. *Weed Sci.* **45**:746–749.

Wanamarta, G., J. J. Kells, and D. Penner. 1993. Overcoming antagonistic effects of Na-bentazon on sethoxydim absorption. *Weed Technol.* **7**:322–325.

Weed Science. 1982. Symposium proceedings on herbicide adjuvants. *Weed Sci.* **14**:764–825.

Whitford, F., A. Martin, and J. Boyer. 1997. *Pesticides and Their Proper Storage*, ed. by A. Blessing. Purdue Pesticide Programs Bulletin PPP-26, Purdue University Extension Service, West Lafayette, IN.

Whitford, F., A. Martin, D. Weisenberger, A. Boger, Z. Reicher, B. Wolf, and D. Huth. 1999. *Landscape Pesticide Application Equipment: A Guide to Selection and Calibration of Liquid Sprayers*, ed. by A. Blessing. Purdue Pesticide Programs Bulletin PPP-47. Purdue University Extension Service, West Lafayette, IN.

Wolf, R. E. 1997. Herbicide application technology with an emphasis on reduced drift. In *Weed Management in Horticultural Crops*, ed. by M. E. McGiffen Jr., Chapter 7. ASHS Press, Alexandria, VA.

WEB SITES

Agriculture Chemical Guide, North Carolina State University
http://ipmwww.ncsu.edu/agchem/agchem.html

Drift Control

University of Missouri http://www.muextension.missouri.edu/xplor/agguides/agengin/g01886.htm
University of Nebraska
http://ianrwww.unl.edu/pubs/pesticides/g1001.htm

Herbicide Surfactants and Adjuvants

Colorado State University http://www.ext.colostate.edu/pubs/crops/00559.html
North Dakota State
http://www.ext.nodak.edu/extnews/weedpro/add

Herbigation/Chemigation

Mississippi State University
http://msucares.com/pubs/pub1551.htm
University of Minnesota http://www3.extension.umn.edu/distribution/cropsystems/DC6122.html
Archive of Chemigation
http://www.chemigation.com/infoupdt.htm

Incorporation of Soil Applied Herbicides (Canola Congress)
http://www.canola-council.org
Click Growers Manual, select Canola Pests, select Weeds.

Nozzles

Spray Systems TeeJet Nozzles
http://www.teejet.com

Specialized Sprayers

Patchen Sprayer
http://www.stanislaus-implement.com/patchen_sprayers.htm

Mowers

Burch Wet Blade Mower
http://www.wetblade.com

For chemical use, see the manufacturer's or supplier's label and follow these directions. Also see the Preface.

PART II
Herbicides

8 Chemistry and Classification of Herbicides by Mechanism of Action

Herbicides have been classified in several different ways, including mode of action, chemical structure, and method of application. We have chosen to classify herbicides by *mechanism of action* because we believe it is the most useful and easiest to understand for students and practitioners. Mode of action means the entire sequence of events from introduction of a herbicide into the environment to the death of the plant. The mechanism of action is the primary biochemical or biophysical event that the herbicide directly effects. For example, the imidazolinone and sulfonylurea herbicides inhibit the synthesis of certain essential amino acids, which is the mechanism of action, whereas the mode of action is the subsequent chain of responses that lead to plant death.

A short discussion of organic chemistry and how it relates to herbicides is presented prior to explaining the mechanism of action classification system used in this book. Knowledge of herbicide chemistry is important to understand how particular herbicides should be used to obtain optimum weed control and their fate in the environment. In Chapters 5, 6, and 7 various chemical attributes of herbicides were discussed in relation to movement in the plant, activity, persistence in the soil, and how they are formulated for effective application and performance. Because most commercial herbicides are organic chemicals, a knowledge of organic chemistry basics allows the practitioner and scientist to understand the nature of herbicides, the various chemical constituent groups that make up a herbicide, and how these affect herbicide performance.

BASICS OF ORGANIC CHEMISTRY

Organic compounds are the basis of life on this planet. All organic chemicals contain one or more atoms of carbon (C). Carbon can bond (share electrons) with another carbon and form chains that can be straight, branched, or formed into rings. A carbon atom can share electrons with four other atoms at a time (Figure 8-1).

The three most frequent atoms found in organic compounds are carbon, oxygen, and hydrogen, with nitrogen, phosphorous, sulfur and the various halogens (chlorine, fluorine, iodine, and bromine) being the next most common, and except for herbicides, many organic compounds contain certain metallic atoms such as iron, manganese, and cobalt.

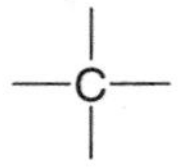

Figure 8-1. A carbon atom with four free electrons for bonding with other atoms.

Hydrocarbons are organic compounds containing carbon and hydrogen. These hydrocarbons can be *saturated* (e.g., the carbons are linked by a single covalent bond) *unsaturated* (e.g., carbons are linked by one or more double or triple bonds between adjacent carbon atoms). Saturated hydrocarbons are called *alkanes*, the simplest being methane with 1 carbon attached to 4 hydrogens; ethane has 2 carbons with 6 hydrogens, propane has 3 carbons with 8 hydrogens, and butane has 4 carbons with 10 hydrogens (see Figure 8-2).

Alkanes with more than four carbons are named with the Greek prefix for the number of carbons; for example, pentane = 5 carbons, hexane = 6 carbons, heptane = 7 carbons, octane = 8 carbons and so on. Unsaturated hydrocarbons with a double bond are called *alkenes* (also olefins), and those with triple bonds are called *alkynes* (also acetylenes). Naming of alkenes and alkynes follows the same scheme as for alkanes (e.g., a 5-carbon alkene is named pentene, and a 5-carbon alkyne is named pentyne).

There can also be substitution on the end of the molecule with a radical group of some type (Figure 8-3). The addition of an OH group results in an alcohol, and addition of a carboxyl group, COOH, results in an acid.

In the case of alkanes, the name is then altered from the -ane suffix to an -yl suffix. Naming of alkenes and alkynes with a radical results in adding the -ene or -yne suffix after the -yl suffix (e.g., methylene or methylyne, respectively).

Isomers are chemicals that have the same composition in regard to the number and types of atoms but differ in configuration. Some herbicides are composed of isomers, one being active and the other being inactive or less active. Chemical companies often eliminate or greatly reduce the inactive isomer from the final formulated product.

Alkene-type hydrocarbons can form 6 carbon rings called "Benzene" rings or "Aromatics." The benzene ring, common to many herbicides, is symbolized in one of the three ways, as shown in Figure 8-4.

The simplified structure is shown in *b* and *c*, where the individual carbons and hydrogens are omitted. A single covalent bond is indicated by a double short dash line, as in *a* and *b*. In *c* the circle inside the ring indicates that the free valence electrons of

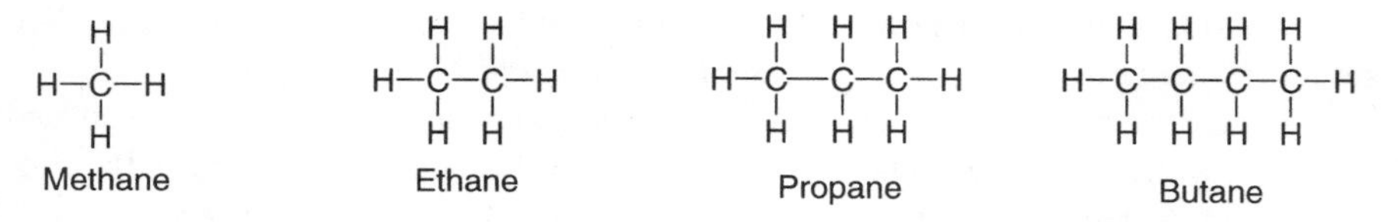

Figure 8-2. Structures of the alkanes: Methan, Ethane, Propane, and Butane.

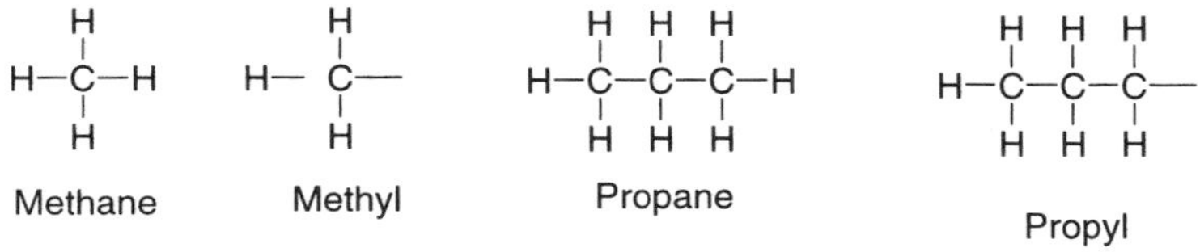

Methane Methyl Propane Propyl

Figure 8-3. The ability for substitution on the end of an alkane. The removal of a H atom results in the formation of a radical. Substitution with a OH atom results in an alcohol, e.g. methyl alcohol (methanol.)

each carbon atom are not attached to any particular carbon atom in the ring but are free to roam around the ring. The benzene ring can react with many atoms, resulting in various substitutions on the ring. The benzene ring with one hydrogen removed is called a *phenyl ring* and can accept many substitutions, including an OH group or other common groups, such as COOH, CH_3, NH_3, and so forth.

The numbering system for identifying the locations for substitutions on the ring is shown in Figure 8-5; however, the number 1 position is not always consistent and is sometimes based on the most reactive or important substitution group. If there are two chemical group substitutions on the ring, their location is noted by the prefix ortho-, meta-, or para-. Ortho- indicates that the substitutions are on adjacent carbon atoms, meta- indicates that the substitutions are separated by one carbon position, and para- indicates that they are separated by two carbon positions (Figure 8-5).

There are many substitution groups (radicals), linkages, and functional groups that occur frequently in various herbicides. Some of the more common of these are listed in Figure 8-6.

The naming of organic chemicals is based on the fundamentals previously described, but does vary. The same chemical can be named in more than one way, but from a chemist's point of view the names are technically correct.

The *Herbicide Handbook* of the Weed Science Society of America (1994,1998) has a listing of specific information for each herbicide that includes the following:

1. *Nomenclature* (common name, chemical family, chemical name, manufacturer)

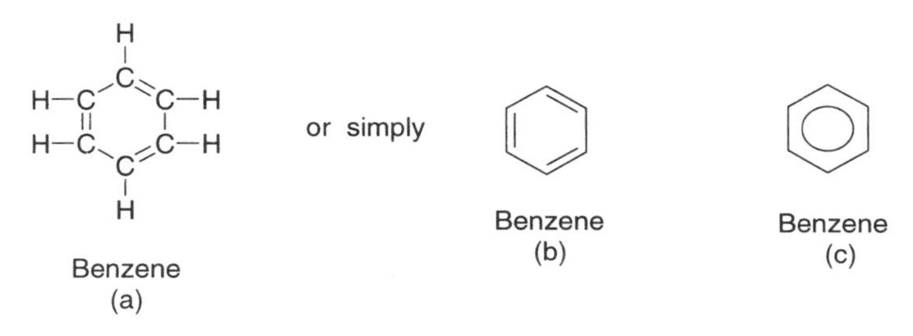

Figure 8-4. The three methods of symbolizing a benzene ring.

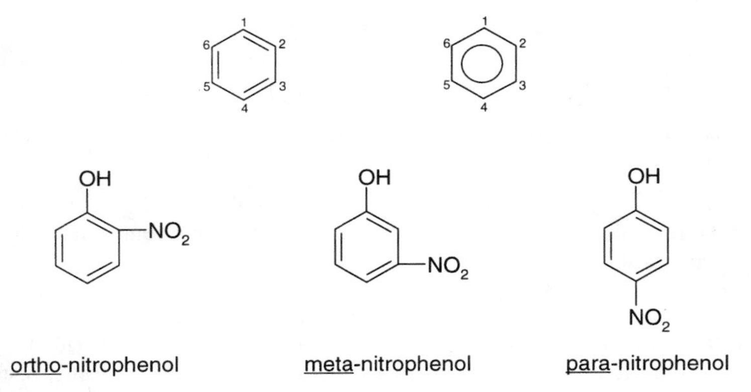

Figure 8-5. a. The numbering system for identifying the locations on the benzene ring for substitutions. b. The *ortho-*, *meta-* and *para-* substitutions on the benzene ring are shown.

2. *Chemical and physical properties of the pure chemical* (chemical structure, molecular formula, molecular weight, description, density, melting and boiling points, vapor pressure, solubility, stability, pK_a, and K_{OW})
3. *Herbicide uses*
4. *Use precautions*
5. *Behavior in plants* (symptomology, absorption and translocation, mechanism of action, metabolism, nonherbicidal biological properties, and mechanism of resistance in weeds)
6. *Behavior in soil* (sorption (K_{OC}, K_d) persistence (carryover potential, half-life), mobility, volatilization, and formulation effects)
7. *Toxicological properties*: acute, including oral and dermal LD_{50}, inhalation $LC_{50,\ eye}$, and skin irritation and skin sensitization; subchronic, including 90-day dietary for mouse, rat, and dog; chronic, including 12-, 18-, and 24-month feeding studies for mouse, rat, and dog, respectively; teratogenicity on rats and rabbits; reproduction on rats; mutagenicity, including gene mutation and structural chromosome aberration; and wildlife effects, on birds, fish, insects, and selected microorganisms use classification
8. *Synthesis and analytical methods*
9. *Sources of additional information*

Herbicides have three names with which all users should be familiar. The first is the *chemical name*, the second is the *common name*, and the third is the *trade name*. There are more than 200 chemicals classified as herbicides. For standardization purposes, this text uses the listing of common and chemical names of herbicides as approved by the Weed Science Society of America and provided in *Weed Science* (1999, Vol. 47, pp. 764–769). This listing includes many herbicides that are no longer manufactured or are not sold in the United States. The chemical names used are those preferred by the Chemical Abstracts Service (CAS)

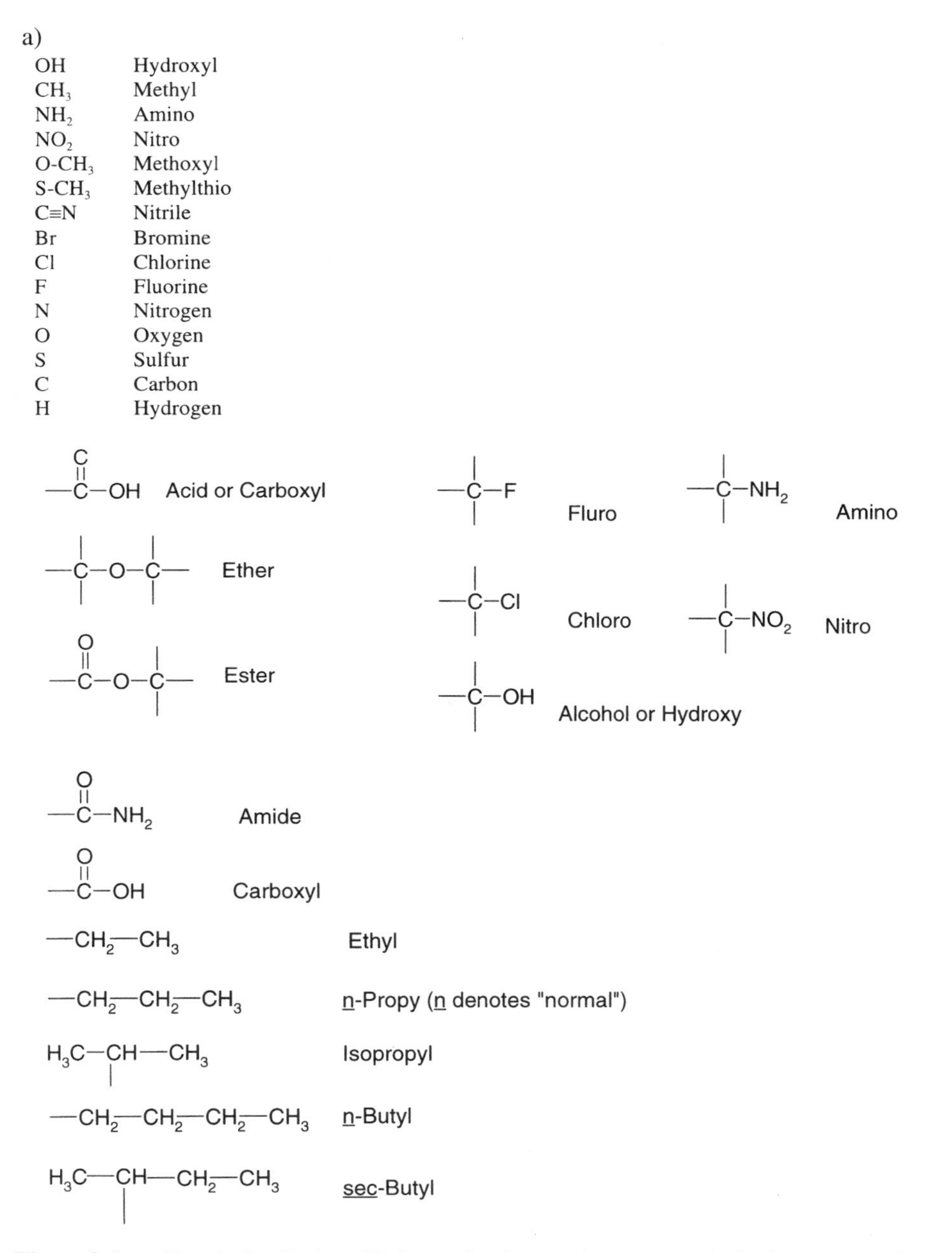

Figure 8-6. a. Chemical radicals and linkages that frequently occur as substitutions on organic herbicides. b. Common parent forms of organic herbicides.

b)

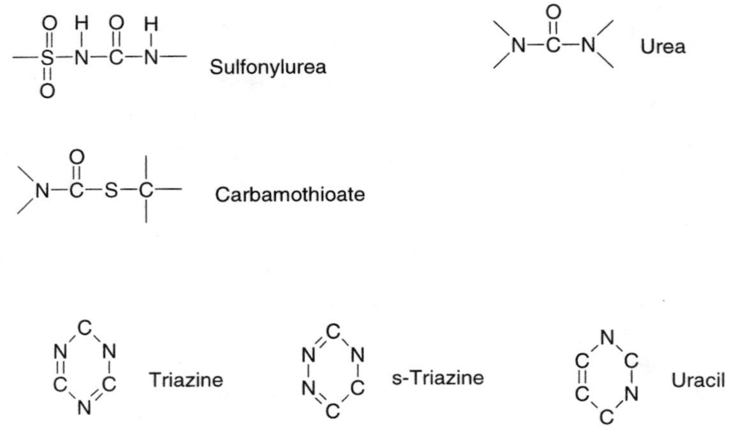

Sulfonylurea

Urea

Carbamothioate

Triazine

s-Triazine

Uracil

Figure 8-6. (continued)

according to their system of nomenclature in effect since 1972. A chemical name is based on the structural components of the chemical and will sometimes vary slightly, depending on who named the chemical originally and the source from which the name was obtained.

The common name is defined as "a coined name that applies to the 100% pure pest control chemical or to a technical pest control chemical of known composition where this composition is the result of chemical reactions occurring during manufacture" (American National Standards Institute, Inc., (ANSI), 1430 Broadway, New York, NY 10018; ANSI/ASC K62.1-1985, paragraph 2.2). A herbicide's common name is not synonymous with the active ingredient of a commercial formulation as identified on the product label, but in essence refers to the parent chemical. The common name will not vary for the parent chemical of a herbicide regardless of where it is sold in the world. Because the common name does not vary and is listed on every herbicide label, users' familiarity with the common name for a particular herbicide can be helpful in its proper use and can avoid the confusion created by its several trade names.

One herbicide can have several trade names for its various uses. This is true for herbicides sold by more than one company, and especially true for herbicides whose patents have expired, and for commercial products containing more than one herbicide in the formulation.

MECHANISM OF ACTION CLASSIFICATION

For the purposes of clarity and ease of use in this text, we do not classify herbicides by chemical similarities but by their mechanism of action. In the table that follows, herbicides are classified on the basis of their apparent type or mechanism of action with the specific chemical family grouping included as a subheading. There are often many diverse chemical families within the same grouping, and to discuss each group on the basis of its chemistry would be confusing and in many cases redundant. Although we have a tremendous amount of specific information about most herbicides and how they result in plant death, it should be recognized that the exact mechanism of action of some herbicides is not known. In spite of the lack of specific information for these particular herbicides, enough is known about their effects on plants to enable us to place them in closely related mechanism of action groups. Moreover, for some herbicides, there may be more than one mechanism of action. Therefore, the classification used is based on current knowledge related to the primary type or mechanism of action for each chemical and grouping. In the following table below, various chemical groups within a mechanism of action and the specific herbicides within each group follow the mechanism of action heading. Each specific herbicide is identified by its common name with the trade name(s) in parenthesis. Not all trade names are included, because some of the older herbicides whose patents have expired are sold under many trade names. At the end of each mechanism of action grouping, a short paragraph relating to the general mode of action is provided. More specific in-depth information about each mechanism of action group and uses worldwide for each herbicide are provided in Chapters 9 to 17.

Classification of Herbicides by Primary Type of Mechanism of Action

	Photosynthesis Inhibitors	
s-triazines		
Chloro	*Methoxy*	*Methylthio*
Atrazine (Aatrex, Atrazine) Simazine (Princep) Cyanazine (Bladex)	Prometon (Pramitol)	Ametryn (Evik) Prometryn (Caparol, Cotton-Pro)
Other Triazines	Substituted Ureas	Uracils
Hexazinone (Velpar) Metribuzin (Sencor, Lexone)	Diuron (Karmex) Fluometuron (Cotoran) Linuron (Lorox) Tebuthiuron (Spike)	Bromacil (Hyvar) Terbacil (Sinbar)
Benzothiadiazole	Benzonitrile	Phenylcarbamates
Bentazon (Basagran)	Bromoxymil (Buctril)	Desmedipham (Betanex) Phenmedipham (Spin-aid)
Pyridazinone	Phenylpyridazine	Other
Pyrazon (Pyramin)	Pyridate (Tough)	Propanil (Stam, Stampede)

This group includes herbicides that are applied only to the soil, some that are applied only postemergence to plant foliage, and some that can be applied both as soil and as foliar treatments. These herbicides inhibit electron transport in photosystem II of the photosynthetic reaction in plants, resulting in the formation of free radicals (potent biological oxidants) that attack and destroy the integrity of cell membranes. When soil applied, weed seeds germinate, their roots absorb the herbicide and translocate it in the xylem to the leaves, and the plant slowly dies as photosynthesis is inhibited. When applied postemergence, the action is contact, requiring complete wetting of the foliage for complete kill. Susceptible plants turn yellow, then die from the bottom to the top. Leaves yellow between the veins and then turn brown from the base and outer leaf edges toward the center, eventually falling off the plant and leaving only a stem with an apical bud.

	Pigment Inhibitors	
No Chemical Family Recognized	Pyridazinone	Isoxazole
Amitrole (Amitrol T) Clomazone (Command) Fluridone (Sonar)	Norflurazon (Zorial, Evital, Solicam, Predict)	Isoxaflutole (Balance)
		Triketones
		Sulcotrione (Mikado) Mesotrione (Callisto)

Pigment inhibitors are mostly applied as preplant or preemegence treatments, with the exception of amitrole (foliar) and fluridone in aquatics. These herbicides inhibit different enzymes in the carotenoid pigment biosynthetic pathway in the plant. Carotenoid pigments are important accessory plant pigments that protect chlorophyll from photooxidation. When carotenoids are absent, chlorophyll is destroyed in the light and plants slowly die. Injury caused by these herbicides is a bleached white to translucent appearance of the leaves. Sometimes the bleaching is not complete on the entire leaf but will be interveinal with pink or red highlights along the margins.

Cell Membrane Disruptors and Inhibitors

A. Direct Effect on Membranes
   Dilute sulfuric acid
   Monocarbamide dihydrogen sulfate (Enquick)
   Herbicidal oils
B. Induce Lipid Peroxidation
   1. Photosynthesis involved

Bipyridyliums

Diquat (Diquat, Reward)
Paraquat (Gramoxone Extra)

   2. Photosynthesis not involved

Diphenylethers	Oxidiazole
Acifluorfen (Blazer)	Oxadiazon (Ronstar)
Fomesafen (Flexstar, Reflex)	
Lactofen (Cobra)	
Oxyfluorfen (Goal)	

N-phenylheterocycles
Carfentrazone (Aim)
Flumiclorac (Resource)
Sulfentrazone (Authority, Spartan)

N-phenylphthalimide
Flumioxazin (Valor)

C. Inhibition of Glutamine Synthetase
Glufosinate (Liberty, Rely)

There is a diversity of chemistry in this group of compounds, with most being applied as postemergence contact herbicides; however, oxidiazon, sulfentrazone, and oxyfluorfen have important uses as soil-applied preemergence herbicides. Although the specific site of inhibition in the plant varies among the cell membrane disruptor groups—group B1 (photosystem I), group B 2 [inhibition of protoporphyrinogen oxidase (PROTOX)], and group C (glutamine synthase)—plant death from these herbicides is rapid when they are applied to foliage. Complete coverage of the leaf is important for best activity, and the rate of plant death is more rapid under high light and warm environmental conditions. Injury symptoms include an initial appearance of water soaked tissue, followed by desiccation of leaf tissue caused by a disruption of cell membranes. Membranes are degraded by free radicals (potent biological oxidants) that form within plants as a result of the action of these herbicides.

Cell Growth Disruptors and Inhibitors

A. Mitotic Disruptors

Dinitroanilines	Pyridine
Benefin (Balan)	Dithiopyr (Dimension)
Ethalfluralin (Sonalan, Curbit)	Thiazopyr (Visor)
Oryzalin (Surflan)	
Pendimethalin (Prowl, Pendulum, Pentagon, others)	**Amide**
Prodiamine (Barricade, Endurance, Factor)	Pronamide (Kerb)
Trifluralin (Treflan, Trifluralin, and several others)	

Other
DCPA (Dacthal)

B. Inhibitors of Shoots of Emerging Seedlings

Carbamothioates (Thiocarbamates)	
EPTC (Eptam, Eradicane, Eradicane Extra)	Butylate (Sutan +)
Cycloate (Ro-Neet)	Molinate (Ordram)
Pebulate (Tillam)	Thiobencarb (Bolero, Abolish)
Triallate (Far-Go, Avadex BW)	

C. Inhibitors of Roots Only of Seedlings

Amide	Phenylurea	Other
Napropamide (Devrinol)	Siduron (Tupersan)	Bensulide (Prefar, Betasan, Bensumec)

D. Inhibitors of Roots and Shoots of Seedlings

Chloroacetamides	Oxyacetamide
Acetochlor (Harness, Surpass, Topnotch)	Flufenacet (Axiom, Epic, Domain)
Alachlor (Lasso, Micro-Tech, Partner)	
Dimethenamid (Frontier)	
Metolachlor (Dual, Pennant)	
Propachlor (Ramrod)	
Butachlor (Machete)	

Herbicides in this classification inhibit root and/or shoot growth of emerging seedlings and are applied to the soil either as preemergence or preplant incorporated treatments. The mitotic disrupters inhibit the early steps in plant cell division responsible for chromosome separation and cell wall formation in plants, the root/shoot inhibitors deplete long-chain fatty acids from the plasma membranes of plants, whereas the specific mechanism shoot inhibitors is not known. *Mitoic disrupters* inhibit shoot elongation when effective, and susceptible weeds never see the light of day. Root inhibition is observed as root pruning, and roots can be swollen and expanded at the tip (club-shaped). The underground portion of the stem can be thickened and shortened, and stems often have callus growth thickenings at the soil surface and become brittle. Inhibitors of roots only (C in this listing) or roots and shoots of seedlings (D) result in root pruning and growth inhibition but no root swelling. The inhibition of shoots by the carbamothioates (B) and the chloroacetamides and oxyacetamides (D) results in lack of seedling shoot emergence. If shoots do emerge, they tend to be twisted and leaves are tightly rolled, with stems sometimes rupturing and new growth protruding from the ruptured tissue.

Cellulose Biosynthesis Inhibitors

Nitrile	Benzamide	Quinolinecarboxylic Acid
Dichlobenil (Casoron, Dyclomec)	Isoxaben (Gallery)	Quinclorac (Facet)

The cellulose biosynthesis inhibitor herbicides are a diverse group of chemically unrelated compounds. The common herbicidal effect is either a direct or indirect inhibition of cellulose biosynthesis, which in effect leads to a lack of cell structure integrity. In most cases these herbicides are used for preemergence control and result in the inability of weed seedlings to grow (Sabba and Vaughn, 1999). Symptoms include stunted growth and root swelling. Dichlobenil and isoxaben are used preemergence and are most effective against dicots, and quinclorac is used both preemergence and postemergence. Quinclorac as a cellulose biosynthesis inhibitor is most active against monocots, although it has a proposed second mechanism against dicots as a growth regulator.

Growth Regulators

Phenoxy Acetic Acids	Phenoxy Propionic Acids	Phenoxy Butyric Acids
2,4-D MCPA	Dichlorprop (2,4-DP) Mecoprop (MCPP)	2,4-DB MCPB

Benzoic Acids	Picolinic Acid and Related Compounds
Dicamba	Picloram (Tordon) Triclopyr (Garlon) Clopyralid (Lontrel, Reclaim, Stinger, Transline) Quinclorac (Facet)

Growth regulator herbicides can be absorbed from the soil by plant roots; however, most of these compounds are applied as postemergence treatments. Translocation can be in both the xylem and phloem to active growth regions, but their action tends to be localized on the shoot system. They selectively kill broadleaf weeds but can injure grass crops if applied at the wrong time. In the case of perennial weeds, many of these herbicides translocate to below-ground portions of the plant for systemic kill. Initial symptomology is quickly apparent on newly developing leaves and shoot regions as a twisting and epinasty of the shoot, cupping and crinkling of leaves, elongated leaf strapping (sometimes called "buggy whip") with parallel veins, stem swelling, and a disruption of phloem transport. Secondary effects can be a fusion of brace roots, such as observed with corn. Root injury is expressed as a proliferation or clustering of secondary roots and inhibition of overall root growth. The specific site of herbicide inhibition is not known for this group, although there appear to be multiple sites that disrupt hormone balance, nucleic acid metabolism, and protein synthesis, resulting in alteration of auxin activity in plants, producing weakened cell walls, rapid cell proliferation (unproductive growth), and plant death within several days or weeks.

Lipid Biosynthesis Inhibitors (Grass-Specific Herbicides)

Aryloxyphenoxypropionates	Cyclohexanediones
Clodinofop-propargyl (Discover, Horizon)	Alloxydim-sodium (Fervin, Kusagard)
Diclofop-methyl (Hoelon, Hoegrass)	Butroxydim (Falcon)
Fenoxaprop-ethyl (Horizon)	Clethodim (Envoy, Prism, Select)
Fenoxaprop-p-ethyl (Acclaim,Whip 360, others)	Clefoxydim (Aura)
Fluazifop-p-butyl (Fusilade DX 2000)	Cycloxydim (Focus, Laser, Stratus)
Haloxyfop (Edge, Verdict, Gallant, Plus, Torpedo)	Sethoxydim (Poast, Poast, Ultims, Vantage)
Quizalofop-p (Assure II)	Tepraloxydim (Equinox)
Isoxapyrifop (HOK-1566)	Tralkoxydim (Achieve)
Cyhalofop-butyl (XDE-537, Clincher)	
Fenthioprop	
Propaquizafop (Agil, Shogun)	

These herbicides have specific activity against grass species only. Dicots and nongrass monocots are tolerant. Some of these herbicides have shown minimal soil activity; however, the main activity occurs after postemergence application to emerged grass. Activity occurs on both annual and perennial grass species but varies depending on the particular herbicide. Translocation of these herbicides can occur in both the xylem and the phloem, and all generally require the addition of an adjuvant to improve leaf coverage and absorption. These herbicides are most effective when applied to unstressed, rapidly growing grasses. Death of the grass is slow, requiring a week or more for complete kill. Symptoms include rapid cessation of shoot and root growth, pigment changes (purpling or reddening) on the leaves occurring within 2 to 4 days, followed by a progressive necrosis beginning at meristematic regions and spreading over the entire plant. These herbicides inhibit the enzyme acetyl-CoenzymeA carboxylase (ACCase) in the biosynthetic pathway leading to lipid biosynthesis in plants, preventing fatty acid formation, which is essential for plant lipid synthesis. Lack of lipids results in loss of cell integrity of membranes and no new growth.

Inhibitors of Amino Acid Synthesis

A. Inhibitors of Aromatic Amino Acid Synthesis

Amino Acid Type
Glyphosate (Roundup, several others)
Sulfosate (Touchdown)

B. Inhibitors of Branched-Chain Amino Acid Synthesis

Sulfonylureas	
Bensulfuron (Londax)	Chlorimuron (Classic)
Chlorsulfuron (Glean, Telar)	Cyclosulfamuron (Invest)
Ethametsulfuron (Muster)	Flupyrsulfuron (Lexus)
Halosulfuron (Permit, Manage-Turf)	Metsulfuron (Ally, Escort)
Nicosulfuron (Accent)	Oxasulfuron (Dynam, Expert)
Primisulfuron (Beacon)	Prosulfuron (Peak)
Rimsulfuron (Matrix, Elim, Prism)	Sulfometuron (Oust)
Sulfosulfuron (Maverick, Outrider)	Thifensulfuron (Pinnacle)
Triasulfuron (Amber)	Tribenuron (Express)
Triflusulferon (Upbeet)	

Imidazolinones	Triazolopyrimidines	Pyrimidinyloxybenzoates
Imazamethabenz (Assert)	Cloransulam (FirstRate)	Pyrithiobac (Staple)
Imazamox (Raptor)	Diclosulam (Strongarm)	
Imazapic (Cadre, Contend)	Flumetsulam(Broadstrike)	
Imazapyr (Arsenal, Chopper, Stalker)		
Imazaquin (Scepter, Image)		
Imazethapyr (Pursuit)		

Sulfonylaminocarbonyltriazolinone
Flucarbazone (Everest)

These herbicides, although differing in chemical structure, all inhibit amino acid synthesis in plants. Group A herbicides inhibit 5-enolpyruvylshikimate-3-phosphate synthase (EPSPS) in the shikimic acid pathway, resulting in limited production of the aromatic amino acids phenylalanine, tryptophan, and tyrosine and many important secondary compounds. Group B herbicides inhibit the enzyme acetolactate synthase [ALS, also called acetohydroxyacid synthase (AHAS)] in the branch chain amino acid pathway, resulting in limited production of isoleucine, leucine, and valine.

These herbicides are potent inhibitors of plant growth and are effective on both dicots and monocots. Glyphosate and sulfosate (Group A) have only foliar activity (no soil activity), and the ALS inhibitors (Group B) have members with foliar, soil, or both soil and foliar activity. Treated plants stop growing almost immediately after application. In the case of EPSPS inhibitors, plants may show a small amount of bleaching around new growth areas; plants die slowly (1 to 2 weeks) and turn a uniform harvest brown color. For ALS inhibitors, 2 to 4 days after treatment, the growing point (apical meristem) becomes chlorotic and later necrotic. Plants may also have shortened internodes, reduced root growth ("bottle brushing"), and pigment changes, including yellowing, purpling, or reddening. Plant death begins at the growing point and gradually spreads to the entire plant, with death occurring within 7 to 10 days.

Miscellaneous Herbicides	
Inhibitors of Auxin Transport	
Naphtylphthalamic acid	Semicarbone
Naptalam (Alanap)	Diflufenzopyr (Distinct)

Naptalam is soil applied, and while diflufenzopyr is foliar applied. These herbicides inhibit auxin (IAA) transport and/or action in plants, resulting in lack of plant growth due to reduced growth hormone for cell expansion. A common symptom of these herbicides, in addition to reduced plant growth, is the upward turning of the root tip.

Mechanism of Action Not Clear		
Organic Arsenicals	Unclassified	
DSMA	Asulam (Asulox) Endothall (Accelerate, Aquathol, Hydrothol)	Acrolein (Magnacide)
MSMA	Ethofumesate (Prograss) Fosamine (Krenite)	Methyl Bromide Metham (Vapam)
Cacodylic acid		Dazomet (Basamid)
	Difenzoquat (Avenge) TCA (Nata) Pelargonic acid (Scythe)	Borates Sodium Chlorate

The unclassified group contains a variety of herbicides in which the specific mechanism of action is not clearly understood. Nor is the specific site of action within the plant understood. Specifics regarding symptomology and plant death are discussed in Chapter 17.

LITERATURE CITED AND SUGGESTED READING

Anderson, W. P. 1983. *Weed Science Principles*. 2d ed., Chapter 6, pp. 205–303. West Publishers, New York.

Herbicide Action. 2000. *An Intensive Course on Activity, Selectivity, Behavior, and Fate of Herbicides in Plants and the Environment*. Purdue University, West Lafayette, IN.

Herbicide Handbook 7th ed. 1994, Weed Science Society of America, Lawrence, KS.

Herbicide Handbook Supplement to 7th ed. 1998. Weed Science Society of America, Lawrence, KS.

Hess, F. D. 2000. Light-dependent herbicides. *Weed Sci*. **48**:160–170.

Ross M. A., and C. A. Lembi. 1999. Applied Weed Science. 2d ed., Chapter 5, pp. 76–96. Prentice-Hall, Upper Saddle River, NJ.

Sabba, R. P., and K. C. Vaughn. 1999. Herbicides that inhibit cellulose biosynthesis. *Weed Sci*. **47**:757–763.

WEB SITES

Herbicide Mode of Action Summaries

Purdue University

http://www.agcom.purdue.edu

Select "On-line Publications," scroll to "Botany and Plant Pathology," select "Weeds and Weed Management," go to "Herbicide Mode of Action WS23."

University of Minnesota

http://www.extension.umn.edu

Scroll down listing to "Herbicide Mode of Action."

Iowa State University

http://www.weeds.iastate.edu

Scroll to "Mode of Action Chart."

North Carolina State University Agriculture Chemical Guide

http://ipmwww.ncsu.edu

Weed Science Society of America

http://www.wssa.net

For chemical use, see the manufacturer's or supplier's label and follow these directions. Also see the Preface.

9 Photosystem II Inhibitors

The mechanism of action of herbicides classified as photosystem II (PS II) inhibitors involves the inhibition of photosynthesis. Photosynthesis inhibitors include many compounds in each of several chemical groups (ureas, uracils, triazines, benzothiadiazoles, benzonitriles, phenylcarbamates, pyridazinones, and phenylpyridazines). This group comprises a large number of our most important herbicides, which are some of the oldest yet most widely used for weed management. These herbicides are generally applied to the soil and are absorbed by the roots of the plant, moving with the flow of water into the foliage through the xylem. Movement in the phloem does not occur with most of these herbicides. There are also herbicides in this group that are active only when applied to plant foliage. When used postemergence, their action is through contact; complete wetting of the foliage is necessary and requires an adjuvant in the spray tank. The foliage and stems of the plant are injured but the root system is not.

Plants exposed to soil applications of most of these herbicides will germinate and emerge, absorb the herbicide from the soil through their roots, and translocate the herbicide to the leaves, where photosynthesis is inhibited. Plants turn yellow as a result of chlorophyll breakdown. As injury proceeds, the plant will turn yellow between the veins, dying from the tip toward the base and from the outer edge toward the center. Leaves then fall off the plant, leaving only the stem. Death resulting from a foliar application follows the same leaf injury pattern.

COMMON CHARACTERISTICS

Common characteristics of the photsystem II inhibiting herbicides are provided in the following list:

1. Rate of CO_2 fixation declines within a few hours in all treated plants (except those that are resistant because of the lack of a herbicide binding site). In tolerant plants the rate of photosynthesis does not go as low as in sensitive plants, and it returns to normal within a few days. In susceptible plants, the rate drops to near zero within 1 or 2 days and does not recover. Symptoms of injury develop on the leaves of treated plants after a few days.
2. These herbicides have no direct effect on root growth at recommended use rates.
3. Apparently, roots can absorb all of the compounds, and leaves absorb most. However, leaf absorption varies greatly between compounds. Because of

variable behavior in the soil, translocation from roots to leaves, uptake from foliar sprays, and so forth, some of these herbicides are of practical value only when applied to the soil, some only when applied to the foliage, and some are effective in applications of both types.

4. All herbicides in this classification move primarily in the xylem. Therefore, perennials are killed only by soil applications.
5. When postemergence sprays are used, thorough coverage of the foliage is important because there is little basipetal translocation and the action is of a "contact" rather than "systemic" type. Surfactants or oils are often added to increase foliar action.
6. Plants are most susceptible to postemergence sprays when low light intensity occurs during the few days before spraying and high light intensity occurs after spraying.
7. The dose response curve is very sharp. For this reason, along with the contact rather than systemic action on the foliage, the drift problem for sensitive crops is less serious with these compounds than with growth regulator and other systemic herbicides.
8. Resistance has developed in several weed species following repeated applications for several years with some of these herbicides.
9. In general, movement of these compounds in the soil is low to moderate, but this varies with the compound, soil, and rainfall.
10. Persistence in the soil varies from less than 1 month to more than 2 years, depending on the herbicide, amount applied, climate, and soil.
11. Repeated applications of soil-active photosynthesis inhibitor herbicides have not resulted in any increase in the rate of breakdown in the soil.
12. Synergistic interaction often occurs when the photosynthesis inhibitor herbicides are applied at or near the same time as cholinesterase inhibitor insecticides.
13. All the compounds in this group have low mammalian toxicity.

Selectivity may be due to one or more of the following:

1. Placement in the soil (depth protection)
2. Localized application of activated carbon
3. Directed sprays
4. Differential uptake by roots and/or leaves
5. Differential translocation from roots to foliage
6. Adsorption at inactive sites in the plants
7. Differential metabolism in roots and/or leaves
8. Differential age of crops and weeds, including annual weeds in perennial crops
9. Within a variety, increased tolerance of larger seeds when these herbicides are applied to the soil
10. Lack of herbicide binding to a protein in the chloroplast membrane

Each chemical grouping is discussed in the following sections below regarding common, trade, and chemical names; chemical structure; general crop uses throughout the world; type of application; and response to soil influences. Specific information relating to chemical properties is available in the WSSA *Herbicide Handbook* (1994 and 1998 Supplement), as well as specific rates for use and timing of applications; the user should always refer to the product label for presently registered uses, application recommendations, precautions, and other information.

TRIAZINE HERBICIDES

The triazine herbicides inhibit plant growth, but this is considered to be a secondary effect caused by an inhibition of photosynthesis. At herbicidal concentration, triazine herbicides cause foliar necrosis, followed by death of the leaf (Figure 9-1). Other leaf effects include loss of membrane integrity and chloroplast destruction.

Triazine herbicides are absorbed by leaves, but translocation from the leaves is essentially nil. The amount of foliar absorption varies for the herbicides in this group. Simazine is poorly absorbed by leaves, whereas ametryn and prometryn are easily absorbed. All triazines are readily absorbed by roots and readily translocated throughout the plant in the xylem. The distribution of herbicide within a species reflects the relative susceptibility and degree of degradation, as shown in Figure 9-2, with ^{14}C simazine-treated to roots of hydroponically grown cucumber, corn, and cotton. Cucumber is quite sensitive to triazines, and accumulation of herbicide along the leaf margins is typical. The distribution of herbicide in tolerant corn suggests extensive metabolism, and intermediately susceptible cotton shows an accumulation of label in the lysigenous glands.

Specific characteristics of the triazine herbicides are discussed in the following paragraphs.

Simazine

Simazine (6-chloro-*N*-*N*′-diethyl-1,3,5-triazine-2,4-diamine) is a white crystalline solid with a vapor pressure of 6.1×10^{-9} mm Hg at 20°C, a low water solubility of 3500 mg/l (ppm) at 20°C, a soil half-life of 60 days, and an oral LD_{50} (rat) >5000 mg/kg. Simazine is subject to UV (ultaviolet) photodecomposition.

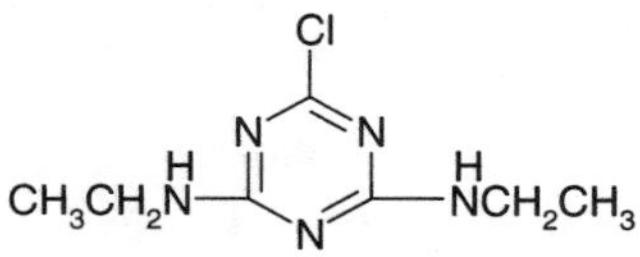

Cl
N N
CH_3CH_2NH — N — $NHCH_2CH_3$

Simazine

Uses Many formulations are available, including WP, DF, WDG, G, and WG (see Chapter 7 for explanation of formulations), and selectivity is usually obtained by placement or as a result of crop metabolism. Simazine has many trade names and is registered for use as a preplant incorporated, preemergence, postemergence, or

Figure 9-1. Leaf chlorosis induced by triazine herbicides. *Upper*: Chlorotriazines (e.g., simazine) almost always show interveinal chlorosis. *Lower*: Methylthiotriazines (e.g., ametryn) usually show veinal chlorosis (C. L. Elmore, University of California.)

postemergence directed application in corn, pome fruit, stone fruit, citrus, nut trees, bush fruits, strawberries, olives, pineapples, field beans, french beans, peas, sweet corn, asparagus, hops, alfalfa, lupins, oilseed rape, artichokes, sugar cane, cocoa, coffee, rubber, oil palms, tea, turf, woody ornamentals, and in forestry and noncrop situations; it also controls many broadleaf and grass weeds.

Atrazine

Atrazine (6-chloro-*N*-ethyl-*N′*-(1-methylethyl)-1,3,5-triazine-2,4-diamine) is a white crystalline solid with a vapor pressure of 2.9×10^{-7}mm Hg at 25°C, a moderate water

Figure 9-2. Autoradiographs showing distribution of [^{14}C] simazine and/or ^{14}C-labeled degradation products 4 days after root treatment via culture solution in (*A*) susceptible cucumber, (*B*) moderately susceptible cotton, and (*C*) tolerant corn. (Davis et. al., 1959.)

solubility of 33 mg/l (ppm) at 22°C, a soil half-life of 60 days, and an oral LD_{50} (rat) of 3090 mg/kg. Atrazine is subject to UV photodecomposition.

Cl

$(CH_3)_2CHN(H)$—[triazine ring: N, N, N]—$N(H)CH_2CH_3$

Atrazine

Uses The main formulations are L and WG, and atrazine is sold under many trade names. Atrazine is registered for use as a preplant incorporated, preemergence, and postemergence treatment in corn, sorghum, sugarcane, forestry, turfgrass, macadamia nuts, guava, chemical fallow, and noncrop applications; it controls many broadleaf and grass weeds. Plant roots readily absorb atrazine, and foliar applications require the use of an adjuvant to improve coverage.

Cyanazine

Cyanazine (2-[[4-chloro-6-(ethylamino)-1,3,5-triazin-2-yl]amino]-2-methylpropane-nitrile) is a colorless crystal with a vapor pressure of 1.6×10^{-9} mm Hg at 20°C, a moderate water solubility of 160 mg/l (ppm) at 23°C, a soil half-life of 14 days, and an oral LD_{50} (rat) of 182 to 334 mg/kg. Cyanazine is relatively stable and not subject to UV photodecomposition.

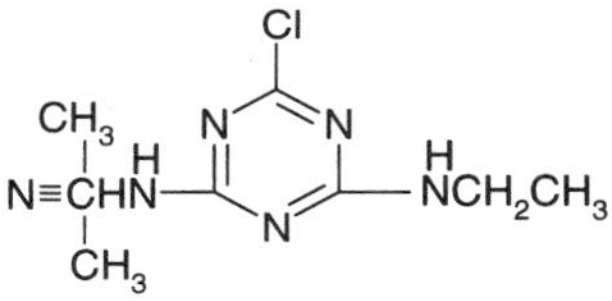

Cyanazine

Uses Cyanazine is formulated as an SC, L, DF, WP, and G and sold as Bladex for use as a preplant incorporated, preemergence, or postemergence application in corn (all types), cotton, broad beans, peas, barley, wheat, oilseed rape, forestry, potatoes, soybeans, and sugarcane and controls many broadleaf and grass weeds. Cyanazine will no longer be manufactured after 2000 but can be used until supplies are depleted.

Prometon

Prometon (6-methoxy-*N,N'*-bis(1-methylethyl)-1,3,5-triazine-2,4-diamine) is a white crystalline solid with a vapor pressure of 2.3×10^{-6} mm Hg at 20°C, a relatively high water solubility of 720 mg/l (ppm) at 22°C, a soil half-life of 500 days, and an oral LD_{50} (rat) of 4345 mg/kg. Prometron is relatively stable but is subject to UV photodecomposition.

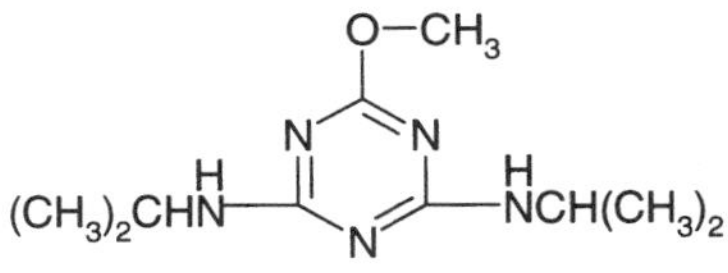

Prometon

Uses Prometon is formulated as an EC and WP and sold as Pramitol. Prometon is registered for use as a preemergence and postemergence treatment in noncrop situations, on industrial sites, in total vegetation control, and in and under asphalt for the control of many broadleaf and grass weeds. The mobility of this compound has caused the death of ornamentals and trees adjacent to sites of application when the compound has leached to their root systems. .

Prometryn

Prometryn (*N,N′*-bis(1-methylethyl)-6-(methylthio)-1,3,5-triazine-2,4-diamine) is a white crystalline solid with a vapor pressure of 1.0×10^{-6} mm Hg at 20°C, a relatively moderate water solubility of 33 mg/l (ppm) at 22°C, a soil half-life of 60 days, and an oral LD_{50} (rat) of 4550 mg/kg. Prometryn is relatively stable but is subject to UV photodecomposition.

SCH_3
N N
H H
$(CH_3)_2CHN$ N $NCH(CH_3)_2$

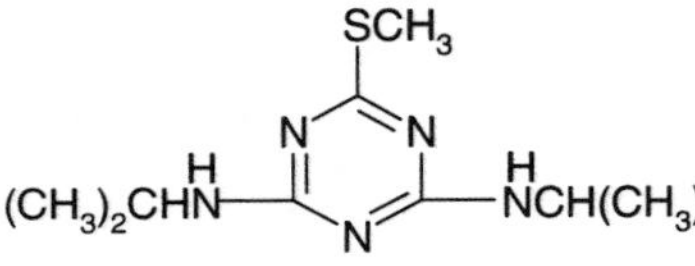

Prometryn

Uses The main formulations are L, SC, and WG, and it is sold as Caporal, Gessagard, Cotton-Pro, Efmetryn, and Prometrex for application as a preemergence or postemergence-directed application in several crops, including cotton, sunflower, peanuts, celery, potatoes, carrots, peas, beans, leeks, and pigeon peas for the control of many broadleaf and grass weeds. Selectivity is due to placement and metabolism and/or sequestration (especially in cotton).

Ametryn

Ametryn (*N*-ethyl-*N′*-(1-methylethyl)-6-(methylthio)-1,3,5-triazine-2,4-diamine) is a white crystalline solid with a vapor pressure of 8.4×10^{-7} mm Hg at 20°C, a relatively high water solubility of 200 mg/l (ppm) at 22°C, a soil half-life of 60 days, and an oral LD_{50} (rat) of 1160 mg/kg. Ametryn is relatively stable but is subject to UV photodecomposition.

SCH_3
N N
H H
$(CH_3)_2CHN$ N NCH_2CH_3

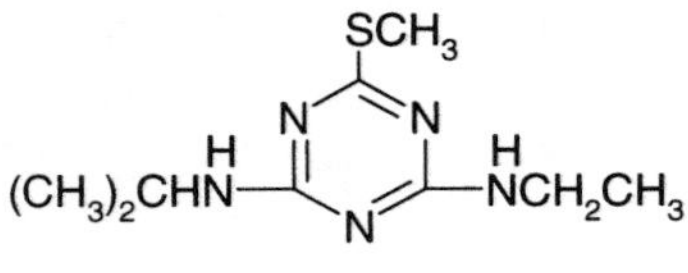

Ametryn

Uses The main formulations are WP, SC, EC, and WG, and it is sold as Evik, Gesapex, Amesip, Ametrex, and Metatryne for use as a preemergence or postemergence-directed application in corn (all types), pineapple, sugarcane, bananas, plantains, citrus, cassava, coffee, tea, sisal, cocoa, oil palms, and noncrop situations for control of many broadleaf and grass weeds.

Metribuzin

Metribuzin (4-amino-6-(1,1-dimethylethyl)-3-(methylthio)-1,2,4-triazin-5(4H)-one) is a white crystalline solid with a vapor pressure of 1.2×10^{-7} mm Hg at 20°C, a water solubility of 1100 mg/l (ppm) at 20°C, which makes it mobile in the soil, a soil half-life of 30 to 60 days, and an oral LD_{50} (rat) of 1090 mg/kg. Metribuzin is relatively stable and not subject to UV photodecomposition.

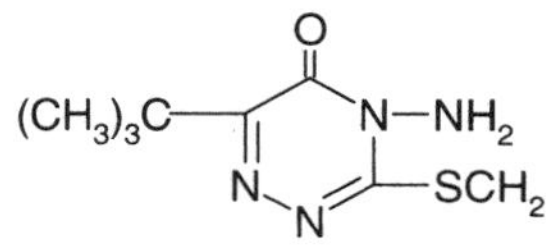

Metribuzin

Uses Metribuzin is formulated as a DF, WG, SC, and L and sold as Sencor, Lexone, and Mistral for use as a preplant incorporated, preemergence, postemergence, and postemergence-directed application in a wide variety of crops, including alfalfa, sainfoin, asparagus, bermudagrass turf, potatoes, soybeans, sugarcane, tomatoes, barley, lentils, peas, wheat, corn, and in noncrop situations and in chemical fallow. Metribuzin controls many broadleaf weeds.

Hexazinone

Hexazinone (3-cyclohexyl-6-(dimethylamino)-1-methyl-1,3,5-triazine-2,4(1*H*3*H*)-dione) is a white crystalline solid with a vapor pressure of 2×10^{-7}mm Hg at 25°C, a water solubility of 33,000 mg/l (ppm) at 25°C, which makes it mobile in the soil, a soil half-life of 90 days, and an oral LD_{50} (rat) of 1690 mg/kg.

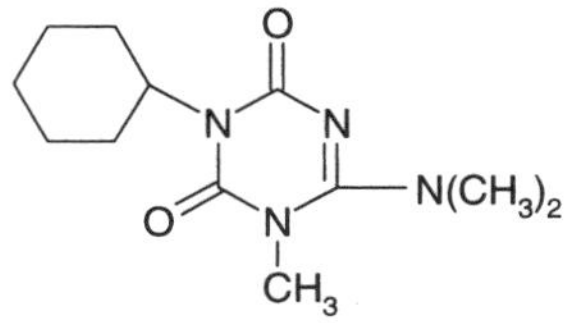

Hexazinone

Uses Hexazinone is formulated as a SC, SP, WG, and G and is sold as Velpar for use as a preemergence and postemergence application in Christmas trees, forestry, alfalfa, sugarcane, pineapple, industrial turf, pastures (bermudagrass and bahiagrass), and noncrop situations, and for brush control in pastures and rangeland, for control of many annual and perennial broadleaf and grass weeds.

Soil Influences

Triazine herbicides are reversibly absorbed by clay and organic colloids. They are not subject to excessive leaching in most soil types. In a study of five triazine herbicides on 25 soil types, adsorption almost always increased in the following order: propazine > atrazine > simazine > prometon > prometryn (Talbert and Fletchall, 1965). Correlation analysis indicated that adsorption of the methythio- (prometryn) and methoxy- (prometon) triazines was more highly related to clay content, whereas adsorption of chlorotriazines (simazine, atrazine, propazine) was more highly related to organic matter. The following relative leachability in Lakeland fine sand has been reported: propazine > atrazine > simazine > ametryn > prometryn (Rodgers, 1968).

Note that the order of triazines common to these two studies is the same, indicating that the leachability of triazine herbicides is directly related to their adsorption to soil colloids. These two studies also indicate that adsorption and leachability have little or no relationship to water solubility of the compounds. A reduction in phytotoxicity of triazine herbicides is associated with increasing amounts of clay and organic matter in soil (Weber, 1970).

Triazine herbicides vary widely in their persistence in soils. Soil type and environmental conditions have considerable influence on the actual period of persistence. Methoxytriazines are generally more persistent than methylthio- or chlorotriazines. Prometon, the most persistent, can remain at phytotoxic levels for several years. Atrazine and simazine are less persistent but can still injure sensitive plants the next season. Ametryn and prometryn are usually even less persistent, but may last from 6 to 9 months. Cyanazine and metribuzin appear to be the least persistent of the triazines; their half-life is 2 to 4 weeks under most conditions. At these rates of disappearance, less than 10% of that applied would remain after 2 to 4 months. A monograph prepared by Gunther (1970) contains several review papers on the interaction of triazine herbicides and soil.

UREA HERBICIDES

Phytotoxic symptoms of urea-type herbicides can be largely seen in the leaves (Figure 9-3). They are readily absorbed by roots and translocated by the xylem throughout the plant.

Figure 9-3. Diuron-induced chlorosis in peaches is usually veinal but can sometimes be interveinal. *Left to right*: Untreated to increasing rates. (C. L. Elmore, University of California.)

Diuron

Diuron (*N′*-(3,4-dichlorophenyl)-*N*,*N*-dimethylurea) is a white crystalline solid with a vapor pressure of 6.9×10^{-8} mm Hg at 25°C, a moderate water solubility of 42 mg/l (ppm) at 25°C, a soil half-life of 90 days, and an oral LD_{50} (rat) of 3400 mg/kg.

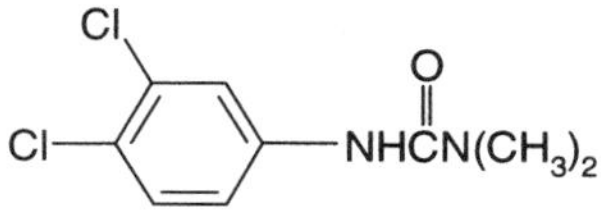

Cl
Cl — [benzene ring] — NHCN(CH$_3$)$_2$ (C=O)

Diuron

Uses Diuron is stable and formulated as an L, SC, WP, WG, F, and G and sold under several names. Diuron is registered as a preplant incorporated, preemergence, postemergence, and postemergence-directed application in a wide variety of crops, including alfalfa, artichokes, asparagus, corn, winter wheat and barley, birdsfoot trefoil, oats, red clover, sorghum, cotton, sugarcane, tree fruit, bush fruits, citrus, vining fruits, olives, nuts, grapes, pineapple, bananas, plantains, papayas, mint, grass seed crops, and tree plantings and in noncropland, for control of a variety of annual and perennial weeds, depending on the use rate.

Linuron

Linuron (*N′*-(3,4-dichlorophenyl)-*N*-methoxy-*N*-methylurea) is a white crystalline solid with a vapor pressure of 1.7×10^{-5} mm Hg at 20°C, a water solubility of 75 mg/l (ppm) at 25°C, a soil half-life of 60 days, and an oral LD_{50} (rat) of 1254 mg/kg.

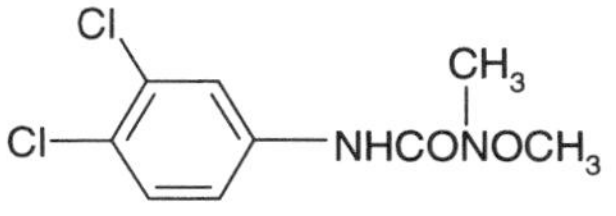

Cl
Cl — [benzene ring] — NHCONOCH$_3$ (N–CH$_3$)

Linuron

Uses Linuron is stable and formulated as an DF,WG, L, and SC and sold as Lorox, Afalon, Linex, and other products for preemergence, postemergence, and postemergence-directed use in asparagus, artichokes, carrots, celery, parsnips, parsley, fennel, celeric, onions, leeks, garlic, peas, corn, cotton, flax, sunflowers, sugarcane, potatoes, soybeans, sorghum, bananas, cassava, coffee, tea, rice, peanuts, hybrid poplar, ornamental trees and shrubs, and in noncrop situations, for control of annual broadleaf weeds and some grasses.

Fluometuron

Fluometuron (*N*,*N*-dimethyl-*N′*-[3-(trifluoromethyl)phenyl]urea) is a white crystalline soild with a vapor pressure of 5×10^{-7} mm Hg at 20°C, a water solubility of 110 mg/l (ppm) at 22°C, a soil half-life of 85 days, and an oral LD_{50} (rat) of 6416 mg/kg.

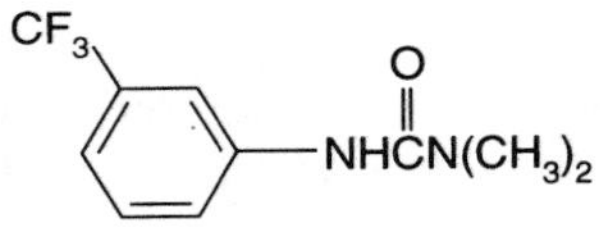

$$CF_3-C_6H_4-NHC(=O)N(CH_3)_2$$

Fluometuron

Uses Fluometuron is stable and formulated as a DF, WG, L, and SC and sold as Cotoran, Cotogard, and Cottonex for preplant incorporated, preemergence, or postemergence treatments in cotton and sugarcane for control of many broadleaf and grass weeds.

Tebuthiuron

Tebuthiuron (*N*-[5-(1,1-dimethylethyl)-1,3,4-thiadiazol-2-yl]-*N*,*N'*-dimethylurea) is a colorless solid with a vapor pressure of 1×10^{-7} mm Hg at 25°C, a water solubility of 2500 mg/l (ppm) at 25°C, a soil half-life of 12 to 15 months, and an oral LD_{50} (rat) of 644 mg/kg.

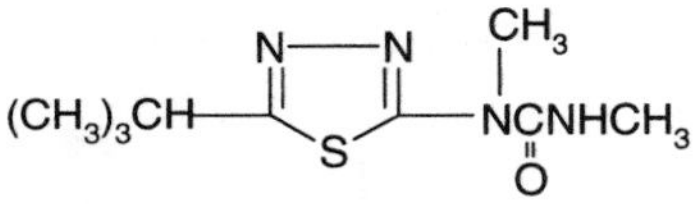

$$(CH_3)_3CH-C_2N_2S-N(CH_3)C(=O)NHCH_3$$

Tebuthiuron

Uses Tebuthiuron is stable and formulated as a WP, P, WG, and G and sold as Spike, Bushwacker, and Tebusan for preemergence use in noncrop situations, rangeland, rights-of-way, industrial sites, and wildlife habitat for control of broadleaf weeds and brush species.

Soil Interaction of Urea Herbicides

As a class, urea-type herbicides are relatively persistent in soils. Under favorable moisture and temperature conditions with little or no leaching, most can be expected to remain phytotoxic for 6 months at the lower selective rates and 24 months or more at higher nonselective rates.

Fluometuron is the least persistent of this group, with a half-life of about 30 days. Linuron and siduron (which is not a PS II Inhibitor) have half-lives of 2 to 5 months and <1 year, respectively. At selective rates, diuron has a half-life of <1 year, but at nonselective rates its half-life is >1 year. Tebuthiuron is very persistent, with a half-life of 12 to 15 months in areas receiving 40 to 60 inches of rainfall and considerably greater in areas of low rainfall.

Principal factors affecting the persistence of ureas in the soil are microbial decomposition, leaching, adsorption on soil colloids, and photodecomposition. The latter is important only when the herbicide remains on the soil surface for an extended period of time. Researchers believe that volatility and chemical decomposition are of minor importance in the persistence of these herbicides.

TABLE 9-1. Water Solubilities and Adsorption on Soil of Urea and Uracil Herbicides

Herbicide	Solubility in Water (ppm)	Adsorption on Keyport Silt Loam[a]
Fenuron	3850	0.3
Tebuthiuron	2300	—
Bromacil	815	1.5
Terbacil	710	1.7
Monuron	230	2.6
Fluometuron	90	—
Linuron	75	5.5
Diuron	42	5.2
Siduron	18	2.5
Neburon	5	16.0

[a]Expressed as ppm (active ingredient) present on the soil in equilibrium with 1 ppm in soil solution.
Note: Monuron is no longer made as a herbicide; fenuron and neburon are sold only outside the United States; and siduron, although a phenylurea herbicide, does not inhibit photosynthesis and is discussed in Chapter 12.
From Wolf et al., 1958.

Decomposition by microorganisms is the most important factor in the loss of most urea-type herbicides. Although tebuthiuron has also been shown to be degraded by microorganisms, this does not appear to be the major cause of its loss from soils. Conditions such as moderate moisture and temperature with adequate aeration, favoring microorganisms, would also favor decomposition. Therefore, under dry, cold, or very wet conditions (poor aeration), the chemicals normally persist for a long time.

The adsorptive forces between the chemical and the soil colloids directly affect the chemical's rate of leaching; its solubility is a less important factor.

In a study using four urea-type herbicides, leachability was correlated with adsorption and water solubility (Wolf et al., 1958) (see Table 9-1). Fenuron was leached the most, followed in order by monuron, fluometuron, linuron, diuron and siduron, and neburon. The water solubilities and available adsorption values for Keyport silt loam of the other urea and uracil (discussed in the next section) herbicides are given in Table 9-1. Some of the urea-type herbicides are no longer available in the United States but are included here because they help in understanding the soil interactions of the entire class.

URACIL HERBICIDES

Bromacil

Bromacil (5-bromo-6-methyl-3-(1-methylpropyl)-2,4(1*H*,3*H*)pyrimidinedione) is a white crystalline solid with a vapor pressure of 3.1×10^{-7} mm Hg at 25°C, a moderate water solubility of 815 mg/l (ppm) at 25°C, a soil half-life of 60 days, and an oral LD_{50} (rat) of 5175 mg/kg.

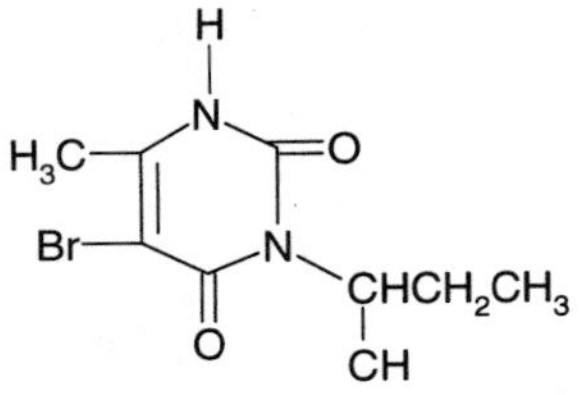

Bromacil

Uses Bromacil is formulated as an L and G and is sold as Hyvar X, Rokar XL, and Urgan for preemergence, postemergence, and postemergence directed applications for use in citrus, pineapple, and noncrop situations for control of annual broadleaf and grass weeds, and at higher rates for brush species.

Terbacil

Terbacil (5-chloro-3-(1,1-dimethylethyl)-6-methyl-2,4-(1*H*,3*H*)-pyrimidinedione) is a white crystalline solid with a vapor pressure of 3.1×10^{-7} mm Hg at 25°C, a moderate water solubility of 710 mg/l (ppm) at 25°C, a soil half-life of 120 days, and an oral LD_{50} (rat) of 1255 mg/kg.

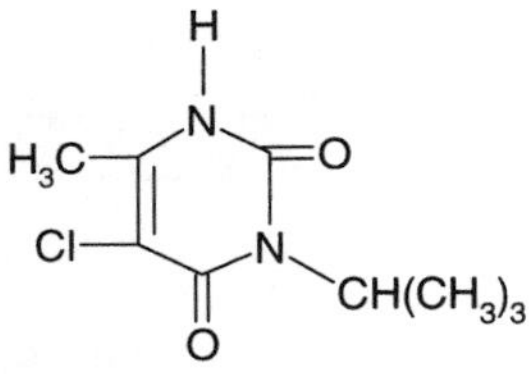

Terbacil

Uses Terbacil is formulated as a WP and sold as Sinbar for use as a preemergence, postemergence, and postemergence directed application in blueberries, cranberries, strawberries, alfalfa, mint, sugarcane, asparagus, apples, peaches, and citrus for control of a wide variety of annual broadleaf weeds.

Soil Interaction of Uracil Herbicides

Bromacil and terbacil are adsorbed less on soil colloids than the urea-type herbicides monuron, diuron, and neburon, but more tightly than fenuron (see Table 9-1). Therefore, they are leached more readily than most ureas. Bromacil and terbacil have a half-life of about 5 to 6 months when applied at 4 lb/acre, but at sterilant rates they persist for more than one season. This loss is apparently the result of microbial degradation. Soil diphtheroids, *Pseudomonas*, and *Penicillium* species have been shown to be able to degrade bromacil (WSSA, *Herbicide Handbook*, 1994).

OTHER PS II INHIBITORS

Propanil

Propanil (*N*-(3,4-dichlorophenyl)propanamide) is a brown to black crystalline solid with a vapor pressure of 4×10^{-5} mm Hg at 20°C, a moderate water solubility of 500 mg/l (ppm) at 25°C, a soil half-life of 1 day, and an oral LD_{50} (rat) of 1080 mg/kg.

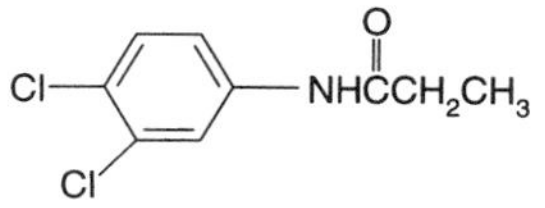

Propanil

Uses Propanil is formulated as an EC, SC, or WG, sold as Stam or Propanil, and registered as a postemergence application in rice, spring barley, oats, durum and hard red spring wheat for control of many grasses and a few broadleaf weeds.

Pyrazon

Pyrazon (5-amino-4-chloro-2-phenyl-3(2*H*)-pyridazinone) is in the Pyridazinone family. Pyrazon is a yellow brown crystalline solid with a vapor pressure of $<7.6 \times 10^{-8}$ mm Hg at 56.5°C, a water solubility of 400 mg/l (ppm) at 20°C, a soil half-life of 21 days, and an oral LD_{50} (rat) of 2200 mg/kg.

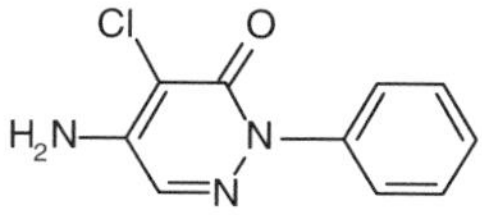

Pyrazon

Uses Pyrazon is formulated as a DF, WG, FL, and SC, sold as Pyramin and registered as a preplant incorporated, preemergence, and postemergence application for control of broadleaf weeds in sugar beets and red beets.

Bromoxynil

Bromoxynil (3,5-dibromo-4-hydroxybenzonitrile) is a member of the Benzonitrile chemical family. Bromoxynil is a light buff to creamy powder (acid) with a vapor pressure of 4.8×10^{-6} mm Hg at 25°C, a water solubility of 130 mg/l (ppm) at 20 to 25°C, a soil half-life of 7 days, and an oral LD_{50} (rat) of 440 mg/kg.

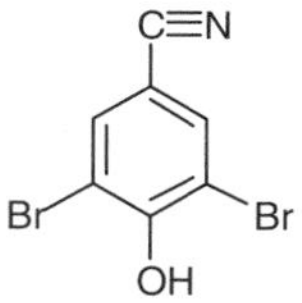

Bromoxynil

Uses Bromoxynil is formulated as an EC, sold as Buctril and under several other tradenames for postemergence control of broadleaf weeds in corn, sorghum, flax, garlic, onions, mint, grasses grown for seed or sod production, nonresidential turfgrass, and noncropland and industrial sites.

Bentazon

Bentazon (3-(1-methylethyl)-(1*H*)-2,1,3-benzothiadiazin-4(3*H*)-one 2,2-dioxide) is a member of the benzothiadiazole chemical family. Bentazon is a white crystalline solid with a vapor pressure of 7.5×10^{-9} mm Hg at 20°C, a water solubility of 500 mg/l (ppm) at 20°C, a soil half-life of 20 days and an oral LD_{50} (rat) of 1100 mg/kg.

Bentazon

Uses Bentazon is formulated as an L and sold as Basagran for postemergence control of broadleaf weeds in soybeans, dry or succulent beans and peas, peanuts, corn, sorghum, rice, peas, mint, and turf. Bentazon has excellent activity against emerged yellow nutsedge (Figure 9-4).

Phenylcarbamates

Phenmedipham Phenmedipham (3-[(methoxycarbonyl)amino]phenyl(3-methylphenyl)carbamate) is a colorless crystalline solid vapor pressure of 1×10^{-11} mm Hg

Figure 9-4. Effectiveness of bentazon applied postemergence for yellow nutsedge control in corn. *Left*: Treated; *Right*: Untreated.

at 25°C, a water solubility of >10 mg/l (ppm) at 20°C, a soil half-life of 25 to 30 days, and an oral LD_{50} (rat) of >4000 mg/kg.

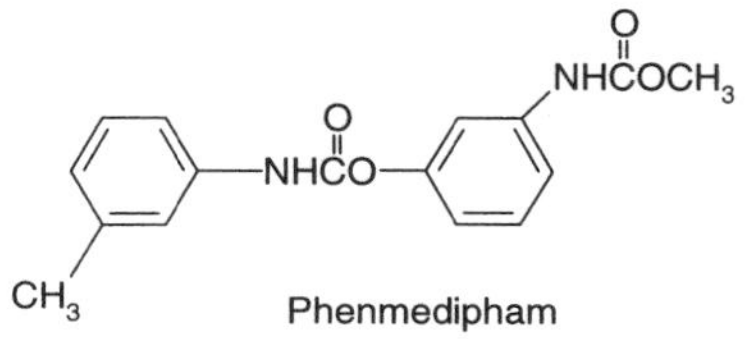

Phenmedipham

Uses Phenmedipham is formulated as an EC and sold as Betanal and Spin-aid for postemergence broadleaf weed control in sugar beets, red beets, and spinach.

Desmedipham

Desmedipham (ethyl [3-[[(phenylamino)carbonyl]oxy]phenyl]carbamate) is a colorless or light yellow crystalline solid with a vapor pressure of 3×10^{-9} mm Hg at 25°C, a water solubility of 7 mg/l (ppm) at 20°C, a soil half-life of <30 days, and an oral LD_{50} (rat) of >10,000 mg/kg.

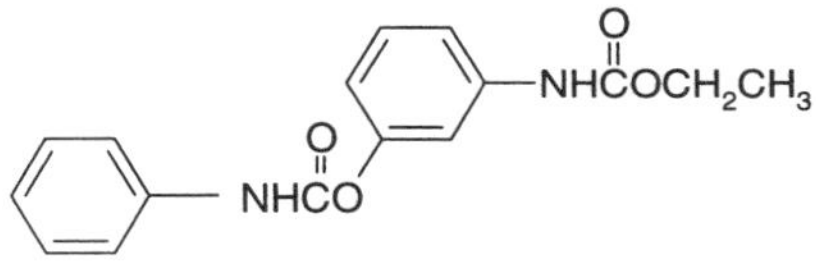

Desmedipham

Uses Desmedipham is formulated as an EC and sold as Betanex for postemergence broadleaf weed control in sugar beets.

Pyridate

Pyridate (*O*-(6-chloro-3-phenyl-4-pyridazinyl) *S*-octyl carbonothioate), a member of the phenylpyriddazine chemical family, is a white crystalline solid with a vapor pressure of 1.01×10^{-7} mm Hg at 20°C, a water solubility of 1.5 mg/l (ppm) at 20°C, a soil half-life of 7 to 21 days, and an oral LD_{50} (rat) of >4690 mg/kg.

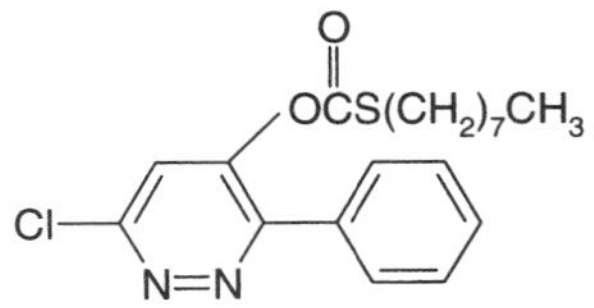

Pyridate

Uses Pyridate is formulated as an EC and sold as Tough and Lentagran for postemergence broadleaf weed control in peanuts, corn, cereals, mint, and rice.

Soil Influence

Propanil, bromoxynil, bentazon, phenmedipham, desmedipham, and pyridate are all foliar-applied herbicides, and soil has no influence on their activity. They tend to be rapidly broken down in soils under warm-moist conditions and therefore present no residual problem for subsequent crops. Most are strongly adsorbed to soils, which reduces leaching potential. These herbicides have the following half-lives: propanil, 1 to 3 days; bromoxynil, 7 days (with some suggestion of marginal short-term soil activity); bentazon, 20 days, with rapid adsorption to soil colloids; phenmedipham, 25 to 30 days, with strong adsorption to soil; desmedipham, <1 month, with strong adsorption to soil; and pyridate, 7 to 21 days, with strong adsorption to soil.

GENERAL CONSIDERATIONS

Metabolism

The various photosynthesis inhibitors are inactivated by a wide variety of reactions. In many cases, rapid inactivation is associated with tolerance of the plant to the herbicide. Apparently, the ring is not split in any of the metabolic reactions in the plant. The three most common inactivation reactions for triazines are (1) dechlorination, demethoxylation, or demethylthiolation and subsequent hydroxylation of the site, (2) dealkylation of the alkyl side chains, and (3) conjugation to glutathione. The most common reactions for substituted ureas are demethylation and/or demethoxylation. These mechanisms are discussed and described in the section "Herbicide Selectivity" in Chapter 5.

Bioassays

Bioactivity of many photosynthesis inhibitors in soil has been measured by fresh weight of seedlings 2 to 3 weeks after planting. Oats have commonly been used for triazines, but cucumbers are also satisfactory and sugar beets are more sensitive. Cabbage is a good test species for terbacil, and several plants have been used for the substituted ureas. A cotyledon disc bioassay is very sensitive for detecting these herbicides (Silva et al., 1976), and algae bioassays also are sensitive. A new method using an aquatic plant appears promising (Selim et al., 1989). Another method measures, in the presence of light, the alleviation by photosynthesis inhibitors of electrolyte leakage caused by paraquat (Yanase et al., 1990).

Environmental Concerns

Herbicides in the *s*-triazine class have come under increasing scrutiny because of their identification in many surface water supplies during the year. Peaks in these levels have been shown to occur during the early part of the growing season and reduced levels at other times of the year. There is a great deal of research and debate concerning whether the detection of *s*-triazines is a safety concern and whether restrictions in uses are necessary. Many articles concerning these issues are cited at the end of this chapter for further reading on this matter.

MODE OF ACTION OF PHOTOSYNTHESIS INHIBITORS

The light reaction of photosynthesis occurs within two reaction centers [photosystem I (PS I) and photosystem II (PS II)] located in chloroplasts within plant cells. PS II is composed of at least a dozen proteins and is the site of action for photosynthesis inhibitor herbicides (see Figures 9-5 and 9-6).

The main purpose of photosystem II and photosystem I is to process light energy and produce reducing power and metabolic energy (ATP) for use in the dark reactions of photosynthesis to produce the carbohydrates necessary for plant growth.

The process of photosynthesis begins in photosystem II. Light energy from the sun is utilized by the light-harvesting pigments (chlorophylls and carotenoids) of green plants to produce reducing power and O_2. Initially, the light energy that is captured by pigments catalyzes reactions resulting in the release of an electron and O_2 from water. Second, the captured light energy excites the available electron and, through a series of interactions with specialized light-harvesting pigments, produces the reducing power necessary for use in plant growth in PS I and the dark reactions of photosynthesis.

The actual process is shown in Figure 9-6. The excited electron is transferred from P_{680} to pheophytin and then to a plastoquinone molecule Q_A. Q_A passes two electrons (one at a time) to Q_B (a protein-bound plastoquinone). Once two electrons are passed from Q_A to Q_B, the fully reduced QB molecule becomes protonated (two hydrogen

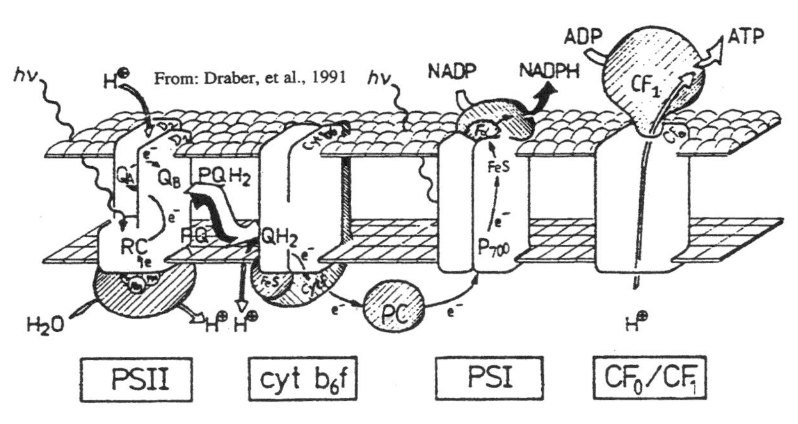

Figure 9-5. The process of photosynthesis is illustrated. Photosynthesis begins in photosystem II (PS II) where light energy ($\lambda\nu$) from the sun is utilized by light-harvesting pigments (chlorophylls and carotenoids) of green plants to produce reducing power and O_2. Initially, the light energy that is captured by pigments catalyzes reactions, resulting in the release of an electron and O_2 from water. Next, the captured light energy excites the available electron, and through a series of interactions with specialized light-harvesting pigments (in PS II and PS I) and various quinones (QA, QB, PQ) and cytochromes (PC), produces reducing power in PS I e.g. reduced nicotinamide adenine dinucleotide phosphate (NADPH) that is used in the dark reactions of photosynthesis (CF_0/CF_1) to create the energy necessary for plant growth.

ions are added from the stroma) to form a bound plastohydroquinone (PQH_2) molecule. PQH_2 has a lowered binding affinity for its protein binding site, so it is easily displaced by an oxidized Q_B (PQ), and the process of PS II is repeated. PQH_2 can now transfer its electrons to the cytochrome b_6f complex, and eventually the electrons are transferred to PS I via plastocyanin.

Herbicides that inhibit photosystem II bind to a protein on the binding niche for Q_B, called the D-1 protein (Figure 9-6). These herbicides compete with Q_B, for the binding niche in the D-1 protein. This competition can lead to displacement of the Q_B and stop electron flow through PS II so that no reduced Q_BH_2 (PQH_2) forms and therefore no reducing power is generated in photosynthesis.

There are two families of PS II inhibiting herbicides that bind to the Q_B binding site. The "classical" family of photosynthesis inhibitors includes the triazines, ureas, and carbamates. This family is also termed the "$serine_{264}$" or "amide-type" family. The binding of classical photosynthesis inhibitors to the Q_B niche not only stops

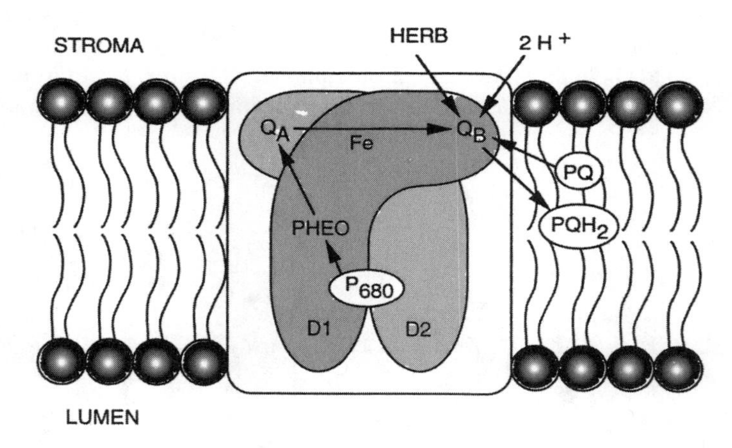

Figure 9-6. The actual process of electron transfer in PS II in which photosynthesis-inhibiting herbicides act. In PS II, an electron excited by the captured light energy is transferred from P_{680} to pheophytin (PHEO) and then to a plastoquinone molecule Q_A bound to the D2 protein. Q_A passes two electrons (one at a time) to Q_B (a protein-bound plastoquinone) on the D1 protein. Once two electrons are passed from Q_A to Q_B, the fully reduced Q_B molecule becomes protonated (two hydrogen ions are added from the stroma) to form a bound plastohydroquinone (PQH_2) molecule, which can then be further processed through photosynthesis to PS I (see Figure 9-5). PQH_2 has a lowered binding affinity for its protein binding, so an oxidized Q_B (PQ) easily displaces it and the process of PS II is repeated. PQH_2 can now transfer its electrons to the cytochrome b_6f complex, and eventually the electrons are transferred to PS I via plastocyanin. Herbicides, that inhibit photosystem II bind to a protein on the binding niche for Q_B called the D1 protein. These herbicides compete with Q_B for the binding niche in the D1 protein. This competition can lead to displacement of the Q_B, which stops electron flow through PS II so that no reduced Q_BH_2 (PQH_2) forms and therefore no reducing power is generated in photosynthesis.

electron flow but also slows D-1 protein turnover, which increases their herbicidal activity. The D-1 protein is slowly damaged by light, and it must be continually produced or the integrity (efficiency) of photosystem II is reduced.

The second family of photosynthesis inhibitors also binds at the Q_B niche. These herbicides, primarily the substituted phenols, are called the "nonclassical" family (sometimes termed the "$histidine_{215}$" family) of inhibitors. Well-known examples are pyridate, bromoxynil, and ioxynil. These herbicides also bind to the Q_B niche but in a slightly different orientation than the classical inhibitors. The nonclassical inhibitors bind to different amino acids in the Q_B binding site, but, for a yet unknown reason, these herbicides apparently do not slow the turnover of the D-1 protein.

How do plants die from PSII inhibitors? Early reports suggested that plants died by "starving to death" as a result of the inhibition of the light reaction of photosynthesis. However, plants die faster if sprayed with photosynthesis inhibitors and placed in the light than if sprayed and placed in the dark. This proves that something other than photosynthesis inhibition is responsible for the observed herbicidal effect.

The leaf chlorosis that develops after treatment is thought to be due to membrane damage caused by lipid peroxidation. When chlorophyll accepts light energy, it changes from a ground energy state to a singlet energy state (^{1}CHL). This singlet energy is normally transferred to the P_{680} reaction center, and the chlorophyll molecule returns to the ground state (Figure 9-7). When electron flow is blocked by herbicide binding in the Q_B pocket of the D-1 protein, the singlet chlorophyll energy cannot be transferred to the PS II reaction centers. The singlet energy state of the chlorophyll molecule is transformed to a more reactive triplet energy state (^{3}CHL). This triplet energy state is normally dissipated by carotenoids; however, because of the mass of triplet chlorophyll molecules produced by blocking electron flow through PS II, the carotenoid system is overloaded. The excess triplet chlorophyll causes lipid peroxidation, resulting in the breakdown of plant cell membranes and eventual plant death as cells become leaky, losing their contents and ability to function.

What is lipid peroxidation and why is it important in herbicide action? Lipid peroxidation, for the purpose of this book, refers to the breakdown of cell membranes and plant death as the ultimate result of the inhibitory action of a herbicide. This process is involved in the mechanism of a surprisingly large number of our most commonly used herbicides. Herbicide mechanisms causing lipid peroxidation fall into several categories. These herbicides include those that inhibit electron flow in photosystem II in the photosynthesis light reaction (e.g., triazines, phenyl ureas, uracils, etc.; this chapter), directly or indirectly affect carotenoid biosynthesis (e.g., norflurazone, isoxaflutole; Chapter 10), capture electrons in PS I in the photosynthesis light reaction (e.g., paraquat and diquat; Chapter 11), inhibit protoporhyrinogen oxidase (PROTOX) during chlorophyll biosynthesis (e.g., diphenyl ethers; Chapter 11), inhibit glutamine synthase in the nitrogen assimilation pathway (glufosinate; Chapter 11), and possibly other herbicides such as glyphosate; Chapter 16).

The mechanism of membrane disruption for these herbicides is essentially the same, the main difference being the factor produced that initiates the process of lipid

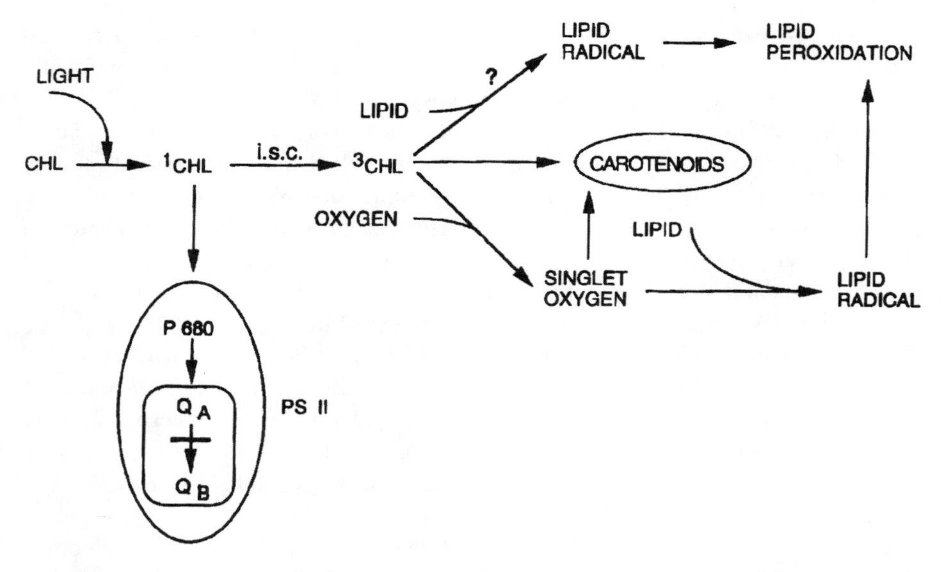

Figure 9-7. How do plants die after treatment with a photosynthesis-inhibiting herbicide? The leaf chlorosis that develops after treatment is thought to be due to membrane damage caused by lipid peroxidation. When chlorophyll accepts light energy, it changes from a ground energy state to a singlet energy state ^{1}CHL. This singlet energy is normally transferred to the P_{680} reaction center and the chlorophyll molecule returns to the ground state (Figure 9-5), and photosynthesis proceeds. When electron flow is blocked, by herbicide binding in the Q_B pocket of the D-1 protein, the singlet chlorophyll energy cannot be transferred to the PS II reaction centers. The singlet energy state of the chlorophyll molecule is transformed to a more reactive triplet energy state (^{3}CHL). This triplet energy state is normally dissipated by carotenoid pigments; however, because of the mass of triplet chlorophyll molecules produced by blocking electron flow through PS II, the carotenoid system becomes overloaded and the excess energy becomes destructive to cell membranes. The excess triplet chlorophyll then causes lipid peroxidation (through various reactions with oxygen and lipids), which results in the breakdown of plant cell membranes. When cell membranes are disrupted, cells within the plant become leaky, lose their contents, and quit functioning, which leads to eventual plant death (i.s.c. = intersystem conversion).

peroxidation and membrane disruption. The destruction of cell membranes is a three-step process involving initiation, propagation, and termination, as shown in Figure 9-8. The first step (*initiation*) is the production of unstable high-energy factors as a result of the primary inhibitory action of the herbicide. There are several types of initiating factors produced by herbicides, including *triplet chlorophyll* by PS II inhibiting herbicides and glufosinate, *singlet oxygen* by PROTOX inhibiting herbicides, and *hydroxy radicals* from PS I inhibiting herbicides. Carotenoid biosynthesis-inhibiting herbicides reduce the amount of pigments available to dissipate these high-energy factors (triplet chlorophyll and singlet oxygen). The process by which the various herbicides result in these initiating factors is described in this chapter and in Chapters 10, 11, and 12.

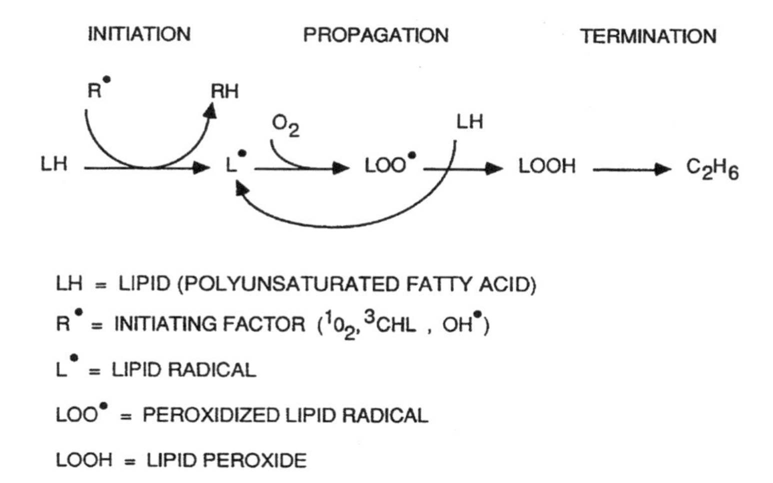

Figure 9-8. The process of lipid peroxidation is illustrated. Lipid peroxidation refers to the breakdown of cell membranes and plant death as the ultimate result of the inhibitory action of a herbicide. The first step (*initiation*) is the production of an unstable high-energy initiation factor (R°) which results from the primary inhibitory action of a herbicide. Initiation begins when R° interacts with a lipid membrane (LH, cell membrane), extracting an electron and becoming reduced (RH). In this process a lipid radical (L°) forms. In propagation, the lipid radical then interacts with oxygen (O_2), forming a highly reactive peroxidized lipid radical (LOO°). The LOO° reacts with another LH cell membrane, forming another L°, and the process is repeated. The formation of initiating factors and their interaction with lipids within cell membranes leads to a rapid peroxidation of the membranes, which results in leaky cells and plant death (termination). There are several types of initiating factors produced by herbicides, including *triplet chlorophyll* by PS II-inhibiting herbicides and glufosinate; *singlet oxygen* by PROTOX-inhibiting herbicides, and *hydroxy radicals* from PS I-inhibiting herbicides. Carotenoid biosynthesis-inhibiting herbicides reduce the amount of pigments available to dissipate these high-energy factors (triplet chlorophyll and singlet oxygen). The process by which the the various herbicides result in these initiating factors is described in Chapters 9, 10, 11, and 12.

Although the initiating factor can vary depending on the particular herbicide, the steps in lipid peroxidation are the same. Destabilization of the polyunsaturated fatty acids results in holes in the membrane and loss of cell integrity so that constituents leak out, causing loss of cell function and plant death. For a more complete description of this process, refer to Figure 9-8 or works by Girotti (1998) and Hippeli et al. (1999). Many of the free radicals formed can also disrupt protein and enzyme function in the chloroplast (e.g., the D-1 protein and RUBP carboxylase).

Mechanism of Resistance to Triazine Herbicides

More than 60 weed species worldwide have been shown to have high levels of resistance to triazine herbicides. Resistance in most cases is caused by differential activity at the site of action (chloroplast membrane); however, there are two examples of resistance due to increased herbicide metabolism in velvetleaf and rigid ryegrass (Anderson and Gronwald, 1991; Burnet et al., 1991, 1993). In susceptible weeds, triazines bind to the Q_B binding niche protein and effectively block electron transport, whereas in resistant weeds the herbicide cannot bind to the protein. The protein is only slightly different, because urea herbicides (e.g., diuron) still bind to the D1 protein in triazine-resistant biotypes, block electron flow, and have normal herbicidal effects. In all resistant populations identified in nature, there is a one amino acid change at position 264 in the protein sequence of the D1 protein. In all but one case, the amino acid change is from a serine to a glycine. Serine is more polar than glycine, which may account for the large change in the triazine binding constant.

This change in the amino acid present at position 264 results in decreased photosynthesis efficiency in triazine-resistant biotypes, as most mutant and modeling studies show that serine$_{264}$ is important for plastoquinone binding at Q_B (hydrogen bonding between serine$_{264}$ and plastoquinone). One theory is that in glycine$_{264}$ biotypes, electrons reside slightly longer on Q_A before being transferred to Q_B, reducing the rate of electron transfer. Another theory suggests that decreased photosynthesis results from increased sensitivity to photoinhibition (light stress) and may be due to a combination of increased photoinhibitory damage, leading to an increased turnover rate of the D1 protein, and the lowering of photosynthetic efficiency (electron transfer from Q_A to Q_B). Whatever the mechanism(s), resistant plants are significantly less fit (lower growth and reproductive ability) as compared with susceptible biotypes. The only time that resistant biotypes are more fit is when a triazine herbicide is used. The resistance gene has been transferred into crop plants (e.g., canola), and the reduced fitness is expressed, resulting in a "yield penalty" (Beversdorf et al., 1988).

Triazine-resistant plants selected from field populations still show sensitivity to urea herbicides such as diuron. This occurs because ureas bind slightly differently in the Q_B binding niche than triazines. Triazines use hydrogen bonding to serine$_{264}$, and when serine is changed to glycine, binding is lost. Diuron (a urea) binding is not affected by a serine$_{264}$ change, because it does not use hydrogen bonding at position 264 when binding to the Q_B binding site. However, if serine$_{264}$ is changed to an alanine or threonine by site-directed mutagenesis, binding of diuron is reduced because of overall conformational changes in the Q_B binding pocket (Mackay and O'Malley, 1993). A purslane biotype identified in nature that is resistant to linuron (a urea) was shown to have the serine$_{264}$ on the D-1 protein changed to threonine (Masabni et al., 1999). This biotype was cross-resistant to atrazine, and, as is the case in the other changes at serine$_{264}$, less fit than the wild type.

Triazine resistance is due to natural selection of resistant biotypes for populations of weeds. Substantial research has concluded that these biotypes are present as a small component of the natural population and are "selected for" when the susceptible biotypes are killed by triazine application. The occurrence of these natural biotypes can be minimized by not using herbicides with the same mode of action year after year. More information on herbicide resistance is given in Chapter 18.

LITERATURE CITED AND SUGGESTED READING

Anderson, M. P., and J. W. Gronwald. 1991. Atrazine resistance in a velvetleaf (*Abutilon theophrasti*) biotype due to enhanced glutathione s-transferase activity. *Plant Physiol.* **96**:104–109.

Beversdorf, W. D., D. J. Hume, and M. J. Donnelly-Vanderloo. 1988. Agronomic performance of triazine-resistant and susceptible reciprocal spring canola hybrids. *Crop Sci.* **28**:932–934.

Böger, P. H., and G. Sandmann, eds. 1989. *Target Sites of Herbicide Action*. CRC Press, Boca Raton, FL.

Burnet, M. W. M., O. B. Hildebrand, J. A. M. Holtum, and S. B. Powles. 1991. Amitrol, triazines, substituted ureas, and metribuzin resistance in a biotype of rigid ryegrass (*Lolium rigidum*). *Weed Sci.* **39**:317–323.

Burnet, M. W. M., B. R. Loveys, J. A. M. Holtum, and S. B. Powles. 1993. Increased detoxification is a mechanism of simazine resistance in *Lolium rigidum*. *Pestic. Biochem. Physiol.* **46**:207–218.

Davis, D. E., H. H. Funderburk, Jr., and N. G. Sansing. 1959. The absorption and translocation of C^{14} labeled simazine by corn, cotton and cucumber. Weeds **7**:300–309.

Dodge, A. D. 1982. The role of light and oxygen in the action of photosynthesis inhibitor herbicides. *Amer. Chem. Soc. Symp. Series* **181**:58–77.

Dwivedi, U., and R. Bhardwaj. 1995. D1 protein of photosystem II: The light sensor in chloroplasts. *J. Biosci.* **20**:35–47.

Girotti, A. W. 1998. Lipid hydroperoxide generation, turnover, and effector action in biological systems. *J. Lipid Res.* **39**:1529–1542.

Gunther, F. A., ed. 1970. The triazine herbicides. *Residue Rev.* **32**:1–413.

Hankamer, B., J. Barber, and E. J. Boekema. 1997. Structure and membrane organization of photosystem II in green plants. *Annu. Rev. Plant Physiol. Plant Mol. Biol.* **18**:641–671.

Herbicide Handbook. 7th ed. 1994. Weed Science Society of America. Lawrence, KS.

Hess, F. D. 2000. Light dependent herbicides—An overview. *Weed Sci.* **48**:160–170.

Hippeli, S., I. Heiser, and E. F. Elstner. 1999. Activated oxygen and free oxygen radicals in pathology: New insights and analogies between animals and plants. *Plant Physiol. Biochem.* **37**:167–178.

Huppatz, J. L. 1996. Quantifying the inhibitor-target site interactions of photosystem II herbicides. *Weed Sci.* **44**:743–748.

LeBaron, H. M., and J. Gressel, eds. 1982. *Herbicide Resistance in Plants*. John Wiley & Sons, Inc., New York.

Mackay, S. P., and P. J. O'Malley. 1993. Molecular modeling of herbicide interactions with the D1 protein of photosystem II. Part 1. The effect of hydrophobicity of inhibitor binding. *Proc. Brighton Crop Protection Conf., Weeds*, pp. 525–532.

Masabni, J. G., and B. H. Zandstra. 1999. A serine to threonine mutation in linuron-resistant *Portulaca oleraceae*. *Weed Sci.* **47**:393–400.

Naber, J. D., and J. J. S. van Rensen. 1991. Activity of photosystem II herbicides is related with their residence times at the D1 protein. *Z. Naturforsch.* **46c**:575–578.

Niyogi, K. K. 1999. Photoprotection revisited: Genetic and molecular approaches. *Annu. Rev. Plant Physiol. Plant Mol. Biol.* **50**:333–359.

Powles, S. B., and J. A. M. Holtum. 1994. *Herbicide Resistance in Plants: Biology and Biochemistry*. CRC Press, Boca Raton, FL.

Rodgers, E. G. 1968. Leaching of seven *s*-triazines. *Weed Sci.* **16**:117.

Selim, S. A., S. W. O'Neal, M.A. Ross, and C.A. Lemib. 1989. Bioassay of photosynthetic inhibitors in water and aqueous soil extracts with Eurasian watermilfoil. *Weed Sci.* **37**:810–814.

Silva, J. F., R. O. Fadayoni, and G. F. Warren. 1976. Cotyledon disc bioassay for certain herbicides. *Weed Sci.* **24**:250–252.

Takahashi, M., M. Takano, T. Shiraishi, K. Asada, and Y. Inoue. 1995. Cooperation of photosynthetic reaction center of photosystem II under weak light as shown by light intensity-dependence of the half–effective dose of atrazine. *Biosci. Biotech. Biochem.* **59**:796–800.

Talbert, R. E., and O. H. Fletchall. 1965. The adsorptrion of some s-triazines in soils. *Weeds* **13**:46.

Weber, J. B. 1970. Mechanisms of adsorption of *s*-triazines by clay colloids and factors affecting plant availability. *Residue. Rev.* **32**:93.

Wolf, D. E., R. S. Johnson, G. D. Hill, and R. W. Varner. 1958. Herbicidal Properties of Neburon. *NCWCC Proc.* **15**:7.

Yanase, D. A., A. Andoh, and N. Yasudomi. 1990. A new simple bioassay to evaluate photosynthetic electron-transport inhibition utilizing paraquat phytotoxicity. *Pestic. Biochem. Physiol.* **38**:92–98.

For chemical use, see the manufacturer's or supplier's label and follow these directions. Also see the Preface.

10 Pigment Inhibitors

HISTORY

Pigment inhibitors are mostly applied as preplant or preemegence treatments, with the exception of amitrole (foliar) and fluridone in aquatics. These herbicides inhibit different enzymes in the carotenoid pigment biosynthetic pathway in the plant. Carotenoid pigments are important accessory plant pigments that protect chlorophyll from photooxidation. When carotenoids are absent, chlorophyll is destroyed in the light and plants slowly die (Figure 10-1). Injury caused by these herbicides is evidenced by a bleached white to translucent appearance of the leaves. Sometimes the bleaching is not complete on the entire leaf but is interveinal with pink or red highlights along the margins.

Figure 10-1. Typical symptomology caused by pigment-inhibitor herbicides. Cabbage transplants planted into clomazone-treated soil show the bleaching effect after root uptake.

Amitrole (3-amino-1,2,4-triazole), or aminotrizole, is the oldest among the currently registered carotenoid biosynthesis inhibitors. The American Chemical Paint Company introduced it as a herbicide in 1954, and it was widely investigated as a selective herbicide in the 1950s and early 1960s. Because of its limited selectivity, its use today is primarily for control of several woody species and perennials in noncrop situations. Five additional pigment inhibitors have subsequently been introduced and registered for use in the United States: norflurazon (1970s), fluridone (1970s), clomazone (1980s), isoxaflutole (1990s), and mesotrione (2001).

CHARACTERISTICS, USES, AND SELECTIVITY

Amitrole

Amitrole (1*H*-1,2,4-triazol-3-amine) is an off-white crystalline powder that is nonvolatile, with a vapor pressure of 4.4×10^{-7} mm Hg at 25°C, a water solubility of 280,000 mg/l (ppm) at 25°C, a soil half-life of 14 days, and an oral LD_{50} (rat) of > 5000 mg/kg. Amitrole is a foliar-applied compound with minimum soil activity.

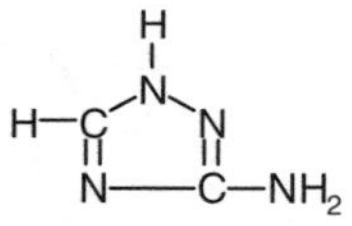

Amitrole

Uses Amitrole is formulated as an SL and sold as Amitrol-T. Amitrole is generally considered nontoxic to mammals, birds, fish, and bees. The compound is highly xylem and phloem mobile in most plants and is readily conjugated in many plants. However, caution should be exercised when it is used around desirable species because plants that are contacted will likely display bleaching symptoms. Amitrole is applied to foliage for the control of certain perennial plants, most commonly for the control of poison ivy and poison oak. Selectivity is minimal, with injury occurring to most plants that come in contact with the herbicide.

Soil Influences Because amitrole is a foliar-applied herbicide, soil has no effect on its performance. Microbial decomposition is the principal path of amitrole dissipation in the soil and is very rapid.

Norflurazon

Norflurazon (4-chloro-5-(methylamino)-2-(3-[trifluoromethyl]phenyl)-3(2*H*)-pyridazinone) is a white to grayish brown crystalline powder with a vapor pressure of 2×10^{-8} mm Hg at 20°C, a water solubility of 28 mg/l (ppm) at 25°C, a soil half-life of 45 to 180 days, and an oral LD_{50} (rat) > 7500 mg/kg. Norflurazon is moderately susceptible to photodegradation, but is nonvolatile.

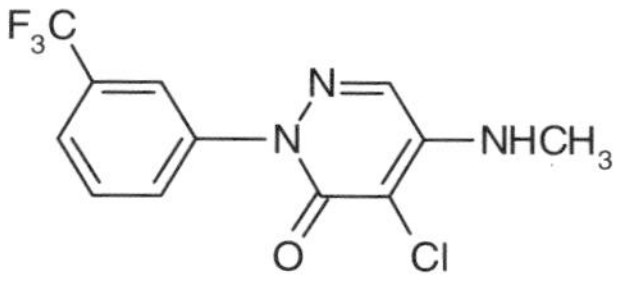

Norflurazon

Uses Norflurazon is formulated as granules and sold as Evital or as water-dispersible granules and sold as Solicam and Zorial.

Norflurazon is xylem mobile in most plants and is generally soil applied and absorbed by plant roots. In tolerant plants norflurazon is demethylated to a biologically inactive metabolite, and environmental stresses such as drought, suboptimum temperatures, and the like, may increase the chances for crop injury. Zorial is used preemergence, either to the surface or incorporated, for the control of annual grasses and small-seed dicot weeds in cotton, soybean, and peanut. It is particularly effective for the control of tropic croton and several malvaceous weeds such as prickly sida, Venice mallow, velvetleaf, and spurred anoda. It also has activity on yellow nutsedge. Selectivity among these crops is apparently achieved via application rate, placement, and differential translocation and metabolism. Apparently, cotton having gossypol glands is more tolerant of both norflurazon and fluridone. Zorial is also being used for annual grass control in bermudagrass hayfields in the southeastern United States. Solicam is used in tree fruits, nuts, caneberries, grapes, and asparagus for the same spectrum of weed control as is Zorial. Typical use rates for Solicam are higher than for Zorial. Selectivity is aided by the fact that norflurazon has limited soil mobility and therefore does not reach the roots of most tree crops. The granular formulation (Evital) is used in cranberries, where selectivity is achieved by differential metabolism.

Soil Influences Norflurazon is readily adsorbed by the clay and organic matter in soil. Its mobility and persistence in the soil is largely governed by soil sorptive properties; however, it is not leached appreciably through most soils. Typical half-lives (45 to 145 days), depend on soil type, soil microorganisms, the colloidal content of the soil, and environmental conditions. Microbial decomposition is apparently an important factor in persistence, although microbial enhancement due to previous applications has not been reported. The persistence of norflurazon can affect the growth of subsequent rotational crops. Soil microorganisms degrade norflurazon, but this is considered to be only partially responsible for its disappearance from soil. Volatilization and photodecomposition can contribute to its loss when it remains on the soil surface.

Fluridone

Fluridone (1-methyl-3-phenyl-5-[3-(trifluoromethyl)phenyl]-4(1*H*)-pyridinone) is a yellow solid with a vapor pressure of $< 10^{-7}$ mm Hg at 25°C, a water solubility of 12 mg/l (ppm) at 25°C, an oral LD_{50} (rat) of > 10,000 mg/kg, and a half-life of 21 days in water. Fluridone is very sensitive to photochemical degradation when used in aquatic weed control; it is formulated as an aqueous suspension or as slow-release pellets and sold as Sonar. Limited foliar absorption occurs in terrestrial plants, but is a very important avenue of entry in aquatic plants.

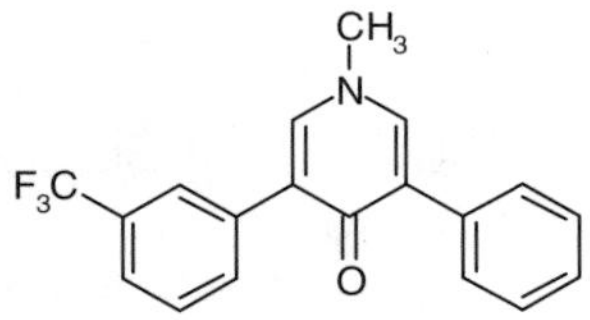

Fluridone

Uses Fluridone is sold as Sonar for the selective control of certain submerged and emerged aquatic weeds such as fanwort, coontail, parrot feather, water milfoil, pondweed, and hydrilla.

Soil and Water Influences Fluridone is strongly adsorbed to organic matter in soil and is highly immobile in soil because of its limited water solubility. It is moderately persistent in soil, where microbial decomposition is the major dissipation path. In water it is also subject to microbial decomposition but the major route of decomposition is via photodecomposition. Its half life in water is about 21 days and in hydrosoil about 90 days.

Clomazone

Clomazone (2-[(2-chlorophenyl)methyl]-4,4-dimethyl-3-isoxazolidinone) is a pale yellow liquid at room temperature with a vapor pressure of 1.44×10^{-4} at 25°C, a water solubility of 1100 mg/l (ppm) at 25°C, a soil half-life of 24 days, and an oral LD_{50} (rat) of 2077 mg/kg.

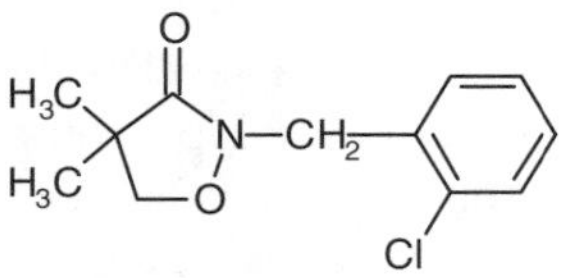

Clomazone

Uses Clomazone is formulated as an ME and sold as Command. Clomazone is generally more efficacious when it is not incorporated. An unbroken, concentrated barrier of the herbicide appears to provide much better weed control than when the herbicide is diluted by incorporation. However, because of volatility and the potential for off-site movement, incorporation is often used (Figure 10-2). The ME formulation is a reduced-volatility microencapsulated formulation developed to reduce the potential for off-site movement.

Figure 10-2. Off-target movement of a herbicide (clomazone) *Foreground*: No injury to an adjacent wheat field; Background: Field on right treated with clomazone while wheat field on left is exhibiting injury from herbicide movement.

Clomazone shows little foliar absorption; however, off-site injury and phytotoxicity to tobacco suggest that foliar absorption does occur. The primary site of uptake is the plant root, and the herbicide is translocated in the xylem. The mechanism of action is not fully understood, but the resulting effects of the herbicide clearly show involvement of carotenoid biosynthesis inhibition.

Clomazone is used for preemergence weed control in soybeans, cotton, tobacco, pumpkins, peppers, and squash. In soybeans, clomazone is apparently metabolized. In cotton, selectivity is achieved by the use of an organophosphate insecticide at planting as a crop safener. Clomazone was first registered for use in cotton in 1993, requiring the addition of either phorate or disulfoton in-furrow at planting. Tobacco is sufficiently tolerant of clomazone to allow the use of either pre- or posttransplant applications of the herbicide. Differential tolerance occurs among various cultivars of pepper, squash, and pumpkins. For example, bell and chili peppers are more tolerant than banana peppers, and ornamental jack-o'-lantern pumpkins are much more sensitive than processing pumpkins. In all of these crops clomazone is used for the control of annual grasses and a few annual dicot weeds such as velvetleaf, morning glory, and Florida beggarweed.

Soil Influences Clomazone has low mobility in sandy loam, silt loam, and clay loam soils and intermediate mobility in fine sand soils. It is subject to volatility losses and vapor drift when applied to the surface of moist or wet soils and may cause significant off-site damage to sensitive species in the event of an inversion. Clomazone is relatively immobile in the soil, and the principal path of breakdown is microbial decomposition. Its degradation is more rapid in sandy loam soils than in silt loam and clay loam soils. Degradation appears to proceed via binding to the soil matrix and mineralization to carbon dioxide, and its half-life ranges from 24 to 80 days, depending on microbiological activity and environmental conditions.

Isoxaflutole

Isoxaflutole (5-cyclopropyl-4-(2-methylsulphonyl-4-trifluoromethyl-benzoyl)isoxazole) is a relatively stable white to pale yellow solid with a vapor pressure of 7.5×10^{-9} at 25°C, a water solubility of 6.2 mg/l (ppm), at 25°C, a soil half-life of 2 months, and an oral LD_{50} (rat) >5000 mg/kg.

O SO$_2$CH$_3$ N O CF$_3$

Isoxaflutole

Figure 10-3. Corn injury caused by soil uptake of isoxaflutole. (D. Bridges, University of Georgia.)

Uses Isoxaflutole is sold as Balance and is formulated as a WG or SC. Isoxaflutole is a broad-spectrum grass and dicot herbicide used primarily as a soil-applied, preemergence herbicide. It is taken up by the roots and transported apoplastically to the shoot. Soil moisture conditions that are favorable for rapid crop growth are also conducive to maximum herbicidal activity.

Isoxaflutole is used primarily for preemergence weed control in corn. It is effective on most small-seeded dicot weeds, including velvetleaf, but ineffective on common cocklebur and wild buckwheat. Balance has excellent activity on woolly cupgrass, providing more consistent control of woolly cupgrass than currently available preemergence herbicides (metolachlor, acetochlor, etc.). Corn injury has been observed in some field trials (Figure 10-3).

Soil Influences Because of its water solubility, isoxaflutole is potentially mobile in the soil, especially with intense or excessive rainfall and it has been shown to be moderately persistent in soil. However, field studies have shown relatively little movement from the surface horizons, attributed to dissipation.

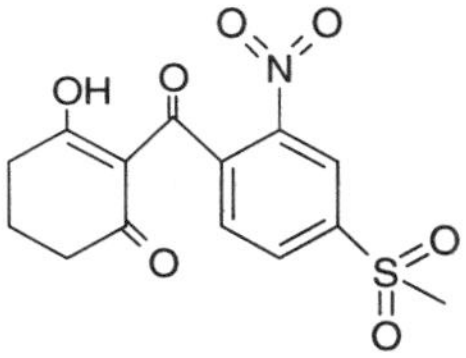

Mesotrione

Mesotrione

Mesotrione (2-[4-(methylsulfonyl)-2-nitrobenzoyl]-1,3-cyclohexanedione is a beige to tan liquid with a vapor pressure of 4.2×10^{-8} mm Hg at 25°C, an LD_{50} (rat) of >5000 mg/kg and a soil 1/2 life of 9 days under field conditions. Mesotrione is soluble in water.

Uses Mesotrione is sold as Callisto and is formulated as a SC for use as a preemergence and postemergence broad-spectrum broadleaf weed herbicide in field corn. Mesotrione is a systemic preemergence and postemergence herbicide for selective contact and residual weed control. It is especially active against velvetleaf, cocklebur, smooth and redroot pigweeds, waterhemps, common lambsquarters, sunflower and nightshades. Weed death may take up to 2 weeks but postemergence foliar applications cause immediate cessation of growth in susceptible species. Mesotrione should not be applied postemegence to field corn previously treated with chlorpyrifos or terbofos and never tank-mixed with organophosphate or carbamate insecticides as corn injury may occur.

Soil Influences Mesotrione has a moderate soil life and is broken down by microbes. Herbicide activity is influenced by soil pH and organic matter content with increased activity at high pH and decreased activity with high soil organic matter levels. Most nonlabeled crops can be replanted onto treated soil the spring following initial application.

MECHANISM OF ACTION OF CAROTENOID BIOSYNTHESIS INHIBITORS

The most striking symptom resulting from treating plants with herbicides that inhibit carotenoid biosynthesis is the totally white foliage produced following treatment (Figure 10-1), which is sometimes termed "albino growth." The white foliage is the result of a primary inhibition of carotenoid biosynthesis coupled with a secondary destruction (photooxidation) of chlorophyll as it is formed and, to some extent, an inhibition of chlorophyll biosynthesis. These herbicides are also called "bleaching herbicides" or "bleachers." Growth does continue for a time after treatment, but without production of green photosynthetic tissue, the growth of affected plants can not be maintained. Growth ceases and necrosis then begins. Plant tissues formed before treatment do not show typical albino symptoms, as these herbicides do not affect preexisting carotenoids. There is a turnover of carotenoid pigments; thus, tissue formed prior to treatment will eventually show some chlorosis and then necrosis.

The loss of chlorophyll in plants treated with carotenoid biosynthesis inhibitors is primarily the result of the destruction of chlorophyll by light (photooxidation). An important role of carotenoids is to protect chlorophyll from photooxidation. After chlorophyll is synthesized and becomes functional, some of the chlorophyll, which has been excited by light photons, is transformed from the singlet stable form to the longer lived, but more reactive and unstable, triplet form. Carotenoids act to protect chlorophyll from photoxidation by transferring the excitation energy of triplet chlorophyll to lower, less destructive energy states. When carotenoids are not present, these triplet chlorophyll states initiate degrading reactions, among which is chlorophyll and membrane destruction by lipid peroxidation (see Chapter 9). Thus, without carotenoids plant cells cannot survive in high light.

If plants are treated with carotenoid synthesis inhibitor herbicides and then grown in very low light intensities (to eliminate photooxidation), new growth contains only about 70% of the chlorophyll present in nontreated plants grown under the same conditions. This newly formed chlorophyll will not be destroyed by photooxidation if plants are maintained in low light, but 80% is destroyed if returned to high light.

Carotenoid biosynthesis inhibitors also disrupt the normal developmental sequence of chloroplasts. When developing shoot tissues (e.g., cotyledons) are grown in the dark in the presence of carotenoid biosynthesis inhibitors, the number of thylakoids, particularly the grana thylakoids, is significantly reduced (Wrischer et al., 1998). The pigment-protein complexes required for assembly of photosystem II (PS II) in the thylakoids of chloroplasts are also reported to be inhibited (Moskalenko and Karapetyan, 1996).

The biosynthesis pathway of carotenoids is shown in the Figure 10-4 and the pathway enzymes inhibited by commercially available carotenoid biosynthesis inhibiting herbicides are listed in Table 10-1. The first series of reactions results in the

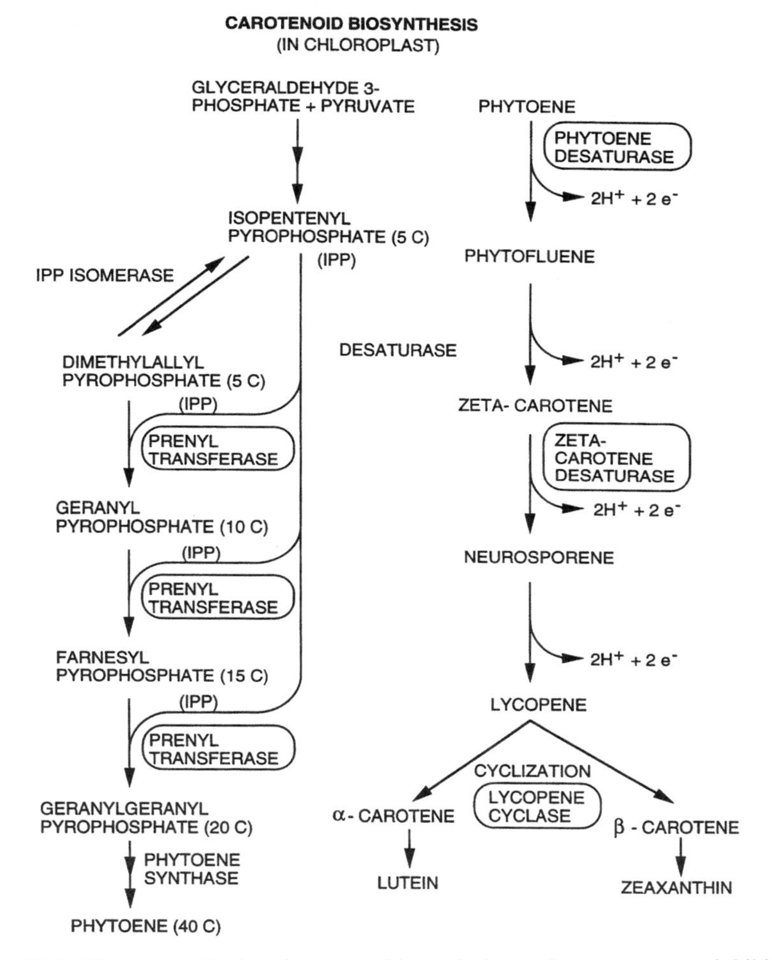

Figure 10-4. The biosynthesis of carotenoids and the pathway enzymes inhibited by norflurazon and fluridone (phytoene desaturase) and amitrol (zeta-carotene desaturase). See text for full explanation.

formation of a 20-carbon intermediate, geranylgeranyl pyrophosphate (GGPP). These reactions take place in the stroma of the chloroplast or on the thylakoid membranes. The critical starting material for the biosynthesis of carotenoids is isopentenyl pyrophosphate (IPP). IPP is an important precursor for many important carotenoids and as a component of chlorophylls and plastoquinone. After GGPP is formed, two GGPP molecules combine to form the 40-carbon intermediate phytoene, which undergoes a series of desaturation reactions (hydrogen abstraction), then cyclization to form α carotene and eventually lutein, β carotene and zeaxanthin. The site of action of noflurazon and fluridone is the inhibition of phytoene desaturase so that phytofluene is not produced. Inhibition causes a large accumulation of phytoene in

TABLE 10-1. Carotenoid Biosynthesis Pathway Enzymes Inhibited by Commercial Herbicides

Enzyme	Herbicide
Phytoene desaturase	Norflurazon, fluridone
Zeta carotene desaturase	Amitrole
Lycopene cyclase	Amitrole
4-hydroxyphenylpyruvate	Isoxaflutole
Dioxygenase	Sulcotrione

From Sandman and Böger, 1992.

treated plants. This accumulation is the most commonly reported proof for herbicide action at the phytoene desaturase site. Phytoene desaturase is also involved in reactions forming zeta carotene, and as a result phytoene epoxide and hydroxy derivatives of phytoene also accumulate.

Amitrole inhibits at 2 points, either the enzyme zeta-carotene desaturase, which results in the accumulation of zeta-carotene or the enzyme lycopene cyclase at the cyclization step following lycopene synthesis, which results in an accumulation of lycopene (Table 10-1). Reviews of chemicals that inhibit at these sites are provided in Sandman and Böger (1992) and Böger (1996).

Clomazone (Command) appears to have a unique site of action in carotenoid biosynthesis, although the specific site at which it inhibits is not known. We do know that treatment of plants with clomazone results in loss of pigments and the appearance of white tissue, which strongly suggests an effect on some aspect of carotenoid biosynthesis. Although much research has investigated possible sites in the pathway, the specific mechanism of action for clomazone remains a mystery.

An additional "new" site of action for carotenoid synthesis inhibitors was discovered in 1993. The herbicide sulcotrione (ICIA-0051, also published as SC-0051), a triketone herbicide used commercially in Europe (trade name, Mikado), produces carotenoid synthesis inhibition symptoms (new growth is white). However, until 1993 its enzyme site of action in carotenoid biosynthesis could not be located, even though there was an accumulation of phytoene suggesting inhibition of phytoene desaturase. Sulcotrione was shown to inhibit the enzyme 4-hydroxyphenylpyruvate dioxygenase (HPPD), which converts 4-hydroxyphenylpyruvate to homogentisate (Figure 10-5) (Schulz et al. (1993). The effect of this enzyme inhibition results in a depletion of plastoquinones, which are needed for proper functioning of the phytoene desaturase enzyme; hence, the reason for the accumulation of phytoene. Clomazone (site of action currently unknown) does not interfere with this enzyme.

The new herbicides isoxaflutole (Balance) and mesotrione (Callisto) also inhibit the HPPD enzyme (Pallett et al., 1997). Isoxaflutole is a proherbicide that (in plants and soils) rapidly undergoes ring opening at the isoxazole ring to form the diketonitrile derivative (Pallett, et al., 1997, 1998). This diketonitrile is thought to be the active

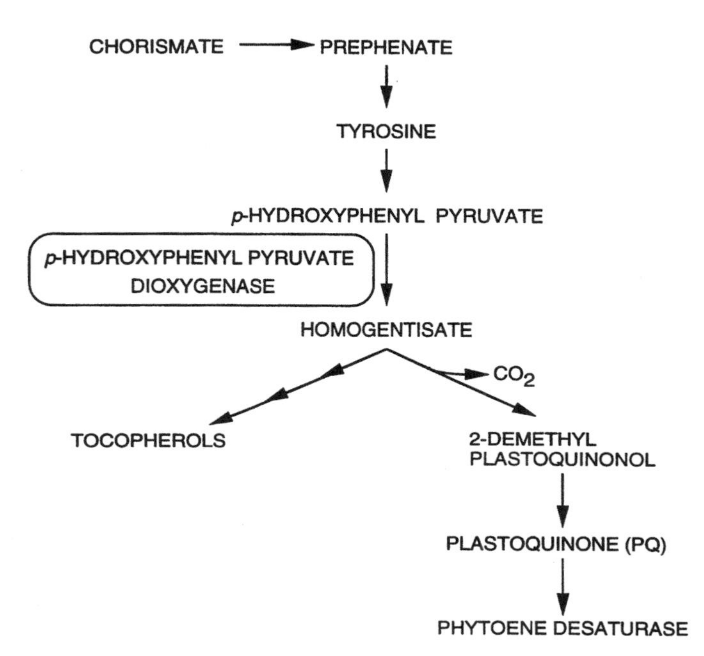

Figure 10-5. Sulcotrione isoxaflutole, and mesotrione inhibit the enzyme 4-hydroxyphenylpyruvate dioxygenase (HPPD), which converts 4-hydroxyphenylpyruvate to homogentisate. The effect of this enzyme inhibition results in a depletion of plastoquinones, which are needed for proper functioning of the phytoene desaturase enzyme, resulting in pigment production inhibition (see text for full explanation).

form of the herbicide, as HPPD is strongly inhibited by this derivative. Selectivity in corn appears to be due to the rapid metabolic degradation of the diketonitrile derivative of isoxaflutole to a benzolic acid derivative.

There are no weeds that have been shown to have resistant biotypes occurring after repeated use of the carotenoid biosynthesis inhibitor herbicides, except for amitrole. Some reports show certain woody perennial species are not currently controlled as well as in the past with amitrole, although the mechanism of resistance is not known.

LITERATURE CITED AND SUGGESTED READING

Achhireddy, N. R., and M. Singh. 1986. Toxicity, uptake, translocation, and metabolism of norflurazon in five citrus rootstocks. *Weed Sci.* **35**:312–317.

Anonymous. 1997. Isoxaflutole. In *The Pesticide Manual,* 11th ed., ed. by C.D.S. Tomlin, British Crop Protection Council. Farnham, Surrey, UK.

Barard, D. F., D. P. Rainey, and C. C. Lin. 1978. Absorption, translocation, and metabolism of fluridone in selected crop species. *Weed Sci.* **26**:252–254.

Bartley, G. E., and P. A. Scolnik. 1995. Plant carotenoids: Pigments for photoprotection, visual attraction, and human health. *The Plant Cell* **7**:1027–1038.

Böger, P. 1996. Mode of action of herbicides affecting carotenogenesis. *J. Pestic. Sci.* **21**:473–478.

Böger, P., and G. Sandmann. 1993. Pigment biosynthesis and herbicide interaction. *Photosynthetica* **28**:481–493.

Keeling, J. W., R. W. Lloyd, and J. R. Abernathy. 1989. Rotational crop response to repeated applications of norflurazon. *Weed Tech.* **3**:122–126.

Loux, M. M., R. A. Liebl, and F. W. Slife. 1989. Absorption of clomazone on soils, sediments, and clays. *Weed Sci.* **37**:440–442.

Lutzow, M., P. Beyer, and H. Kleinig. 1990. The herbicide Command does not inhibit the prenyl diphosphate-forming enzymes in plastids. *Z. Naturforsch.* **45c**:856–858.

Moskalenko, A. A., and N. V. Karapetyan. 1996. Structural role of carotenoids in photosynthetic membranes. *Z. Naturforsch.* **51c**:763–771.

Pallet, K. E., J. P. Little, M. Sheekey, and P. Veerasekaran. 1998. The mode of action of isoxaflutole. I. Physiological effects, metabolism and selectivity. *Pestic. Biochem. Physiol.* **62**: 113–124.

Pallet, K. E., J. P. Little, P. Veerasekaran, and F. Viviani. 1997. Inhibition of 4-hydroxyphenyl-pyruvate dioxygenase: The mode of action of the herbicide RPA 201772 (isoxaflutole). *Pestic. Sci.* **50**:83–84.

Sandmann, G., and P. Böger. 1992. Chemical structure and activity of herbicidal inhibitors of phytoene desaturase. In *Rational Approaches to Structure, Activity, and Ecotoxicology of Agrochemicals*, ed. by W. Draber and T. Fujita, pp. 357–371. CRC Press, Boca Raton, FL.

Sandmann, G., A. Schmidt, H. Linden, and P. Böger. 1991. Phytoene desaturase, the essential target for bleaching herbicides. *Weed Sci.* **9**:474–479.

Schulz, A., O. Ort, P. Beyer, and H. Kleinig. 1993. SC–0051, a 2-benzoyl-cyclohexane-1,3-dione bleaching herbicide, is a potent inhibitor of the enzyme *p*-hydroxyphenylpyruvate dioxygenase. *FEBS Letters* **318**:162–166.

Scott, J. E., L. A. Weston, J. Chappell, and K. Hanley. 1994. Effects of clomazone on IPP isomerase and prenyl transferase activities in cell suspension cultures and cotyledons of solanaceous species. *Weed Sci.* **42**:509–516.

Secor, J. 1994. Inhibition of barnyardgrass 4-hydroxyphenylpyruvate dioxygenase by sulcotrione. *Plant Physiol.* **106**:1429–1433.

Wrischer, M., N. Ljubesic, and B. Salopek. 1998. The role of carotenoids in the structural and functional stability of thylakoids in plastids of dark grown spruce seedlings. *J. Plant Physiol.* **153**:46–52.

For chemical use, see the manufacturer's or supplier's label and follow these directions. Also see the Preface.

11 Membrane Disruptors

There is a diversity of chemistry in the compounds of the membrane disruptor herbicides. Most are applied as postemergence contact herbicides; however, oxidiazon, sulfentrazone, and oxyfluorfen have important uses as soil-applied preemergence herbicides. Although the specific site of inhibition in the plant varies from photosystem I inhibition, to protoporphyrinogen oxidase (PROTOX) inhibition, to glutamine synthase inhibition, plant death from these herbicides is rapid after foliar application. Complete coverage of the leaf is important for best activity, and the speed of plant death is more rapid under high light and warm environmental conditions. Injury symptoms include an initial appearance of water-soaked tissue, followed by desiccation of leaf tissue caused by a disruption of cell membranes (Figure 11-1). Membranes are degraded by free radicals (potent biological oxidants) that form within plants as a result of the action of these herbicides.

Figure 11-1. Typical injury symptomology of tissue necrosis caused by a membrane-disrupter herbicide. Figure illustrates initial injury sometimes seen on soybean after application of a diphenylether. Soybean plants recover from this injury.

PHOTOSYSTEM I INHIBITORS

Diquat and paraquat herbicides are heterocyclic (more than one type of atom in the ring) organic compounds belonging to the bipyridilium class. The ring structures have a positive charge and are formulated as salts. Diquat and paraquat are the most important members of this group. The herbicidal activity of diquat was discovered in England in 1955, and paraquat was discovered a few years later. These are now widely used as herbicides and crop desiccants throughout the world.

General Characteristics

1. Very soluble in water, formulated as dichloride or dibromide salts.
2. Strong cations.
3. Rapidly adsorbed and absorbed by foliage.
4. Plants are killed quickly, usually within 1 or 2 days of application.
5. Action is much more rapid in the light than in the dark.
6. Usually, plants are killed so rapidly that there is very little translocation.
7. Increased tolerance to paraquat has developed in certain weeds after several years of repeated applications.
8. They are strongly adsorbed by inorganic soil colloids, especially by the expanding lattice clays where these herbicides are trapped between the layers. Soil activity is rare.

General Uses

1. Extensive use in land preparation for production of corn, soybeans, and other crops in reduced tillage programs
2. Contact preemergence sprays in slow-germinating crops
3. Directed sprays in corn, sugarcane, soybeans, tree fruits, bush fruits, grapes, and other crops
4. Pasture renovation
5. Aquatic weed control
6. Noncrop weed control
7. Preharvest crop and weed desiccation
8. Selective weed control in peanuts

The specific characteristics and labeled uses for diquat and paraquat are discussed in the following paragraphs.

Diquat

Diquat (6,7-dihydrodipyrido[1,2-α:2′,1′-*c*]pyrazinediium ion) is a yellow crystalline solid with a vapor pressure of $< 1 \times 10^{-8}$, a water solubility of 718,000 mg/l (ppm) at 25°C, and an oral LD_{50} (rat) of 230 mg/kg. Applicators should use caution in handling diquat and avoid breathing spray or contacting the concentrate on skin.

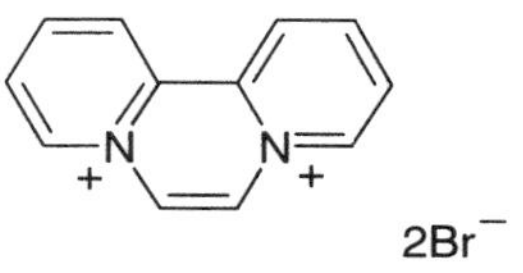

$2Br^-$

Diquat

Uses Diquat is formulated as an SL and is applied as a contact spray to foliage. Primary uses are for preharvest desiccation of foliage in alfalfa, clover, grain sorghum, soybeans (seed crop only), and potatoes; for vegetation control and weed control in nonbearing grapes and noncrop or nonplanted areas around farms, and for aquatic weed control. There are geographical restrictions, including Special Local Need (SLN) Registrations, harvest time, and other limitations on some of the aforementioned uses (see label).

Paraquat

Paraquat (1,1′-dimethyl–4,4′-bipyridinium ion) is a crystalline white solid with a vapor pressure of $<1 \times 10^{-7}$, a water solubility of 620,000 mg/l at 25°C, and an oral LD_{50} (rat) of 138 mg/kg. Absorption through intact skin is minimal but may be facilitated if the skin is damaged. Several poisonings and deaths have been reported as a result of ingesting relatively small amounts of the liquid concentrate. Fatalities resulted from progressive pulmonary fibrosis associated with liver and kidney damage. As a safeguard against oral ingestion, current formulations contain a material that induces vomiting when ingested.

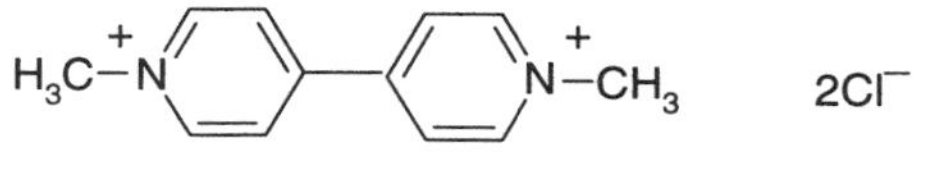

H_3C-N^+ N^+-CH_3 $2Cl^-$

Paraquat

Uses Paraquat is formulated as an SL in the dichloride salt form, which is very soluble in water, and is applied as a contact herbicide to emerged vegetation. In general, it can be used in any crop utilizing techniques that keep sprays off leaves and succulent stems of the crop plant. These techniques include preplant, preemergence, or directed spray treatments or use in crops when they are dormant or as a chemical fallow. Paraquat is nonselective, and crops are usually injured by the spray or spray drift when it comes in contact with the crop foliage. Perennial weed growth is suppressed by foliar desiccation but soon recovers, and annual weeds emerging after paraquat application are not controlled. Paraquat is mixed with soil active herbicides to kill existing weeds and obtain residual control. Generally, paraquat is used as a site prep herbicide prior to planting, as a directed spray to avoid contact with the crop or foliage, or as a harvest aid. The following can be treated, depending on label recommendations, with paraquat:

Alfalfa, new seeding	Asparagus	Dry beans
Cacao	Corn (field, sweet, pop)	Cotton
Easter lilies	Grasses (for seed)	Guar
Lentils	Clover and other legumes	
Mint	Onions	
Garlic	Pineapple	Potato
Peanuts	Pines	Rice
Safflower	Small fruits	Strawberries
Small grains (barley, wheat)	Grain sorghum	Soybeans
Sugarcane	Sunflower	Taro (dryland)
Sugar beets	Guava	Hops
Passion fruit	Tyfon	Native pastures
Vegetables (wide variety)	Pastures (reseeding)	
Fruit, nut, and ornamental trees and vines		

Selectivity

There are differences in selectivity between species, but most crops lack sufficient tolerance to these herbicides to make over-the-top applications safe. However, strains of perennial ryegrass have been bred for tolerance to paraquat and can be used where this herbicide is applied for pasture renovation. In the past few years peanuts have been shown to have considerable tolerance to paraquat, and starting in 1988 it has been sold for this crop in the southeastern United States.

Soil Influences

An important and unique property of both diquat and paraquat is their rapid inactivation in soil. The inactivation results from a reaction between the positively charged herbicide ion and the negatively charged clay minerals. The presence of the double positive charge on the herbicide molecules causes them to become tightly adsorbed within the clay lattice. Therefore, these herbicides are essentially nonphytotoxic in most soils. Some phytotoxicity, however, has been demonstrated at high rates in very sandy soils high in organic matter that contain little or no clay. Although nonphytotoxic, the bound diquat and paraquat may persist in soils for some period of time. Because of the tight binding to soil, these herbicides do not leach. They are highly persistent because of their strong binding to clay and unavailability to microbes but are not taken up by plants, so no residual toxicity occurs.

Metabolism

These herbicides are not degraded in higher plants in the usual sense. Rather, they are reversibly converted from the ion form to the free radical within the plant (Duke, 1985). However, any of the compound that remains on the leaf surface may be degraded by photodecomposition.

Mechanism of Action

The first visible damage caused by the bipyridilium-type herbicides is the appearance of many dark green spots on the leaf (often termed "water soaking"), followed by wilting and death (necrosis) of the tissues, often within a few hours (Figure 11-2). High light intensities increase the rate of development of the phytotoxic symptoms. Conversely, cloudy conditions can interfere with good activity of these herbicides. Best results in the field have often been obtained by a late-afternoon application, rather than a morning or midday application. This appears to allow some internal translocation during the night, before the development of acute phytotoxicity induced by light, which would limit movement. Translocation following foliar application appears to be solely via the apoplastic system when it does occur.

Diquat and paraquat (di-cations) have the ability to accept an electron from photosystem I (PS I) during electron flow in photosynthesis and become free radicals (mono-cations). The formation of these free radicals stops electron transport to oxidized nicotinamide adenine dinucleotide phosphate (NADP) and effectively inhibits normal functioning of PS I. In PS I, plastocyanin transfers its electron through a series of steps to ferrodoxin, and finally on to NADP. In paraquat- or diquat-treated plants, the herbicide binds near the ferrodoxin binding site and accepts electrons, becoming a free radical (Figure 11-3). The difference between a paraquat or diquat free radical and reduced ferrodoxin is that the herbicide free radical does not pass its accepted electron to NADP, but instead initiates a series of reactions leading to cell membrane disruption and plant death.

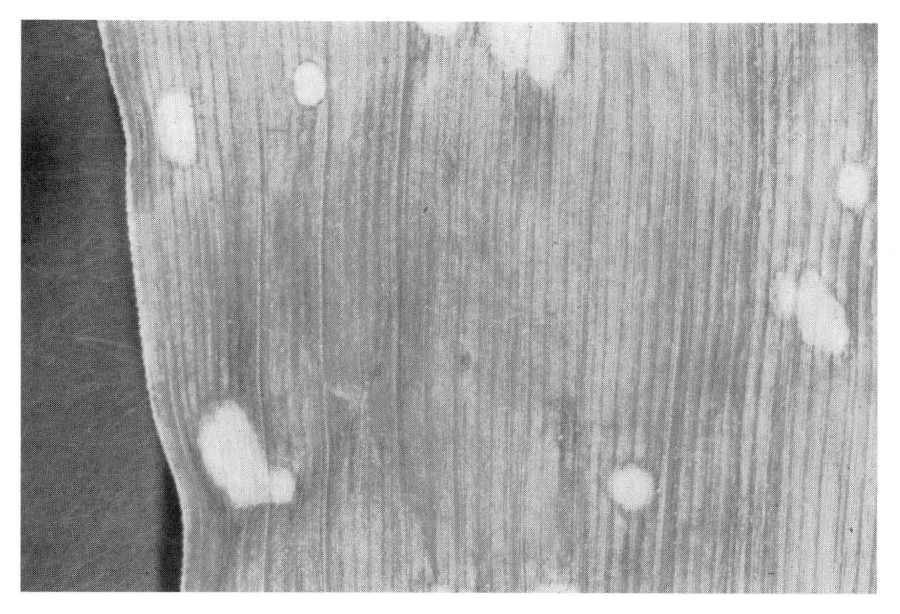

Figure 11-2. Typical tissue necrosis observed on plants after application of paraquat. Under bright, sunny conditions tissue necrosis can occur within an hour.

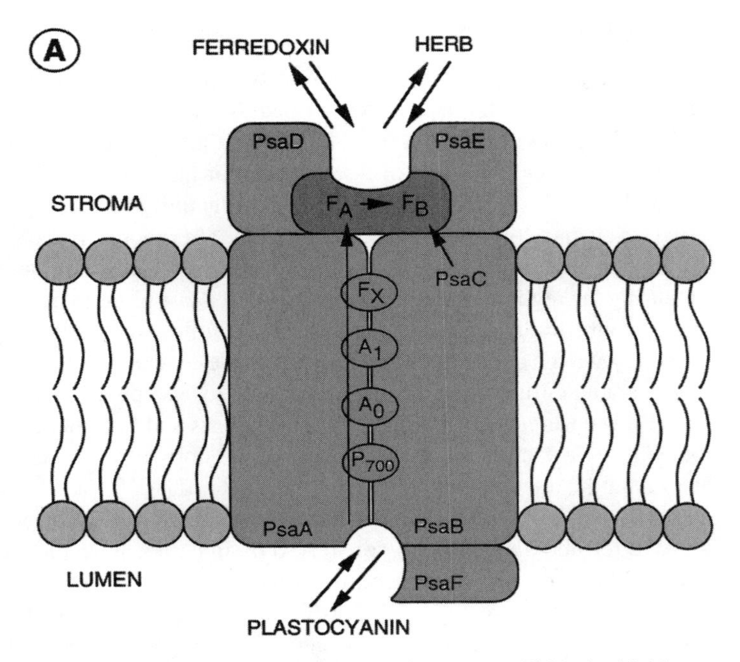

Figure 11-3. Mechanism of action of diquat and paraquat. Either herbicide can accept an electron from photosystem I (PS I) during electron flow in photosynthesis and become a free radical (mono-cation). The formation of such free radicals stops electron transport to NADP and effectively inhibits normal functioning of PS I. In PS I, plastocyanin transfers its electron through a series of steps (P_{700}, A_O, A_1, F_X, F_A) to ferrodoxin and finally on to NADP. In paraquat or diquat treated plants, the herbicide binds near the ferrodoxin binding site in PS I and accepts electrons, becoming a free radical. The difference between a paraquat or diquat free radical and reduced ferrodoxin is that the herbicide free radical does not pass its accepted electron to NADP, but instead initiates a series of reactions leading to cell membrane disruption and plant death.

The paraquat or diquat free radicals (mono-cations) are not the agents causing the tissue damage. These free radicals are unstable and rapidly undergo "auto-oxidation" back to the parent compound, where they can accept another electron from PS I and form another herbicide free radical. During auto-oxidation the herbicide free radical reacts with water and oxygen, forming superoxide and the parent herbicide (Figure 11-4). Superoxide then reacts with the enzyme superoxide dismutase to form hydrogen peroxide, which in turn forms hydroxy radicals (see Figure 11-4). Hydroxy radicals are the most potent biological oxidants known, and they quickly and effectively initiate membrane disruption through lipid peroxidation (as discussed in Chapter 9). Under normal conditions with no herbicide, there are mechanisms that dissipate the low levels of the oxidative stress molecules such as superoxides and peroxides that form, so little membrane damage occurs. These scavenging enzymes and antioxidants include superoxide dismutase (SOD), ascorbate peroxidase, and

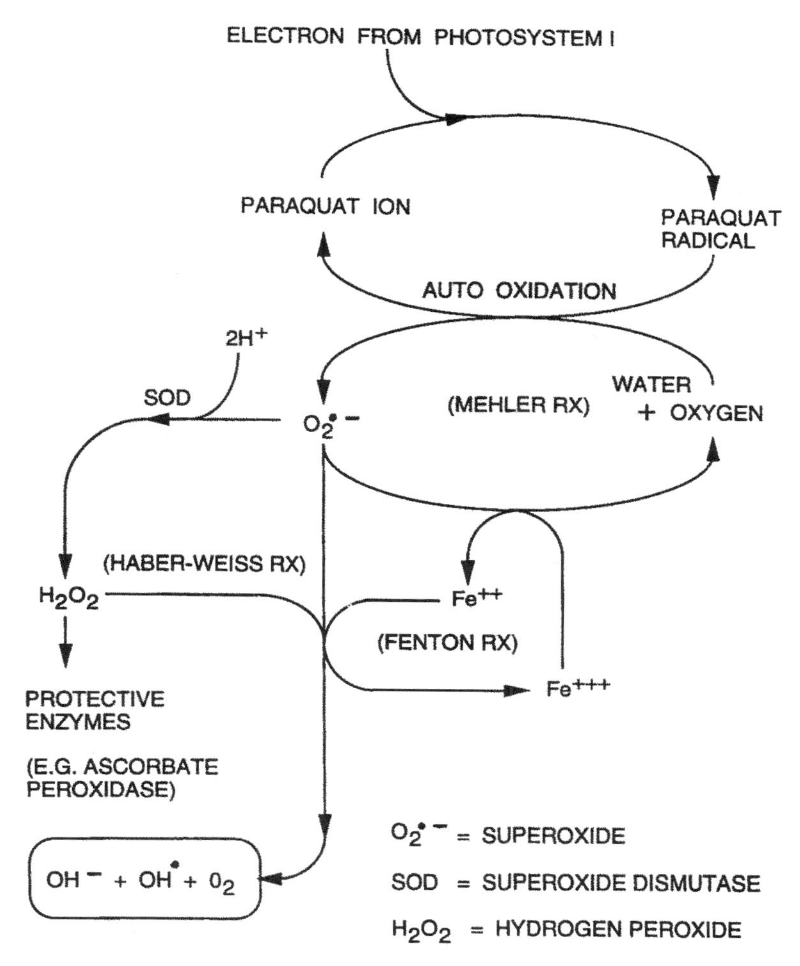

Figure 11-4. Formation of free radicals in plants after treatment with paraquat or diquat. The paraquat or diquat free radicals (monocations) formed (as shown in Figure 11-3) are not the agents causing the tissue damage and plant death. These free radicals are unstable and rapidly undergo "auto-oxidation" back to the parent compound, where they can accept another electron from PS I and form another herbicide free radical. During auto-oxidation the herbicide free radical reacts with water and oxygen, forming superoxide and the parent herbicide. Superoxide then reacts with the enzyme superoxide dismutase to form hydrogen peroxide, which in turn forms hydroxy radicals. Hydroxy radicals, the most potent biological oxidants known, quickly and effectively initiate membrane disruption through lipid peroxidation (as discussed in Chapter 9). Under normal conditions when no herbicide is present, there are cellular mechanisms that dissipate the low levels of oxidative stress molecules such as superoxides and peroxides that form, so little membrane damage occurs. These scavenging enzymes and antioxidants include superoxide dismutase (SOD), ascorbate peroxidase, and glutathione reductase, which act to detoxify active biological oxidants. However, the large amounts of free radicals formed in plants treated with paraquat or diquat completely overload the plant protective systems, and membrane disruption careens out of control.

glutathione reductase, which act to detoxify active biological oxidants. However, the large amount of free radicals formed in plants treated with paraquat or diquat completely overloads the plants' protective systems, and membrane disruption careens out of control.

Paraquat has a mechanism of action in mammalian systems that can lead to severe tissue damage and death. Paraquat is reduced from a di-cation to a mono-cation by cytochrome *c* reductase from mitochondrial respiration. As in plants, the mono-cation reacts with oxygen to form superoxide. This active oxygen species is a precursor for hydroxy radical formation. As in plants, hydroxy radicals initiate lipid peroxidation, which causes cell membrane damage and cell death in tissue coming in contact with paraquat.

An interesting approach to improve paraquat activity is to treat weeds with low rates of photosynthesis inhibitors plus paraquat. Such a treatment will delay the appearance of the phytotoxic symptoms associated with paraquat and allows an applicator to achieve complete kill on difficult-to-control perennial and biennial weeds. The effectiveness of such a treatment is probably due to a slowing of the rapid cell membrane damage (and cell death) caused by paraquat and allowing some translocation of paraquat to occur. The slowed membrane damage is the result of reduced electron flow into PS I, because the photosynthesis-inhibiting herbicide is blocking some electron flow in PS II. This reduces the rate at which paraquat captures electrons from PS I, thus reducing the levels of superoxide formed during auto-oxidation of the paraquat radical and slowing plant death. There is a commercially available product containing paraquat and diuron.

Weed Resistance

There are at least 25 species of weeds that have been reported to have resistance to paraquat. The first reported resistant weed was annual bluegrass, but now resistance is also known in several *Conyza* spp., *Hordeum glaucum* (wall barley), and *Solanum* spp. (black nightshade). Resistance has appeared in populations of weeds that have been repeatedly treated with paraquat within a season and over seasons. For example, *Conyza bonariensis* growing in citrus and vine plantations in Egypt developed resistance when paraquat was applied eight times annually for 9 years in succession (Turcsányi, 1994).

The exact mechanism of resistance in these weeds is not known, although resistance at the site of action in PS I is not responsible. Five possible mechanisms of resistance have been suggested (Feurst and Vaughn, 1990), which include (1) restricted cuticular penetration, (2) inactivation of the herbicide by metabolic processes, (3) an altered site of action, (4) enzymatic detoxification of the active oxygen species generated from the herbicide free radical by SOD, ascorbate peroxidase, and glutathione reductase (as described earlier), and (5) a sequestration of the herbicide in the cell wall, cytoplasm, or vacuole, which prevents the herbicide from reaching its action site in the cell.

There is no evidence to support reduced cuticular penetration, herbicide metabolism, or an altered site of action within the plant as a mechanism of resistance.

The best research evidence supports sequestration of the paraquat molecule in the cell wall away from its site of action as the primary mechanism for resistance (Norman et al., 1994). Enhanced detoxification of the active oxygen species formed after paraquat treatment (Shaaltiel and Gressel, 1986), would provide only short-term survival.

PROTOX INHIBITORS

The PROTOX inhibitor group includes the *diphenylethers*: acifluorfen (Blazer), fomesafen (Reflex), lactofen (Cobra), oxyfluorfen (Goal); an *oxidiazole*: oxadiazon (Ronstar); *N-phenylheterocycles*: flumiclorac (Resource), carfentrazone (Aim), and sulfentrazone (Authority) and the *N-phenylphtalimide*: flumioxazin(Valor).

Characteristics (unless as noted for a particular herbicide)

1. PROTOX inhibitors can enter the roots, stems, or leaves of young plants. Most of these compounds are applied to the foliage; however, sulfentrazone is applied preplant incorporated or preemergence; oxyfluorfen, and flumioxazin can be applied preemergence.
2. There is usually little or no translocation in the plant, although when translocation occurs it is in the xylem (sulfentrazone).
3. Light is required for activity.
4. Plant parts exposed to the herbicide and light become chlorotic, then desiccated and necrotic, and die rapidly (within a day or two).
5. Youngest expanded leaves of tolerant crops (such as soybeans) may also show chlorosis and necrosis (foliar bronzing) but recover with no reduction in yield.
6. All compounds in this group inhibit the enzyme protoporphyrinogen oxidase, which results in membrane disruption.
7. These herbicides are generally strongly adsorbed by soil organic matter and are highly resistant to leaching.
8. Because of the preceding characteristics, when these herbicides are applied preemergence, the action takes place near the soil surface during seedling emergence.
9. Soil incorporation greatly reduces activity.
10. There have been no reports of the development of weed resistance resulting from repeated application of herbicides in this group.
11. Residual life in the soil varies considerably within this group, but little injury has been reported on rotational crops, except on occasion with late-season applications of fomesafen on soybeans causing injury the following spring to corn planted on the site.

Specific characteristics and labeled uses for PROTOX inhibiting herbicides are provided in the following paragraphs.

Diphenylethers

Acifluorfen

Acifluorfen (5-[2-chloro-4-(trifluoromethyl)phenoxy]-2-nitrobenzoic acid) is a light tan to brown solid, whereas the Na salt is a light yellow solid with a vapor pressure of $< 7.6 \times 10^{-8}$ mm Hg at 25°C, a water solubility of 120 mg/L (ppm) at 23 to 25°C for the acid and 250,000 mg/L (ppm) for the sodium salt, a soil half-life of 14 to 60 days, and an oral LD_{50} (rat) of 4790 mg/kg (acifluorfen Na). Acifluorfen is subject to photodegradation and is readily degraded to nonphytotoxic products.

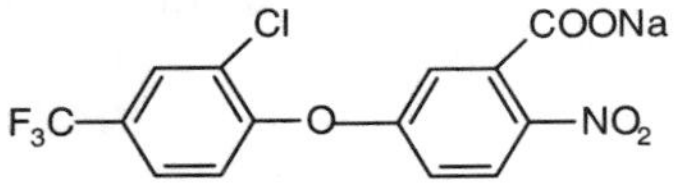

Cl COONa
F_3C O NO_2

Aciflourfen - Sodium

Uses Acifluorfen is formulated as the sodium salt as an SL and sold under the trade name Blazer for control of broadleaf weeds and some grasses in soybeans, peanuts, and rice. Acifluorfen is mainly applied to plant foliage and requires the addition of an adjuvant such as nonionic surfactant (NIS) or crop oil concentrate (COC) or urea ammonium nitrate (UAN) (in soybeans) to achieve consistent weed control. Acifluorfen can be tank mixed with several other herbicides (see label) and is often mixed with bentazon and/or sethoxydim in commercial formulations. Soil characteristics have little influence on its herbicidal effectiveness; however, it is strongly adsorbed on soil and not subject to leaching. There are 18-month replant restrictions for certain root crops such as carrots, turnips, and sweet potatoes.

Fomesafen

Fomesafen (5-[2-chloro-4-(trifluoromethyl)phenoxy]-*N*-(methylsufonyl)-2-nitrobenzamide) is a white crystalline solid with a vapor pressure of 1×10^{-7} mm Hg at 50°C, a water solubility of 50 mg/l (ppm) at 25°C as the acid and 600,000 mg/l (ppm) as the Na salt, a soil half-life of 100 days, and an oral LD_{50} (rat) of 8160 mg/kg.

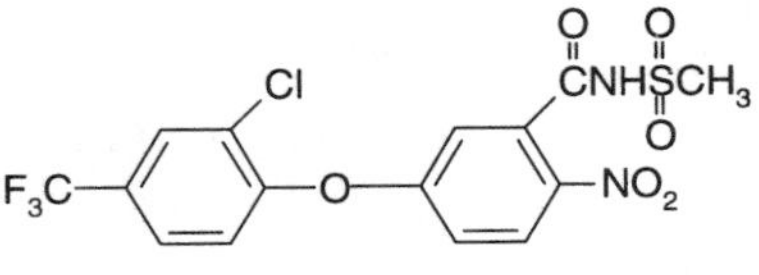

O O
Cl CNHSCH$_3$
O
F_3C O NO_2

Fomesafen

Uses Fomesafen is formulated as the sodium salt and sold under the trade name Reflex, primarily as a selective postemergence herbicide for broadleaf weed control in soybeans. Weed control is best when weeds are young, actively growing, and not under stress. Certain weeds may be controlled by soil residual activity if rainfall occurs soon after application. Fomesafen may be applied alone or tank mixed with a variety of herbicides (see label) and must be applied with an adjuvant such as COC, NIS, methylated seed oil (MSO), or others as specified on the label. Limitations include geographic location, the rate used to control specific weeds at various stages of growth, and replanting restrictions (see label).

Lactofen

Lactofen ((±)-2-ethoxy-1-methyl-2-oxoethyl 5-[2-chloro-4-(trifluoromethyl)phenoxy]- 2 nitrobenzoate) is a dark brown to tan material with a vapor pressure of 4×10^{-9} mm Hg at 20°C, an extremely low water solubility of <1 mg/l (ppm) at 22°C, and an oral LD_{50} (rat) of 5.0 mg/kg.

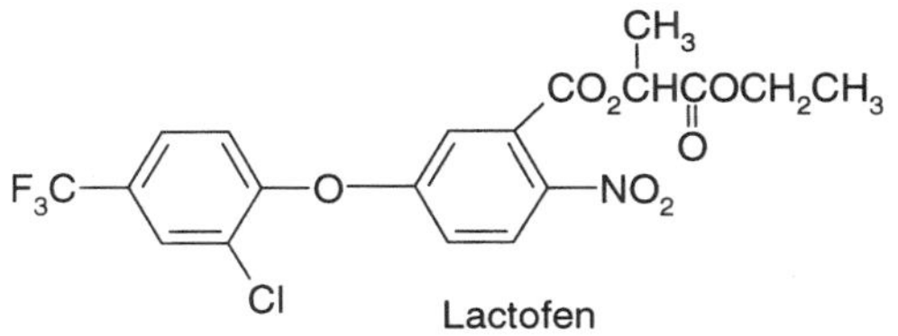

Lactofen

Uses Lactofen is formulated as an emulsifiable concentrate sold under the trade name Cobra and is used as a selective postemergence herbicide to control many broadleaf weeds in soybeans, cotton (directed spray when cotton is at least 6 inches tall, with restrictions), and Southern pine seedbeds including Eastern white pine, loblolly, sand pine, shortleaf pine, slash pine, and Virginia pine after stand emergence. Adjuvants and additives such as COC, NIS, or ammonium sulfate fertilizer (AMS) can be added to enhance activity, and numerous tank mixture combinations are registered (see label). There are limitations in regard to grazing and use of forage for animals (see label).

Oxyfluorfen

Oxyfluorfen (2-chloro-1-(3-ethoxy-4-nitrophenoxy)-4-(trifluoromethyl)benzene) is a dark red-brown to yellow semisolid with a vapor pressure of 2×10^{-6} mm Hg at 25°C, a very low water solubility of 0.1 mg/l (ppm) at 25°C, and an oral LD_{50} (rat) of >500 mg/kg. Oxyfluorfen is readily adsorbed on soil, not readily desorbed, and shows negligible leaching. It undergoes detoxification in soils, with a half-life of about 30 to 40 days under normal field conditions, but microbial degradation does not appear to be a major factor.

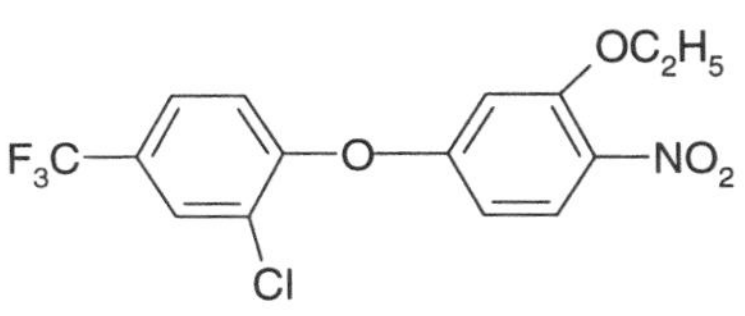

Oxyfluorfen
(Goal)

Uses Oxyfluorfen is formulated as an emulsifiable concentrate, sold under the trade name Goal, which has utility as a postemergence or a preemergence treatment for control of broadleaf weeds in many crops. When applied to the soil, activity is due to contact of the herbicide with the emerging shoot. Oxyfluorfen can be applied alone or in combination with other herbicides and is sold in premixes for use to control weeds in various ornamentals (see label). Oxyfluorfen has registrations in the following crops (check label for specific application instructions):

Artichokes—post-directed
Tree fruit/nut/vines—when crop is dormant
Broccoli, cabbage, cauliflower—pre to transplants
Cocoa—pre- and post
Coffee—pre- and post
Conifers—seedbeds, transplants, and container stock—pre- and post
Citrus—nonbearing, Cotton—post-directed
Cottonwood—pre- and post
Eucalyptus—pre- and post
Fallow beds—ground and aerial
Guava—pre- and post
Horseradish—pre
Jojoba—pre- and post
Mint—dormant
Onions—post
Papaya—pre
Taro—pre
Numerous ornamentals—pre- and post

Oxadiazon

Oxadiazon (3-[2,4-dichloro-5-(1-methylethoxy)phenyl]-5-(1,1-dimethylethyl)-1,3,4-oxadiazol-2-(*3H*)-one) is a white crystalline solid with a vapor pressure of 7.76×10^{-7} mm Hg at 25°C, a low water solubility of about 0.7 mg/l (ppm) at 20°C, and an oral acute LD_{50} (rat) of 5000 mg/kg. Oxadiazon is strongly adsorbed on soil colloids and humus, and very little leaching occurs. Its soil half-life is 60 days. Young seedlings absorb Oxadiazon as they emerge through the soil, and there appears to be limited movement to growing points.

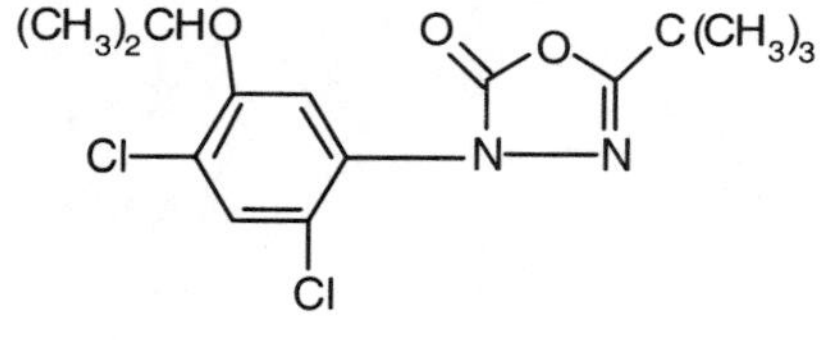

$(CH_3)_2CHO$ O O $C(CH_3)_3$
Cl N N
Cl

Oxadiazon

Uses Oxadiazon is formulated as a water-soluble packet (WSP) (50%) or granule (2%) and sold under the trade name Ronstar for use as a preemergence herbicide for control of certain annual grasses and broadleaf weeds in turf and ornamentals. In lawns and turf, the WSP can be applied to established bermudagrass, St. Augustinegrass, zoysia grass, selected ornamental shrubs, vines, and trees, and conifer nurseries. The granular formulation can be applied to established bermudagrass, perennial bluegrass, perennial ryegrass, St. Augustinegrass, seashore paspalum, buffalo grass, tall fescue, bentgrass and zoysia turf and to bermudagrass, zoysiagrass, seashore paspalum, and tropic lalo during establishment from sprigs. In ornamental nurseries, oxadiazon granular is used in newly transplanted and established ornamental shrubs, vines, and trees, in nursery containers, and in selected forestry nursery species. Check label for rates, timing, and sensitive species.

N-phenylheterocycles

Flumiclorac

Flumiclorac ([2-chloro-4-fluoro-5-(1,3,4,5,6,7-hexahydro-1,3-dioxo-2*H*-isoindol-2-yl)-phenoxy]acetic acid) is a beige powdered solid with a vapor pressure of 10^{-7} mm Hg at 22°C, a water solubility of 0.189 mg/l (ppm) at 25°C, and an oral LD_{50} (rat) of >5000 mg/kg. Flumiclorac is strongly adsorbed to both clay and organic matter, with a short soil half-life of <1 to 6 days, and it does not leach. Flumiclorac is applied to plant foliage and is readily absorbed into leaves with little or no translocation occurring. Soybeans and corn readily degrade flumiclorac.

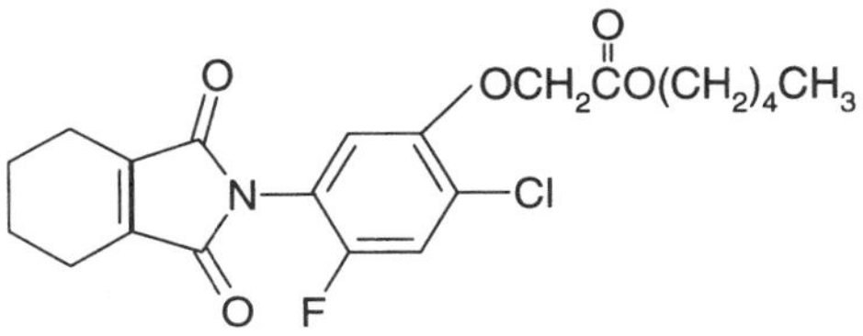

Flumiclorac

Uses Flumiclorac is formulated as an EC and sold under the trade name Resource for postemergence control of certain broadleaf weeds such as velvetleaf, lambsquarters, common ragweed, pigweed species, and spotted spurge in soybeans and corn (field). Flumiclorac can be applied with certain adjuvants (COC, NIS, MSO) or additives (AMS) for enhanced activity and tank mixed with a variety of herbicides for broader-spectrum control (see label). Flumiclorac has no soil residual activity and has no replant restrictions when applied alone.

Carfentrazone

Carfentrazone (α,2-dichloro-5-[4-(difluoromethyl)-4,5-dihydro-3-methyl-5-oxo-1*H*-1,2,4-triazol-1-yl]-4-fluorobenzenepropanoic acid, ethyl ester) is a viscous yellow liquid with a vapor pressure of 1.2×10^{-7} mm Hg at 25°C, a water solubility of 12 μg/ml (ppm) at 20°C, a soil half-life (in the laboratory) of <10 days, making it nonpersistent in the soil, and an oral LD_{50} (rat) of 5143 mg/kg. Carfentrazone is readily absorbed by foliage, and although translocation is minimal, some symplastic movement can occur.

Uses Carfentrazone is formulated as an EC and sold as Aim and Affinity for postemergence broadleaf weed control in corn, wheat, and rice. It has excellent activity on velvetleaf, kochia, and a wide range of winter annual weeds.

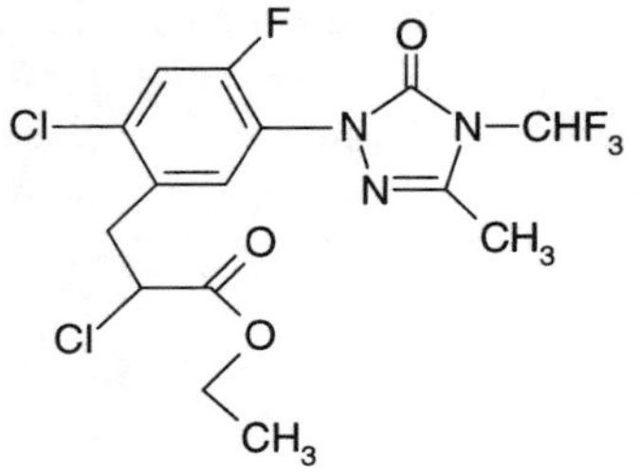

Carfentrazone

Sulfentrazone

Sulfentrazone (*N*-[2,4-dichloro-5-[4-(difluoromethyl)-4,5-dihydro-3-methyl-5-oxo-1*H*-1,2,4- triazol-1-yl]phenyl]methanesulfonamide) is a tan solid with a vapor pressure of 1 $\times 10^{-6}$ mm Hg at 25°C, a water solubility of 0.11 mg/g (ppm) at 25°C, making it moderately mobile in soil, a soil half-life of 110 to 280 days, making it relatively persistent, and an LD_{50} (rat) of 2689 mg/kg. Sulfentrazone is taken up by roots and foliage of treated plants. Soil applications result in root uptake, with some limited translocation assumed to occur inasmuch as it has activity against perennial nutsedge.

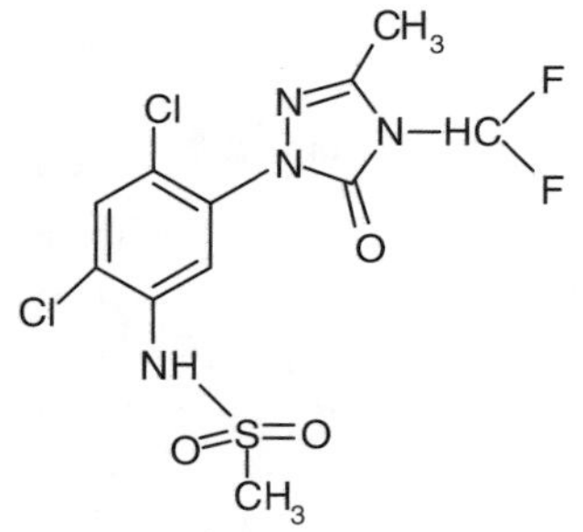

Sulfentrazone

Uses Sulfentrazone is formulated as a 75% DF and sold under the trade names Authority and Spartan or in various premixes. Sulfentrazone can be applied preemergence or preplant-incorporated for control of certain annual broadleaf and grass weeds and sedges in soybeans, sugarcane, peas, and tobacco. Replant restrictions vary from 3 months for many grain crops up to 30 months for canola and sugarbeet (check label for specifics).

Flumioxazin (2-[7-fluoro-3,4-dihydro-3-oxo-4-(2-propynyl)-2*H*-1,4-benzoxazin-6-yl]-4,5,6,7-tetrahydro-1*H*-isoindole-1,3(2*H*)-dione, is a *N*-phenylphthalimide herbicide. Flumioxazin is a tan granule with a vapor pressure of 2.41×10^{-6} mm Hg at 22°C, a water solubility of 1.78 mg/l (ppm) at 25°C, a soil half-life of 12–17 days and an oral LD_{50} (rat) > 2250 mg/kg. Flumioxazin is soil applied and will cause emerging plants to turn necrotic and die shortly after exposure to sunlight.

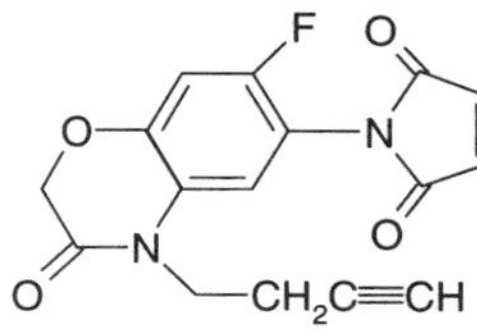

Flumioxazin

Uses Flumioxazin is formulated as a 51% water dispersible granule and sold as Valor for preemergence and as a component of burndown treatments for control of a wide variety of broadleaf weeds in soybeans and peanuts.

Selectivity

Compounds in this group are most active on broadleaf weeds and are generally applied to the leaf foliage. The exceptions are oxyfluorfen, sulfentrazone, azafenidin, and flumioxazin, which can be applied preemergence. Sulfentrazone also has activity against sedges. When in contact with nondormant crop foliage, selectivity of herbicides in this group is always somewhat marginal. In the case of acifluorfen, lactofen, and fomesafen on soybeans, the crop metabolizes the herbicides, but because of the rapid killing action there is always some foliage injury (see Figure 11-1). Acifluorfen and fomesafen are formulated as the sodium salts, which penetrate leaves more slowly than the esters that cause much more crop injury. Another factor in soybean tolerance is this crop's excellent capacity to recover from early-season damage without yield loss. Onion tolerance to oxyfluorfen increases with age and is probably partly due to less wetting and leaf penetration as compared with many susceptible weeds. Resistance to wetting and penetration is also probably a factor in the tolerance of conifer seedlings to oxyfluorfen. In many crops, contact with foliage is avoided by directed sprays (oxyfluorfen under dormant fruit and nut trees and grapes) or granular applications (oxadiazon on woody ornamentals and turf). Thus, differential wetting and penetration of foliage, metabolism, ability to recover from injury, and selective placement are important factors in selectivity for herbicides in this group.

Metabolism

Several metabolites have been reported to be formed in plants following application of one or more of the diphenylether herbicides. Reduction of the $-NO_2$ group to $-NH_2$ and cleavage of the ether linkage are frequently reported. Subsequently, glucosides and other conjugates may be formed. Acifluorfen is rapidly metabolized in plants, with an initial conjugation with glutathion, which subsequently forms the N-malonylcystein conjugate.

Mechanism of Action of PROTOX Inhibitors

The activity of diphenylethers (DPE), oxadiazon, N-phenylheterocycles and flumioxazin is expressed as foliage necrosis after 4 to 6 hours of sunlight following herbicide application. The first symptoms are a water-soaked appearance (dark green spots on the foliage), followed by necrosis of the water-soaked areas. Water soaking is indicative of membrane damage and cellular substances leaking into intercellular spaces, which changes the refractive index of the tissue. With preemergence applications, tissue is damaged by contact with the herbicide as the plant emerges above the soil surface. As with postemergence applications, the damage is tissue necrosis. Following absorption and movement to the site of action, light is an obligate requirement for herbicidal activity.

Researchers agree that the membrane damage observed after treatment with these herbicides is the result of lipid peroxidation of polyunsaturated fatty acids in cell membranes, but the actual lipid peroxidation-initiating factor produced after treatment with these herbicides has been elusive. The initial work was done with DPE herbicides; however, all herbicides in this class, including oxadiazon and the N-phenylheterocycles, act by the same mechanism. First it was shown that treating plant chloroplasts with DPE herbicides caused a large amount of singlet oxygen to be formed (Haworth and Hess, 1988). As discussed in Chapter 9, singlet oxygen ($^1O_2$) is known to be an efficient initiating factor of lipid peroxidation. Data in 1988 publications by Lydon and Duke; Matringe et al. (1989a); and Witkowski and Halling (1989) suggested that an accumulation of the tetrapyrrole, protoporphyrin IX, was central to the mechanism of action of these herbicides (see Figure 11-5). Protoporphyrin IX is a precursor in the chlorophyll biosynthesis pathway and was shown to accumulate to high concentrations in tissue treated with diphenylether herbicides (Lehnen et al., 1990). It is known that oxygen and light can interact with protoporphyrin IX to produce singlet oxygen, which then initiates lipid peroxidation and membrane damage, as discussed in Chapter 9.

In 1989 two groups (Matringe et al., Witkowski and Halling, 1989) reported the inhibited enzyme to be protoporphyrinogen oxidase (abbreviated PROTOX). PROTOX is a protein located in the chloroplast, where it is involved in chlorophyll and heme synthesis (see Figure 11-5), and in mitochondria, where it is involved in nonplastidic heme synthesis. This enzyme converts protoporphyrinogen IX to protoporphyrin IX.

Considering that PROTOX is the enzyme prior to protoporphyrin IX, how can its inhibition by these herbicides lead to an accumulation of protoporphyrin IX? This was perplexing for several years, but it was found that as protoporphyrinogen IX molecules accumulate in the chloroplast after herbicide treatment, they diffuse from the chloroplast into the cytoplasm. Once in the cytoplasm, they are oxidized to protoporphyrin IX and readily interact with oxygen and light forming singlet oxygen, which initiates lipid peroxidation and cell membrane disruption (as described in Chapter 9), which rapidly leads to plant death.

The very rapid accumulation of protoporphyrin IX in herbicide-treated plants is due to a deregulation of the entire pathway resulting from a decrease in heme levels.

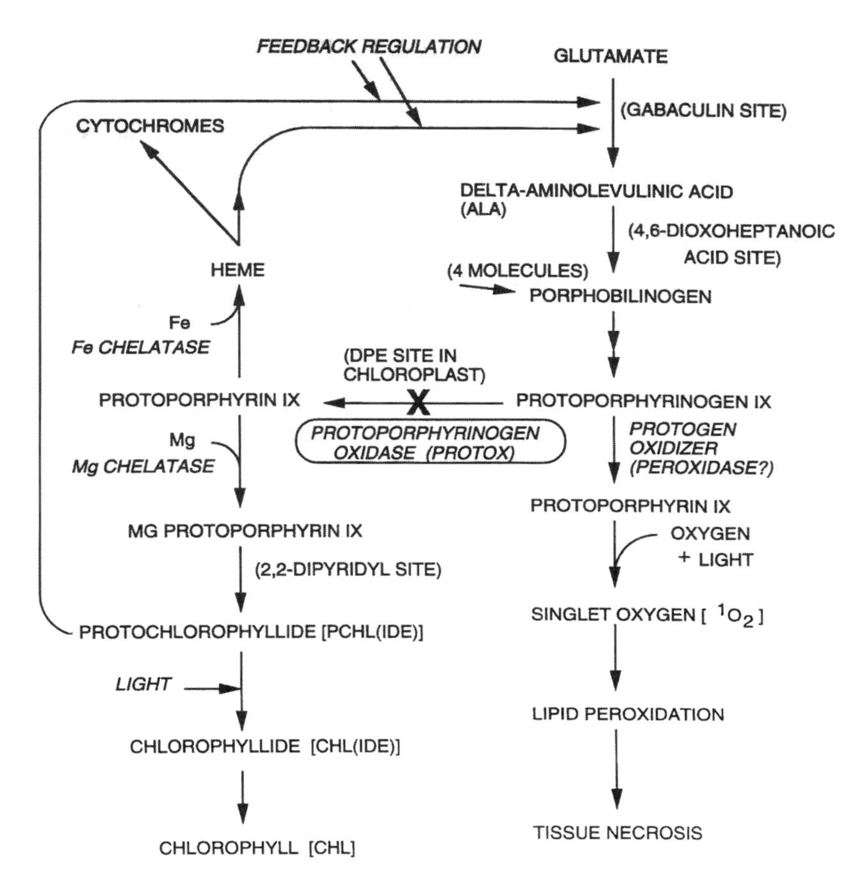

Figure 11-5. Mechanism of action of PROTOX inhibitors. Protoporphyrinogen oxidase (PROTOX) is a protein located in the chloroplast, where it is involved in chlorophyll and heme synthesis, and in mitochondria, where it is involved in nonplastidic heme synthesis. This enzyme converts protoporphyrinogen IX to protoporphyrin IX, which is involved in the biosynthesis of chlorophyll. When PROTOX is inhibited (X) by a herbicide, protoporphyrin IX cannot be formed and protoporphyrinogen IX accumulates. The protoporphyrinogen IX that accumulates then leaks out of the chloroplast into the cytoplasm. Once in the cytoplasm, it is oxidized to protoporphyrin IX. Protoporphyrin IX then reacts with oxygen and light and forms singlet oxygen. The singlet oxygen rapidly reacts with lipids in cell membranes, causing lipid peroxidation and membrane disruption, and this leads to plant death.

In the absence of a herbicide, the heme levels would keep the entire pathway in a balanced production of products. However, in the presence of a herbicide heme levels decrease and are a signal for increased pathway activity and result in high levels of protoporphyrinogen IX formation. This protoporphyrinogen IX then leaks into the cytoplasm where it is oxidized to protoporphyrin IX, which forms free radicals that lead to membrane disruption.

Until this research on the DPE mechanism of action was done, little was known about the mode of action of oxadiazon other than there is a rapid loss of membrane integrity (lipid peroxidation) that is light dependent after herbicide treatment. In publications by Duke et al. (1989) and Matringe et al. (1989b) it was shown that the mode of action of oxadiazon is identical to that of the DPEs. In addition to the diphenylethers and oxadiazon, the N-phenyl heterocycles and flumioxazin are also known to inhibit PROTOX and result in plant membrane damage.

INHIBITORS OF GLUTAMINE SYNTHASE

Glufosinate

Glufosinate (2-amino-4-(hydroxymethylphosphinyl)butanoic acid) is a white to light yellow crystalline powder with a solubility of 1,370,000 mg/l (ppm), 0 mm Hg vapor pressure at 25°C, and a soil half-life of 7 days. It is foliar applied and has no soil activity.

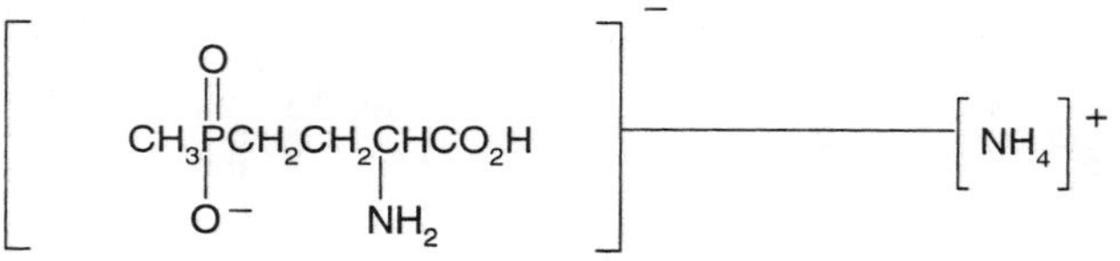

$$\left[CH_3\overset{O}{\underset{O^-}{P}}CH_2CH_2\underset{NH_2}{C}HCO_2H \right]^- \text{———} \left[NH_4 \right]^+$$

Glufosinate

Uses Glufosinate is formulated as an SL and sold under the trade name Finale for use in noncropland, Rely for use in apples, grapes, and tree nuts, and Liberty for use in soybeans and corn resistant to glufosinate. It is nonselective and active only when applied postemergence, with best activity on annuals and control of the above ground foliage of perennials. Xylem and phloem transport are poor, thus thorough spray coverage is required for complete kill of targeted weeds. The symptoms include chlorosis, which is then followed by necrosis, usually beginning to develop within 3 to 5 days after treatment. The symptoms are somewhat like those of membrane-disrupting herbicides (e.g., diphenylethers and paraquat); however, the speed of membrane disruption (necrosis) is slower than in other herbicides having a direct membrane disruption mode of action (e.g., paraquat). If treated plants are placed in the dark immediately after treatment, symptoms develop but at a greatly reduced rate.

Although glufosinate is the preferred name of the synthesized herbicide, a common name often used in the literature for the chemical structure is phosphinothricin (PPT). Glufosinate is a mixture of the D- and L- isomers of PPT, but only the L-isomer is active. PPT is also the active ingredient in the biological herbicide bialaphos. Bialaphos, a tripeptide, is not herbicidal; however, it is rapidly hydrolyzed to PPT in plants.

Mechanism of Action Glufosinate is known to directly inhibit the glutamine synthetase (GS) enzyme in the nitrogen assimilation pathway of plants. GS is the enzyme responsible for converting glutamate plus ammonia to glutamine (Figure 11-6). Glufosinate competes with glutamate for binding to the GS enzyme, and once bound to GS, glufosinate effectively blocks the GS enzyme from forming glutamine. Glufosinate inhibition of GS results in decreased levels of glutamine, which in turn results in decreased levels of several other important plant amino acids (e.g., glutamate, aspartate, asparagine, alanine, and serine), whose ultimate synthesis is dependent on the presence of glutamine.

After glufosinate application, ammonia levels in leaves, which are usually very low, increase dramatically and within 4 hours of treatment are about ten times higher than in nontreated leaves. After 1 day, ammonia levels can be as much as 100 times higher in glufosinate-treated tissue than in nontreated tissue. This accumulation of

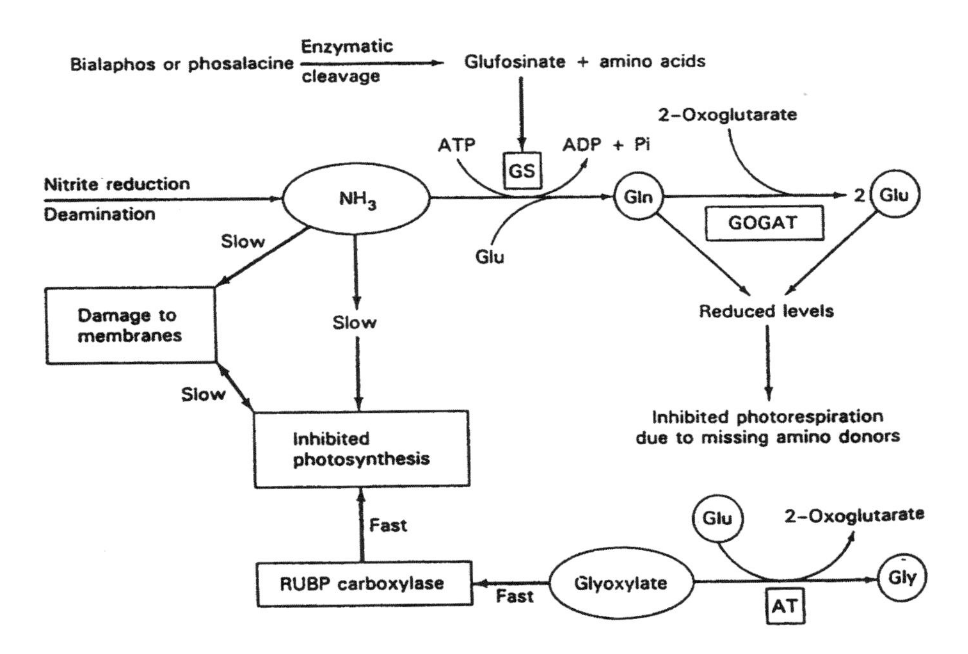

Figure 11-6. Glufosinate mechanism of action. Glufosinate is known to directly inhibit the glutamine synthetase (GS) enzyme in the nitrogen assimilation pathway of plants. GS is the enzyme responsible for converting glutamate (glu) plus ammonia (NH_3) to glutamine (Gln) resulting in decreased levels of Gln, which in turn results in decreased levels of several other important plant amino acids (e.g., glutamate (Glu), aspartate, asparagine, alanine, and serine), whose ultimate synthesis is dependent on the presence of gln. Glufosinate disrupts many important nitrogen metabolism (nitrogen assimilation) synthetic reactions in plants by inhibiting Gln and Glu formation. As a result, electron flow in photosynthesis is indirectly inhibited through the decrease in amino donors (from glu) for glyoxylate. Glyoxylate accumulates to a level that reduces carbon fixation in the Calvin cycle, which leads to an inhibition of the light reaction in photosynthesis. In the presence of light, inhibition of electron flow in photosynthesis causes induction of lipid peroxidation (membrane damage) from a buildup of triplet chlorophyll (see lipid peroxidation on page 219, Chapter 9).

ammonia in glufosinate-treated plants is known to be due to the direct inhibition of the GS enzyme by glufosinate. Although early research suggested that ammonia accumulation resulted in plant death, recent data show that ammonia is not directly responsible for the toxic effects of glufosinate.

So, What Kills the Plant? The sequence of events leading to plant death after glufosinate treatment can be summarized as follows: Glufosinate disrupts many important nitrogen metabolism (nitrogen assimilation) synthetic reactions in plants by inhibiting glutamine and glutamate formation. The loss of both of these important substrates is due to inhibition of GS. Electron flow in photosynthesis is indirectly inhibited through the decrease in amino donors (from glutamate) for glyoxylate. Glyoxylate accumulates to a level that reduces carbon fixation in the Calvin cycle, which leads to an inhibition of the light reaction in photosynthesis (Wild and Wender, 1993). In the light, inhibition of electron flow in photosynthesis causes induction of lipid peroxidation (membrane damage) from a buildup of triplet chlorophyll (see lipid peroxidation, Chapter 9). *Ammonia accumulation is not associated with the lethal effect of glufosinate.* The glufosinate mechanism of action is a wonderful example of how the metabolic pathways of plants are interconnected and how inhibition of an enzyme in one pathway can have a cascading effect in other pathways and lead to plant death.

Crop Resistance Crop plants have been made resistant to glufosinate by genetic engineering. The resistance is due to detoxification of the glufosinate. A gene encoding phosphinothricin acetyltransferase (PAT) activity isolated from *Streptomyces hygroscopicus* (*bar* gene) and *Streptomyces viridochromogenes* (*pat* gene) was engineered into several crop species. When this enzyme is produced in plants, it converts glufosinate to a nonherbicidal acetylated form. In greenhouse trials 0.4 kg/ha killed nontransformed plants, whereas 4.0 kg/ha did not affect transformed plants. The trait is inherited as a dominant Mendelian trait. In transgenic plants with high levels of resistance, the only metabolite detected after glufosinate treatment was *N*-acetyl-glufosinate (Dröge-Laser et al., 1994), and this metabolite was the final stable product in plants.

LITERATURE CITED AND SUGGESTED READING

Di Tomaso, J. M., J. J. Hart, and L. V. Kochian. 1993. Compartmentation analysis of paraquat fluxes in maize roots as a means of estimating the rate of vacuolar accumulation and translocation to shoots. *Plant Physiol.* **102**:467–472.

Dodge, A. D. 1989. Herbicides interacting with photosystem I. *Soc. Exp. Biol. Seminar Series* **38**:37–50.

Dröge-Laser, W., U. Siemeling, A. Pühler, and L. Broer. 1994. The metabolites of the herbicide L-phosphinothricin (glufosinate). *Plant Physiol.* **105**:159–166.

Duke, S. O., J. Lydon, and R. N. Paul. 1989. Oxadiazon activity is similar to that of p-nitro-diphenyl ether herbicides. *Weed Sci.* **37**:152–160.

Fuerst, E. P., and K. C. Vaughn. 1990. Mechanisms of paraquat resistance. *Weed Technol.* **4**:150–156.

Halliwell, B. 1991. Oxygen radicals: Their formation in plant tissues and their role in herbicide damage. In *Herbicides*, ed. by N. R. Baker and M. P. Percival. pp. 87–129. *Topics in Photosynthesis*, Vol. 10. Elsevier, New York.

Harvey, B. M. R., J. Muldoon, and D. B. Harper. 1978. Mechanism of paraquat tolerance in perennial ryegrass. I. Uptake, metabolism and translocation of paraquat. *Plant Cell Envir.* **1**:203–209.

Haworth, P., and F. D. Hess. 1988. The generation of singlet oxygen ($^1O_2$) by the nitrodiphenyl ether herbicide oxyfluorfen is independent of photosynthesis. *Plant Physiol.* **86**:672–676.

Herbicide Handbook. 7th ed. 1994. Weed Science Society of America. Lawrence, KS.

Herbicide Handbook Supplement to 7th ed. 1998. Weed Science Society of America. Lawrence, KS.

Hess, F. D. 2000. Light-dependent herbicides—An overview. *Weed Sci.* **48**:160–170.

Kocher, H. 1989. Inhibitors of glutamine synthetase and their effects on plants. *Monograph, British Crop Prot. Council* **42**:173–182.

Lehnen, L. P., Jr., T. D. Sherman, J. M. Becerril, and S. O. Duke. 1990. Tissue and cellular localization of acifluorfen-induced porphyrins in cucumber cotyledons. *Pestic. Biochem. Physiol.* **37**:239–248.

Lydon, J., and S. O. Duke. 1988. Porphyrin synthesis is required for photobleaching activity of the p-nitrosubstituted diphenyl ether herbicides. *Pestic. Biochem. Physiol.* **31**:74–83.

Matringe, M., J. M. Camadro, P. Labbe, and R. Scalla. 1989a. Protoporphyrinogen oxidase as a molecular target for diphenyl ether herbicides. *Biochem. J.* **260**:231–235.

Matringe, M., J. M. Camadro, P. Labbe, and R. Scalla. 1989b. Protoporphyrinogen oxidase inhibition by three peroxidizing herbicides: Oxadiazon, Ls 82-556, and M & B 39279. *Febs Letters* **245**:35–38.

Norman, M. A., R. J. Smeda, K.C. Vaughn, and E. P. Fuerst. 1994. Differential movement of paraquat in resistant and sensitive biotypes of *conyza*. *Pestic. Biochem. Physiol.* **50**:31–42.

Shaaltiel, Y., and J. Gressel. 1986. Multienzyme oxygen radial detoxifying system correlated with paraquat resistance in *Conyza bonariensis*. *Pestic. Biochem. Physiol.* **26**:22–28.

Turcsányi, E., G. Surányi, E. Lehoczki, and G. Borbély. 1994. Superoxide dismutase activity in response to paraquat resistance in *Conyza canadensis* (L.) *Cronq*. *J. Plant Physiol.* **144**:599–606.

Wild, A., and C. Wendler. 1993. Inhibitory action of glufosinate on photosynthesis. *Z. Naturforsch.* **48c**:369–373.

Witkowski, D. A., and B. P. Halling. 1989. Inhibition of plant protophyrinogen oxidase by the herbicide acifluorfen-methyl. *Plant Physiol.* **90**:1239–1242.

WEB SITE

International Survey of Herbicide Resistant Weeds
http://www.weedscience.org

For chemical use, see manufacturer's or supplier's label and follow the directions. Also see the Preface.

12 Cell Growth Disrupters and Inhibitors

Herbicides classified as cell growth disrupters and inhibitors inhibit the root and/or shoot growth of emerging seedlings and are applied to the soil as either preemergence or preplant incorporated treatments. The herbicide groups include the mitotic disrupters, inhibitors of shoots and/or roots, inhibitors of roots, and inhibitors of shoots. They have a variety of modes and mechanisms of action, with the common characteristic that they all inhibit either the shoot or the root of the emerging seedling in some manner.

General Properties of Cell Growth Disrupter and Inhibitor Herbicides

- These herbicides inhibit the growth of roots and shoots of seedlings.
- Established annuals and perennials are killed only in a few special cases.
- There is little or no translocation of these herbicides in plants.
- There is little or no activity on the foliage of established plants.
- These herbicides are moderately to highly selective among species.
- They are moderately to highly resistant to leaching in the soil.

MITOTIC DISRUPTERS

The *dinitroaniline* class of chemicals includes some of the most important soil-applied herbicides in the world, which are widely used in agronomic, vegetable, and tree crops, as well as in turf. These compounds were discovered in the mid-1950s. Early compounds were active postemergence and preemergence and generally exhibited little selectivity. In the early 1960s, amino alkyl substitution in the synthesis of these compounds yielded selectivity directed toward grass control and usefulness as preemergence and preplant incorporated applications.

The dinitroanilines (DNAs) inhibit early seedling growth but do not prevent germination. Roots show initial injury: increased root diameter and swelling near the tip (Figure 12-1). Primary roots become stubby, and lateral root development is inhibited. Absorption of the herbicide by developing seedlings occurs through roots and shoots. Very little translocation occurs from the site of uptake.

Surface preemergence treatments may result in stem swelling, brittleness, and breakage at the soil surface in some dicot species. Such results do not occur when these herbicides are incorporated into the soil before planting.

Figure 12-1. Inhibition of lateral root formation by trifluralin, which is a typical effect of the dinitroaniline class of herbicides on susceptible species. (T. N. Jordan, Purdue University.)

These herbicides are especially effective on grass weeds and small-seeded broadleaf weeds. With deep incorporation at high doses, trifluralin controls johnsongrass from both seed and rhizomes. The DNAs are generally combined with other herbicides to broaden the spectrum of weeds controlled.

Selectivity

There are large differences in tolerance to DNAs among species. Carrot is extremely tolerant of dinitroaniline herbicides, and work by Vaughan and Vaughn (1987) showed that this resistance is at the site of action and due to differences in tubulin, as compared with that of susceptible plants. Dinitroaniline-resistant goosegrass biotypes in North and South Carolina have altered tubulin. In addition to a resistant site of action, metabolism has been implicated as a mechanism of selectivity.

Positional placement is used to obtain selectivity where the crop does not have sufficient true tolerance. In some cases the herbicide is applied to the surface after planting, as with oryzalin on several crops and pendimethalin on corn. In others it may be shallowly incorporated above a deep-planted crop, as with trifluralin on potatoes or wheat. Certain dinitroanilines are widely used to control germinating annual grasses in established turf.

Trifluralin

Trifluralin (2,6-dinitro-*N*,*N*-dipropyl-4-(trifluoromethyl)benzenamine) is an orange crystalline solid with a vapor pressure of 1.1×10^{-4} mm Hg at 25°C, an extremely low water solubility of 0.3 mg/l (ppm) at 25°C, a field soil half-life of 45 days, and an oral LD_{50} (rat) of >5000 mg/kg.

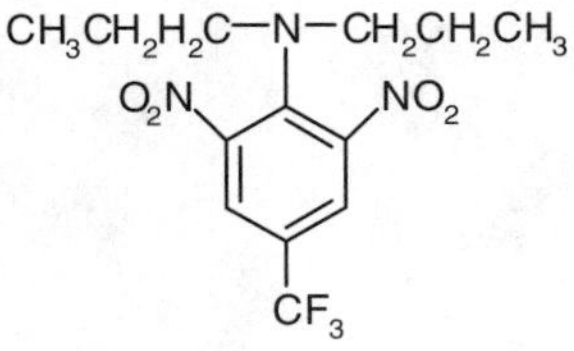

Trifluralin

Uses Trifluralin is formulated as an EC or a 5 or 10% granule and sold as Treflan, Trifluralin, Preen, or Trilin. Trifluralin was the first DNA registered for use on food crops and is widely used as a preemergence soil-incorporated herbicide in soybean, cotton, peanuts, alfalfa, canola, wheat, dry beans, oilseed rape, sunflowers, safflowers, numerous vegetable crops, including beans, carrots, cole crops, cucurbits, tomatoes, greens, okra, peas, and potatoes, sugarcane, sugar beets, small fruits, grapes, tree fruits, citrus, pecans, forestry, noncropland, and various types of nursery stock. In tomatoes, sugar beets, cantaloupes, cucumber, and watermelon it is too toxic to be applied in the direct seeded crop but can be used on transplants or established plants.

Pendimethalin

Pendimethalin (*N*-(1-ethylpropyl)-3,4-dimethyl-2,6-dinitrobenzeneamine) is a crystalline orange-yellow solid with a vapor pressure of 9.4×10^{-6} mm Hg at 25°C, a low water solubility of 0.275 mg/l (ppm) at 25°C, a soil half-life of 44 days, and an oral LD_{50} (rat) of >5000 mg/kg.

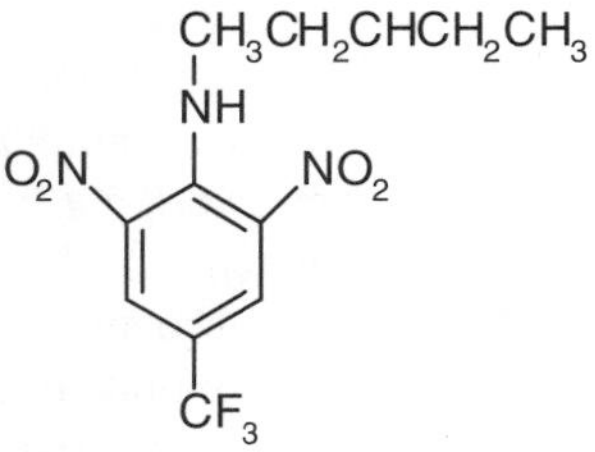

Pendimethalin

Uses Pendimethalin is formulated as an EC, WG, WP, DG or G and sold as Prowl, Pentagon, Pendulum, and other products and used as a preemergence, early postemergence, preplant soil-incorporated, or postemergence soil-incorporated treatment, depending on the crop. Pendimethalin is used in cereals, onions, leeks, garlic, fennels, corn, sorghum, rice, soybeans, peanuts, brassicas, carrots, celery, black salsify, field beans, peas, lupines, evening primrose, tulips, potatoes, cotton, hops, pome fruit, stone fruit, berry fruit, citrus, lettuce, capsicums, established turf, a variety of ornamentals, and in transplanted tomatoes, sunflowers, and tobacco.

Benefin

Benefin (*N*-butyl-*N*-ethyl-2,6-dinitro-4-(trifluoromethyl)benzeneamine) is a yellow-orange crystalline solid with a vapor pressure of 7.8×10^{-5} mm Hg at 25°C, an extremely low water solubility of 0.1 mg/l (ppm) at 25°C, a soil half-life of 30 to 60 days, and an oral LD_{50} (rat) of >5000 mg/kg.

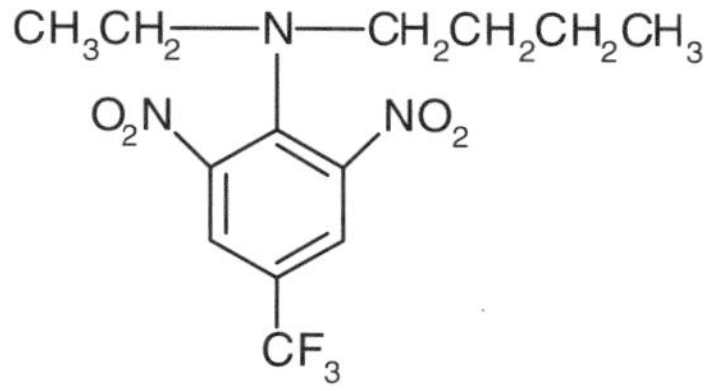

CH_3CH_2—N—$CH_2CH_2CH_2CH_3$
O_2N NO_2
CF_3

Benefin

Uses Benefin is formulated as an EC, WG, or G and sold as Balan or Benefex for use in preplant incorporated or surface (G) treatments in turf, alfalfa, clovers, peanuts, lettuce, cucumbers, chicory, endive, field beans, lentils, clovers, trefoils, and tobacco.

Ethalfluralin

Ethalfluralin (*N*-ethyl-*N*-(2-methyl-2-propenyl)-2,6-dinitro-4-(trifluoromethyl)benzeneamine) is a yellow-orange crystal with a vapor pressure of 8.2×10^{-5} mm Hg at 25°C, an extremely low water solubility of 0.3 mg/l (ppm) at 25°C, a soil half-life of 60 days, and an oral LD_{50} (rat) of >10,000 mg/kg.

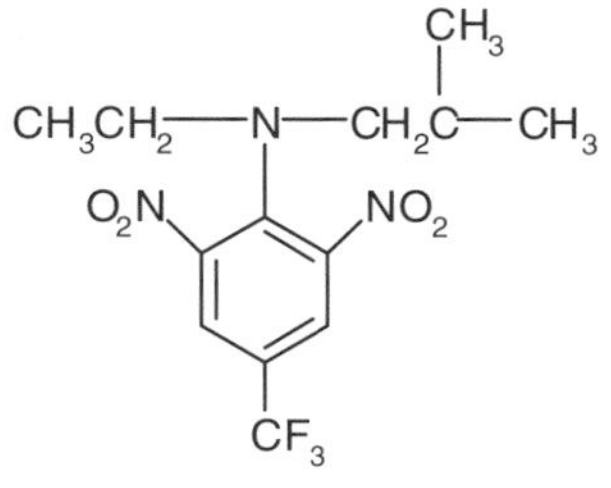

CH_3
CH_3CH_2—N—CH_2C—CH_3
O_2N NO_2
CF_3

Ethalfluralin

Uses Ethalfluralin is formulated as an EC or G and sold as Sonolan and Edge for use in cotton, soybeans, dry beans, dry peas, lentils, peanuts, safflowers, and sunflowers and as Curbit for use in direct-seeded cucurbits prior to crop emergence.

Oryzalin

Oryzalin (4-(dipropylamino)-3,5-dinitrobenzenesulfonamide) is a bright orange crystalline powder with a vapor pressure of $<10^{-8}$ mm Hg at 25°C, a low water solubility of 2.6 mg/l (ppm) at 25°C, a soil half-life of 20 to 128 days, depending on use rate, and an oral LD_{50} (rat) of >5000 mg/kg.

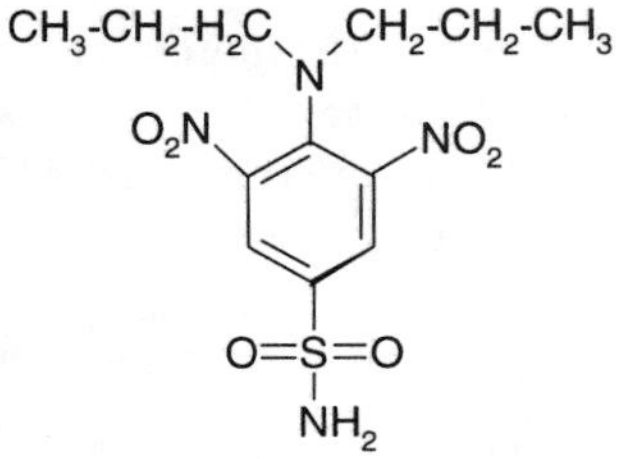

Oryzalin

Uses Oryzalin is formulated as an AS or G and sold as Surflan for use in tree fruits and nuts, grapes, various woody and herbaceous ornamentals and turf, soybeans, cotton, peanuts, oilseed rape, sunflowers, alfalfa, peas, mint, and noncrop areas. Unlike most DNAs, oryzalin is relatively nonvolatile and weakly acidic and can be applied to the soil surface, although rainfall or irrigation is required to optimize soil activity.

Prodiamine

Prodiamine (2,4-dinitro-N^3,N^3-dipropyl-6-(trifluoromethyl)-1,3-benzenediamine) is a yellow-orange powdered solid with a vapor pressure of 2.51×10^{-8} mm Hg at 25°C, a water solubility of 0.013mg/l (ppm) at 25°C, a soil half-life of 120 days, and an oral LD_{50} (rat) of >5000 mg/kg.

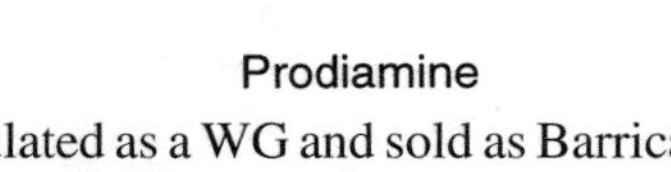

Prodiamine

Uses Prodiamine is formulated as a WG and sold as Barricade, Endurance, and Factor for preplant and preemergence use in cotton, soybeans, alfalfa, nonbearing fruit trees, vines, nuts, turf, ornamentals, and wildflowers.

Soil Influences

The dinitroanilines (DNAs) are commonly incorporated in the soil before planting because of the potential for vapor losses and photodecomposition and their very low water solubility. However, the less volatile oryzalin (Surflan) is usually applied to the soil surface after planting and pendimethalin (Prowl) is surface-applied when used on corn. In established turf, dinitroanilines are applied to the surface in granular formulations.

DNAs are strongly adsorbed to soil clay and organic matter. As a consequence, leaching is not a problem. Strong adsorption is another reason (in addition to volatility and photolysis) that mechanical incorporation of the herbicide in soil is necessary. Dissipation in soil is primarily microbial, but volatility losses also contribute. The persistence of the dinitroaniline herbicides in soil is relatively long and must be taken into consideration for the subsequent planting of sensitive crops. At usual application rates and under normal field conditions, benefin persists 4 to 5 months and the others less than 12 months or 1 growing season (Herbicide Handbook, 1994, 1998).

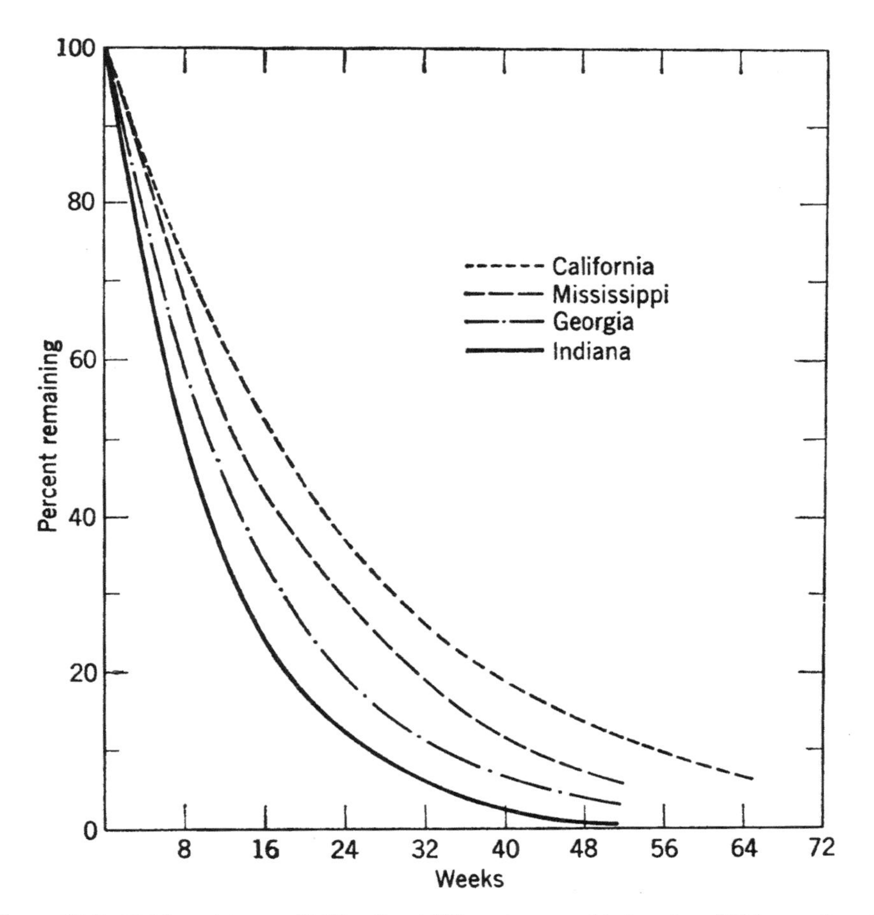

Figure 12-2. Field persistence of trifluralin at different geographic locations (different climatic conditions). In most locations, almost all of the trifluralin disappeared within 48 weeks. However, it persisted longer under dry land conditions of California. (Dow Agro Sciences.)

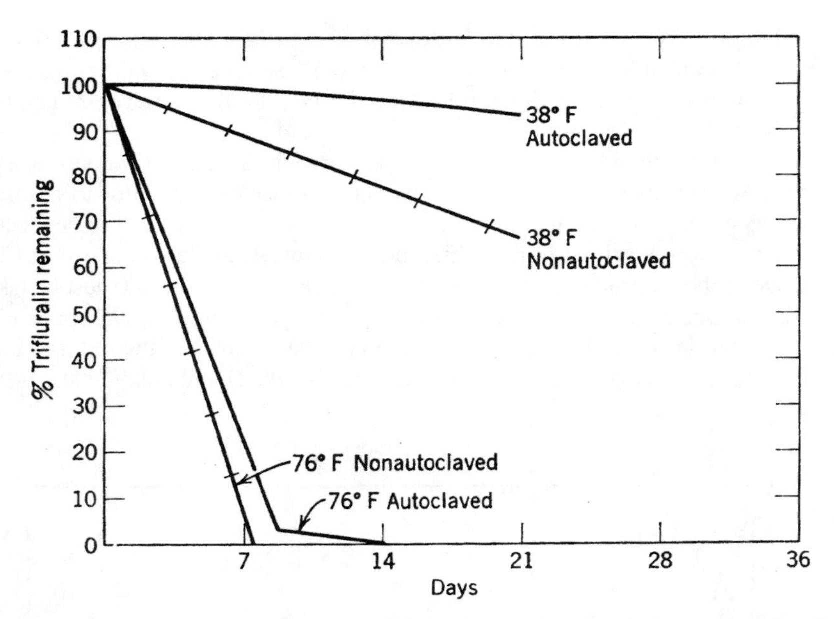

Figure 12-3. Anaerobic degradation of trifluralin in autoclaved and nonautoclaved soil as a function of temperature in soil saturated with water (200% field capacity). (Probst et al., 1967.)

Trifluralin has been shown to persist longer in an arid environment than in areas receiving more rainfall (Figure 12-2). Trifluralin has also been shown to be degraded rapidly under anaerobic conditions (Figure 12-3) (Probst et al., 1967).

Metabolism

DNAs are *slowly* degraded in higher plants. N-dealkylation and reduction of the nitro have been reported.

OTHER MITOTIC DISRUPTERS

DCPA

DCPA (dimethyl 2,3,5,6-tetrachloro-1,4-benzenedicarboxylate) is a white crystalline solid with a vapor pressure of 2.5×10^{-6} mm Hg at 25°C, an extremely low water solubility of 0.5 mg/l (ppm) at 25°C, a soil half-life of 60 to 100 days, and an oral LD_{50} (rat) of >10,000 mg/kg.

DCPA (Dacthal)

Uses and Selectivity DCPA is formulated as a WP or F and sold as Dacthal; it is applied preemergence (including post-plant after establishment) and preplant incorporated for the control of annual grasses but will kill some other weeds, including dodder. The primary use of this herbicide is in vegetable and fruit crops and turf. Selectivity is often by differential placement in the soil, but there are large differences in species tolerance even when this herbicide is in the zone of germination, suggesting differences at the site of action. DCPA stops growth of both root tips and seedling shoot growing points and can cause callus growth on stems (Figure 12-4). DCPA is slowly metabolized in plants.

Soil Influences DCPA, because of its very low water solubility (0.5 ppm), has low mobility in both the plant and the soil. It is active only in the soil. The low solubility and mobility of DCPA requires that it be present at a high concentration (4.5 to 9.0 lb/acre) in the zone of uptake. DCPA is adsorbed to organic matter in

Figure 12-4. Effect of misapplication of DCPA (application at transplanting) on stem enlargement and callus formation on muskmelon.

the soil. Degradation is via microbes, but the long half-life (100 days) results in long soil persistence, which can affect rotation crop-planting decisions.

Pronamide

Pronamide (3,5-dichloro(*N*-1,1-dimethyl-2-propynyl)benzamide) is an off-white solid with a vapor pressure of 8.5×10^{-5} mm Hg at 25°C, a low water solubility of 15 mg/l (ppm) at 25°C, a soil half-life of 60 days (but persistence can vary from 2 to 9 months, depending on soil type and climatic conditions), and an oral LD_{50} (rat) of 16,000 mg/kg.

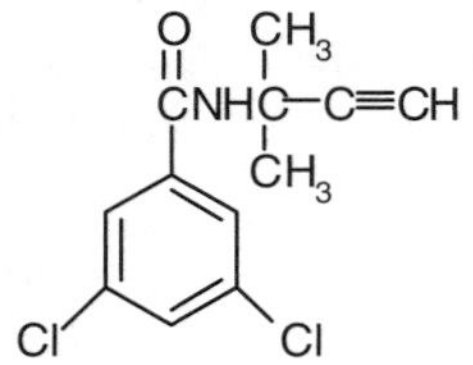

Pronamide

Uses Pronamide is formulated as a WP and sold as Kerb for soil application (preemergence and preplant incorporated) to control germinating grasses and several broadleaf weeds in small-seeded legumes, bermudagrass, fallow land, lettuce, endive, escarole, artichokes, rhubarb, brambles, grapes, blueberries, woody ornamentals, nursery stock, and Christmas trees. Pronamide is excellent for control of quackgrass and winter annuals. As with the other mitotic disrupters, pronamide inhibits early seedling growth and there are large differences between species in tolerance. Pronamide is readily absorbed by roots with some apoplastic translocation. Little metabolism occurs in plants.

Soil Influences This herbicide is used primarily in cooler climatic zones or during cool periods of the year because of its short residual life resulting from volatility losses under conditions of high soil temperature. Pronamide is readily adsorbed on organic matter and other colloidal exchange sites and therefore leaches very little in most soils. It has intermediate persistence in soil, ranging from 2 to 9 months.

Dithiopyr

Dithiopyr (*S,S*-dimethyl 2-(difluoromethyl)-4-(2-methylpropyl)-6-(trifluoromethyl)-3,5-pyridinedicarbothioate) is a colorless crystal with a vapor pressure of 4×10^{-6} mm Hg at 25°C, a low water solubility of 1.4 mg/l (ppm) at 20°C, a soil half-life of 17 days (with a range of 3 to 49 days, depending on the conditions) and an oral LD_{50} (rat) of >5000 mg/kg.

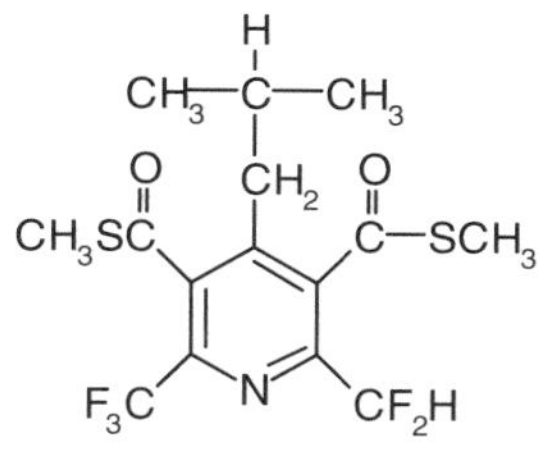

Dithiopyr

Uses Dithiopyr is formulated as an EC and a G and sold as Dimension for use as a preemergence or early postemergence treatment on turf (established lawns, ornamental turf, and putting greens) and is a relatively new herbicide in this mode of action group. It is especially effective on crabgrass. Thiazopyr (Visor) is another pyridine herbicide that is being tested for potential use in cotton, peanuts, and tree crops.

Soil Influences Dithiopyr is strongly adsorbed to soil organic matter. Like other cell growth disrupters, dithiopyr has low water solubility, exhibits little movement in soil, and is degraded by microbes.

Mechanism of Action

Many preemergence herbicides inhibit growth shortly after seed germination, thus preventing emergence of the weed above the soil surface. Even if the herbicide causes no other effect on the weed, control is achieved. Although, there may be a plentiful supply of carbohydrates and lipids in the seed for maintaining metabolic reactions prior to emergence from the soil, the survival period is finite. The growth of plant roots and shoots is a combination of cell division and cell enlargement that results in an irreversible increase in size. An inhibition or disruption of either or both of these processes inhibits growth. There are herbicides that disrupt, as well as inhibit, the growth process in plants. This section focuses on herbicides that disrupt the cell division and cell enlargement process.

Mitosis, cell division, during weed seedling growth takes place primarily in the meristems located at the root and shoot tips. Specialized cells in the meristem are in a continuous cycle of events.

Mitosis is a sequence of events that forms two cells, each containing one complement of chromosomes, from a single cell. Each stage of mitosis has been given a name (prophase, metaphase, anaphase, and telophase). During prophase the nuclear envelope breaks down and chromosomes are more or less randomly distributed around the cell. During metaphase the chromosomes become aligned in the middle of the cell (metaphase plate), during anaphase one set of chromosomes is moved to each end of the cell, and during telophase a new cell wall (cell plate) is formed between the two sets of chromosomes and the nuclear envelope reforms around each chromosome set (Figure 12-5).

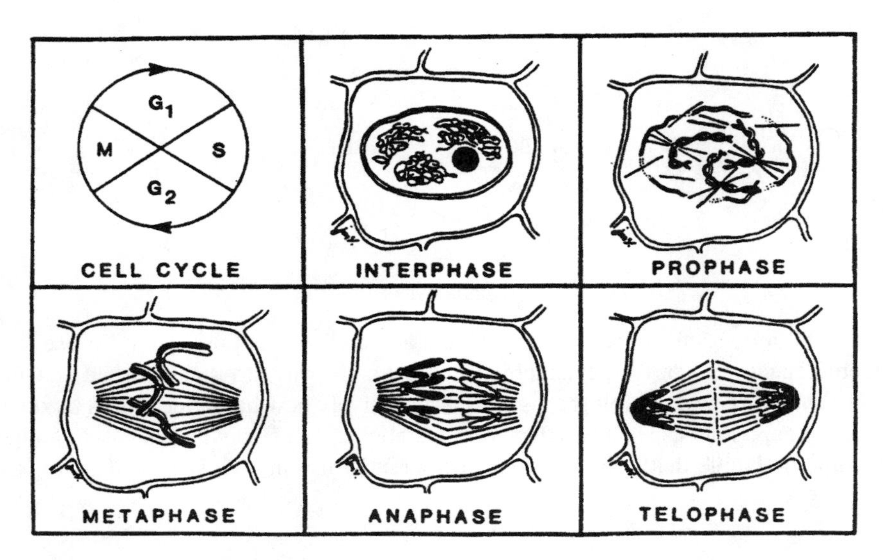

Figure 12-5. The mitotic sequence in cell division. Cell division during weed seedling growth takes place primarily in the meristems located at the root and shoot tips. Specialized cells in the meristem are in a continuous cycle of events. Mitosis is a sequence of events that forms two cells, each containing one complement of chromosomes, from a single cell. Each stage of mitosis has been given a name (prophase, metaphase, anaphase, and telophase). During prophase the nuclear envelope breaks down and chromosomes are more or less randomly distributed around the cell. During metaphase the chromosomes become aligned in the middle of the cell (metaphase plate), during anaphase one set of chromosomes is moved to each end of the cell, and during telophase a new cell wall (cell plate) is formed between the two sets of chromosomes and the nuclear envelope reforms around each chromosome set. The dinitroanilines, DCPA, pronamide, and dithiopyr, all inhibit growth shortly after seed germination by interfering with the mitotic process and prevent emergence of the weed seedling above the soil surface. If the herbicide causes no other effect on the weed, control has been achieved. Even with a plentiful supply of carbohydrates and lipids in the seed for maintaining metabolic reactions prior to emergence from the soil, the survival period is finite. Growth of plant roots and shoots is a combination of cell division and cell enlargement, which results in an irreversible increase in size. An inhibition or disruption of either or both of these processes inhibits growth.

The spindle apparatus, which is composed of protein structures called microtubules, is the framework responsible for moving chromosomes during the various stages of mitosis. Herbicides known to interfere with the movement of chromosomes do so by interfering with the spindle apparatus (microtubules) and their assembling and disassembling which prevents their proper functioning—thus, inhibition of cell division and plant growth inhibition.

Dinitroanilines and Pronamide These herbicides do not inhibit the onset of mitosis, but rather disrupt the mitotic sequence once initiated. All of these compounds interfere with the normal movement of chromosomes during the mitotic sequence through causing the loss of microtubules during mitosis by slowing or preventing their assembly, which results in the disappearance of microtubules. The prophase sequence appears normal; however, without the presence of a spindle apparatus (microtubules) the chromosomes are unable to move to the metaphase configuration, the daughter chromosomes cannot migrate to their respective poles (anaphase), and cell wall formation does not occur at telophase. After a time in the prophase state, the chromosomes concentrate in the middle of the cell and the nuclear envelope reforms, causing a polyploid nucleus. Without the production of new cells, plant growth will eventually stop.

The mechanism of action of these herbicides is that they bind to tubulin proteins, which are the proteins that form the microtubules that lead to spindle fiber formation. The herbicide binds in a manner that slows or prevents microtubule formation and thus effects a stoppage of mitosis. Vaughan and Vaughn (1987) reported a slight difference between the dinitroanilines and pronamide in regard to their mechanism of inhibition of tubulin action. With pronamide treatment, very short microtubules actually form near the chromosomes, but these short microtubules cannot function properly in the mitotic process.

DCPA The precise mechanism of action of DCPA is unknown. Mitosis is disrupted, and there is abnormal cell division in the root tip meristem area. Similar to the type seen after trifluralin treatment. In addition to affecting chromosome movement, DCPA causes a significant disruption of cell wall formation. New cell walls form during mitosis, but their direction is random within the cell, rather than producing the usual straight walls formed between the two daughter nuclei (Holmsen and Hess, 1984). Microscopic observations show that some microtubules are present in disrupted mitotic cells. Thus, DCPA may inhibit microtubule function as well as microtubule formation.

Propham and Chlorpropham The carbamate herbicides propham and chlorpropham (no longer sold in the United States) are known to cause the spindle microtubules to function improperly. Again, prophase is normal; however, rather than the microtubules moving the chromosomes to two distinct locations (poles) at opposite ends of the cell, chromosomes are moved to several locations in the cell. This has been termed a “multipole spindle apparatus.”

Other Mechanisms

Two other herbicides (dithiopyr and thiazopyr) are known to have sites of action associated with the disruption of mitosis. When annual grass species are treated with dithiopyr or thiazopyr, the symptoms are the same as with dinitroaniline treatment. Root growth is inhibited, and the root tip becomes abnormally enlarged. Cell division is disrupted at metaphase because of the absence of microtubules. In protein binding studies, dithiopyr was found to bind to protein other than tubulin (Armbruster et al.,

1991). It was suggested that dithiopyr altered the microtubule's ability to form and function.

Cause of Root Tip Swelling Dinitroaniline and many carbamate herbicides, as well as DCPA and dithiopyr, cause characteristic root tip swelling in susceptible plants (Figure 12-6). This swelling is due to abnormal cell enlargement in the zone of elongation. In this zone, cells enlarge to a spherical shape rather than to the normal cylindrical shape (Hess, 1982). When herbicides cause the loss of microtubules, the microfibril orientation is random and the wall has no preferential support in any direction; thus, cell enlargement is spherical and the root tip area becomes swollen (Hess, 1982).

Weed Resistance

There are four dinitroaniline-resistant weeds reported to have been selected from the environment after repeated use of these herbicides (for review see Smeda and Vaughn, 1994). The first report of weed resistance to the dinitroaniline herbicides was for goosegrass in the southeastern United States. Populations of green foxtail (Canada), johnsongrass, and Palmer amaranth have since been reported to be resistant to dinitroaniline herbicides. Two cases of dinitroaniline resistance are reported for weeds

Figure 12-6. Root tip swelling caused by dinitroaniline herbicides.

that have multiple resistances (rigid ryegrass in Australia and slender foxtail in the United Kingdom). The exact mechanism for resistance in each of these weeds is not known; however, in a highly resistant goosegrass biotype, the resistance is apparently due to the reduced ability of the herbicide to interfere with microfiber formation and function; thus, mitosis can occur in the presence of the herbicide.

INHIBITORS OF SHOOTS AND/OR ROOTS

CHLOROACETAMIDES/OXYACETAMIDES

The chloroacetamide herbicides have been among the most widely used herbicide groups. Since the introduction of CDAA (Randox, 2-chloro-*N*,*N*-di-2-propenyl-acetamide) in 1954, the use of these herbicides has continued to grow.

In the United States this is attributable to extensive use on both corn and soybeans. Metolachlor, acetochlor, flufenacet, and dimethenamid are widely used because they have excellent weed control activity, long soil residual, and are formulated with protective safeners that ensure crop safety. Safeners include dichlormid and MON 4660 in acetochlor, and benoxacor in metolachlor.

Chloroacetamide herbicides are neutral/nonionic, with low to moderate water solubility and moderate to low vapor pressures. These herbicides control germinating seeds, very small emerged seedlings of many annual grasses, and a few small-seeded broadleaf species such as lambsquarters and pigweed. Treated seeds usually germinate, but the seedlings either do not emerge from the soil or emerge and exhibit abnormal growth.

Grasses show inhibition of primary leaf emergence from the coleoptile, and nutsedge shows shoot inhibition. If leaves do emerge, they often do not unroll completely, trapping the tip of the next developing leaf and causing it to loop (Figure 12-7). The primary anatomical sites of action are the developing leaves beneath the coleoptile and the apical and intercalary meristems near the coleoptilar node.

Root inhibition has also been reported, but roots are generally less sensitive than shoots to these herbicides. In the case of cotton, roots are more sensitive to alachlor than the shoots. Thus, the relative activity on roots versus shoots varies with species.

Chloroacetamides apparently are absorbed by both the roots (especially dicots) and shoots (especially monocots). Because chloroacetamide herbicides affect germinating seeds or very small seedlings, it has been difficult to study translocation patterns. Available studies indicate very limited translocation.

Chloroacetamides are applied as both preemergence and preplant incorporated treatments at 0.8 to 3.5 lb/acre, with the use of either spray or granular applications. Studies with alachlor on yellow nutsedge indicate that best results are achieved from placement around or just above the tubers. In field studies, nutsedge control has been much better with soil incorporation than with surface application. Soil activity is approximately 4 to 5 weeks. Crops on which one or more of these herbicides have been used include corn, sorghum, soybeans, dry beans, peanuts, peas, cotton, many vegetable crops, some fruit crops, and ornamentals.

Figure 12-7. Shoot inhibition and effect on leaf development in corn caused by a chloroacetamide herbicide (T. N. Jordan, Purdue University.)

Selectivity

Selectivity appears to be related to the rate of metabolism; tolerant plants rapidly metabolize chloroacetamide herbicides as compared with susceptible plants. Protection of sorghum from chloroacetamides can be obtained by treating seeds with any of several compounds (see the section on the use of safeners, Chapter 5). The exact mechanism of protection (safening) is not understood but may be related to increased synthesis of glutathione after treatment or improved specificity of glutathione for the herbicide. Good correlation has been observed between tolerance and glutathione levels in several species.

Alachlor

Alachlor (2-chloro-*N*-(2,6-diethylphenyl)-*N*-(methoxymethyl)acetamide) is a cream to wine-red colored solid, with a vapor pressure of 1.6×10^{-5} mm Hg at 25°C, a water solubility of 200 mg/l (ppm) at 20°C, a soil half-life of 21 days, and an oral LD_{50} (rat) of >2000 mg/kg, depending on formulation.

CH_2CH_3
$N(COCH_2Cl)(CH_2OCH_3)$
CH_2CH_3

Alachlor

Uses Alachlor is formulated as an EC, WDG, and G and sold as Lasso, Micro Tech, Partner, and other products for use as a preplant incorporated, preemergence, and directed postemergence treatment in corn, soybeans, peanuts, dry beans, lima beans, cotton, brassicas, oilseed rape, radish, sugarcane, and certain woody ornamentals. Check label for specific crop uses and restrictions. There is a concern about movement of alachlor into groundwater. It has been detected in many well samples over the past decade, and this problem has led to less use in agriculture.

Acetochlor

Acetochlor (2-chloro-*N*-(ethoxymethyl)-*N*-(2-ethyl-6-methylphenyl)acetamide) is a thick, oily, light amber to violet liquid with a vapor pressure of 3.4×10^{-8} mm Hg at 25°C, a water solubility of 223 mg/l (ppm) at 25°C, an oral LD_{50} (rat) of 2148 mg/kg; it provides 8 to 12 weeks of weed control, so soil carryover is not a problem.

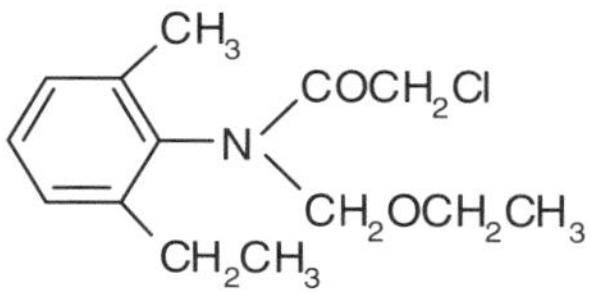

Acetochlor

Uses Acetochlor is formulated as an EC and sold as Harness and Harness Xtra for preplant incorporated or preemergence use in corn, peanuts, soybeans, cotton, potatoes, and sugarcane.

Dimethenamid

Dimethenamid (2-chloro-*N*-[(1-methyl-2-methoxy)ethyl]-*N*-(2,4-dimethyl-thien-3-yl)-acetamide) is a dark brown viscous liquid with a vapor pressure of 2.76×10^{-4} mm Hg at 25°C, a water solubility of 1174 mg/l (ppm) at 25°C, a soil half-life of 20 days, and an oral LD_{50} (rat) of 2400 mg/kg.

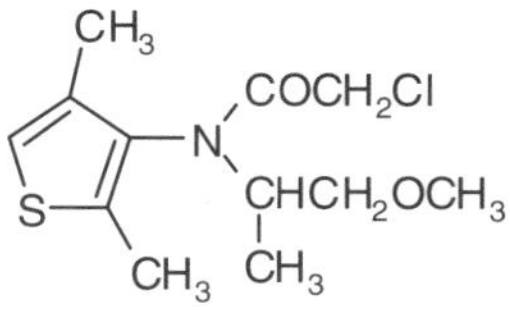

Dimethenamid

Uses Dimethenamid is formulated as an EC and sold as Frontier and Outlook for preplant incorporated or preemergence use in corn and soybeans.

Flufenacet

Flufenacet (*N*-(4-fluorophenyl)-*N*-(1-methylethyl)-2-[[5-trifluoromethyl)-1,3,4-thiadiazol-2- yl]oxy]acetamide) is a white solid with a vapor pressure of 9×10^{-7} mm Hg at 20°C, a water solubility of 56 mg/l (ppm) at 25°C, a soil half-life of 29 days, and an oral LD_{50} (rat) of 2347 mg/kg.

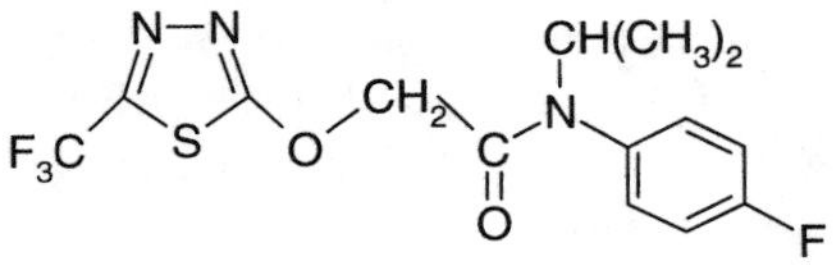

Flufenacet

Uses Flufenacet is formulated as a DF and sold as Axiom for use as a preplant incorporated, preplant, or preemergence treatment for corn and soybeans.

Metolachlor

Metolachlor (2-chloro-*N*-(2-ethyl-6-methylphenyl)-*N*-(2-methoxy-1-methylethyl)-acetamide) is a white to tan liquid with a vapor pressure of 1.3×10^{-5} mm Hg at 20°C, a water solubility of 448 mg/l (ppm) at 20°C, a soil half-life of 90 to 150 days, and an oral LD_{50} (rat) of > 2500 mg/kg, depending on formulation.

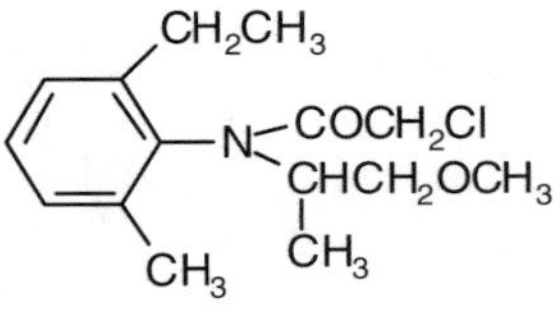

Metolachlor

Uses Metolachlor is formulated as an EC, DF, WG, and G and sold under many trade names, including Dual, Dual II, Dual Magnum, Dual II Magnum, and Pennant, for preplant incorporated or preemergence use in field corn, soybeans, sugar beets, sugarcane, peanuts, various vegetables such as potatoes and tomatoes, fruit and nut trees, sorghum, dry beans, safflower, sunflowers, and numerous ornamentals.

Propachlor

Propachlor (2-chloro-*N*-(1-methylethyl)-*N*-phenylacetamide) is a light tan solid with a vapor pressure of 7.9×10^{-5} mm Hg at 25°C, a water solubility of 580 mg/l (ppm) at 25°C, a soil half-life of 7 days, and an oral LD_{50} (rat) of 3269 mg/kg.

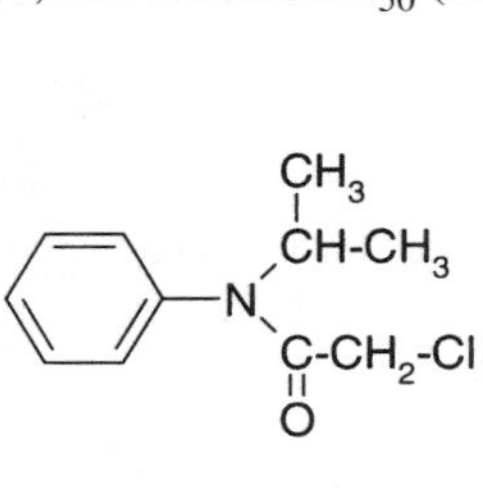

Propachlor

Uses Propachlor is formulated as an SC, WP, and G and sold as Ramrod for preemergence and preplant incorporated use in corn and grain sorghum.

Soil Influences

Herbicides in this group tend to be adsorbed by clay and organic matter and not generally susceptible to excessive leaching in most soils. The exception is alachlor, as noted earlier. There is not generally a problem with soil persistence, as the compounds are degraded to nontoxic forms by soil microorganisms. Persistence can be longer in soils high in organic matter.

Metabolism

Chloroacetamides are metabolized in plants, and glutathione conjugation appears to be the mechanism of metabolism. Glutathione conjugates are commonly found in plants treated with chloroacetamide herbicides.

Mechanism of Action

Although the chloroacetamides have been examined extensively for many years, their primary biochemical mechanism of action was only recently shown to be related to the depletion of very-long-chain fatty acids from the plasma membranes of plant cells (Böger, 2001). The action is due to an inhibition of the biosynthesis of very-long-chain fatty acids (important constituents of phosphatidylcholin, phosphatidylethanolamine, and cerebrosides) that are important constituents of the plasma membrane. The loss of these very-long-chain fatty acids stops the biosynthesis and function of the plasma membrane and, because cell integrity is lost, leads to the death of the plant.

Resistance

There have been three weeds reported to exhibit resistance to the chloroactemide herbicides. These include barnyardgrass in China, a barnyardgrass biotype in Thailand, and a rigid ryegrass biotype in Australia.

INHIBITORS OF ROOTS

Napropamide

Napropamide (*N,N*,-diethyl-2-(1-Naphthalenyloxy)propanamide) is a brown solid with a vapor pressure of 4×10^{-6} mm Hg at 25°C, a water solubility of 73 mg/l (ppm) at 20°C, a soil half-life of 70 days, and an oral LD_{50} (rat) of >5000 mg/kg.

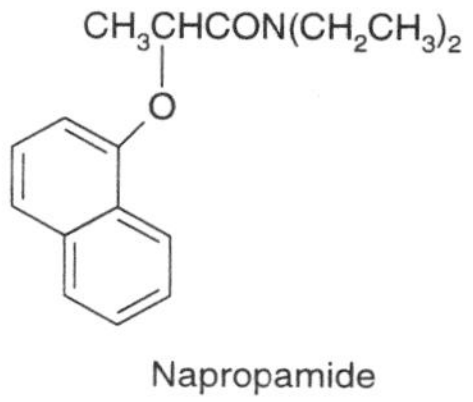

$CH_3CHCON(CH_2CH_3)_2$

Napropamide

Uses Napropamide is formulated as a WP, EC, DF, and G and sold as Devrinol for use as a preplant or preplant incorporated treatment in many vegetable crops, oilseed rape, tobacco, sunflowers, safflowers, olives, figs, mint, turf, strawberries, grapes, several kinds of fruit and nut trees, and woody ornamentals. Annual grasses are best controlled; however, a few broadleaf weeds are also sensitive.

Soil Influences Because of photodecomposition and low mobility in soil, napropamide is usually incorporated mechanically. Napropamide is resistant to leaching in most mineral soils and is slowly decomposed by soil microorganisms. This herbicide is quite persistent in the soil, being active for more than 6 months. Fall-planted cereals are often damaged by a napropamide application used the previous spring. In addition, alfalfa, sorghum, corn, and lettuce should not be seeded until 12 months after napropamide use.

Metabolism Napropamide is rapidly metabolized in fruit trees and tomatoes to water-soluble metabolites, the major metabolites being hexose conjugates of 4-hydroxynapropamide.

Mechanism of Action The exact mode of action of napropamide is not known. The growth of susceptible seedling roots is stopped without malformation almost immediately after exposure to the herbicide (Figure 12-8). The effect on roots is localized, as roots that are not in direct contact with the herbicide are not affected, and no shoot inhibition has been observed. Napropamide has been shown to be rapidly absorbed by tomato roots and readily translocated throughout the stem and leaves.

Figure 12-8. Complete inhibition of strawberry root development caused by napropamide. *Left:* Treated root; *Right*: Untreated root.

However, upward translocation from root absorption in corn was much slower. The distribution pattern suggests that it is primarily translocated in the apoplast.

Bensulide

Bensulide (*O*,*O*-bis(1-methylethyl)*S*-[2-(phenylsulfonyl)amino]ethyl]phosphorodithioate) is a white crystalline solid with a vapor pressure of 8×10^{-7} mm Hg at 25°C, a water solubility of 25 mg/l (ppm) at 20°C, a soil half-life of 120 days, and an oral LD_{50} (rat) of 770 mg/kg.

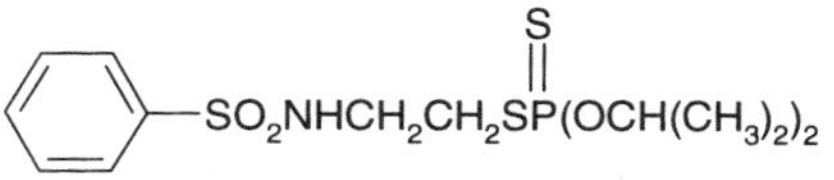

$$C_6H_5-SO_2NHCH_2CH_2SP(=S)(OCH(CH_3)_2)_2$$

Bensulide

Uses Bensulide is formulated as an EC and G and is sold as Betasan and Prefar for preemergence control of crabgrass and other weeds in established grass and dichondra lawns, cucurbits, cole crops, eggplant, peppers, lettuce, garlic, onions, shallots, grass seed, and cotton. In general, preplant applications require soil incorporation, and preemergence applications are used only on crops that are irrigated to promote emergence.

Soil Influences Bensulide is tightly bound to organic matter in soil and is inactive in soils containing high amounts of organic matter. It is not subject to significant leaching in any soil type, and soil microorganisms slowly degrade it. Bensulide has a soil half-life of about 4 months in a moist loam soil at 70 to 80°F, and about 6 months in a moist loamy sand soil. Treated areas should not be planted to nonregistered crops within 18 months (12 months for soybeans) or turf species for 4 months.

Mechanism of Action Bensulide inhibits the growth of roots and partially inhibits cell division and root elongation. Binucleated cells were observed in treated tissue, suggesting that bensulide may inhibit some stage of mitosis. It is adsorbed on root surfaces, and the root absorbs a small amount. Little, if any, however, is translocated upward to the leaves. Bensulide appears to be degraded by higher plants.

Siduron

Siduron (*N*-(2-methylcyclohexyl)-*N*′-phenylurea) is a white crystalline solid with a vapor pressure of 4×10^{-9} mm Hg at 25°C, a water solubility of 18 mg/l (ppm) at 25°C, a soil half-life of 90 days, an oral LD_{50} (rat) of >7500 mg/kg.

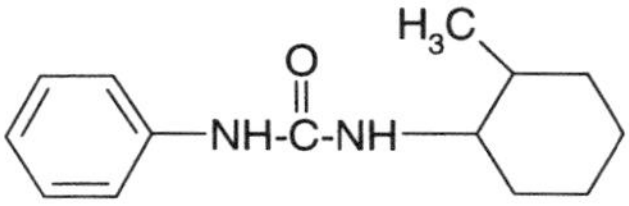

$$C_6H_5-NH-C(=O)-NH-C_6H_{10}(CH_3)$$

Siduron

Uses Siduron is formulated as a WP and sold as Tupersan, a specialty herbicide for preemergence control of smooth and hairy crabgrass, foxtail, and barnyardgrass in newly seeded or established turf; however, it does not control annual bluegrass, clovers, or most broadleaf weeds. Most turfgrass species are tolerant to siduron even when germinating from seed. Although most turfgrasses are tolerant to siduron, bermudagrass and certain bentgrass strains are subject to injury by siduron.

Soil Influence Siduron is resistant to leaching and in most respects is similar to the other phenylurea herbicides in its soil response (see Chapter 9).

Mechanism of Action Siduron, unlike the rest of the phenylurea herbicides, does not inhibit photosynthesis. It is not known exactly what the mode of action of siduron is; however, it inhibits some aspect of cell division.

INHIBITORS OF SHOOTS

CARBAMOTHIOATES (ALSO CALLED THIOCARBAMATES)

Carbamothioates are soil-applied herbicides with high vapor pressures. They require mechanical incorporation into the soil to reduce loss by volatilization and ensure adequate herbicidal activity. Soil incorporation as a standard practice was principally developed to overcome the ineffectiveness of surface applications of EPTC. However, these herbicides have been successfully applied as a layer or lines below the soil surface and are sometimes applied as granules to the soil surface or metered into irrigation water.

These herbicides are most active on annual grasses but control several other annual weeds, and some are used to suppress nutsedge (*Cyperus* species) and quackgrass. They act by inhibiting shoot growth of emerging weeds and have no direct effect on root growth. The first demonstration of shoot uptake of soil-applied herbicides was with EPTC. They enter roots readily, but must be translocated to the shoot growing point to be active. In emerging grasses, translocation from roots is limited, so the herbicides are ineffective through root exposure alone. The placement of these herbicides in the soil has been used to provide selectivity for certain grass crops while inhibiting weed growth.

Because of the requirement for soil incorporation of carbamothioates and the changing patterns of crop management (reduced or no-till), the use of this group of herbicides has decreased dramatically in recent years, and many of these compounds are disappearing from the commercial marketplace.

EPTC is still important in field beans, potatoes, small-seeded legumes (alfalfa), and sweet corn. Other crops in which one or more of these herbicides might be used include corn, soybeans, flax, safflower, tobacco, peanuts, rice, barley, wheat, sugar beets, several other vegetables and fruits, and nursery crops.

Selectivity and Safeners

There are large inherent differences among species as to susceptibility. In general, dicots are more tolerant than grasses, but there are wide differences in tolerance within dicots and within grasses. Safeners have been useful in protecting corn from EPTC, vernolate, and butylate injury (Figure 12-9). The nonherbicidal compounds dichlormid (2,2-dichloro-*N,N*-di-2-propenylacetamide) and R-29148 [3-(dichloro-acetyl)-2,2,5-trimethyloxazolidine] significantly increase the tolerance of corn to these herbicides (Chang et al., 1972). The mechanism of increased tolerance in corn is proposed to be the result of increased herbicide inactivation by conjugation to glutathione. Dichlormid stimulates glutathione production. Conjugation occurs between the sulfoxide form of carbamothioate herbicide and glutathione.

Butylate

Butylate (*S*-ethyl bis(2-methylpropyl)carbamothioate) is a colorless liquid with a vapor pressure of 1.3×10^{-2} mm Hg at 25°C, a water solubility of 44 mg/l (ppm) at 20°C, a soil half-life of 13 days, and an oral LD_{50} (rat) > 3500 mg/kg.

Butylate

Figure 12-9. Protection of corn from EPTC injury by an antidote dichlormid. *Left*: Untreated control; *Center*: EPTC with no safeners; *Right*: EPTC + safeners. (M. V. Hickman, Purdue University.)

Uses Butylate is formulated as an EC with a safener and as a G and sold as Sutan+ for preplant incorporated treatments in corn and pineapple for control of annual grass weeds and nutsedge suppression.

Cycloate

Cycloate (*S*-ethyl cyclohexylethylcarbamothioate) is a colorless liquid with a vapor pressure of 6.2×10^{-3} mm Hg at 25°C, a water solubility of 85 mg/l (ppm) at 20°C, a soil half-life of 30 days, and an oral LD_{50} (rat) of 3200 mg/kg.

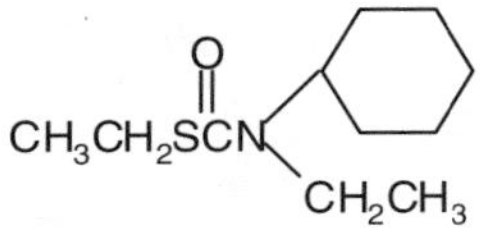

$CH_3CH_2SC(=O)N(C_6H_{11})CH_2CH_3$

Cycloate

Uses Cycloate is formulated as an EC and G and sold as Ro-Neet for preplant incorporated applications in sugar beets, fodder beets, table beets, and spinach.

EPTC

EPTC (*S*-ethyl dipropyl carbamothioate) is a light yellow liquid with a vapor pressure of 3.4×10^{-2} mm Hg at 25°C, a water solubility of 370 mg/l (ppm) at 20°C, a soil half-life of 6 days, and an oral LD_{50} (rat) of 1325 to 1500 mg/kg.

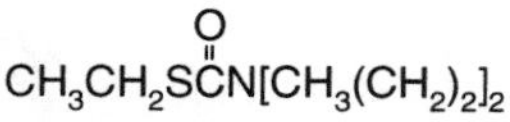

$CH_3CH_2SC(=O)N[CH_3(CH_2)_2]_2$

EPTC

Uses EPTC is formulated as an EC and G and sold as Eptam or, with a safener, as Eptam+ or Eradicane for use preplant incorporated as a surface application in potatoes, beans, peas, forage legumes, sugar beets, table beets, alfalfa, trefoil, clover, cotton, corn, flax, sweet potatoes, dry beans, castor beans, corn, flax, safflowers, sunflowers, citrus, almonds, walnuts, pineapples, strawberries, pine nurseries, and nursery stock.

Molinate

Molinate (*S*-ethyl hexahydro-1*H*-azepine-1-carbothioate) is a clear, bright orange liquid with a vapor pressure of 5.6×10^{-3} mm Hg at 25°C, a water solubility of 970 mg/l at 20°C, a soil half-life of 21 days, and an oral LD_{50} (rat) of 955 mg/kg.

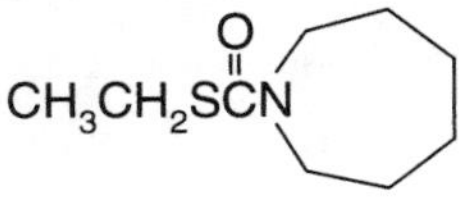

$CH_3CH_2SC(=O)N$ (azepane ring)

Molinate

Uses Molinate is formulated as an EC and G and sold under the trade name Ordam for preplant incorporated applications before planting to water-seeded rice or shallow soil-seeded rice, or postflood, postemergence use in other types of rice culture.

Pebulate

Pebulate (*S*-propyl butylethylcarbamothioate) is a yellow liquid with a vapor pressure of 8.9×10^{-3} mm Hg at 25°C, a water solubility of 60 mg/l (ppm) at 20°C, a soil half-life of 14 days, and an oral LD_{50} (rat) of 1400 mg/kg.

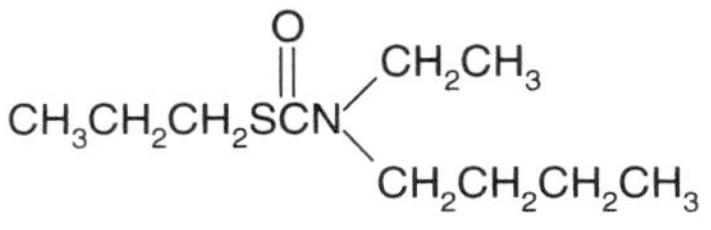

$CH_3CH_2CH_2SC(=O)N(CH_2CH_3)(CH_2CH_2CH_2CH_3)$

Pebulate

Uses Pebulate is formulated as an EC and G and sold under the trade name Tillam for preplant incorporated applications in tobacco and sugar beet and preplant incorporated or post-transplant use in tomatoes.

Thiobencarb

Thiobencarb (*S*-[(4-chlorophenyl)methyl]diethylcarbamothioate) is a light yellow to brownish yellow liquid with a vapor pressure of 1.476×10^{-6} mm Hg at 20°C, a water solubility of 30 mg/l (ppm) at 25°C, a soil half-life of 30 to 90 days, and an oral LD_{50} (rat) of 1033 mg/kg.

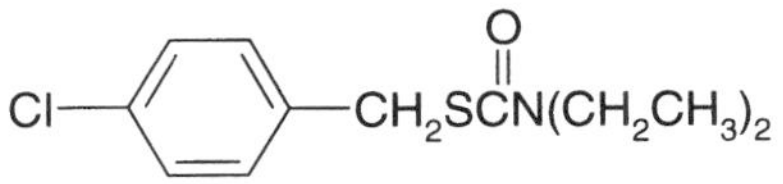

$Cl-C_6H_4-CH_2SC(=O)N(CH_2CH_3)_2$

Thiobencarb

Uses Thiobencarb is formulated as an EC and G and sold as Bolero for preemergence or early postemergence use in direct-seeded or transplanted rice. Some barnyardgrass biotypes are resistant to thiobencarb.

Triallate

Triallate (*S*-(2,3,3-trichloro-2-propenyl)bis(1-methylethyl)carbamothioate) is an amber-colored oily liquid with a vapor pressure of 1.1×10^{-4} mm Hg at 25°C, a water solubility of 4 mg/l (ppm) at 20–25° C, a soil half-life of 82 days and an oral LD_{50} (rat) of 2193 mg/kg.

$$[(CH_3)_2CH_2]N\overset{O}{\overset{\|}{C}}SCH_2\overset{Cl}{\overset{|}{C}}{=}CCl_2$$

Triallate

Uses Triallate is formulated as an EC and G and sold as Avadex-BW and Far-Go for preplant incorporated use to control wild oats and some grasses in wheat, barley, rye, field beans, peas, lentils, beets, oilseed rape, maize, flax, alfalfa, clover, vetches, sainfoil, safflower, sunflowers, and certain vegetables. Some wild oat biotypes are resistant to triallate.

Soil Influences

In general, carbamothioate herbicides are adsorbed by the clay and organic matter in soils and are not readily leached (*Herbicide Handbook*, 1994). However, adsorption and leaching do vary somewhat within this class of herbicides. The relative leachability of some of these compounds is molinate > EPTC > pebulate > cycloate > butylate. Quantitative data are generally lacking on the leachability of these compounds; however, it has been reported that cycloate is leached 3 to 6 inches with 8 inches of water in a loamy sand soil (*Herbicide Handbook*, 1994). Vernolate is less subject to leaching than EPTC, but data on its relationship to other compounds is not available. Likewise, the relative leachability of thiobencarb and triallate with respect to the other carbamothioates is unknown.

The carbanothioate herbicides compete with moisture for the adsorption sites on soil particles. Therefore, they are readily adsorbed on dry soil but poorly adsorbed on wet soil. Soil persistence is relatively short for these herbicides, and dissipation is mostly by volatilization, especially from wet soil.

Degradation of these herbicides is primarily by microorganisms and is a major cause of their disappearance (*Herbicide Handbook*, 1994). They have a relatively short period of persistence in soils under aerobic conditions, with loss of phytotoxicity within 6 weeks or less, depending on the specific herbicide. The half-life of six of these compounds ranges from 1 to 6 weeks under standard conditions (Table 12-1).

After repeated use of carbamothioate herbicides, some soils become enriched with microorganisms that break down the herbicide so rapidly that the period of weed control is inadequate to protect the crop from weeds. EPTC was the first herbicide whose effectiveness was restored by an additive after microbial enrichment had created a situation wherein the herbicide had become ineffective (Obrigawitch et al., 1982). The nonherbicidal compound dietholate (*O, O*-diethyl *O*-phenyl phosphorothioate), which protects carbamothioates from rapid microbial decomposition, is called a "herbicide extender." In Eradicane Extra, a crop protectant, R29148, and a herbicide extender, dietholate, are formulated together with EPTC. Eradicane Extra is used for weed control in corn where the herbicide is subject to rapid decomposition because of soil enrichment.

TABLE 12-1. Half-Life of Six Carbamothioate Herbicides in Moist Loam Soil at 70 to 80°F

Herbicide	Weeks
EPTC	1.0
Vernolate	1.5
Pebulate	2.0
Butylate	3.0
Molinate	3.0
Cycloate	3.5

Metabolism

The carbamothioates are rapidly metabolized in most plants. The first practical use of herbicide antidotes or safeners was with EPTC and butylate in the formulation to expand their use in marginally tolerant crops such as corn (Chang et al., 1972) (see the section "Selectivity and Safeners").

Mechanism of Action

The mechanism of action of these compounds is not known, but the site of action in germinating grasses and nutsedge is in the developing leaves and the shoot growing point. The mechanism of action of carbamothiates is probably closely related to that of the chloroacetamides. In general, root tissue is extremely tolerant.

In using carbamothiates in seedling grasses, as with chloroacetamides, a common symptom is distortion of the first foliar leaf and restriction of its emergence from the coleoptile. In treated fields, susceptible grasses often emerge but remain very small, with severely distorted young leaves, and eventually the plant dies. Research results indicate that effects on gibberellin may be involved (Donald et al., 1979).

The sprouting of nutsedge tubers is stimulated, but the sprouts are stunted and the tips enlarged. The tubers are not killed, and new sprouts emerge and grow after the compound dissipates sufficiently in the soil. Wax development on leaves of plants is reduced (Flore and Bukovac, 1976), and several carbamothioate herbicides have been reported to stimulate weed seed germination.

Alfalfa tolerates exposure to massive rates of EPTC in close proximity to the planted seed because the germinating seed and emerging seedling are extremely tolerant of the herbicide. Alfalfa becomes susceptible to EPTC injury only after the seedling has emerged, the hypocotyl has unhooked, and the cotyledons have diverged. By this time, the EPTC has diffused (probably as vapors) and the seedling is never exposed to extreme concentrations after it reaches the susceptible stage (Dawson, 1983).

Weed Resistance

In spite of a long history of use, weeds resistant to carbamothioates have appeared only recently. Wild oat resistance to triallate has been reported in many fields in Canada, Montana, and Idaho. Late watergrass biotypes in California are resistant to thiobencarb and molinate; rigid ryegrass in Australia, annual bluegrass in the United States, and Chilean needlegrass in New Zealand are resistant to one or more of the carbamothioates.

LITERATURE CITED AND SUGGESTED READING

Mitotic Disrupters

Armbruster, B. L., W. T. Molin, and M. W. Bugg. 1991. Effects of the herbicide dithiopyr on cell division in wheat root tips. *Pestic. Biochem. Physiol.* **39**:110–120.

Herbicide Handbook. 7th ed. 1994. Weed Science Society of America. Lawrence, KS.

Herbicide Handbook. Supplement to 7th ed. 1998. Weed Science Society of America. Lawrence, KS.

Hess, F. D. 1982. Determining causes and categorizing types of growth inhibition induced by herbicides. *Amer. Chem. Soc. Symp. Ser.* **181**:208–230.

Hoffman, J. C., and K. C. Vaughn. 1994. Mitotic disrupter herbicides act by a single mechanism but vary in efficacy. *Protoplasm* **179**:16–25.

Holmsen, J. D., and F. D. Hess. 1984. Growth inhibition and disruption of mitosis by DCPA in oat (*Avena sativa*) roots. *Weed Sci.* **32**:732–738.

Jasieniuk, M., A. L. Brule-Babel, and I. N. Morrison. 1994. Inheritance of trifluralin resistance in green foxtail (*Setaria viridis*). *Weed Sci.* **42**:123–127.

Peter, C. J., and J. B. Weber. 1985. Adsorption and efficacy of trifluralin and butralin as influenced by soil properties. *Weed Sci.* **33**:861–867.

Probst, G. W., T. Golab, R. J. Herberg, F. J. Halzer, S. J. Parka, C. van der Shans, and J. B. Tepe. 1967. Fate of trifluralin in soils and plants. *J. Agr. Food Chem.* **15**:592–599.

Smeda, R. J., and K. C. Vaughn. 1994. Resistance to dinitroaniline herbicides. In *Herbicide Resistance in Plants*, ed. by S. B. Powles and J. A. M. Holtum, pp. 215–228. CRC Press, Boca Raton, FL.

Vaughan, M. A., and K. C., Vaughn. 1987. Pronamide disrupts mitosis in a unique manner. *Pestic. Biochem. Physiol.* **28**:182–193.

Vaughn, K. C., and L. P. Lehnen Jr. 1991. Mitotic disrupter herbicides. *Weed Sci.* **39**:450–457.

Weber, J. B. 1990. Behavior of dinitroaniline herbicides in soils. *Weed Technol.* **4**:394–406.

Chloroacetamides

Böger, P., and B. Matthes. 2001. Chloroacetamides deplete the plasma membrane of very long-chain fatty acids. *WSSA Abstr.* **41**:70.

Fuerst, E. P. 1987. Understanding the mode of action of the chloroacetamide and thiocarbamate herbicides. *Weed Tech.* **1**:270–277.

Carbamothioates

Bean, B. W., F. W. Roeth, A. R. Martin, and R. G. Wilson. 1988. Influence of prior pesticide treatments on EPTC and butylate degradation. *Weed Sci.* **36**:70–77.

Chang, F. Y., J. B. Bandeen, and G. R. Stephenson. 1972. A selective antidote for prevention of EPTC injury in corn. *Can. J. Plant Sci.* **52**:707–714.

Dawson, J. H. 1983. Tolerance of Alfalfa (*Medicago Sativa*) to EPTC. *Weed Sci.* **31**:103–108.

Donald, W. W., R. S. Fawcett, and R. G. Harvey. 1979. EPTC effects on corn (*Zea mays*) growth and endogenous gibberellins. *Weed Sci.* **27**:122–127.

Flore, J. A., and M. J. Bukovac. 1976. Pesticide effects on the plant cuticle: II. EPTC effect on leaf cuticle morphology and composition in *Brassica oleracea* L. *J. Amer. Soc. Hort. Sci.* **101**:586–590.

Gray, R. A., and A. J. Weierich. 1969. Importance of root, shoot, and seed exposure on the herbicidal activity of EPTC. *Weed Sci.* **17**:223–229.

Moorman, T. B. 1988. Populations of EPTC-degrading microorganisms in soils with accelerated rates of EPTC degradation. *Weed Sci.* **36**:96–101.

Obrigawitch, T., F. W. Roeth, A. R. Martin, and R. G. Wilson Jr. 1982. Addition of R-33865 to EPTC for extended herbicide activity. *Weed Sci.* **30**:417–422.

Root Inhibitors

Di Tomaso, J. M., T. L. Rost, and F. M. Ashton. 1988. The comparative cell cycle, and metabolic effects of the herbicide napropamide on root tip meristems. *Pestic. Biochem. Pysiol.* **31**:166–174.

Romanowski, R. A., and A. Borowy. 1979. Soil persistence of napropamide. *Weed Sci.* **27**:151–153.

WEB SITE

International Survey of Herbicide Resistant Weeds
http://www.weedscience.org

For chemical use, see the manufacturer's or supplier's label and follow these directions. Also see the Preface.

13 Cellulose Biosynthesis Inhibitors

The cellulose biosynthesis inhibitor herbicides are a diverse group of chemically unrelated compounds. The common herbicidal effect is either a direct or indirect inhibition of cellulose biosynthesis, which in effect leads to a lack of cell structure integrity. In most cases these herbicides are used for preemergence control and result in the inability of weed seedlings to grow (Sabba and Vaughn, 1999). Symptoms include stunted growth and root swelling. Dichlobenil and isoxaben are used preemergence and are most effective against dicots, whereas quinclorac is used both preemergence and postemergence. Although quinclorac is an auxinic type of herbicide, it has a second mechanism of action on monocots as a cellulose biosynthesis inhibitor with excellent activity in most monocots.

CELLULOSE BIOSYNTHESIS INHIBITORS

Nitrile	Benzamide	Quinolinecarboxylic Acid
Dichlobenil (Casoron, Dyclomec, Norosac)	Isoxaben (Gallery)	Quinclorac (Facet)

Dichlobenil

Dichlobenil (2,6-dichlorobenzonitrile) is a white to slightly yellow crystal with a vapor pressure of 5.5×10^{-4} mm Hg at 20°C, making it quite volatile (thus the granular formulation and soil incorporation), a water solubility of 20.5 mg/l (ppm) at 25°C, a soil half-life of 30 to 180 days, depending on soil type, and an oral LD_{50} (rat) of 4460 mg/kg.

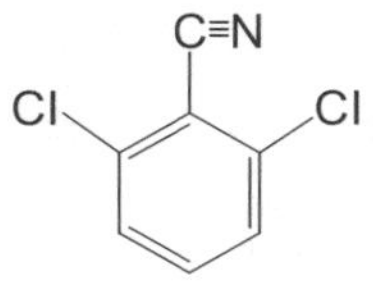

Dichlobenil

Uses Dichlobenil is formulated as a G (4 or 10%) and sold as Casoron, Dyclomec, Norosac, and Silbenil for application in fruit and nut orchards and nurseries, cranberries, vineyards, bush fruit, woody ornamentals, shelterbelts, hybrid cottonwood and poplar plantations and stoolbeds, and in noncropland.

Dichlobenil was the first herbicide reported to have an effect on cell wall biosynthesis (Hogetsu et al., 1974). It acts on growing points and root tips and is the simplest chemically of this group, being in the nitrile family. Other herbicide members of this family (hydroxylated bezonitriles) such as bromoxynil do not inhibit cellulose biosynthesis but inhibit electron transport in photosystem II (see Chapter 9).

Dichlobenil (DCB) has been marketed since the early 1960s and is used as a preemergence herbicide to control a broad range of annual broadleaf and grass weeds and certain biennials and perennials when it comes in contact with actively growing organs in the soil. Dichlobenil can be selectively used in several established crops whose roots do not come in contact with the herbicide that is located in the upper soil layers. Roots and leaves readily absorb DCB, with movement most common in the xylem and slow or no movement in the phloem.

Soil Influences Dichlobenil is tightly adsorbed on soil colloids, particularly organic matter. Therefore, it is not subject to leaching in most agricultural soils. Its volatility and codistillation with water can cause its rapid loss from the soil surface. This loss is accelerated under high temperatures, in wet soil, and with low relative air humidity. The loss is minimized when the herbicide is applied to dry soil just prior to rainfall, overhead irrigation, or mechanical soil incorporation. In laboratory studies, dichlobenil was slowly degraded (hydrolyzed) by microbes in the soil to 2,6-dichlorobenzimide, then to 2,6-dichlorobenzoic acid, and eventually to CO_2 and other breakdown products with a soil half-life of 60 days.

Isoxaben

Isoxaben (*N*-[3-(1-ethyl-1-methylpropyl)-5-isoxazolyl]-2,6-dimethoxybenzamide is a white crystalline solid with a vapor pressure of $< 3.9 \times 10^{-7}$ mm Hg at 25°C, a very low water solubility of 1 mg/l (ppm) at 25°C, a soil half-life of 50 to 120 days, and an oral LD_{50} (rat) >10,000 mg/kg.

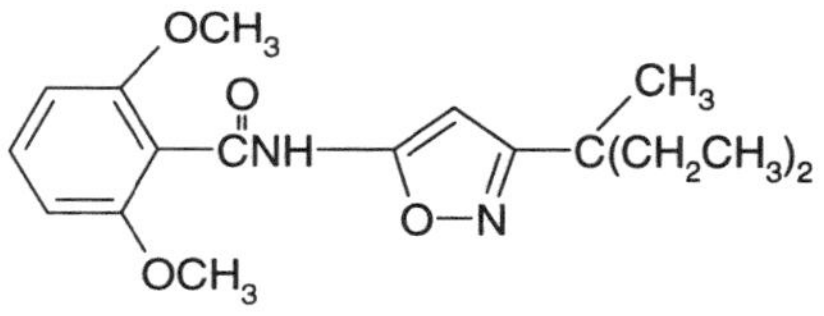

Isoxaben

Uses Isoxaben is formulated as a DF and sold as Gallery, as well as in a variety of herbicide mixtures and with fertilizer for use in established turf, ornamentals, nursery stock, nonbearing fruit and nut trees, nonbearing vines, Christmas trees, and noncropland areas.

Isoxaben is chemically more complex than DCB and provides excellent control of broadleaf weeds such as bittercress, common chickweed, clover, dandelion, henbit, prostrate knotweed, plantain, and spurge. Roots of plants readily absorb isoxaben, and it is translocated to stems and leaves in the xylem. The primary effect of this herbicide is on seedling growth; it also affects root and shoot development. Germination is not prevented, but once germination occurs, growth is inhibited. Isoxaben is a highly active molecule with an I_{50} concentration for growth inhibition in a model plant, *Arabidopsis* of 4.5 nM (Heim et al., 1989). Isoxaben is metabolized by hydroxylation of the alkyl side-chain in some species, but this is not a mechanism of selectivity.

Soil Influences Isoxaben is strongly adsorbed to soil colloids and is subject to little leaching. Degradation by microbes is the primary means of loss from soils. Dissipation studies indicate a soil half-life of 50 to 120 days, resulting in a moderate to long residual life in the field and effective weed control for 5 to 6 months at labeled rates.

Quinclorac

Quinclorac (3,7-dichloro-8-quinolinecarboxylic acid) is a colorless crystal with a vapor pressure of $< 10^{-7}$ mm Hg at 25°C, a water solubility of 62 mg/l, (ppm) at 20°C, and an oral LD_{50} (rat) of >2160 mg/kg.

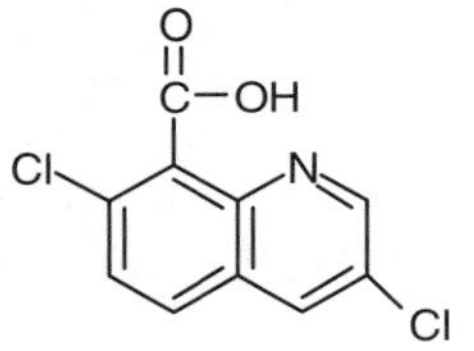

Quinclorac

Uses Quinclorac is formulated as a WP, WG, and DF and sold as Facet for preemergence and postemergence control of barnyardgrass, foxtail species, and some broadleaf weeds in direct seeded and transplanted rice.

Quinclorac is a herbicide active on many grass and some broadleaf weeds and originally was classified as a "auxin agonist" herbicide similar to 2,4-D and dicamba. This classification is largely based on the morphological response of dicot plants (stem curvature) after treatment with quinclorac and the observed increase in ethylene evolution in susceptible plants. However, the ethylene evolution induced by quinclorac is substantially lower than that induced by 2,4-D. Koo et al. (1994) reported that the activity of quinclorac in sensitive grass species did not appear to be auxinlike and that a different mechanism may be operating in these species than in broadleaf species. In grass species, quinclorac inhibited the growth of roots and shoots, induced electrolyte leakage in young root and shoot tissue, and produced a necrotic band in the elongation zone of roots and shoots. In susceptible grasses, quinclorac causes chlorosis and eventual necrosis in expanding leaves. Quinclorac is absorbed by the emerging shoots and roots of weeds and can be translocated in both the phloem and xylem, and it is not readily metabolized in susceptible plants.

Soil Influences Quinclorac has a soil residual activity of up to 1 year for susceptible plants. Its mobility is dependent on soil type, amount of organic matter, and soil percolation rate, although it can have moderate movement based on its solubility. Soil breakdown is not well characterized, but microbes degrade it, with 3-chloro-8-quinolinecarboxylic acid being a major metabolite.

MECHANISM OF ACTION

The most characteristic symptom of cellulose biosynthesis inhibiting herbicides is a swelling of the apical regions of the germinating seedling. Histological analyses of root tip meristems show a progressive disappearance of the meristematic zone. Cells in the meristematic zone become enlarged and sometimes are almost devoid of cytoplasm (Lefebvre et al., 1987).

The cellulose biosynthesis inhibiting herbicides are a diverse group of chemistries that all inhibit cellulose biosynthesis in plants. No definitive experimental data exist describing a single specific site(s) of inhibition for these herbicides in the cellulose biosynthesis pathway; however, recent evidence suggests that the three previously described commercial herbicides act at different points in the biosynthetic pathway. All of this evidence is indirect, but it indicates that the final effect of herbicide inhibition is a lack of cellulose biosynthesis and thus no cell wall development. Therefore, there is a lack of cellular integrity, which leads to arrested or abnormal growth, resulting in plant death.

Cellulose is a linear polymer of glucose arranged as a β-1,4-glucan that is synthesized in the plasma membrane of plants and provides structural organization to the plant. Cellulose biosynthesis is presently very poorly understood. There is a tremendous amount of active research in this area, and new information on the overall biosynthetic process and possible sites of herbicide action is becoming available. Detailed information regarding the current knowledge of herbicide action at such sites is provided by Sabba and Vaughn (1999), and information on cellulose biosynthesis is provided by Delmer (1999).

Dichlobenil is known to have a site in the cellulose synthase enzyme complex, and it apparently acts to inhibit the shunting of glucose molecules necessary for building the cellulose molecule. Vaughn et al. (1996) have provided ultrastructural evidence that dichlobenil inhibits cellulose biosynthesis. These authors show that during telophase in onion root tips, dichlobenil inhibited the stiffening and straightening of the plate stage of cell wall formation, which is associated with the accumulation of callose in the newly forming cell wall (Figure 13-1).

Isoxaben is thought to inhibit the conversion of sucrose into UDP-glucose, which in turn inhibits cell wall completion due to a lack of cellulose. This inhibition results in reduced substrate for cellulose formation and is at an earlier site in the biosynthesis pathway than dichlobenil. (See Figure 13-2 for a model of dichlobenil and isoxaben action.) In isoxaben-treated plants, enlarged cells were observed in the meristematic zone in root tips similar to those seen in plants inhibited by mitotic disrupting herbicides like trifluralin. However, there was no effect on the various stages of mitosis in treated tissue as compared with control tissue; thus, isoxaben is not a mitotic inhibitor nor a mitotic disrupter. Grasses are generally more tolerant to isoxaben than

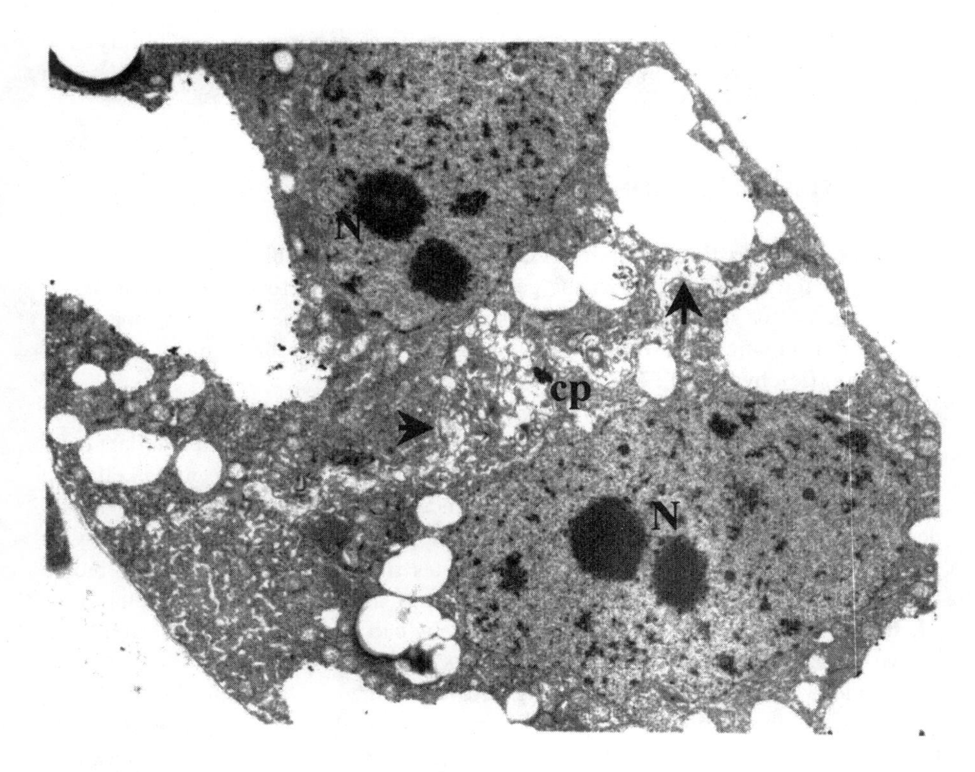

Figure 13-1. Electron micrograph of a BY-2 tobacco cell treated for 4 hr with 1.0 μM dichlobenil. The cell plate (cp) forming between daughter nuclei of a recent cell division is opaque, thick, and irregular and spread haphazardly throughout the cytoplasm (arrows). Magnification ×6300. (Sabba and Vaughn, 1999, published with permission.)

dicots. A study by Heim et al. (1993) showed that the tolerance level of *Agrostis palustris* (var. Penncross) was explained by no inhibition of glucose formation or use in the synthesis of cellulose. Considering that such natural tolerance (selectivity) may be at the site of isoxaben action, the development of weed resistance in the field after repeated use of isoxaben could be an issue. Thus, management practices should be used that alternate isoxaben use with herbicides having different mechanisms of action so as to minimize the potential development of resistant weed populations.

Quinclorac, as mentioned, appears to inhibit cellulose biosynthesis in grasses. Koo et al. (1996) have shown that in grass species (corn) quinclorac inhibits the incorporation of glucose into cellulose within 3 hours after treatment. Inhibition of glucose incorporation into hemicellulose was also inhibited, but at a higher concentration. Because of the inhibition of hemicellulose, these authors concluded that the specific target site is somewhat different from that of dichlobenil and isoxaben, which do not inhibit the synthesis of hemicellulose. Quinclorac has been a

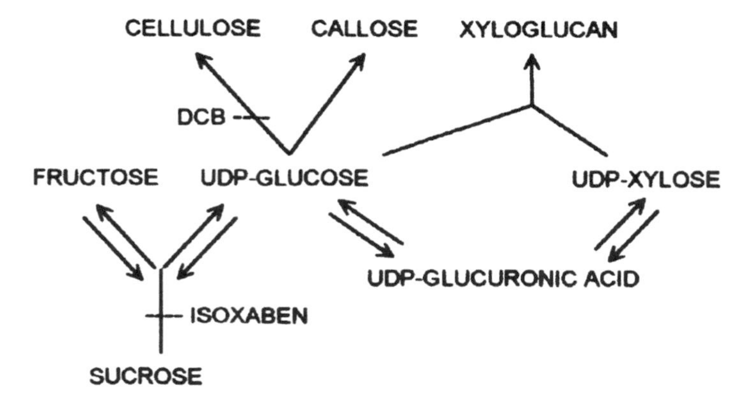

Figure 13-2. Proposed sites of inhibition in the cellulose biosynthesis pathway by isoxaben and dichlobenil (DCB). Isoxaben appears to inhibit the production of UDP-glucose from sucrose, which results in an inhibition of the synthesis of cellulose, and callose and xyloglucan synthesis is inhibited. DCB inhibits the conversion of UDP-glucose to cellulose as aftertreatment, UDP-glucose is shunted toward callose and xyloglucan synthesis. (Sabba & Vaughn, 1999, published with permission.)

very successful herbicide for control of barnyardgrass in rice. Although this unique site of action was thought to be highly unlikely to result in the development of resistant biotypes, there are now at least two examples of barnyardgrass biotypes that are resistant to quinclorac in the United States and Brazil.

LITERATURE CITED AND SUGGESTED READING

Delmer, D. P. 1999. Cellulose biosynthesis: Exciting times for a difficult field of study. *Ann. Rev. Plant Physiol. Plant Mol. Biol.* **50**:245–276.

Delmer, D. P., and Y. Amor. 1995. Cellulose biosynthesis. *Plant Cell* **7**:987–1000.

Heim, D. R., J. L. Roberts, P. D. Pike, and I.M. Larrinua. 1989. Mutation of a locus of *Arabidopsis thaliana* confers resistance to the herbicide isoxaben. *Plant Physiol.* **90**:146–150.

Heim, D. R., J. L. Roberts, P. D. Pike, and I. M. Larrinua. 1990. A second locus *Ixr* B1 in *Arabidopsis thaliana*, that confers resistance to the herbicide isoxaben. *Plant Physiol.* **92**:858–861.

Heim, D. R., L. A. Bjelk, J. James, M. A. Schneegurt, and I. M. Larrinua. 1993. Mechanism of isoxabon tolerance in *Agrostis palustris* var. penncross. *J. Exper. Botany* **44**:1185–1189.

Herbicide Handbook, 7th ed. 1994. Weed Science Society of America, Lawrence, KS.

Herbicide Handbook Supplement to 7th ed. Weed Science Society of America, Lawrence, KS.

Hogetsu, T., H. Shibaoka, and M. Shimokoriyama. 1974. Involvement of cellulose synthesis in actions of gibberellin and kinetin on cell expansion: 2,6-Dichlorobenzonitrile as a new cellulose synthesis inhibitor. *Plant Cell Physiol.* **15**:389–393.

Koo, S. J., J. C. Neal, and J. M. DiTomaso. 1994. Quinclorac-induced electrolyte leakage in seedling grasses. *Weed Sci.* **42**:1–7.

Koo, S. J., J. C. Neal, and J. M. DiTomaso. 1996. 3,7-Dichloroquinolinecarboxylic acid inhibits cell-wall biosynthesis in maize roots. *Plant Physiol.* **112**:1383–1389.

Lefebvre, A., D. Maizonnier, J. C. Gaudry, D. Clair, and R. Scalla. 1987. Some effects of the herbicide EL-107 on cellular growth and metabolism. *Weed Res.* **27**:125–134.

Sabba, R. P., and K. C. Vaughn. 1999. Herbicides that inhibit cellulose biosynthesis. *Weed Sci.* **47**:757–763.

Schneegurt, M. A., D. R. Heim, and I. M. Larrinua. 1994. Investigation into the mechanism of isoxaben tolerance of dicot weeds. *Weed Sci.* **42**:163–167.

Vaughn, K. C., J. C. Hoffman, M. G. Hahn, and L. A. Staehelin. 1996. The herbicide dichlobenil disrupts cell plate formation: Immunogold chracterization. *Protoplasma.* **194**:117–132.

For chemical use, see the manufacturer's or supplier's label and follow these directions. Also see the Preface.

14 Growth Regulator Herbicides

The growth regulator class of herbicides is used on more land area worldwide than any other group. Some are used extensively on the three leading world crops (wheat, rice, and corn), and there is substantial use on brush, rangeland, turf, and other grass crops. Historically, 2,4-D and MCPA are important because they helped provide the stimulus for the development of our agricultural chemical industry.

Growth regulator herbicides can be absorbed from the soil by plant roots; however, most of these compounds are applied as postemergence treatments. Translocation can be in both the xylem and phloem to active growth regions, but their action tends to be localized on the shoot system. They selectively kill broadleaf weeds but can injure grass crops if applied at the wrong time. In the case of perennial weeds, many of these herbicides translocate to below ground portions of the plant for systemic kill. Initial symptomology is quickly apparent on newly developing leaves and shoot regions as a twisting and epinasty of the shoot, cupping and crinkling of leaves, elongated leaf strapping (sometimes called "buggy whip") with parallel veins, stem swelling, and a disruption of phloem transport. Secondary effects can be a fusion of brace roots, such as observed with corn. Root injury is expressed as a proliferation or clustering of secondary roots and overall root growth inhibition. The specific site of herbicide inhibition is not known for this group. There appear to be multiple sites of action that disrupt hormone balance, nucleic acid metabolism, and protein synthesis. The herbicide action alters auxin activity in plants, resulting in weakened cell walls, rapid cell proliferation (unproductive growth), and plant death within several days or weeks.

HISTORY

The introduction of 2,4-D and MCPA in the mid-1940s, immediately after World War II, revolutionized weed control. They demonstrated that synthetic compounds could be developed and used to selectively control weeds in crops economically. Following their introduction, the chemical industry began major synthesis and evaluation programs, which led to the development of the wide array of herbicides that are available today,

The discovery of the phytotoxic properties of the phenoxy herbicides came directly from basic research on plant growth regulators, which began in the 1930s with the discovery that indole-3-acetic acid (IAA) and 2-naphthoxyacetic acid (NAA) promoted cell growth. In the early 1940s, the phenoxyacetic acids 2,4-D, MCPA, and 2,4,5-T were discovered. These compounds were found to be more active than IAA at affecting cell growth and not as readily metabolized in plant cells. This early research on these herbicides was not reported as it progressed because of

World War II security regulations; see the *Botany Gazette*, Volume 107 (1946), for several papers on this research. Slade et al. (1945) reported that the plant growth regulator α-napthaleneacetic acid controlled yellow charlock (wild mustard) in oats with only slight injury to the crop.

The phenoxy herbicides are often referred to as auxinlike herbicides because they induce twisting and curvature (epinasty) of the petioles and stems of broadleaf plants, reminiscent of plant response to high application doses of the native auxin, IAA. Benzoic acids were first reported to have growth regulating properties in plants in the 1940s (Zimmerman and Hitchcock, 1942), and dicamba was tested and introduced in the early 1960s. Dow researchers in 1963 reported the discovery of picloram, the first herbicidal pyridine derivative, and the last member of this group, quinclorac, was developed by BASF in 1984.

Characteristics

1. The growth regulator compounds affect plant growth in a similar way and appear to act at the same site as the natural plant auxin, IAA. However, all are much more active than IAA and persist in the plant longer.
2. All growth regulator herbicides are weak acids with pK_a values ranging from 2 to 4. Water solubility is highly influenced by formulation—high for salts, moderate for acids, and low for esters.
3. Volatility is formulation dependent, with esters being most volatile and amines being less volatile.
4. The compounds are used primarily to control broadleaf weeds in cereals, corn, and other grass crops and in noncropland.
5. Their effect on the plant is "systemic" rather than "contact."
6. These herbicides produce profound effects on the growth and structure of plants, including malformed leaves, epinastic bending and swelling of stems, deformed roots, and tissue decay. They cause parenchyma cells to divide rapidly, often producing callus tissue, excessive vascular tissue in young leaves, plugging of the phloem, and root growth inhibition. Meristematic tissues are more affected than mature tissues, with cambium, endodermis, pericycle, and phloem parenchyma being particularly sensitive.
7. Physical properties (weak acids) of growth regulator herbicides are consistent with phloem transport, which results in good control of perennial weeds.
8. These herbicides have a flat dose response, and they induce plant symptoms at concentrations well below the lethal dose, which creates potential problems with spray drift to susceptible crops/plants.

PHENOXYS

Six phenoxy herbicides (2,4-D, MCPA, MCPB, 2,4-DB, dichlorprop, and mecoprop) are currently used in the United States. In addition to the phenoxy "mainframe," all have a chlorine atom on the 4-position of the ring and an aliphatic acid attached to the oxygen atom (Table 14-1). The aliphatic acids are acetic, butyric, and proprionic acid.

TABLE 14-1. Common Name, Chemical Name, and Chemical Structure of Several Phenoxy Herbicides

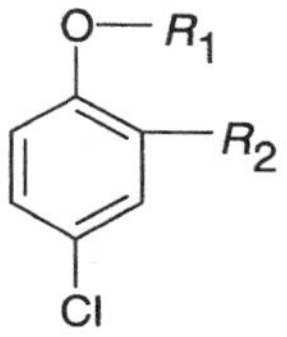

Common Name	Chemical Name	R_1	R_2
2,4-D	(2,4-Dichlorophenoxy) acetic acid	$-CH_2-COOH$	$-Cl$
MCPA	(4-Chloro-2-methylphenoxy)acetic acid	$-CH_2-COOH$	$-CH_3$
2,4-DB	4-(2,4-Dichlorophenoxy) butanoic acid	$-(CH_2)_3-COOH$	$-Cl$
MCPB	4-(4-Chloro-2-methylphenoxy) butanoic acid	$-(CH_2)_3-COOH$	$-CH_3$
Dichlorprop (2,4-DP)	(±)-2-(2,4-Dichlorophenoxy)propanoic acid	$-CH(CH_3)-COOH$	$-Cl$
Mecoprop (MCPP)	(±)-2-(4-Chloro-2-methylphenoxy)-propanoic acid	$-CH(CH_3)-COOH$	$-CH_3$

Each of these acids has a chlorine atom or methyl group on the 2-position of the ring. Although these six phenoxy herbicides have many characteristics in common, each has its own unique selective and phytotoxic uses. In general, the acetic forms are used in grass crops, lawns and turf, and noncropland, and the butyric acid forms are used in legume crops. The proprionic form dichlorprop (2,4-DP) is used for woody plant control, and the proprionic form mecoprop (MCPP) is primarily used in lawns and turf.

2,4-D

2,4-D [(2,4-dichlorophenoxy) acetic acid] is a white crystalline solid with a vapor pressure of 1.4×10^{-7} mm Hg at 25°C, which can be higher, depending on form (Table 14-2). The acid has a water solubility of 900 mg/l (ppm), at 25°C and other molecular forms of 2,4-D vary in their water solubilities—for example, butoxyether ester = 100 mg/l (ppm); dimethylamine salt = 796 mg/l (ppm), and the isooctyl ester = 0.0324 mg/l (ppm). The soil half-life is 10 days. The acute oral LD_{50} (rat) for the acid is 746 mg/kg, and ranges up to > 1000 mg/kg for other formulations.

In 1942, Zimmerman and Hitchcock of the Boyce Thompson Institute first described the use of 2,4-D as a plant growth regulator. In 1944, Marth and Mitchell of the USDA reported that 2,4-D killed dandelion, plantain, and other weeds in a bluegrass lawn.

Common forms of 2,4-D include the parent acid, amine salts, and esters. These various forms involve the substitution of another chemical group for the terminal hydrogen atom of the acetic side chain of the parent molecule. These substitutions

TABLE 14-2. General Characteristics of Different Forms of 2,4-D

Form	Solubility in Water	Solubility in Oil	Appearance When Mixed with Water	Precipitates Formed with Water	Volatility Hazard[a]
Acid	Low	Low	Milky	Yes	Low
Amine salts					
Water-soluble	High	Low	Clear	Yes	None
Oil-soluble	Low	High	Milky	Yes	None
Esters					
Low-volatile	Low	High	Milky	No	Medium
High-volatile	Low	High	Milky	No	High

[a]The tendency to form volatile fumes or gases that can injure susceptible plants.

alter the physical and biological characteristics of the parent molecule and thereby facilitate the use and/or increase the effectiveness of 2,4-D in the field. The relative effectiveness of the various forms usually refers to their different degrees of phytotoxicity at equal rates of application. Increased effectiveness of a particular form is usually associated with increased absorption, but volatility of the compound usually also increases with increased absorption. The general characteristics of these different forms are given in Table 14-2.

However, regardless of the substitution, it is the parent molecule that acts as the herbicide at its site of action in the plant. Some forms may contain more than one molecular form of 2,4-D in a given formulation—for example, two different amines or ester plus acid. The concentration of essentially all phenoxy herbicide formulations is expressed as acid equivalent in pounds per gallon. *Acid equivalent* refers to that part of the formulation that theoretically can be converted to the acid. Recommendations are also made on this basis.

Acid forms are not commonly used because they are only moderately soluble in water, slightly volatile, and relatively expensive to formulate. Other less expensive formulations are equally effective for many purposes. However, acid forms are more effective on certain hard-to-kill weeds than amine forms. They are available as emulsifiable concentrates, alone, or in combination with other forms of 2,4-D or other herbicides. One formulation contains a 2,4-D ester and is particularly effective in control of field bindweed, Russian knapweed, Canada thistle, leafy spurge, cattails, tules, and nutsedge.

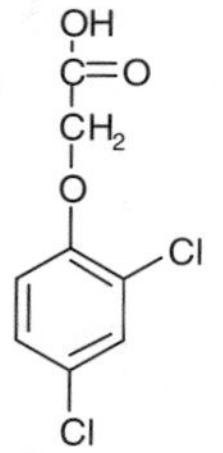

OH
C=O
CH_2
O
Cl
Cl

2,4-D

Amines are produced by reacting the 2,4-D acid with an amine, forming an amine salt of 2,4-D. There are two types of 2,4-D amines, namely, water-soluble and oil-soluble amines that are distinctly different in their physical and biological properties. Water-soluble amines are the most commonly used form of 2,4-D because of their high water solubility, very low volatility, ease of handling in the field, and overall cost. They are formulated as water-soluble concentrates. They are somewhat less effective than most other forms but provide effective weed control for many purposes at minimal cost.

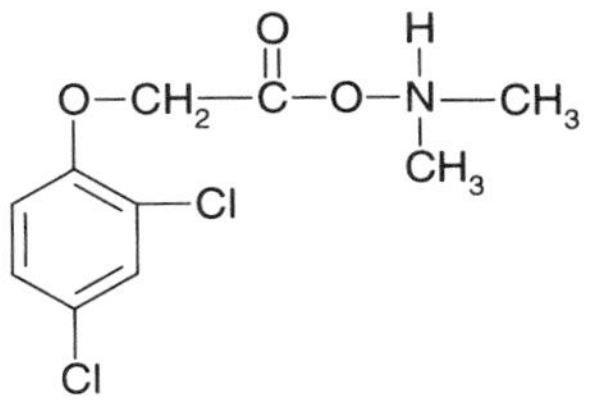

Dimethylamine Salt of 2,4-D (water soluble)

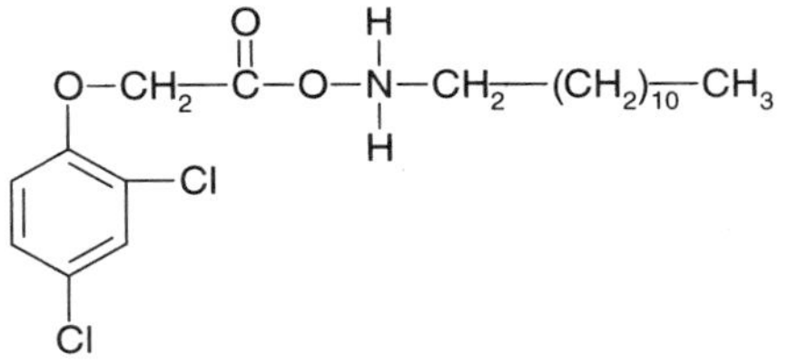

Dodecylamine Salt of 2,4-D (oil soluble)

Oil-soluble amines are essentially insoluble in water and are used as emulsifiable concentrates. Their major advantage is that their effectiveness approaches that of low-volatile esters of 2,4-D with minimal volatility hazard, especially at high temperatures.

Esters are produced by reacting the 2,4-D acid with an alcohol, and a number of different esters are used. Increasing the length of the alcohol side chain reduces the volatility of the compound and generally reduces its leaf absorption. In general, however, esters are absorbed more readily than any of the other forms of 2,4-D. There are three types of 2,4-D esters: low-volatile esters, high-volatile esters, and invert esters. They are distinctly different in their physical and biological properties.

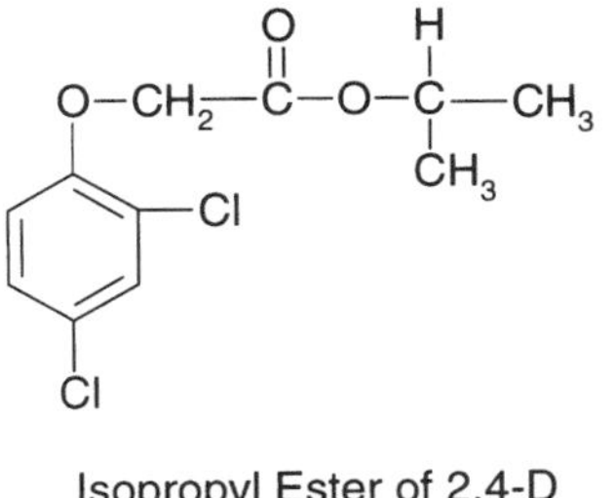

Isopropyl Ester of 2,4-D (a volatile form)

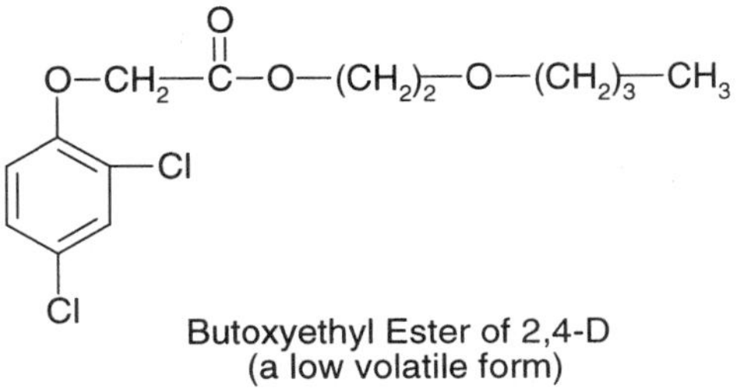

Butoxyethyl Ester of 2,4-D (a low volatile form)

Low-volatile esters are essentially insoluble in water and used as emulsifiable concentrates; for some uses they are dissolved in kerosene or diesel oil. They are somewhat volatile and present a volatility hazard, particularly under hot conditions. Low-volatile esters are more effective than amines for controlling certain hard-to-kill weeds, such as bindweed, thistles, smartweed, wild garlic, curled dock, tansey ragwort, and wild onion.

High-volatile esters are essentially insoluble in water and used as emulsifiable concentrates. They are very volatile and present a serious volatility hazard; therefore, they are used only in isolated areas where volatility drift will not cause injury to desirable species. Their use is prohibited in many areas.

Invert esters are unique formulations that produce an invert emulsion (water-in-oil, W/O) when mixed with water. This is in contrast to the oil-in-water (O/W) type of emulsion commonly used in herbicidal sprays. A detailed discussion of emulsions (emulsifiable concentrates) is presented in Chapter 7. Invert emulsions produce a more viscous solution than oil-in-water emulsions and are therefore less subject to spray drift.

Uses 2,4-D is formulated as an EC, SL, SP, SL, and G under a wide variety of trade names and used primarily as a postemergence treatment to control annual and perennial herbaceous and woody weeds. It is registered for use in corn, small grains, grain sorghum, rice, sugarcane, orchards (pome and stone fruit), cranberries, strawberries, asparagus, turf, pastures and rangeland, conifer release, aquatic situations, and noncrop areas in the United States and throughout the world. 2,4-D also has some herbicidal activity via the soil and is used preemergence in corn after planting, but before crop emergence on high organic matter soils. Injury to the corn can occur if the compound is soil applied on coarse-textured and low organic matter soils. It is used in soybeans as an early preplant soil application at least 30 days ahead of no-till planting. Symptoms of 2,4-D response are shown in Figures 14-1 and 14-2.

Figure 14-1. A common burdock plant twisted and curled following treatment with 2,4-D.

Figure 14-2. Abnormality of corn brace roots induced by a high rate of 2,4-D applied during a susceptible stage of growth.

MCPA

MCPA [(4-chloro-2-methylphenoxy) acetic acid] is a light brown solid, with a water solubility of 825 mg/l (ppm) at 25°C for the acid, 866,000 mg/l (ppm) for the dimethylamine salt, 5 mg/l (ppm) for the isooctyl ester, and 270,000 mg/l (ppm) for the sodium salt. MCPA acid has a vapor pressure of 1.5×10^{-6} mm Hg at 20°C for the acid, and the dimethylamine salt has a negligible vapor pressure. The soil half-life is 5 to 6 days and the acute oral LD_{50} (rat) is 1160 mg/kg for the acid.

O—CH_2C(=O)—OH
CH_3
Cl

MCPA

Uses MCPA is formulated as an SL, SP, and EC and sold throughout the world under numerous MCPA trade names for the various amine, ester, and inorganic salt forms. MCPA was one of the first hormone-type herbicides discovered in England. It is used as a postemergence herbicide, with characteristics similar to those of 2,4-D, except that it is more selective on cereals, legumes, and flax at equal rates and may be more effective than 2,4-D on certain broadleaf weed species. MCPA is used in the United States in flax, peas, wheat, barley, oats, rye, alfalfa, birdsfoot trefoil, clovers (alsike, red, and ladino), pastures, rangelands, conservation reserve programs (CRP) and set-aside acres, in other parts of the world, in rice, vines, potatoes, and under fruit trees. The rates used vary, depending on formulation (amine, ester, or inorganic salt), weed type (annual, biennial, or perennial), stage of crop growth, and variety.

MCPB

MCPB [4-(4-chloro-2-methylphenoxy)butanoic acid] is a white crystalline solid which as the acid is essentially water insoluble and as the sodium salt has a water solubility of 200,000 mg/L (ppm) at 25°C. The vapor pressure is not reported for the sodium salt. The soil half-life is 14 days for the sodium salt and the oral LD_{50} (rat) is 690 mg/kg.

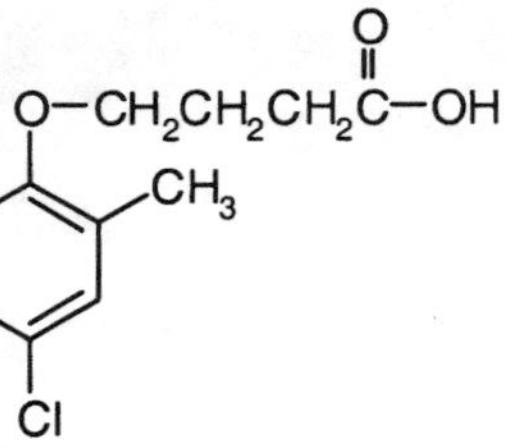

$O-CH_2CH_2CH_2C(=O)-OH$
CH_3
Cl

MCPB

Uses MCPB is formulated as the sodium salt as an SL and sold as Thistrol, Bellmac Straight, Madek, and Topotox for use in controlling annual and perennial broadleaf weeds (very effective on thistles) in peas (in the United States), and in cereals, clovers, sainfoil, peas, peanuts, and pastures in other parts of the world.

2,4-DB

2,4-DB [4-(2,4-dichlorophenoxy)butanoic acid] is a white crystalline solid that has a negligible vapor pressure and a water solubility of 46 mg/l (ppm) for the acid, 709,000 mg/l (ppm) for the dimethylamine salt, and 8 mg/l (ppm) for the low-volatile butoxyethyl ester at 25°C. The soil half-life is 5 days, 10 days, and 7 days, respectively, for the acid, amine, and ester forms and the LD_{50} (rat) for the acid is 1960 mg/kg.

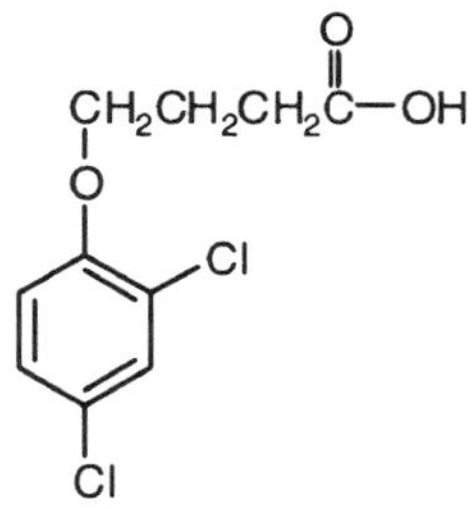

2,4-DB

Uses 2,4-DB is formulated as an SL and WG in the amine form and the acid forms, respectively, and sold as Butyrac or 2,4-DB. It is used for early postemergence weed control in seedling and established alfalfa, in birdsfoot trefoil, seedling alsike, ladino and red clover, and peanuts, for postemergence or postemergence directed treatments in soybeans in the Unitd States and in grasslands in other parts of the world. The ester form has been discontinued.

The 2,4-DB mechanism of action is described in a later section; however, it is not highly phytotoxic per se. Once applied to the plant, it undergoes β-oxidation (Chapter 5) in plants and soil to form 2,4-D, which is phytotoxic. This reaction is more rapid in susceptible plants than in tolerant plants (e.g., small-seeded legumes). Therefore, many broadleaf weeds are controlled by 2,4-DB, unlike the small-seeded legumes, which are less subject to injury.

Dichlorprop

Dichlorprop (also called 2,4-DP) [(±)-2-(2,4-dichlorophenoxy)propanoic acid] is a white to tan crystalline solid with a vapor pressure of 3×10^{-6} mm Hg at 20°C, a water solubility of 710 mg/l (ppm) at 28°C for the acid and 50 mg/l (ppm) at 25°C for the ester, a soil half-life of 10 days, and an LD_{50} (rat) of 800 mg/kg.

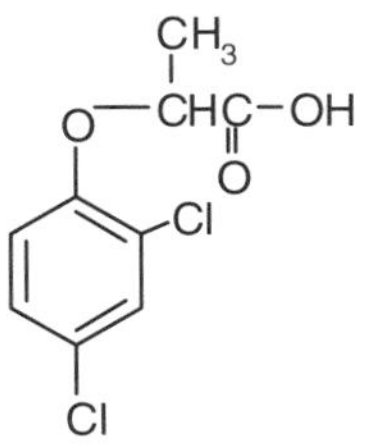

Dichlorprop

Uses Dichlorprop is formulated as an EC and SL and is generally sold only in mixtures with other herbicides under various trade names for control of broadleaf weeds in small grains and turf and in brush control in noncropland in the United States. It is used in cereals, grasslands, and in aquatic weed control in other parts of the world and is especially good for control of chickweed, smartweed, and woody species.

Mecoprop

Mecoprop (also called MCPP) [(±)-2-(4-chloro-2-methylphenoxy)propanoic acid] is a colorless crystalline solid with a moderate water solubility of 620 mg/l (ppm) at 20°C, a soil half-life of 21 days, and an oral LD_{50} (rat) of 650 mg/kg.

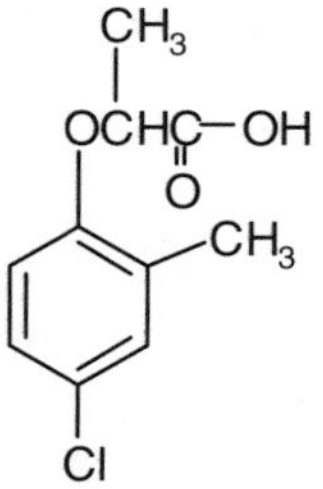

Mecoprop

Uses Mecoprop is formulated as an SL and EC and sold under various MCPP trade names to control 2,4-D-tolerant weeds such as chickweed, clover, plaintain, knotweed, and ground ivy in lawns and turf and in small grains. There are some use restrictions for certain grass species, environmental conditions, and time of mowing. Check product labels for specific uses and restrictions.

Soil Influences

The soil influences on phenoxies are similar for 2,4-D, MCPA, MCPB, 2,4-DB, mecoprop, and dichlorprop. Soil type and product formulation influence leaching in soil. For example, 2,4-D is adsorbed on soil colloids, and less leaching occurs in clay and organic soils than in sandy soils. Microorganisms are of major importance in the disappearance of 2,4-D (and other phenoxies) from soil. 2,4-D persists at phytotoxic levels from 1 to 4 weeks in warm, moist loam soil at usual application rates and does not persist into the next growing season. Although 2,4-D does not generally reduce the total number of microorganisms in the soil, it may reduce nodulation of legume species (Payne and Fults, 1947).

BENZOICS

Dicamba

Dicamba (3,6-dichloro-2-methoxybenzoic acid) is a white crystalline solid with a vapor pressure of 9.24×10^{-6} mm Hg at 25°C, a water solubility of 4500 mg/l (ppm) at 25°C for the acid, 720,000 mg/l (ppm) at 25°C for the dimethylamine, and 400,000 mg/l (ppm) for the sodium salt; the newer glycolamine form is also highly soluble. The soil half-life is 4.4 days, and the oral LD_{50} (rat) is 1707 mg/kg.

Dicamba

Uses Dicamba is formulated as an SL and G and sold under several trade names, such as Banvel in the dimethylamine salt and sodium salt forms and Clarity and Vanquish in the diglycolamine salt form, for selective broadleaf weed control in corn, grain sorghum, small grains, cotton, perennial grass seed crops (including fescues, bluegrass, and ryegrass) and established turf and for control of undesirable brush and trees, cut surface treatment, and annual and perennial broadleaf control in fallow land, pastures, noncropland, and aquatics. As with 2,4-D use on corn, sorghum, and small grains, postemergence and postemergence directed applications must be made at the appropriate growth stage to avoid injury. See label for details. Although similar to 2,4-D in its general weed control spectrum, dicamba is outstanding for the control of Polygonacae weeds, wild buckwheat, pigweed, jimsonweed, chickweed, purslane, prickly sida, smartweed, and black nightshade. Dicamba is combined with 2,4-D and other herbicides to broaden the spectrum of weeds controlled. Dicamba can drift to sensitive crops and cause significant damage, especially to high-value horticulture crops (Fig. 14-3).

Figure 14-3. Modification of leaf morphology of cucumber induced by dicamba.

Soil Influence Dicamba is relatively mobile in the soil and the degree of leaching is dependent on the amount of rainfall. Leaching of dicamba to the roots of certain ornamental plants can cause injury and death. Taxus are especially sensitive to dicamba soil residues. Soil microorganisms degrade this compound, and the rate of degradation is most rapid under warm, moist soil conditions and in slightly acid soils. Under cool, dry soil conditions dicamba persists up to several months.

PICOLINIC ACIDS

Picloram

Picloram (4-amino-3,5,6-trichloro-2-pyridinecarboxylic acid) is a white powder with a vapor pressure of 6.16×10^{-7} mm Hg at 35°C, a water solubility for the acid of 430 mg/l (ppm) and 200,000 mg/l (ppm) for the amine salt at 25°C, a soil half-life of 90 days and an oral LD_{50} (rat) of >5000 mg/kg for the acid and the amine. Picloram is a restricted-use pesticide that is highly active and very mobile, yet very persistent, in the soil, often persisting into the next growing season. Very small amounts can kill or injure many broadleaf plants. Extreme care must be taken when applying it to prevent its escape from the target site.

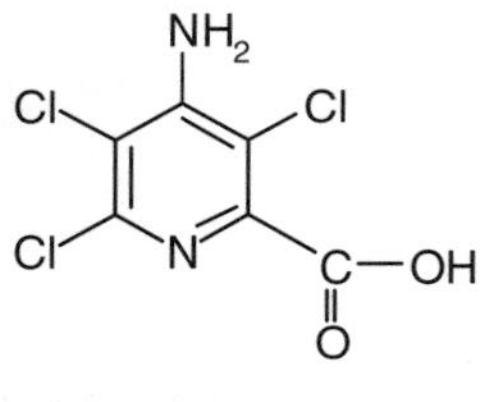

Picloram

Uses Picloram is formulated as an SL and sold under the trade name Tordon for effective control of many perennial broadleaf weeds and brush in noncropland, rights-of-way, range areas, permanent grass pastures, and in small grains. Picloram is also sold in combination with 2,4-D for use in permanent grasslands and rangeland. It is applied to both the foliage and the soil and is also used as a basal or cut surface treatment for unwanted tree control.

Soil Influences Organic matter and certain clays adsorb picloram. Picloram is readily leached through sandy and montmorillonite clay soils low in organic matter, but not through soils high in organic matter or lateritic soils. Salts of picloram appear to be leached more readily than the parent acid form.

Picloram is very persistent in soils, which is one of the reasons it is a restricted-use herbicide. Microorganisms slowly degrade it. Conditions that favor microbial growth, such as warm, moist soil and organic matter, reduce its period of persistence. The application rate also influences its period of phytotoxicity in soils. Phytotoxicity may often be detected well over 1 year after application.

Triclopyr

Triclopyr [(3,5,6-trichloro-2-pyridinyl)oxy]acetic acid) is a fluffy white solid with a vapor pressure of 1.26×10^{-6} mm Hg at 25°C, a water solubility of 430 mg/l (ppm) for the acid, 23 mg/l (ppm) for the ester, and 2,100,000 mg/l (ppm) for the amine salt at 25°C, a soil half-life of 30 days, and an LD_{50} (rat) of 713 mg/kg. Triclopyr is closely related to picloram but has greater selectivity.

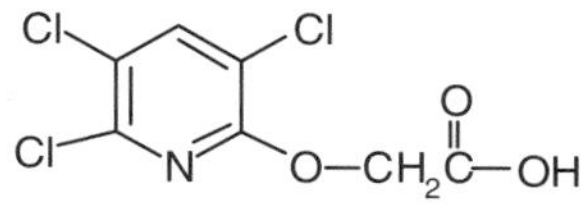

Triclopyr

Uses Triclopyr is formulated as an SL and is sold under various trade names in the amine and ester forms, either as a stand-alone product or in combination with clopyralid or 2,4-D, for control of many woody and broadleaf weeds. It controls ash, oaks, and other root sprouting species better than other auxin-type herbicide. Most grass species are tolerant. Triclopyr stand-alone products include Tuflon Ester, Remedy (ester), Garlon (ester or amines), Grandstand (amine), and combination products include Redeem and Confront (+ clopyralid) and Crossbow (+ 2,4-D). These products are registered for use in various applications, including turf, sod farms, permanent grass pastures, rangeland, certain ornamentals, noncropland, industrial sites, rights-of-way, forest and wildlife openings (including grazed sites), rice, CRP, and nonirrigation-ditch banks. Check product labels for specific uses.

Soil Influences Organic matter content and pH influence triclopyr adsorption, but this herbicide is not considered to be strongly adsorbed on soil colloids. Some leaching may occur in light soils under high rainfall conditions. It is degraded in soils by microorganisms at a rate that is considered to be relatively rapid.

Clopyralid

Clopyralid (3,6-dichloro-2-pyridinecarboxylic acid) is an off-white crystalline solid with a vapor pressure of 1.3×10^{-6} mm Hg at 25°C, a water solubility of 1000 mg/l (ppm) for the acid and 300,000 mg/l (ppm) for the amine at 25°C, a soil half-life of 12 to 70 days, depending on soil conditions and an LD_{50} (rat) of 4300 mg/kg. Clopyralid, like triclopyr, is closely related to picloram but has much greater selectivity.

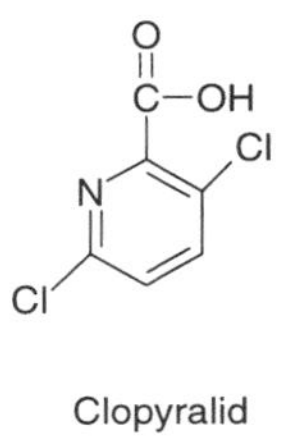

Clopyralid

Uses Clopyralid is formulated as an SL under various trade names in the amine and acid forms as a stand-alone product or in combination with tryclopyr, 2,4-D, or flumetsulam. It is used as a postemergence herbicide applied to the foliage of plants, but it can also affect susceptible species by root uptake. Clopyralid controls many annual and perennial broadleaf weeds and certain woody species (e.g., mesquite and associated species). It is particularly effective on members of the Umbelliferae, Polygonaceae, Asteraceae, and Leguminosae families, but does not control grasses. Clopyralid stand-alone products include Translin (amine), Lontrel Turf and Ornamental (amine), and Stinger (amine); combination products include Curtail M (+ 2,4-D), Hornet (+ flumetsulam), Reedem, and Confront (+ triclopyr). These products are registered for various uses, including wheat, oats, and barley not underseeded with legumes, fallow cropland, grasses grown for seed, rangeland, permanent grass pastures, CRP, noncropland, rights-of-way, industrial sites, wildlife openings including grazed sites, established turf, sod farms, field corn, asparagus, Christmas tree plantations, mint, sugar beets, selected ornamentals, and tree plantations. Clopyralid is used in European cereals to extend the spectrum of weed control with 2,4-D or MCPA. It is also available in Canada for use in canola, particularly for the control of Canada thistle.

Soil Influences Clopyralid is not strongly adsorbed by soil colloids. It exists in the soil primarily in the salt form and is therefore subject to leaching. It is degraded by microorganisms at a medium to fast rate in a wide range of soils. No injury to susceptible broadleaf crops was observed the year following a field application of 0.5 lb/acre (0.56 kg/ha).

Quinclorac

Quinclorac (3,7-dichloro-8-quinolinecarboxylic acid) is a colorless crystal with a vapor pressure of $< 10^{-7}$, a water solubility of 62 mg/L (ppm) at 20°C, and an oral LD_{50} (rat) of 2610 mg/kg. The soil half-life is not reported, but soil residual amounts may injure certain susceptible species for up to 1 year after application (WSSA, *Herbicide Handbook*, 1994, 1998).

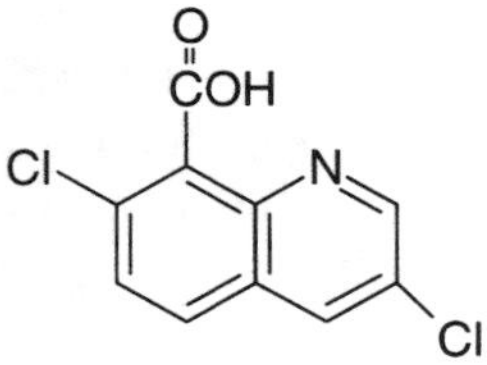

Quinclorac

Uses Quinclorac is formulated as a WP and sold under the trade name Facet. It is registered for preemergence control of annual grasses (barnyardgrass is especially well controlled) and certain broadleaf weeds in rice in several countries, but not in Japan. Its mechanism of action is discussed in the next section of this chapter and in Chapter 13, and it apparently has two different mechanisms that result in plant death.

Soil Influences Quinclorac is only slightly adsorbed by the soil, but soils with organic matter and clay will adsorb some of this herbicide. It is relatively mobile in soil, especially in lighter soils with low organic matter, and leaching increases with greater amounts of rainfall. Microorganisms in the soil degrade quinclorac, and the water regimes used in rice paddies can affect the rate of disappearance.

SELECTIVITY

The selectivity of growth regulator herbicides does not seem to be the result of a single factor, but is determined by the sum of many plant reactions to the herbicide. Potential selectivity mechanisms include the following:

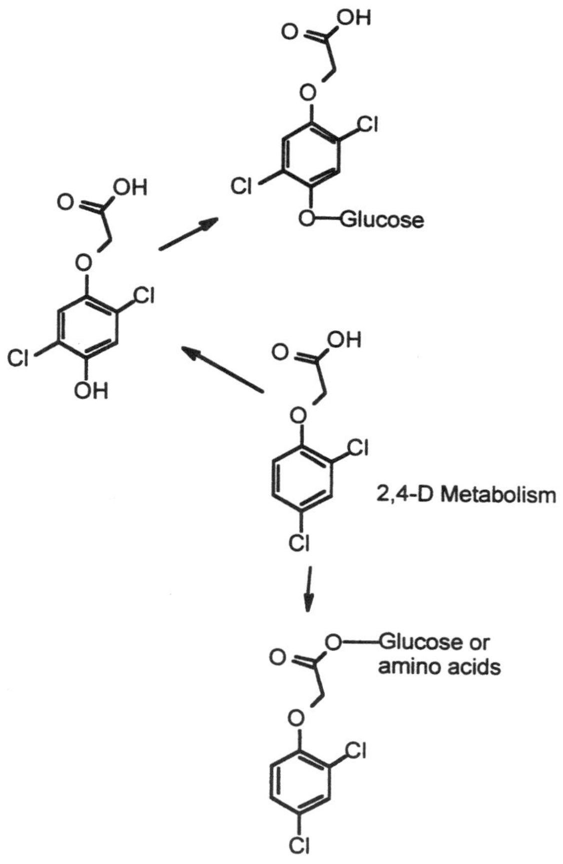

Figure 14-4. A major mechanism of metabolism of 2,4-D in plants is by aryl hydroxylation to 2,5-dichloro-4-hydroxyphenoxyacetic acid and 2,3-D-4-OH. This metabolism results in a loss of auxin activity.

1. The arrangement of vascular tissue in scattered bundles surrounded by protective sclerenchyma tissue in grasses (monocotyledons) may prevent destruction of the phloem by disorganized growth caused by growth regulator herbicides.
2. Metabolism of 2,4-D by aryl hydroxylation of 2,4-D to 2,5 dichloro-4-hydroxy-phenoxyacetic acid and 2,3-D-4-OH is a major pathway for 2,4-D metabolism (Figure 14-4). Aryl hydroxylation of 2,4-D results in the loss of auxin activity. In addition to aryl hydroxylation and subsequent glycosylation, conjugation of 2,4-D with amino acids has been reported in many species. However, amino acid conjugates of 2,4-D are biologically active and therefore may not represent a major mechanism of detoxification. Metabolism reactions serve to reduce the amount of herbicide within the plant.
3. Some plants can excrete or release herbicides through the root system.
4. Altered affinity for an auxin-binding site on the plasmalemma may modify sensitivity.

PROBLEMS

Because of the flat dose response curve and extreme sensitivity of certain plants (i.e., grapes, tomatoes, redbuds, and cotton to 2,4-D and soybeans to dicamba), drift from treated fields can cause serious problems. Such problems can be reduced by the following practices:

1. Avoid use of volatile formulations.
2. Use high volumes of spray if practical.
3. Use low pressure.
4. Avoid spraying when wind is blowing toward susceptible crops.

MECHANISM OF ACTION OF GROWTH REGULATORS

Epinasty is among the most obvious effects of all growth regulator herbicides on broadleaf plants. These plants usually develop grotesque and malformed leaves and stems when treated (Figure 14-1). Brace roots of corn also develop abnormally (Figure 14-2). The herbicides concentrate in young embryonic or meristematic tissues that are growing rapidly, and these tissues are more sensitive than mature or relatively inactive young tissue.

Histological studies with red kidney beans showed that the cambium, endodermis, embryonic pericycle, phloem parenchyma, and phloem rays were grossly altered by 2,4-D. The cortex and xylem parenchyma showed little response, and the epidermis, pith, mature xylem, mature sieve tubes, and differentiated pericycle showed no response. These results suggest that active cell division is essential for the development of 2,4-D toxicity symptoms. The types of tissue affected by 2,4-D in field bindweed and sow thistle were much the same as those in beans.

Leaves readily absorb nonpolar forms (acids, esters, oil-soluble amines) of 2,4-D, whereas polar forms (inorganic salts, water-soluble amines) are absorbed more slowly. The use of surfactants usually increases foliar absorption, and absorption increases with increasing temperature and humidity. Plant stems and roots also absorb all these compounds. Rainfall shortly after application may decrease effectiveness, but rainfall 6 to 12 hours later has little effect. Nonpolar forms have a tendency to resist removal by rainfall.

After absorption by plant foliage, all these herbicides are translocated throughout the plant in the phloem. With 2,4-D, it was shown to move from the leaves (source) with the photosynthate in the phloem, but more slowly than the photosynthate itself. 2,4-D accumulated in the sink areas of the plant (e.g., developing organs and meristems). Limited translocation occurred in grasses relative to broadleaf plants (see the section "Selectivity"), which may partially explain grass tolerance to 2,4-D (Ashton, 1958).

Translocation of a foliar-applied herbicide to underground roots and rhizomes is essential for the control of perennial weeds. Therefore, periods of maximum growth and photosynthate accumulation in those underground organs and minimum growth of the aboveground organs favor the control of perennial weeds. This usually occurs in the fall for most perennial species but may occur at other times of the year for certain perennial species. Excessive rates of application may damage the phloem and reduce translocation; therefore, sequential low rates usually give better control of perennial weeds than a single high rate of application.

Plant age and associated rate of growth influence susceptibility; however, this will vary with different growth regulator herbicides. In general for 2,4-D, younger plants are more susceptible than older plants of the same species. However, some plants are tolerant while small, and others never gain more than slight tolerance. Some plants may develop a second period of susceptibility. For example, small grains are very susceptible to 2,4-D in the germinating and small seedling stages but become tolerant in the fully tillered stage, susceptible again in the jointing, heading and flowering stages, and tolerant again in the "soft-dough" stage (see Figure 19-3). The periods of susceptibility coincide with periods of rapid growth. At this time, the cells of the meristems are dividing rapidly, have a high level of metabolic activity, and are very susceptible to 2,4-D.

The biochemical and metabolic changes in plants reported to be induced by growth regulator herbicides are numerous. Studies suggest that nucleic acid metabolism and the metabolic aspects of cell wall plasticity are most relevant to the mode of action of growth regulator herbicides. Early work showed stimulated synthesis of RNA and DNA in IAA-treated tobacco pith cells. A few years after the introduction of the phenoxy herbicides, similar results were found in a wide variety of plant tissues exposed to these herbicides. 2,4-D appears to be acting in a manner similar to the native auxin (IAA). However, IAA has endogenous control mechanisms that maintain its concentration within the appropriate physiological range. Growth regulator herbicides have no such control mechanisms. It is well known that 2,4-D can stimulate or inhibit cell growth, depending on the concentration present in the tissue. In general, low concentrations stimulate growth and high concentrations inhibit growth. In fact, low concentrations of 2,4-D are a common component of tissue culture growth media and are used to promote cell growth. The level of the growth regulating herbicide in the meristem and developing organs of the intact treated plant increase with time after application; the level is initially low and later high. Thus, there is first a stimulation of cell metabolic processes, resulting in uncontrolled growth, and later an inhibition of these processes and plant death.

It is well established that soon after growth regulator herbicide application, early plant responses are associated with cell wall acidification and changes in gene expression. Auxin and auxinic growth regulating herbicides induce proton efflux through the plasma membrane by stimulation of proton pumping ATPase, which leads to the acidification of the cell wall matrix. Low pH increases cell wall extensibility and activates enzymes (extracellular cellulases) that degrade cell walls. Together, these events weaken the cell wall and enable growth via turgor-driven cell expansion. Auxin also promotes changes in gene expression. Approximately 25 auxin-responsive genes have been identified, however, with the exception of ACC synthase, the precise biochemical action of other auxin-responsive gene products is unknown. ACC synthase is the key regulatory enzyme in ethylene biosynthesis. Ethylene has been suggested to be causally involved in the effects induced (epinasty) in susceptible plants by growth regulating herbicides.

Tissue proliferation induced by a growth regulator herbicide leads to epinasty, stem swelling, and disruption of the phloem, preventing photosynthate movement from the leaves to the root system. This unproductive growth causes death in several days or weeks.

Mechanism of Action of Quinclorac

Differing theories on quinclorac's mode of action of have been proposed, and it is now thought that this herbicide has two mechanisms of action in plants. One of those theories involves an auxinic mode of action in broadleaf weeds and inhibition of cellulose biosynthesis in grass (see Chapter 13). This makes quinclorac somewhat unique because, unlike other growth regulator herbicides, it is very active on grass weeds. Klaus Grossmann (1998) has proposed a mechanism of action based on its auxin activity, which induces ethylene biosynthesis in susceptible species. Quinclorac stimulated the synthesis of 1-aminocyclopropane-1-carboxylic acid (ACC, the immediate precursor of ethylene) in barnyardgrass, a sensitive species. The formation of ethylene from ACC produces cyanide as a byproduct. Grossmann has suggested that the accumulation of endogenous cyanide is related to the phytotoxic symptoms produced by quinclorac.

RESISTANCE

Resistant biotypes of wild mustard to 2,4-D (Hall et al., 1993), yellow starthistle to picloram (Fuerst et al., 1996), and chickweed to MCPA (Coupland et al., 1991) have been reported. Moreover, there are frequent reports on the varying tolerance of biotypes within a species. There are now at least two species of barnyardgrass resistant to quinclorac in the United States and Brazil.

LITERATURE CITED AND SUGGESTED READING

Ashton, F. M. 1958. Absorption and translocation of radioactive 2,4-D in sugarcane and bean plants. *Weeds* **6**:257–262.

Cardenas, J., F. W. Slife, J. B. Hanson, and H. Butler. 1968. Physiological changes accompanying the death of cocklebur plants treated with 2,4-D. *Weed Sci.* **16**:96–100.

Coupland, D., D. T. Cooke, and C. S. James. 1991. Effects of 4-chloro-2-methylphenoxypropionate on plasma membrane ATPase activity in herbicide-resistant and herbicide-susceptible biotypes on *Stellaria media*. *J. Exp. Bot.* **42**:1065–1071.

Fites, R. C., J. B. Hanson, and F. W. Slife. 1969. Alteration of messenger RNA and ribosome synthesis in soybean hypocotyl by 2,4-D. *Bot. Gaz.* **130**:118–126.

Frear, D. S., H. R. Swanson, and E. R. Mansager. 1989. Picloram metabolism in leafy spurge: Isolation and identification of glucose and gentiobase conjugates. *Agric. Food Chem.* **37**:1408–1412.

Fuerst, E. P., T. M. Sterling, M. A. Norman, T. S. Prather, G. P. Irzyk, Y. Wu, N. K. Lownds, and R. H. Callihan. 1996. Physiological characterization of picloram resistance in yellow starthistle. *Pestic. Biochem. Physiol.* **56**:149–161.

Grossman, K. 1998. Quinclorac belongs to a new class of highly selective auxin herbicides. *Weed Sci.* **46**:707–716.

Hall, J. C., S. M. M. Alam, and D. P. Murr. 1993. Ethylene biosynthesis following foliar application of picloram to biotypes of wild mustard (*Sinapis arvensis*) susceptible or resistant to auxinic herbicides. *Pestic. Biochem. Physiol.* **47**:36–43.

Hamner, C. L., and H. B. Tukey. 1944. Selective herbicidal action of midsummer and fall applications of 2,4-D. *Bot. Gaz.* **106**:232–245.

Hartwig, L., M. Claussen, and M. Bottger. 1999. Growth: Progress in auxin research. *Prog. Bot.* **60**:315–340.

Herbicide Handbook. Supplement to 7th ed. 1998. Weed Science Society of America, Lawrence, KS.

Herbicide Handbook. 7th ed. 1994. Weed Science Society of America, Lawrence, KS.

Koo, S. J., J. C. Neal, and J. M. Di Tomaso. 1994. Quinclorac-induced electrolyte leakage in seeding grasses. *Weed Sci.* **42**:1–7.

Krueger, J. P., R. G. Butz, and D. J. Cork. 1991. Aerobic and anaerobic soil metabolism of dicamba. *J. Agric. Food Chem.* **39**:995–999.

Lerner, P., and W. J. Owen. 1990. Selective action of the herbicide triclopyr. *Pestic. Biochem. Physiol.* **36**:187–200.

Lutman, P. J. W., and C. R. Heath. 1990. Variation in the resistance of *Stellaria media* to mecoprop due to biotype, application method and 1-aminobenzotriazole. *Weed Res.* **30**:129–138.

Lym, R. E., and K. D. Moxness. 1990. Absorption, translocation and metabolism of picloram and 2,4-D in leafy spurge (*Euphorbia esula*). *Weed Sci.* **37**:498–502.

MacDonald, R. L., C. J. Swanton, and J. C. Hall. 1994. Basis for the selective action of fluroxpyr. *Weed Res.* **34**:333–344.

Marth, P. C., and J. W. Mitchell. 1944. 2,4-dichlorophenoxyacetic acid as a differential herbicide. *Bot. Gaz.* **126**:224–232.

Moyer, J. R., P. Bergen, and G. B. Schaalje. 1992. Effect of 2,4-D and dicamba residues on following crops in conservation tillage systems. *Weed Technol.* **6**:149–155.

Sabba, R. P., T. M. Sterling, and M. K. Lownds. 1998. Effect of picloram on resistant and susceptible yellow starthistle: The role of ethylene. *Weed Sci.* **46**:297–300.

Slade, R. E., W. G. Templeman, and W. A. Sexton. 1945. Found MCPA controlled charlock in cereals. *Nature* **155**:497–498.

Slife, F. W. 1956. The effect of 2,4-D and several other herbicides on weeds and soybeans when applied as postemergence sprays. *Weeds* **4**:61–68.

Stephenson, G. R., K. R. Solomon, C. S. Bowley, and K. Lyer. 1990. Persistence, leachability and lateral movement of triclopyr in selected Canadian forestry soils. *J. Agric. Food Chem.* **38**:584–588.

Sterling, T. M., and J. C. Hall. 1997. Mechanism of action of natural auxins and the auxinic herbicides. In *Herbicide Activity: Toxicology, Biochemistry and Molecular Biology*, ed. by R. M. Roe. IOS Press, Amsterdam.

Walker, L., and M. Estelle. 1998. Molecular mechanisms of auxin action. *Curr. Opin. Plant Biol.* **1**:434–439.

Wall, D. A. 1994. Potato (*Solanum tuberosum*) response to simulated dicamba drift. *Weed Sci.* **42**:110–114.

White, R. H., R. A. Liebl, and T. Hymowitz. 1990. Examination of 2,4-D tolerance in perennial *Glycine* species. *Pestic. Biochem. Physiol.* **38**:153–161.

Zimmerman, P. W., and A. E. Hitchcock. 1942. Substituted phenoxy and benzoic acid growth substances and the relation of structure to physiological activity. *Contr. Boyce Thompson Inst.* **12**:321–344.

For chemical use, see the manufacturer's or supplier's label and follow these directions. Also see the Preface.

15 Lipid Biosynthesis Inhibitors

Compounds in the lipid biosynthesis inhibitor (LBI) group are used mostly for postemergence control of grasses. They were first introduced in 1975, and new compounds within this group are continuing to be developed. The LBI herbicides are classified under two general chemical groupings, the aryloxyphenoxy-propionates (AOPP) and the cyclohexanediones (CHD).

These herbicides have specific activity against grass species only and are commonly referred to as graminicides. Dicots and nongrass monocots are tolerant. Some of these herbicides have shown minimal soil activity; however, the main activity occurs after postemergence application to emerged grass. Activity occurs on both annual and perennial grass species but varies, depending on the particular herbicide. Translocation of these herbicides can occur in both the xylem and the phloem, and all generally require the addition of an adjuvant to improve leaf coverage and absorption. These herbicides are most effective when applied to unstressed, rapidly growing grasses. Death of the grass is slow, requiring a week or more for complete kill. Symptoms include rapid cessation of shoot and root growth and pigment changes (purpling or reddening) on the leaves within 2 to 4 days of treatment, followed by a progressive necrosis beginning at meristematic regions and spreading over the entire plant. These herbicides inhibit the enzyme acetyl-CoenzymeA carboxylase (ACCase) in the biosynthetic pathway leading to lipid biosynthesis in plants, and prevent fatty acid formation, which is essential for plant lipid synthesis. Lack of lipids results in the loss of cell integrity of membranes, no new growth, and plant death.

General Characteristics

1. Used in postemergence control of annual and perennial grasses.
2. Selectivity occurs within grass weed and grass crop species.
3. Nongrass species are resistant.
4. Readily absorbed by plant foliage. Translocation may vary among species but occurs both in the xylem and the phloem.
5. Usually requires the addition of a surfactant or other spray additive to the spray solution for maximum activity.
6. These herbicides are most effective when applied to unstressed, rapidly growing grasses and are less effective if the grass is under stress.
7. Death of susceptible species is slow, requiring a week or more for complete death. Symptoms include a rapid cessation of shoot and root growth, with pigment changes on the leaves occurring within 2 to 4 days, followed by a

TABLE 15-1. Examples of Reported Interactions of Herbicide Mixtures of Lipid Biosynthesis Inhibitors (LBI) and Other Herbicides

I. Interactions That Result in Antagonism of LBI	
Sulfonylureas	2,4-DB
Imidazolinones	Metribuzin
MCPA	Dicamba
Bentazon	Bromoxynil
2,4-D	Acifluorfen
Carfentrazone	Pyrithiobac
II. LBI Herbicides Reported to Be Affected	
Diclofop-methyl	Sethoxydim
Haloxyfop-methyl	Fluazifop
Fenoxaprop-methyl	Fenoxaprop
	Tralkoxydim

progressive necrosis beginning at meristematic regions and spreading over the entire plant.

8. Rapidly degraded in soil.
9. Under normal use rates, most of these herbicides have insufficient soil activity to control grass weeds. Diclofop is the only herbicide within this group that has a soil application label.
10. Antagonism has been observed when these herbicides are tank-mixed with some postemergence broadleaf herbicides such as 2,4-D, acifluorfen, or bentazon (Table 15-1).
11. There are more than 20 species of grass that have developed resistance to this group.

GENERAL USES

Lipid biosynthesis inhibitor herbicides are effective for control of many annual and perennial grass species. Because they have no effect on broadleaf species and nongrass monocots, these herbicides are used widely in many cropping systems. Some of the herbicides within this group also show selectivity within grass species, which allows their use in certain grass crops and turf.

The list of crop and noncrop registrations for the herbicides within this group is extensive. A few examples of registrations will allow a greater appreciation of their overall utility for controlling grass weeds. Registrations have been granted for these herbicides in most fruit and vegetable crops, rice, cotton, peanut, soybean, canola, flax, wheat, barley, turf, ornamentals, forestry, rights-of-way, and other noncrop situations. A possible new use of nonselective grass herbicides (e.g., haloxytop, clethedim, sethoxydim) is control of volunteer Roundup Ready cereals in no-tillage and reduced tillage production systems.

Antagonism has been reported when the postemergence grass compounds are applied as mixtures with postemergence broadleaf herbicides (Table 15-1). Antagonism has also been reported when certain LBI herbicides are mixed with growth regulator type herbicides (2,4-D and dicamba), sulfonylureas (Devine and Rashid, 1993) and other ALS inhibitors (Ferreira and Coble, 1994), and bentazon and acifluorfen. Antagonism generally results in reduced grass control with little or no reduction in broadleaf weed control.

The antagonism is not at the ACCase site of action (Aguero-Alvarado et al., 1991). For bentazon, the antagonism is thought to be due to bentazon reducing the absorption of the graminicide (e.g., sethoxydim) across the cuticle and/or plasma membrane. For example, if droplets of the two herbicides are applied next to each other, there is no antagonism; whereas when the droplets are mixed together, there is antagonism (Jensen and Caseley, 1990). Thelen et al. (1995), using NMR (Nuclear Magnetic Resonance) spectroscopy, have shown that the sodium cation from the bentazon formulation associates with sethoxydim, perhaps making it less able to absorb through the cuticle. Adding ammonium sulfate overcomes the antagonism, and these authors have shown that the ammonium sulfate prevents the sodium in the bentazon formulation from complexing with sethoxydim.

One possible cause of the antagonism is a reduced uptake and translocation of the grass herbicide component. Tribenuron reduced basipital translocation of diclofop in wild oat, perhaps as a result of antagonism of dichlofop-methyl absorption by tribenuron (separating applications in time reduced antagonism) (Baerg et al., 1996). Devine and Rashid (1993) showed a reduced uptake of tralkoxydim when it was applied with metsulfuron-methyl; however, this reduction was transient, suggesting that the antagonism may be due to some other mechanism(s).

The antagonism resulting from mixing graminicides with growth regulator herbicides (2,4-D and dicamba) is at least partly due to decreased translocation of the grass herbicide to the site of action. Auxin agonist herbicides are known to cause a general reduction in phloem translocation. However, the effect may also be that auxin agonist herbicides reduce the disruption of the membrane proton gradient caused by diclofop methyl (Shimabukuro et al., 1989). Another suggestion, by Barnwell and Cobb (1993) was that AOPP herbicides inhibit growth by being auxin antagonists, and that by being mixed with auxin herbicides their inhibitory effect was reduced.

Applying each herbicide in the mixture sequentially can reduce antagonism. The longer the time interval between the application of each herbicide, the greater the chances to lessen or eliminate antagonism. Mixing the compounds together in the spray tank does not cause antagonism, but it is probably caused by an interaction between the herbicides at the leaf surface, which reduces the grass herbicide absorption and subsequent translocation. Some commercial mixtures of grass-specific and broadleaf herbicides for wheat take advantage of this antagonism and improve crop safety of fenoxaprop. Examples include Tiller, which is a mixture of fenoxaprop, 2,4-D, and MCPA, and Dakota, which is a mixture of fenoxaprop and MCPA.

Some of the lipid biosynthesis inhibitor herbicides, such as sethoxydim, have been shown to be quite susceptible to photodecomposition, which reduces their overall efficacy. Sethoxydim is presently being marketed as Poast Plus, which includes an additive that reduces or prevents photodecomposition of sethoxydim on the leaf surface and increases the time in which the active herbicide molecule can penetrate the leaf.

ADDITIVES

Nonionic surfactants, crop oil concentrates, and nitrogen solutions (e.g., ammonium sulfate and urea ammonium nitrate [UAN]) are added to most grass herbicides to improve weed control. Some LBI herbicides require the addition of a specific additive to the spray solution (see product labels for specific use guidelines).

SELECTIVITY

Selectivity of these compounds in several grass species has been shown to be due to differential susceptibility at the site of action (acetyl-coenzymeA carboxylase [ACCase]) or due to metabolism of the herbicide molecule. Grass-specific herbicides kill susceptible species by inhibiting the action of ACCase, a key enzyme in the lipid biosynthetic pathway. This enzyme in grasses is generally sensitive to inhibition by these herbicides, whereas ACCase in nongrass monocots and dicots is either not sensitive or has a very low sensitivity. This differential sensitivity at the site of action allows their widespread use in nongrass crops. In situations where these herbicides are used in grass crops or turf, the mechanisms of selectivity include both differential susceptibility at the site of action (fescues) and the ability of the crop (barley or wheat) to metabolize the herbicide.

Differential susceptibility occurs in several types of grasses. For example, fine-leafed fescues and annual bluegrass are relatively tolerant. Among important perennial weedy grasses, johnsongrass is most sensitive, quackgrass is moderately susceptible, and bermudagrass is the most difficult to control. Annual grass weeds all vary in sensitivity. Wild cane, giant foxtail, wild proso millet, and volunteer corn are very sensitive, but higher rates are generally required for crabgrass and green, yellow, and robust foxtails. The mechanisms for this variation in response of weedy grass species are not specifically known. These differences in susceptibility have allowed certain of these compounds to be registered for the turf market. Certain varieties of corn have been selected that are resistant to sethoxydim (Poast-tolerant corn lines; see the section on sethoxydim).

SOIL INFLUENCES

The efficacy of these compounds is not dependent on soil type, inasmuch as they are generally postemergence herbicides. There is, in some cases, soil activity at high use rates that prevents grass germination immediately after application. Soil persistence of these herbicides is short in all cases, and they have low to moderate potential for leaching. Soil breakdown by microorganisms is rapid, and residues of these compounds disappear within 30 to 60 days of application. All of these compounds have short soil half-lives (see specifics for each herbicide in the following section).

ARYLOXYPHENOXY-PROPIONATES

Diclofop-Methyl

Diclofop-methyl [(±)-2-[4-(2,4-dichlorophenoxy)phenoxy]propanoic acid] is a colorless solid with a vapor pressure of 2.58×10^{-7} mm Hg at 20°C, a water solubility of 0.8 mg/l (ppm), a soil half-life of 30 days at pH 7.0, and an oral LD_{50} (rat) of 557 to 580 mg/kg. Its use form is the methyl ester, and it is formulated as an emulsifiable concentrate.

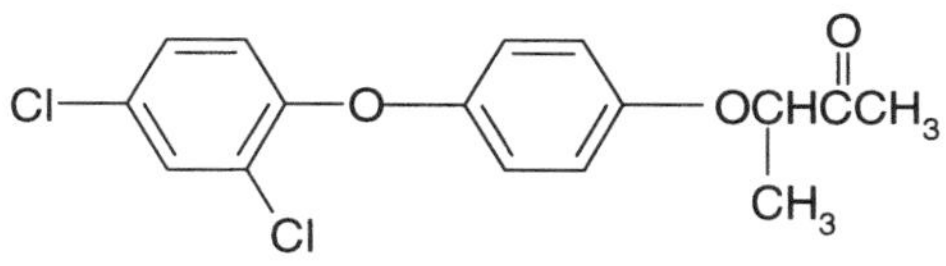

Diclofop - Methyl

Uses Diclofop-methyl is sold as Hoelon, Hoegrass (in Canada), and Illoxan for control of emerged annual grasses (especially wild oats), in small grains, and on golf courses; however, it does not control perennial grass. Best results are obtained when most wild oat and annual grasses are in the one- to three-leaf stage. It also has phytotoxicity to annual grasses via soils and is used this way in winter wheat.

Soil Influences Studies have shown that diclofop-methyl does not leach downward or move laterally in soils. Dissipation is rapid in soil under aerobic conditions, and under anaerobic conditions it disappears even faster, with up to 85% of the parent compound being metabolized within 2 days.

Fluazifop-p-Butyl

Fluazifop-p-butyl [(*R*)-2-[4-[[5-(trifluoromethyl)-2-pyridinyl]oxy]phenoxy]propanoic acid] is the active isomer. It is a light straw-colored, odorless liquid with a vapor pressure of 2.5×10^{-7} mm Hg at 20°C, a water solubility of 1.1 mg/l (ppm) at 25°C, a soil half-life of 15 days, and an oral LD_{50} (rat) of 4096 mg/kg.

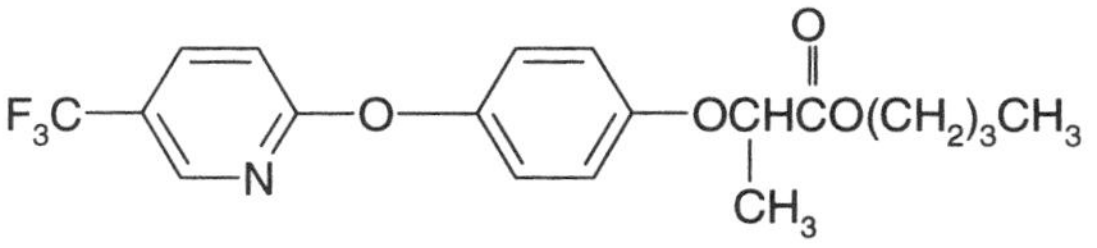

Fluazifop - Butyl

Uses Fluazifop-p-butyl is formulated as an EC and sold as Fusilade DX and in mixtures with fenoxaprop and fomesafen. It is registered for use in soybeans, cotton, several vegetables, fruits, and many ornamentals for selective postemergent control of annual and perennial grasses.

Soil Influences Fluazifop-p-butyl has low water solubility and is not subject to leaching in the soil. Microbes in moist soils rapidly degrade it. Residual soil life depends on soil type, rainfall, and rates used, but susceptible crops can normally be planted 60 days after its application.

Fenoxaprop-p-Ethyl

Fenoxaprop-p-ethyl[(*R*)-2-[4-[(6-chloro-2-benzoxazolyl)oxy]phenoxy]propionic acid] is a white, odorless solid with a vapor pressure of 1.4×10^{-7} mm Hg at 20°C, a water solubility of 0.7 mg/l (ppm) at 20°C, a soil half-life of 9 days, and an oral LD_{50} (rat) of 3310 mg/kg.

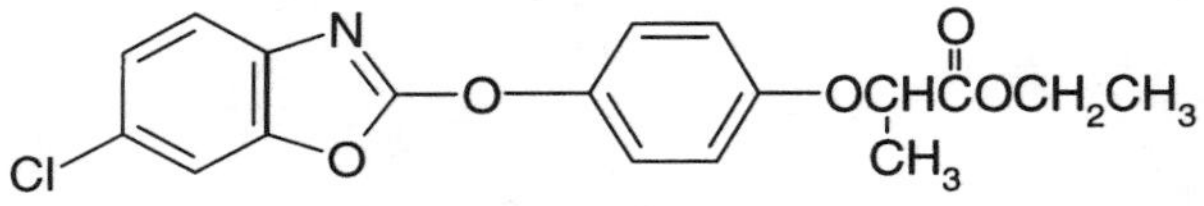

Fenoxaprop - Ethyl

Uses Fenoxaprop-p-ethyl is formulated as an EC and sold alone under the trade names of Silverado for use in winter wheat, Puma for use in wheat (not durum) and barley, Whip 360 for use in rice, Horizon for use in certain grass seed crops, and Acclaim Extra for use in ornamentals. Fenoxaprop-p-ethyl is also sold in combinations with growth regulator herbicides, as mentioned earlier (in the section "General Uses"), under the trade names Tiller (+ 2,4-D and MCPA) for use in spring wheat (not durum), winter wheat, and spring barley; Cheyenne (+ MCPA) for use in wheat (not durum); and Dakota (+ MCPA) for use in spring and winter wheat.

Soil Influences Soil has little influence on its activity, as fenoxaprop is applied to plant foliage. It has low mobility in silt loam and silty clay soils, with rapid microbial degradation under most conditions. Fenoxaprop has no soil residual activity with normal use rates.

Quizalofop-p-Ethyl

Quizalofop-p-ethyl [(*R*)-2-[4-[(6-chloro-2-quinoxalinyl)oxy]phenoxy]propanoic acid] is a colorless crystal with vapor pressure of 3×10^{-7} mm Hg at 20°C, a water solubility 0.3 mg/l (ppm) at 20°C, a soil half-life of 60 days, and an oral LD_{50} (rat) of 1670 mg/kg.

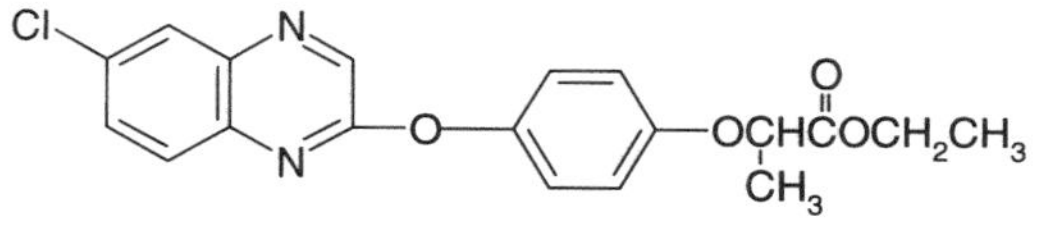

Quizalofop - Ethyl

Uses Quizalofop-p-ethyl is formulated as an EC and sold under the trade name Assure II. Assure II provides control of emerged annual and perennial grass in a number of crops, including canola, crambe, cotton, dry beans, dry and succulent peas, lentils, spearmint and peppermint, snap beans, soybeans, sugar beets and in noncrop areas.

Soil Influences Because quizalofop-p-ethyl is applied to emerged grass, soil has little effect on its efficacy. It is moderately adsorbed on sandy loam soils and is strongly adsorbed to silt loam soils. Quizalofop-p-ethyl is rapidly degraded by microbes under aerobic and anaerobic conditions; it has a moderate soil residual life but poses no problem to following crops.

Other Aryloxyphenoxy-propionates sold outside the United States or in Development

There are several other aryloxyphenoxy-propanoates that are sold outside the United States or are in commercial development. These compounds include clodinafop-propargyl, sold as Discover; cyhalofop-butyl, sold as Clincher; haloxyfop sold as Edge, Verdict, and Gallant; propaquizafop sold as Agil and Shogun; and fenthioprop and isoxapyrifop, which are in development. Commercial products are sold for annual and, in some cases, perennial emerged grass control in a wide variety of crops.

CYCLOHEXANEDIONES

Sethoxydim

Sethoxydim ±(2-[1-(ethoxyimino)butyl-5-[2-(ethylthio)-propyl]-3-hydroxy-2-cyclohexen-1-one) is an amber-colored, oily, odorless liquid with a vapor pressure of 1.6×10^{-7} mm Hg at 25°C, a water solubility of 257 mg/l (ppm) at 25°C, a soil half-life of 5 days, and an oral LD_{50} (rat) of 2676 to 3124 mg/kg.

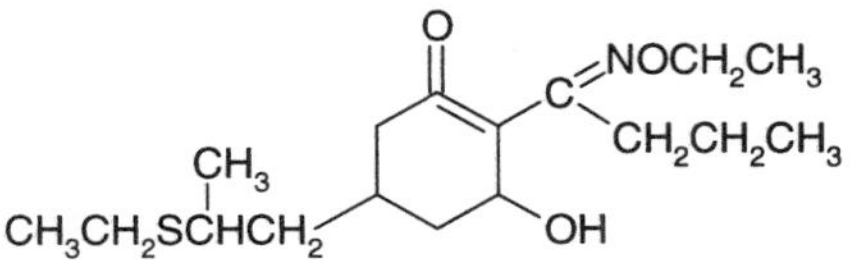

Sethoxydim

Uses Sethoxydim is formulated as an EC and sold under the trade names Poast, Poast Plus, Conclude G, and Vantage. Poast is registered for use in most soybeans and other dicot agronomic crops and most vegetables and fruits. Poast Plus is registered for use in alfalfa, birdsfoot trefoil, sainfoil, citrus, clover, clover hay, cotton, peanuts, soybeans, and Poast-resistant corn. Conclude G is registered for use in soybeans as a component mix with bentazon. Vantage is registered for use in a wide variety of ornamental herbaceous and woody species (ground covers, bushes, and trees). Sethoxydim is applied postemergence for control of a wide variety of annual and some perennial grass species (Figure 15-1). Sethoxydim is also sold as Conclude Xtra, which is a three-component tank mixture of sethoxydim, acifluorfen, and bentazon, for use in soybeans for broadspectrum postemergence grass and broadleaf weed control.

Soil Influences The efficacy of sethoxydim is not dependent on soil type inasmuch as it is a postemergence herbicide, but soil activity reduces emergence of germinating grasses immediately after application. Soil adsorption varies with organic matter, and soil persistence, although short, is pH dependent with a 4- to 5-day half-life at pH 6.8 and an 11-day half-life at pH 7.4.

Clethodim

Clethodim [(E,E-(±)-2-[1-[[3-chloro-2-propenyl)oxy]-imino]propyl]-5-[2-(ethylthio)-propyl]-3-hydroxy-2-cyclohexen-1-one] is a clear amber viscous liquid with a vapor

Figure 15-1. Sethoxydim control of giant foxtail in soybean. *Left*: Unsprayed control; *Right*: Sprayed.

pressure of $<10^{-7}$ mm Hg at 20°C, a water solubility that is pH dependent (increasing as pH increases), a soil half-life of ~ 3 days, and an oral LD_{50} (rat) of 1630 mg/kg.

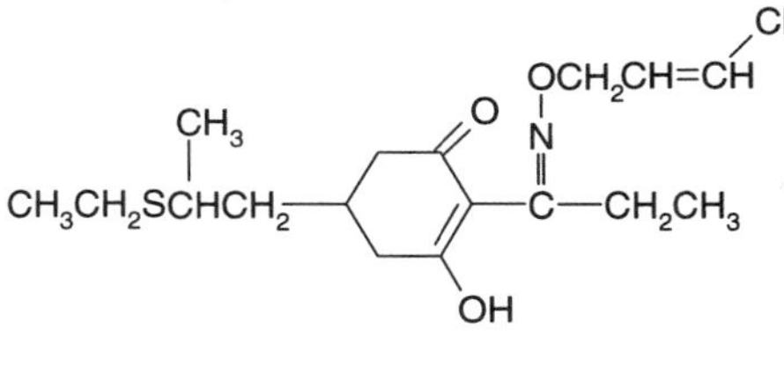

$CH_3CH_2SCH(CH_3)CH_2$–[cyclohexenone ring with =O and –OH]–$C(=N–OCH_2CH=CHCl)–CH_2CH_3$

Clethodim

Uses Clethodim is formulated as an EC and sold under the trade names Select, Prism, and Conclude Xtra G for use in controlling emerged annual and perennial grasses in a wide variety of crops, including soybeans, cotton, sugar beets, dry bulb (onions, garlic and shallots), tomatoes, alfalfa, dry beans, peanuts, nonbearing food crops, and in fallow land and noncropland. It is also sold as Envoy for use in ornamental flowers, trees, ground covers, shrubs, nonbearing fruits and nuts, and noncropland.

Soil Influences Clethodim is weakly adsorbed to soils and rapidly degraded by both hydrolysis and photodegradation. The soil residual life is very short, and there is no problem of injury to subsequent crops.

Tralkoxydim

Tralkoxydim (2-[1-(ethoxyimino)propyl]-3-hydroxy-5-(2,4,6-trimethylphenyl)-2-cyclohexen-1-one) is a colorless, odorless solid with a water solubility of 5 mg/l (ppm) at pH 5 at 20°C (increasing with increasing pH), a soil half-life of 3 days, and an oral LD_{50} (rat) of 1324 mg/kg.

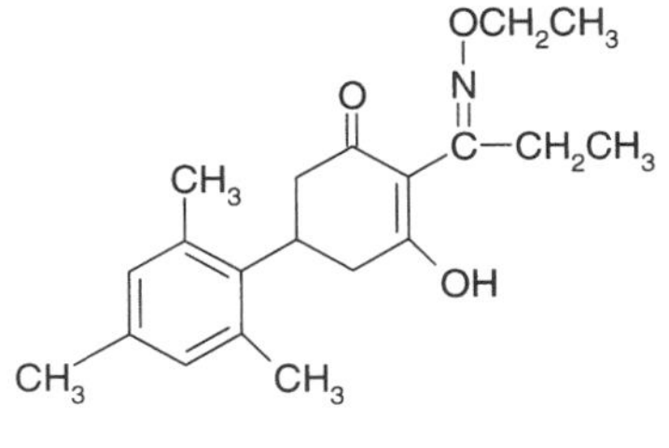

[2,4,6-trimethylphenyl (CH_3, CH_3, CH_3)]–[cyclohexenone ring with =O and –OH]–$C(=N–OCH_2CH_3)–CH_2CH_3$

Tralkoxydim

Uses Tralkoxydim is formulated as a DG and sold under the trade name Achieve for control of emerged wild oats, giant and yellow foxtail, annual rye (Italian), and Persian darnel in wheat and barley.

Soil Influences Soil type does not influence its activity. It degrades rapidly under aerobic conditions but slower under anaerobic or flooded conditions.

Other Cyclohexanediones not sold in the United States or in Development

Several cyclohexanediones are sold outside the United States. These compounds include alloxydim-sodium sold as Fervin and Kusagard; cycloxydim sold as Focus, Laser, and Stratos, butroxydim sold as Falcon, clefoxydim sold as Aura, and tepraloxydim sold as Equinox. All are used for annual and, in some cases, perennial emerged grass control in a wide variety of crops.

MECHANISM OF ACTION

Many studies on the mechanism of action of the grass-specific herbicides have been conducted with the aryloxyphenoxy-propionates (AOPPs) herbicide diclofop-methyl. It serves as the model of how these herbicide groups are absorbed, translocated, and act in killing grassy weeds. Diclofop-methyl is rapidly absorbed through the cuticle and into leaf cells, where it is deesterified by esterase enzymes. The acid is then translocated to meristematic areas of the shoots and roots. The amount translocated is low in relation to the amount applied; however, translocation occurs in both the xylem and the phloem. Roots and shoots of grass plants cease growth within a few hours after treatment, long before visual signs of injury appear. Plants then develop red coloration in the leaves (anthocyanin formation) and necrotic areas in the nongreen meristematic tissue just above the growing point and at nodes (Figure 15-2). This tissue begins to change color and consistency, with individual cell disruption observed as soon as 8

Figure 15-2. Effect of lipid biosynthesis-inhibiting herbicides on the meristematic regions of the grass shoot and root. *Left*: Unsprayed control; *Right*: Necrotic shoot meristem of grass and lack of root development treated 10 days prior with fluazifop.

hours after treatment. This time period corresponds closely to the translocation time required to reach the affected area. Respiratory activity declines, with a resulting increase in sugar and anthocyanin levels, and mitosis is affected as a result of an inhibition of cell wall formation; however, these are thought to be secondary effects. Overall cells appear as if they are proceeding through a rapid senescence.

Inhibition of lipid biosynthesis could explain the reduction of growth (a lipid-requiring process), the reported increase in membrane permeability, and the

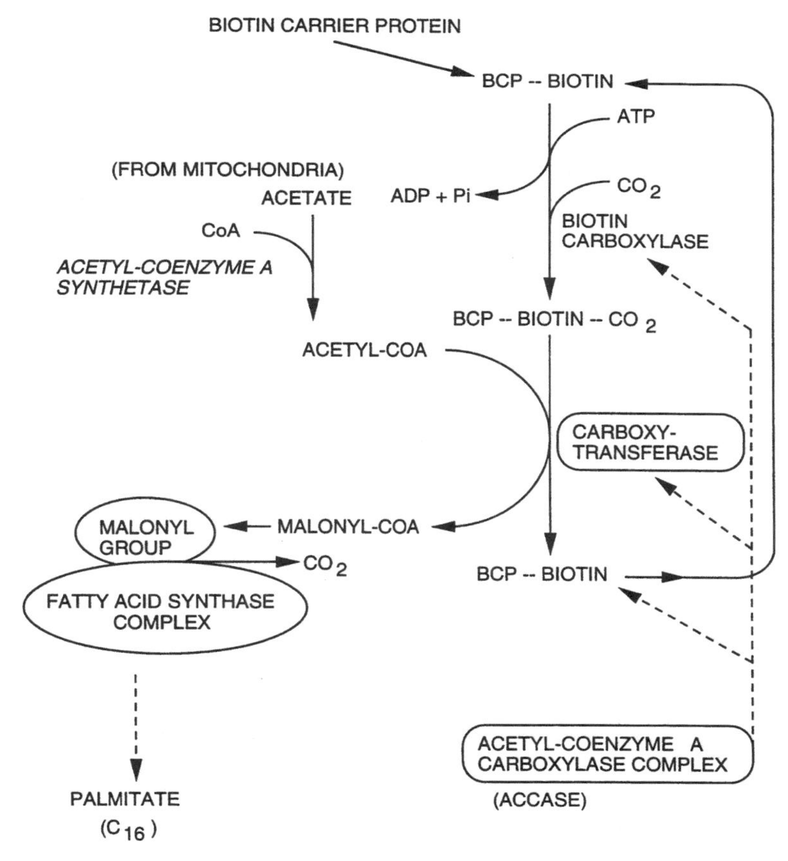

Figure 15-3. Site of action of lipid biosynthesis inhibitors in the lipid biosynthesis pathway of grasses. The herbicide inhibits acetyl-coenzyme A carboxylase (ACCase). ACCase converts acetyl-CoA to malonyl-CoA by the addition of CO_2 to acetyl CoA. This is a key early biosynthetic reaction in lipid biosynthesis and is thought to be the rate-limiting step in lipid biosynthesis.

ultrastructural effects commonly observed after herbicide treatment. While researchers were evaluating the influence of these herbicides on lipid synthesis, they also studied their effects on many other processes. In the factors evaluated (protein synthesis, nucleic acid synthesis, sterol synthesis, cellular ATP levels, free sugar content, glycolytic intermediates, cell wall material, and photosynthesis), no effects were observed at concentrations and treatment times where lipid synthesis was inhibited.

In 1987 research showed that the site of herbicide action was on acetyl-Coenzyme A carboxylase (ACCase), a key early enzyme in the lipid biosynthesis pathway (Figure 15-3). The step catalyzed by ACCase is thought to be the rate-limiting step in lipid biosynthesis. Burton et al. (1987) found sethoxydim- and haloxyfop-inhibited ACCase isolated from chloroplasts of corn seedling with I_{50} concentrations of 2.9 and 0.5 μM, respectively, and that the same enzyme isolated from pea chloroplasts was not inhibited. Focke and Lichtenthaler (1987) reported that cycloxydim, sethoxydim, and clethodim inhibited fatty acid biosynthesis in a chloroplast enzyme preparation from barley. Many other publications have implicated ACCase as the target site for AOPP and CHD herbicides. ACCase was found to be the site of action for the acid forms of diclofop, fenoxaprop, fluazifop, and haloxyfop (Kobek et al., 1988).

Chloroplastic ACCase from dicot species is not sensitive to these herbicides, which is due to the fact that dicots have a different form of the enzyme in their chloroplasts than grasses. This may also be the case for certain nongrass insensitive monocots. No significant differences in absorption, translocation, or metabolism of these herbicides have been reported between dicot and grass species. For more detail on the two enzyme forms, refer to papers by Konishi and Sasaki (1994) and Alban et al. (1994).

CROP RESISTANCE

Perhaps the best method of proving whether an identified site of herbicide action is its only site of action, is to obtain mutants with a herbicide-resistant site of action. The site of action can be characterized in the resistant mutant, and the genetics of the mutant can then be evaluated to determine whether the resistance can be segregated as a single trait. In 1990 a sethoxydim-resistant corn mutant was recovered from tissue culture cell selection (Parker et al., 1990). Sethoxydim was added to corn tissue culture cell lines, and cells were selected that had a greater than 40-fold resistance to sethoxydim. This line also exhibited a 20-fold cross-resistance to haloxyfop. ACCase isolated from plants regenerated from the resistant mutant cell lines was significantly less sensitive to sethoxydim (>100-fold) and haloxyfop (10-fold), demonstrating that resistance was at the site of action. These data strongly support a single site of action at ACCase for the CHD and AOPP herbicides.

WEED RESISTANCE

The ease with which resistant mutants were recovered from corn tissue culture is a significant concern in regard to how easily resistance to these herbicides may occur. In grass species there are forms of ACCase not inhibited by these herbicides (see

Tardif et al., 1993), suggesting a significant potential for selection of resistant weeds in the environment. There are more than 20 grass species reported to be resistant to these herbicides. Resistance has developed in annual ryegrass (*Lolium rigidum*), Italian ryegrass (*Lolium multiflorum*), perennial ryegrass (*Lolium perenne*), wild oat (*Avena fatua* and *A. sterilis*), giant foxtail (*Setaria faberi*), green foxtail (*Setaria viridis*), johnsongrass (*Sorghum halepense*), goosegrass (*Eleusine indica*), and large crabgrass (*Digitaria sanguinalis*). Many reports are available regarding weed populations with resistance to 1 or more of the CHD and AOPP herbicides.

In nearly every case of weeds with resistance, resistance is explained by the presence of an insensitive ACCase enzyme. However, not all tolerances for these herbicides are due to differences at the site of action. Wheat tolerance to diclofop is due to hydroxylation (33% in 6 hours) of the diclofop by a cytochrome P-450 monoxygenase (McFadden et al., 1989).

The most disconcerting report is of annual ryegrass in Australia that was shown to be resistant to diclofop-methyl in 1982. This resistant biotype was observed after diclofop-methyl had been used in wheat for only 4 years. Research has shown that certain ryegrass biotypes are not only resistant to diclofop-methyl, but are cross-resistant to other grass specific herbicides and have resistance to a wide variety of other herbicide classes. The mechanism of resistance in these biotypes varies considerably.

Apparently, the Australian biotype does not have a resistant ACCase and resistance is related to a differential response of the susceptible and resistant biotypes relating to the stability of the plasma membrane after herbicide exposure.

In growth experiments with LBI-resistant biotypes, the fitness of resistant and susceptible biotypes appears to be similar (Wiederholt and Stoltenberg, 1996a, 1996b), and in most biotypes analyzed, resistance is governed by a single, partially dominant, nuclear gene (Murray et al., 1995). Considering the foregoing information on resistance, it is important that LBI herbicides be carefully used in weed control programs to minimize the occurrence of resistance.

LITERATURE CITED AND SUGGESTED READING

Aguero-Alvarado, R., A. P. Appleby, and D. J. Armstrong. 1991. Antagonism of haloxyfop activity in tall fescue *(Festuca arundinacea)* by dicamba and bentazon. *Weed Sci.* **39**:1–5.

Alban, C., P. Baldet, and R. Douce. 1994. Localization and characterization of two structurally different forms of acetyl-CoA carboxylase in young pea leaves, of which one is sensitive to aryloxyphenoxypropionate herbicides. *Biochem. J.* **300**:557–565.

Baerg, R. J., J. W. Gronwald, C. V. Eberlein, and R. E. Stucker. 1996. Antagonism of dichlofop control of wild oat (*Avena fatua*) by tribenuron. *Weed Sci.* **44**:461–468.

Barnwell, P., and A. H. Cobb. 1993. An investigation of aryloxyphenoxypropionate antagonism of auxin-type herbicide action on proton-efflux. *Pestic. Biochem. Physiol.* **47**:87–97.

Burton, J. D., J. W. Gronwald, D. A. Somers, J. A. Connelly, B. G. Gengenbach, and D. L. Wyse. 1987. Inhibition of plant acetyl-coenzyme A carboxylase by the herbicides sethoxydim and haloxyfop. *Biochem. and Biophy. Res. Comm.* **148**:1039–1044.

Devine, M. D., and A. Rashid. 1993. Antagonism of tralkoxydim activity in *Avena fatua* by metsulfuron methyl. *Weed Res.* **33**:97–104.

Ferreira, K. L., and H. D. Coble. 1994. Effect of DPX-PE350 on the efficacy of graminicides. *Weed Sci.* **42**:222–226.

Focke, M., and H. R. Lichtenthaler. 1987. Inhibition of the acetyl-CoA carboxylase of barley chloroplasts by cycloxydim and sethoxydim. *Z. Naturforsch.* **42c**:1361–1363.

Holtum, J. A. M., J. M. Matthews, R. E. Häusler, D. R. Liljegren, and S. B. Powles. 1991. Cross-resistance to herbicides in annual ryegrass (*Lolium rigidum*). III. On the mechanism of resistance to diclofop-methyl. *Plant Physiol.* **97**:1026–1034.

Jensen, K. I. N., and J. C. Caseley. 1990. Antagonistic effects of 2,4-D amine and bentazon on control of *Avena fatua* with tralkoxydim. *Weed Res.* **30**:389–395.

Kobek, K., M. Focke, and H. K. Lichtenthaler. 1988. Fatty acid biosynthesis and acetyl-CoA carboxylase as a target of diclofop, fenoxaprop, and other aryloxyphenoxy-propionic acid herbicides. *Z. Naturforsch.* **43c**:47–54.

Konishi, T., and Y. Sasaki. 1994. Compartmentalization of two forms of acetyl-CoA carboxylase in plants and the origin of their tolerance toward herbicides. *Proc. Natl. Acad. Sci. USA* **91**:3598–3601.

McFadden, J. J., D. S. Frear, and E. R. Mansager. 1989. Aryl hydroxylation of diclofop by cytochrome P450 dependent monooxygenase from wheat. *Pestic. Biochem. Physiol.* **34**:92–100.

Murray, B. G., I. N. Morrison, and A. L. Brûlé-Babel. 1995. Inheritance of acetyl-CoA carboxylase inhibitor resistance in wild oat (*Avena fatua*). *Weed Sci.* **43**:233–238.

Parker, W. B., D. A. Somers, D. L. Wyse, R. A. Keith, J. D. Burton, J. W. Gronwald, and B. G. Gengenbach. 1990. Selection and characterization of sethoxydim tolerant maize tissue cultures. *Plant Physiol.* **92**:1220–1225.

Powles, S. B., and P. D. Howat. 1990. Herbicide-resistant weeds in Australia. *Weed Technol.* **4**:178–185.

Shimabukuro, R. H., W. C. Walsh, and J. P. Wright. 1989. Effect of diclofop methyl and 2,4-D on transmembrane proton gradient: A mechanism for their antagonistic interaction. *Physiol. Plant.* **77**:107–114.

Shimabukuro, R. H., and B. L. Hoffer. 1992. Effects of diclofop on the membrane potentials of herbicide-resistant and -susceptible annual ryegrass root tips. *Plant Physiol.* **98**:1415–1422.

Stoltenberg, D. E., and R. J. Wiederholt. 1995. Giant foxtail (*Setaria faberi*) resistance to aryloxyphenoxypropionate and cyclohexanedione herbicides. *Weed Sci.* **43**:527–535.

Tardif, F. J., J. A. M. Holtum, and S. B. Powles. 1993. Occurrence of a herbicide-resistant acetyl-coenzyme A carboxylase mutant in annual ryegrass (*Lolium rigidum*) selected by sethoxydim. *Planta* **190**:176–181.

Thelen, K. D., E. P. Jackson, and D. Penner. 1995. Characterizing the sethoxydim-bentazon interaction with proton nuclear magnetic resonance spectrometry. *Weed Sci.* **43**:337–341.

Wiederholt, R. J., and D. E. Stoltenberg. 1996a. Absence of differential fitness between giant foxtail (*Setaria faberi*) accessions resistant and susceptible to acetyl-coenzyme A carboxylase inhibitors. *Weed Sci.* **44**:18–24.

Wiederholt, R. J., and D. E. Stoltenberg. 1996b. Similar fitness between large crabgrass (*Digitaria sanguinalis*) accessions resistant or susceptible to acetyl-coenzyme A carboxylase inhibitors. *Weed Technol.* **10**:42–49.

For chemical use, see the manufacturer's or supplier's label and follow these directions. Also see the Preface.

16 Inhibitors of Amino Acid Biosynthesis

The herbicides discussed in this chapter, although differing in chemical structure, all inhibit amino acid synthesis in plants. The first herbicide described is glyphosate herbicide, which inhibits an enzyme in the aromatic amino acid biosynthetic pathway. The second group of herbicides discussed includes the imidazolinones, pyrimidyl-oxy-benzoates, sulfonylureas, triazolopyrimides, and a sulfonylaminocarbonyl-triazolinone, which all inhibit an enzyme in the branch-chain amino acid biosynthetic pathway.

GLYPHOSATE—AN INHIBITOR OF AROMATIC AMINO ACID BIOSYNTHESIS

History

Glyphosate was discovered and developed as a herbicide by Monsanto Chemical Company. *N*-(phosphonomethyl) glycine, the active ingredient in glyphosate, is a derivative of the amino acid glycine and phosphonic acid. In 1964, Stauffer Chemical Company patented a series of phosphonic and phosphinic acids to be used as industrial cleaners. In 1969 and 1971, Monsanto obtained patents for the use of phosphonic acids as growth regulators and selective herbicides. Then, in 1974, Monsanto received a patent for the use of phosphonic acid derivatives as nonselective herbicides.

In the early 1980s, Stauffer Chemical Company released for development a related herbicide, sulfosate, which is now sometimes referred to as glyphosate-trimesium (trimethylsulfonium salt). It is derived from the same parent acid, *N*-(phosphonomethyl)glycine. Today both herbicides are marketed in the United States.

Chemical and Physical Characteristics

N-(phosphonomethyl)glycine is a white solid and a weak organic acid derivative of phosphonic acid and the amino acid glycine. It has a vapor pressure of 1.84×10^{-7} mm Hg at 45°C, a water solubility of 900,000 mg/l (ppm) at 25°C, a soil half-life of 47 days, and an oral LD_{50} (rat) of >5000 mg/kg. Glyphosate is generally formulated as salts; hence, it is water soluble, and it is mildly to moderately corrosive to iron and galvanized materials. The formulated herbicides are very temperature stable under normal conditions (–20 to 40°C) and are essentially nonvolatile and photostable.

$$\left[HO_2CCH_2NHCH_2\overset{O}{\overset{\|}{P}}(OH)-O \right]^-$$

N- (phosphonomethyl) glycine

$$\left[CH_3\underset{NH_3}{CH}CH_3 \right]^+$$

{isopropylamine} glyphosate

$$\left[CH_3\underset{CH_3}{S}CH_3 \right]^+$$

trimesium {sulfosate}

Use and Selectivity

The various salts of glyphosate are essentially nonselective postemergence-applied herbicides that control a vast range of annual and perennial weeds. Glyphosate is sold under many different trade names—for example, Accord, Rodeo, Roundup Ultra, Roundup Pro, Polado, and Touchdown, to name a few—for use in a variety of cropping and noncrop situations. These various products are effective on most herbaceous annual and perennial species at rates ranging from 0.5 to 1.0 lb/acre (0.56 to 1.12 kg/ha). In general, the control of herbaceous perennials, perennial vines, and woody species requires higher rates. Because glyphosate is inherently nonselective, selectivity has often been achieved via placement and timing—for example, as a preplant or preemergence herbicide for the control of existing vegetation in no-till systems and for turfgrass renovation, where existing sod must be killed before reseeding, sprigging, or sodding operations.

Glyphosate is used as a postharvest treatment or during off-season for the control of perennial weeds. For example, late summer or fall applications after the harvest of corn, grain sorghum, or cotton are common for the control of johnsongrass. In some cases glyphosate has been successfully used as a harvest aid. Some selectivity may be achieved by adjusting rates and timing of application, as has been accomplished with alfalfa and broadbean.

Novel application equipment such as the rope wick and other "wiper" applicators, selective placement recirculating sprayers, and, most recently, "hooded" sprayers have permitted the selective use of these otherwise nonselective herbicides. Glyphosate is also widely used for spot treatment, as an alternative to hoeing, in many noncrop sites for the control of undesirable species and in forestry.

Some of the most important developments in the use of these herbicides have occurred in the past 5 years; for instance, through the use of bioengineering, crop tolerance to glyphosate has been developed. This technology has already yielded soybeans, cotton, corn, and canola that are tolerant of glyphosate at rates that will allow for relatively broad-spectrum weed control in these crops. It is anticipated that the use of Roundup Ready crops will continue to increase in the near future.

Glyphosate is a safe and widely used aquatic herbicide. Special formulations that do not include surfactants (which can be toxic to fish) have been developed for aquatic weed control. The high water solubility of these formulations makes them ideal aquatic herbicides because they are easy to disperse in standing water. The combination of environmental and human safety and the herbicide's efficacy against a number of important aquatic weed species has resulted in its becoming an important aquatic herbicide.

Injection systems have been developed that allow for the direct injection of glyphosate into trees. Likewise, one company now markets pruning sheers that are fitted with a small container and injector that applies a small amount of concentrated herbicide when the user snips woody vines such as poison ivy, greenbriar, and the like.

It is noteworthy that formulation does significantly affect the activity, and sometimes the selectivity, of glyphosate. For instance, a special formulation of glyphosate, Polado, is used as a plant growth regulator to enhance sugar content in sugarcane.

Soil Influences

Glyphosate is applied to the aboveground parts of weeds, and soil type has little effect on its performance. It is strongly bound to soil clay particles; it does not leach, nor is it herbicidally active because of its strong adsorption to soil. Therefore, even though the salts of glyphosate are very water soluble, they generally do not leach.

Factors Affecting Performance

Environmental Conditions Because glyphosate is applied as a foliar spray, foliar absorption is required for activity. Rapidly growing nonstressed plants of most species are most sensitive to glyphosate and show symptoms of injury within 3 to 10 days after application. Severe environmental conditions, extreme heat or cold or drought, will both slow and reduce the effectiveness of glyphosate.

Physical and Chemical Conditions The uptake, translocation, and activity of glyphosate can be affected by the addition of other herbicides. For example, herbicides that result in rapid disruption of membranes or significantly impair normal metabolic processes can reduce the effectiveness of glyphosate. Glyphosate activity can also be greatly reduced when the water carrier contains high levels of iron or salts. The inclusion of various spray adjuvants has been widely investigated with glyphosate. The use of adjuvants may enhance the activity of glyphosate under some conditions. The use of ammonium sulfate as an additive increases activity in some species and is permitted on the product label. Glyphosate is a weak organic acid, and some reports indicate that activity may be enhanced at lower pH values and decreased under high pH conditions. Reducing carrier volumes, which results in more concentrated spray droplets, also appears to enhance activity.

Mechanism of Action

The first detectable symptom after glyphosate treatment is growth inhibition, followed by a noticeable yellowing (chlorosis) of treated tissue (Figure 16-1). Symptoms are slow to develop, and if the environment is cool and cloudy after treatment, symptoms develop at a very slow pace. Five to 10 days after treatment, the chlorosis turns into necrosis and the plants begin to die. Glyphosate is xylem and phloem mobile; however, phloem mobility is not as great as with some other phloem-mobile herbicides. Such lower mobility is thought to be due to glyphosate's being less well trapped in the phloem during movement of the phloem sap than weak acid herbicides such as 2,4-D.

The slow rate of kill is an advantage in allowing maximum phloem mobility to roots and perennial organs prior to the death of treated leaves. Glyphosate is one of the few herbicides that have been shown to cross the plasma membrane using a carrier protein (see Chapter 5).

Reports of early mechanism-of-action research with glyphosate indicated that levels of the aromatic amino acids phenylalanine and tyrosine were reduced in treated

Figure 16-1. Glyphosate symptomology on shoot tips of a treated tomato plant. Note the chlorosis at the meristematic regions.

tissue. A reduction in amino acids can reduce protein synthesis and subsequently cause an inhibition of plant growth. Additional research showed that feeding phenylalanine and tyrosine to glyphosate-treated plants reversed glyphosate-induced growth inhibition in some test systems.

Further research showed that levels of shikimate were increased in glyphosate-treated tissue. Shikimate accumulation was shown to be due to glyphosate inhibiting the chloroplastic enzyme 5-enolpyruvylshikimate-3-phosphate synthase (EPSPS). EPSPS is the penultimate enzyme in the aromatic amino acid biosynthetic pathway that occurs in the chloroplast and converts shikimate-3-phosphate (S-3-P) to enolpyruvylshikimate-3-phosphate (EPSP) and eventually leads to the production of the amino acids phenylalanine, tyrosine, and tryptophan (Amrhein et al., 1980). Shikimate builds up in glyphosate-treated plants because S-3-P cannot be converted to EPSP, and because S-3-P is unstable, it is rapidly converted to the more stable shikimate, which accumulates (Figure 16-2).

How Do Glyphosate-Treated Plants Die?

No one knows for sure exactly how glyphosate-treated plants die. As with all of the herbicides discussed in this book, the initial inhibitory site results in a cascade of biochemical reactions that lead to the eventual death of the plant. As is often the case, some investigators believe that the single site of action at EPSPS cannot explain all of the phytotoxicity observed after glyphosate treatment (Shieh et al., 1991). The large accumulation of shikimate and the depletion of products arising from the shikimate acid pathway (aromatic amino acids) have not been shown to be herbicidal. There is strong support that EPSPS is the only target site for glyphosate, inasmuch as high levels of resistance are achieved when a glyphosate-resistant EPSPS gene is engineered into crop plants. Some investigators suggest that the herbicidal activity of glyphosate is actually the result of a carbon drain in the carbon reduction cycle, due to deregulation of the shikimate acid pathway (Figure 16-2). In the presence of glyphosate the activity of 3-deoxy-D-arabino-heptulosonate-7-phosphate synthase is increased, perhaps because of decreased regulation by aroginate (Pinto et al., 1988). Even without deregulation by glyphosate, more than 20% of the carbon fixed by leaves passes through the aromatic amino acid pathway and up to 30% of the plant dry weight is aromatic molecules derived from this pathway! Schultz et al. (1990) reported that shikimate and S3P can account for up to 16% of the dry weight of sink tissue after glyphosate treatment. In addition to the carbon drain and loss of aromatic amino acid biosynthesis, this pathway is a precursor to numerous important plant products (e.g., auxin, folic acid, lignin, plastoquinone, flavonoids, phenolics, and alkaloids).

Therefore, it appears that a block in the action of the EPSPS enzyme by glyphosate results in plant death in a number of ways. First is the reduced production of important amino acid and secondary product building blocks for plant growth, and second is a carbon drain into this pathway that reduces the overall functioning of other biochemical pathways of the plant. These effects would account for the slow death of the treated plant, as well as the effectiveness of this herbicide on most annual and perennial plants due to translocation to all regions of the plant.

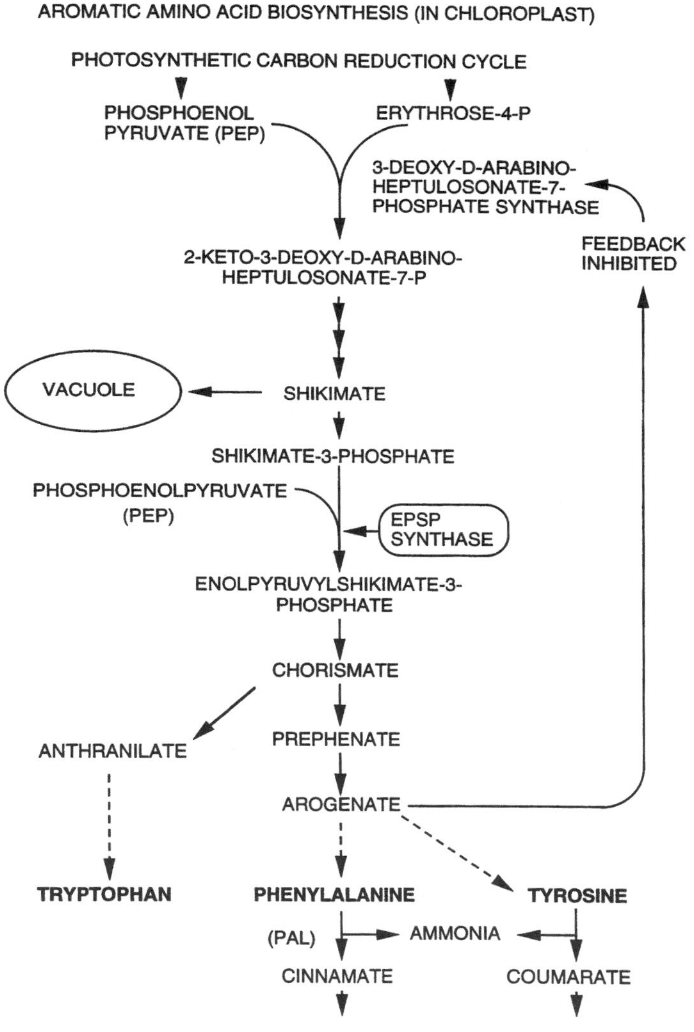

AROMATIC AMINO ACID BIOSYNTHESIS (IN CHLOROPLAST)
PHOTOSYNTHETIC CARBON REDUCTION CYCLE
PHOSPHOENOL PYRUVATE (PEP)
ERYTHROSE-4-P
3-DEOXY-D-ARABINO-HEPTULOSONATE-7-PHOSPHATE SYNTHASE
FEEDBACK INHIBITED
2-KETO-3-DEOXY-D-ARABINO-HEPTULOSONATE-7-P
VACUOLE
SHIKIMATE
SHIKIMATE-3-PHOSPHATE
PHOSPHOENOLPYRUVATE (PEP)
EPSP SYNTHASE
ENOLPYRUVYLSHIKIMATE-3-PHOSPHATE
CHORISMATE
ANTHRANILATE
PREPHENATE
AROGENATE
TRYPTOPHAN
PHENYLALANINE
TYROSINE
(PAL)
AMMONIA
CINNAMATE
COUMARATE

Figure 16-2. Site of action of glyphosate in the aromatic amino acid biosynthetic pathway. In glyphosate-treated tissue the levels of shikimate are increased. Shikimate accumulation is due to glyphosate inhibiting the chloroplastic enzyme 5-enolpyruvylshikimate-3-phosphate synthase (EPSPS). EPSPS is the penultimate enzyme in the aromatic amino acid biosynthetic pathway that occurs in the chloroplast and converts shikimate-3-phosphate (S-3-P) to enolpyruvylshikimate-3-phosphate (EPSP) and eventually leads to the production of the amino acids phenylalanine and tyrosine, as well as tryptophan. Shikimate builds up in glyphosate-treated plants because S-3-P cannot be converted to EPSP, and because S-3-P is unstable, it is rapidly converted to the more stable shikimate, which accumulates.

Crop Resistance

During the past several years research seeking to genetically engineer crops resistant to glyphosate has been successful. The greater part of this work occurred at Monsanto; however, Calgene, Inc. was active in the initial research involving genes for resistant EPSPS enzymes. The most prominent method to obtain resistant crops has been to insert resistant EPSPS genes from bacteria into crop plants [see Bradshaw et al. (1997) for a review of EPSPS genes obtained]. The best resistance gene identified to date, termed CP4, came from an *Agrobacterium* species (strain CP4) of bacteria. The resistant EPSPS gene exhibits a high level of glyphosate resistance and has a similar efficiency to the native plant EPSPS.

Soybean plants engineered to be resistant to glyphosate have been extensively field-tested, and to date the field performance has been acceptable (Delannay et al., 1995). Commercial introduction of glyphosate-resistant soybeans occurred in 1996. Glyphosate-resistant corn and cotton have also been developed and were commercially introduced in 1998. Other glyphosate-resistant crops in development include wheat, sugar beets, lettuce, potatoes, and many more (see Chapter 18).

A second mechanism for generating glyphosate-resistant crops being explored is glyphosate degradation. An amine oxidase enzyme [termed glyphosate oxidoreductase (GOX) by Monsanto] that converts glyphosate to aminomethyl phosphonate + glyoxylate has been isolated from bacteria, and this gene is then coupled to the resistant CP4 EPSPS gene for insertion into plants (Zhou et al., 1995). Glyphosate-resistant sugar beet currently in development contains both the GOX gene and the resistant CP4 EPSPS gene (Mannerlöf et al., 1997).

Weed Resistance

Even though glyphosate has been used extensively and repeatedly throughout the world since 1975, no cases of glyphosate-resistant weeds were identified in the environment until 1996. Bradshaw et al. (1997) have reviewed the reasons for the very low frequency of glyphosate resistance in weeds.

Australia, a country known for frequent problems with herbicide-resistant weeds, reported a glyphosate-resistant annual ryegrass population (Sindel, 1996). This population was located on a farm in northern Victoria where glyphosate had been used repeatedly over a 15-year period. Since that initial report, other glyphosate-resistant ryegrass types have been identified in Australia and California. The mechanism of resistance is not known for these types; however, Lorraine-Colwill et al. (2001) have shown that differences in translocation patterns between the resistant and susceptible types may play a role. Simarmata et al. (2001) have shown that the California ryegrass from an almond orchard where glyphosate had been applied repeatedly over a 10-year period, has resistance to 9.96 kg/ha of glyphosate, which is eight times the normally toxic dose. A goosegrass biotype from Maylasia is resistant to glyphosate because of an alteration in the EPSPS enzyme. Recently, a *Conyza canadensis* biotype with elevated resistance to glyphosate has been reported in Delaware (VanGessel, 2001).

Possibly more important than the development of classical resistance is the fact that because of the degree to which naturally occurring interspecific and instraspecific differential tolerance occurs, heavy reliance on glyphosate as a primary defense against weeds will lead to shifts toward the more tolerant populations and species. Gressel (1996) warns that the intra- and interspecific variability often reported in the literature for quantitative levels of glyphosate resistance should not be ignored. For example, a resistant isozyme of EPSPS was isolated from a corn cell line (Forlani et al., 1992). This corn cell line (black Mexican sweet) was reported to have two EPSPS isozymes, one of which was resistant to glyphosate. Corn tissue culture has shown some somaclonal variation with respect to glyphosate sensitivity (Racchi et al., 1995). Westwood and Weller (1997) report that different biotypes of bindweed show different levels of glyphosate sensitivity and that multiple mechanisms are responsible for the increased tolerance. Boerboom et al. (1991) showed that a several-fold difference in glyphosate tolerance was observed between different birdsfoot trefoil selections. Specific activity differences of EPSPS among the selections assayed were positively correlated with plant tolerance level, providing further evidence of differential tolerance, especially intraspecific variability, which indicates the potential for future control failures. More detailed examples of resistance and crop genetic engineering are presented in Chapter 18.

INHIBITORS OF BRANCH CHAIN AMINO ACID BIOSYNTHESIS

Imidazolinones

American Cyanamid Company developed the imidazolinone class of herbicides. These herbicides are used for the control of many broadleaf and grass weeds in cereals, soybeans, corn, alfalfa, peanuts, peas, and noncrop areas. The various compounds can be applied preemergence or postemergence, and treated plants stop growing almost immediately after application. Two to 4 days after treatment, the growing point (apical meristem) becomes chlorotic and later necrotic. Plant death begins in the growing point and gradually spreads to the entire plant, and within 7 to 10 days after treatment the plant becomes chlorotic and eventually dies. These compounds are also potent root inhibitors and can cause root pruning. All the imidazolinones are translocated in the phloem, as the herbicides are weak acids, which is consistent with phloem transport. These herbicides are inhibitors of the enzyme acetolactate synthase (ALS, also known as AHAS) in branched chain amino acid biosynthesis that leads to the production of the amino acids leucine, isoleucine, and valine.

The imidazolinones are weak acids. At low pH, the imidazolinones are neutral and therefore fat loving, whereas at pH 7 they are ionized. Ionization increases water solubility and decreases soil binding. The vapor pressures of the imidazolinones are very low.

Imazamethabenz

Imazamethabenz [(±)-2-[4,5-dihydro-4-methyl-4-(1-methylethyl)-5-oxo-1*H*-imidazol-2-yl]-4(and 5)-methylbenzoic acid], is commercially formulated as the methyl ester in a 3:2 mixture of the para- and meta-methyl isomers. The herbicide is an off-white solid with a vapor pressure of 1.1×10^{-8} mm Hg at 25°C, a water solubility of 1370 mg/l (ppm) for the *m*-isomer and 857 mg/l (ppm) at 25°C for the *p*-isomer, a soil half-life of 25 to 36 days, and an oral LD_{50} (rat) of >5000 mg/kg. The soil influence on imazamethabenz and all the other imidazolinones is presented later in this chapter.

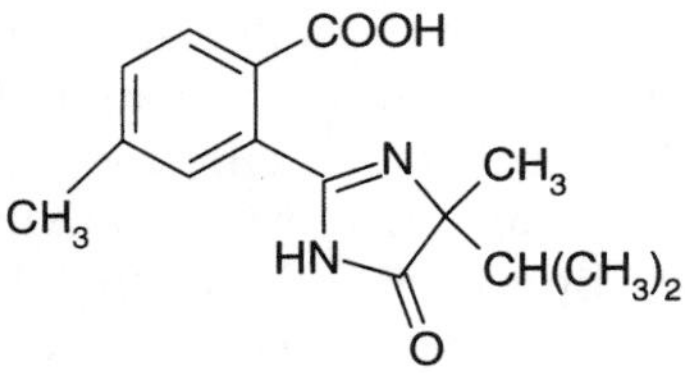

Imazamethabenz

Uses Imazamethabenz is formulated as an SC or LC and sold as Assert for postemergence control of wild oats and mustard species in winter and spring wheat and spring barley. Wild oat control is most effective at the one- to two-leaf stage. Rotational crops such as corn, alfalfa, sunflower, and edible beans can be planted the next cropping season following application of Assert. Although the half-life for imazamethabenz is reported to be 25 to 36 days, certain very susceptible crops can be injured at extremely low herbicide soil concentrations. Sugar beets cannot be planted for 20 months after imazamethabenz has been applied, and many other crops (such as vegetables) have a 15-month rotation interval.

Imazapyr

Imazapyr [2-[4,5-dihydro-4-methyl-4-(1-methylethyl)-5-oxo-1*H*-imidazol-2-yl]-3-pyridinecarboxylic acid] is formulated as the water-soluble isopropylamine salt for herbicide use, but application rates and concentrations in the formulation refer to the acid equivalent. The acid form is a white to tan solid, and the isopropylamine salt is a pale yellow to dark green liquid. The isopropylamine salt has a vapor pressure of $<10^{-7}$ mm Hg at 45°C, a very high water solubility of 11,272 mg/L (ppm) at 25°C, a soil half-life of 25 to 142 days, depending on soil type and environmental conditions, and an oral LD_{50} (rat) of >5000 mg/kg.

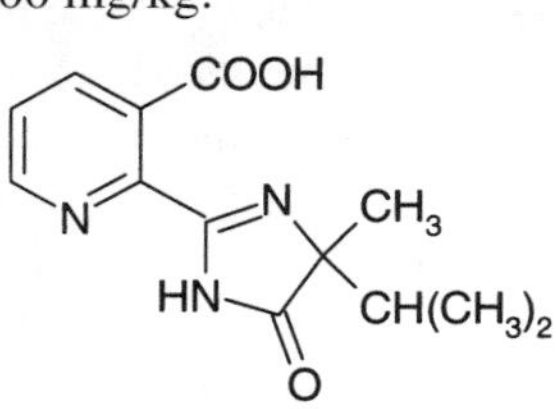

Imazapyr

Uses Imazapyr is formulated as an SL or G and sold as Arsenal, Chopper, and Contain as an industrial/noncrop vegetation control material. It provides broad-spectrum foliar and soil activity on many weed species, including sedges, field bindweed, johnsongrass, woollyleaf bursage, and other herbaceous and woody species such as vines, brambles, brush species, and deciduous trees. Postemergence applications are generally superior, especially for perennial weeds. The preemergence activity of imazapyr provides residual control of most weed species following a postemergence application. For best control, the weeds should be actively growing at the time of application.

In forestry, imazapyr is used for site preparation and release of loblolly pine stands. Planting of loblolly pine should be delayed for 3 months following a site preparation application. Broadcast applications for loblolly pine release should not be used before the conifer is 3 years old. Treatments applied during active growth may cause some minor growth inhibition in the conifer. Treatments made after formation of final conifer resting bud formation in the fall minimize potential conifer injury. Directed spray applications may be made at all ages of loblolly pine.

For woody plant control, imazapyr can be applied as a cut stump, tree injection, frill girdle, or low-volume basal bark treatment. The treatment is applied with diesel oil or a penetrating oil. Imazapyr provides excellent control of unwanted hardwood species in forests and has the advantage of controlling grassy weeds as compared with some growth regulator herbicides used in these situations.

Biological activity of imazapyr in soils is from 3 to 12 months, and there is little movement of imazapyr in soil.

Imazapic

Imazapic [(±)-2-[4,5-dihydro-4-methyl-4-(1-methylethyl)-5-oxo-1*H*-imidazol-2-yl]-5-methyl-3-pyridinecarboxylic acid] is an off-white to tan powder with a vapor pressure of $<10^{-7}$ mm Hg at 60°C, a water solubility of 2150 mg/l (ppm) at 25°C, a soil half-life of 120 days, and an oral LD_{50} (rat) of >5000 mg/kg.

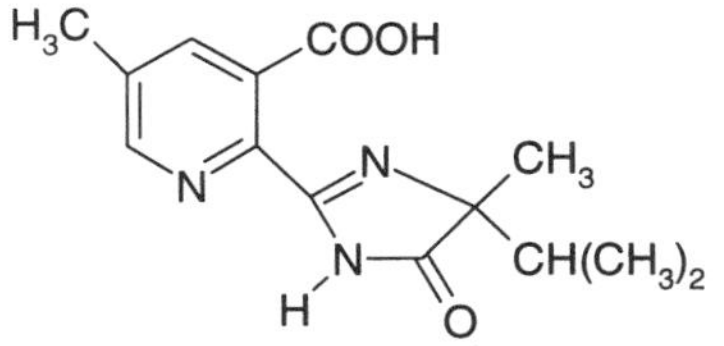

Imazapic

Uses Imazapic is formulated as a DG and sold as Cadre and Plateau for use in peanuts and noncropland for control of sicklepod, hemp sesbania, nutsedge, and many grass weeds, including *Panicum* spp., johnsongrass, foxtail, and crabgrass. Cadre and Plateau are also used for weed control and suppression of bahiagrass, bermudagrass, and centipedegrass. Plateau is used for prairie grass renovation as well.

Imazaquin

Imazaquin ±(2-[4,5-dihydro-4-methyl-4-(1-methylethyl)-5-oxo-1*H*-imidazol-2-yl]-3-quinolinecarboxylic acid) is a tan solid with a vapor pressure of $<2 \times 10^{-8}$ mm Hg at 45°C, a water solubility of 60 mg/l (ppm) at 25°C, a soil half-life of 60 days, and an oral LD_{50} (rat) of >5000 mg/kg.

COOH
N
N
CH3
HN
CH(CH3)2
O

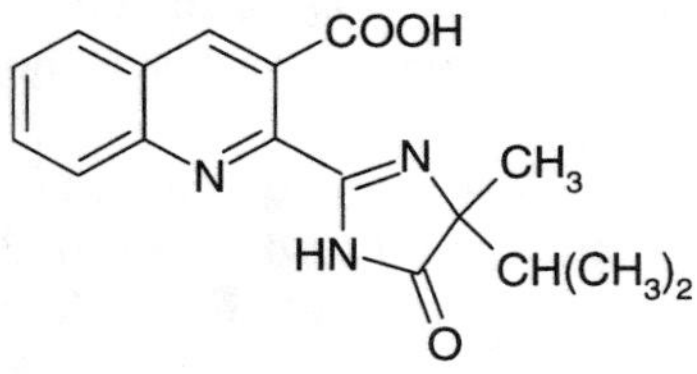

Imazaquin

Uses Imazaquin is formulated as an SL or WG under the trade name Scepter as a selective soybean herbicide with excellent broadleaf weed activity. Scepter is applied preplant incorporated, preemergence, and postemergence and controls cocklebur, pigweed, prickly sida, velvetleaf, black nightshade, smartweed, ragweed, jimsonweed, morning glory, and volunteer corn. Soil persistence and rotational crop sensitivity impact the cropping interval for sugar beets and many vegetables (26 months), cotton (18 months), corn, and sorghum (11 months) in some areas. Corn is more sensitive to imazaquin residues than sorghum. Scepter was the first imidazolinone herbicide commercialized when it was registered for use on soybeans in the southeastern United States in 1986. Imazaquin sold as Image is also registered for use on established warm-season turfgrasses such as bermudagrass, St. Augustinegrass, centipedegrass, and zoysiagrass.

Imazethapyr

Imazethapyr ±[2-[4,5-dihydro-4-methyl-4-(1-methylethyl)-5-oxo-1*H*-imidazol-2-yl]-5-ethyl-3-pyridinecarboxylic acid] is an off-white to tan solid with a vapor pressure of $<10^{-7}$ mm Hg at 60°C, a water solubility of 1400 mg/l (ppm) at 25°C, a soil half-life of 60 to 90 days, and an oral LD_{50} (rat) of >5000 mg/kg.

CH3H2C
COOH
N
N
CH3
HN
CH(CH3)2
O

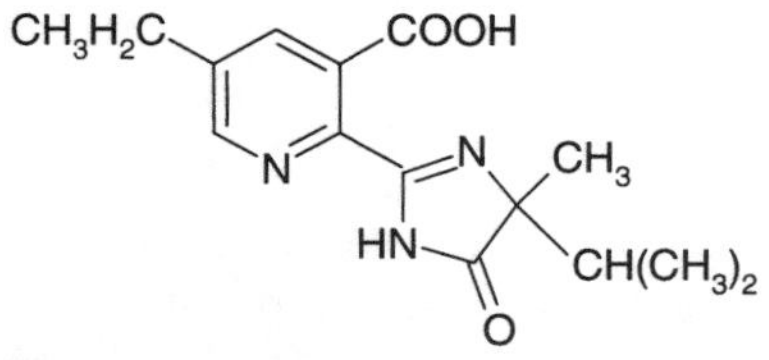

Imazethapyr

Uses Imazethapyr is formulated as the ammonium salt for herbicide use in the SL or WG form and sold as Pursuit for use in soybeans, peas, lentils, alfalfa, peanuts, and imidazolinone-resistant corn. Imazethapyr has similar weed control attributes as Scepter, but Pursuit has better postemergence activity on velvetleaf and grasses. Imazethapyr is applied preplant incorporated, preemergence, and postemergence. There is less concern about carryover of imazethapyr as compared with imazaquin. The greater flexibility in application timing, good control of many troublesome weeds, and reduced soil carryover have resulted in widespread use of imazethapyr, particularly in the northern corn belt. In 1994 registration was obtained for the use of Pursuit on imidazolinone (IMI)-resistant/tolerant corn. Imazethapyr-containing products for use in corn include Lightning (imazethapyr + imazapyr), Resolve (imazethapyr + dicamba), and Contour (imazethapyr + atrazine).

Imazamox

Imazamox (2-[4,5-dihydro-4-methyl-4-(1-methylethyl)-5-oxo-1*H*-imidazol-2-yl]-5-(methoxymethyl)-3-pyridinecarboxylic acid) is an odorless powdered solid with a vapor pressure of $<10^{-7}$ mm Hg at 20°C, a moderate water solubility, a soil half-life of 20 to 30 days, and an oral LD_{50} (rat) of >5000 mg/kg.

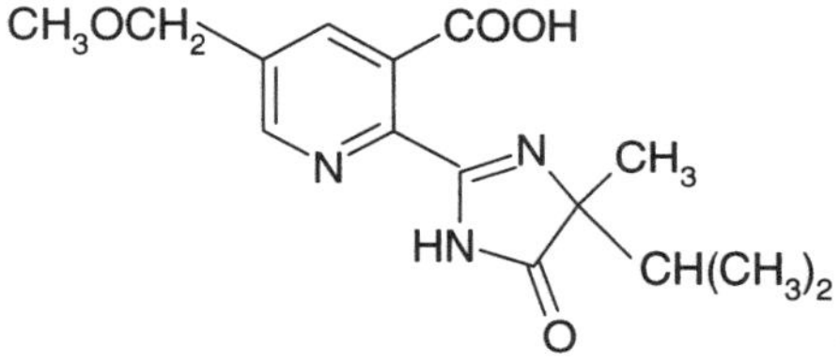

CH_3OCH_2 COOH N N CH_3 HN $CH(CH_3)_2$ O

Imazamox

Uses Imazamox is formulated as an EC in the ammonium salt form and sold as Raptor for use in soybeans. It is similar to imazapyr, but imazamox has better grass activity and shorter soil residual. The addition of an adjuvant and a fertilizer in the spray tank mixture is required to obtain maximum weed efficacy with this herbicide.

Soil Interaction

Imidazolinone herbicides are moderately to highly persistent in soil, lasting from several weeks to several months under temperate conditions. Greater adsorption, and therefore persistence, occurs with decreasing soil pH and increasing soil organic matter. Dissipation in soil is via microbial degradation. Under drier soil conditions, more herbicide is bound to clay and organic matter and less is available for degradation or plant uptake. Although degradation in soils with 5.0 and 7.0 pH is relatively slow, they undergo rapid hydrolysis at pH 9.0. Imidazolinone herbicides will degrade in light, but little photolysis has been observed on soil or plant foliage under field conditions. Little leaching has been documented under field conditions, even though some laboratory studies have shown moderate mobility.

Selectivity

The primary mechanism of natural selectivity to the imidazolinones is metabolism to nontoxic metabolites; however, uptake and translocation can also impact tolerance. Routes of metabolic degradation depend on the specific imidazolinone herbicide under investigation. For detailed information on imidazolinone metabolism in plants, see Shaner and Conner (1991).

Pyrimidyl-oxy-benzoates

The pyrimidyl-oxy-benzoates (POBs) were discovered by the Kumiai chemical industry. The POBs have the same mode of action (inhibition of ALS) as the imidazolinone herbicides. Only one POB herbicide (pyrithiobac) is presently on the market in the United States.

Pyrithiobac

Pyrithiobac (2-chloro-6-[(4,6-dimethoxy-2-pyrimidinyl)thio]benzoic acid, sodium salt) is a yellowish white solid with a water solubility of 760 mg/l (ppm) at 10°C, an oral LD_{50} (rat) of 1000 to 3000 mg/kg. It has a moderate soil persistence, with photodecomposition and microbial breakdown both involved in degradation.

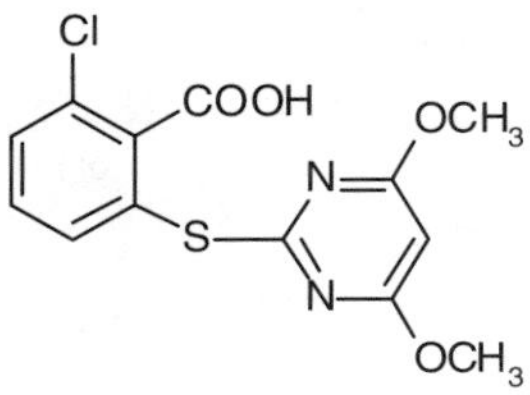

Pyrithiobac

Uses Pyrithiobac is formulated as an SP and sold under the trade name Staple for broadleaf weed control in cotton. Pyrithiobac is generally applied postemergence and provides good control of morning glory, pigweed, common cocklebur, velvetleaf, prickly sida, hemp sesbania, nightshades, and johnsongrass. Its selectivity in cotton is due to crop metabolism of the herbicide. Pyrithiobac is used on approximately 16% of the cotton acreage in the United States. Most horticultural crops cannot be planted on treated soil for at least 12 months.

Sulfonylureas, Triazolopyrimides, and the Sulfonylaminocarbonyltriazolinones

Sulfonylurea, triazolopyrimide, and sulfonylaminocarbonyltriazolinone herbicides move in the xylem and phloem in plants and prevent biosynthesis of branched-chain amino acids. They are potent and rapid inhibitors of plant cell division. Inhibition of growth is rapid in the growing tips of both the roots and shoots of sensitive plants. Devine et al. (1990) found that chlorsulfuron quickly reduces assimilate translocation, thus limiting its own translocation. The specific mechanisms of action for these herbicides are described later in this chapter.

Sulfonylureas were first commercialized for wheat and barley in 1982. These herbicides have high levels of activity at low application rates. Currently there are 26 sulfonylurea active ingredients on the market or under development. Slight changes in the basic structure can have great effects on the selectivity of these herbicides from one crop to another. These compounds are active at low rates of application, with some compounds active at 0.06 oz/acre. They provide excellent control of many important broadleaf and grass weeds in most major crops (Figure 16-3). Activity occurs through shoot and root uptake, as these herbicides are readily translocated to active growing points in plants. Selectivity in plants is due to differential metabolism. Examples of crops tolerant (due to various natural processes) to one or more sulfonylurea herbicides include wheat and other small grains, soybean, rice, oilseed rape/canola, flax, corn, potato, tomato, sugar beet, cotton, blueberries, turf, plantation crops, and conifers. Sulfonylurea herbicides are also used to control vegetation in industrial and right-of-way sites.

Triazolopyrimide herbicides were first commercialized in 1993, and presently five are registered or are under development. Corn, cereals, and soybean are tolerant to one or more of these herbicides, with use rates ranging from 0.25 to 1.50 oz ai/acre.

Sulfonylaminocarbonyltriazolinones are a new class of ALS-inhibitor herbicides. Currently, flucarbazone herbicide (Everest) is approved for use in wheat in the United States. Weeds resistant to sulfonylurea herbicides are usually resistant to

Figure 16-3. Chlorsulfuron weed control in wheat. Note excellent weed control in center of photo. (A. Appleby, Oregon State Univ.)

triazolopyrimide herbicides and cross-resistance in certain weeds between sulfonylurea and sulfonylaminocarbonyltriazolinone herbicides has been reported in Oregon.

SULFONYLUREAS

Chlorsulfuron

Chlorsulfuron (2-chloro-*N*-[[(4-methoxy-6-methyl-1,3,5-triazin-2-yl)amino]carbonyl] benzenesulfonamide) is an odorless white crystalline powder with a vapor pressure of 2.3×10^{-11} mm Hg at 25°C, a water solubility of 587 mg/l (ppm) at pH 5 and 31,800 mg/l (ppm) at pH 7, a soil half-life of 40 days, and an oral LD_{50} (rat) of >5000 mg/kg.

OCH_3 N N $SO_2NHCONH$ N Cl CH_3

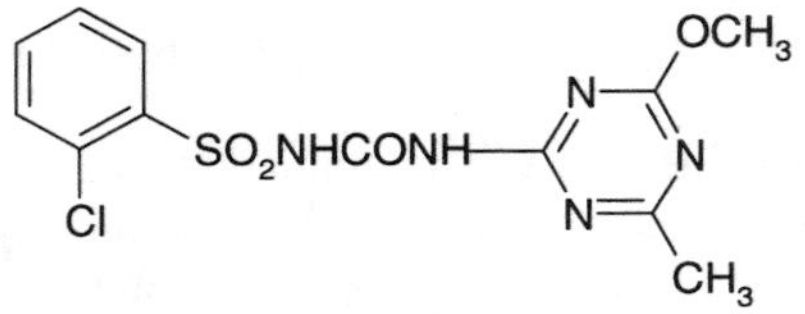

Chlorsulfuron

Uses Chlorsulfuron is formulated as a WG, sold as Glean, and is registered for use in wheat, barley, and spring oats. Other crops that are somewhat tolerant include triticale and flax. The trade name for the noncrop and industrial-site weed control formulation is Telar, and for turf, sod, and seed production, it is sold as TFC. It controls a wide range of annual broadleaves at label rates of 0.188 to 0.75 oz ai/acre depending on the situation.

Sulfometuron

Sulfometuron (methyl 2-[[[[(4,6-dimethyl-2-pyrimidinyl)amino]carbonyl]amino] sulfonyl]benzoate) is an odorless white solid with a vapor pressure of 5.5×10^{-16} mm Hg at 25°C, a water solubility of 10 mg/l (ppm) at 25°C at pH 5 and 300 mg/l (ppm) at pH 7, a soil half-life of 20 to 28 days, and an oral LD_{50} (rat) of >5000 mg/kg.

CO_2CH_3 CH_3 N $SO_2NHCONH$ N CH_3

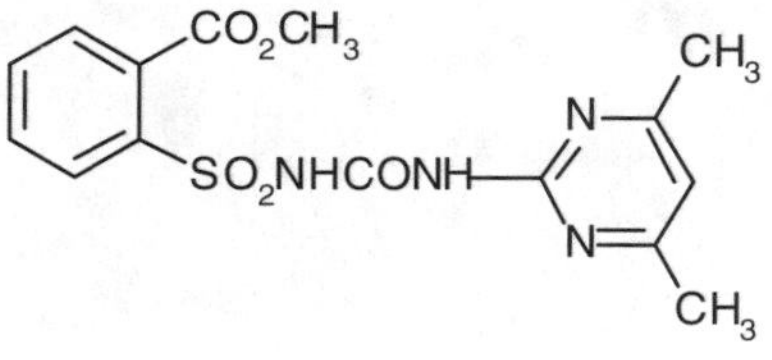

Sulfometuron

Uses Sulfometuron is formulated as a WG and sold as Oust. Oust has a wider range of activity than chlorsulfuron. It has little selectivity in grass-type crops and is used mostly in noncrop weed control with some use in industrial, unimproved turf, hardwood plantations, and conifer site preparation and release and, when combined with hexazinone, is sold as Oustar for forestry uses.

Metsulfuron

Metsulfuron (methyl 2-[[[[(4-methoxy-6-methyl-1,3,5-triazin-2-yl)amino]carbonyl]-amino]sulfonyl]benzoate) is a white to pale white solid with a sweet ester like odor, a vapor pressure of 2.5×10^{-12} mm Hg at 25°C, a water solubility of 548 mg/l (ppm) at 25°C at pH 5 and 2790 mg/l (ppm) at pH 7, a soil half-life of 30 days, and an oral LD_{50} (rat) of >5000 mg/kg.

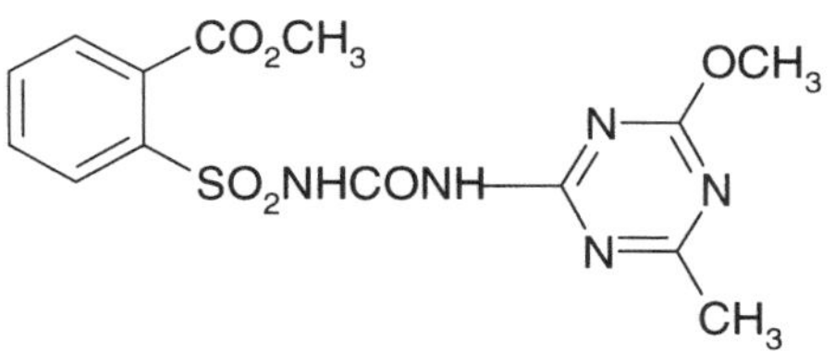

Metsulfuron

Uses Metsulfuron is formulated as a WG and sold as Ally, Escort, or Manor. It is used for selective weed control in wheat, barley, pastures, rangeland, conifer plantations, site preparation, noncropland, and turf. Metsulfuron has a smaller margin of selectivity, especially in barley, than chlorsulfuron; however, it is slightly more active on many broadleaf weeds. Metsulfuron is also registered for brush control and may be slightly less persistent than chlorsulfuron in the soil.

Chlorimuron Ethyl

Chlorimuron ethyl (ethyl 2-[[[[(4-chloro-6-methoxy-2-pyrimidinyl)amino]carbonyl]-amino]sulfonyl]benzoate) is a white solid with a vapor pressure of 4×10^{-12} mm Hg at 25°C, a water solubility of 11 mg/l (ppm) at 25°C at pH 5 and 1200 mg/l (ppm) at pH 7, a soil half-life of 40 days, and an oral LD_{50} (rat) of 4102 mg/kg.

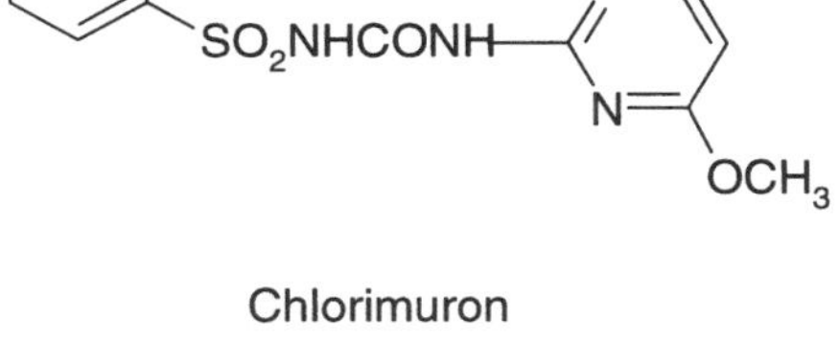

Chlorimuron

Uses Chlorimuron is formulated as a WG and sold as Classic, with selectivity in soybean, peanut, and noncropland. It is particularly effective against small broadleaf weeds, but yellow nutsedge is susceptible and grass weeds are suppressed. Chlorimuron may be used preplant incorporated, preemergence, or postemergence. Chlorimuron is sold as a combination with metribuzin as Canopy and Canopy SP and with sulfentrazone as Canopy XL for use in soybeans, and with thifensulfuron as Reliance STS, Reliance STS SP, and Synchrony STS, STS DF, and STS SP for use in STS soybeans (sulfonylurea-tolerant soybeans).

Bensulfuron

Bensulfuron (methyl 2-[[[[[(4,6-dimethoxy-2-pyrimidinyl)amino]carbonyl]amino]-sulfonyl]methyl]benzoate) is a white to pale white odorless solid with a vapor pressure of 2.1×10^{-14} mm Hg at 25°C, a water solubility of 3 mg/l (ppm) at 25°C at pH 5 and 120 mg/l (ppm) at pH 7, a soil half-life of 4 to 8 weeks, and an oral LD_{50} (rat) of >5000 mg/kg.

CO_2CH_3 OCH_3 N $CH_2SO_2NHCONH$ N OCH_3

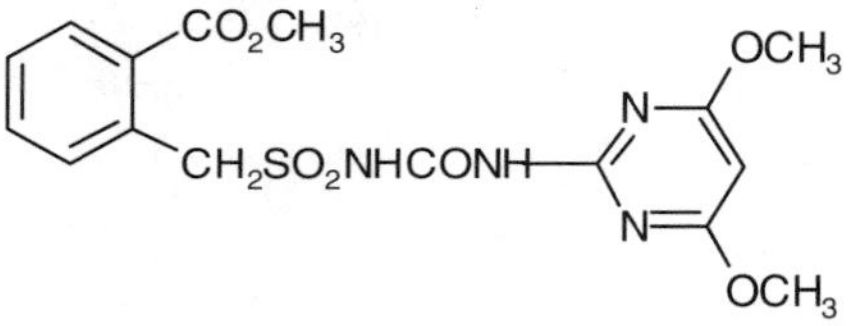

Bensulfuron

Uses Bensulfuron is formulated as a WG and sold as Londax for weed control in rice, and as Duet combined with propanil for rice. It is particularly effective against broadleaves and sedges, although grasses may be suppressed.

Thifensulfuron

Thifensulfuron (methyl 3-[[[[(4-methoxy-6-methyl-1,3,5-triazin-2-yl)amino]-carbonyl]amino]sulfonyl]-2-thiophenecarboxylate) is an odorless white solid with a vapor pressure of 1.3×10^{-10} mm Hg at 25°C, a water solubility of 223 mg/l (ppm) at 25°C at pH 5 and 2240 mg/l (ppm) at pH 7, a soil half-life of 12 days, and an oral LD_{50} (rat) of >5000 mg/kg.

OCH_3 N $SO_2NHCONH$ N N CH_3 S CO_2CH_3

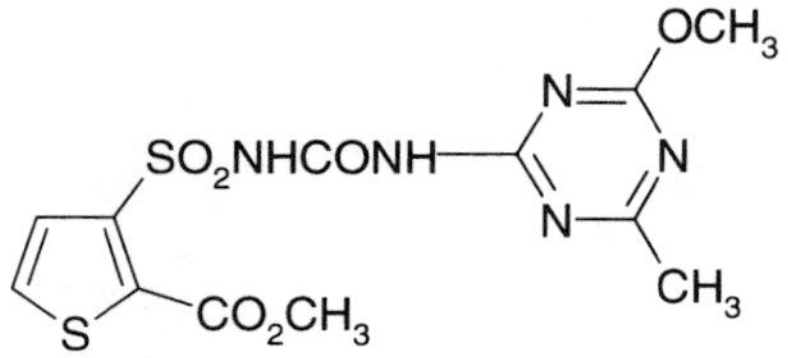

Thifensulfuron

Uses Thifensulfuron is formulated as a WG and sold as Pinnacle for use in soybeans, as Harmony GT for use in cereals, as Upbeet for use in sugar beets, in mixture with tribenuron as Harmony Extra in small grains, as Canvas (a mixture of thifensulfuron, tribenuron, and metsulfuron) for use in wheat, barley, and fallow, and under various names for use in soybeans when combined with chlorimuron, as stated earlier. Its soil persistence is generally less than 1 month, and it is somewhat less injurious to some varieties of wheat than tribenuron.

Tribenuron

Tribenuron (methyl 2-[[[[(4-methoxy-6-methyl-1,3,5-triazin-2-yl)-methylamino]-carbonyl]amino]sulfonyl]benzoate) is a light brown solid, with a vapor pressure of 3.9×10^{-10} mm Hg at 25°C, a water solubility of 48 mg/l (ppm) at 25°C at pH 5 and 2040 mg/l (ppm) at pH 7, a soil half-life of 10 days, and an oral LD_{50} (rat) of >5000 mg/kg.

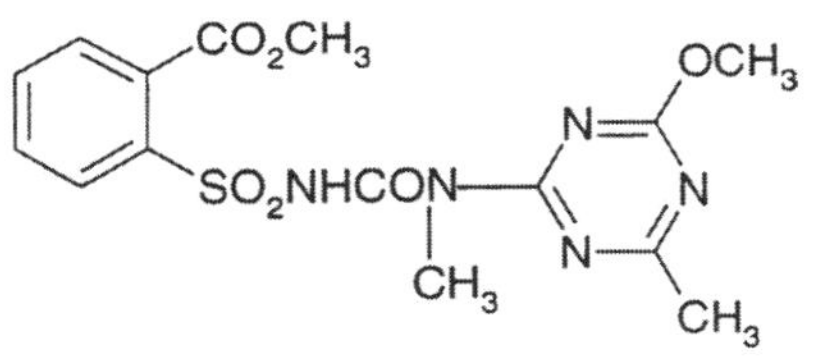

Tribenuron

Uses Tribenuron is formulated as a WG and sold under the trade name Express for selective weed control in small grains, controlling a slightly different spectrum of broadleaves than thifensulfuron—for example, it is more effective against Canada thistle. It has been formulated as a mixture with thifensulfuron, which is called Harmony Extra. It is very short-lived in the soil.

Triasulfuron

Triasulfuron (2-(2-chloroethoxy)-*N*-[[(4-methoxy-6-methyl-1,3,5-triazin-2-yl)amino) carbonyl]benzenesulfonamide) is a colorless and odorless crystalline solid with a vapor pressure of $<1.5 \times 10^{-8}$ mm Hg at 25°C, a water solubility of 32 mg/l (ppm) at 25°C at pH 5 and 815 mg/l (ppm) at pH 7, a soil half-life that can vary from 11 to 95 days, depending on soil type and environmental conditions, and an oral LD_{50} (rat) of >5000 mg/kg.

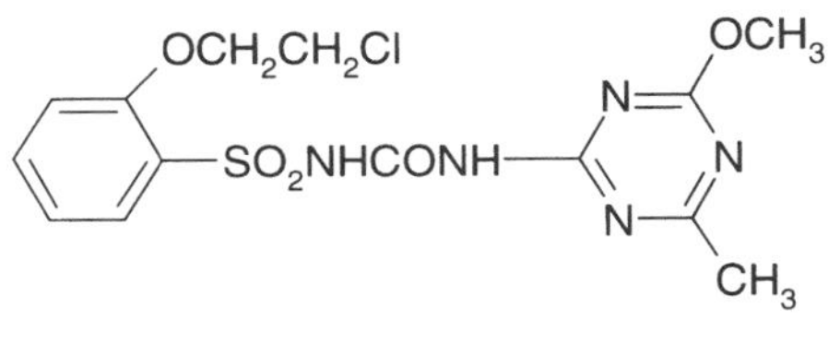

Triasulfuron

Uses Triasulfuron is formulated as a WG and sold under the trade name Amber, for use in wheat, barley, pastures, rangelands, and CRP, and as Rave, combined with dicamba, for wheat. It is similar to Glean and Ally in selectivity and soil persistence; however, it is somewhat more effective against vetch and bedstraw.

Primisulfuron

Primisulfuron (methyl 2-[[[[[(4,6-bis(difluoromethoxy)-2-pyrimidinyl)amino]carbonyl]-amino]sulfonyl]benzoate) is a colorless to yellow odorless crystalline solid with a vapor pressure of 3.75×10^{-8} mm Hg at 25°C, a water solubility of 3.3 mg/l (ppm) at 20°C at pH 5 and 5280 mg/l (ppm) at pH 7, a soil half-life of 30 days, and an oral LD_{50} (rat) of >5050 mg/kg.

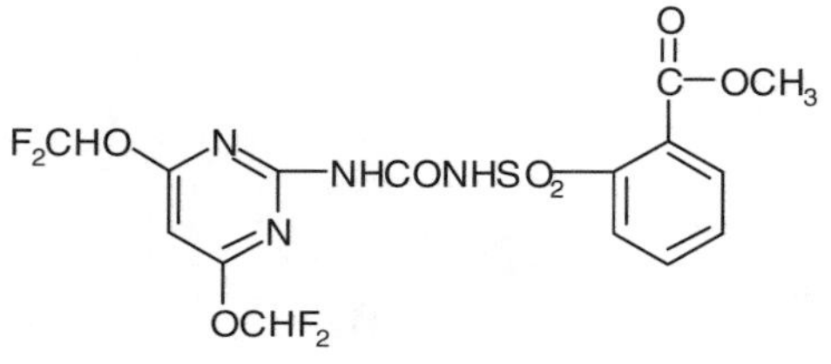

Primisulfuron

Uses Primisulfuron is formulated as a WG and SG and sold as Beacon for selective weed control in field corn, popcorn, and Kentucky bluegrass and for control of both broadleaves and grasses when applied postemergence. It is used in grass seed crops for control of quackgrass and annual grass weeds (e.g., downy brome, wild oat). Exceed is a mixture of primisulfuron and prosulfuron, and NorthStar is a combination with dicamba for uses in field corn.

Nicosulfuron

Nicosulfuron (2-[[[[(4,6-dimethoxy-2-pyrimidinyl)amino]carbonyl]amino]-sulfonyl]-*N,N*-dimethyl-3-pyrimdinecarboxamide) is a white solid with a phenolic odor, with a vapor pressure of 1.2×10^{-16} mm Hg at 25°C, a water solubility of 360 mg/l (ppm) at 25°C at pH 5 and 12,200 mg/l (ppm) at pH 7, a soil half-life of ~21 days, and an oral LD_{50} (rat) of >5000 mg/kg.

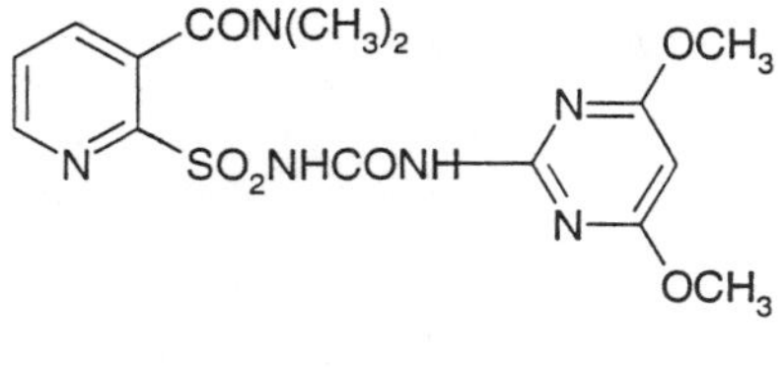

Nicosulfuron

Figure 16-4. Grass control in corn by nicosulfuron applied postemergence to the grass and the crop; all the grass was killed with no damage to the corn.

Uses Nicosulfuron is formulated as a WG and sold under the trade name Accent for selective postemergence weed control of annual and some perennial grasses, as well as some broadleaves, in field corn and popcorn (Figure 16-4). Nicosulfuron is also sold in combination with rimsulfuron, flumetsulam, and clopyralid as Accent Gold, with rimsulfuron and atrazine as Basis Gold, with dicamba as Celebrity Plus, for field corn, and with dicamba and diflufenzopyr as Celebrity for field corn and sugarcane.

Ethametsulfuron

Ethametsulfuron (methyl 2-[[[[[(4-ethoxy-6-(methylamino)-1,3,5-triazin-2-yl]-amino]carbonyl]amino]sulfonyl]benzoate) is a white crystalline solid with a vapor pressure of 5.8×10^{-15} mm Hg at 25°C, a water solubility of 1.7 mg/l (ppm) at 25°C at pH 5 and 410 mg/l (ppm) at pH 7, a short soil persistence, and an oral LD_{50} (rat) of >5000 mg/kg.

CO_2CH_3 — $SO_2NHCONH$ — OCH_2CH_3 — $NHCH_3$

Ethametsulfuron

Uses Ethametsulfuron is formulated as a WG and sold under the trade name Muster for selective weed control in oilseed rape/canola. It is sold in Canada, but it is not registered for use in the United States except as a Section 18 emergency exemption (see Chapter 4).

Rimsulfuron

Rimsulfuron (N-[[(4,6-dimethoxy-2-pyrimidinyl)amino]carbonyl]-3-(ethylsulfonyl)-2-pyridinesulfonamide) is an off-white to tan odorless solid, with no reported vapor pressure, a water solubility of <10 mg/l (ppm) at 25°C in unbuffered distilled water and 7300 mg/l (ppm) at pH 7, a soil half-life of 1.7 to 4.3 days, and an oral LD_{50} (rat) of >5000 mg/kg.

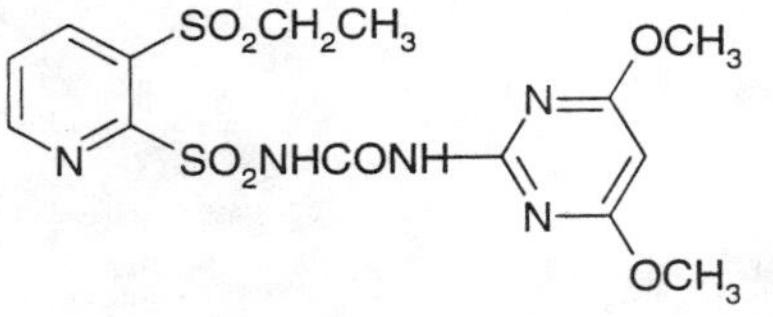

Rimsulfuron

Uses Rimsulfuron is formulated as a DF and sold as Matrix and Shadeout for preemergence and postemergence control of grasses and broadleaf weeds in potatoes and tomatoes, respectively, and as Basis with rimsulfuron and thifensulfuron for use in field corn.

Triflusulfuron

Triflusulfuron (methyl 2[[[[[4-(dimethylamino)-6-[2,2,2-trifluoroethoxy)-1,3,5- triazin-2-yl]amino]carbonyl]amino]sulfonyl]-3-methylbenzoate) is an off-white crystalline solid with a vapor pressure of $<10^{-7}$ mm Hg at 25°C, a water solubility of 1 mg/l (ppm) at 25°C at pH 5 and 110 mg/l (ppm) at pH 7, a soil half-life of 2 to 4 days, and an oral LD_{50} (rat) of >5000 mg/kg.

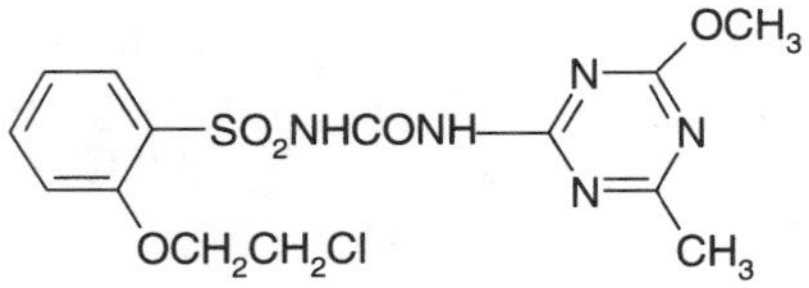

Triflusulfuron

Uses Triflusulfuron is formulated as a WG under the trade names Safari, Upbeet, and Debut for early postemergence control of both broadleaves and grasses in sugar beet and fodder beet.

Cinosulfuron

Cinosulfuron (*N*-[[(4,6-dimethoxy-1,3,5-triazin-2-yl) amino]carbonyl]- 2-(2-methoxyethoxy)-benzenesulfonamide) is a colorless crystalline solid with a low vapor pressure, a water solubility of 120 mg/l (ppm) at 25°C at pH 5 and 4000 mg/l (ppm) at pH 7, a soil half-life of 3 days, and an oral LD_{50} (rat) of >5000 mg/kg.

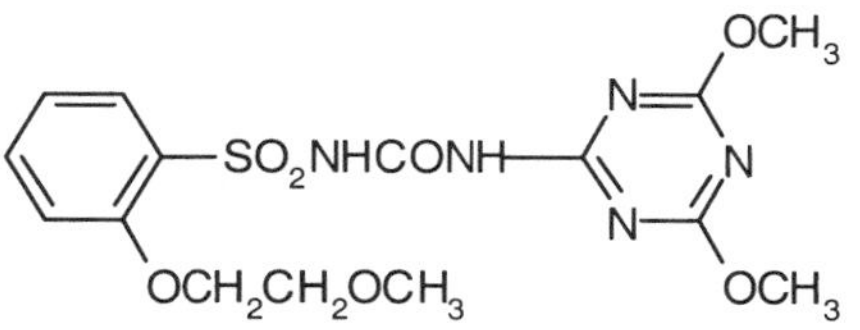

Cinosulfuron

Uses Cinosulfuron is formulated as a WG and sold under the trade name Setoff for preemergence and early postemergence broadleaf and grass weed control in rice and tropical plantation crops.

Prosulfuron

Prosulfuron (N-[[3-(4-methoxy-6-methyl-1,3,5-triazin-2-yl)-amino-carbonyl]-2-(3,3,3-trifluoropropyl)benzene sulfonamide) is an odorless and colorless crystalline solid with a vapor pressure of $<2.6 \times 10^{-8}$ mm Hg at 25°C, a water solubility of 30 mg/l (ppm) at 25°C at pH 5 and 4000 mg/l (ppm) at pH 7, an oral LD_{50} (rat) of 986 mg/kg, and a soil half-life of 8 to 20 days.

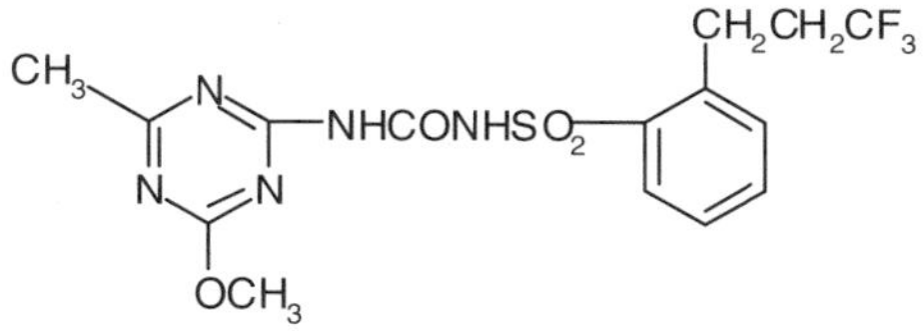

Prosulfuron

Uses Prosulfuron is formulated as a WG and sold as Peak for preemergence and postemergence broadleaf weed control in grain sorghum, wheat, barley, rye, oats, triticale, and proso millet, and as Exceed mixed with primisulfuron for use in field corn.

Halosulfuron

Halosulfuron (3-chloro-5-[[[[(4,6-dimethoxy-2-pyrimidinyl) amino] carbonyl]amino]-sulfonyl]-1-methyl-1*H*-pyrazole-4-carboxylate) is a white powder with a vapor pressure of 2.8×10^{-12} mm Hg at 25°C, a water solubility of 15 mg/l (ppm) at 25°C, a soil half-life of 33 to 35 days, and an oral LD_{50} (rat) of 8865 mg/kg.

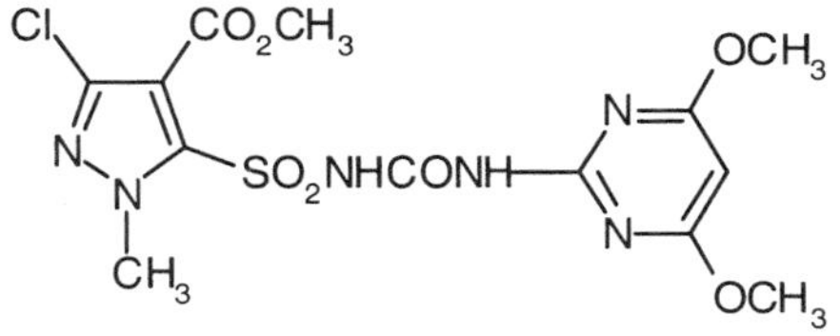

Halosulfuron

Uses Halosulfuron is formulated as a WG and sold under the trade name Permit for preemergence and postemergence control of broadleaf weeds and some grasses and sedges in field and sweet corn, grain sorghum, and fallow, as Manage in industrial and roadside weed control, and as Sempra in corn and sugarcane.

Sulfosulfuron

Sulfosulfuron [*N*-[[4,6-dimethoxy-2-pyrimidinyl)amino]carbonyl]-2-(ethylsulfonyl)-imidazo[1,2-α] pyridine-3-sulfonamide] is an odorless white solid with a vapor pressure of 6.61×10^{-10} mm Hg at 25°C, a water solubility of 17.6 mg/l (ppm) at 20°C at pH 5 and 1627 mg/l (ppm) at pH 7, a soil half-life of 14 to 75 days, and an oral LD_{50} (rat) of >5000 mg/kg.

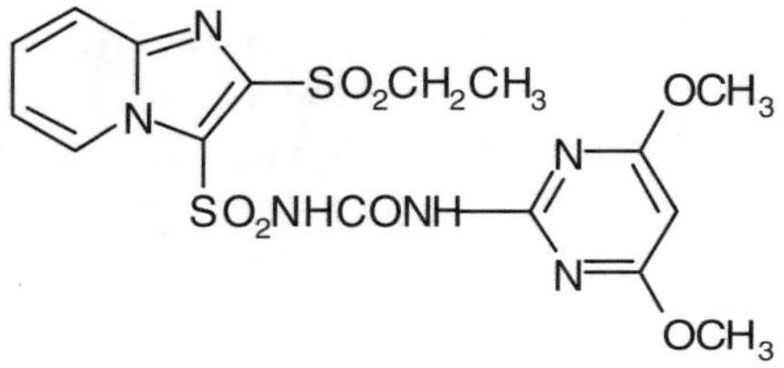

Sulfosulfuran

Uses Sulfosulfuron is formulated as a WG and sold as Maverick for preemergence or postemergence broad-spectrum weed control in wheat, and as Outrider for industrial and roadside weed control. Sensitive grass weed species include annual bromes, quackgrass, volunteer barley, and wild oat.

TRIAZOLOPYRIMIDES

Flumetsulam

Flumetsulam (*N*-(2,6-difluorophenyl)-5-methyl[1,2,4]triazolo[1,5-α]pyrimidine-2-sulfonamide) is an off-white to light tan solid with a vapor pressure of 2.8×10^{-15} mm Hg at 25°C, a water solubility of 49 mg/l (ppm) at 25°C at pH 5 and 5600 mg/l (ppm) at pH 7, a soil half-life of 1 to 3 months, and an oral LD_{50} (rat) of >5000 mg/kg.

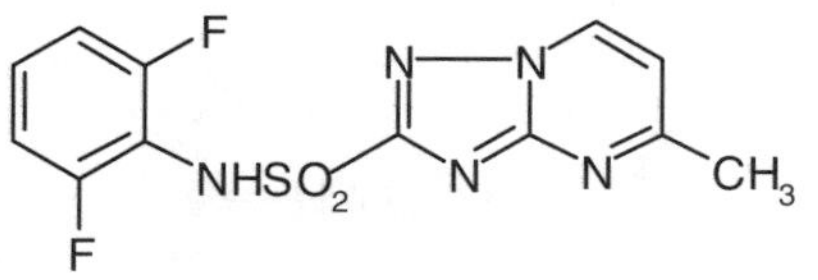

Flumetsulam

Uses Flumetsulam is formulated as an SC and sold as Python for selective weed control in soybeans, and sold in combination with trifluralin as Broadstrike + Treflan for soybeans, with metolachlor as Broadstrike + Dual for field corn and soybeans, and with clopyralid as Hornet for field corn to control broadleaf weeds and annual grasses. Cotton, sugar beet, and canola can be planted 18 to 26 months after application.

Cloransulam-Methyl

Cloransulam-methyl (methyl 3-chloro-2-[[(5-ethoxy-7-fluoro[1,2,4]triazolo[1,5-*c*]-pyrimidin-2 yl)sulfonyl]amino]benzoate) is an off-white powdered solid with a vapor pressure of 3×10^{-16} mm Hg at 25°C, a water solubility of 3 mg/l (ppm) at 25°C at pH 5 and 184 mg/l (ppm) at pH 7, a soil half-life of 13 to 28 days, and an oral LD_{50} (rat) of >5000 mg/kg.

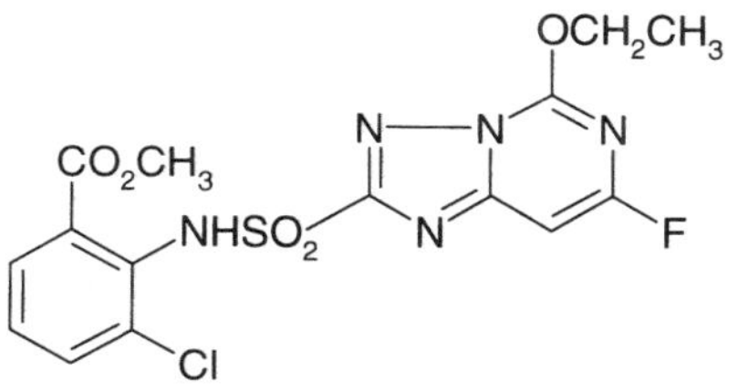

Cloransulam Methyl

Uses Cloransulam is formulated as a WSP and sold under the trade names First Rate for use in soybeans, combined with flumetsulam as Frontrow for soybeans, and with sulfentrazone as Gauntlet for soybeans in preemergence and postemergence broadleaf weed control. Weeds controlled include cocklebur, velvetleaf, ragweed, morning glory, and sunflower.

Diclosulam

Diclosulam (*N*-(2,6-dichlorophenyl)-5-ethoxy-7-fluoro[1,2,4]triazolo-[1,5-*c*]pyrimidine-2-sulfonamide). No physical data are available.

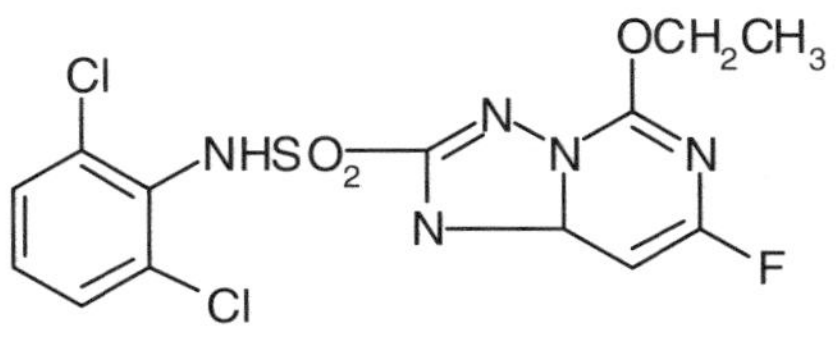

Diclosulam

Uses Diclosulam is a new herbicide sold as StrongArm for preemergence and early postemergence broadleaf weed control in soybeans and peanuts.

SULFONYLAMINOCARBONYLTRIAZOLINONE

Flucarbazone-Sodium

Flucarbazone-sodium (4,5-Dihydro-3-methoxy-5-oxo-*N*-[[2-(trifluoromethoxy)-phenyl)-1*H*-1,2,4-triazole-1-carboxamide, sodium salt) is a tan solid with a vapor pressure of 1×10^{-10} mm Hg at 20°C, a water solubility of 30,000 mg/l (ppm), and an oral LD_{50} (rat) of >5000 mg/kg.

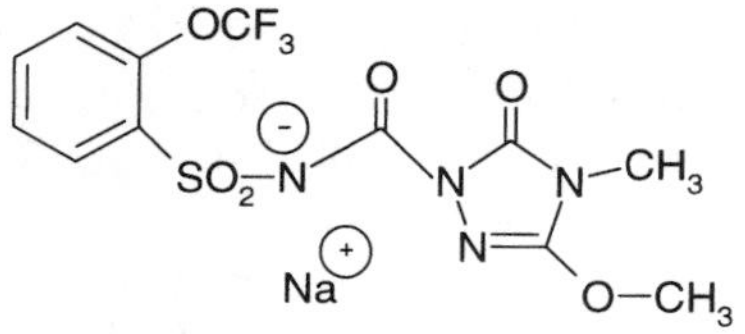

Flucarbazone - Sodium

Uses Flucarbazone is formulated as a dispersible granular (DG) and sold as Everest for postemergence control of wild oat, green foxtail, and Italian ryegrass in winter and all types of spring wheat, durum wheat, and winter wheat.

SELECTIVITY

For sulfonylurea and triazolopyrimide herbicides, the single most important mechanism of selectivity is rapid conversion to inactive compounds in tolerant crops; little or no metabolism is measured in sensitive plants. Sweetser et al. (1982) have found that wheat, oat, and barley can add an –OH group to the phenyl ring of chlorsulfuron, after which the herbicide conjugates with a carbohydrate to form an inactive compound. Some broadleaves, such as flax, nightshades, and leafy spurge, are also tolerant to sulfonylureas (Swisher and Wiemer, 1986). This tolerance is apparently based on the ability to add an –OH group to the N-containing ring and attach a carbohydrate to it. An interesting observation has been the relatively flat response curve often seen with chlorsulfuron. Often, even low rates can cause observable effects, but much higher rates do not cause proportionately greater injury. Brewster and Appleby (1983) found that 560 g/ha of chlorsulfuron did not reduce wheat yields significantly below those at 35 g/ha, a range of 16-fold. The range of sensitivity among plants is unusually great. For example, wheat tolerates nearly 100 ppb of chlorsulfuron, whereas sugar beet growth can be negatively affected at 0.1 ppb (*DuPont Agricultural Products Technical Bulletin*).

The relative sensitivity of plants to triazolopyrimidine herbicides is directly related to the metabolism rate in each plant species. For example, the degradation half-life of chloransulam-methyl in soybean (tolerant), velvetleaf, ivyleaf morning glory, and pitted morning glory (susceptible) is 4.9, 62.1, 69.3, and 165.3 hours, respectively (Swisher et al., 1991).

Sulfonylaminocarbonyltriazolinone herbicides are used to selectively control many different grass weeds and some broadleaf weeds in wheat, rye, and triticale. Like the sulfonylurea and triazolopyrimidine herbicides, sulfonylaminocarbonyltriazolinones are ALS-inhibitor herbicides. The variation in the mechanisms of selectivity between susceptible and tolerant species has not been reported, but likely is caused by a differential rate of metabolism among species. Herbicide uptake is enhanced by addition of a nonionic surfactant.

Application of certain soil- or foliage-applied organophosphate insecticides can reduce the selectivity of some sulfonylureas in crops. For example, certain sulfonylurea-containing herbicides can be used (e.g., Accent and Beacon) with banded applications of Counter CR, but others cannot be used (e.g., Basis and Basis Gold). Di-Syston or malathion applied shortly before or after chlorsulfuron (Glean), chlorsulfuron + metsulfuron (Finesse), and probably triasulfuron (Amber), can cause injury in wheat.

SOIL INTERACTION

Sulfonylureas are the best-studied types in this herbicide grouping. They can be mobile in the soil, depending somewhat on their specific chemical structure. The herbicides move more readily in high pH soils than in low pH soils. For example, the solubility of chlorsulfuron in water (25°C) at pH 5 and 7 is 587 and 31,800 mg/L, respectively. Chloransulam-methyl, a triazolopyrimide herbicide, exhibits low soil mobility. Its water solubility (25°C) at pH 5 and 7 is 3 and 184 mg/L, respectively. This indicates that some of these herbicides have properties of chemicals that can move through soil and potentially reach groundwater. However, groundwater contamination has not been detected, because these herbicides are applied at low use rates and usually dissipate quickly. The water solubility (25°C) of flumetsulam at pH 2.5 and 7 is 49 and 5,600 mg/L, respectively. However, lysimeter studies indicate that risk of groundwater contamination is negligible (Feucht et al., 1999).

In general, all sulfonylurea herbicides degrade in the soil by both chemical hydrolysis and microbial breakdown. Because they are active at such low concentrations, carryover injury to rotation crops is a major concern with some sulfonylurea herbicides. This is particularly true in areas with diversified cropping systems. For example, chlorsulfuron, metsulfuron, triasulfuron, sulfometuron, chlorimuron, nicosulfuron, primisulfuron, and ethametsulfuron have rotational crop planting restrictions that exceed 120 days. Bensulfuron, thifensulfuron, and tribenuron have rotational crop restrictions of about 45 to 120 days.

Both microbial breakdown and chemical hydrolysis occur in acid soils. Sulfonylurea herbicides are weak acids. Thus, hydrolysis occurs much faster in acid soils. In high pH soils, chemical hydrolysis is drastically reduced and degradation is primarily by microbe action. On soils with pH of 7.0 to 7.5, a 48-month interval or more may be required before certain sensitive crops can be grown. Several sulfonylurea herbicides, such as chlorsulfuron, triasulfuron, and metsulfuron, should not be used in soils above pH 7.9. In general, sulfonylurea herbicides degrade quickly in warm, moist, acid soils and more slowly in cool, dry, alkaline soils.

A wide range of crop plants are sensitive to recommended rates of sulfonylurea herbicides. Differences in sensitivity among crop species to sulfonylurea herbicides are listed in Table 16-1, and a ranking of sensitivity of rotation crops to triasulfuron and chlorsulfuron is shown in Table 16-2 (Carda et al., 1991). When these crops are grown in rotation with persistent sulfonylurea herbicides, possible damage will depend on the amount of herbicide persisting into the following season, which, in turn, is influenced by soil pH , moisture, temperature, and such factors. Moreover, the same crop will respond differently to the same level of a particular sulfonylurea herbicide, depending on soil and environmental factors. This makes predicting the chances of carryover problems difficult. The presence of sensitive weeds in previously treated areas has proven not to be a good indicator of when crops can be planted safely.

Triazolopyrimide herbicides are metabolized primarily by common soil microorganisms. Laboratory and field studies show that the half-life of chloransulam-methyl is 13 to 28 days for 16 agricultural soils incubated at 25°C. Degradation is biphasic, with a rapid initial degradation followed by slower subsequent degradation. Two of the degradation products, chloransulam and sulfonamide, have about 10 and 1000 times less herbicidal activity, respectively, than the parent chloransulam-methyl. The rotational crop intervals for chloransulam-methyl are 3 months for wheat; 9 months for alfalfa, corn, cotton, peanuts, rice, and sorghum; and 30 months for sugar beet, sunflower, tobacco and all other crops.

Soil pH and organic matter affect availability of flumetsulam in soil, and activity increases as pH increases and organic matter decreases. Rotational crop intervals for flumetsulam (70 g/ha) plus trifluralin (0.95 kg/ha) applied to soybeans varies greatly, depending on the crop. For example, the interval is 1 month for peanuts; 3 months for wheat, barley, field corn, and rice; 12 months for grain sorghum and sunflower; and up to 22 months for cotton, sugar beet, and canola.

Reported soil half-lives of flucarbazone-sodium ranges from 9 days in northern Europe (Feucht et al., 1999) to about 80 days in the semiarid Pacific Northwest region

TABLE 16-1. Relative Sensitivity of Rotation Crops to Chlorsulfuron Residues

Tolerant------------------------------				------------------------------Sensitive
Wheat	Barley	Safflower Dry beans Flax Fava beans Sorghum Red clover Peas	Garbanzo beans Potatoes	Corn Lentils Sunflowers Alfalfa Sugar beets Mint

Estimated from several sources (Burkhart et al., 1984; Dyer and Fay, 1983).

TABLE 16-2. Rotation Crop Sensitivity Ranking

Crop	Sensitivity Ranking[a]	
	Triasulfuron	Chlorsulfuron
Lentils	1	1
Sugar beets	2	2
Alfalfa	3	3
Sunflower	4	7
Safflower	5	10
Flax	6	5
Durum	7	8
Pinto bean	8	6
Barley	9	9
Oats	10	11
Corn	11	4

[a]1 = most sensitive; 11 = least sensitive
(From Carda et al., WSWS, 1991).

of the United States (Rainbolt et al., 2000). Sensitive rotational crops include sweet corn, oat, barley, canola, and mustard.

MECHANISM OF ACTION

The five chemical classes of commercial herbicides (imidazolinones, sulfonylureas, triazolopyrimidines, sulfonylaminocarbonyltriazolinone, and the pyrimidinyl-oxy-benzoates) discussed earlier have their site of action at a single enzyme in the branched-chain amino acid biosynthesis pathway that leads to the production of isoleucine, leucine, and valine. This biosynthetic pathway occurs in the chloroplasts of higher plants. Acetolactate synthase (ALS), also known as acetohydroxyacid synthase (AHAS), an early enzyme in the pathway, is the target site for these herbicides (Figure 16-5). In higher plants ALS is feedback regulated by valine and leucine. Work by various researchers has shown that all these herbicides are potent inhibitors of ALS, showing slow-tight binding to the ALS enzyme complex. In the following paragraphs, each group is briefly discussed regarding specific plant responses after herbicide application.

The *imidazolinone* herbicides are active in both grass and dicot weed species. Crop selectivity is due to differential herbicide degradation between weed and crop. The action of the compounds is quite slow, with death of weeds occurring only several weeks after treatment. The meristematic tissue appears to be the first tissue affected. When corn tissue cultures (suspension cultures) were treated with an imidazolinone herbicide, most amino acids increased or remained constant in concentration. However, the levels of valine, leucine, and isoleucine decreased. Adding these three

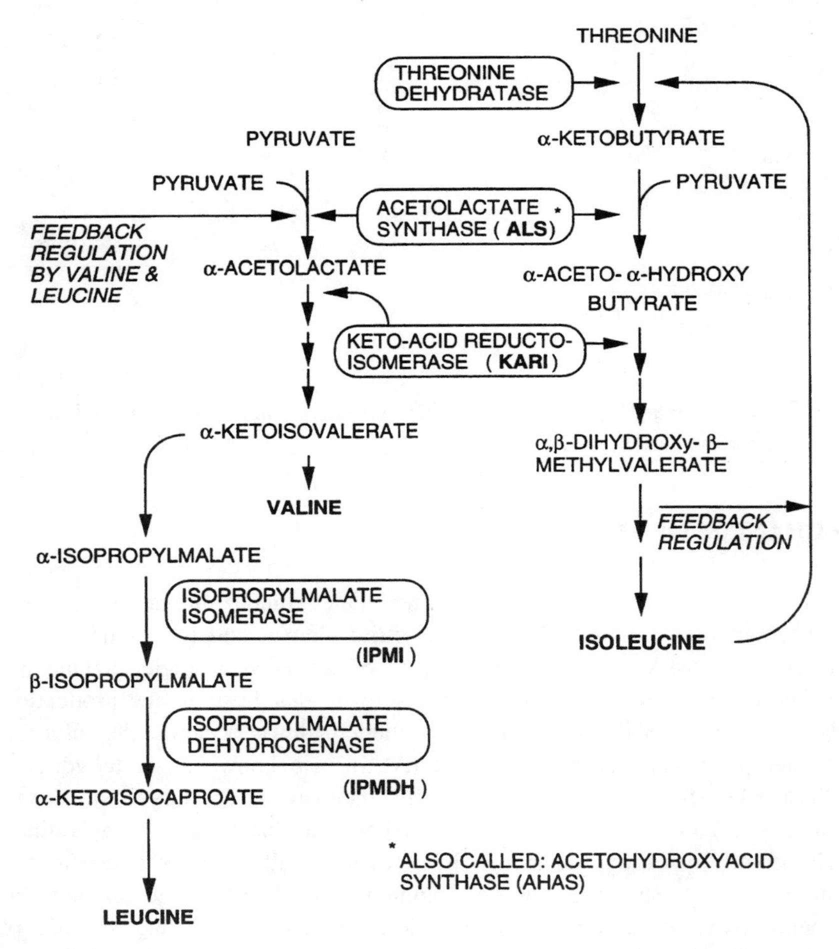

Figure 16-5. Site of action of imidazolinones, sulfonylureas, triazolopyrimidines, sulfonylaminocarbonyltriazolinones, and pyrimidinyl-oxy-benzoates) in the branch chain amino acid biosynthetic pathway that leads to the production of isoleucine, leucine, and valine. All these herbicides have their site of action at acetolactate synthase (ALS), also known as acetohydroxyacid synthase (AHAS), an early enzyme in the pathway, and all these are potent inhibitors of ALS. This biosynthetic pathway occurs in the chloroplasts of higher plants.

amino acids to the tissue culture medium prevented the phytotoxic effects. Protection was also achieved when these amino acids were supplied to seedling corn plants treated with the imidazolinone herbicide imazapyr. For a review of this mechanism of action, see Shaner and O'Conner (1991).

The *sulfonylurea* herbicides control dicots and some grass weeds in many crops. Selectivity is due to herbicide degradation in tolerant crops. Chlorsulfuron, which is the best-studied herbicide within this group, has been shown to be a potent inhibitor of weed growth by inhibiting cell division with little effect on cell enlargement. The first ultrastructural abnormalities observed inside root cells after chlorsulfuron treatment occur in the nuclei, mitochondria, and chloroplasts (Stoynova et al., 1997). There is some evidence that root growth inhibition (a partial herbicidal effect) may be responsible for reports that sulfonylureas aggravate symptoms of zinc deficiency in wheat in marginal soils (Figure 16-6) (Rengel and Wheal, 1997).

The *triazolopyrimidines* are a relatively new chemical class of amino acid biosynthesis inhibitors (e.g., flumetsulam and cloransulam from Dow AgroSciences). Names other than those given here have appeared in the literature for this herbicide class (e.g., 1,2,4 triazolo [1,5-*a*] pyrimidines; sulfonamides; and triazolo [1,5-*a*] pyrimidines). Reports show activity to be higher in dicots than in grasses because of differences in rate of degradation of the parent herbicide in grasses and dicots.

The pyrimidinyl-oxy-benzoates, sometimes termed the "pyrimidinyl ethers," contain one registered herbicide—pyrithiobac. Pyrithiobac controls broadleaf weeds in cotton under the trade name Staple. General information about this chemistry is available in Rheinheimer et al. (1994).

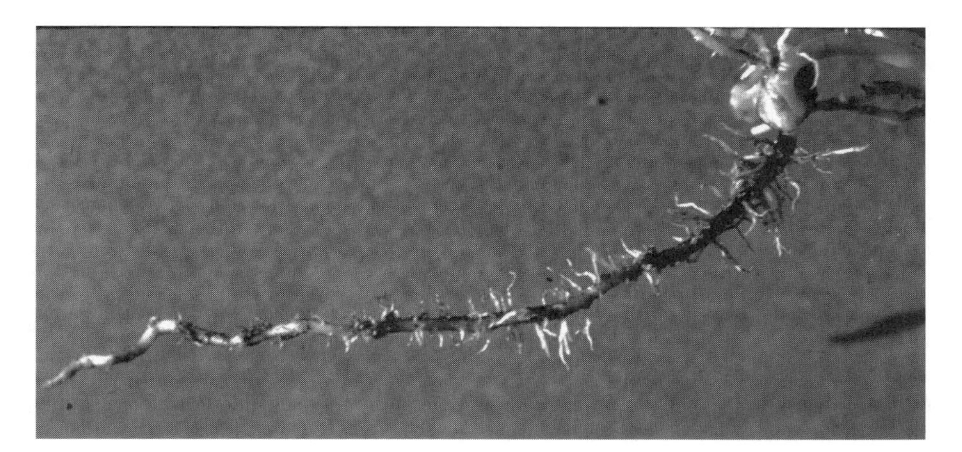

Figure 16-6. Root inhibition caused by branched-chain amino-acid-inhibiting-herbicides. There is a lack of lateral root development, and the plant root looks like a "bottle brush." (T. N. Jordan, Purdue University.)

HERBICIDAL ACTIVITY OF ALS INHIBITORS

The question arises as to whether the depletion of branched-chain amino acids is the sole cause of plant death for herbicides that inhibit ALS (AHAS). Inhibition of amino acid biosynthesis will lead to growth inhibition, inasmuch as new amino acid production is necessary to sustain the protein synthesis required for plant growth. Feeding exogenous branched-chain amino acids to plants treated with ALS-inhibiting herbicides in many cases significantly reduces injury, suggesting that depletion of these amino acids is related to phytotoxicity. Analysis of plant tissue (duckweed) after chlorsulfuron treatment showed a large accumulation of α-aminobutyrate (Rhodes et al., 1987), and it was originally thought that this compound was the toxic agent. However, later studies showed α-aminobutyrate was not involved in the herbicide toxicity response. Some evidence suggests that singlet oxygen accumulates and is involved in the mechanism of action of ALS-inhibitor herbicides (Durner et al., 1994). The ALS enzyme has an oxygen-consuming side reaction, which may generate singlet oxygen if its normal action is inhibited. To date, however, the possible involvement of singlet oxygen in the overall toxic mechanism for these herbicides has not been absolutely confirmed.

HERBICIDE RESISTANCE IN WEEDS AND CROPS

Weed Resistance

A major problem has arisen with herbicides that inhibit ALS. There are more than 60 weed species, including prickly lettuce, kochia, Russian thistle, biotypes of Amaranthacea (including common and tall waterhemp), rigid ryegrass, common chickweed, cocklebur, and others that have developed resistance to ALS-inhibitor herbicides in all the classes discussed. This has led to withdrawing registration entirely in some states and their use on fallow land in other states. Unfortunately, resistance is an incompletely dominant trait, increasing the rate at which resistant populations can build up in a normally susceptible population (Mallory-Smith et al., 1990).

Cross-resistance can occur between different classes of acetolactate-synthase-inhibiting herbicides. Downy brome resistance to primisulfuron (Beacon) in the Pacific Northwest region of the United States is an example. Primisulfuron-resistant downy brome was selected in a Kentucky bluegrass seed production field following two annual applications of the herbicide over 3 years (Mallory-Smith et al., 1990). Sulfosulfuron and imazamox effectively control downy brome in winter wheat, which is grown in rotation with Kentucky bluegrass. However, these ALS-inhibitor herbicides do not control primisulfuron-resistant downy brome (Ball and Mallory-Smith, 2000). Caution and an understanding of this potential resistance problem are necessary with the use of these herbicides.

In addition, common chickweed, perennial ryegrass, and Russian thistle biotypes have shown resistance to five sulfonylureas, one imidazolinone, and one triazolopyrimidine herbicides (Saari et al., 1992). These resistances were all due to a decrease in sensitivity of the ALS enzyme. Horak and Peterson (1995) found biotypes of Palmer amaranth and common waterhemp that were resistant to imazethapr and thifensulfuron. In Australia, a rigid ryegrass biotype resistant to diclofop-methyl is cross-resistant to chlorsulfuron. This resistance has been found (Cotterman and Saari, 1992) to be due to rapid metabolism of chlorsulfuron (glucose conjugate of hydroxy-chlorsulfuron). Tall morning glory tolerance to pyrithiobac was also shown to be due to metabolism (Sunderland et al., 1995). In 1994 there was documented ALS resistance in at least 17 weed species at more than 1200 sites in seven countries (Saari and Cotterman, 1994). Studies on the competitiveness of the resistant weed biotypes have shown that they have a similar growth rate as susceptible biotypes.

The reasons for such resistance are the various mutations on the ALS enzyme. These mutations result in the inability of the herbicide to inhibit the ALS enzyme. The basis of the resistance is interesting inasmuch as mutations can occur in four primary sites on the ALS gene and can result in different levels of resistance to the various ALS herbicides.

Crop Resistance

Laboratory researchers have shown that it is relatively easy to obtain crop plants resistant to both imidazolinone and sulfonylurea herbicides. This has been done by both induced mutations and selection under herbicide pressure. Soybean lines resistant to sulfonylureas and corn lines resistant to the imidazolinones that have been obtained using these methods are currently on the market. Resistance in these lines is due to a single-gene dominant trait. Interestingly, an imidazolinone-resistant (IR) corn line from Pioneer is resistant to all four chemical classes of ALS inhibitors. Because IR corn is thought to have a single relevant mutation site [tryptophan$_{542}$ to leucine (Bernasconi et al., 1995)], the ALS herbicides must share some portion of a common binding site on the enzyme. The resistance in the crops is due to an altered ALS enzyme not being sensitive to the herbicides, and no detrimental characteristics in the crops have been reported.

ANTAGONISM

There are some reports of antagonism when ALS inhibitors are applied in mixtures with other herbicides. Mixes of sulfonylureas plus MCPA were less potent on dicot weeds than either herbicide applied alone. Sequential applications and selective placement studies show that antagonism occurred only when the two herbicides were in physical contact. Antagonism of grass species control was observed when pyrithiobac was mixed with ACCase inhibitors (Ferreira et al., 1995). Sequential applications eliminated the antagonism. Antagonism was thought to be due to decreased translocation of the ACCase inhibitor in the presence of pyrithiobac. Kim and Vanden Born (1996) clearly showed that chlorsulfuron treatment decreased the export of sucrose out of herbicide-treated leaves more than 50% within 24 hours after application, thus supporting antagonism's being due to decreased transport.

Interestingly, Kim and Vanden Born (1996) also reported a 50% decrease in sucrose movement into developing leaves that were importing sucrose (sinks) when the importing leaves were treated with chlorsulfuron.

SYNERGISM

Adding organophosphate insecticides (e.g., terbufos) to the furrow during planting results in an increased sensitivity to postemergence-applied sulfonylureas in corn (e.g., Diehl et al., 1995) and pyrimidinyl-oxy-benzoates in cotton (Allen and Snipes, 1995). Diehl et al. (1995) studied the interaction of terbufos + nicosulfuron and found that terbufos, and to a greater extent its oxidative metabolites (terbufos-sulfoxide and terbufos-sulfone), inhibited cytochrome P_{450}-mediated hydroxylation of nicosulfuron. Organophosphate insecticides are known to inhibit P_{450} reactions, and nicosulfuron is known to be selective in corn because of ring hydroxylation (pyrimidine-OH). Kwon and Penner (1995) found that adding safeners known to induce P_{450} (e.g., naphthalic anhydride) reduced corn injury resulting from the combination of terbufos + nicosulfuron. Adding P_{450} inhibitors (e.g., piperonyl butoxide) increased injury resulting from sufonylureas (Kwon and Penner, 1996). As would be expected, there is no increase in sulfonylurea injury when terbufos or piperonyl butoxide is applied to IR corn.

LITERATURE CITED AND SUGGESTED READING

Aromatic Amino Acid Biosynthesis Inhibitors—Glyphosate

Amrhein, N., B. Deus, P. Gehrke, and H. C. Steinrücken. 1980. The site of the inhibition of the shikimate pathway by glyphosate. II. Interference of glyphosate with chorismate formation *in vivo* and *in vitro*. *Plant Physiol.* **66**:830–834.

Boerboom, C. M., N. J. Ehlke, D. L. Wyse, and D. A. Somers. 1991. Recurrent selection for glyphosate tolerance in birdsfoot trefoil. *Crop Sci.* **31**:1124–1129.

Bradshaw, L. D., S. R. Padgette, S. L. Kimball, and B. H. Wells. 1997. Perspectives on glyphosate resistance. *Weed Tech.* **11**:189–198.

Delannay, X., T. T. Bauman, D. H. Beighley, M. J. Buettner, and others. 1995. Yield evaluation of a glyphosate-tolerant soybean line after treatment with glyphosate. *Crop Sci.* **35**:1461–1467.

Forlani, G., E. Nielsen, and M. L. Bacchi. 1992. A glyphosate resistant 5-*enol*-pyruvyl-shikimate-3-phosphate synthase confers tolerance to a maize cell line. *Plant Sci.* **85**:9–15.

Gressel, J. 1996. Fewer constraints than proclaimed to the evolution of glyphosate-resistant weeds. *Resist. Pest Manag.* **8**:2–5.

Lorraine-Colwill, D. F., S. B. Powles, T. R. Hawks, and C. Preston. 2001. Inheritance of evolved glyphosate resistance in *Lolium rigidum*. *Theor. and Appl. Gen.* **102**:545–550.

Mannerlöf, M., S. Tuvesson, P. Steen, and P. Tenning. 1997. Transgenic sugar beet tolerant to glyphosate. *Euphytica*. **94**:83–91.

Nalewaja, J. D., and R. Matysiak. 1991. Salt antagonism of glyphosate. *Weed Sci.* **39**:622–628.

Pinto, J. E. B. P., W. E. Dyer, S. C. Weller, and K. M. Herrman. 1988. Glyphosate induces 3-deoxy-D-arabino-heptulosonate-7-phosphate synthase in potato (*Solanum tuberosum* L.) cells grown in suspension culture. *Plant Physiol.* **87**:891–893.

Racchi, M. L., M. Rebecchi, G. Todesco, E. Nielsen, and G. Forlani. 1995. Glyphosate tolerance in maize (*Zea mays* L.). 2. Selection and characterization of a tolerant somaclone. *Euphytica.* **82**:165–173.

Schultz, A., T. Munder, H. Hollander-Czytko, and N. Amrhein. 1990. Glyphosate transport and early effects on shikimate metabolism and its compartmentation in sink leaves of tomato and spinach plants. *Z. Naturforsch.* **45c**:529–534.

Shieh, W. J., D. R. Geiger, and J. C. Servaites. 1991. Effects of *N*-(phosphonomethyl)glycine on carbon assimilation and metabolism during a simulated natural day. *Plant Physiol.* **97**:1109–1114.

Simarmata, M., J. E. Kaufmann, and D. Penner. 2001. Progress in determining the origin of the glyphosate-resistant ryegrass in California. *WSSA Abstr.* **41**:95.

Sindel, B. 1996. Glyphosate resistance discovered in annual ryegrass. *Resist. Pest Manag.* **8**:5–6.

VanGessel, M. J. 2001. Personal communication.

Westwood, J. H., and S. C. Weller. 1997. Cellular mechanisms influence differential glyphosate sensitivity in field bindweed (*Convolvulus arvensis*) biotypes. *Weed Sci.* **45**:2–11.

Zhou, H., J. W. Arrowsmith, M. E. Fromm, C. M. Hironaka, and others. 1995. Glyphosate-tolerant CP4 and GOX genes as a selectable marker in wheat transformation. *Plant Cell Reports* **15**:159–163.

Inhibitors of Branch-Chain Amino Acid Biosynthesis

Allen, R. L., and C. E. Snipes. 1995. Interactions of foliar insecticides applied with pyrithiobac. *Weed Technol.* **9**:512–517.

Ball, D. A., and C. A. Mallory-Smith. 2000. Sulfonylurea herbicide resistance in downy brome. *Proc. Western Soc. Weed Sci.* **53**:41–42.

Bernasconi, P., A. R. Woodworth, B. A. Rosen, M. V. Subramanian, and D. L. Siehl. 1995. A naturally occurring point mutation confers broad range tolerance to herbicides that target acetolactate synthase. *J. Biol. Chem.* **270**:17381–17385.

Brewster, B. D., and A. P. Appleby. 1983. Response of wheat and rotation crops to chlorsulfuron. *Weed Sci.* **31**:861–865.

Burkhart, D. C., W. E. Dyer, and P. K. Fay. 1984. The effect of chlorsulfuron residues on seven crops in Montana. *Proc. Western Soc. Weed Sci.* **37**:188–190.

Carda, K. M., D. Mulageta, P. K. Fay, and E. S. Davis. 1991. The residual properties of triasulfuron in Montana. *Proc. Western. Soc. Weed Sci.*, pp. 80–81.

Cotterman, J. C., and L. L. Saari. 1992. Rapid metabolic inactivation is the basis for cross-resistance to chlorsulfuron in diclofop-methyl-resistant rigid ryegrass (*Lolium rigidum*) biotype SR4/84. *Pest. Biochem. Physiol.* **43**:182–192.

Devine, M. D., H. D. Bestman, and W. H. Vanden Born. 1990. Physiological basis for the different phloem mobilities of chlorsulfuron and clopyralid. *Weed Sci.* **38**:1–9.

Diehl, K. E., E. W. Stoller, and M. Barrett. 1995. *In vivo* and *in vitro* inhibition of nicosulfuron metabolism by terbufos metabolites in maize. *Pestic. Biochem. Physiol.* **51**:137–149.

DuPont Agricultural Products Company. Glean Weed Killer. Technical Bulletin E-43612. 12 pp.

Durner, J., O. C. Knörzer, and P. Böger. 1994. Ketol-acid reductoisomerase from barley (*Hordeum vulgare*): Purification, properties and specific inhibition. *Plant Physiol.* **103**:903–910.

Dyer, W. E., P. W. Chee, and P. K. Fay. 1993. Rapid germination of sulfonylurea-resistant *Kochia scoparia* L. accessions is associated with elevated seed levels of branched chain amino acids. *Weed Sci.* **41**:18–22.

Ferreira, K. L., J. D. Burton, and H. D. Coble. 1995. Physiological basis for antagonism of fluazifop-P by DPX-PE350. *Weed Sci.* **43**:184–191.

Feucht, D., K. H. Muller, A. Wellman, and H. Santel. 1999. BAY MKH 6561—A new selective herbicide for grass control in wheat, rye, and triticale. *Proc. Brit. Crop Prot. Conf.—Weeds.* **1**:53–58.

Horak, M. J., and D. E. Peterson. 1995. Biotypes of palmer amaranth (*Amaranthus palmeri*) and common waterhemp (*Amaranthus rudis*) are resistant to imazethapyr and thifensulfuron. *Weed Technol.* **9**:192–195.

Kim, S., and W. H. Vanden Born. 1996. Chlorsulfuron decreases both assimilate export by source leaves and import by sink leaves in canola (*Brassica napus* L.) seedlings. *Pestic. Biochem. and Physiol.* **56**:141–148.

Kwon, C. S., and D. Penner. 1995. The interaction of insecticides with herbicide activity. *Weed Technol.* **9**:119–124.

Kwon, C. S., and D. Penner. 1996. The effect of piperonyl butoxide and adjuvants on sulfonylurea herbicide activity. *Weed Technol.* **10**:127–133.

Mallory-Smith, C. A., D. C. Thill, and M. J. Dial. 1990. Identification of sulfonylurea herbicide-resistant prickly lettuce (*Lactuca serriola*). *Weed Technol.* **4**:163–168.

Rainbolt, C. R., D. C. Thill, and D. A. Ball. 2001. Response of rotational crops to BAY MKH 6561. *Weed Technol.* **15** (in press).

Rengel, Z., and M. S. Wheal. 1997. Herbicide chlorsulfuron decreases growth of fine roots and micronutrient uptake in wheat genotypes. *J. Exp. Bot.* **48**:927–934.

Rheinheimer, J., U. Vogelbacher, E. Baumann, K. O. Westphalen, and M. Gerber. 1994. Pyrimidyl ethers, potent ALS-inhibiting herbicides. Eighth IUPAC International Congress of Pestic. Chem. Washington, DC. Abstr. 134.

Rhodes, D., A. L. Hogan, L. Deal, G. C. Jamieson, and P. Haworth. 1987. Amino acid metabolism of *Lemna minor.* II. Responses to chlorsulfuron. *Plant Physiol.* **84**:775–780.

Saari, L. L., and J. C. Cotterman, 1994. Survey of acetolactate synthase (ALS) insensitivity to ALS-inhibitors among resistant *Kochia scoparia* populations in North America. Eighth IUPAC International Congress of Pestic. Chem. Washington, DC. Abstr. 175.

Saari, L. L., J. C. Cotterman, W. F. Smith, and M. M. Primiani. 1992. Sulfonylurea herbicide resistance in common chickweed, perennial ryegrass, and Russian thistle. *Pest. Biochem. Physiol.* **42**:110–118.

Shaner, D. L., and S. L. O'Conner. 1991. *The Imidazolinone Herbicides*. 290 pages. CRC Press, Boca Raton, FL.

Stoynova, E., P. Petrov, and S. Semerdjieva. 1997. Some effects of chlorsulfuron on the ultrastructure of root and leaf cells in pea plants. *J. Plant Growth Reg.* **16**:1–5.

Sunderland, S. L., J. D. Burton, H. D. Coble, and E. P. Maness. 1995. Physiological mechanism for tall morningglory (*Ipomoea purpurea*) resistance to DPX-PE350. *Weed Sci.* **43**:21–27.

Sweetser, P. B., G. S. Schow, and J. M. Hutchison. 1982. Metabolism of chlorsulfuron by plants: Biological basis for selectivity of a new herbicide for cereals. *Pestic. Biochem. Physiol.* **17**:18–23.

Swisher, B. A., and M. R. Weimer. 1986. Comparative detoxification of chlorsulfuron in leaf disks and cell cultures of two perennial weeds. *Weed Sci.* **34**:507–512.

Swisher, B. A., G. L. deBoer, D. Ouse, T. C. Geselius, J. J. Jachetta, and V. W. Miner. 1991. Metabolism of selected triazolo (1,5-c) pyridine sulfonamides in plants. Internal report of DowAgro Sci. (cited in Dorich and Schultz, 1997, *Down to Earth* **52**(2):1–10).

For chemical use, see the manufacturer's or supplier's label and follow these directions. Also see the Preface.

17 Miscellaneous Herbicides

Many herbicides are not specifically classed by mechanism of action inasmuch as the specific site within the plant where they act is not known. In this chapter, the miscellaneous herbicides are discussed in regard to their physical and chemical properties and their uses.

ALIPHATICS

Aliphatic means no rings in the structural formula. The aliphatic herbicides are the chlorinated aliphatic acid (TCA), the organic arsenicals (cacodylic acid, MSMA, and DSMA), acrolein, fosamine, methyl bromide, metham, and pelargonic acid. Except for the aliphatics, the other herbicides discussed in this chapter have chemical structures containing one or more rings.

TCA

TCA trichloroacetic acid is a chlorinated aliphatic acid, white deliquescent solid with a vapor pressure of 5 mm Hg at 77°C, a water solubility of 83,000 mg/l (ppm) at 25°C, a soil half-life of 21 to 90 days, and an oral LD_{50} (rat) of 400 mg/kg. Only the sodium salt is used as a herbicide, and the following comments refer to this form. Depending on purity, it may be a white solid or a pale yellow to amber liquid.

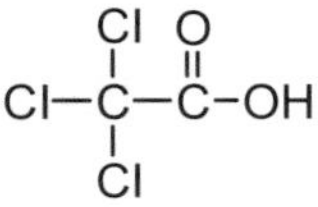

$$Cl-\underset{Cl}{\overset{Cl}{|}}{C}-\overset{O}{\overset{\|}{C}}-OH$$

TCA, Acid

Uses TCA is formulated as a water-soluble liquid concentrate, granule, or powder. It is hygroscopic, and when exposed to 90 to 95% relative humidity at 70°F, the chemical will absorb its weight in water in 8 to 10 days. Therefore, it must be stored in moisture-proof containers. When handling and using sodium TCA, avoid contact with skin and eyes and avoid breathing spray mist. It is irritating to the skin, eyes, nose, and throat. Wear protective goggles and clothing, impervious gloves, and a respirator. The major use of sodium TCA is as a grass killer. It is primarily active through the soil but has some foliar contact activity at nonselective rates. It has proven useful as a nonselective treatment of perennial weedy grasses such as johnsongrass, common bermudagrass, and quackgrass. Sodium TCA is also used as a selective preemergence treatment in sugar beets to control annual grasses. In sugarcane, one preemergence or early postemergence treatment controls seedling grasses, including johnsongrass seedlings. It has also been used selectively in oilseed rape. Throughout the world, TCA is used in numerous field agronomic, field vegetable, and noncropland situations.

Soil Influence TCA is not tightly adsorbed to soil and is subject to leaching, especially by heavy rain. It is slowly degraded by microorganisms. It persists 3 to 10 weeks, depending on rate applied and soil moisture and temperature.

Mode of Action When applied to foliage, TCA often causes rapid necrosis by contact action. It inhibits the growth of both shoots and roots and causes leaf chlorosis and formative effects, especially in the shoot apex. TCA is readily absorbed by leaves and roots (Blanchard, 1954). It is translocated throughout the plant from the roots, but only small amounts are translocated from leaves. Therefore, it is primarily translocated via the apoplastic system. Perhaps its rapid contact action prevents symplastic movement. TCA is degraded slowly, if at all, by higher plants.

It has been suggested that TCA acts by precipitating proteins, inasmuch as chemists commonly use it for this purpose. However, this theory has not been generally accepted. Perhaps it modifies sulfhydryl or amino groups of enzymes or induces conformational changes in enzymes.

Organic Arsenicals

The organic arsenical herbicides include three similar compounds—cacodylic acid, MSMA, and DSMA—and others.

Cacodylic Acid

Cacodylic acid dimethylarsinic acid is a colorless crystalline solid with no reported vapor pressure, a very high water solubility of 102,000 mg/l (ppm) at 25°C, a soil half-life of 50 days, and an oral LD_{50} (rat) of 2,756 mg/kg. Like other organic arsenical herbicides, it has much lower mammalian toxicity than elemental arsenic. It was one of the first organic arsenical herbicides introduced.

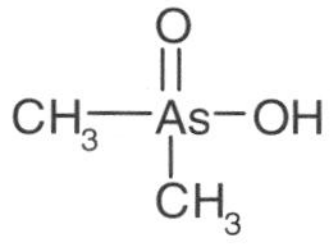

$$CH_3-\overset{\overset{O}{\|}}{\underset{\underset{CH_3}{|}}{As}}-OH$$

Cacodylic Acid

Uses Cacodylic acid is formulated as an SL and as the acid form, the sodium salt, and combinations of the acid and sodium salt, and sold as Cotton-Aide, Monter, and Acme as a general contact nonselective spray to desiccate and defoliate a wide variety of plant species. It is used primarily to control emerged annual weeds in lawn-turf seedbeds, lawn renovation areas, and noncrop areas and provides some control of certain perennial grasses. Treated forage should not be fed to livestock and livestock should not graze in treated areas. Some formulations are also used as a directed spray in nonbearing citrus orchards. For cotton desiccation and defoliation, it is applied 7 to 10 days prior to harvest, when more than 50% of the bolls are open.

MAA

MAA methylarsonic acid is usually formulated as monosodium methylarsonate (MSMA) or as disodium methane arsonate (DSMA). Purified MAA white crystalline solid is used in these formulations. MSMA is a clear, odorless liquid with a vapor pressure of 16 mm Hg at 20°C, a high water solubility of 1,040,000 mg/l (ppm) at 25°C, a soil half-life of 180 days, and an oral LD_{50} (rat) of 2833 mg/kg. DSMA has a vapor pressure of $<10^{-7}$ mm Hg at 20°C, a water solubility of 269,000 mg/l (ppm) at 25°C, a soil half-life of 180 days, and an oral LD_{50} (rat) of 1935 mg/kg.

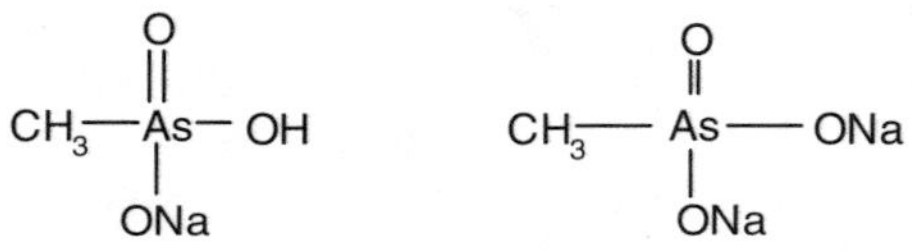

$CH_3-As(=O)(ONa)-OH$ $\qquad$ $CH_3-As(=O)(ONa)-ONa$

MSMA $\qquad$ DSMA

Uses There are several suppliers and many trade names for these herbicides and their various formulations. MSMA is formulated as an SL and DSMA is formulated as an SP. The first major use of these compounds was for postemergence control of crabgrass, dallisgrass, and other weedy grasses in lawns and turf. MSMA and DSMA are used to control many annual and perennial grasses and nutsedges on noncrop areas, in lawns and turf, in citrus (except in Florida), and in cotton. MSMA is used preplant in cotton, whereas DSMA is used as a directed spray in emerged cotton, citrus, noncropland, and turf. These compounds are particularly effective on johnsongrass, dallisgrass, crabgrass, and nutsedges. There is an organic arsenical resistant biotype of common cocklebur that has been identified in Mississippi, although the mechanism of resistance is not known (Nimbal et al., 1995).

Mode of Action Cacodylic acid desiccates and defoliates many species of plants, while MSMA and DSMA induce foliar chlorosis followed by gradual tissue browning and, finally, necrosis. These are general growth inhibitors, inhibiting the sprouting of rhizome and tuber buds and causing aberrant cell division. Cacodylic acid is primarily translocated in the apoplast, and significant symplastic transport is probably prevented by its rapid contact action, which kills the phloem. MSMA and DSMA are translocated in both the phloem and the xylem.

These compounds appear to be metabolized in plants by conjugation with sugars, organic acids, and amino acids. However, only the conjugated amino acid forms have been widely reported. The carbon-arsenic bond of these compounds appears to be very stable in higher plants.

Acrolein

Acrolein acrylaldehyde, or 2-propenal is a colorless liquid with a vapor pressure of 210 mm Hg at 20°C, a very high water solubility of 250,000 mg/l (ppm), a soil half-life of 150 hours, and an acute oral LD_{50} of 29 mg/kg for rats, which makes it very toxic to most organisms. Acrolein is a flammable liquid subject to explosive reactions under certain conditions; therefore, it may be applied only by licensed applicators.

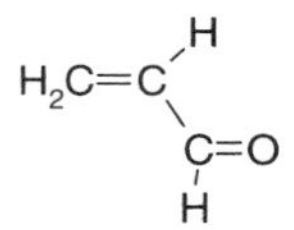

$H_2C{=}C(H){-}C(H){=}O$

Acrolein

Uses and Mode of Action Acrolein is formulated as an SL and is sold as Magnacide for injection in water to control submersed and floating aquatic weeds. It is a contact herbicide and a general cell toxicant and acts on enzyme systems through its destructive sulfhydryl reactivity.

Fosamine

Fosamine ethyl hydrogen (aminocarbonyl) phosphonate, is a white crystalline solid with a vapor pressure of 4×10^{-6} mm Hg at 25°C, a water solubility of 1,790,000 mg/l (ppm) at 25°C, and an oral LD_{50} (rat) of 24,400 mg/kg.

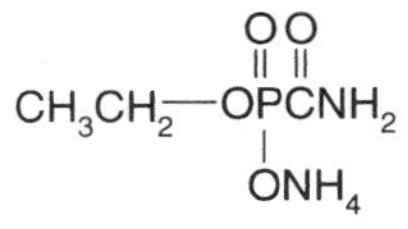

$CH_3CH_2{-}O{-}P(=O)(ONH_4){-}C(=O)NH_2$

Fosamine

Uses Fosamine is formulated as an S or SL and sold as Krenite for nonselective control of many brush and woody species on noncropland. It also controls blackberries, brackenfern, and field bindweed. It is applied as a complete-coverage spray to the leaves, stems, and buds for most effective control. Best control of deciduous woody species is obtained when it is applied within the 2 months before the foliage changes color in the fall.

Soil Influence Because fosamine is applied to the aboveground parts of weeds, soil type has little effect on its performance. It is rapidly degraded in soils by microorganisms and has a half-life of 8 days. This rapid degradation precludes significant leaching of fosamine under field conditions.

Mode of Action When fosamine is applied to susceptible deciduous woody plants in the fall, little or no phytotoxic symptoms are observed until the following spring. In the spring, bud development does not occur or only small malformed leaves develop. In contrast, certain species (e.g., pines and field bindweed) may develop phytotoxic symptoms soon after the herbicide treatment. Some suppression of terminal growth of many species may occur; therefore, spray drift onto desirable plants should be avoided.

Little research on the absorption, translocation, and metabolism of fosamine has been published (Kitchen et al., 1980; Mann et al., 1986; Müller, 1981; *Herbicide Handbook*, 1994). Fosamine is readily absorbed by leaves, stems, and buds. Retention and penetration are less when the herbicide is applied to rough or hairy leaves than to leaves with smooth surfaces. Translocation from treated areas appears to be relatively slow and limited. This supports the field observations that have shown that complete coverage of the plant is required for control. Translocation studies indicate greater transport in susceptible species than in tolerant species and may explain the differential selectivity among species. The reported translocation patterns suggest that the herbicide is transported in both the symplastic and apoplastic systems. Metabolism studies with [^{14}C] fosamine in higher plants indicates that degradation is relatively rapid, with a half-life of about 2 to 3 weeks.

Methyl Bromide

Methyl bromide bromomethane has many trade names. It is a colorless, nearly odorless liquid or gas. At 1 atmosphere of pressure and 5°C, the liquid boils and becomes a gas. The gas is 3.2 times heavier than air at 12°C. It is slightly soluble in water and very soluble in alcohol or ether. It is generally considered nonflammable and nonexplosive. However, mixtures containing between 13.5 and 14.5% of the gas in air may be exploded by a spark. Methyl bromide gas is poisonous to humans and animals, and the effects of exposures are cumulative. The acute oral LD_{50} is 100 mg/kg, and the inhalation LD_{50} is 3150 mg/l for rats. For humans, the inhalation LD_{50} is 60,000 mg/l for 2 hours. Because methyl bromide is nearly odorless, 2% chloropicrin (teargas) is often added as a warning agent. See product label for protective clothing requirements and respiratory protection.

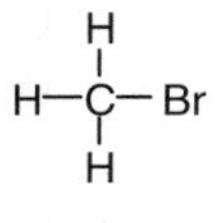

```
      H
      |
   H—C—Br
      |
      H
```

Methyl Bromide

Uses Methyl bromide is used as a temporary soil sterilant. It kills plant tissues and most seeds, insects, and disease organisms by acting as a potent inhibitor of respiration in living tissue. It is an excellent preplant soil treatment for seedbeds or propagating beds of tobacco, flowers, vegetables, turf, and tree seedlings. It is also used preplant for strawberry beds and to sterilize potting soil or compost mixtures (Figure 17-1 and Figure 17-2). Because methyl bromide treatments are expensive ($1000 to $1500/acre), they are used only in high-value crops for serious pest problems. Formulations of methyl bromide (45%) and chloropicrin (55%) give increased control of certain plant pathogens.

Figure 17-1. Methyl bromide fumigation applicator that injects the fumigant into the soil and then places a sealed plastic tarp over the treated soil to contain the gas in the soil atmosphere. (D. Sanders, North Carolina State University.)

Figure 17-2. Strawberries growing on polyethylene black plastic in a field that was fumigated prior to planting with methyl bromide. (D. Monks, North Carolina State University.)

The volatile nature of methyl bromide requires that a plastic tarpaulin confine it for about 48 hours to be effective. Planting can usually be safely done 72 hours after removal of the tarpaulin. The herbicide is injected into the soil and covered with a plastic tarpaulin in a single operation at a rate of 250 to 300 lb/acre for large areas. For small areas or soil volumes, it is merely released under a plastic tarpaulin at a rate of 1 to 2 lb/100 ft^2 of surface or 100 ft^3 of soil.

Although methyl bromide kills most weed seeds, certain seeds with hard seed coats are not killed. These include prickly sida, redstem filaree, certain morning glories, malvas, and clovers. However, most vegetative propagules, including bermudagrass and johnsongrass rhizomes and nutsedge tubers, are killed by methyl bromide fumigation.

Methyl bromide may kill beneficial microorganisms as well as disease microorganisms. This may result in an inhibition of the normal decomposition of organic matter into ammonia, then nitrite, and finally nitrate. If the microorganisms responsible for one or more of these processes are killed, phytotoxic levels of ammonia or nitrite may accumulate following methyl bromide fumigation. This seldom occurs in low-organic-matter soils; however, ammonium forms of nitrogen fertilizers should not be used. Usually, the addition of peat, sawdust, or other organic matter to potting or compost mixtures after fumigation will provide the microorganism inoculum necessary to prevent this problem. Methyl bromide uses are severely restricted, and by 2005 it will no longer be available for any uses in the United States because its effect in depleting the earth's ozone.

Metham

Metham methylcarbamodithioic acid, in its pure form is a white crystalline solid. Because the acid is unstable, the commercial formulation is a stable concentrated 32.7% aqueous solution of the sodium salt. The commercial product has a vapor pressure of 24 mm Hg at 25°C, a water solubility of 722,000 mg/l (ppm), a soil half-life of 7 days, and an oral LD_{50} (rat) of 812 mg/kg.

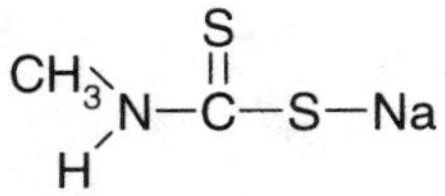

S
||
CH_3 N—C—S—Na
H

Metham

Uses Metham is a temporary soil fumigant sold as Vapam for use in controlling nematodes, garden centipedes, soil-borne disease organisms, and most germinating weed seeds and seedlings. It has also been used to control certain shallow perennial weeds (e.g., nutsedges) and kill roots in sewers. This herbicide is used in the field and for potting soil. When used in the field, the soil should be cultivated before application to allow diffusion of the gaseous toxicant.

Metham may be applied in various ways, depending on the size of the area to be treated and the equipment available. For small areas, a sprinkling can or hose proportional diluter may be used. For larger areas, soil injection, spray application with immediate soil incorporation, or application through a sprinkler-irrigation system may be used.

Metham is most effective when it is possible to confine the vapors with a plastic tarp; however, the water seal method (saturating the top 2 inches of soil with water) may also be used. When using a tarp, the treated area should be covered for 48 hours or longer. Seven days after treatment the area should be cultivated to a depth of 2 inches. At least 14 to 21 days should pass after application before the treated area is seeded to a crop.

Pelargonic Acid

Pelargonic acid is a fatty acid derivative natural product herbicide, which is a water-white liquid with a waxy fatty-acid-like odor similar to that of crayons. It has a vapor pressure of 20 mm Hg at 25°C, a very high water solubility, and an oral LD_{50} (rat) of > 5000 mg/kg. Pelargonic acid has no persistence in soil and no soil activity.

$$CH_3(CH_2)_7COOH$$

Pelargonic Acid

Pelargonic acid is formulated as an EC and sold as Scythe for use as a foliar-applied contact nonselective herbicide in various landscape, noncropland, and site prep situations for burndown of annual and perennial plant foliage.

Dazomet

Dazomet tetrahydro-3,5-dimethyl-2*H*-1,3,5-thiadiazine-2-thione, is a white, odorless crystalline solid with a vapor pressure of 3×10^{-6} mm Hg at 20°C, a water solubility of 2000 mg/l (ppm) at 20°C, a soil half-life of 7 days, and an oral LD_{50} (rat) of 650 mg/kg.

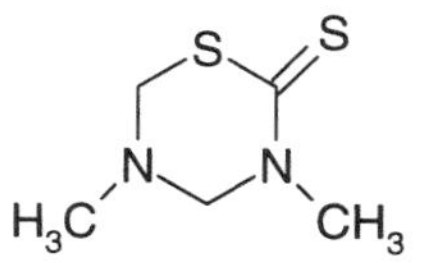

S S N N H_3C CH_3

Dazomet

Uses Dazomet is formulated as a granular and sold as Basamid as a soil fumigant to control annual and perennial weeds, nematodes, soil fungi, and certain soil insects. It is used on seedbeds of ornamentals, tobacco, and lawns and turf. Seedbeds should be well prepared and have adequate moisture for good plant growth. Planting of the crop should be delayed until dazomet and its toxic degradation products have disappeared, which usually occurs within 10 to 30 days after application.

Soil Influence Dazomet undergoes chemical degradation into methylisothiocyanate, formaldehyde, hydrogen sulfide, and monomethylamine in moist, warm soil. Soil moisture is essential for its biological activity. An approved respirator, gloves, and protective clothing are recommended during handling and application.

Naptalam

Naptalam 2-[(1-naphthalenylamino)carbonyl]benzoic acid, is a purple crystalline solid with no reported vapor pressure, a water solubility of 200 mg/l (ppm) for the acid and 230,800 mg/l (ppm) for the sodium salt, a soil half-life of 14 days, and an oral LD_{50} (rats) of > 8200 mg/kg for the acid and 1800 mg/kg for the sodium salt.

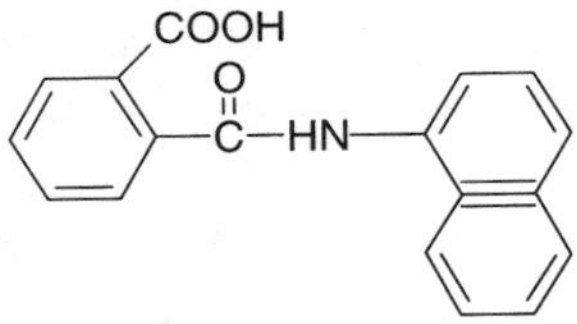

Naptalam

Uses Naptalam is formulated as the sodium salt in water (2lb/gal) and sold as Alanap for preemergence use to control broadleaf weeds in cucurbits, peanuts, soybeans, and woody ornamentals. Combinations with other herbicides are also used to broaden the spectrum of weeds controlled in cucurbits and soybeans.

Soil Influence Naptalam is subject to extensive leaching in porous soils. If heavy rains occur shortly after seeding and herbicide application, both crop injury and poor weed control may result. It is relatively nonpersistent and presents no soil-residual problem. Weeds are usually controlled at 3 to 8 weeks after application of the herbicide.

Mode of Action Naptalam has the unique property of acting as an antigeotropic agent; growing shoots and roots have a tendency to lose their ability to grow up or down, respectively. This may effect be associated with its herbicidal action (Ashton and Crafts, 1981).

Naptalam is an auxin (IAA) antagonist; it inhibits the polar transport of IAA and binds to the plasma membrane and is thought to inhibit the efflux of auxin. At submicromolar concentrations, naptalam stimulates IAA absorption (Duke, 1985). The latter results from inhibition of the efflux of IAA from cells and has been proposed to explain the inhibition of polar transport. This may be related to naptalam's binding to sites on plant cell membranes.

Difenzoquat

Difenzoquat 1,2-dimethyl-3,5-diphenyl-1*H*-pyrazolium, is a white to off-white crystalline solid with a vapor pressure of $< 10^{-7}$ mm Hg at 25°C, a water solubility of

817,000 mg/l (ppm) at 25°C, a soil half-life of <4 weeks, and an oral LD_{50} (rat) of 617 mg/kg.

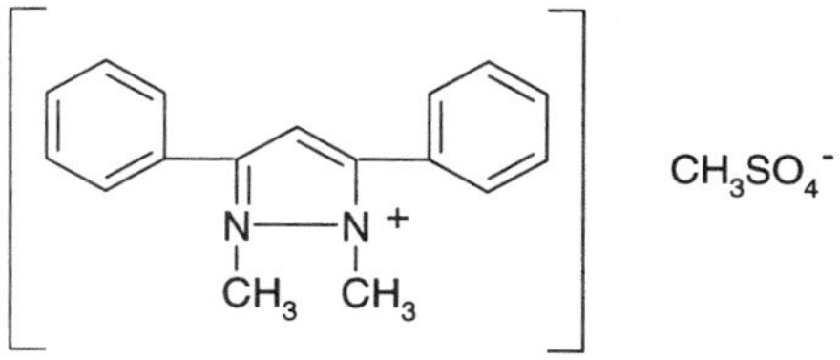

$CH_3SO_4^-$

Difensoquat

Uses Difenzoquat is formulated as the methyl sulfate salt as an AS and sold as Avenge for use as a selective postemergence herbicide to control wild oats in barley and wheat. It is applied when the wild oats are in the three- to five-leaf stage, and it can be tank mixed with 2,4-D, MCPA, bromoxynil, chlorsulfuron, or MCPA plus bromoxynil to broaden the spectrum of weeds controlled. Check the product label for crop variety restrictions.

Soil Influence Difenzoquat is strongly adsorbed to soil particles and therefore not subject to leaching. It is not readily metabolized by soil microorganisms. It is readily demethylated photolytically to the relatively volatile monomethyl pyrazole. Difenzoquat residues disappear from soil at a moderate rate, and rotation to other crops can be made the year following application.

Mode of Action Difenzoquat causes chlorosis and necrosis in leaves. It is readily absorbed by leaves, but translocation is limited. It is not significantly metabolized by plants. It has been suggested that difenzoquat may act at the molecular level, like diquat and paraquat (Fedtke, 1982; Halling and Behrens, 1983). Cohen and Morrison (1982) indicated that difenzoquat interferes with active ion transport across the plasma membrane.

Asulam

Asulam methyl[(4-aminophenyl)sulfonyl]carbamate, is a white crystalline solid with a vapor pressure of 10^{-7} mm Hg at 25°C, a high water solubility of 5000 mg/l (ppm) at 20 to 25°C, a soil half-life of 7 days, and an oral LD_{50} (rat) of 5000 mg/kg.

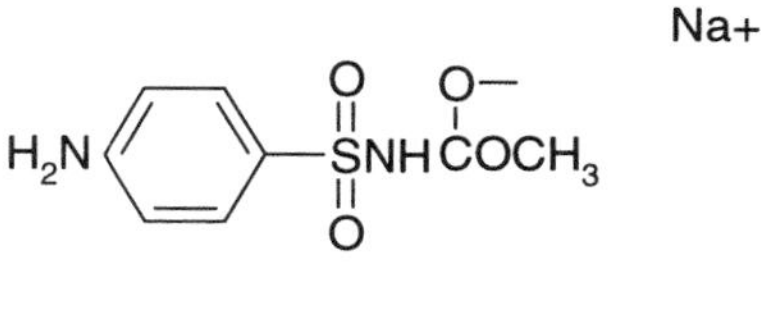

Na+

H_2N SNH COCH$_3$

Asulam

Uses Asulam is formulated as an SL of the sodium salt and sold as Asulox for postemergence control of several perennial grasses (including johnsongrass), annual grasses, brackenfern, tansy ragwort, and certain broadleaf weeds. Asulam is used in sugarcane; field-grown conifer, juniper, and yew nurseries; Christmas tree plantings; conifer release in reforestation areas; noncropland; alfalfa; flax; chickory; poppies; and in certain tolerant St. Augustinegrass and bermudagrass turf. The addition of surfactants to the spray solution increases the effectiveness of asulam for some of these uses but also reduces its selectivity for some uses. Because asulam is applied to the foliage of the target species, soil type does not affect its performance.

Mode of Action Asulam inhibits meristem growth. This involves an interference with cell division and is related to its effect on microtubule assembly or function (Fedtke, 1982). It is readily absorbed by susceptible species, but absorption is increased by the addition of certain surfactants in some, but not all, species (Catchpole and Hibbitt, 1972; Babiker and Duncan, 1975). It appears to be translocated in both the symplast and the apoplast, but the degree of translocation seems to be species dependent.

The inhibition of folic acid biosynthesis is considered to be the major site of action of asulam (Stephen et al., 1980; Veerasekaran et al., 1981a,b.) It inhibits 7,8-dihydropteroate synthetase, an enzyme involved in folic acid synthesis. Folic acid is required for biosynthesis of purine nucleotides, which are components of both DNA and RNA. This concept supports the previous findings that asulam inhibits nucleic acid and protein synthesis.

Endothall

Endothall 7-oxabicyclo[2.2.1]heptane-2,3-dicarboxylic acid is a white crystalline solid with no reported vapor pressure, a water solubility of 100,000 mg/l (ppm), a soil half-life of 7 days, and an oral LD_{50} (rat) 38 to 51 mg/kg.

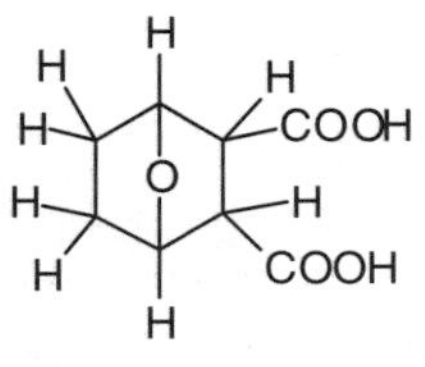

H H H H-COOH H-O-H H COOH H H

Endothall

Uses Endothall is formulated as an SL and in granular forms, and various salts are available, such as disodium, dipotassium, and amine, and sold under several trade names. Endothall is used to control certain annual grass and broadleaf weeds in sugar beets and established lawns and turf. It is also used to control aquatic weeds in still and moving water and in rice. In cotton, potatoes, and legumes for seed production, it is utilized as a preharvest crop desiccant or harvest aid.

In sugar beets, endothall is applied to the soil as a preemergence or a shallow (<1.5 inches) preplant soil-incorporated treatment or as a postemergence treatment. To broaden the spectrum of weeds controlled by a postemergence application of endothall, desmedipham, or phenmedipham, can be added to the spray solution. The sugar beets should be in the four- to six-leaf stage at the time of application for these combination treatments.

In established lawns and turf, endothall is applied when the temperature and soil moisture favor good growth. Endothall is used after sowing for aquatic weed control in rice. In still water (lakes, ponds), a combination of the dipotassium salt of endothall and mixed copper-ethanolamine complexes is used for aquatic weed control. Endothall is subject to considerable leaching in soils; however, it is rapidly degraded in both soil and water.

Mode of Action Endothall is absorbed readily by leaves and roots. It is translocated to a limited extent from roots to shoots of plants via the xylem, but it is not phloem mobile and is thus not translocated from leaves to other plant parts. Its action appears to be via contact, causing rapid desiccation to germinating seedlings, browning of the foliage, or both.

Ethofumesate

Ethofumesate (±)-2-ethoxy-2,3-dihydro-3,3-dimethyl-5-benzofuranyl methanesulfonate is a white crystalline solid with a vapor pressure of 6.45×10^{-7} mm Hg at 25°C, a moderate water solubility of 110 mg/l (ppm) at 25°C, a soil half-life of 5 to 14 weeks, and an oral LD_{50} (rat) of technical ethofumesate of 6400 mg/kg.

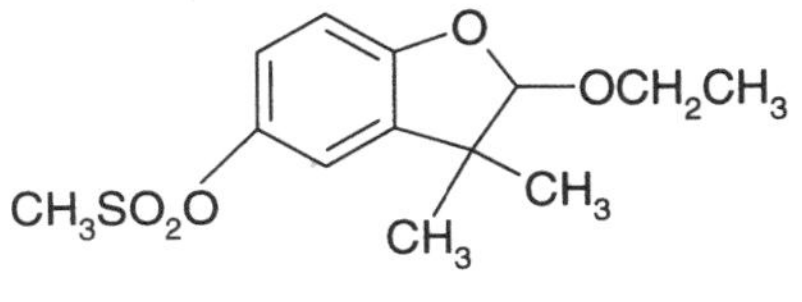

CH_3SO_2O — O — OCH_2CH_3 — CH_3 — CH_3

Ethofumesate

Uses Ethofumesate is formulated as an EC and sold as Nortron and Prograss for use as a selective herbicide in sugar beets, established lawns and turf, and established grass seed crops for control of annual grass and broadleaf weeds. Ethofumesate is primarily a soil-applied herbicide but has some postemergence activity on young weeds. Postemergence activity may be increased when it is used in combination with certain other herbicides in sugar beets.

Soil Influence Ethofumesate appears to be adsorbed to organic matter in soils. Field studies have demonstrated that it is not readily leached below 6 inches, and laboratory studies have shown that it does not leach in soils having an organic matter content greater than 1% (*Herbicide Handbook*, 1994). The effective application rate varies with soil type. Ethofumesate is biologically degraded in soils. The half-life ranges from less than 5 weeks under warm, moist conditions to more than 14 weeks under cold, dry conditions. The label states that crops other than sugar beets or ryegrass should not be planted within 12 months following application.

Mode of Action Ethofumesate is absorbed by the emerging shoots and roots of most plants (*Herbicide Handbook*, 1994). Foliar absorption is reduced as leaves mature and their cuticles develop. A preemergence application of this herbicide decreased epicuticular wax formation on sugar beet leaves (Duncan et al., 1982). Ethofumesate is translocated to the foliage following emerging shoot or root absorption but is not translocated from treated leaves. This suggests that it is translocated in the apoplast. It is also translocated more in at least some susceptible species than in tolerant sugar beets (Duncan et al., 1982).

The tolerant species sugar beets and ryegrass metabolize ethofumesate into a major and a minor conjugated metabolite (*Herbicide Handbook*, 1994). These data suggest that selectivity involves both translocation and metabolism of the herbicide.

BORATES

Boron, the phytotoxic element of the borate herbicides, is an essential minor element for plant growth. In areas that are deficient in boron, small amounts of boron must be applied for optimum growth. However, in large quantities boron is toxic to plants.

Borax

Borax sodium tetraborate ($Na_2B_4O_7$) and its hydrated forms, sodium tetraborate pentahydrate ($Na_2B_4O_7{\cdot}5H_2O$) and sodium decahydrate ($Na_2B_4O_7{\cdot}1OH_2O$). Their water solubilities are very high: 25,600 mg/l (ppm), 38,200 ppm, and 59,300 ppm, respectively, at 20°C. The acute oral LD_{50} (rat) of these compounds ranges from 2000 to 5600 mg/kg.

Uses The various sodium borate salts are seldom used alone for weed control today. In the past, they were extensively used alone or in combination with sodium chlorate for total vegetation control on noncropland—railroads, farms, and industrial sites—and in the paving industry to prevent weeds from growing though asphalt. They are currently used in package mix combinations with sodium chlorate as a harvest aid and certain triazine herbicides for total vegetation control on noncropland. Atratol contains atrazine, borax, and sodium chlorate; Pramitol 5PS contains prometon, simazine, borax, and sodium chlorate. In these combinations, the borates not only act as herbicides but also function as a fire retardant for sodium chlorate and an inhibitor of microorganism growth. The latter delays the degradation of the triazine herbicides by soil microorganisms. In the recent past, package mixes of borax plus bromacil (Borocil) and borax plus monuron (Ureabor) were available. Borax is also used as a fire retardant for sodium chlorate when it is used as a harvest aid (see the following discussion of sodium chlorate).

Soil Influence Borax is moderately adsorbed by inorganic components of the soil and subject to slow leaching—in contrast to sodium chlorate, which is readily leached. This difference plus the fact that borax acts as a fire retardant for sodium chlorate explains why borax and sodium chlorate are often used together for weed control. The overall advantage of a borate-chlorate combination is that they both are nonselective and that borax controls shallow-rooted weeds and sodium chlorate controls deep-rooted weeds for total vegetation control on noncropland. The sodium borate salts are relatively persistent, usually lasting 1 or more years. The period of persistence depends on soil type and rainfall. They are less persistent in acid soils and areas of high rainfall.

Mode of Action Herbicidal rates of borates cause plant desiccation beginning with necrosis of leaf margins, which progressively continues throughout the leaves. They are principally absorbed by roots and translocated through the xylem to all parts of the plant, accumulating in the leaves. The herbicide is most effective on young and tender plants. Therefore, treatment should be applied early enough to allow the material to be leached into the root-absorption zone by the time weed growth is just beginning. Virtually nothing has been published on the mechanism of action of borate as a herbicide (Brian, 1976). However, Crafts (1964) suggested that boron compounds apparently "tie up" calcium in the plant inasmuch as the injury symptoms resulting from excessive boron use resemble those of calcium deficiency.

$NaClO_3$

Sodium Chlorate

Sodium chlorate $NaClO_3$ is a white crystalline solid when pure; commercial forms are white to pale yellow. It has a very high water solubility of 1,000,000 mg/L (ppm) at 20°C. The oral LD_{50} (rat) of sodium chlorate is 5000 mg/kg.

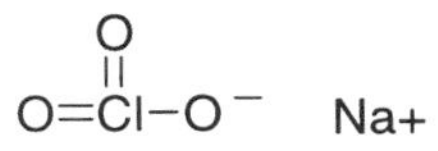

$$O{=}\overset{\overset{\displaystyle O}{\|}}{Cl}{-}O^{-} \quad Na+$$

Sodium Chlorate

Sodium chlorate has a salty taste. "Salt-hungry" animals may eat enough of treated plants to be poisoned. One pound of this chemical per 1000 pounds of animal is considered lethal. An additional hazard is that certain poisonous plants that are ordinarily avoided by livestock become palatable when treated with sodium chlorate. Sodium chlorate is 30 to 50 times more toxic to higher plants than common table salt, NaCl.

Fire Danger Sodium chlorate is a strong oxident; contact with combustible materials such as clothing, leather, wood, and plants may cause a fire. Sodium chlorate is dangerously flammable and has been ignited by the sun's rays, clothing friction, or shoes scraping a rock. Numerous precautions related to this herbicide in regard to application, spillage, storage, and container disposal are given on the label and in the *Herbicide Handbook* (1994). Among these precautions are (1) wear rubber boots and apron, (2) remove contaminated clothing promptly and immediately wash with water, and (3) apply only in dry form to dry vegetation for weed control.

Uses Sodium chlorate is seldom used alone for weed control today. In the past, it was used alone as a foliar contact spray or as crystals for a soil treatment even when emerged weeds were present. The foliar-contact treatment has been largely discontinued, mainly because of the fire hazard presented when the foliage dries. When dry crystals are applied to dry vegetation, there is usually no fire hazard because most of the material falls to the ground. However, sodium chlorate-sodium borate combinations as soil treatments usually gave superior total vegetation control than either product alone, for the reasons presented previously; see the earlier section "Borates."

Currently, sodium chlorate is used in package mixes of borates plus various triazine herbicides for total vegetation control on noncropland; see the section "Borates" for details. Sodium chlorate plus a fire retardant is widely used as a harvest aid to desiccate weeds in the mature crop and/or to reduce the moisture content of seeds and seed heads of many crops. These crops include cotton, corn (field, sweet, and popcorn), flax, gaur beans, peppers (chili, processing), rice, safflower, sorghum, soybeans, and sunflower.

Soil Influence Sodium chlorate is not adsorbed to soils to a significant degree and is subject to leaching. Soil microorganisms degrade sodium chlorate to sodium chloride. This occurs most rapidly in moist soils at temperatures above 70°C. Its phytotoxicity may persist 5 years or longer in areas of low rainfall and low microbial activity in the soil. In areas of high rainfall and high microbial activity in the soil, its phytotoxicity may be lost in 12 months in heavy soils and 6 months in sandy soils. Heavy rains or irrigation soon after a sodium chlorate application may leach the chemical from the upper 2 to 3 inches of the soil, rendering this zone free of the herbicide. This allows weed seeds to germinate and shallow-rooted weeds to continue growth. See the section "Borates" for the basis of the advantages of a sodium chlorate-sodium borate combination for total vegetation control, related to their differential behavior in soils.

Mode of Action Sodium chlorate desiccates foliage of plants quickly and is injurious to the roots and other organs and living tissues that it contacts. It penetrates the cuticle and comes in direct contact with the living cells (Loomis et al., 1933). Because it kills living cells rapidly, translocation from the leaves to the rest of the plant via the living phloem is minimal. However, it is rapidly translocated from the roots to the shoots through the nonliving xylem.

In addition to the contact action of sodium chlorate, which probably is related to an alteration of cell membranes, it has been reported to have an effect on certain metabolic processes. Plants treated with sodium chlorate are rapidly depleted of their food reserves (Crafts, 1935), mainly carbohydrates, apparently by an increased rate of respiration (Wort, 1964). However, Gorenflot (1947) found that sodium chlorate inhibited respiration and photosynthesis as well as protoplasmic streaming of leaf cells of *Elodea canadensis*, but that the effect was reversible if treatment was not too prolonged. Sodium chlorate has also been reported to decrease catalase activity (Neller, 1931), which could include an increase in hydrogen peroxide, which is toxic to plants. Presumably many other effects on normal plant metabolism could be demonstrated from a sodium chlorate treatment, considering its high chemical reactivity.

LITERATURE CITED AND SUGGESTED READING

Ashton, F. M., and A. S. Crafts. 1981. *Mode of Action of Herbicides*. John Wiley & Sons, Inc. New York.

Babiker, A. G. T., and H. J. Duncan. 1975. Penetration of bean leaves by asulam as influenced by adjuvants and humidity. *Pestic. Sci.* **6**:655–664.

Blanchard, F. A. 1954. Uptake distribution and metabolism of carbon-14 labeled trichloroacetate in corn and pea plants. *Weeds* **3**:274–278.

Brian, R. C. 1976. The history and classification of herbicides. In *Herbicides*, ed. by L. J. Audus, p. 41. Academic Press, New York.

Catchpole, A. H., and C. J. Hibbitt. 1972. Studies on the retention, penetration and translocation of asulam in some perennial weed species. *Proc. 11th British Crop Prot. Council*, pp. 77–83.

Cohen, A. S., and I. N. Morrison. 1982. Differential inhibition of potassium ion absorption by difenzoquat in wild oat and cereals. *Pestic Biochem Physiol.* **18**:174–179.

Crafts, A. S. 1935. Physiological problems connected with the use of sodium chlorate in weed control. *Plant Physiol.* **10**:699–711.

Crafts, A. S. 1964. Herbicide behavior in the plant. In *The Physiology and Biochemistry of Herbicides*, ed. by L. J. Audus, p. 93. Academic Press, New York.

Duke, S. O. 1985. *Weed Physiology. Vol. II. Herbicide Physiology*, CRC Press, Boca Raton, Fl.

Duncan, D. N., W. F. Meggett, and D. Penner. 1982. The basis for selectivity of root-applied ethofumasate in sugarbeet and three weed species. *Weed Sci.* **30**:191–194.

Duncan, D. N., W. F. Meggett, and D. Penner. 1982b. Basis for increased activity from herbicide combinations ethofumasate applied on sugarbeet. *Weed Sci.* **30**:195–200.

Fedtke, C. 1982. *Biochemistry and Physiology of Herbicide Action*. Springer-Verlag, New York.

Goatley, J. M., Jr., and R. E. Schmidt. 1994. Crabgrass control and dollar spot suppression in creeping bentgrass with DSMA. *HortSci.* **29**:884–886.

Gorenflot, R. 1947. *Relation entre les variations de la cyclose, de l'assimilation chlorophyllienne et de la respiration de l'Elodes canadensis L. sous l'influence du chlorate de sodium. Rev. Gen. Bot.* **54**:153–185.

Halling, B. P., and R. Behrens. 1983. Effect of difenzoquat on photoreactions and respiration in wheat and wild oat. *Weed Sci.* **31**:693–699.

Herbicide Handbook. 7th ed. 1994. Weed Science Society of America, Lawrence, KS.

Jordan, D., M. McClelland, A. Kendig, and R. Frans. 1997. Monosodium methanearssonate influence on broadleaf weed control with selected postemergence-directed cotton herbicides. *J. of Cotton Sci.* **1**:72–75.

Kidd, B. R., N. H. Stephen, and H. J. Duncan. 1982. The effect of asulam on purine biosynthesis. *Plant Sci. Lett.* **19**:211–217.

Killmer, J. L., J. M. Widholm, and F. W. Slife. 1980. Antagonistic effect of p-aminobenzoate or folate on asulam inhibition of carrot cultures. *Plant Sci. Lett.* 203–208.

Kitchen, L. M., C. E. Rieck, and W. W. Witt. 1980. Absorption and translocation of ^{14}C fosamine by three woody species. *Weed Res.* **20**:285–289.

Latshaw, W. L., and J. W. Zahnley. 1927. Experiments with sodium chlorate and other chemicals as herbicides for field bindweed. *J. Agric. Res.* **35**:757–767.

Lawrie, J., and D. V. Clay. 1994. Tolerance of some broadleaved and coniferous forestry trees species to herbicides with potential for bracken control (asulam). *J. Soc. For. Great Britain* **67**:237–244.

Leek, G., and D. Penner. 1980. Activity of CGA-43089 [α-(cyanomethoximino)-benzacetonitrile] as a potent antidote for weed control in sorghum. *Abstract Weed Sci. Soc. Am.* #182.

Loomis, W. E., E. V. Smith, R. Bissey, and L. E. Arnold. 1933. The absorption and movement of sodium chlorate when used as an herbicide. *J. Am. Soc. Agron.* **25**:724–739.

Mann, R. K., W. W. Witt, and C. E. Rleck. 1986. Fosamine absorption and translocation in multiflora rose (*Rosa multiflora*). *Weed Sci.* **34**:830–833.

Müller, F. 1981. Studies on the behavior of fasamine in young forest plants. Mitleilungen-Biologische Bundesanetalt fur land-und forstwirtechaft. **203**:264–265.

Neller, J. R. 1931. Effect of chlorates upon the catalase activity of the roots of bindweed. *J. Agric. Res.* **43**:183–189.

Nimbal, C. I., G. D. Wills, S. O. Duke, and D. R. Shaw. 1995. Uptake, translocation, and metabolism of 14C-MSMA in organic arsenical-resistant and –susceptible Mississippi biotypes of common cocklebur. *Weed Sci.* **43**:540–554.

Stephen, N. H., G. T. Cook, and H. J. Duncan. 1980. A possible mechanism of action of asulam involving folic acid biosynthesis. *Ann. Appl. Biol.* **96**:227–234.

Subramanian, M. V., S. A. Brunn, P. Bernasconi, B. C. Patel, and J. D. Reagan. 1997. Revisiting auxin transport inhibition as a mode of action for herbicides (naptalm). *Weed Sci.* **45**:621–627.

Veerasekaran, P., R. C. Kirkwood, and E. W. Parnell. 1981a. Studies of the mechanism of action of asulam in plants. Part I. Antagonistic interaction of asulam and 4-aminobenzoic acid. *Pestic. Sci.* **12**:325–329.

Veerasekaran, P., R. C. Kirkwood, and E. W. Parnell. 1981b. Studies of the mechanism of action of asulam in plants. Part II: Effect of asulam on the biosynthesis of folic acid. *Pestic. Sci.* **12**:330–338.

Wort, D. J. 1964. Effects of herbicides on plant composition and metabolism. In *The Physiology and Biochemistry of Herbicides*, ed. by L. J. Audus, p. 312. Academic Press, New York.

For chemical use, see the manufacturer's or supplier's label and follow these directions. Also see the Preface.

18 Herbicide Resistance in Weeds and Crops

There were no confirmed reports of weed resistance to any herbicide until 1968, when triazine resistance was reported in common groundsel (*Senecio vulgaris*) in western Washington. Since that time there have been increasing reports of herbicide resistance to several classes of herbicides, which has resulted in a more cautious approach to herbicide use in weed management programs.

Herbicides are a valuable tool for effective and economical control of weeds and have resulted in improved crop yields and overall production efficiency. In weed control, the use of plant selectivity is the major factor in effective herbicide use (Chapter 5). Selectivity is the differential effect of herbicides on various plant species based on the particular plant being treated, the specific herbicide, the environment, and the farmer. The loss of this selectivity due to the development of herbicide resistance in previously susceptible weed species will dramatically affect our ability to produce crops with high yields economically and is of concern to all people involved in agriculture. Research and development programs are necessary to provide a better understanding of the mechanisms of herbicide resistance, the ecological and management factors that influence its development, and to design strategies that minimize or eliminate its influence in agriculture.

There are three main plant responses to herbicide application. *Susceptibility* is demonstrated when the treated plant dies as a result of a herbicide used at a normal rate. *Tolerance* is the inherent ability of a plant to survive and reproduce after herbicide treatment. This implies that there is no selection or genetic manipulation but that the plant is naturally tolerant—for example, grass tolerance to 2,4-D or carrot tolerance to trifluralin. *Resistance*, as defined by the Weed Science Society of America, is the inherited ability of a plant to survive and reproduce following exposure to a dose of herbicide normally lethal to the wild type. Resistance may occur naturally or be induced by selection pressure.

Within resistance there are three subcategories used to describe particular types of resistance. *Cross-resistance* occurs when a plant selected for resistance to a specific herbicide is also resistant to other herbicides within a similar chemical group (such as resistant to atrazine and to other s-triazines as well). *Multiple resistance* occurs when a plant is resistant to herbicides that are chemically unrelated and that have different modes of action (such as a weed resistant to both sulfonylurea and aryloxyphenoxypropanate herbicides). *Negative cross-resistance* occurs when a biotype resistant to a particular herbicide is more susceptible to other classes of herbicides than the susceptible biotype—for instance, an atrazine-resistant redroot pigweed (*Amaranthus retroflexus*) that is more susceptible to fluometuron (cotoran/meturon) than the triazine-susceptible biotype.

HERBICIDE-RESISTANT WEEDS

Herbicide resistance was first noted in 1968 with the identification of a biotype of common groundsel that was resistant to simazine (Princep). There has been a rapid increase in the development of resistance in weeds to herbicides since about 1977. Internationally (45 countries), there are now about 233 species resistant to more than 16 herbicide chemical families (Heap, 2000).

Resistance can occur when a particular herbicide is applied repeatedly for as few as 3 to 5 years to sensitive species. The use of soil-residual herbicides usually increases selection pressure, and such practices result in continuous high selection pressure for resistant biotypes. This naturally occurring mutation can be passed on to progeny. Over time, resistant biotypes become the dominant biotypes in a population. For this to occur, one or more alleles for resistance must be present in the field population of the weed. Typically, monogene phenotypes are present at frequencies ranging from 10^{-5} for dominant genes to 10^{-11} for recessive genes. Chloroplastic genes are thought to be present at even lower frequencies (Debreuil et al., 1996). Resistance of a particular weed species is generally detectable when the resistant types compose approximately 30% of the population.

Within resistant species there are two primary mechanisms responsible for resistance. The mechanism is either metabolism of the herbicide or an altered site of action. An altered site of action has been the most common mechanism observed in resistant biotypes. Herbicide-resistant weed biotypes have been reported for most identified herbicide sites of action. Weed biotypes with an altered site of action are usually resistant to all herbicides within the same family (e.g., a biotype resistant to all sulfonylurea herbicides). They may also be resistant to other herbicides that attack the same target site (e.g., a sulfonylurea-herbicide-resistant weed biotype that is resistant to imidazolinone or triazolopyrimidine herbicides). This type of cross-resistance among herbicide families is not predictable and must be determined for each biotype and herbicide. Metabolism-based resistance may result in multiple resistance to herbicides. For example, wheat (*Triticum aestivum*), unlike many other crops, seems to have only one mechanism to degrade herbicides—a monooxygenase system that oxidizes several selective herbicides (Gressel, 1988). Likewise, multiple resistance to herbicides in different families may occur in a herbicide-resistant weed with an altered monooxygenase system. If the altered system was similar to that of wheat, a weed would likely be resistant to all herbicides used selectively in wheat, whether or not the herbicide had been applied previously to the weed.

Triazine resistance was the first resistance reported and is the most thoroughly studied. Since 1968 more than 60 *s*-triazine-resistant species (or biotypes) have been identified (40 dicot and 20 grass species in 35 genera) in the world. These biotypes occur in 31 states in the United States and 18 countries worldwide. Most of these species are resistant because of an altered herbicide target site (Figure 18-1), although, herbicide metabolism is the mechanism in a velvetleaf (*Abutilon theophrastis*) biotype.

Figure 18-1. Atrazine-resistant jimsonweed. *Right*: Resistant plant sprayed with 2.2 kg/ha atrazine with no injury, whereas, *Left*: Control plant was killed.

Resistance has also occurred in many other herbicide groups. In addition to resistance to triazine herbicides, widespread herbicide resistance has been reported for the bipyridiliums, substituted urea, sulfonylureas, dinitroanilines, aryloxyphenoxy propanates, cyclohexanediones, synthetic auxins, and glyphosate. The exact mechanisms involved in all these resistances are not known. Resistance to the bipyridiliums is due to sequestration. Resistance to the dinitroanilines is due to an altered binding site at the active site. In the case of the sulfonylureas, aryloxyphenoxy propanates, and cyclohexanediones, resistance depends on the biotype being studied and can include an altered target site or herbicide metabolism. Mechanisms of resistance to synthetic auxin herbicides have not been determined. For glyphosate, resistance is known to be related to an altered binding site in a goosegrass biotype from Malaysia; however, for field bindweed and rigid ryegrass biotypes, the mechanism appears to be related to multiple mechanisms; however, an altered binding site is not responsible.

The recent identification of species with multiple resistance is of special concern for the continued effective use of herbicides. Annual ryegrass (*Lolium rigidum*) biotypes identified in Australia rapidly developed resistance to diverse groups of herbicides, and some of these biotypes are resistant to many herbicides within ten different chemical classes. These biotypes may possess multiple mechanisms of resistance. Weeds with multiple resistance are of great concern because such weeds would be almost impossible to control with any herbicide. Other species with multiple resistance to herbicides include blackgrass (*Alopecurus myosuroides*) and hood canarygrass (*Phalaris paradoxa*), resistant to triazines and diclofop-methyl (Hoelon), and wild oat (*Avena fatua*) resistant to triallate (Fargo) and difenzoquat (Avenge).

MANAGEMENT STRATEGIES FOR HERBICIDE-RESISTANT WEEDS

Knowledge of how herbicide resistance develops is important in designing production systems that prevent or decrease its occurrence. Important factors to consider in the design of resistance management programs are weed characteristics, herbicide characteristics, and cultural practices. A poor understanding of any or all of these factors in the design of a weed management program results in an increase in selection pressure and greater potential for the occurrence of herbicide-resistant weeds. See Mallory-Smith et al. (1999) and Powles and Shaner (2001) for more details on resistance management strategies.

Certain weedy characteristics are conducive to the rapid development of resistance to a particular herbicide: annual growth habit, high seed production, high percentage of viable seed germinating each year (i.e., little seed dormancy), several reproductive generations per growing season, and extreme susceptibility to a particular herbicide.

Herbicide characteristics include a single site of action, high rate of effectiveness on a particular species (this leads to heavy selection pressure), long residual activity in the soil (paraquat and glyphosate are exceptions), high use rate relative to the amount needed for weed control, and high frequency of use.

Cultural practices include continuously growing the same crop, little or no cultivation or tillage for weed control or no elimination of weeds that escape herbicide control, and continuous or repeated use of a single herbicide or several herbicides that have the same mode of action.

Taken together, the presence of these factors allows certain weed biotypes that contain resistance alleles to become the prominent members of a population that is put under heavy selection pressure. The important point is that poor cropping and herbicide use patterns will result in the development of herbicide-resistant weeds.

Many strategies and models have been proposed to delay or prevent the development of resistance. At the present time none of these has been tested thoroughly enough to be defined as the preferred method for managing weed resistance to herbicides. Most farmers faced with the problem of controlling weeds resistant to a particular herbicide are using good management practices, as summarized here:

1. Cultivate as part of a plan to control weeds.
2. Rotate herbicide modes of action as crops are rotated.
3. Rotate weed control herbicide modes of action in any continuous crop rotation.
4. Use tank mixes or sequential mixes of herbicides with different modes of action to control specific weeds.

5. Use herbicides that have short-term soil activity to avoid high selection pressure on weeds.

As should be the case for all weed control, the existence of single- or multiple-herbicide-resistant weeds requires an involved approach that includes a well-designed, integrated weed management system in which the use of herbicides is only one of the control methods employed. The use of herbicides should be based on the biological and evolutionary realities of the weeds involved. These realities are not defined at the present time. However, with more research focused on weed ecology, genetics, and biology, we will be able to understand the mechanisms responsible for and the management strategies necessary to prevent the widespread occurrence of weeds developing resistance to herbicides.

A tremendous amount of research is focused on resistance, addressing most of the topics discussed here. As more information becomes available concerning the fitness of resistant biotypes and the processes that govern the development of multiple resistances, along with a more complete knowledge of weed biology, ecology, and genetics, our ability to deal with resistance will improve. Weed resistance to herbicides is a major area of concern. It must be dealt with in a timely manner if we are to avoid major loss of effective herbicides for use in our agricultural cropping systems. Great care must be taken by farmers, university researchers, and chemical companies in designing systems to control weeds that avoid the common practices that have led to the development of resistance in the past.

It is important to note that the U.S. Environmental Protection Agency (EPA) considers resistance and resistance management in its decision to register and regulate pesticides (Matten, 1997). EPA believes in implementing effective resistance management strategies. Thus, EPA created the Pesticide Resistance Management Workgroup of the Office of Pesticide Programs (PRMW), in part to examine the agency's role in managing pest resistance and to provide policy options for regulating pesticides to reduce selection pressure for resistance.

HERBICIDE-RESISTANT CROPS

The use of biotechnology to develop crop plants with new resistance to herbicides, notably broad-spectrum herbicides such as glyphosate, has had a significant impact on weed control practices. This will continue as the ability to manipulate herbicide tolerance using these methods expands: More genes for herbicide resistance are being developed, and the number of crops that can be genetically modified in this way is increasing. Indeed, it is possible that this approach will change the manner in which herbicides are developed. Rather than identifying a herbicide with selectivity, genes for resistance to specific herbicides (including new ones) may be incorporated into specific crops. The science that underlies this technology is discussed in the following sections.

There are three principal requirements for manipulating herbicide resistance in plants through biotechnology:

1. Identification of a gene that can provide resistance to a herbicide
2. Modification of this gene so that it can be expressed in the target crop plant
3. Transfer of the modified gene into the crop plant

Each of these components has been developed over the last 15 years as our understanding of plant biochemistry, physiology, and molecular biology has increased. The application of genetic engineering to herbicide resistance is the first major commercial success in agricultural biotechnology. However, there is tremendous potential for genetic manipulation of crop plants to impact crop production practices as well as the downstream uses of agricultural products.

Identification of Genes for Herbicide Resistance

Plants use a number of mechanisms to tolerate herbicides, including reduced uptake of the herbicide, increased metabolism, and a target site that is not affected by the herbicide. Conceptually, genes could be identified that would alter any of these steps to provide resistance to a specific herbicide. In practice, the two mechanisms that have been favored for genetic engineering are modification of the target site and metabolism of the herbicide. The *revolutionary* aspect of this new technology is that genes for such traits need not come from the traditional germ plasm resources used by plant breeders. Rather, genes for herbicide resistance may be found in any organism, including microbes. Once identified, these genes can be isolated, modified to function appropriately in a plant, and transferred to the crop.

The strategy used to identify genes for herbicide resistance depends on the nature of the herbicide and its mode of action. An approach that has proven effective in many cases is to identify microbes that either survive in the presence of the herbicide or use the herbicide as a food source. There are a number of advantages to screening for herbicide resistance genes in bacteria:

- Large numbers of bacteria can be tested through simple procedures, such as screening for bacteria that grow on agar medium supplemented with the herbicide.
- Large collections of bacteria can be obtained simply from soil samples or other sources.
- Genes that are responsible for herbicide resistance can be isolated rapidly because bacterial genomes are small and molecular genetic techniques facilitate this process.

A number of examples are presented here to illustrate how bacteria have been used to identify genes for herbicide resistance.

Glyphosate (Roundup) inhibits the enzyme EPSP synthase, an enzyme required for the synthesis of one class of amino acids. This enzyme is found in both plants and bacteria, and both groups of organisms are susceptible to glyphosate. Bacteria that are able to grow in the presence of glyphosate frequently have an altered form of EPSP synthase that is no longer inhibited by glyphosate. The first demonstration that plants could be engineered to be glyphosate resistant used a gene for EPSP synthase from a glyphosate-resistant strain of the bacterium *Salmonella typhimurium*. The gene used in the first Roundup Ready soybeans was similarly derived from another glyphosate-resistant bacterium.

Bromoxynil (Buctril) is a photosynthetic inhibitor. There are a number of reasons that selecting for bromoxynil-resistant bacteria would not aid in the development of plants with resistance to this herbicide. However, bacteria that grow using bromoxynil as the only source of carbon and nitrogen must be capable of metabolizing the herbicide. A gene encoding the enzyme responsible for this metabolic process was identified from a bacterium that could grow on bromoxynil. This gene was used to develop plants with resistance to bromoxynil.

Glufosinate (Liberty) inhibits the activity of glutamine synthetase. Glufosinate is chemically synthesized but was originally discovered as a metabolite produced by a soil microorganism, *Streptomyces viridochromogenes*. The bacteria that produce this compound protect themselves from it by producing a second enzyme that inactivates glufosinate within the bacteria. The gene encoding this detoxifying enzyme was isolated and used to develop plants with resistance to glufosinate.

In addition to these examples illustrating the use of bacteria to identify genes for herbicide resistance, there are other instances in which genes for herbicide resistance have been identified in plants. Microbes provide an invaluable source of genes for agricultural biotechnology, not only in the area of herbicide resistance but also for insect resistance, manipulation of metabolism, and other processes.

Modification of Genes for Expression in Plants

Genes carry two essential pieces of information (Figure 18-2). First, they carry information to produce proteins. The sequence of bases in the DNA is decoded by the cell to direct the synthesis of a specific protein. The machinery of the cell first transcribes the sequence of bases in DNA into a similar sequence of bases in RNA. This is then translated into a sequence of amino acids in the protein. The DNA code for proteins is essentially the same in all organisms, from bacteria to plants and animals. Consequently, a gene transferred from a bacterium to a plant will encode the same protein if the gene is expressed.

In addition to carrying the code for a specific protein, each gene also carries instructions about where, when, and to what level that gene should be expressed—that is, activated or switched on. These regulatory features are critical for the function of all organisms. The regulatory information is also carried in the sequence of bases in the DNA and is referred to as the *promoter*. The promoter of a gene determines where and when that gene is expressed and whether the protein will be made in abundance or at a low level. However, unlike the codes for proteins, regulatory instructions are not the same from one organism to another. Consequently, genes transferred between organisms will not be expressed unless the promoters are tailored to work in the host into which this gene is being moved. For example, a gene isolated from *Salmonella*

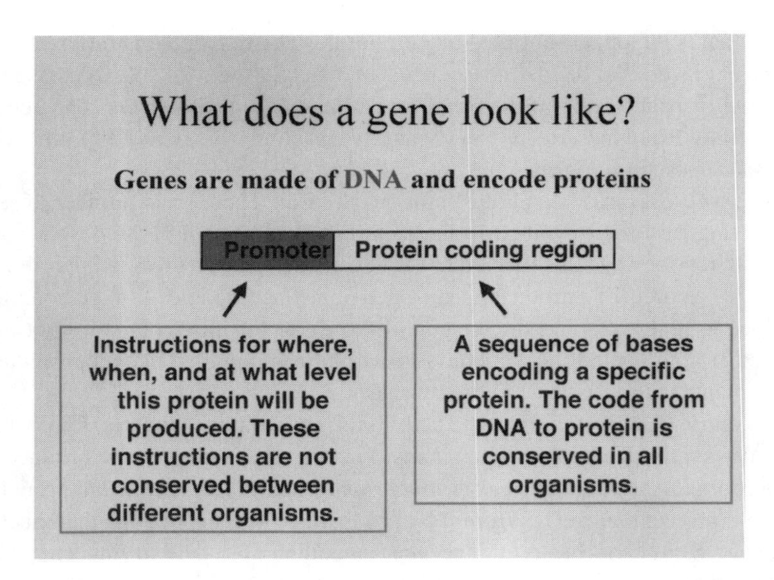

Figure 18-2. What is a gene? A gene contains the promoter and the protein coding region. (P. B. Goldsbrough, Purdue University.)

typhimurium, a bacterium, will not be expressed in a plant unless the promoter that controls its expression is altered to function in plants.

As our understanding has increased in regard to how gene expression in plants is regulated, promoters with different properties have been identified. Promoters that are available for genetic engineering include those that function in all tissues (providing constitutive expression), as well as promoters that are expressed in specific tissues such as leaves, roots, or meristems.

Genes that are used to provide tolerance to herbicides through biotechnology are therefore comprised of two essential components:

1. A DNA sequence encoding a specific protein that will either metabolize the herbicide or provide a target that is not inhibited by the herbicide
2. A promoter sequence that will direct the expression of this protein in particular tissues of the plant

Transfer of Genes into Crop Plants

Development of plants with novel traits, such as herbicide resistance, depends on the ability to transfer genes into these plants. Although this has been a routine technique for several years with many plants that are used in research, (e.g., tobacco, petunia, and *Arabidopsis*), the majority of crop plants have proven to be less amenable to gene transfer. However, the combination of persistence, skill, and large research expenditures has resulted in workable, but in most cases still inefficient, methods to move genes into crop plants such as corn and soybean. The process of gene transfer is commonly referred to as *transformation*. Cells that have taken up a gene by this method, and plants that are regenerated from these cells, are described as *transformed* or *transgenic*. The term *genetically modified organism* (GMO) is also used to describe transgenic organisms, including transformed plants. Genes that are transferred into plants are known as *transgenes*.

The various techniques used to transform plants include a number of common steps:

- A tissue is selected from which fertile plants can be regenerated through tissue culture.
- The desired gene(s) are transferred into as many cells as possible in this target tissue.
- Selection is used to identify transformed cells.
- Plants are regenerated from transformed cells.
- Transgenic plants are analyzed to confirm that the genes have been transferred, are being expressed as expected, and are having the desired effect on the phenotype of the plant.

Transformation of most plants has used either *Agrobacterium tumefaciens* or particle bombardment to move genes into the plant. *Agrobacterium tumefaciens* is a soil bacterium that has the unique ability to transfer DNA from the bacterium into plant cells. In nature the bacterium uses this ability to modify plant cells to produce compounds that the bacterium can use for growth. For genetic manipulation of plants, *Agrobacterium* can be used to transfer specific genes into plants without the deleterious effects normally associated with this bacterium. Until recently, *Agrobacterium* could not be used for transformation of most crop plants. Consequently, other methods were developed to transfer genes into these species, the most widely used being particle bombardment. Metallic particles are coated with the genes (pieces of DNA) to be transferred into the plant, and these DNA-coated particles are literally shot into the plant cells. Once inside a cell, the DNA is released from the particle, moves into the nucleus, and incorporates into the genome. Particle bombardment can transfer DNA into essentially any species, as it relies on physical methods to move DNA, unlike the biological system used in *Agrobacterium*-mediated transformation.

A general feature of transformation procedures is that they are inefficient; that is, very few plant cells are actually transformed. One of the genes normally transferred into cells during transformation is a selectable marker gene encoding resistance to an antibiotic or herbicide. This allows the rare transformed cells to be identified, because they can grow in the presence of the antibiotic or herbicide whereas the nontransformed cells die. Another requirement for successful transformation is that these transformed cells be able to be regenerated into fertile plants. This is accomplished through tissue culture, where the medium is modified first to stimulate production of shoots, followed by the generation of roots. Each plant produced through this process is genetically unique, because the DNA inserts at random in the plant genome and because of variation resulting from the tissue culture process.

The final step is to confirm that the transgene is performing as expected. For herbicide tolerance, this involves demonstrating that transgenic plants have tolerance to field applications of the herbicide while maintaining suitable agronomic properties. The transgene must also be stably expressed in all plants and from generation to generation.

The time required to progress from the initial transformation of plant cells to release of a new variety is likely to be at least 5 years. However, once the new herbicide tolerance gene is incorporated into elite breeding material, it can be handled simply as another trait that must be followed during variety development.

HERBICIDE-RESISTANT CROPS ON THE MARKET OR IN DEVELOPMENT

Herbicide-resistant crops that have been released or are being developed are listed in Table 18-1. As shown, most major crops and several minor crops are being developed for resistance to many widely used herbicides.

The acreage of the major herbicide-resistant crops (Table 18-2) has shown a tremendous increase since 1996, and world acreage projections (Table 18-3) suggest that increased production of these crops will continue in the future. Farmers have accepted this new technology with great excitement, especially in the United States, because of its effectiveness, ease of use, and ability to reduce input costs (Figure 18-3). However, as discussed in the following section, many issues concerning herbicide-resistant crops and genetically modified organisms in general must be

TABLE 18-1. Major Herbicides for Which Transgenic Herbicide-Resistant Cultivars Have Been or Are Being Developed

Herbicide	Herbicide-Resistant Plant Species
Bromoxynil	Cotton, potato, tobacco[a]
Glyphosate	Sugar beet,[a] corn, cotton, lettuce,[a] canola,[a] potato, oilseed rape, soybean, tobacco, tomato, strawberry, winter and spring wheat, barley, poplar, rice, bentgrass, perennial ryegrass, Kentucky bluegrass
Glufosinate	Alfalfa,[a] *Arabidopsis*, sugar beet, corn, barley, lupin, flax, creeping bentgrass, melon,[a] peanut, rice,[a] canola, sugarcane, soybean, tomato, wheat,[a] durum wheat, rye, oilseed rape, asparagus, papaya, pea
Sulfonylureas	Corn, cotton, tobacco, tomato (flax, soybean, sugar beet[b])
Imidazolinones[b]	Corn, canola, wheat, rice, sugar beet, sunflower, sugarcane, soybean
2,4-D	Potato, sweetgum,[a] cotton[b]
Sethoxydim[b]	Corn
Triazines[b]	Canola

[a]Can hybridize with wild relatives
[b]Nontransgenic

TABLE 18-2. Herbicide-Tolerant Crop Areas, 1995–1998

	Area (million acres)			
Herbicide-Resistant Crop	1996	1997	1998	1999
Corn				
IMI corn	4.0	4.0	6.1	
Sethoxydim-resistant corn	0.1	0.7	2.0	
Liberty Link corn	–	0.7	6.5	
Roundup Ready corn	–	–	1.0	
Canola				
IMI canola	1.0	1.5	2.0	
Liberty Link canola	.4	2.2	2.5	
Roundup Ready canola	0.06	0.5	3.0	
Soybean/Cotton				
STS soybean	3.0	5.0	6.0	
Roundup Ready soybean	1.75	12.5	29.0	41.0
Cotton				
Roundup Ready cotton	–	0.8	2.5	1.9
BXN cotton	0.04	0.28	0.8	
Stacked Gene Crops				
Bt/Roundup Ready cotton	–	0.06	1.1	1.5
Bt/LibertyLink corn	–	–	0.1	
Total area (million acres)	10.35	28.24	62.6	

addressed in a systematic manner for this technology to become fully accepted by the general public.

ISSUES RELATED TO HERBICIDE-RESISTANT CROPS

Herbicide-Resistant Plants

Widespread plantings of herbicide-resistant crops will likely result in the development of weeds that are resistant/tolerant to the herbicides used in this management system. This is especially true where a limited number of herbicides are applied over an extensive area. For example, there are already cases of weeds that have become resistant/tolerant to herbicides used in the Roundup Ready^{TM} soybean system in the United States. The incorporation of herbicide resistance traits in other crops, especially in those within the same cropping system, increases the likelihood of resistance developing in the targeted weed population.

TABLE 18-3. Planted Acres of Herbicide-Resistant Crops in 1998 and Forcasted Acres for 2003

Region	Crop	Acres (millions)	
		1998	2003
United States	*Herbicide-Resistant Crops*		
	IMI corn	6.0	2.0
	Poast protected corn	2.0	1.0
	Liberty Link corn	6.5	10.0
	Roundup Ready corn	1.0	8.0
	STS soybean	6.0	4.0
	Roundup Ready soybean	23.0	44.0
	Liberty Link soybean		10.0
	Roundup Ready cotton	2.5	3.0
	BXN Cotton	0.8	1.0
	Stacked Gene Crops		
	Bollgard/Roundup Ready cotton	1.0	2.0
	Bt/BXN cotton	0.1	0.5
	Roundup Ready/bt corn		22.0
	Bt/Liberty Link corn	0.1	18.0
Canada	IMI Canola	2.0	0.5
	Liberty Link canola	2.5	4.0
	Roundup Ready canola	3.0	5.0
	Roundup Ready soybean		1.0
	Bt/Liberty Link corn		0.1
Latin America	IMI corn	0.1	2.0
	Roundup Ready soybean	6.0	15.0
	Bt/Roundup Ready corn		45.0
	Roundup Ready cotton	0.1	0.8
Rest of World	Roundup Ready cotton		4.0
	Bt/Roundup Ready corn		10.5
	Herbicide-resistant sugarbeet		4.0
	Herbicide-resistant canola		5.0

The criticism of herbicide-resistant crops relates primarily to four potential areas of concern (Zimdahl, 1999): public health, environmental, social, and weed control issues. There is a concern on the part of some that the development of herbicide-resistant plants will result in increased use of herbicides and that this may compromise public health. If a herbicide is not completely metabolized in plants, humans and other organisms can ingest it. However, this does not necessarily mean that health-related problems will develop if a herbicide or its byproduct is consumed. Yet problems could develop where people are constantly exposed to a herbicide over

Figure 18-3. Corn resistant to glyphosate as a result of genetic engineering. Glyphosate was sprayed over the top of the corn and killed the weeds but not the crop.

a long period of time. There are also issues related to a potential increase in allergic reactions due to direct and indirect exposure to these herbicides. Extensive use of a product could increase the likelihood of this happening.

Environmental concerns as a result of the extensive use of this technology are common. The extensive use of a small number of products could compromise naturally occurring nontarget plant species, aquatic plant systems, potable water, and so on. It could also lead to the selection of species that are unaffected by the herbicide, which could threaten the balance of nature. Social concerns related to the high use of herbicide-resistant crops reflect a fear that this technology will favor large farms and result in the loss of small farming operations. The cost of the technology, which is passed on to the farmer in the form of increased seed costs, could put additional pressure on small farmers as they try to compete with larger operations. Moreover, if the price of the technology is carried through to the final product, it could mean higher food costs for the consumer.

In addition, there are concerns about weed control. It has already been shown that weed resistance to herbicides can and will occur. The properties and use pattern of a herbicide in a cropping system and the intrinsic properties of the weed are the factors important in the selection of herbicide-resistant weeds—not whether the crop is resistant or tolerant to a herbicide. The risk of selecting herbicide-resistant weeds in a herbicide-resistant crop is no greater than using a selective herbicide in a naturally tolerant crop. In either case, repeated use of the same herbicide or herbicides with the same mode of action will eventually select for herbicide-resistant weed biotypes or for weeds never susceptible to the herbicide (weed species shifts). This can be prevented or greatly delayed by using effective integrated weed management strategies in all parts of the crop production system. Most important is to avoid using the same herbicide or herbicides with the same site of action routinely in any cropping system. The widespread use of a single product repeatedly would increase the likelihood of a resistant-weed situation developing, and such use patterns must be avoided.

There is also concern about transferring resistance from one plant to another through gene flow. A good example of a situation in which this has already occurred is the case of herbicide-resistant canola. Genes from the canola moved to a nonweedy relative in the mustard family, then to a wild mustard. This is especially a risk where closely related plant species are in close proximity. The probability of introgression between each herbicide-resistant crop and weedy relatives should be known before the crop is grown commercially, especially if weedy relatives are present in the farming region. If gene transfer can occur, specific management strategies will be required to mitigate introgression.

Another concern is the possibility of herbicide-resistant plants becoming weeds after being successfully grown in cultivation. The weediness of herbicide-resistant and existing crop cultivars should be similar in the absence of a specific herbicide selection pressure. There are no reported cases of herbicide-resistant crops having greater fitness than existing cultivars. In fact, sometimes they are less fit (Mallory-Smith, 1996). Control of volunteer herbicide-resistant and nonresistant crop plants should be the same, except that herbicide choice will be more limited with the herbicide-resistant crop. This will be especially true when one herbicide-resistant crop is followed immediately with another crop resistant to the same herbicide (e.g., glyphosate-resistant corn followed by glyphosate-resistant soybean). Control of volunteer herbicide-resistant crops could become more difficult with gene stacking (multiple herbicide resistance)—for example, control of Garst Seed Company's corn hybrid with resistance to both imidazolinone and glufosinate herbicide in the subsequent soybean crop. It may be necessary to control volunteer herbicide-resistant crop plants in both arable and nonarable sites to prevent introgression with weedy relatives and existing susceptible crop cultivars. This will be especially important for outcrossing crops such as canola, and perhaps wheat. It will also be necessary to prevent gene flow through seed movement to areas outside the field seeded to herbicide-resistant crops. Although this has not specifically been shown to occur with herbicide-resistant plants, wild proso millet (*Panicum miliaceum* L.) is an example of a cultivated plant that became a weed.

An additional concern is the suppression of gene expression by other genes, as well as a single gene exerting simultaneous effects on more than one character. In nature, when one changes or modifies an organism, there is rarely just one outcome. Although specific outcomes are intended, unplanned events are not uncommon.

Ethics

Issues of the ethics of GMOs are as big and broad as the world itself. Some consider the moving of genes from one organism to another as an unethical act and contrary to their beliefs and the system of nature. Others feel that this results from the human ability to advance technology and improve life.

Markets and Plant Identity Preservation

Another issue of concern relative to GMOs is their acceptance in foreign markets. It is still unclear as to how this will be resolved, but some markets are, and possibly will be, closed to GMO-derived raw and processed commodities. Moreover, because of foreign market exclusions and restrictions, farmers, elevators, terminals, processors, and shippers are requiring that the identity of genetically modified seed in commerce be known throughout the selling and distribution chain.

Public perceptions and opinions related to these GMOs, as well as the regulations associated with the buying and selling of agricultural commodities, bring a degree of uncertainty to the future international trade of GMOs.

Registration and Regulation

In the United States, GMOs are regulated by the EPA and the Food and Drug Administration (FDA). Issues related to the registration and regulation of GMOs are considered by these agencies through Scientific Advisory Panels (SAP) which establish periods for public comment before final rulings are made.

GMOs have affected and will continue to change our lives. One can only hope that the changes will be positive and that these organisms will be developed and used responsibly. It is imperative that they be adequately tested and evaluated prior to their release and use. It is also important that there be a free flow of dialogue between those in favor of the technology and those who oppose or are concerned about it.

A FINAL THOUGHT

Biotechnology and the development of genetically modified organisms (GMOs) hold great promise for the human race and our planet. However, like any technology, this has potential for being misused or abused and must be closely watched and regulated to ensure that its negative aspects are minimized. The potential output to be derived from this technology in our agricultural system is barely a blip on the screen at this time. There are endless possibilities of things that can be derived from GMOs. It is almost beyond comprehension. In a workshop report published in *Agricultural Biotechnology & the Public Good*, Harrwood and DeWitt (1994) noted that "biotechnology should be directed toward solution of problems within a context of integrated systems, and that such technologies should then be applied within the context of sustainable systems."

LITERATURE CITED AND SUGGESTED READING

Herbicide-Resistant Weeds

Caseley, J. C., G. W. Cussans, and R. K. Atkin, eds. 1991. *Herbicide Resistance in Weeds and Crops*. Butterworth-Heinmann, Stoneham, MA.

Feng, P. C. C., J. Pratley, and J. A. Bohn. 1999. Resistance to glyphosate in *Lolium rigidum*. II. Uptake, translocation, and metabolism. *Weed Sci.* **47**:412–415.

Fischer, A. J., C. M. Ateh, D. E. Bayer, and J. E. Hill. 2000. Herbicide-resistant *Echinochloa oryzoides* and *E. phyllopogon* in California *Oryza sativa* fields. *Weed Sci.* **48**:225–230.

Gadamske, G., D. Clarka, J. Gressel, and S. W. Gawronski. 2000. Negative cross-resistance in triazine-resistant biotypes of *Echinochloa crus-galli* and *Conyza canadensis*. *Weed Sci.* **48**:176–180.

Gressel, J. 1988. Wheat Herbicides: The Challenge of Emerging Resistance. Biotechnology Affiliates, CPL Scientific Ltd. Checkendon, Reading, U.K.

Herbicide resistance: Mechanisms and diagnostics. 1999. *Proc. Brighton Crop Protection Conf—Weeds* **1**:153–202.

Herbicide resistance symposium. 1996. In *Proc. of the Second International Weed Control Congress*, Vol. II, pp. 377–479. Copenhagen, Denmark.

Herbicide resistant weeds: What's new? 1999. *Proc. Brighton Crop Protection Conf—Weeds*. **3**:769–802.

Mallory-Smith, C. A., D. C. Thill, and D. W. Morishita. 1999. *Herbicide-Resistant Weeds and Their Management*. Pacific Northwest Extension Bulletin 437(rev.).

Powles, S. B., and J. A. M. Holtum. 1994. *Herbicide Resistance in Plants: Biology and Biochemistry*. Lewis Publications-CRC Press, Boca Raton, FL.

Powles, S. B., C. Preston, I. B. Bryan, and A. R. Jutsum. 1997. Herbicide resistance: Impacts and management. *Advances Agron*. **58**:57–93.'

Herbicide-Resistant Crops

Askew, S. D., and J. W. Wilcut. 1999. Cost and weed management with herbicide programs in glyphosate-resistant cotton (*Gossypium hirsutum*). *Weed Technol.* **13**:308–313.

Ateh, C. M., and R. G. Harvey. 1999. Annual weed control by glyphosate in glyphosate-resistant soybean (*Glycine max*). *Weed Technol.* **13**:394–398.

Bergelson, J., C. B. Purrington, and G. Wichmann. 1998. Promiscuity in transgenic plants. *Nature* **395**:25.

Beringer, J. E. 2000. Releasing genetically modified organisms: Will any harm outweigh any advantage? *J. Applied Ecol.* **37**:207–214.

Brown, J., and A. P. Brown. 1997. Gene transfer between canola (*Brassica napus* L. and *Brassica campestris* L.) and related weed species. *Ann. Appl. Biol.* **129**:513–522.

Culpepper, A. S., and A. C. York. 1999. Weed management and net returns with transgenic, herbicide-resistant, and nontransgenic cotton (*Gossypium hirsutum*). *Weed Technol.* **13**:411–420.

Debreuil, D. J., L. F. Fiegen, and I. N. Morrison. 1996. Growth and seed return of auxin-type herbicide-resistant wild mustard (*Brassica Kaber*) in wheat. *Weed Sci.* **44**:871–878.

Duke, S. O., ed. 1996. *Herbicide-Resistant Crops: Agricultural, Economic, Environmental, Regulatory, and Technical Aspects*. Lewis Publications, Boca Raton, FL.

Dyer, W. E. 1994. Herbicide-resistant crops: A weed scientist's perspective. *Phytoprotection* **75**(supp.):71–77.

Field Testing Genetically Modified Organisms—Framework for Decisions. 1989. National Academy Press, Washington DC (Chapters 3, 4, 5, and 6).

Gelvin, S. B. The introduction and expression of transgenes in plants. 1998. *Current Opinion in Biotechnology* **9**:227–232. An up-to-date review of technical advances in plant transformation technology.

Global development in herbicide tolerant corps. 1997. *Proc. British Crop Protection Conference*. Papers 9B1-9B4, pp. 929–963.

Herbicide tolerant crops: Their value to world agriculture. 1999. *Proc. Brighton Crop Protection Conf—Weeds* **2**:637–660.

Harwood, R. and J. DeWitt. 1994. Workshop Report. In Agricultural Biotechnology and the Public Good, ed. by J. F. MacFarland, pp. 195–199. National Agricultural Biotechnology Council, Ithaca, NY.

Madsen, K. H., and J. C. Streibig. 1999. *Modeling of Herbicide Use in Genetically Modified Herbicide Resistant Crops. 2. Description of Models and Model Output*. Ministry of Environment and Energy, Danish Environmental Protection Agency, Copenhagen, Denmark.

Madsen, K. H., E. R. Poulsen, and J. C. Streibig. 1999. *Modeling of Herbicide Use in Genetically Modified Herbicide Resistant Crops*. 1. *Content and Input for Models*. Environ. Proj. 346, Ministry of Environment and Energy, Danish Environmental Protection Agency, Copenhagen, Denmark.

Mallory-Smith, C. A., and C. V. Eberlein. 1996. Potential pleiotropic difficulties with herbicide resistant crops. In *Herbicide-resistant crops: Agricultural, Economic, Environmental, Regulatory, and Technological Aspects*. ed. by S. D. Duke, Chap. 13:201–210, Lewis Publishers, Boca Raton, FL.

Matten, S. R. 1997. Pesticide resistance management activities by the U.S. Environmental Protection Agency. *Resistant Plant Mgmt.* **9**:3–5.

Miflin, B. J. 2000. Crop biotechnology. Where now? *Plant Physiol.* **123**:17–27.

Parrott, W. 2000. Forum on genetically modified crops. *Agron. J.* **92**:792–806.

Rieger, M. A., C. Preston, and S. B. Powles. 1999. Risks of gene flow from transgenic herbicide-resistant canola (*Brassica napus*) to weedy relatives in southern Australian cropping systems. *Aust. J. Agric. Res.* **50**:115–128.

Seefeldt, S. S., R. Zemetra, F. L. Young, and S. S. Jones. 1998. Production of herbicide-resistant jointed goatgrass (*Aegilops cylindrica*) x wheat *Triticum aestivum*) hybrids in the field by natural hybridization. *Weed Sci.* **46**:632–634.

Zimdahl, R. L. 1999. *Fundamentals of Weed Science*, 2nd ed. Academic Press, New York.

WEB SITES

Heap, I. M. 2000. *International Survey of Herbicide-Resistant Weeds*. WSSA/HRAC. http://www.weedscience.org.

Agricultural Biotechnology Information

www.aphis.usda.gov/biotechnology/

The Alliance for Better Food

www.betterfoods.org

Ag Biotech InfoNet

http://www.biotech-info.net

Council for Biotechnology Information

www.whybiotech.com

DNA Learning Center. An excellent site that features outstanding explanations of the basic concepts of molecular biology.

http://vector.cshl.org

For chemical use, see the manufacturer's or supplier's label and follow these directions. Also see the Preface.

PART III
Practice

19 Small Grains and Flax

The small grains discussed here include wheat, oats, barley, rye, and rice. Flax is also included because the cultural practices used in growing flax and its weed problems are similar to those of small grains.

There are winter- and spring-planted varieties of many small grains. Winter varieties are planted in the fall, live through the winter, and are harvested the following summer. Spring varieties are planted in early spring and harvested in mid to late summer of the planting year. As an average, small grain tolerance to cold is in this order: rye > wheat > barley > oats > rice. Therefore, winter rye is grown in far northern areas and rice is normally grown in the United States only in southern areas (e.g., California, Florida, Louisiana, and Texas).

In general, winter plantings are most often infested with winter-annual weeds, and to a lesser extent, by summer annuals that germinate in early spring. Spring varieties are primarily infested by summer annuals that germinate in the early spring, and perennial weeds can be troublesome in all areas.

EFFECT OF WEEDS ON YIELD

Weeds compete with small-grain and flax crops for light, carbon dioxide, and soil moisture and nutrients. Certain weeds have also been shown to reduce yields by allelopathic effects (Chapters 1 and 2).

Bridges (1992) reported on crop losses due to weeds in small grains and flax seed in the United States grown with and without the use of herbicides (Table 19-1). The mean values in this table indicate that yield reductions due to weeds were similar in all grain crops where herbicides were used. Estimates in production systems without herbicides showed that loss due to weeds would be greatest in rice (56%). The full report shows great differences in crops losses due to weeds for a given crop over the various geographic areas (e.g., 1 to 20% in wheat). Yield reductions for a flax seed crop are also considerable, at about 11.7% when grown with the use of herbicides and 41.7% in flax grown without herbicides.

Weed competition early in the season reduces yields more than late-season competition. Although yields are not greatly reduced by late-season weeds, such weeds may cause difficulty in harvesting. Weeds also lower crop quality and may reduce the protein content of the grain.

TABLE 19-1. Estimated Annual Average Yield Reduction Due to Weeds in Small Grains and Flax in the United States Grown with Best Management Practices (BMP) Including (+) or Excluding (−) Herbicides[a]

	Yield Return			
	BMP (+) Herbicides		BMP (−) Herbicides	
Crop	Mean %	Range %	Mean %	Range %
Barley	6.3	1.0–25.0	20.4	2.0–60.0
Flax	11.7	10.0–15.0	41.7	35.0–50.0
Oat	6.8	3.0–15.0	19.4	3.0–40.0
Rice	6.8	1.0–15.0	56.6	10.0–90.0
Rye	5.2	0.0–15.0	13.9	2.0–40.0
Wheat	5.8	1.0–20.0	20.0	0.0–40.0

[a]From Bridges, 1992.

WEED CONTROL METHODS

Weed control in small grains and flax includes using “clean seed,” crop rotation, good seedbed preparation, crop competition, and application of herbicides.

Clean Seed

Crop seed free of weed seeds is commonly referred to as “clean seed.” The importance of using clean seed in small-grain crops can hardly be overemphasized. Most commercial crop seed currently available is relatively free of weed seed, but this aspect should be confirmed by checking the label (also discussed in Chapter 3).

Crop Rotation

When a small grain or flax is continuously grown in the same field year after year, weeds that grow during the same season tend to adapt to the crop-management programs and flourish. Their populations increase with time as a result of an increasing seed bank and/or vegetative propagules when management is poor.

Small grains and flax can usually be grown for several years on “new land” before weeds become a serious problem. However, as with all continuous monoculture systems over time, it was noted with this practice that weeds became more of a problem, and farmers were forced to use a wider array of effective methods of weed control.

Crop rotation is a practical weed control practice, as small grains and flax can be rotated with a number of crops. Crop rotation achieves variations in cultural practices that break the monoculture cycle favoring the dominance of certain weed species. In some areas, the rotation of cotton, corn, and small grain with lespedeza illustrates this point. The variations in time of seedbed preparation, time of cultivation, and periods of growth of these four crops are detrimental to certain weed species. Herbicides can be combined with crop rotation to broaden the spectrum of weeds controlled. These crops allow the use of many different herbicides that control different weed species. Consequently, a more effective weed control program can be developed by utilizing both crop rotation and herbicide rotation in a concerted effort rather than using either type of rotation alone.

Seedbed Preparation

Weed control is one of the principal purposes of seedbed preparation. Because small grains cannot be effectively cultivated after sowing, the importance of controlling weeds before sowing is obvious. A presowing cultivation can eliminate most annual weed seedlings that appear before sowing and often reduces early competition from perennial weeds. Summer fallow can reduce weed seed production and the growth of perennial weeds.

Crop Competition

Dense stands of fast-growing small grains provide considerable competition to weeds (Figure 19-1). Maximal crop competition can be obtained by using high seeding rates,

Figure 19-1. Competition from the crop crowds out many weeds. (North Carolina State University.)

well-adapted varieties, and proper planting dates. Adequate soil moisture and fertility may also increase crop competition. Under less favorable conditions, weeds can flourish and not only reduce yields but also produce seeds and vegetative propagules that increase the weed problems for subsequent years.

Flax is only partially effective in competing with weeds, because its initial growth is slow and the leaf surface area is small. Nevertheless, thick stands of flax are somewhat helpful for weed control.

Chemical Control

Even with the use of clean seed, crop rotation, seedbed preparation, and crop competition, weeds can often be a serious problem in small grains and flax. Some weeds thrive under the same management practices as these crops do.

Crop competition is often increased with the use of herbicides, as effective herbicides control weed growth while causing little or no reduction in the growth of the crop plant. This gives the crop plant a competitive advantage over the weeds.

The performance of herbicides often varies under different climatic and edaphic conditions. Therefore, their use may vary or be restricted in specific geographic areas; see labels. It is also advisable to consult with local public authorities, agricultural consultants, and/or company field representatives about their proper use.

Barley, Oats, Wheat, and Rye Many herbicides are registered for use in these crops, more in barley and wheat than in oats and in rye (Table 19-2). Each of these herbicides has specific uses in regard to time and method of application and type of weeds controlled. For example, paraquat can be applied preplant or preemergence to control most emerged weeds. Triallate can be applied preplant or preemergence as a soil-incorporation treatment for wild oat control. 2,4-D can be applied postemergence to control many broadleaf weeds, and difenzoquat can be applied postemergence to control wild oats in barley and wheat. Most postemergence herbicides used in these crops must be applied at specific stages of crop and weed growth to obtain optimal selectivity (Figure 19-2). Combinations of certain herbicides can also be used to broaden the spectrum of weeds controlled. A detailed discussion of all of these herbicides, the combinations, and their uses is not possible because of space limitations; however, 2,4-D and related compounds are covered because of their wide use and importance.

2,4-D This herbicide is widely used in small grains because it is relatively inexpensive and controls a broad spectrum of broadleaf weeds. In 1997, 2,4-D was applied to more than 35% of the wheat acreage in the United States. It has been said that more acres of small grain are treated with 2,4-D than any other crop with any other herbicide.

Good weed control without crop injury usually depends on proper timing in the application of herbicides. This is particularly true for 2,4-D in small grains.

The time of *greatest susceptibility* of small grains to 2,4-D is during periods of rapid growth, which are at the zero- to four-leaf stage and the boot stage (or jointing

TABLE 19-2. Small Grains Herbicides[a]

Wheat and Barley				
No-till Preplant	Postemergence[b]	Postemergence (continued)	Postemergence (continued)	Preharvest
Paraquat	Tralkoxydim	Bromoxynil + MCPA	Propanil	Glyphosate
Glyphosate	Metsulfuron	Bromoxynil + 2,4-D	Fluroxypyr	Glyphosate + 2,4-D
Sulfosate	Metsulfuron + 2,4-D	Fenoxaprop + MCPA +	Fluroxypyr + MCPA	2,4-D
Dicamba	Metsulfuron + MCPA	thifensulfuron + tribenuron	Fluroxypyr + 2,4-D	2,4-D + dicamba
Dicamba +	Triasulfuron	Clopyralid + 2,4-D	Fenoxaprop + 2,4-D +	Metsulfuron + 2,4-D
glyphosate	Imazamethabenz	Clopyralid + MCPA	MCPA	
Pelargonic acid	Difenzoquat	Clopyralid	Picloram + 2,4-D	
	Dicamba	Fenoxaprop + MCPA	2,4-D	
Preemergence	Dicamba + metsulfuron	Fenoxaprop + 2,4-D	MCPA	
	Trifensulfuron + tribenuron	Tribenuron	Bromoxynil	
Triallate	Trifensulfuron + tribenuron	Chlorsulfuron + metsulfuron	Dicamba	
Trifluralin	+ metribuzin	Chlorsulfuron + 2,4-D	Chlorsulfuron	
Triallate + trifluralin	Triasulfuron + dicamba	Diclofop-methyl	Chlorsulfuron +	
	Bromoxynil	Prosulfuron	metsulfuron + metribuzin	
		Metribuzin	Carfentrazone	
			Flucarbazone	

TABLE 19-2. Continued

Oat Herbicides[a]			
Preplant	Preemergence	Postemergence	Postemergence (continued)
Paraquat	None	Dicamba	Fluoxypyr
Glyphosate		Dicamba + MCPA	Fluoxypyr + MCPA
		Bromoxynil	Picloram + 2,4-D
		Bromoxynil + MCPA	Picloram + MCPA
		Bromoxynil + 2,4-D	2,4-D
		Trifensulfuron + tribenuron	MCPA
		Prosulfuron	Carfentrazone
			Chlorsulfuron

Rye		
Preplant	Preemergence	Postemergence
Paraquat	None	Prosulfuron
Glyphosate		2,4-D
		Bromoxynil
		MCPA

[a]Always read and follow all instructions on the herbicide label.
[b]Durum—no picloram, no fenoxaprop, or fenoxaprop combinations with MCPA, MCPA + 2,4-D, or MCPA + trifensulfuron + tribenuron.

Figure 19-2. Blue mustard control in wheat with chlorsulfuron. Untreated area in center of picture. (A. Appleby, Oregon State University.)

stage) through flowering (Figure 19-3). Applications during the zero- to four-leaf stage will usually cause malformations of the head and "onion" leaves and a general stunting of the plant, which leads to reduced yields (Figure 19-4). Applications from the beginning of the boot stage through flowering usually reduce yields. During the boot stage the internodes elongate rapidly. Injury during these susceptible periods is associated with high meristematic activity (active sinks), resulting in increased symplastic translocation of the herbicide to these sites.

Periods of *least susceptibility* of small grains to 2,4-D are at the four-leaf to boot stage and the soft-dough-grain stage to maturity (Figure 19-3). The most desirable time to apply 2,4-D is at the four-leaf stage to just before the boot stage, including the fully tillered stage. At this time, the grain is relatively tolerant; the weeds are usually small and easily controlled. Such small weeds have not yet caused any serious competitive damage to the crop, and ground-spray equipment causes only slight mechanical damage to the grain. Although grain is very tolerant to phenoxy herbicides from the soft-dough-grain stage to maturity, treatment at this time is not usually recommended because weed competition has already had its major effect and ground-spray equipment can damage the grain. Late application can also result in possible residues of 2,4-D in the harvested grain. In general, wheat varieties are most tolerant to 2,4-D, with barley having intermediate tolerance and oats being the least tolerant.

MCPA and *dicamba* are other auxin-type herbicides used as postemergence treatments to control broadleaf weeds in small grains. MCPA is similar to 2,4-D in both chemical structure and action. It is less injurious to small grains, especially oats, but is not as active on most broadleaf weeds. However, it is more effective on a few

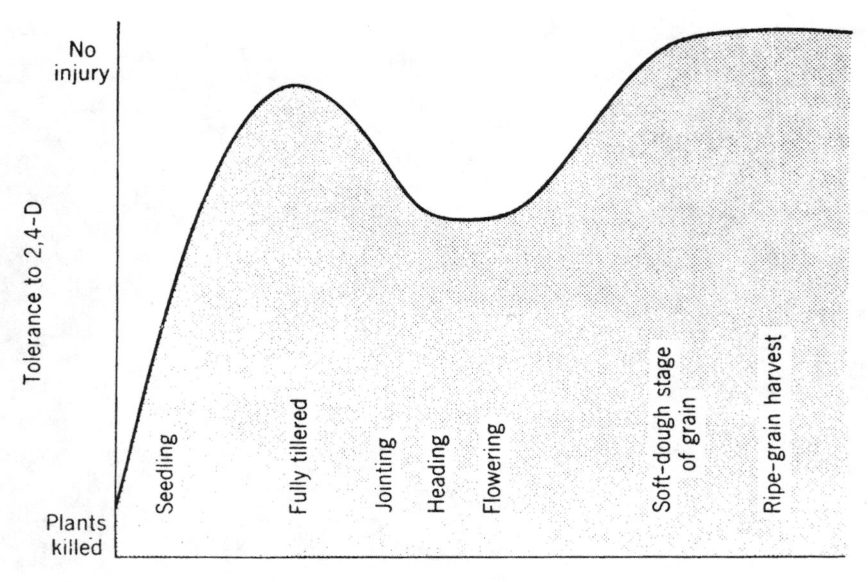

Figure 19-3. Stages of small-grain growth and the degree of tolerance to 2,4-D. (North Carolina State University.)

Figure 19-4. Wheat injured by premature treatment with 2,4-D. The wheat was about 3 inches tall when treated, and 60 to 70% of the heads were abnormal. (J. B. Harrington, University of Saskatchewan.)

weeds (e.g., hempnettle and Canada thistle) and it can be applied somewhat earlier than 2,4-D without causing injury to the small grain. MCPA usually costs more than 2,4-D in the United States. In contrast, dicamba is more injurious than 2,4-D to small grains but controls several broadleaf weeds better than 2,4-D (e.g., Canada thistle, chickweeds, field bindweed, and wild buckwheat).

Wild Oat Wild oat is a serious problem weed in spring wheat, spring barley, spring oats, and flax. It is less of a problem in winter cereals. Yield losses in North America have been estimated as high as $1 billion per year. Crop yield losses are due to both allelopathy and competition. More than 50 different strains of wild oat give the weed a wide area of adaptation, and many of the seeds are dormant, making eradication impossible. Most seeds germinate with soil temperatures of 50 to 60°F, which results in emergence with the small grain crop seed.

Crop rotation and summer fallow provide only limited control. Delayed spring seeding, with effective spring cultivation just before seeding, is reasonably effective. However, delayed seeding may result in a serious drop in crop yield even with good wild oat control.

Several herbicides can be used to control wild oats in barley and wheat: diclofop, fenoxaprop, tralkoxydim, difenzoquat, glyphosate, imazamethabenz, triallate, and trifluralin. Glyphosate can also be used in oat crops and is applied as a preplant or preemergence treatment to emerged wild oats. Triallate is applied to the soil before or after seeding the crop and requires soil incorporation.

Trifluralin is applied to the soil preplant, and crop seed must be planted below the depth of soil incorporation. Diclofop can be used preplant soil-incorporated, preemergence, or postemergence. Difenzoquat and imazamethabenz are used postemergence to both grain and wild oats. Specific stages of growth of both crop and wild oats are essential for all postemergence treatments to obtain good control of wild oats without crop injury especially for diclofop and tralkoxydim. See labels for detailed information on rates, time and methods of application, and restrictions.

Tillage Substitutes Although small grains are grown in many areas of the United States, the major production area is the Great Plains. This is particularly true for wheat. The region is subject to high winds and severe thunderstorms, which cause considerable soil erosion if sufficient plant cover is not present (Wicks, 1986). In addition, rainfall for maximum crop production is frequently limited (Wicks, 1986). These factors require soil and moisture conservation practices for a sustainable agriculture. Such practices include stubble mulch, chemical fallow, ecofarming, and no-till (Wicks, 1986). Each system has been adapted to specific areas in the Great Plains.

Stubble mulch uses shallow tillage with V sweeps and a rod weeder to control weeds during the fallow year between wheat crops, leaving much of the previous wheat residue on the soil surface. *Chemical fallow* uses herbicides to control weeds during the fallow period. *Ecofarming* is a system of controlling weeds and managing crop residues throughout a crop rotation with herbicides and minimum tillage (e.g., winter wheat-sorghum or corn-fallow). *Low-till* or *no-till* practices are usually used where rainfall is sufficient for continuous winter wheat with no fallow year. Herbicides are applied as needed with minimal cultivation for planting and crop emergence. The objective of these practices is to reduce soil erosion caused by wind and rainfall and conserve soil moisture. Comprehensive coverage of this topic is available in *No-Tillage and Surface-Tillage Agriculture*, edited by Sprague and Tripplett (1986).

Rice Rice culture utilizes many of the same weed control practices common to wheat, barley, and oats. However, some of these are discussed further because of the unique cultural practices used for rice. Rice is grown under lowland (paddy flooded with water) or upland (dry land) conditions. *Lowland* rice indicates crops grown in fields that are inundated with water for at least part of the time between establishment and harvest, and *upland* rice indicates the cultivation of "rainfed" rice on well-drained nonirrigated fields. The cultural practices, weed species present, and control methods are somewhat different in these two systems. Upland rice is dibble, broadcast, or row seeded, and the most common weed control method is by hand or with the use of animals, with few purchased inputs such as herbicides. A wide variety of weed species, including many pantropical grasses (*Echinochloa colona*, *Elusine indica*, *Digitaria* spp., *Paspalum* spp., *Rottboellia cochinchinensis*) and broadleaf species (*Commelina* spp., *Ageratum conzyoides*, *Portulaca oleracea*, *Amaranthus* spp., *Euphorbia* spp.) infest upland rice (D. E. Johnson, National IPM Network Website).

Lowland rice is either seeded or transplanted, and fields must be relatively level to allow flooding to be used. Water management is a very important aspect of weed control in lowland rice (Bayer, 1984). Flooding prevents the germination and growth of many weeds, and the rice plant is more tolerant of the soil conditions and anaerobic environment in flooded fields. In direct-seeded fields flooding cannot be used until crop establishment, so postemergence weed control is essential. The most serious weeds in lowland rice include a variety of sedges, fringerushes, bulrushes, grasses (barnyardgrass, saramollagrass, sprangletop, red rice, knotgrass, itchgrass), and broadleaves (primrose, eclipta, gooseweed, *Euphorbia* spp., dayflower), which tend to be well adapted to the aquatic environment, grow rapidly, reproduce prolifically, and compete well with rice. Many of these weeds mimic rice (red rice and barnyardgrass species), especially at early growth stages, which results in difficulty of hand removal. Several species have developed resistance to widely used rice herbicides, as discussed later in this section.

The management of weeds in rice requires an integrated approach that includes good land preparation, good water management, judicious herbicide use, and a competitive crop. Tillage provides a good seedbed and is an effective method to control weeds, especially because tillage will expose underground stems and tubers of cattails, bulrush, sedges, and the like. If sufficient drying of these reproductive organs occurs on the soil surface in the spring, their densities in the rice field will be reduced.

The flooding of rice fields is the most effective cultural practice for weed control. A continuous water depth of 3 to 6 inches prevents the germination of most weed seeds and kills the majority of any emerged weed seedlings, while improving the effectiveness of rice herbicides and allowing crop competition to aid in weed management. Maintaining a water depth of 6 to 8 inches for 21 to 28 days after planting can provide partial control of barnyardgrass. In seeded beds, fields are flooded immediately after planting for approximately 48 hours to encourage uniform rice germination and emergence, which allows better rice competition against weeds. The permanent flood is then reapplied within 3 to 6 weeks of planting, depending on the rice growth rate, and must be carefully managed to avoid killing the rice plants with extended periods of total submergence. Rice can tolerate total submergence for approximately 48 hours.

The objective of herbicide use is to kill or stunt the weeds while allowing the rice to grow and gain a competitive advantage. Maintaining an effective height differential between the rice and the weeds allows flooding to control the weeds by keeping them submerged while the rice grows above the water. If flooding is practiced well, herbicide application can be minimized. Many of the herbicides used in rice are not used in other small grains (Table 19-3). Nonselective herbicides (e.g., glyphosate) are sometimes used before establishment to control especially difficult weeds such as perennials and red rice. These include herbicides primarily for grass control such as

TABLE 19-3. Rice and Flax Herbicides

	Rice Herbicides[a]		
Preplant	Postplant Preemergence		Postemergence
Molinate Pendimethalin	Molinate Thiobencarb Bensulfuron Triclopyr Fenoxaprop Pendimethalin Quinclorac		Propanil Triclopyr Fenoxaprop $CuSO_4$ Bentazon Acifluorfen Halosulfuron Carfentrazone 2,4-D MCPA Quinclorac
	Flax Herbicides[a]		
Preplant Incorporated	Postemergence		Desiccant
Trifluralin EPTC Propachlor	Bromoxynil MCPA Sethoxydim	Diclofop	Sodium chlorate

[a]Always read and follow instructions on the herbicide label.

fenoxaprop; herbicides for broadleaf and grass control such as propanil, molinate, butachlor, thiobencarb, and pendimethalin, and herbicides primarily for broadleaf and sedge weed control such as 2,4-D, MCPA, triclopyr, halosulfuron, bentazon, oxyfluorfen, and bensulfuron. Most herbicides used in rice are selective, and the time of application relative to the growth stage of the crop and the weeds, along with proper water management, is critical for crop safety and good weed control. Each herbicide has its own specific requirements, and the product label must be referred to prior to use.

With foliar-applied herbicides such as MCPA, bentazon, 2,4-D, and propanil, the weed foliage must be exposed to the herbicide spray, which may require lowering the water level. To allow time for herbicide absorption, the water level should not be raised over the weeds for 24 to 48 hours. Small submerged weeds not sprayed with the herbicide will not be controlled. Molinate can be applied as a preplant or preflood soil incorporation treatment, or either preemergence or postemergence postflood. Pendimethalin is used only in combination with propanil, as propanil controls certain emerged weeds and pendimethalin provides residual control of other weeds. Pesticide combinations or sequential applications may cause rice damage (e.g., propanil with a carbamate or organophosphate insecticide). Because glyphosate is nonselective, it must not contact the crop, and rice fields and levees must not be treated when the fields contain water. There are many other limitations and precautions for the use of rice herbicides; see labels.

Resistance to rice herbicides has occurred in red rice, several barnyardgrass biotypes, and some broadleaf weeds and sedges. Grassy-weed-resistant biotypes have been reported for propanil, molinate, thiobencarb, fenoxaprop, and quinclorac, with barnyardgrass biotypes showing the greatest diversity of resistant types. Resistance to bensulfuron has been reported in smallflower umbrella sedge, ricefield bulrush, California arrowhead, and redstem. Management of these resistant weeds, which is critical to prevent their spread, includes preventing their growth and reproduction, preventing movement of propagules to unifested fields, and the use of herbicides with different mechanisms of action for ensuring effective control and minimizing selection pressure.

Flax Flax weeds can be a serious problem in flax if left uncontrolled. Nonchemical methods of weed control in flax are essentially the same as in barley, oats, and wheat. Seed should be sown on relatively weed-free land, especially free of quackgrass. Use of postharvest tillage and herbicides in the previous season to suppress perennials and stimulate annual weed seed germination will reduce weeds in the flax crop. Some growers delay flax planting to allow the first flush of spring weeds that can be killed prior to planting. The herbicides used in flax are listed in Table 19-3. Bromoxynil and MCPA are postemergence herbicides used to control broadleaf weeds. Diclofop and sethoxydim are postemergence herbicides used to control wild oats and other annual grasses. Sethoxydim also controls certain perennial grasses. Propachlor is a preemergence herbicide used to control annual broadleaf and grass weeds, but not wild oats. EPTC and trifluralin are preplant soil-incorporated herbicides that control many grass and broadleaf weeds. EPTC is applied in the late fall before the ground freezes, and trifluralin is applied just before planting. EPTC controls wild oats, but trifluralin does not control this weed. Appropriate combinations of some of these herbicides are used to broaden the spectrum of weeds controlled. Limitations and precautions for the use of these flax herbicides are given on the labels.

LITERATURE CITED AND SUGGESTED READING

Bayer, D. E., J. E. Hill, and D. E. Seaman. 1984. Rice. *Principles of Weed Control in California*. Thomson Publications, Fresno, CA.

Bridges, D. C., ed. 1992. *Crop Losses Due to Weeds in the United States, 1992*. Weed Science Society of America, Lawrence, KS.

Crop Protection Reference, 17 ed. 2001. Chemical and Pharmaceutical Press, New York.

Donald, W. N., and E. F. Eastin. 1995. Weed management systems for grain crops. In *Handbook of Weed Management Systems*, ed. by A. E. Smith, Chapter 11. Marcel Dekker, New York.

Herbicide Handbook. 1994. Weed Science Society of America, Lawrence, KS.

Herbicide Handbook Supplement to 7th ed. Weed Science Society of America, Lawrence, KS.

Sprague, M. A., and G. B. Triplett, eds. 1986. *No-Tillage and Surface-Tillage Agriculture*. John Wiley & Sons, Inc., New York.

Wicks, G. A. 1986. Substitutes for tillage on the Great Plains. In *No-Tillage and Surface-Tillage Agriculture*, ed. by M. A. Sprague and G. B. Triplett, pp. 183–196. John Wiley & Sons, Inc., New York.

WEB SITES

Flax Council: *Weed Control in Flax*

http://www.flaxcouncil.ca

Search "Weed Control."

National IPM Network: *Weed Management in Small Holder Rice Production in the Tropics*. 1996. D. E. Johnson. In Radcliffe's IPM World Textbook Homepage. University of Minnesota, National IPM Network.

http://www.ipmworld.umn.edu/chapters/johnson.htm

North Carolina State University:

North Carolina Agricultural Chemicals Manual

http://ipmwww.ncsu.edu/

Select Pesticide Recommendations

University of Florida: *Weed Management in Small Grains*. 2001. J. A. Tredaway and B. J. Brecke. Publication SS-AGR-07. Cooperative Extension Service, Institute of Food and Agricultural Sciences.

http:edis.ifas.ufl.edu

Select "Pest Management," then select "Pest Management Guides," select "Weed Management in Field Crops and Pastures," select "Weed Management in Small Grains."

University of Florida: Weed Management in Rice. 2001. J. A. Dusky and J. A. Tredaway. Publication SS-AGR-10. Cooperative Extension Service, Institute of Food and Agricultural Sciences.

http://edis.ifas.ufl.edu

Select "Pest Management Guides," select "Weed Management Guides," select "Weed Management in Field Crops and Pastures Grasses," select "Weed Management in Rice."

University of California: *Rice Integrated Weed Management*. 1999. J. Hill, M. Carriere, A. Fischer, D. E. Bayer, J. F. Williams and S. C. Scardaci. Publication 3339. University of California Davis, Agriculture and Natural Resources Statewide Integrated Pest Management Project.

University of California: *Small Grains Integrated Weed Management*. 1999. J. P. Orr, W. M. Canevari, R. N. Vargas, S. D. Wright, C. E. Bell, H. Carlson, L. W. Mitich, M. J. Smith, J. F. Williams, T. E. Kearey, D. W. Cudney, L. Jackson, and J. E. Hill. Publication 3339. University of California Davis, Agriculture and Natural Resources Statewide Integrated Pest Management Project.

http://www.ipm.ucdavis.edu

Select "How to Manage Pests," select "Pests of Agriculture Crops" and select "Rice" or "Small Grains."

University of Minnesota: *Cultural and Chemical Weed Control Guide in Field Crops*

http://www.extension.umn.edu

Select "Crops," then select "Weed Control," select "Cultural and Chemical Weed Control in Field Crops."

AgChemical Label Information, Crop Data Management Systems, Inc., Marysville, CA

http://www.cdms.net/manuf/manuf.asp

For chemical use, see the manufacturer's or supplier's label and follow these directions. Also see the Preface.

20 Field Crops Grown in Rows

Several field crops are grown in rows to facilitate their culture, in contrast to other field crops, such as small grains, flax, and small-seeded legumes, that are not grown in rows. The nature of the individual crop plant usually determines whether it is grown in rows. In general, crops grown in rows have larger seeds, develop into larger and more vigorous plants, and require more space per plant. These characteristics permitted earlier farmers to utilize "horse-hoeing" in the development of the row crop concept. However, with the replacement of the horse with machinery and other advances in agriculture, the rows have gradually become closer and yields have increased. Some of these advances include cultural implements, new varieties, and weed control practices, especially the use of herbicides.

Weed control in most annual crops can be divided into early-season and late-season phases. Early-season weeds usually have a greater effect on crop yields than late-season weeds. Late-season weeds may make harvesting difficult, reduce the quality of the crop (e.g., grass in cotton), and reinfest the soil with weed seeds.

WEED CONTROL METHODS

Weed control in field crops grown in rows involves an integrated approach utilizing the methods discussed in Chapter 3. These include mechanical means, competition, crop rotation, and chemical methods. Biological control by predators and diseases has not been perfected for weed control in cultivated crops, other than by the development of certain disease-resistant cultivars and crops genetically modified for insect resistance (i.e., Bt corn and cotton). Weed control is important, as shown by the crop yield losses with the varying levels of weed control efficiency (Table 20-1).

Monitoring

When choosing the most effective weed management practices involving herbicide choice and rotational crops, growers should survey each field and adjacent fields for weeds present, past cropping history, and soil type. Such information is vital to establishing the best management practices for weed control. Farmers should also keep records of weeds within fields and how they change within crops and over seasons with the use of various weed management practices.

TABLE 20-1. Estimated Annual Average Yield Reduction Due to Weeds in the United States Grown with Best Management Practices (BMP) Including (+) Herbicides and Excluding (–) Herbicides

	Yield Reduction			
	BMP (+) Herbicides		BMP (–) Herbicides	
Crop	Mean %	Range %	Mean %	Range %
Corn	6.2	1.0–25	30.2	10.0–60.0
Cotton	8.0	2.0–20.0	51.7	20.0–70.0
Grain Sorghum	7.3	5.0–15.0	34.1	15.0–90.0
Peanuts	7.1	4.0–22.0	54.0	30.0–80.0
Soybeans	7.8	2.0–20.0	40.0	15.0–90.0
Sugar Beets	6.7	1.0–10.0	30.5	15.0–70.0
Sugarcane	9.2	1.0–23.0	37.0	13.0–45.0
Tobacco	5.9	0–15.0	16.5	0.0–40.0

From Bridges, 1992.

Mechanical Means

Mechanical weed control in row crops primarily involves cultivation. Some crops may benefit from one early cultivation to loosen the soil if it becomes hard and packed when dry. On other soils, cultivation is of little value if the weeds are controlled by other means.

Cultivation has some disadvantages. Some crops grow slowly, and weeds may get ahead of the crop before the field can be cultivated. Cultivation frequently fails to control weeds in the crop row, may injure roots, and may result in slower crop growth and, ultimately, yield reductions. Heavy weed growth may develop after the last cultivation, and repeated cultivation, especially in wet soils, injures the physical condition of the soil. Cultivation also consumes costly fuel, and large tractors are an expensive investment.

Competition

Providing conditions favoring crop competition is paramount for good weed control in field crops grown in rows. These conditions include all cultural and environmental factors that provide rapid seed germination, vigorous seedling growth, and subsequent maximum crop growth. For most field crops grown in rows, the crop canopy covers the row middles within 7 to 8 weeks, providing major competition to late-emerging weeds. A herbicide that keeps the crop weed-free until crop canopy closure provides an integrated weed control approach. In the absence of good crop competition, other integrated methods of weed control often fail to reach their optimum effectiveness. Dense, vigorous crop stands provide maximum competition to weeds (Figure 19-1).

Crop Rotation

Crop rotation is an essential component of any integrated weed control program. A row crop can be rotated with a number of crops that allow variations in cultural practices. Such rotations break the monoculture cycle, which favors the dominance of certain weed species, and reduces selection pressure for herbicide-resistant weed development by allowing herbicide rotation. Variations in time of seedbed preparation, time of cultivation, and periods of growth of various crops are detrimental to certain weed species. Herbicide rotation combined with crop rotation broadens the spectrum of weeds controlled, allowing the use of herbicides with a variety of mechanisms of action. Consequently, a more effective weed control program can be developed by utilizing both crop rotation and herbicide rotation in a concerted action, rather than using either type of rotation alone.

Chemical Weed Control

Chemical weed control in field crops grown in rows should be integrated with other weed control methods for optimum effectiveness, as presented in Chapter 3. The following chemical weed control methods have proven effective in various parts of the United States and Canada, but not necessarily in all areas. Herbicides are applied at different times in row crops: (1) preplant—before planting, (2) preemergence—after planting but before crop emergence, (3) postemergence—after emergence of the crop and/or weeds.

Preplant treatments can be applied to emerged weeds or the soil surface, or incorporated into the soil. Preplant treatments to emerged weeds with nonselective, nonpersistent herbicides (e.g., glyphosate, pelargonic acid, sulfosate, or paraquat) can be used in most field crops grown in rows. Soil-surface applications of preemergence-type herbicides must be followed by rain or sprinkler irrigation to be most effective. Overhead water and soil incorporation puts the herbicide into the area of the soil where the weed seeds germinate. In general, soil incorporation is also required for certain volatile herbicides such as the carbamothioates (e.g., EPTC). Crop seeds must be tolerant to the herbicide, or, with some herbicides, the seeds can be placed below the herbicide-treated layer (e.g., cotton with trifluralin) (Figure 6-1).

With soil incorporation, the herbicide is applied to the soil surface and incorporated to a depth of 2 to 3 inches with appropriate equipment. Incorporation equipment includes power-driven rotary hoes with L-shaped knives, ordinary rotary hoes, discs, sweeps, drag-harrows, and other similar soil tillage implements (Chapter 7). The equipment must break up large clods and mix the herbicide evenly throughout the treated soil profile. Overhead water is not needed with soil incorporation; therefore, the herbicide seldom fails to be effective. Other preplant herbicide treatments are discussed shortly, in the section on minimum and no-till.

Preemergence treatments are applied to the soil surface after planting but before the crop emerges and, depending on the herbicide, can be made before or after weed emergence. Rain or sprinkle irrigation is required to move the herbicide into the soil. The discussion related to surface applications of herbicides and overhead water in preplant treatments also applies to preemergence treatments.

Postemergence treatments are made any time after the crop emerges. Highly selective herbicides may be applied as a broadcast spray over the top of the crop and the weeds. The crop is not injured, and the susceptible weeds are controlled. Moderately selective or nonselective herbicides can also be applied postemergence to the crop with the use of methods that apply the chemical to the weeds but minimize or essentially avoid application of the chemical to the crop. Moderately selective herbicides can also be applied as a directed spray that minimizes crop exposure, (e.g., 2,4-D) to small broadleaf weeds in a taller corn crop. Drop nozzles are often used for this type of application. Nonselective herbicides can be applied with a recirculating sprayer or rope-wick applicator that essentially avoids crop exposure (e.g., glyphosate applied to johnsongrass that is taller than cotton [Chapter 7]), or they can be applied to herbicide-resistant crops (Chapter 18).

Tillage Systems

Various tillage systems are used in row crops and other crops for weed control and optimum crop growth. Choices depend on the farmer's preference, the crop grown, the area of the country, and soil conservation regulations.

Conventional tillage generally involves mechanical disturbance of the soil by field cultivation equipment in the fall, in the spring, and during the season. Often referred to as "clean cultivation," these techniques leave little plant or crop residue on the soil surface. All types of herbicide application timings are possible in these cultivation systems.

Minimum-Till and No-Till The concept of minimum-till or no-till evolved after the development of herbicides that could replace certain cultivation practices. Many of the sequential tillage operations were performed primarily for weed control. Minimum-till and no-till methods result in reduced need for cultivation for weed control and the retention of plant residues as a mulch cover on the soil surface. This reduces soil erosion from wind and rain, increases water conservation, and improves mineral nutrition (Sprague and Triplett, 1986). Other advantages of these methods include more efficient use of labor, fuel, land, fertilizers, pest control, and water. The preplant incorporated herbicides used in conventional systems are not used in no-till and infrequently used in reduced-till.

Ridge-Till A ridge-till system (Chapter 3) generally leaves the soil undisturbed after fall harvest and over the winter. Crops are planted on ridges formed the previous season, which are rebuilt annually by cultivation in the spring. During planting, the ridge is cleared of crop residue and most weed seed; often a preemergence herbicide is applied to the ridge or a timely cultivation occurs after planting to eliminate emerged weeds. Sometimes an early preplant herbicide is applied in early spring to eliminate winter annual and early summer annual weeds.

Ecofarming

Ecofarming is a technique used in grain-growing regions to assist in soil and water conservation. This system is based on using winter wheat stubble and good weed control to manage soil moisture and to prevent weed seed production. The general technique is to apply herbicides after grain harvest to prevent weed growth and seed production, to leave the field in fallow during the winter, and to no-till plant the following spring, using herbicides early preplant or at planting (see Guide for Weed Management in Nebraska, University of Nebraska Web Site).

CORN

Modern weed control in corn utilizes an integrated program that includes field selection, crop rotation, crop competition, cultivation, and herbicides. Weeds reduce yields by competing with the crop for essential growth factors and they can interfere with harvest and produce seed that is reinvested in the soil seed bank. Yields are usually reduced when weeds are not controlled before the corn is 6 to 8 inches high.

In the selection of cultivation methods, the growth pattern of the crop (e.g., root distribution) must be considered (Figure 20-1). Because corn roots are relatively shallow, corn should be cultivated only deeply enough to remove or cover the weeds, so as to minimize root pruning. Rotary weeders, finger weeders, and harrows are used for cultivation when the corn is small. When it is 3 to 4 inches tall, shovel or sweep

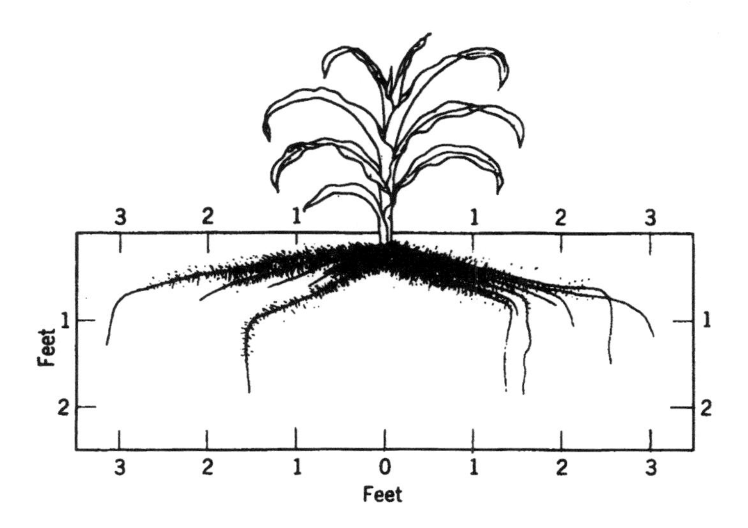

Figure 20-1. Corn root distribution. Cultivation should be shallow to avoid root damage, especially after the plant is 15 inches tall. (Adapted from Nebraska Research Bulletin No. 161, 1949.)

cultivation can be started to remove weeds from the interrow area and throw soil into the crop row to bury small weeds. Cultivation is repeated as often as needed until the corn is 20 to 24 inches tall. Such cultivation is usually performed three or four times when cultivation is the only method of weed control used. The last cultivation is often referred to as a "Layby" cultivation.

Much of a crop's root system will be pruned off by a cultivation within 6 inches of the stem to a depth of 6 inches (Figure 20-1). Serious crop wilting may follow a deep cultivation in dry weather. Late-season deep cultivation will nearly always reduce corn yields; therefore, shallow sweeps should be used if a late cultivation is necessary. Roots on the cultivator shank indicate that the cultivation is too deep.

The use of herbicides reduces the number of postplant cultivations required. A program of reduced cultivation with the use of herbicides can provide very effective, season-long weed control in corn.

Chemical Weed Control in Corn

The knowledge of weeds within a field is important in determining which herbicides to choose. Anticipation of the potential weed problems, based on observation and information gained from weed maps, will assist in choosing the appropriate herbicides.

Since 2,4-D was introduced in the mid-1940s to control broadleaf weeds in corn, many other herbicides have been developed for corn (Table 20-2). In fact, more herbicides have been developed for corn and soybeans than for any other crops. Most corn herbicides are used to control annual weeds; however, some also control or suppress the growth of perennial weeds. Combinations of two or more herbicides are often used as a tank mix or in sequential treatments to extend the period of control and/or broaden the spectrum of weeds controlled. In addition, corn cultivars are available that are tolerant to certain herbicides that provide broad-spectrum weed control (Table 20-2). Herbicides should be selected primarily on the basis of weed species present, stage of crop growth, and succeeding crop rotation. Certain limitations relative to soil type, geographic location, and other factors must be considered in the use of many herbicides; see labels. Depending on their specific properties, corn herbicides can be applied preplant, preemergence, or postemergence to the crop. Many, but not all, of the herbicides listed in Table 20-2 can also be used in non-field corn crops (e.g., silage, sweet corn, popcorn, and seed corn).

Corn has both susceptible and tolerant stages of development relative to 2,4-D. This herbicide is applied as an overall postemergence spray, beginning at the three-leaf stage until the corn is just less than 10 inches tall. Thereafter, until tassel initiation, directed sprays with drop nozzles are used. Corn is most susceptible to 2,4-D injury during or after tasseling to the dough stage, and it should not be applied during this period (Figure 20-2).

TABLE 20-2. Field Corn Herbicides[a]

No-Till or Minimum-Till (all preemergence to crop)	Preplant Incorporated	Preplant or Preemergence
Atrazine	EPTC + acetochlor	Acetachor
Atrazine + alachlor	EPTC + atrazine	Acetachlor + atrazine
Atrazine + glyphosate	Butylate	Acetachlor + dicamba
Atrazine + metolachlor		Acetachlor + pendimathalin
Atrazine + paraquat		Alachlor
Atrazine + acetochlor		Alachlor + atrazine
Atrazine + dicamba		Alachlor + dicamba
Atrazine + dimethenamid		Alachlor + atrazine + dicamba
Atrazine + flufenacet + metribuzin		Atrazine
Atrazine + pendimethalin		Atrazine + simazine
Atrazine + 2,4-D		Atrazine + s-metolachlor
Glyphosate		Atrazine + pendimethalin
Glyphosate + alachlor		Atrazine + dimethenamid
Glyphosate + alachlor + atrazine		Atrazine + nicosulfuron + thifensulfuron
Glyphosate + dicamba		Dicamba
Glyphosate + 2,4-D		Dicamba + dimethenamid
Glyphosate + acetochlor		Dicamba + metolachlor
Glyphosate + atrazine		Dicamba + pendimethalin
Glyphosate + atrazine + metolachlor		Dicamba + atrazine
Glyphosate + bromoxynil		Isoxaflutole
Glyphosate + diphenamid		Dimethenamid
Glyphosate + flumetsulam + clopyralid		Dimethenamid + pendimethalin
Glyphosate + flufenacet + metribuzin		Flufenacet + metribuzin
Glyphosate + metolachlor		Flufenacet + metribuzin + atrazine
Glyphosate + acetochlor + atrazine		Flufenacet + metribuzin + dicamba
Sulfosate		Flufenacet + metribuzin + flumetsulam
Sulfosate + alachlor		Flufenacet + metribuzin + flumetsulam + clopyralid

TABLE 20-2. Field Corn Herbicides, Continued

No-Till or Minimum-Till (all preemergence to crop)	Preplant Incorporated	Preplant or Preemergence
Sulfosate + acetochlor		Flufenacet + metribuzin + pendimethalin
Sulfosate + atrazine		Flumetsulam
Sulfosate + acetochlor + atrazine		Flumetsulam + clopyralid
Sulfosate + dicamba		Flumetsulam + s-metolachlor
Sulfosate + diphenamid		s-Metolachlor
Sulfosate + flufenacet + metribuzin		s-Metolachlor + atrazine
Sulfosate + metolachlor		s-Metolachlor + dicamba
Sulfosate + pendimethalin		Pendimethalin
Sulfosate + 2,4-D		Cyanazine
Paraquat		Cyanazine + atrazine
Paraquat + atrazine		Mesotrione
Paraquat + atrazine + alachlor		Mesotrione + labelled mixtures
Paraquat + atrazine + metolachlor		
Paraquat + dicamba		
Paraquat + flumetsulam + clopyralid		
Paraquat + flufenacet + metribuzin		
Paraquat + pendimethalin		
Paraquat + 2,4-D		
Paraquat + acetochlor		
Pelargonic acid		

TABLE 20-2. Field Corn Herbicides, Continued

Postemergence		
Acetachlor + atrazine	Dicamba + s-metolachlor	Metribuzin
Alachlor + atrazine	Dicamba + nicosulfuron	Metribuzin + atrazine
Atrazine	Dicamba + pendimethalin	Metribuzin + bentazon
Atrazine + dimethenamide-p	Dicamba + primsulfuron	Metribuzin + bentazon + atrazine
Atrazine + s-metolachlor	Dicamba + 2,4-D	Metribuzin + bromoxynil
Atrazine + nicosulfuron	Dicamba + diflufenzopyr + nicosulfuron	Metribuzin + dicamba
Atrazine + primsulfuron	Dicamba + diflufenzopyr	Metribuzin + dicamba + atrazine
Atrazine + pendimethalin	Flumetsulam + clopyralid	Metribuzin + 2,4-D
Bentazon	Flumetsulam + clopyralid + 2,4-D	Nicosulfuren
Bentazon + atrazine	Flumiclorac	Nicosulfuren + atrazine
Bromoxynil	Flumiclorac + atrazine	Nicosulfuren + bromoxynil + atrazine
Bromoxynil + atrazine	Flumiclorac + bromoxynil	Nicosulfuren + dicamba + atrazine
Bromoxynil + clopyralid	Flumiclorac + dicamba	Nicosulfuren + dicamba + primisulfuron
Bromoxynil + dicamba	Flumiclorac + dicamba + atrazine	Paraquat
Bromoxynil + nicosulfuron	Flumiclorac + nicosulfuron	Primisulfuron
Bromoxynil + primsulfuron	Flumiclorac + primsulfuron	Primisulfuron + bromoxynil + atrazine
Bromoxynil + 2,4-D	Flumiclorac + primsulfuron + prosulfuron	Primisulfuron + dicamba + atrazine
Carfentrazone	Flumiclorac + 2,4-D	Primisulfuron + nicosulfuron
Clopyralid	Halosulfuron	Primisulfuron + 2,4-D
Dicamba	Halosulfuron + atrazine	Prosulfuron + primisulfuron
Dicamba + acetochlor	Halosulfuron + bromoxynil	Prosulfuron + primisulfuron + atrazine
Dicamba + alachlor	Halosulfuron + bromoxynil + atrazine	Prosulfuron + primisulfuron + bromoxynil
Dicamba + atrazine	Halosulfuron + dicamba	Prosulfuron + primisulfuron + bromoxynil + atrazine
Dicamba + atrazine + bentazon	Halosulfuron + dicamba + atrazine	Prosulfuron + primisulfuron + dicamba
Dicamba + clopyralid + flumetsulam	Halosulfuron + nicosulfuron	Prosulfuron + primisulfuron + dicamba + atrazine
Dicamba + clopyralid	Halosulfuron + primisulfuron	
Dicamba + dimethenamid-p	Halosulfuron + 2,4-D	

TABLE 20-2. Field Corn Herbicides, Continued

Postemergence

Prosulfuron + primisulfuron + nicosulfuron
Prosulfuron + primisulfuron + 2,4-D
Prosulfuron + primisulfuron + pyridate
Pyridate
Pyridate + atrazine
Pyridate + dicamba
Pyridate + nicosulfuron
Pyridate + primisulfuron
Pyridate + 2,4-D
Rimsulfuron + nicosulfuron + atrazine
Rimsulfuron + nicosulfuron + atrazine + dicamba
Rimsulfuron + nicosulfuron + atrazine + flumetsulam + clopyralid
Rimsulfuron + nicosulfuron + atrazine + nicosulfuron
Rimsulfuron + nicosulfuron + atrazine + pyridate
Rimsulfuron + nicosulfuron + flumetsulam + clopyralid
Rimsulfuron + nicosulfuron + dicamba
Rimsulfuron + nicosulfuron + nicosulfuron
Rimsulfuron + thifensulfuron
Rimsulfuron + thifensulfuron + atrazine
Rimsulfuron + thifensulfuron + dicamba
Rimsulfuron + thifensulfuron + dicamba + atrazine
2,4-D
Mesotrione
Mesotrione + labelled mixtures

Chesterfield Corn Hybrids Only

Imazethapyr
Imazethapyr + alachlor
Imazethapyr + atrazine
Imazethapyr + bromoxynil
Imazethapyr + dicamba
Imazethapyr + dimethenamid-p
Imazethapyr + s-metolachlor
Imazethapyr + pendimethalin
Imazethapyr + imazapyr

Roundup Ready Corn Hybrids Only

Glyphosate
Glyphosate + acetochlor
Glyphosate + alachlor
Glyphosate + atrazine
Glyphosate + halosulfuron

Glufosinate-Tolerant Corn Hybrids Only

Glufosinate
Glufosinate + atrazine
Glufosinate + bromoxynil
Glufosinate + flumetsulam + clopyralid
Glufosinate + halosulfuron
Glufosinate + primisulfuron

Sethoxydim-Resistant Corn Hybrids Only

Sethoxydim
Sethoxydim + atrazine
Sethoxydim + bentazon
Sethoxydim + 2,4-D
Sethoxydim + dicamba

Lay-by

s-Metolachlor
Pendimethalin
Trifluralin
Ametryn
Linuron
Dimethenamid

Directed Postemergence

Ametryn
Paraquat

Preharvest

Glyphosate
2,4-D
Sodium chlorate

Spot-Treatment

Glyphosate
Sulfosate

Postharvest

2,4-D + dicamba

[a]Always read and follow all instructions on the herbicide label.

Figure 20-2. Injury to corn caused by 2,4-D applied over the top after corn was more than 12 inches tall.

GRAIN SORGHUM

Nonchemical weed control in grain sorghum (milo) is similar to that used in corn, including an integrated program of field selection, crop rotation, and cultivation. These methods are further integrated with chemical methods.

Chemical Weed Control in Grain Sorghum

In general, grain sorghum is less tolerant to most herbicides than corn. The development of herbicides to give season-long weed control with minimum tillage has allowed the rows to be planted closer together, increasing their competitiveness to weeds. The herbicides used in grain sorghum are listed in Table 20-3.

Like corn, grain sorghum has susceptible and tolerant stages of development relative to 2,4-D (Figure 20-3). It is most tolerant at the 4- to 12-inch stage, and treatment with 2,4-D at the seedling to four-leaf stage and after the 12-inch stage (head starts to develop rapidly) to the soft-dough stage is not recommended, inasmuch as significant injury can occur. When the grain is half-formed, grain sorghum again becomes tolerant; however, by this time weeds have already produced most of their adverse effects.

SOYBEANS

A well-planned integrated weed control program involving field analysis and a combination of cultural and chemical practices is the best approach to soybean weed management. This program depends on the weeds present, the soil, and the tillage system, as well as the crop rotation and row-spacing practices used.

TABLE 20-3. Sorghum Herbicides[a]

No-Till	Conventional Tillage	Postemergence		Lay-by	Harvest Aid
		Over the Top	*Directed*		
Glyphosate	Atrazine	Atrazine	Linuron	Pendimethalin	Diquat
Paraquat	Alachlor	2,4-D	Paraquat	Trifluralin	Glyphosate
2,4-D + dicamba	Dimethenamid	2,4-D + atrazine	Linuron + 2,4-D		Sodium chlorate
Atrazine	s-Metolachlor	Dicamba			
Atrazine + paraquat	Alachlor + atrazine	Bentazon			
Atrazine + glyphosate + alachlor	Dimethenamid + atrazine	Bentazon + atrazine			
Atrazine + glyphosate + s-metolachlor	s-Metolachlor + atrazine	Bromoxynil + atrazine			
Atrazine + s-metolachlor	Propachlor + atrazine	Dicamba + atrazine			
Atrazine + dimethenamid-p	Propachlor	Prosulfuron			
Pelargonic acid	Propazine	Bromoxynil			
		Carfentrazone			
		Quinclorac			
		Halosulfuron			

[a]Always read and follow all instructions on the herbicide label.

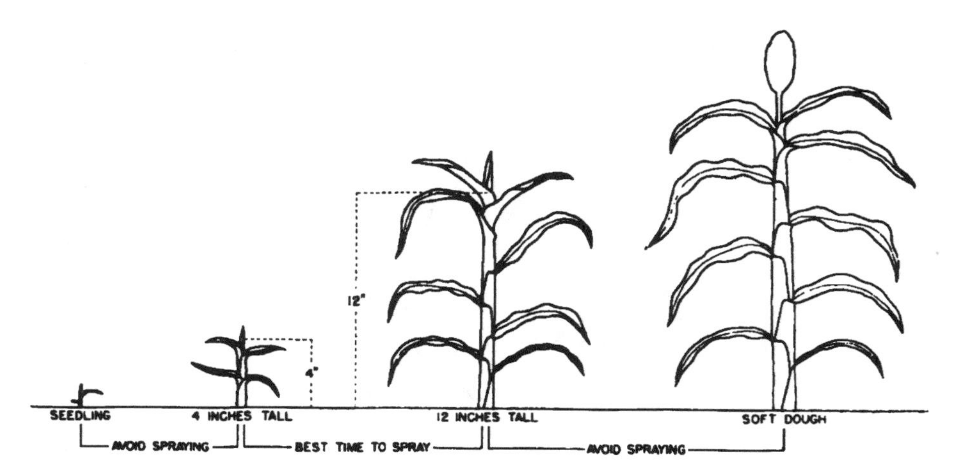

Figure 20-3. Sorghum tolerance to 2,4-D at various stages of development. (W.M. Phillips, Kansas Agricultural Experiment Station and USDA.)

Weeds usually germinate and emerge with the soybeans, so it is important to effect early-season weed management to prevent crop loss. Any shading from weeds or reduction in moisture due to competition can severely reduce soybean growth; this can be minimized by obtaining good weed control during the first 3 to 4 weeks after planting. Many growers plant soybeans in narrow rows and encourage vigorous early-season growth. Thorough seedbed preparation, use of vigorous cultivars, and soil moisture management allow the soybean canopy to close quickly, which shades small weed seedlings, reducing their growth. If weeds grow over the closed canopy, they can be eliminated with selective herbicides. This is the general technique used with herbicide-tolerant soybean cultivars that results in excellent weed control and minimizes the number of herbicide applications necessary. In wide row culture, weeds must be managed until the canopy closes, and some tillage is generally used prior to planting and during the early growth periods.

The rotary hoe is an effective and economical means of weed control in soybeans (Figure 20-4). For best results, it should be used when the soil surface is dry and

Figure 20-4. Control of green foxtail in soybeans by rotary hoeing. *Left*: Before rotary hoeing. *Right*: After rotary hoeing; note the lack of injury on the soybeans.

Figure 20-5. Trifluralin and Imazethapyr applied preplant incorporated in soybeans. *Left*: Untreated *Right*: herbicide treatment.

Figure 20-6. Effectiveness of postemergence acifluorfen for weed control in soybeans.

slightly crusted and when the weeds are just emerging, but not more than $\frac{1}{4}$ inch tall. The rotary hoe is also effective on these small weeds when the soil surface is moist. It is important to cultivate only deeply enough to kill the weeds without disturbing the soybean roots. This technique does not bring additional weed seeds to the soil surface or disturb the layer of soil previously treated with a herbicide.

Chemical Weed Control in Soybeans

Many herbicides have been developed for the control of weeds in soybeans. A chemical weed control program in soybeans usually begins with either a preplant soil-incorporated herbicide treatment or a surface-applied preemergence herbicide treatment after planting (Figure 20-5). If necessary, this may be followed by a rotary hoeing and one or two cultivations. The use of postemergence treatments is widespread in soybean culture (Figure 20-6). Frequently, the decision regarding their use depends on the number of weeds that have survived earlier herbicide treatments. The herbicides used in soybeans are listed in Table 20-4. The recent introduction of herbicide resistant soybeans (see Chapter 18) has greatly altered herbicide use patterns in this crop.

DRY BEANS

Weed control in dry beans is similar to that described for soybeans. The main difference is that fewer herbicides are registered and nonchemical approaches to weed management are more critical to achieving satisfactory weed control. Herbicides registered for use in dry beans are listed in Table 20-5.

PEANUTS

The peanut root grows very fast after planting and germination, even though leaves emerge slowly. Under favorable conditions, the root may be 2, 6, or 15 inches long in 4, 6, or 12 days, respectively. In contrast, the first leaves usually do not emerge until 7 to 12 days after planting. This unique growth pattern allows the use of both mechanical and chemical weed control methods.

The rotary weeder, flexible-spike weeder, and cultivator are effective tools for weed control in peanuts. The rotary weeder can be used just before and again just after peanuts emerge to control small weed seedlings. Before a wide array of herbicides were available for use in peanuts, a timely use of the rotary weeder was considered to reduce cultivating time by 25% and hand hoeing by 50%. Cultivating two to four times has been common, but the need for cultivation has been reduced by the use of herbicides. Cultivation is now generally used if effective weed control is not obtained with herbicides; otherwise, cultivation provides little benefit to peanuts. When cultivating, it is important not to throw soil on the peanut vines, as this slows growth.

TABLE 20-4. Soybean Herbicides[a]

No-Till or Minimum-Till	Preplant Incorporated	Preplant or Preemergence
Glyphosate	Clomazone	Alachlor
Glyphosate + alachlor	Clomazone + alachlor	Alachlor + clomazone
Glyphosate + alachlor + metribuzin	Clomazone + ethalfluralin	Alachlor + imazethapyr
Glyphosate + flumetsulam	Clomazone + s-metolachlor	Alachlor + metribuzin
Glyphosate + flumioxazin	Clomazone + pendimethalin	Alachlor + trifluralin
Glyphosate + imazethapyr	Clomazone + trifluralin	Clomazone + alachlor
Glyphosate + imazethapyr + alachlor	Ethalfluralin	Clomazone + ethalfluralin
Glyphosate + imazethapyr + pendimethalin	Ethalfluralin + alachlor	Clomazone + s-metolachlor
Glyphosate + s-metolachlor + metribuzin	Ethalfluralin + s-metolachlor	Clomazone + metribuzin
Glyphosate + s-metolachlor + imazethapyr	Ethalfluralin + metribuzin	Cloransulam + sulfentrazone
Glyphosate + 2,4-DB	Ethalfluralin + metribuzin + alachlor	Dimethenamid
Glyphosate + 2,4-D	Ethalfluralin + metribuzin + clomazone	Dimethenamid + clomazone
Sulfosate	Flumetsulam + trifluralin	Dimethenamid + imazethapyr
Sulfosate + alachlor	Flumetsulam + trifluralin + metribuzin	Dimethenamid + metribuzin
Sulfosate + clomazone	Imazethapyr	Flufenacet + metribuzin
Sulfosate + dimethenamid-p	Imazethapyr + alachlor	Flumetsulam
Sulfosate + metribuzin	Imazethapyr + s-metolachlor	Flumetsulam + s-metolachlor
Sulfosate + s-metolachlor	Imazethapyr + pendimethalin	Flumetsulam + trifluralin
Sulfosate + flumioxazin	Imazethapyr + trifluralin	Flumioxazin
Metribuzin	Pendimethalin	Lactofen
2,4-D	Pendimethalin + alachlor	s-Metolachlor
2,4-DB	Pendimethalin + s-metolachlor	s-Metolachlor + clomazone

Paraquat	Pendimethalin + metribuzin	s-Metolachlor + ethalfluralin
Paraquat + alachlor	Trifluralin	s-Metolachlor + imazethapyr
Paraquat + imazethapyr	Trifluralin + alachlor	s-Metolachlor + metribuzin
Paraquat + s-metolachlor	Trifluralin + s-metolachlor	s-Metolachlor + trifluralin
Paraquat + metribuzin	Trifluralin + metribuzin	s-Metolachlor + cloranulam
Paraquat + pendimethalin	Trifluralin + metribuzin + clomazone	s-Metolachlor + flumetsulam
Paraquat + 2,4-DB	Trifluralin + imazethapyr	s-Metolachlor + pendimethalin
Paraquat + 2,4-D	Trifluralin + cloransulam	s-Metolachlor + chlorimuron-ethyl
Paraquat + alachlor + metribuzin		Metribuzin
Paraquat + s-metolachlor + metribuzin		Metribuzin + s-metolachlor
Paraquat + imazethapyr		Metribuzin + pendimethalin
Paraquat + imazethapyr + alachlor		Metribuzin + trifluralin
Paraquat + imazethapyr + pendimethalin		Metribuzin + chlorimuron-ethyl
Paraquat + s-metolachlor + imazethapyr		Sulfentrazone
Paraquat + sethoxydim + 2,4-D		Sulfentrazone + clomazone
Paraquat + flumioxazin		Sulfentrazone + clorasulam
		Sulfentrazone + s-metolachlor
		Dimethenamid-p
		Imazaquin
		Imazaquin + pendimethalin
		Imazaquin + dimethenamid-p

[a]Always read and follow all instructions on herbicide label.

TABLE 20-4. Soybean Herbicides, Continued

Postemergence		
Acifluorfen	Clethodim + bentazon	Fluazifop + bentazon
Acifluorfen + bentazon	Clethodim + cloransulam	Fluazifop + chlorimuron
Acifluorfen + fluazifop-	Clethodim + chlorimuron	Fluazifop + imazethapyr
p-butyl	Clethodim + fomesafen	Fluazifop + fenoxaprop
Acifluorfen + imazethapyr	Clethodim + flumiclorac	Fluazifop + fomesafen
Acifluorfen + sethoxydim	Clethodim + imazamox	Fomesafen
Acifluorfen + thifensulfuron	Clethodim + imazethapyr	Fomesafen + thifensulfuron
Acifluorfen + bentazon +	Clethodim + lactofen	Fomesafen + imazethapyr
sethoxydim	Cloransulam methyl	Imazamox
Acifluorfen + 2,4-DB	Cloransulam methyl +	Imazaquin
Bentazon + quizalofop-p-ethyl	acifluorfen	Imazethapyr
Bentazon + cloransulam	Cloransulam + fluazifop +	Imazethapyr + lactofen
Bentazon + fomesafen	fenoxaprop	Imazethapyr +
Bentazon + imazamox	Cloransulam + fomesafen	quizalofop-p-ethyl
Bentazon + imazethapyr	Cloransulam + imazethapyr	Imazethapyr + sethoxydim
Bentazon + sethoxydim	Cloransulam + lactofen	Imazethapyr + thifensulfuron
Bentazon + thifensulfuron	Cloransulam + sethoxydim	Lactofen
Bentazon + aciflurofen +	Cloransulam + trifensulfuron	Lactofen + glyphosate
sethoxydim	Cloransulam + quizalofop	Lactofen + imazamox
Bentazon + lactofen +	Flumiclorac	Lactofen + quizalofop-p-ethyl
acifluorfen + thifensulfuron	Flumiclorac + bentazon	Lactofen + thifensulfuron
+ sethoxydim	Flumiclorac + chlorimuron	Paraquat
Chlorimuron	Flumiclorac + cloransulam	Quizalofop-p-ethyl
Chlorimuron + lactofen	Flumiclorac + fomesafen	Quizalofop + thifensulfuron
Chlorimuron +	Flumiclorac + glyphosate	Sethoxydim
quizalofop-p-ethyl	Flumiclorac + imazethapyr	Thifensulfuron
Chlorimuron + thifensulfuron	Flumiclorac + lactofen	2,4-DB
Chlorimuron + fomasafen	Flumiclorac + thifensulfuron	
Clethodim	Fluazifop	
Clethodim + acifluorfen	Fluazifop + lactofen	

Sulfonylurea Tolerant Soybeans[b]	Roundup Ready Soybeans[c]	Postemergence-directed
Thifensulfuron + chlorimuron	Glyphosate	Linuron
Thifensulfuron + chlorimuron	Glyphosate + aciflurofen	Linuron + 2,4-DB
+ cloransulam methyl	Glyphosate + bentazon	Metribuzin
Thifensulfuron + chlorimuron	Glyphosate + chlorimuron	2,4-DB
+ fomesafen	Glyphosate + flumioxazin	
Thifensulfuron + chlorimuron	Glyphosate + imazethapyr	
+ quizalofop-p-ethyl	Sulfosate	

[a]Always read and follow all instructions on the herbicide label.

[b]Sulfonylure tolerant soybeans are not injured by applications of chlorimuron.

[c]Roundup Ready soybeans are resistant to applications of glyphosate herbicide. Use only glyphosate products registered for use in Roundup Ready soybeans.

TABLE 20-5. Dry Edible Bean Herbicides[a]

Preplant Incorporated or Preemergence	Postemergence	Desiccants
Alachlor	Quizalofop-p-ethyl	Sodium chlorate
EPTC	Bentazon	Paraquat
Ethalfluralin	Sethoxydim	
s-Metolachlor	Clethodim	
Pendimethalin	Imazethapyr	
Trifluralin		
Imazethapyr		
Dimethenamid-p		
Pelargonic acid		

[a]Always read and follow all instructions on the herbicide label.

Chemical Weed Control in Peanuts

The use of herbicides is the most effective method of weed control in peanuts, which includes preplant soil-incorporated, preemergence, and postemergence treatments, with herbicide combinations and sequential applications being common. The herbicides available for use in peanuts are listed in Table 20-6.

Preplant soil-incorporated treatments are usually applied as part of the final seedbed preparation and planting operation. In one pass through the field, the herbicide is applied to the soil surface and incorporated into the soil, the seed planted, and the seedbeds shaped. Premergence herbicides are applied after planting but before weed emergence. This may be an application immediately after planting, a delayed application, or an application at cracking time. A delayed application is performed about 6 days after planting to allow time for the root tip to grow deep into the soil and reduce the risk of injury by certain herbicides. A cracking time treatment is also a delayed preemergence application, performed when the emerging peanut shoot cracks the soil surface but has not yet emerged. This treatment is usually applied 6 to 10 days after planting. However, both of these delayed applications may reduce the herbicide's effectiveness if the weed seedlings are beyond their optimal control period.

COTTON

Cotton was once planted thickly and thinned to stand during the first hand hoeing. This required considerable hand labor and cultivation. Even so, cotton was often still weedy at harvest time. With mechanical harvesting, weed control became increasingly important because weedy trash stained and contaminated the cotton lint. Herbicides have replaced most hand labor and reduced cultivation. Now cotton is usually planted at a final stand spacing of about 45,000 plants/acre.

TABLE 20-6. Peanut Herbicides[a]

Preplant Incorporated	Preemergence	Cracking	Postemergence	Lay-by
Alachlor	Alachlor	Paraquat	Acifluorfen	Dimethenamid-p
Ethalfluralin	Dimethenamid-p	Alachlor	Acifluorfen + 2,4-DB	s-Metolachlor
Pendimethalin	Flumioxazin	Dimethenamid-p	Bentazon	
Dimethenamid-p	Metolachlor	s-Metolachlor	Bentazon + acifluorfen	
s-Metolachlor	Diclosulam	Imazethapyr	Bentazon + acifluorfen + 2,4-DB	
Diclosulam	Diclosulam + s-metolachlor	2,4-D	Bentazon + 2,4-DB	
Diclosulam + pendimethalin	Diclosulam + dimethenamid-p	Bentazon	Imazapic	
Diclosulam + ethalfluralin	Imazethapyr	Pyridate + 2,4-DB	Imazethapyr	
Diclosulam + s-metholacholor	Norflurazon	Acifluorfen	Pyridate + 2,4-DB	
Diclosulam + dimethenamid-p		Diclosulam	2,4-DB	
			Paraquat	
			Paraquat + bentazon	
			Paraquat + acifluorfen	
			Chlorimuron	
			Clethodim	
			Sethoxydim	

[a]Always read and follow all instructions on the herbicide label.

Good production methods are needed to allow for the best possible growth and yields of cotton. Among these are the use of well-prepared seedbeds, high-quality seed, and well-adapted varieties; adequate fertility; and proper control of diseases, insects, and weeds. Planting should be delayed until the soil temperature has reached at least 60°F, which usually provides the desired stand. A vigorous crop of cotton helps control weeds, especially late-season weeds, through competition.

Flaming

The use of fire by means of a flame directed toward the base of the cotton stem can control many small weed seedlings. The flat-type burner usually uses volatile gases such as butane (Figure 20-7). The specialized equipment required is called a flame cultivator. It may be used when cotton stems are $\frac{1}{4}$ inch in diameter. At this stage of growth, the cotton stem is tolerant to the small amount of heat required to control small weed seedlings. However, the current high cost of fuel has greatly reduced the use of this practice.

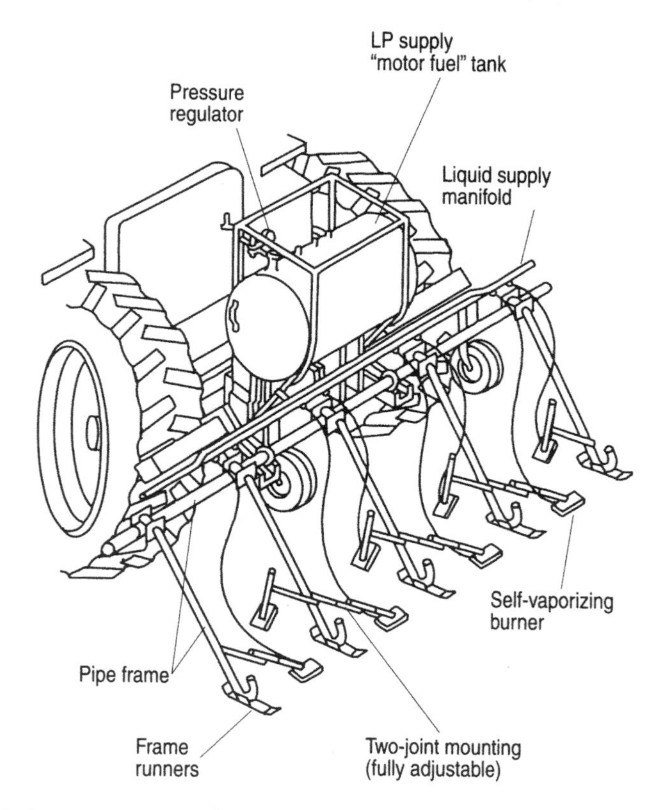

Figure 20-7. A tractor mounted flame weeder. Reprinted with permission from *Sustainable Vegetable Production from Start-Up to Market* (NRAES–104), published by the National Resource, Agriculture and Engineering Service, Cooperative Extension, 152 Riley-Robb Hall, Ithaca, NY 14853–5071. Adapted with permission from illustrations by John Gist in *Steel in the Field: A Farmers Guide to Weed Management Tools* (SAN-2).

Cultivation

Mechanical removal of weeds is still used in cotton production. In areas using winter fallow beds, sectioned rolling cultivators can be effective on young weeds. Cultivation is used preplant to remove existing weeds, to prepare the seedbed, and to assist in control of perennial weeds such as nutsedge. Postplant mechanical cultivation during the time from seeding to canopy closure (lay-by) with equipment (sweeps, bed knives, rolling cultivators, reversed disk-hillers, and bed-weeders), carefully aligned to follow the seeded rows as close to the plants as possible, provides excellent weed control.

Hand removal of weeds within the row or removal of perennial weeds with hoes is effective but not common because of the costs involved. In some regions, geese (three to five per acre) are used to control grass and nutsedge. If geese are used, the field must be temporarily fenced and the farmer must provide supplemental food, water, and protection (from dogs and coyotes) for the geese.

Newer weed control technology (herbicide-tolerant crops, better selection of available herbicides) has reduced the amount of cultivation used. Less cultivation results in fewer weed flushes, reduced loss of moisture from the soil, and less root damage to the cotton plants.

Chemical Weed Control in Cotton

Herbicides used in cotton can be applied preplant, preplant soil-incorporated, preemergence, or postemergence. Preplant treatments to emerged weeds with nonselective, nonpersistent herbicides (e.g., glyphosate or paraquat) can be used in cotton. The other herbicides applied in cotton are listed in Table 20-7. Most preplant soil-incorporated or preemergence herbicides are applied just before or just after planting, or often at the same time as planting, with the application and planting equipment attached to the same tractor. Certain relatively persistent preplant soil-incorporated herbicides (e.g., trifluralin) can be applied in the fall or at any time up to the date of planting.

Several postplant herbicides provide control of both broadleaf and grass weeds after cotton planting. Herbicide-tolerant cotton varieties also are available that allow application of bromoxynil or glyphosate for broad-spectrum weed control, which allows growers wider options and eliminates the need for early postplant directed herbicide applications (Figure 20-5).

SUGAR BEETS

Sugar beet "seed" is actually a fruit and may contain more than one seed. However, monogerm seed, often containing only one seed, is usually planted. If more than one plant develops from such seeds, they should be thinned to one plant for maximum production. The seed is usually planted at a high rate and thinned to the desired stand mechanically or by hand. Some farmers use precision planting and thereby achieve lower seeding rates. The weeds that escape the usual previous herbicide treatment can also be removed during the thinning operation.

TABLE 20-7. Cotton Herbicides[a]

Early Preplant Burndown	Preplant Incorporated	Preemergence	Postemergence
Glyphosate	Pendimethalin	Fluometuron	Over The Top
Glyphosate + thifensulfuron + tribenuron	Trifluralin	Pyrithrobac-sodium	MSMA
Glyphosate + 2,4-D	Trifluralin + clomazone	Norfluorazon	Fluometuron
Paraquat	Norflurazon	Norfluorazon + fluometuron	Pyrithiobac-sodium
Paraquat + cyanazine		Pendimethalin	Clethodim
Paraquat + oxyfluorfen		Pendimethalin + fluometuron	Fluazifop-p-butyl
Paraquat + thifensulfuron + tribenuron		Clomazone	Fluazifop-p-butyl + fenoxaprop-p-ethyl
Paraquat + 2,4-D		Diuron	Quizalofop-p-ethyl
Pelargonic acid			Sethoxydim
Sulfosate			
Sulfosate + dicamba			
Sulfosate + oxyfluorfen			
Sulfosate + thifensulfuron + tribenuron			
Sulfosate + 2,4-D + dicamba			

[a]Always read and follow instructions on the herbicide label.

TABLE 20-7. Cotton Herbicides, Continued

Postemergence (continued)			
Directed	Lay-by	BXN Cultivars (bromoxynil resistant)	Roundup Ready Cultivars (glyphosate resistant)
DSMA	Fluometuron	Bromoxynil	Glyphosate
MSMA	Pendimethalin	Bromoxynil + MSMA	Glyphosate + fluazifop-p-butyl
Fluometuron + MSMA	Prometryn	Bromoxynil + pyrithiobac-sodium	Glyphosate + pyrithiobac
Fluometuron + MSMA + s-metolachlor	Prometryn + lactofen		Glyphosate + s-metolachlor
Pyrithiobac-sodium + MSMA			Glyphosate + fluometuron
Cyanazine + MSMA			
Dimethenamid-p			
Dimethenamid-p + MSMA			
Diuron			
Lactofen + MSMA			
Lactofen + prometryn			
MSMA + prometryn			
Oxyfluorfen			
Oxyfluorfen + cyanazine			
Oxyfluorfen + MSMA			
Prometryn + s-metolachlor			

[a] Always read and follow instructions on the herbicide label.

TABLE 20-8. Sugarbeet Herbicides[a]

Preplant	Preplant Incorporated	Postplant Preemergence	Postemergence	Lay-by (after cultivation)
Paraquat	Pyrazon	Pyrazon	Endothall	Trifluralin
Glyphosate	Ethofumesate	Endothall	Desmedipham	EPTC
	Cycloate		Pyrazon	Cycloate
	Pebulate		Phenmedipham + desmedipham	
			Sethoxydim	
			Phenmedipham + desmedipham + ethofumesate	
			Clopyralid	
			Triflusulfuron	
			Quizalofop-p-ethyl	
			Clethodim	

[a]Always read and follow all instructions on the herbicide label.

Sugar beet weed management must include several methods, because no one method can provide complete weed control. Sugar beet is not competitive, and uncontrolled weeds can reduce yield by more than 90%. Dense weeds make weed removal by hand, mechanically, or with herbicides difficult and complicate harvest.

The major weed problem in sugar beets is annual weeds, both grasses and broadleaf weeds. They are particularly troublesome at emergence through thinning time and after lay-by. Favorable production methods, discussed in regard to the aforementioned crops, induce a vigorous crop that helps control the weeds through competition; however, because sugar beets are low-growing plants, growers cannot allow any prolonged period of weed interference to occur.

Chemical Weed Control in Sugar Beets

Preplant and preemergence herbicides used in sugar beets are degraded fairly rapidly in the soil and control weeds for only about 4 to 6 weeks. Therefore, additional herbicide applications are required later in the season to obtain weed control and acceptable yields. The herbicides used in sugar beets are listed in Table 20-8.

SUGARCANE

Typically, a sugarcane field is planted every 2 to 4 years and the perennial crop is harvested at 12- to 24-month intervals. Following the initial planting, two to four sequential harvests are made before productivity declines and replanting is necessary. The crop is reproduced vegetatively by planting ~20-inch pieces of the stem containing two to four buds, with one bud at each node. These vegetative propagules

TABLE 20-9. Sugarcane Herbicides[a]

Preplant	Preemergence	Postemergence
Glyphosate	Atrazine	Metribuzin
Pelargonic acid	Diuron	Atrazine
	Pendimethalin	Asulam
	Glyphosate	Ametryn
		Halosulfuron
		Glyphosate (hooded sprayer)

[a]Always read and follow all instructions on the herbicide label.

are referred to as *seed pieces*. The initial planting is referred to as *plant cane* and subsequent crops as *ratoon* or *stubble cane*.

Sugarcane requires a tropical or semitropical climate with a continuous or long growing season, high soil fertility, and abundant rainfall or irrigation. Under these conditions, all types of weeds flourish—annuals, biennials, and perennials. After planting, weeds must be controlled with cultivation and herbicides until the row middles are covered by sugarcane foliage. Depending on environmental conditions, this usually takes 4 to 7 months for plant cane. After a field has been harvested, growers maintain it weed free while the second crop of stalks (ratoon) grows from the old stubble. The period required from initial plant growth to canopy closure is 2 to 5 months for a ratoon crop. Weeds developing after "close-in" are usually controlled by competition of the sugarcane plants.

Chemical Weed Control in Sugarcane

A chemical weed control program for sugarcane usually requires at least two herbicide applications, one immediately after planting but before crop emergence, and another postemergence to the crop but before close-in. Glyphosate is used preplant or as a spot treatment within the crop, usually for perennial weed control. However, it should not be applied to vegetation in or around ditches, canals, or ponds containing water to be used for irrigation. Glyphosate can also be used to control undesirable sugarcane plants. The herbicides used in sugarcane are listed in Table 20-9.

TOBACCO

Tobacco has very small seed, and, initially, the emerging seedlings grow very slowly. Therefore, the plants are started from seed in plant beds and later transplanted to the field. Different weed control programs are required in each of these situations.

Seedbeds

Tobacco seedbeds are not as important as they once were, because the majority of tobacco transplants are now produced in greenhouses in plug trays using a float system.

Soil-borne diseases, insects, and nematodes, as well as weeds, may create a problem in tobacco seedbeds. A soil fumigant that controls all of these plant pests is typically used, such as methyl bromide, metham, or dazomet. The seedbed should be well prepared, without clods, and porous so that the fumigant thoroughly permeates the soil. Adequate soil moisture is required to convert metham and dazomet into their toxic degradation products. Adequate soil moisture also initiates seed germination, which makes all three of these fumigants more effective. Dormant seeds and seeds with impermeable seed coats (hard seed) may not be killed by such fumigants.

Methyl bromide is usually injected into the soil and the soil covered with a gasproof plastic tarp in a single operation. After at least 48 hours the tarp can be removed, and planting must be delayed for at least an additional 72 hours.

The label for metham recommends that it be applied in the fall if possible. It can be applied using either a tarp or a drench method. In the tarp method, metham is diluted with water (40 gal/100 yd^2), applied to the soil surface, and immediately covered with a plastic tarp for no less than 1 day but not more than 2 days. Seven days later the soil should be "loosened" to a depth of 2 inches; planting should be delayed until at least 21 days after the metham application. In the drench method, metham is diluted with water (150 to 200 gal/100 yd^2). Application may be made with sprinklers, sprayers and nozzles, or any suitable equipment. Additional procedures are the same as for the tarp method.

Dazomet is applied as a preplant soil-incorporation treatment using conventional equipment. The application is made 3 weeks before seeding in the fall or summer, or 4 weeks before seeding in the early spring.

Field Transplants

The influence of cultivation on tobacco transplants in the field was studied in North Carolina. Cultivation did not increase yields on sandy soils if the weeds were otherwise controlled. On loam or clay-loam soils, however, increased yields were observed with one or two cultivations; additional cultivations did not further increase yields.

Herbicides are commonly used in field-transplanted tobacco (Table 20-10). Most are applied either as a soil treatment prior to transplanting or as an over-the-top posttransplant treatment. These herbicides include benefin, napropamide, pebulate, pendimethalin, clomozone, and sulfentrazone. Napropamide and pendimethalin can also be applied at lay-by.

SUNFLOWERS

The success of a weed control program in sunflowers depends greatly on the choice of hybrid, in regard to vigor and size, and the field management practices employed. Because few herbicides are labeled for use in sunflowers, cultural practices are

TABLE 20-10. Tobacco Herbicides[a]

Plant Beds	Pre-Transplant	Post-Transplant	Lay-by
Methyl bromide	Pebulate	Napropamide	Napropamide
Dazomet	Napropamide	Clomazone	Pendimethalin[c]
Metham sodium	Pendimethalin	Sethoxydim	
Sethoxydim	Clomazone		
Pelargonic acid	Sulfentrazone		
	Benefin[b]		

[a]Always read and follow all instructions on the herbicide label.
[b]Burley only.
[c]Flue-cured only.

important for reducing weed competition. Field selection based on previous good weed management is important; fields with heavy weed infestations of annuals and/or perennial weeds are thereby avoided. Close row spacing of 12 inches is gaining in popularity, although this limits cultivation options. Because there are no postemergence herbicides for broadleaf weed control in sunflowers, wider rows of 20 to 30 inches are frequently used to facilitate cultivation.

Chemical Weed Control in Sunflowers

Common practice is to use a burndown herbicide immediately prior to planting, followed by preemergence herbicides incorporated into the soil. Many of the preplant incorporated herbicides can be applied in early spring up to 30 days before planting. Postemergence herbicides for control of emerged grass and wild mustard are available. Herbicides used in sunflowers are listed in Table 20-11.

SAFFLOWER

Safflower requires many of the same cultural considerations as sunflowers. Safflower seedlings grow slowly, remaining in a rosette stage for 3 to 4 weeks after emergence, and weeds can easily become established and reduce crop yield if not controlled. The best weed control practices involve selecting nonweedy fields, using clean seed, and combining timely cultivation with cultural practices that encourage rapid crop growth.

Chemical Weed Control in Safflower

Herbicides available for safflower are listed in Table 20-11. Preplant applications of paraquat or glyphosate control weeds prior to seeding, and trifuralin or ethalfluralin can be soil applied at planting. Additional cultivation is necessary to control

TABLE 20-11. Sunflower and Safflower Herbicides[a]

Burndown	Preplant Incorporated or Preemergence	Postemergence	Harvest Aid
	Sunflower		
Glyphosate	EPTC	Sethoxydim	Paraquat
Pelargonic acid	EPTC + trifluralin	Imazethabenz	Sodium chlorate
	EPTC + ethalfluralin	Clethodim	
	Pendimethalin		
	Trifluralin		
	Paraquat		
	Sulfentrazone + pendimethalin		
	Safflower		
Glyphosate	Trifluralin		
Paraquat	Ethalfluralin		
Pelargonic acid			

[a]Always read and follow all instructions on the herbicide label.

late-season emerging weeds, as no postemergence herbicides are available for use in safflower.

LITERATURE CITED AND SUGGESTED READING

Bridges, David C. 1992. *Crop Losses Due to Weeds in the United States, 1992*. Weed Science Society of America, Lawrence, KS.

Gist, J. 1996. *Steel in the Field: A Farmers Guide to Weed Management Tools*. Natural Resource, Agriculture and Engineering Service (NRAES), Cornell University, Ithaca.

Grubinger, V.P. 1999. *Sustainable Vegetable Production from Start-Up to Market*. Natural Resource, Agriculture and Engineering Service NRAES Publication 104, Cornell University, Ithaca.

Sprague, M. A., and G. B. Triplett, eds. 1986, *No-Tillage and Surface-Tillage Agriculture*. John Wiley & Sons, Inc., New York.

Smith, A. E. 1995. *Handbook of Weed Management Systems*, Marcel Dekker, New York.

WEB SITES

University of Minnesota: *Cultural and Chemical Weed Control Guide in Field Crops*

http://www.extension.umn.edu

Under "Topics," select "Crops," select "Weed Control," and select "Cultural and Chemical Weed Control Guide in Field Crops."

University of Nebraska: *Guide for Weed Management in Nebraska*

http://www.ianr.unl.edu

Select "Publications," then select "Find Information," and select "Weeds."

University of Florida: *Weed Management in Field Crops and Pasture Grasses*

http://edis.ifas.ufl.edu

Select "Pest Management Guides," then select "Weed Management Guide," then select "Crop of Choice (Corn, Cotton, etc.)"

North Carolina State University: *North Carolina Agricultural Chemicals Manual*

http://ipmwww.ncsu.edu

Select "Pesticide Recommendations"

University of California: *Pest Management Guidelines*

http://www.ipm.ucdavis.edu

Select "How to Manage Pests," select "Management of Pests of Agricultural Crops," then select "the Agronomic Crop of Choice"

AgChemical Label Information: Crop Data Management Systems Inc., Marysville, CA

http://www.cdms.net/manuf/manuf.asp

For chemical use, see the manufacturer's or supplier's label and follow these directions. Also see the Preface.

21 Small-Seeded Legumes

Several small-seeded legumes are used for hay, pasture, and soil improvement. Alfalfa, birdsfoot trefoil white clover, and ladino clover are perennial crops usually grown for hay and pasture. Red clover, alsike clover, crimson clover, sweet clover, and lespedeza are hay and soil improvement crops.

Weeds in small-seeded legumes can reduce yields, lower quality, cause premature loss of stand, and present harvesting problems. Common chickweed, a winter annual, can form thick mats in a forage stand and reduce yield by up to 30%. Weeds can also increase disease and insect problems in these crops. Certain weeds in hay or pastures can adversely effect animals and/or their products. Some weeds, such as spurge, fiddleneck, and common groundsel, are toxic to livestock when consumed in large amounts (Figure 1-4). Spiny weeds, such as thistles and downy brome, are irritating to livestock. Other weeds, such as wild onion and wild garlic, cause off-flavors in milk. When these small-seeded legume crops are grown for certified seed production, they must be essentially free of weeds.

Weed control in these crops during the seedling and establishment stage requires a different approach than from that used in weed management in row crops. A significant portion of the weed control program in an established healthy forage comes from crop competition. Maintenance of a relatively weed-free forage, grown with proper fertilization, cutting management, insect control, the use of disease-resistant varieties, and herbicide application, is necessary to keep the forage healthy and competitive. These conditions increase the vigor of the crop and make it more competitive at all stages of growth.

SEEDLING STAGE

Because small-seeded legume seedlings do not grow as vigorously as many common weeds, it is essential that control measures be taken to establish a strong stand. These control measures include (1) elimination of problem weeds before planting, (2) the use of clean planting seed, (3) weed control before seeding, (4) proper date of seeding, and (5) use of chemicals. Mowing weeds that "outgrow" legume seedlings can also be an effective method of control.

Clean Seed

The use of *certified* seed is the best way to avoid sowing weed seeds along with the crop. This type of crop seed is certified by the state as to its high purity in regard to the presence of weed seeds. Less expensive crop seed is frequently contaminated with weed seeds. Once introduced, these weed seeds may develop into serious weed problems that persist for many years and are difficult and expensive to control. This is especially true for the parasitic weed dodder, whose seeds are about the same size as most small-seeded legume seeds.

The seeds of many seriously problematic weeds are very similar to the seeds of small-seeded legumes. Once the legume seeds are contaminated with these weed seeds, they can only partially be cleaned by presently available methods. Many of the mechanical seed cleaners are quite ingenious and take advantage of small physical differences between the seeds to be separated; such as differences in size, shape, weight, seed coat, hairiness, and appendages. An example is the cleaner used to remove dodder seed from small-seeded legume seed, which takes advantage of the fact that legume seed is smooth and waxy and dodder seed is rough and pitted. Consequently, the dodder seeds adhere to felt cloth in the cleaner and legume seeds do not. Adjoining felt-covered rollers rotate in opposite directions, and seeds pass over them. The rollers are slanted so that the legume seeds flow out the lower end, but the dodder seeds do not.

Weed Control Before Seeding

Small-seeded legume seedlings are not competitive with aggressive weeds, as their initial growth is slow. Therefore, it is desirable to eliminate seeds and propagules of both annual and perennial weeds before seeding these crops. Control methods must be appropriate for the weed species present.

Annual weeds are often controlled by two methods. The first method is crop rotation before seeding. For example, a rotation of row crops and small grain with good weed control for 2 or more years usually reduces weed seed populations in the soil. The second method involves killing one or more "crops" of weeds by cultural or chemical methods before, during, or after seedbed preparation, but before seeding the crop. Cultivations to eliminate these small weeds should be shallow so as to prevent bringing deeply buried weed seeds near the soil surface where they may germinate. Deeply buried weed seeds usually remain dormant and present no problem if undisturbed.

Perennial weeds should be controlled by cultural or chemical methods before planting small-seeded legumes. Once these crops have been planted, the options for perennial weed control without crop injury are limited.

Date of Seeding

The date of seeding may determine the weediness of a crop (Klingman, 1970). Most small-seeded legume crops can be planted either in the fall or in the spring. Weediness will depend on whether winter- or summer-annual weeds present a more serious problem. Summer-annual weeds are often more problematic. Thus, alfalfa is usually planted in the fall. With fall planting, the legume crop is well established by spring and competes well with summer-annual weeds. However, there are the hazards of fall drought and winter injury in some areas and excessive rainfall in other areas.

Spring planting may be preferred if winter-annual weeds are the major problem. Spring planting avoids the flush of winter-annual weed growth. However, with spring planting, summer-annual weeds may crowd out legume seedlings before they become established. More efficacious herbicides may make it possible to take advantage of the more desirable seeding conditions usually found in the spring. It is important to keep the forage relatively weed free for the first 60 days after planting.

Chemicals

Herbicide treatment practices for small-seeded legumes at the seedling stage are usually different from those used in the established crop. In general, these crops are more likely to be injured by a given herbicide at a specific rate at the seedling stage than at the established stage. Small-seeded legumes vary in their tolerance to different herbicides. This tolerance and the susceptibility of the weeds are the most important considerations in choosing an appropriate herbicide treatment program.

Herbicides are applied preplant, preplant soil-incorporated, preemergence, or postemergence to the crop for the establishment of a stand from seed. Table 21-1 lists the herbicides that can be used both at the seedling stage and in the established crop. Refer to each herbicide label for complete information concerning application to a crop: the proper stage of growth, the rate, whether herbicide combinations can be used, and restrictions regarding use. See, in the Appendix, Tables A-1 and A-2 for weeds controlled and Table A-4 for trade names.

ESTABLISHED STAGE

Controlling weeds in established forages is generally of most benefit for the first cutting. As the forage ages, weeds often become a serious problem in established stands of the small-seeded legumes, as the vigor of stand decreases as a result of any of a number of environmental conditions or cultural practices. Weeds can reduce the number of crop plants in an established stand by competition. This results in an increasing loss in yield and quality over time. Weed control methods employed in established stands include (1) crop competition, (2) mowing, (3) flaming, (4) cultivation, and (5) the use of chemicals. Before using a herbicide in an established stand, it is important to evaluate the forage to determine whether the cost of the herbicide is justified.

Crop Competition

Although seedlings of small-seeded legumes often do not compete favorably with weeds, a vigorous established stand is very competitive to many weeds. Therefore, maintaining a thick, well-established stand is very important in weed suppression in these crops. The selection of proper adapted varieties, fertilization, drainage, moisture conservation, irrigation, mowing time, and control of diseases, insects, and weeds help to maintain a thick legume stand. These practices also ensure rapid regrowth of the crop after mowing, which increases its competitive ability.

TABLE 21-1. Herbicides Used in Small-Seeded Legumes[a]

	Alfalfa	Birdsfoot Trefoil	Clovers[b]
Preplant Incorporated/Preemergence			
Benefin	X	X	X
EPTC	X	X	X
Trifluralin	X		
Pronamide	X	X	X
Pelargonic acid	X	X	X
Postemergence of Crop/Preemergence of Weeds			
EPTC	X		
Norflurazon	X		
Postemergence			
Bromoxynil	X		
Clethodim	X	X	
Imazethapyr	X		
Sethoxydim	X		
Paraquat	X		
2,4-DB	X	X	X
Dormant			
Diuron	X	X	X
Hexazinone	X		
Imazethapyr	X		
Metribuzin	X		
Norflurazon	X		
Terbacil	X		
MCPA	X		X
Pelargonic acid	X	X	X
Harvest Aid			
Diquat (seed crop only)	X		X
Glyphosate (crop termination)	X		
Pelargonic acid	X	X	X

[a]Always read and follow instructions on the herbicide label.
[b]Some labels list the types of clover, others merely state "Clovers."

Mowing

Many erect annual weeds may be killed and the vigor of erect perennial weeds reduced by mowing. However, mowing does not control prostrate weeds. With repeated mowing, some erect weeds may develop prostrate growth patterns. In fact, at least one erect weed, yellow foxtail, has developed a genetically stable biotype in some areas that is prostrate under mowing regimes. If the crop is mowed too frequently or when it is immature, its vigor will be reduced and weed problems may increase. Mowing controls many weeds especially well when a thick legume stand is maintained.

Flaming

Propane and diesel burners have been used to control weeds in established alfalfa. Winter-annual broadleaf weeds can be controlled with flaming just before the crop resumes growth in the spring. This treatment also suppresses larval populations of the alfalfa weevil. Flaming just after cutting has controlled established dodder plants; however, this usually results in a few days' suppression of alfalfa growth. Flaming has been generally replaced by other methods of weed control because of the currently high price of petroleum.

Cultivation

Although tillage of alfalfa fields to control annual weeds has been recommended and practiced, there is little experimental evidence to support this practice. The early research of Kiesselback and Anderson (1927) showed that cultivation of alfalfa neither increased nor decreased yields. However, in a good stand of alfalfa, one cultivation may control some annual weeds. The use of the spring-tooth harrow is preferred because it kills many weeds without injury to alfalfa crowns. The spike-tooth harrow is effective only on very small weeds, and the disk harrow may cause considerable damage to alfalfa by cutting the crowns.

When these crops are grown for seed production, they are often planted in rows. The areas between the rows can be cultivated as in any row crop.

Chemicals

Some general rules (Pennsylvania State University: PSU Alfalfa Web Home page) should be considered before using a herbicide in an established forage: (1) Thin or irregular stands will not thicken once weeds are removed. A minimum of five alfalfa plants per square foot should be present before considering applying a herbicide. (2) Weeds tolerant to the applied herbicide may invade the space left by the susceptible species, ultimately creating a more severe weed problem. (3) Only well-established, vigorous stands should be treated with herbicides. (4) If the forage stand is at least 2 years old with 25 to 30% of the area being weedy, removing the weeds with a herbicide application is of questionable value. (5) If 50% or more of the stand is weedy, it is time to rotate to another crop.

If a herbicide is to be applied to an established forage, remember that small-seeded legumes vary in their tolerance to herbicides. This tolerance and the susceptibility of the weeds are most important in choosing an appropriate herbicide treatment. Most herbicides applied to established forage are applied in the fall immediately after the last cutting, in the winter, and/or in the early spring. The crop is usually dormant or semidormant at these times. Table 21-1 lists the herbicides that can be used both at the seedling stage and in the established crop. Refer to each herbicide label for complete information on herbicide use.

LITERATURE CITED AND SUGGESTED READING

Bridges, D.C., ed. 1992. *Crop Losses Due to Weeds in the United States, 1992*. Weed Science Society of America, Lawrence, KS.

Crop Protection Reference, 17th ed. 2001. Chemical and Pharmaceutical Press, New York.

Kiesselback, T. A., and A. Anderson. 1927. *Nebraska Expt. St. Bull.*, 222.

Klingman, D. L. 1970. Brush and weed control on forage and grazing lands. *FAO Internet. Conf. on Weed Control*, pp. 401–424. Weed Science Society of America, Lawrence, KS.

Mitich, L. W. 1985. Alfalfa (*Medicago sativa*). In *Principles of Weed Control in California*, pp. 232–237. Thompson Publications, Fresno, CA.

WEB SITES

Pennsylvania State University: Alfalfa Home Page. 1997. M. H. Hall

http://www.cas.psu.edu

Search “Forage,” scroll to and select “Forage Home Page,” select “Pests, then select Weed Management in Alfalfa and Legumes”

North Carolina State University: *North Carolina Agricultural Chemicals Manual*

http://ipmwww.ncsu.edu

Select “Pesticide Recommendations”

Ag Chemical Label Information: Crop Data Management Systems Inc., Marysville, CA

http://www.cdms.net/manuf/manuf.asp

For chemical use, see the manufacturer’s or supplier’s label and follow these directions. Also see the Preface.

22 Vegetable Crops

Weed management in vegetable crops requires a multifaceted approach, built on an understanding of weeds and the crop. Reductions in vegetable crops caused by weeds often mean the difference between profit and loss. Monaco et al. (1981) reported severe reductions in direct-seeded tomatoes with full-season competition of cocklebur, tall morning glory, redroot pigweed, and large crabgrass. Most work with vegetables indicates that weed competition in the first 4 weeks of crop growth can result in severe crop reduction. A review of weed/crop competition by Zimdahl (1980) gives many other examples of serious losses caused by weeds in vegetable crops.

Vegetable crops are generally more vulnerable to weed competition than agronomic crops, because many of them are short-season crops and they are usually weak competitors against weeds. In addition, weeds can reduce the efficiency of protection against disease and insect pests, thereby lowering quality and marketability, and can cause crop losses by interfering with mechanical and hand harvesting.

Past methods of weed control in vegetable crops have centered on the cultivator and the hoe, but hoeing is expensive and labor is in short supply. With increased labor costs, more efficient methods of weed control are necessary if the grower is to make a profit. Tillage (cultivation) is discussed in Chapter 3, and Grubinger (1999) is a good reference for nonchemical approaches to weed control. Experienced growers use a multifaceted approach to weed management and supplement their cultivation and cultural practices with effective selective herbicides to reduce the costs of manual labor and increase the efficiency of their operations.

MANAGEMENT STRATEGIES

In managing weeds in vegetable crops, there are many methods employing various combinations of nonchemical and chemical means. Practices that promote a reduction in herbicide use are of increasing importance because of consumer concerns about pesticide residues and potential environmental contamination from pesticides, and because many useful herbicides are not registered for use in vegetable crops.

Cultural Practices

Farm practices should aim to establish a vigorous crop that competes effectively with weeds. This starts with *land selection*. A general rule is, Do not plant vegetables on land with a history of heavy weed infestation, especially with perennial weeds. *Crop selection* can reduce the effects of weed competition. One criterion in determining whether to plant a vegetable crop should be the weed problems of the field. Plant the most competitive crops in the most weed-infested fields and the least competitive crops in the cleanest ones. Planting heavily infested fields as long-term set-aside acres, or in non-row crops such as alfalfa as a permanent cover, should help prevent the buildup of annual weeds.

Crop rotation is another practice that can reduce weed problems. The characteristics of a crop, the methods used to grow it, and the herbicides used, may inadvertently allow certain weeds to escape control. Rotating crops will improve crop growth and competitiveness. Rotation also affects the weed management tools at your disposal. Related vegetables should not be grown in the same location in successive years (Table 22-1). Once a crop is selected, use adaptive, vigorous varieties resistant to diseases. Disease-infected plants cannot effectively compete with weeds.

Narrower row spacing and proper plant densities help ensure rapid crop closure. A closed canopy shades out later emerging weeds and prevents germination of weed seeds requiring light. Weeds are seldom a problem once canopy closure occurs. Proper row spacing and plant density also allow row cultivation, which is important in most vegetable crops.

Another useful cultural practice to improve crop competitiveness is to plant at the correct time. Crops can be divided into warm- and cool-season plants, depending on the optimum temperature for their growth. Planting date affects the time to emergence and early seedling vigor of a crop, which are important in determining crop competitiveness. Cool-season crops germinate at cooler soil temperatures and thus

TABLE 22-1. Botanically Related Vegetables

Poaceae	*Cucurbitaceae*	*Brassicaceae*	*Leguminosae*
Sweet corn	Cucumber	Rutabaga	Soybean
Dent corn	Winter squash	Kale	Pea
	Summer squash	Broccoli	Snap bean
	Pumpkin	Cauliflower	Lima bean
	Muskmelon	Cabbage	Dry bean
	Watermelon	Brussels sprout	
Amaryllidaceae		Radish	
		Horseradish	
Onion			
Garlic	*Solanaceae*		
	Potato		
	Tomato		
Chenopodiaceae	Pepper		
	Eggplant		
Beet			
Chard			
Spinach			

compete better against early emerging weeds than warm-season crops. Table 22-2 lists crops according to their adaptation to field temperatures. Time plantings so that temperatures are favorable for optimum crop growth.

Adequate fertilization and *appropriate insect and disease management* are also important in establishing a competitive crop. Adequate fertility ensures rapid, uniform germination and good crop growth, which enhance the crop's competitive ability.

Mulching can be useful in managing weeds. Mulches can be classified as either natural (straw, leaves, paper, and compost) or synthetic (plastics). Because natural mulches are difficult to apply over large areas, they are best for small, specialized areas. Natural mulches should be spread evenly at least 1.0 to 1.5 inches thick over the soil to prevent light penetration (Figure 22-1). Natural mulch materials must be free of weed seeds and other pest organisms and be heavy enough that wind or water will not easily displace them. A major advantage of natural mulches is that they add organic matter to the soil and do not have to be disposed of at the end of the season.

Synthetic mulches are easy to apply, control weeds within the row, conserve moisture, and increase soil temperature. There is increased emphasis on the use of plastic mulches in the production of vegetables (Figure 22-2). Weed management in plantings of this type differs from production in bare-ground settings.

When methyl bromide fumigation is used under the plastic, there is no need for a herbicide application under the plastic. Moreover, if black plastic or the newer colored plastics are used with no methyl bromide fumigation, a herbicide is generally not necessary under the plastic. An exception is fields containing high populations of nutsedge. If clear plastic is used without soil fumigation, a herbicide will be required under the mulch. Herbicides applied to the soil surface should be activated by rainfall or overhead irrigation before a plastic mulch is placed over the soil. Herbicides that are safe to use under field conditions for various vegetable crops may be injurious when covered with plastic. Higher temperatures under the plastic enhance herbicide

TABLE 22-2. Classification of Vegetable Crops According to Their Adaptive Field Temperatures

Cool-Season		Warm-Season	
Hardy[a]	Semihardy	Tender	Very Tender
Asparagus	Carrot	Snap bean	Cucumber
Broccoli	Cauliflower	Sweet corn	Eggplant
Cabbage	Chinese cabbage	Tomato	Lima bean
Horseradish	Lettuce		Muskmelon
Onion	Potato		Okra
Pea			Pumpkin
Spinach			Squash
			Watermelon

[a]Hardy crops are most tolerant of cool temperatures and frost, whereas very tender crops are most susceptible to frost and cool temperatures.

Figure 22-1. Use of fall-seeded and spring-desiccated rye as natural mulch on tomato plant beds.

Figure 22-2. Use of black polyethylene mulch in muskmelon. Between-row weeds controlled with cultivation and herbicides. Rye is used as a windbreak between rows to protect muskmelon plants.

volatility, and this can cause increased crop injury. Consult product labels for precautions and specifics.

If weeds are present between plastic strips before planting, and glyphosate, pelargonic acid, or paraquat is registered for the crop under the Stale Bed Provision, such a herbicide can be used to control all existing vegetation. If glyphosate is used as a broadcast application, overhead irrigation or rainfall is required prior to planting to wash the glyphosate off the plastic. Do not use glyphosate as a broadcast spray if holes for the crop have been punched in the plastic.

Only herbicides registered for the specific crop grown can be used for controlling weeds between the plastic strips. Crop roots generally extend beyond the plastic mulch into row middles, and use of a nonregistered herbicide can result in crop damage and/or illegal herbicide residues in the crop. Preemergence herbicides can be applied prior to or after crop planting. However, in either case the surface of the plastic should not receive any herbicide. The herbicide can be concentrated by washing off the plastic into the holes where the crop plants are and thus cause crop damage.

Postemergence herbicides such as paraquat, sethoxydim, and metribuzin can often be applied as directed sprays to row middles where permitted by label. Where paraquat is utilized, physical shielding is recommended to prevent contact with the crop so as to avoid injury.

A disadvantage of plastic mulch is disposal at the end of the season. Many landfills do not accept plastic mulches. Photodegradable plastic mulches have been developed and are commonly used, but their season-long persistence has been a problem, and they degrade into small pieces that can contaminate the environment.

Mechanical Practices

Mechanical weed management relies on primary and secondary tillage implements such as the rotary hoe and the row cultivator. Mechanical weed management starts with seedbed preparation. Few no-till systems have been developed for vegetable crops.

Field cultivation is usually the first step in mechanically managing weeds and is particularly useful in controlling emerged annual weeds. *Rotary hoeing* is often an important second step in mechanically managing weeds in large-seeded vegetable crops (sweet corn, snap bean, lima bean, and pea). Rotary hoeing should be done after the weeds germinate but before they emerge. Rotary hoeing does not control large-seeded weeds such as velvetleaf and shattercane.

Once the crop has emerged or transplants are established, a *row cultivator* can be used to manage emerged weeds. Many new cultivation implements are available to vegetable growers that offer additional options for weed management. These include flex tine harrows, brush hoes, and finger and torsion weeders. *Flex tine harrows* (Figure 22-3) are used broadcast, both over and between crop rows. They are most efficient when used on small newly emerged weeds. The harrows can be used directly over seeded rows and, with proper tine adjustment, after the crop emerges. The *brush hoe* (Figure 22-4) is a power take-off (PTO)-driven plastic bristle brush that rips weeds from the soil. Shields and an additional operator are used to protect the crop

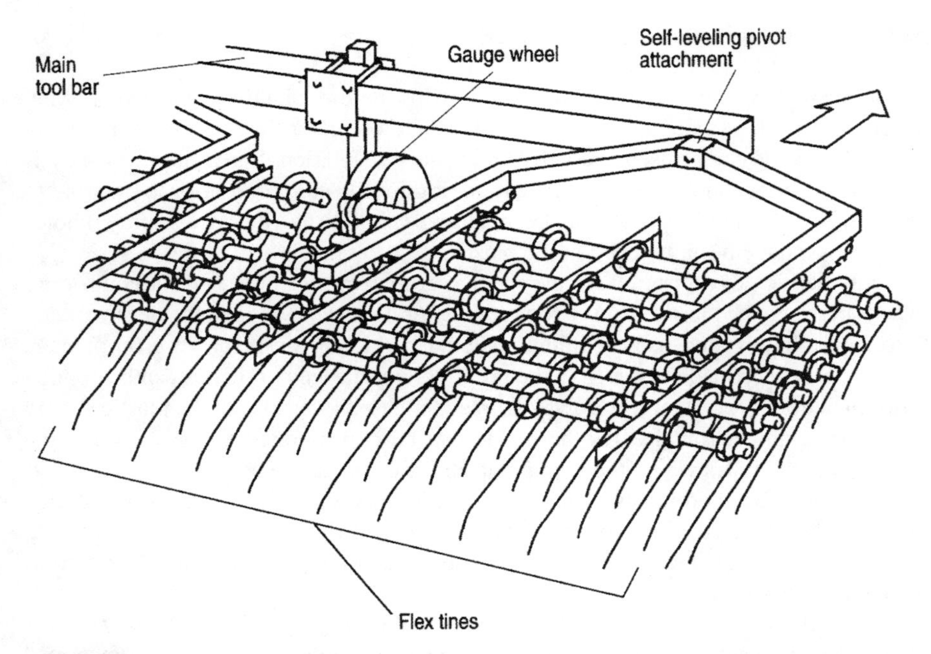

Figure 22-3. Flex tine harrow for cultivation in vegetable crops. Reprinted with permission from *Sustainable Vegetable Production from Start-Up to Market* (NRAES-104), published by the National Resource, Agriculture and Engineering Service, Cooperative Extension, 152 Riley-Robb Hall, Ithaca, NY 14853-5071. Adapted with permission from illustrations by John Gist in *Steel in the Field: A Farmers Guide to Weed Management Tools* (SAN-2).

from damage. Cultivation depth can also be modified to reduce crop damage. The *finger weeder* (Figure 22-5) is designed specifically for in-row cultivation. There are three pairs of ground-driven fingers that uproot and push weeds away from the crop and then move soil back into the row to cover any missed weeds. This machine has to be driven slowly to provide precise cultivation. The *torsion weeder* (Figure 22-6) is mounted on an interrow cultivator and has spring-loaded steel rods on each side of the crop row for undercutting small weeds. Each of these machines offers some advantages over conventional cultivators in their ease of adjustment to minimize crop damage and to provide intra-row weed control. These implements can be used alone or in combination with banded herbicides to allow reduced herbicide use (elimination of broadcast treatments) while providing effective weed control.

Mechanical control has many limitations that must be considered in designing weed management systems. Because mechanical management relies on relatively dry soil, a rainy period may prevent the use of mechanical weed management options and lead to severe weed competition. Relying entirely on mechanical practices to manage weeds is labor intensive, and many growers use herbicides combined with nonchemical approaches to control especially difficult weeds. Among these difficult-to-control weeds are wild proso millet in sweet corn, Canada thistle, hemp dogbane, field bindweed, quackgrass, and johnsongrass. Newly introduced problem

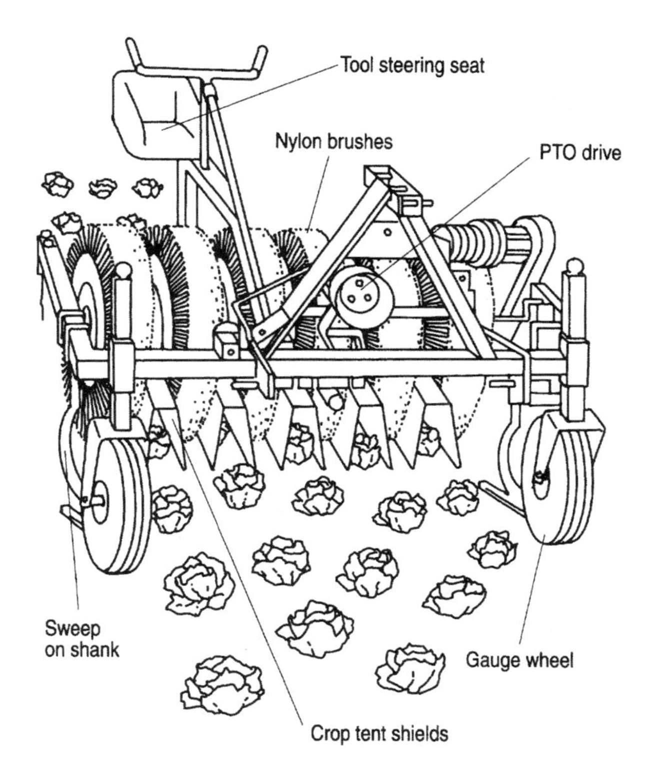

Figure 22-4. Brush hoe is a PTO-driven plastic brush that rips weeds from the soil. A shield protects the crop. Reprinted with permission from *Sustainable Vegetable Production from Start-Up to Market* (NRAES–104), published by the National Resource, Agriculture and Engineering Service, Cooperative Extension, 152 Riley-Robb Hall, Ithaca, NY 14853–5071. Adapted with permission from illustrations by John Gist in *Steel in the Field: A Farmers Guide to Weed Management Tools* (SAN-2).

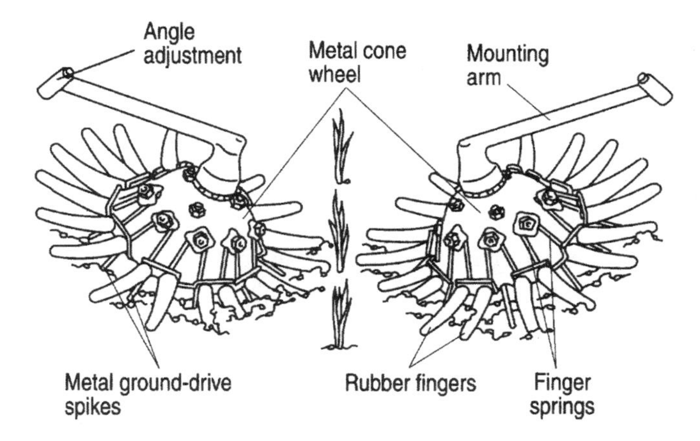

Figure 22-5. Finger weeder is designed specifically for in-row weed removal. Reprinted with permission from *Sustainable Vegetable Production from Start-Up to Market* (NRAES–104), published by the National Resource, Agriculture and Engineering Service, Cooperative Extension, 152 Riley-Robb Hall, Ithaca, NY 14853–5071. Adapted with permission from illustrations by John Gist in *Steel in the Field: A Farmers Guide to Weed Management Tools* (SAN-2).

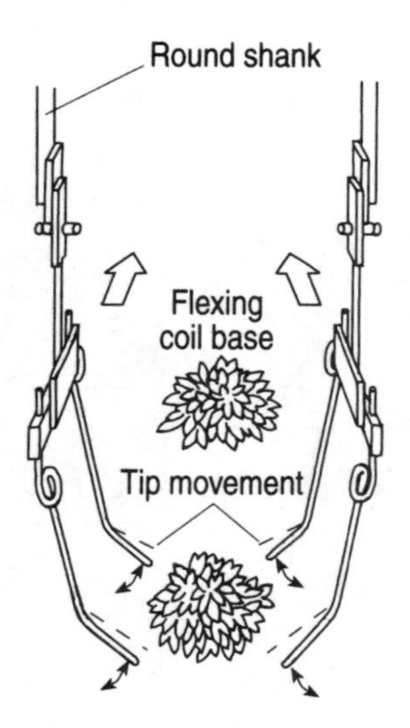

Figure 22-6. Torsion weeder is mounted on an interrow cultivator and has spring-loaded steel rods on each side of the crop row to undercut weeds. Reprinted with permission from *Sustainable Vegetable Production from Start-Up to Market* (NRAES–104), published by the National Resource, Agriculture and Engineering Service, Cooperative Extension, 152 Riley-Robb Hall, Ithaca, NY 14853–5071. Adapted with permission from illustrations by John Gist in *Steel in the Field: A Farmers Guide to Weed Management Tools* (SAN-2).

weeds often show up in scattered patches along headlands and field borders. These are best controlled or eradicated with herbicides before large areas are infested.

Biological Practices

There are currently no management system tools for using insects or diseases to control weeds common in vegetable crops; however, a rust (*Puccinia*) is being developed in Georgia for control of yellow nutsedge in vegetables. Most biological weed management systems to date have been developed to control problem weeds in rangeland areas in the West. One biological system that has potential is the use of cover crops to suppress the development of weeds. These systems are still mostly experimental, but have promise for reducing herbicide use once they are fully developed.

The most promising cover crop system is the use of winter rye, other cereal grains, and certain legumes (Figure 22-7). Winter rye is planted in late summer or early fall and overwintered as a cover crop. In the spring, the rye is killed with a herbicide and/or mowed or rolled 1 week prior to planting the crop. The rye is left as a mulch on the soil surface, and the crop is no-till planted. The system has shown potential to provide early season control of many annual weeds.

Problems encountered in biological management systems include the duration of weed control obtained, the spectrum of weeds controlled, and the requirement for herbicides to initially kill the cover crop. Herbicides may also be needed to provide management of weeds that escape control by the cover during the season.

Figure 22-7. Cabbage grown in a fall-seeded and spring-killed rye and hairy vetch cover crop.

Chemical Weed Management

Vegetables are considered minor crops in reference to pesticide use inasmuch as the acreage of any one vegetable crop is small as compared with that of small grains, corn, soybeans, or cotton. Generally, chemical companies do not consider it economically feasible to develop a herbicide solely for a minor crop if that herbicide is not suitable for a major crop as well. As a result, most herbicides used in vegetable crops were previously developed for a major crop. Moreover, most minor crops are expensive (high-cost) crops. Liability, or risk of lawsuits, and additional registration costs further reduce a company's interest in adding such crops to a label. For these reasons, fewer herbicides are available for use in vegetable crops as compared with major crops. See Chapter 4 for additional information detailing the process for herbicide registration in minor crops and how Interregional Project No. 4 (IR-4) coordinates and supports federal and state research directed toward the development of the information required for registration of a pesticide for use in a minor crop by the Environmental Protection Agency (EPA).

Vegetable crop production is a very specialized business, with crops grown intensively under varying conditions of soil and climate. Several herbicides are often labeled for a particular crop. Scouting fields to determine which weeds are present allows proper selection of a herbicide that will give the best control. Applicators must be aware that a herbicide may produce excellent results under one set of conditions but may injure the crop or fail to control weeds under other conditions. The herbicide label should be checked for specific instructions and precautions. Anyone not experienced in using herbicides on a given crop should proceed cautiously. With experience, the farmer can expand a control program to include a variety of herbicides and integrate other management tactics to obtain acceptable weed control. Extensive listings of herbicides available for use in vegetable production are available from most states' cooperative extension service. Specifics regarding their use in most vegetable crops are provided in the following sections of this chapter.

ARTICHOKES

Artichokes are typically planted from stem portions or transplanted seedlings. They are grown in some regions as an annual crop, but in most regions are grown as a perennial crop for 5 to 10 years. Annual plantings have close in-row and between-row spacing for increased plant density. Most perennial plantings have wide in-row and between-row spacing, although an increasing percentage of the perennial acreage is now grown with closer spacing. Weed control is a major emphasis in both cropping systems, as weeds are a major limitation on yield if not controlled, especially during the crop establishment period and after harvest in perennial plantings.

Preventative weed control prior to planting is critical in artichoke fields, and all perennial weeds should be eliminated prior to planting. Producing a vigorous artichoke stand and maintaining its vigor is the best weed management strategy, especially against annual weeds. Cultural weed control commonly includes cultivation and hand removal in both annual and perennial plantings. In perennial plantings with wide row spacing, bidirectional mechanical cultivation followed by herbicides or hand removal near plants is widely used. When perennial weeds are present in a planting, they must be removed quickly, as they are difficult to eliminate once established.

Herbicides registered for use in artichokes are listed in Table 22-3. Weed management during stand establishment is difficult, but it is essential during the first 8 to 12 weeks or until the crop is established. Because few herbicides are registered for use in new plantings, most growers use both cultural and herbicide methods. When herbicides are used, pronamide is applied prior to weed emergence over the top of newly planted crowns or transplants and irrigated into the soil. Sethoxydim can be applied to control emerged grasses in new plantings.

After artichoke establishment, pronamide or napropamide can be applied over the plant row before or shortly after shoot emergence but before weeds emerge. Other herbicides available for preemergence or postemergence use include simazine, diuron, and oxyfluorfen. Diuron and simazine are used in the fall and are relatively persistent

TABLE 22-3. Artichoke Herbicides[a]

Stand Establishment		
Preemergence	*Postemergence*	
Pronamide	Sethoxydim	Clethodim
Established Plantings		
Preemergence	*Postemergence*	
Napropamide	Diuron	
Simazine	Oxyfluorfen	
Pronamide	Sethoxydim	
Diuron	Clethodim	
Oxyfluorfen		

[a]Always read and follow instructions on the herbicide label.

in soils; rotational restrictions for sensitive crops must be observed. In addition, for crop safety, simazine should not be applied to sandy soils. Oxyfluorfen can be applied as a directed spray for postemergence and preemergence control of annual weeds. Spray contact with the crop will cause injury; therefore, care must be exercised with application of oxyfluorfen. Glyphosate can be used for postemergence control of weeds prior to crop emergence, and sethoxydim and clethodim can be used to selectively control emerged annual grasses and perennial bermudagrass, johnsongrass, and ryegrass.

ASPARAGUS

Asparagus is a perennial crop that may remain in production 10 to 20 years; therefore, year-round weed control must be provided without injuring the crop. There is a wide selection of herbicides for weed control in asparagus (Table 22-4). Herbicide choice(s) depends on many factors, such as crop age, crop growth stage, geographic location, target weed species, and soil type.

Asparagus is established by transplanting crowns that are produced from seed, or seedlings produced in the field or in greenhouses.

Crown Production

For crown production from direct seeding in the field, methyl bromide fumigation of the soil can destroy most weed seeds and many harmful soil organisms prior to planting. If fumigation is not used, there are several herbicides that can aid in reducing weed competition. Paraquat, pelargonic acid, and glyphosate can be applied either prior to seeding or after seeding, but prior to asparagus emergence, to control emerged

TABLE 22-4. Asparagus Herbicides[a]

Seedbeds for Crown Production[b]	
Preemergence	*Postemergence*
Glyphosate	Linuron
Paraquat	Sethoxydim
Pelargonic acid	Fluazifop
Linuron	
Terbacil	
Newly Planted Crowns	
Preemergence[a]	*Postemergence*
Glyphosate	Sethoxydim
Paraquat	Fluazifop
Pelargonic acid	Linuron
Linuron	
Established (1 year or more)	
Preemergence	*Postemergence*
Diuron	Linuron
Norflurazon	2,4-D
Metribuzin	Clopyralid
Terbacil	Dicamba
Napropamide	Fluazifop
	Sethoxydim
	Paraquat
	Glyphosate[c]

[a]Always read and follow instructions on the herbicide label.
[b]Methyl bromide can be applied at least 2 weeks prior to planting.
[c]Preemergence to asparagus, spot treatment immediately after harvest, or broadcast after harvest after all spears are removed. Do not let herbicide contact spears or ferns.

weeds. This is called a *stale bed* application. The stale bed technique consists of preparing the soil for seeding, waiting 7 to 10 days for weed germination, and applying herbicides, followed by seeding. Herbicide application must be followed by minimum soil disturbance to prevent bringing additional weed seeds to the soil surface. This technique is not feasible on soil types that crust severely, interfering with crop seeding.

Because asparagus seeds are slow to germinate, 2 to 3 weeks after planting, paraquat and glyphosate can be applied to emerged weeds during this period. This treatment, which is preemergence to the crop and postemergence to the weeds, is a variation of the stale bed technique.

Preemergence herbicides that can be applied subsequent to seeding include terbacil and linuron. Linuron can also be applied postemergence to weeds when the asparagus fern is 6 to 18 inches tall. Because the selectivity of linuron is based on limited retention of the spray solution by the asparagus fern, no surfactant or crop oil is used in the spray mixture. Sethoxydim and fluazifop can be applied postemergent to the crop for control of emerged grasses.

Newly Planted Crowns

There are only a limited number of herbicides for first-year crown plantings; therefore, it is essential that fields for these plantings are free of noxious and difficult-to-control weeds such as field bindweed, Canada thistle, nutsedges, bermudagrass, and johnsongrass. Methyl bromide fumigation can be utilized to control problem weeds prior to planting, but this is generally too expensive for crown plantings.

Linuron has some regional registrations as a preemergence application to new crown plantings for control of germinating annual broadleaf weeds. Linuron can be applied postemergence when fern growth is 6 to 18 inches tall and weeds are less than 4 inches tall. Fluazifop or sethoxydim applied postemergence will control emerged annual grasses and some perennial grasses such as johnsongrass and bermudagrass. Because there are only regional registrations of linuron, cultivation and the aformentioned postemergence herbicides are used to accomplish weed control in many first-year plantings.

Established Plantings

There are several herbicides registered for established asparagus plantings. Because asparagus is dormant in the winter and crowns are planted relatively deep (8 to 12 inches), growers can control weeds by disking the soil over the crop rows during this period. Herbicides are effectively used during the rest of the year.

Napropamide can be applied early in the season prior to weed and crop emergence. Diuron, terbacil, norflurazon, and metribuzin are preemergence herbicides that can be applied prior to crop emergence and/or immediately after the last harvest. Choice of material and time of application will depend on the type of weeds to be controlled, soil type, economics, and geographic location.

Herbicides that can be applied postemergence to control emerged annual and/or perennial weeds include linuron, 2,4-D, dicamba, paraquat, glyphosate, clopyralid, fluazifop, and sethoxydim. Paraquat, glyphosate, or 2,4-D can be applied to emerged weeds in the spring prior to spear emergence or immediately after the last harvest. Paraquat is the choice for control of annual broadleaf and grass weeds, and glyphosate should be used for perennial weeds and 2,4-D for certain annual and perennial broadleaf weeds. Dicamba or dicamba plus 2,4-D can be used to control annual and perennial broadleaf weeds; however, only one application per season is permitted. Paraquat, glyphosate, 2,4-D, and dicamba will cause crop damage if they contact emerged spears and/or ferns. Linuron can be used as a directed spray to the base of the ferns to control annual broadleaf weeds. Fluazifop and sethoxydim can be applied to plantings of any age for control of emerged annual grasses, and contact with crop foliage does not have to be avoided.

BEANS AND PEAS

Weeds are a major problem for all types of beans and peas. Weeds that are problematic in a particular crop will vary, depending on the location where they are grown. Green beans and lima beans tend to tolerate weed competition more successfully than many other vegetable crops because of their rapid emergence and early growth, although summer annuals can become a problem in mid to late season. English pea is a cool-weather crop, and hence annual cool-season weeds such as chickweed, henbit, sheperds' purse and annual bluegrass are the primary weeds. Unlike English pea, southern pea is a warm-weather crop, and warm-season summer annual broadleaf and grass weeds are the major weeds of concern. Growers must maintain a total weed control program that integrates cultivation and cultural and chemical methods to reduce early-season weed competition and the amount of weeds present at harvest that can interfere with harvest and reduce quality and yields. In all these crops, the most harmful weeds are (1) annual weeds that emerge soon after planting and are not removed and (2) tall weeds that compete for light and other resources. Nutsedges, quackgrass, and Canada thistle can be serious competitors, depending on the geographic location. Preirrigation and planting into moist soil are used in arid areas of the country to minimize early annual weed competition.

There are several herbicides and methods of application that can be used in bean and pea culture. The types of beans and peas grown commercially (Table 22-5) vary considerably in their tolerance to herbicides, and herbicide labels must be checked for instructions on specific use regarding weeds controlled, cultivar or species tolerance, and application timing and rates. There is some confusion in regard to labels of herbicides for beans and peas. Southern pea is a *Vigna* species, which is considered a bean. Therefore, if "bean" appears on the herbicide label, the material can be applied to *Phaseolus* and *Vigna* types, which include snap beans, lima beans, and southern peas. However, if the label says "green bean," it may be applied only to green-colored types, and if the label says "snap bean," it may also be applied to wax types.

Choice of a herbicide(s) for inclusion in a weed management program depends on the weed species to be controlled, geographic limitations, type of bean grown, and soil type. Paraquat, pelargonic acid, or glyphosate can be used in the stale bed technique to nonselectively control emerged annual weeds and, in the case of glyphosate, certain perennial weeds in all these legumes. Paraquat can also be used as a harvest aid in all bean and pea types except snap beans.

Trifluralin and pendimethalin can be applied preemergence or preplant incorporated for annual weed control in all the various beans and peas. Bentazon, sethoxydim, and quizalofop can be applied in all these crops for postemergence control of broadleaf weeds (bentazon) or grasses (sethoxydim and quizalofop).

Certain other herbicides are labeled for use in some, but not all, of these crops in addition to those discussed earlier. Refer to Table 22-5 for specific herbicides labeled

TABLE 22-5. Herbicides for Beans and Peas (Green Beans, Lima, English Peas, Southern Peas)[a]

	Green Beans	Lima	Southern Peas	English Peas
Glyphosate	SB	SB	SB	SB
Paraquat	SB	SB, harvest aid	SB, harvest aid	SB, harvest aid
Pelargonic acid	SB	SB	SB	SB
Clomazone	-	-	-	PPI
Alachlor	-	PPI	-	-
EPTC	PPI, pre	PPI, pre	-	-
s-Metolachlor	PPI, pre		PPI, pre	PPI, pre
Pendimethalin	PPI, pre	PPI, pre	PPI, pre	PPI, pre
Trifluralin	PPI, pre	PPI, pre	PPI, pre	PPI, pre
Imazethapyr	-	PPI, pre post	PPI, pre post	PPI, pre post
Bentazon	post	post	post	post
Quizalofop	post	post	post	post
Sethoxydim Clethodim	post	post	post	post
MCPB	-	-	-	post

[a]Always read and follow instructions on the herbicide label

SB = seed bed prior to crop emergence, PPI = preplant incorporated, pre = preemergence, post = postemergence, and harvest aid = application prior to crop harvest.

for use in each crop. Uses include EPTC for green beans and lima beans; s-metolachlor for snap beans, southern and English peas; imazethapyr in lima beans, southern and English peas; and clomazone and MCPB in English peas. The timing of preemergence, preplant incorporated, and postemergence treatments is guided by crop growth stage, target weed, weed growth stage, and preharvest interval limitations.

CARROT FAMILY (CARROTS, CELERY, DILL, PARSLEY; PARSNIP, TURNIP, AND KOHLRABI ROOTS)

Weed control is difficult in crops of the carrot family because few herbicides are labeled for use and mechanical cultivation within the season is difficult once the crop has emerged. Weed control by necessity involves good site preparation, use of available herbicides, and targeted mechanical cultivation between rows, supplemented with hand weeding within rows. The use of some type of organic mulch within the rows of dill, parsley, and parsnips can provide a measure of additional weed management to reduce the amount of hand weeding necessary. Weed control in carrots is especially important because weeds reduce carrot size and yield, interfere with mechanical harvest, and, if not removed, cause roots to be deformed and unmarketable.

Herbicides registered for weed control in crops of the carrot family are listed in Table 22-6. Weed control in carrots can involve site preparation with paraquat, glyphosate, or pelargonic acid to eliminate emerged weeds prior to planting or prior to seedling emergence. Linuron and trifluralin can be applied preemergence or preplant incorporated (PPI) for annual weed control. Linuron and metribuzin can be applied postemergence to control small emerged broadleaf annual weeds, and sethoxydim, clethodim, and fluazifop will control annual and some perennial grasses when applied postemergence to the weeds. With use of linuron or metribuzin, care must be taken to apply the herbicide at the proper rate and crop growth stage to avoid crop injury.

In celery, site preparation with glyphosate or pelargonic acid can be used to eliminate any emerged weeds prior to or immediately after planting. Preemergence or PPI applications of s-metolachlor, bensulide, thiobencarb, or trifluralin are effective for

TABLE 22-6. Herbicides for Carrot Family (Carrots, Celery, Parsley, Dill, and Parsnips) and Rutabaga, Turnips, and Kohlrabi

	Preemergence	Postemergence
Carrots	Glyphosate Paraquat Pelargonic acid Linuron Trifluralin	Linuron Metribuzin Fluazifop Sethoxydim Clethodim
Celery	Glyphosate Pelargonic acid Bensulide s-Metolachlor Thiobencarb Trifluralin	Prometryn Linuron Sethoxydim Clethodim
Parsley	Glyphosate Pelargonic acid Bensulide Linuron	Prometryn Linuron Sethoxydim Clethodim
Dill	—	Prometryn
Parsnips	Glyphosate Linuron	Linuron Clethodim
Rutabaga	Pelargonic acid	Sethoxydim Clethodim
Kohlrabi	Glyphosate	Clethodim
Turnips	Paraquat Bensulide Trifluralin	Clethodim

control of germinating annual weeds. Prometryn or linuron will control many emerged annual broadleaf weeds, and sethoxydim and clethodim will control emerged grasses.

The herbicides for parsley are similar to those labeled for celery, except that thiobencarb and s-metolachlor are not registered for preemergence or PPI application. Use of linuron should closely follow the label's rate and timing recommendations to avoid crop injury.

Herbicides registered for use in dill, parsnips, turnip, and kohlrabi are limited. There are no herbicides registered for preemergence use in dill, and prometryn and clethodim are registered for postemergence control of emerged annual weeds. Linuron can be used both preemergence and postemergence in parsnips, and glyphosate can be used as a site preparation treatment. In turnips and kohlrabi, pelargonic acid, paraquat, and glyphosate can be used for site preparation and control of emerged weeds before crop emergence; bensulide and trifluralin can be applied preemergence for control of germinating annuals, and sethoxydim and clethodim can be used postemergence to control emerged grasses. In these four crops, weed removal by hand and precision cultivation are widely practiced to reduce weed competition.

COLE CROPS (BROCCOLI, BRUSSELS SPROUTS, CABBAGE, CAULIFLOWER)

Weed control in cole crops is important to reduce early season weed competition and to obtain maximum yields. Weed management is generally a combination of techniques, including site preparation, early season planting for quick crop establishment prior to maximum summer annual weed germination, and use of herbicides along with mechanical cultivation. Multiple mechanical cultivation is not recommended, as yield reductions due to mechanical injury to the crop are common. All cole crops appear to respond similarly to herbicides; however, always check the herbicide label for the registered use for each.

Cole crops may be directly seeded in the field or transplanted. In general, transplanted cole crops are more tolerant of herbicides than direct-seeded crops. However, among the preemergence herbicides registered for the various cole crops (Table 22-7), all except oxyfluorfen can be used for both transplants and direct seeded crops. Selectivity of the herbicides in direct-seeded cole crops is achieved by application timing and, in some cases, rate adjustments.

Glyphosate, pelargonic acid, or paraquat can be applied to emerged weeds prior to seeding or transplanting broccoli, cabbage, brussels sprouts, and cauliflower. The usefulness of these applications is also dependent on minimal soil disturbance during the planting process to avoid bringing new weeds seeds to the soil surface.

Bensulide, trifluralin, napropamide, clethodim and sethoxydim are herbicides that are used in direct-seeded and transplanted broccoli, brussels sprouts, cabbage, and cauliflower. Consult manufacturers' labels for rates, geographic restrictions, and precautions.

Oxyfluorfen is applied to the soil surface prior to transplanting broccoli, cabbage, or cauliflower. Minimal soil disturbance during the transplanting process is essential

TABLE 22-7. Herbicides for Cole Crops (Broccoli, Brussels Sprouts, Cabbage, Cauliflower)[a]

	Broccoli	Brussels Sprouts	Cabbage	Cauliflower
		Preemergence[b]		
Glyphosate	X	X	X	X
Pelargonic acid	X	X	X	X
Paraquat	X	X	X	X
Bensulide	X	X	X	X
Trifluralin	X	X	X	X
s-Metolachlor			X	
Clomazone			X	
Oxyfluorfen[c]	X		X	X
Napropamide	X	X	X	X
		Postemergence		
Pyridate (45WP)[d]			X	
Sethoxydim	X	X	X	X
Clethodim	X	X	X	X

[a]Always read and follow instructions on the herbicide label.
[b]Soil fumigation can be used preplant for site preparation. Allow at least 2 weeks before planting.
[c]Transplants only.
[d]Wettable powder formulation only.

for this treatment to provide the most effective broadleaf weed control. Pyridate in the wettable powder formulation is registered for cabbage for emerged broadleaf weed control; however, the liquid formulation (EC) is *not* registered.

CUCURBIT FAMILY (CUCUMBERS, WATERMELONS, CANTALOUPES, PUMPKINS, AND SQUASH)

Members of the cucurbit family are primarily warm-weather crops; thus, their major weed problems are summer-annual weeds such as barnyardgrass, crabgrass, pigweed, and lambsquarters. An integrated system of proper site preparation, early planting for improved crop competitiveness against weeds, and timely cultivation early in the season, plus the use of herbicides, is essential for good crop growth and yield. Because of the limited number of herbicides available for these crops, many growers use colored or clear polyethylene (plastic) mulch for production of watermelon (Figure 22-8) and muskmelon/cantaloupe to reduce weed competition and promote yield and crop earliness. When clear plastic is used, soil fumigation and/or use of a preemergence herbicide under the plastic is essential to reduce weed pressure. In black (or colored) plastic culture, the main weed problems occur between the rows or in the holes in the plastic where the crop was planted (Figure 22-8).

Figure 22-8. Watermelon grown on black polyethylene mulch. The soil was fumigated with methyl bromide prior to planting. (D. Monks, North Carolina State University.)

Herbicides registered for the various cucurbit crops are listed in Table 22-8. Although there are ten herbicides registered for weed control in cucurbit crops, only seven can be used on all five types of these crops. Paraquat, glyphosate, and pelargonic acid can be used in all cucurbits for site preparation and application before crop emergence. Soil-applied preemergence herbicides for all cucurbits include ethalfluralin and bensulide; clethodim and sethoxydim can be used for emerged grass control. Other preemergence herbicides include naptalm for use in cantaloupes, cucumbers, and watermelons; clomazone (4EC) in cucumbers and pumpkins; and trifluralin in cantaloupes and cucumbers. In some cases, there are regional use restrictions for each of the labeled herbicides in cucurbits. These restrictions are generally based on differences in herbicide selectivity between the cultivars and use practices.

LETTUCE AND GREENS

Unlike cucurbit crops, lettuce and greens are cool-weather plants. Their annual-weed complex generally includes many cool-season annuals, such as annual bluegrass, chickweed, henbit, and common groundsel. The key to weed control in these crops is early planting and quick emergence and stand establishment, so that the crop can outcompete any weeds. Growers generally use cultivation and available herbicides to manage weeds and reduce their negative effects on crop growth and yield. Herbicides labeled for use in these crops are shown in Table 22-9.

TABLE 22-8. Cucurbit Herbicides (Cucumbers, Muskmelons, Watermelons, Pumpkins, Squash)[a]

	Cantaloupes (muskmelon)	Cucumbers	Pumpkins	Squash	Watermelon
	Preplant Preemergence[b]				
Paraquat	X	X	X	X	X
Glyphosate	X	X	X	X	X
Ethalfluralin	X	X	X	X	
Bensulide	X	X	X	X	X
Naptalan	X	X			X
Trifluralin	X	X			X
Clomazone		X	X		
Pelargonic acid	X	X	X	X	X
	Postemergence				
Sethoxydim	X	X	X	X	X
Clethodim	X	X	X	X	X

[a]Always read and follow all instructions on the herbicide label.
[b]Preplant soil fumigation can be used for site preparation. Allow at least 2 weeks before planting.

Lettuce

Although lettuce is usually a cool-weather crop, it may be subjected to high temperatures. In Arizona and southern California, lettuce is planted in late August and September when air and soil daytime temperatures may exceed 100°F. Lettuce will not germinate at temperatures above 90°F. However, the beds are kept wet, so they are cooled by the constant evaporation of water from the bed surface in this arid climate. This cooling allows shallow-planted lettuce seeds to germinate. Under these conditions, a summer annual-weed complex is more common.

In specialized situations, such as lettuce on plastic mulch, methyl bromide is utilized preplant as a soil fumigant to eliminate annual and perennial weeds. Another option for weed management in lettuce is the use of glyphosate, pelargonic acid, or paraquat to control emerged weeds prior to planting or after seeding but before crop emergence.

Preplant soil-incorporated application of benefin (see Figure 22-9), bensulide, or pronamide or preemergence application of pronamide or imazethapyr is used to control annual weeds in lettuce. Pronamide and imazethapyr can also be applied postemergence. Sethoxydim and clethodim can be used postemergence for control of emerged annual and some perennial grasses.

Greens

The greens included in this discussion are chickory, collards, endive, escarole, kale, mustard greens, turnip greens, and spinach. Herbicide tolerances of these crops vary somewhat (Table 22-9); however, collards, kale, mustards, and turnip greens are

Figure 22-9. Annual weed control in lettuce with a preplant soil-incorporated application of benefin. *Center*: Treated band. *Left and right*: Untreated.

closely related members of the mustard family and have similar herbicide tolerances. Endive, escarole, and chickory have similar herbicides registered for use in most cases, but only cycloate and sethoxydim can be used in spinach.

BULB CROPS (ONIONS, GARLIC, LEEK AND SHALLOTS)

Bulb crops include all *Allium* species (onion, garlic, leeks, and shallots) except chives. Depending on the crop, the growing season can be long (dry bulb onions and leeks) or short (green onions). Because herbicides available for use in these crops are limited (Table 22-10) and, by nature, the crops are not competitive against weeds, a comprehensive weed control program is essential. Although the home gardener often grows onions or garlic from bulbs (sets) or transplants, most commercial growers use seed, which makes weeds a particularly serious problem. Because the crop seed germinates and emerges slowly and has a cylindrical upright leaf that does not shade the soil, weed growth within the row is not suppressed. Field selection is critical, so fields that have perennial weeds such as nutsedge or other difficult-to-control weeds must be avoided. Cultivation for weed removal is difficult because these crops are fairly shallow rooted, and extreme care must be taken to avoid root pruning with cultivation, especially during the bulbing period, as crop yield and quality will be decreased. Pulling or hand hoeing is often used during the bulbing period. The cylindrical waxy leaves of *Allium* species provide an advantage when using certain contact herbicides; the spray droplets bounce off the leaves of the crop but remain on leaves of many broadleaf weeds.

Several herbicides used as early postemergence treatments for annual-weed control in dry bulb onions and garlic must be applied only at certain stages of growth to avoid injury to the crop. These stages are classified as loop (crook), flag, one-true-leaf, and two-true-leaf (Figure 22-10).

Glyphosate, pelargonic acid, or paraquat can be used to control annual weeds that emerge before the onions and garlic. These herbicides have no residual soil activity,

TABLE 22-9. Herbicides for Lettuce, Chickory, Collards, Endive, Escarole, Kale, Mustards, Spinach, and Turnip Greens[a]

	Lettuce	Collards	Endive	Escarole	Kale	Mustards	Spinach	Chickory	Turnip Greens
				Preplant					
Glyphosate	X	X			X	X	X		X
Paraquat	X	X			X	X			X
Pelargonic acid	X	X	X		X	X	X		X
Pronamide	X		X	X				X	
Thiobencarb	X		X	X					
Trifluralin		X	X	X	X	X		X	X
Imazethapyr	X		X	X					
Benefin	X								
Bensulide	X	X	X		X	X			X
Cycloate							X		
				Postemergence					
Imazethapyr	X		X	X					
Fluazifop			X	X				X	
Sethoxydim	X	X	X	X	X	X	X	X	X
Clethodim	X							X	

[a]Always read and follow instructions on the herbicide label.

TABLE 22-10. Bulb Crop Herbicides[a]

	Onion		Leek	Garlic	Shallots
	Dry Bulb	Green			
	Preemergence/Preplant Incorporated				
Paraquat	X			X	X
Glyphosate	X		X	X	X
Pelargonic acid	X	X	X	X	X
Bensulide	X			X	
DCPA	X	X			
Pendimethalin	X			X	
	Postemergence				
Fluazifop	X			X	
Sethoxydim	X	X	X	X	X
Clethodim	X	X		X	X
Oxyfluorfen	X				
Bromoxynil	X			X	

[a]Always read and follow instructions on the herbicide label.

and hence a second flush of weeds may appear, particularly if the soil has been disturbed through cultivation. Soil-applied and postemergence herbicides are available to aid in suppressing weeds in onion and garlic plantings. However, these herbicides are specific for the types of onion grown. Green bunching onions are generally short season and nonbulbing, and the entire shoot is eaten; only DCPA and sethoxydim are registered for use. Dry bulb onions and garlic require a longer growing season, and only the bulb is consumed. Herbicides that can be used for dry bulb onions and garlic include glyphosate, pelargonic acid, paraquat, DCPA, pendimethalin, and bensulide preemergence, and sethoxydim, fluazifop, and clethodim can be used to control emerged grass weeds. Oxyfluorfen is registered for use in dry bulb onions only for control of many annual broadleaf weeds when applied postemergence after the onions have at least two fully developed true leaves. Injury to onions and garlic from

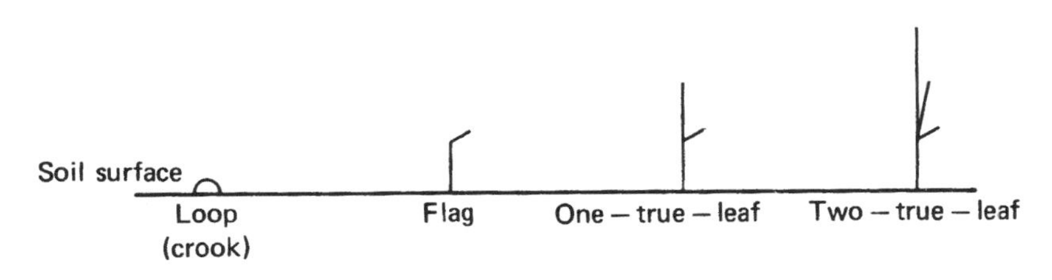

Figure 22-10. Stages of onion growth.

some of the aforementioned herbicides can occur if the products are used incorrectly or under adverse or poor growing conditions. Consult labels for specifics.

For dry bulb onions only, methyl bromide used as a preplant soil fumigant will control many annual and perennial weeds. Because of the expense involved, its use is warranted in only very specialized farming operations.

Few herbicides are registered for green onions (see the preceding discussion), shallots, or leeks. No preemergence soil-applied herbicides are registered for control of germinating annual weeds in leeks or shallots, and only sethoxydim is labeled for emerged grass control. Cultivation and hand removal by pulling or hoeing is commonly used for weed control.

POTATOES

White (Irish)

The potato is the leading vegetable in the world and ranks with the major cereals as one of the leading food crops. The major weed problem in potatoes is annual broadleaf and grass weeds; in some areas perennial weeds such as nutsedges, quackgrass, or johnsongrass can also be serious. Most growers use an integrated approach to weed management including chemical, mechanical, and cultural practices to reduce weed problems. Early-season cultivation is generally very important in potato culture. Rolling cultivators behind the blades used to form hills can uproot many annual weeds that escaped preplant herbicides. Cultivation and hilling is useful for weed control but disrupts the efficacy of previously applied soil herbicides; therefore, many growers apply some herbicide during the hilling operation.

Glyphosate can be applied as a preplant treatment to emerged quackgrass and other perennial weeds. If emerged annuals are present after planting, paraquat or pelargonic acid can also be used. Postplant applications to emerged weeds must be made prior to crop emergence to prevent crop injury.

Preplant treatments are usually incorporated before planting potatoes. EPTC is an effective treatment when incorporated into the soil in preplant and/or lay-by applications to control quackgrass and nutsedge. Postemergence applications of sethoxydim or clethodim will also control quackgrass and johnsongrass.

Many herbicides can be applied to the soil to control annual weeds in potatoes (Table 22-11). Preplant preemergence or PPI treatments include EPTC, pendimethalin, and metolachlor, and delayed preemergence or lay-by treatments can include herbicides such as metribuzin, linuron, and rimsulfuron. Many of the preemergence herbicides are used as tank mix combinations. The majority of these tank mixes include metribuzin combined with herbicides such as pendimethalin or metolachlor (Figure 22-11) to provide broad-spectrum grass and broadleaf weed control. Linuron and rimsulfuron also control annual broadleaf weeds and are frequently tank mixed with the grass-specific herbicides.

Metribuzin or rimsulfuron, and sethoxydim or clethodim can be used postemergence to control annual broadleaf weeds and annual grasses, respectively.

TABLE 22-11. Potato Herbicides (Irish and Sweet)[a]

Irish

PREPLANT INCORPORATED OR PREEMERGENCE	DELAYED PREEMERGENCE[b]	POSTEMERGENCE	VINE KILL
Glyphosate	Metribuzin	Clethodim	Diquat
Paraquat	Linuron	Sethoxydim	Endothall
Pelargonic acid	Rimsulfuron	Metribuzin	Paraquat[c]
s-Metolachlor	s-Metolachlor + linuron	Rimsulfuron	Glufosinate[c]
EPTC	s-Metolachlor + metribuzin		
Pendimethalin	EPTC + pendimethalin		
	Metribuzin + pendimethalin		

Sweet

PREPLANT INCORPORATED OR PREEMERGENCE	POSTEMERGENCE
Glyphosate	Clethodim
Clomazone	Fluazifop
EPTC	Sethoxydim
Napropamide	
Pelargonic acid	

[a]Always read and follow instructions on the herbicide label.
[b]Apply after weeds emerge but before potatoes emerge.
[c]Do not use on potatoes grown for seed or storage.

Figure 22-11. Metolachlor and metribuzin applied preemergence for weed control in Irish potato.

Injury to potato can occur with postemergence applications of metribuzin if several days of cloudy weather precede the applications. In addition, potato varieties differ in metribuzin tolerance. Consult the label for specifics.

Diquat, paraquat, endothall, and glufosinate are registered for vine killing in potatoes (Table 22-11). Vine desiccation is important because tubers harvested from fields containing green vines are more likely to lose skin and bruise during harvest, and the green foliage can interfere with harvesting equipment. Specific instructions on the label must be followed as to proper application timing and storage restrictions for these desiccants.

Sweet Potato

Worldwide, sweet potato is an important vegetable and ranks seventh in production, based on weight, among the food crops of the world. The sweet potato is used for human consumption, animal feed, and industrial purposes. It is a dicotyledonous plant in the morning glory (*Convolulaceae*) family and is not related to the yam, which is a monocotyledonous plant in the family Dioscoreaceae.

The sweet potato is a warm-weather crop having a predominately postrate vine growth habit; the plant expands very rapidly and develops a relatively dense canopy. As with many crops, weed control is critical during the period from planting to canopy closure. Although the sweet potato is an extremely important crop worldwide, relatively few herbicides are registered for use in this crop (Table 22-11).

Sweet potato is generally planted as stem pieces referred to as cuttings, slips, sprouts, or transplants. Bedding roots in the soil and subsequently pulling the sprouts as they emerge produces the transplants. Tissue culture and micropropagation techniques have recently been utilized to develop virus-indexed and improved seed stock. These plants are increased under greenhouse conditions for use in the production of seed potatoes or planted directly into production fields.

Glyphosate and pelargonic acid can be applied to emerged weeds prior to transplanting in the field or prior to crop emergence. Napropamide is the most common preemergence herbicide used on sweet potato plant beds and in fields to control annual weeds. EPTC is used preplant soil-incorporated in certain regions for suppression of nutsedge and control of annual weeds, and clomazone can be applied preemergence for control of annual grasses and certain broadleaf weeds. Generally, preemergence herbicides are applied as banded treatments over the row, and cultivation is relied on for weed control in the row middles. Fluazifop, sethoxydim, and clethodim can be applied postemergence for control of emerged annual grasses, bermudagrass, and johnsongrass. Two applications are generally used for the latter two weeds.

EGGPLANT, PEPPER, AND TOMATO

Eggplant, pepper, and tomato are similar in their tolerance to herbicides. All three crops can be grown commercially from transplants or direct-seeded in the field.

Although annual weeds are the major problem in either planting practice, they are particularly problematic in direct-seeded plantings. Weeds can be removed from direct-seeded plantings of these crops during thinning, but by this time competition may have already delayed the growth of young seedlings. Furthermore, labor costs for thinning may be doubled or tripled by heavy annual weed infestations. Lack of early-season weed control delays crop maturity and, in the case of tomatoes, greatly reduces the efficiency of machine harvesting. Some common annual weeds in these crops are barnyardgrass, crabgrass, foxtails, lambsquarters, purslane, mustard, smartweed, and nightshades, and, in the southern United States, dodder. Nightshades are particularly difficult to control because they are in the same family as the crops (Solanaceae), and most herbicides safe for use in eggplant, peppers, and tomatoes do not control solanaceous weeds.

In general, transplants are more tolerant to herbicides than the direct-seeded crop. However, when the direct-seeded crop is at a transplant size, the plants are generally tolerant to herbicides. The size of the plant, method of application, and other critical factors vary for each herbicide; therefore, detailed information from the herbicide label must be followed.

When these crops are grown on plastic mulch, methyl bromide fumigation immediately prior to laying the plastic is generally used to control annual and perennial weeds, as well as many soil-borne diseases and insects. Bare ground between the plastic can be treated with herbicides listed for the respective crops in Table 22-12.

Eggplant

Relatively few herbicides can be used in eggplant, a vegetable generally handled as a transplanted crop. Prior to transplanting, glyphosate, paraquat, or pelargonic acid can be used to eliminate emerged weeds. Napropamide, triflurailin, or bensulide can be used preplant soil-incorporated or preemergence after planting for annual weed control. Sethoxydim and clethodim are labeled for application to control emerged annual and perennial grasses.

Peppers

Pepper is an important vegetable, and the two types grown are referred to as *bell* and *nonbell*. The bell type is produced primarily for the fresh market, and the nonbell types (banana and chili) are processed (dried, pickled) prior to consumption.

In direct-seeded peppers, applications of glyphosate, pelargonic acid, or paraquat are used to control emerged weeds prior to planting or after seeding but before crop emergence. These materials can also be applied prior to transplanting.

Herbicides used in direct-seeded peppers include bensulide, clomazone and napropamide. They can be used as either preplant soil-incorporation or preemergence treatments. Sethoxydim and clethodim can be applied postemergence to direct-seeded and transplanted pepper plantings for control of annual grasses, as well as quackgrass, bermudagrass, and johnsongrass. In transplants, preplant soil-incorporation treatments include trifluralin as well as bensulide, clomazone and napropamide. There

TABLE 22-12. Tomato, Pepper, and Eggplant Herbicides[a]

Herbicides	Tomato	Pepper	Eggplant
Preplant Incorporated/Preemergence			
Paraquat	X	X	X
Glyphosate	X	X	X
Pelargonic acid	X	X	
Napropamide	X	X	X
Trifluralin[b]	X	X	X
Metribuzin[c]	X		
Pebulate	X		
Bensulide	X	X	X
Trifluralin + metribuzin	X		
Clomazone		X	
Row Middle Application Between Plastic Mulch Applications			
Diquat[c]	X	X	
MCDS	X	X	
Pelargonic acid	X	X	
Paraquat	X	X	
Metribuzin		X	
Postemergence			
Sethoxydim	X	X	X
Clethodim	X	X	X
Metribuzin	X		
Rimsulfuron[d]	X		
Setoxydim + metribuzin	X		

[a]Always read and follow instructions on the herbicide label.
[b]Transplants only.
[c]Vine or plant burndown after harvest.
[d]Processing only.

are several Special Local Needs (Section 18)-labeled herbicides for shielded application to row middles in peppers grown on plastic. These include metolachlor, diquat, MCDS (Enquick), pelargonic acid, and paraquat. These treatments are effective against emerged weeds, and metolachlor will provide preemergence weed control. Such treatments are also useful against dodder present in row middles; however, they cannot be used within the crop row.

Tomatoes

Tomatoes are widely grown for both the fresh market and processing. Most processing acreage is in California, with minor production in the states of Indiana, Ohio,

Michigan, Pennsylvania, and Illinois. Fresh market tomatoes tend to be trellised or staked, whereas processing tomatoes are grown in twin-row beds.

Weed competition in fresh market tomatoes is a problem throughout the life of the planting, because the crop canopy never fully shades the soil within the row, so many growers use black polyethylene mulches (Figure 22-12). In processing tomatoes, crop canopy closure eventually occurs, but early-season weed control is critical to reduce crop/weed competition and prevent yield losses. The noncompetitive nature of tomato necessitates a season-long weed management program involving preplant, preemergence, and postemergence herbicide treatments and mechanical cultivation with some hand hoeing, especially for removal of nightshade weeds.

There are several herbicides registered for use in tomatoes (Table 22-12). As in pepper production, paraquat, pelargonic acid, and glyphosate can be used to control emerged weeds when applied prior to tomato seeding, after seeding but before crop emergence, or prior to transplanting. Paraquat, diquat, and MCDS can also be applied as a directed application (using a shielded sprayer) between rows of established tomato plants grown on plastic, and diquat is labeled for use as a vine desiccant after harvest. Preemergence herbicides include bensulide and napropamide for control of annual weeds in direct-seeded tomatoes. In areas of limited rainfall or to ensure activity, these herbicides are usually incorporated into the soil before planting. In addition to bensulide and napropamide, preplant soil-incorporated herbicides for transplanted tomatoes include metribuzin, pebulate, trifluralin; rimsulfuron is used as a preemergence treatment for processing types only. Various tank mixes of these

Figure 22-12. Weed management in staked fresh market tomatoes with black polyethylene mulch placed on soil surface following methyl bromide soil fumigation. (D. Monks, North Carolina State University.)

products are most commonly used. Metolachlor has been labeled under a Section 18 emergency exemption for nightshade control in tomatoes in several states.

Postemergence herbicides include sethoxydim and clethodim for emerged grass control and metribuzin for emerged broadleaf control. Rimsulfuron can be used postemergence for broadleaf weed control in processing tomatoes.

CORN, SWEET AND POPCORN

Most of the weed control methods described for field corn in Chapter 20 apply to sweet corn and popcorn. The major difference is that the wide variety of sweet corn cultivars available can differ significantly in their response and sensitivity to the many field corn herbicides. The difference in tolerance to herbicides is especially true for the newer *sh*2 (supersweet) cultivars. The sweet corn grower must carefully review each herbicide label for cultivar restrictions and use precautions. Herbicides labeled for use in sweet corn and popcorn, listed in Table 22-13, include many preemergence and preplant incorporated herbicides and postemergence herbicides that can provide effective weed management. Postemergence herbicides must be applied at the appropriate stages of crop and weed growth to avoid significant crop injury. The major difference in available products is that acetochlor, bromoxynil, pyridate, primsulfuron, and nicosulfuron cannot be used in sweet corn and 2,4-D cannot be used in popcorn.

TABLE 22-13. Sweet Corn and Popcorn Herbicides[a]

Preemergence/Preplant Incorporated	Postemergence
Paraquat	Atrazine
Pelargonic acid	Alachlor + atrazine
Glyphosate	Dimethenamid + atrazine
Alachlor	s-Metolachlor + atrazine
Dimethenamid	Bentazon
s-Metolachlor	Halosulfuron
Pendimethalin (processing sweet corn only)	Carfentrazone
Atrazine	2,4-D amine (sweet corn only)
Carfentrazone	Ametryn
Cyanazine	Bromoxynil (popcorn only)
EPTC	Pyridate (popcorn only)
Simazine	Primsulfuron (popcorn only)
Acetochlor (popcorn only)	Nicosulfuron (popcorn only)
Alachlor + atrazine	
Dimethenamid + atrazine	
Metolachlor + atrazine	
s-Metolachlor + atrazine	

[a]Always read and follow instructions on the herbicide label.

MISCELLANEOUS VEGETABLES

There are several vegetable crops that do not fall into clear categories of relatedness to each other. Most of these crops have few herbicides registered for weed control, resulting in a need for intensive cultural and mechanical weed management to reduce crop yield losses. All these crops are mechanically cultivated early in the season and hand hoed as necessary. These crops, listed in Table 22-14, include horseradish, Chinese vegetables, radish, rhubarb, Swiss chard, red beets, okra, and mint.

Horseradish

Horseradish is grown on about 3000 acres in the United States. The crop is propagated from crown sprouts, rhizomes, or occasionally from tissue culture. Propagation stock is planted in the field in early April and harvested the following year. Extensive cultivation is used in most regions to control weeds. Herbicides registered for use include glyphosate and pelargonic acid for site preparation and application prior to crop emergence. Oxyfluorfen is used preemergence for broadleaf weed control, and sethoxydim and clethodim are labeled for postemergence control of emerged grasses. DCPA is also registered as a preemergence application.

Chinese Vegetables

Chinese vegetables include a wide variety of families, including legumes, cole crops, cucurbits, and many others. Some common vegetables in this group are mustard spinach, daikon radish, asparagus bean, napa cabbage, mustard spinach, thick and thin petiole white Chinese leaf cabbage, red and green mustard, bitter melon, winter melon, Japanese greens, Chinese cabbage, edible luffa gourd, and oriental eggplant. Weed control methods must consider cultivation, hand hoeing, and the use of mulching materials, as little is known about most of these crops' responses to herbicides. Presently, only bensulide, clethodim and sethoxydim are labeled for use in any of these crops, and the product labels should be checked for specific crop uses.

Radish

Radish is a cool-season crop that requires about 28 days to mature after planting. The short, early season is a real advantage for weed control, because the crop emerges quickly and most troublesome summer annuals have not germinated this early in the season. The radish seedlings grow quickly and essentially outgrow most weeds that are present. Glyphosate can be used to kill any emerged weeds present prior to radish emergence, and trifluralin can be used as a PPI application prior to planting. Clethodim is registered for emerged grass control.

Rhubarb

Rhubarb is a perennial vegetable whose leaf stalks are harvested and consumed. Clean culture is practiced as much as possible, with high levels of cultivation in the spring

TABLE 22-14. Herbicides for Miscellaneous Vegetables[a]

Preemergence	Postemergence	
Horseradish		
Glyphosate	Sethoxydim	
Pelargonic acid	Clethodim	
DCPA		
Oxyfluorfen		
Chinese Vegetables		
Bensulide	Sethoxydim	Clethodim
Mint (Peppermint and spearmint)		
Terbacil	Terbacil	
Diuron	Bentazon	
Pendimethalin	Bromoxynil	
Oxyfluorfen	Pyridate	
Trifluralin	Sethoxydim	
	Quizalofop	
	Clopyralid	
	MCPB	
	Glyphosate (spot)	
	Paraquat (spot)	
Radish		
Preemergence/Preplant Incorporated	Postemergence	
Trifluralin	Clethodim	
Glyphosate		
Rhubarb		
Pronamide	Sethoxydim	
Paraquat	Clethodim	
Glyphosate		
Swiss Chard		
—	Sethoxydim	Clethodim
Red Beets		
Glyphosate	Pyrazon	Clethodim
Pyrazon	Phenmedipham	
Cycloate	Sethoxydim	
Okra		
Glyphosate		
Pelargonic acid		
Trifluralin		

[a]Always read and follow instructions on the herbicide label.

before emergence and after leaf cutting. Glyphosate and paraquat are labeled for preemergence application and eliminate many troublesome annual and perennial weeds. Pronamide can be used preemergence for annual weed control, and sethoxydim and clethodim are labeled for control of emerged grasses.

Swiss Chard

Swiss chard is seeded, and the leaves are harvested for consumption. The only herbicides registered for use in the crop are sethoxydim and clethodim for postemergence grass control. Preparation of a clean seedbed and the use of both mechanical and hand hoeing are necessary to provide acceptable weed control.

Red Beets

Red or table beets and sugar beets have similar tolerances to herbicides, and the same application methods are used. However, not all sugar beet herbicides (Chapter 20) are registered for use in red beets.

Stale bed application of glyphosate to emerged weeds is useful for eliminating most annual weeds and certain perennial weeds. Cycloate can be applied preplant soil-incorporated, and pyrazon can be applied to the soil surface immediately after planting or postemergence when beets are at the two-leaf stage and before weeds have four leaves. Phenmedipham is labeled for postemergence application to beets past the two-leaf stage and when weeds are in the cotyledon to four-leaf stage. Sethoxydim and clethodim can be applied for control of emerged grass. There are restrictions concerning soil type, growth stage, and environmental conditions for all these herbicides.

Okra

Okra can be seeded or transplanted; however, because few herbicides are registered for use, transplanting provides the crop with some advantage over weeds. Okra can also be grown on plastic to reduce weed competition. Herbicides registered for use in okra are glyphosate and pelargonic acid for preplant kill of emerged weeds, and trifluralin applied PPI before planting. Supplemental mechanical cultivation is essential to remove weeds that are not controlled.

Mint (Peppermint and Spearmint)

Mint is a vegetatively propagated perennial crop that grows from rhizomes (sometimes referred to as roots) that are initially planted into the field in rows. In the first year of the crop the rows canopy over, and in subsequent seasons the mint is grown as a meadow crop. Mint fields are managed on a 3- to 5-year rotational scheme. Mint is harvested as a hay crop, and the hay is distilled to obtain the essential oils widely used in oral hygiene products and candies and for food flavorings.

The two main production areas in the United States are the midwestern states of Indiana, Michigan, and Wisconsin and the northwestern states of Washington, Oregon, Idaho, Montana, South Dakota, and California. In the Midwest mint is fall plowed in most areas, and in the Northwest it is grown as a no-till crop. Weed

Figure 22-13. Terbacil applied in bands in peppermint for preemergence weed control. *Center*: Terbacil banded over peppermint provided complete control of redroot pigweed. *Left and Right*: Untreated peppermint with no control of pigweed.

management by cultivation after the first season is nearly impossible because the mint is grown as a meadow crop, and herbicides are a critical component of any weed management program. In the Midwest some weed control is obtained by fall plowing, and in the Northwest fields are often burned after harvest.

Weed control involves the use of preemergence and postemergence herbicides (Table 22-14); however, winter annuals and perennials tend to be a greater problem in the no-till regions of the Northwest. The most common preemergence herbicide used in the Midwest is terbacil (Figure 22-13), and terbacil, diuron, pendimethalin, oxyfluorfen, and trifluralin are all commonly used in the Northwest. In both regions, postemergence weed control is common. Terbacil and bentazon, along with the grass herbicides quizalofop and sethoxydim, are most commonly used to obtain wide-spectrum postemergence weed control. Pyridate has a 24C label in all mint-producing states and controls emerged pigweed species. MCPB and clopyralid are used where Canada thistle is a problem; clopyralid also controls many emerged winter annual Asteraceae weeds in the northwest. Bromoxynil is used on small emerged annual broadleaf weeds, but caution must be used when applying bromoxynil because severe crop burn can occur. Glyphosate and paraquat have registrations for spot or dormant applications.

PLANT BEDS FOR TRANSPLANT PRODUCTION

Many vegetable transplants are produced in beds both in enclosed structures (greenhouses or cold frames) and in the field. Methyl bromide applied as a soil fumigant can be used in plant beds for the production of any vegetable transplant. Furthermore, any herbicide registered for direct-seeded plantings of the various vegetable crops can also be used in plant beds utilized for transplant production of those vegetables.

LITERATURE CITED AND SUGGESTED READING

Grubinger, V. P. 1999. *Sustainable Vegetable Production from Start-up to Market*. National Resource, Agriculture, and Engineering Service (NRAES), Cornell University, Ithaca, NY.

McGiffin, M., ed. 1997. *Weed Management and Horticulture Crops*. American Society for Horticultural Science (ASHS) Press, Alexandria, VA.

Monaco, T. J., D. C. Sanders, and A. Grayson, 1981. Influence of four weed species on the growth, yield and quality of direct-seeded tomatoes. *Weed Sci*. **29**:394–396.

Smeda, R. J., and L. A. Weston. 1995. Weed management systems for horticulture crops. In *Handbook of Weed Management Systems*, ed. by A. E. Smith, Chapter 14. Marcel Dekker, New York.

Zimdahl, R. L. 1980, *Weed-Crop Competition*. International Plant Protection Center, Oregon State University, Corvallis.

WEB SITES

North Carolina State University: *North Carolina Agricultural Chemicals Manual*

http://ipmwww.ncsu.edu

Select “Pesticide Recommendations.”

University of Florida: *Weed Management in Vegetable Crops*

http://edis.ifas.ufl.edu

Go to “Pest Management Guides,” select “Weed Management Guides,” select “*Weed Management in Vegetable Crops,*” select “Vegetable Crop of Choice.”

Ohio State University: *Ohio Vegetable Production Guide*. 2001. ed. by R. J. Precheur, C. Welty, D. Doohan, R. C. Rowe, R. M. Riedel, and S. Miller. Bulletin 672-01, Cooperative Extension Service, The Ohio State University

http://www.ag.ohio-state.edu

Select “Agriculture and Natural Resources,” scroll to “Ohio Vegetable Production Guide,” and scroll to “Weed Management.”

University of California: *Pest Management Guidelines*

http://www.ipm.ucdavis.edu

Select “How to Manage Pests of Agricultural Crops,” select “Pests of Agricultural Crops,” select “Vegetable Crop of Choice.”

Vegetable Crop Herbicide Label Information

AgChemical Label Information: Crop Data Management Systems, Inc. Marysville, CA

http://www.cdms.net/manuf/manuf.asp

For herbicide use, see the manufacturers or supplier’s label and follow these directions. Also see Preface.

23 Fruit and Nut Crops

Weeds can damage fruit and nut crops in many ways. In newly planted crops, weeds compete directly with young trees for soil moisture, soil nutrients, carbon dioxide, and perhaps light. A review on orchard floor management (Shribbs and Skroch, 1986) indicated that ground covers could vary in their competitive ability. Commonly used ground covers such as orchardgrass, tall fescue, and Kentucky bluegrass were reported to be more competitive with young apple trees than a number of other species. Allelopathy has also been implicated as a detrimental effect of ground covers on the growth of orchard crops. In older plantings, weed competition may be less serious, but still may reduce yields noticeably.

In addition, weeds may harbor plant diseases, insects, and rodents such as field mice or pine voles that girdle trees. Weeds such as poison ivy may interfere at harvest time; when nut crops are harvested from the ground, these and other weeds may seriously interfere with harvesting. The wasteful use of water by weeds is always an important consideration, especially in arid regions.

The crops discussed in this chapter are perennials. They are propagated in nurseries and transplanted to fields. This discussion is limited to field weed control. Although fruit nurseries have serious weed problems, they are not covered here because crop tolerance to most herbicides varies according to several factors, including plant age, soil type, climatic conditions, and geographic location. The herbicide label should be consulted for more specific instructions.

Troublesome perennial weeds, such as quackgrass, johnsongrass, bermudagrass, nutsedge, field bindweed, and Canada thistle, should be brought under control before setting a new orchard or making a small fruit planting. Perennial weeds can be controlled much more easily and at less cost before planting than afterward.

If perennial weeds are absent or controlled before planting the crop, the primary weed problem is annual weeds. However, perennial weeds may invade the area later. This is especially true if perennial weeds are tolerant to the herbicides used. Tolerant perennial and annual weeds flourish when competition from susceptible weeds has been removed by a herbicide.

A total weed control system in fruit and nut crops may combine several methods including *prevention*, *cultivation*, *mowing*, *mulching*, and the *use of herbicides*. Weed control in fruit and nut crops has been reviewed by Smeda and Weston (1995).

TREE FRUITS AND NUTS

Tree fruits (deciduous and citrus) and nuts present a special challenge to the grower regarding weed control. The crops are perennial and remain on the site for many years.

As with any perennial crop, after the initial planting, soil disturbance is minimal. Tillage, although effective for weed control, can result in injury to the trees due to trunk and root damage, and large trees make movement of cultivation equipment difficult. Extensive cultivation in orchards also has a destabilizing effect on the soil, and many orchards are not grown on land conducive to cultivation.

The perennial nature of orchards leads to a typical sequence of succession of plants in the orchard if left unchecked. During the first year many annual weeds are problematic, and into the second and third years annuals slowly give way to biennials and simple perennials. As the site ages beyond the fourth year, more woody perennials appear. The objective of an orchard management program is to maintain the orchard floor so that the weeds present are typical of those found in the first and second year of succession (annuals and biennials). Such a management plan prevents difficult herbaceous and woody perennials from becoming established. This has sometimes been referred to as the "floristic relay" in perennial crops, whereby management delays or prevents floral succession beyond the primary stage.

Ground cover management in orchards can be achieved with cultivation or by use of mowing or herbicides. The specific method depends somewhat on the crop, but perhaps more important are the slope of the land and the soil type. Some kind of ground cover is desirable on hillsides to prevent erosion; mowing and herbicides can be used in this case. On relatively flat terrain, cultivation, mowing, and herbicides can be used, and most often a combination of these methods is employed. Frequently, a herbicide band is used along the tree rows and cultivation or mowing is used between rows. Because mowing or cultivation can be used, the amount of herbicide used is minimized (Figure 23-1).

Cultivation

Cultivation in fruit and nut crops is just as effective and economical as in other tilled crops. The advantages of cultivation are well known by growers. As for the disadvantages, shallow feeder roots are damaged, soil structure is changed, soil erosion increases (especially on hillsides), weeds under trees are difficult to control by mechanized methods, and cultivation brings new weed seeds to the surface, where they may germinate and grow. Cultivation can also spread propagules of perennial weeds throughout the planting, and movement of equipment around trees is difficult. For one or more of these reasons, growers are depending less and less on cultivation alone as the main weed control tactic.

Mowing

Mowing is a popular means of weed control in borders, between rows, and in areas between trees of many fruit and nut crops, especially where soil erosion is a serious threat. These areas can be maintained as short turf, effectively controlling erosion, allowing for drive rows during harvesting, and if properly maintained, keeping weed competition to a minimum. Mowed turf also gives a clean and neat appearance. Disadvantages include competition of the sod with young trees for nutrients and water, and, over time, grass species may shift to those that are lower growing and more

Figure 23-1. Banding herbicides along apple tree rows for weed management. The area between the rows is then mowed to manage the vegetation.

competitive, especially in southern latitudes. Mowing also can spread seeds and propagules if not timed properly. In citrus, air temperatures can be lower by several degrees in groves with sod middles, which can make a critical difference between some or no tree and fruit injury with air temperatures near freezing. The area directly under the tree must be kept weed free, as ground covers near the trunk can harbor rodents that may damage the tree during the winter season.

Cover Crops and Mulching

There is much interest in the use of planted cover crops as a weed suppression tool in orchards and groves. The advantages are similar to those discussed for sod covers. A wide selection of various grain or legume crops can be used as cover crops. These covers can attract beneficial insects, but they may also attract undesirable insects. Disadvantages are that the covers have to be planted and mowed or desiccated with herbicides, which requires additional labor. The covers can compete with the tree for nutrients and water and, in the case of legumes, may release nitrogen into the soil that delays tree hardening-off in the fall. The residues around the tree trunks can be a refuge for rodents that may damage trunks.

Mulching usually controls weeds by depriving them of light. Most mulches also conserve soil moisture. Decomposable organic matter, such as straw, is used in many young fruit tree plantings. Disadvantages include the labor involved in spreading the mulch and initial cost of the material, both of which have limited the use of mulch in tree plantings. In recent years, both black plastic sheets and fabric sheets have been used. The trees are planted through small holes made in these sheets.

Herbicides

Numerous herbicides are registered for use in orchards to control annual and perennial weeds. Tables 23-1 through 23-4 list herbicides available for use in orchard fruits, citrus, and nut crops. In all these crops, many growers maintain a weed-free zone beneath the trees by repeated herbicide applications. The best practices are to use a variety of herbicides and not to rely on one or two specific herbicides over the life of the planting (see Chapter 18 and the following text). Continual use of the same herbicides leads to weed shifts, reduced weed control efficiency, and possibly the development of herbicide-resistant weeds.

Citrus poses the most difficult weed management problems of all the tree crops. The long growing season and the wide variety of potentially problematic weeds result in severe weed competition. The many effective products available make herbicide application an attractive tool for weed control programs. Increasingly, citrus growers are applying herbicides through irrigation systems to increase their efficiency and reduce production costs. Citrus is also an example of a crop in which biological weed management is being practiced. DeVine, a *mycoherbicide* for control of perennial strangler or milkweed vine in citrus groves, is a biological herbicide formulation consisting of live *chlamydospores* of the root rot fungus *Phytophthora palmivora*. The mycoherbicide is applied to the soil surface, which must be moist at time of application to achieve root infection. The fungus will not infect citrus roots.

Generally speaking, the types of herbicides registered for use in orchard, citrus, and nut crops are quite similar. When applying these herbicides, the labels should be consulted and followed for appropriate rates, timing of application, and restrictions based on tree type and age. Several of the herbicides listed in the tables in this chapter can be used only on mature, established trees, and others can be applied only to nonbearing trees or during the dormant season.

There are many soil-applied preemergence herbicides that are effective against annual weeds: isoxaben, napropamide, oryzalin, pendimethalin, and trifluralin. Other soil-applied herbicides, such as bromacil, diuron, norflurazon, dichlobenil, EPTC, metolachlor, oxyfluorfen, simazine, and terbacil, control annual weeds and are effective on certain perennial weeds. Metolachlor is useful for yellow nutsedge control, EPTC suppresses both yellow and purple nutsedge, and bromacil, diuron, norflurazon, dichlobenil, oxyfluorfen, simazine, and terbacil are capable of controlling several perennial weeds. Pronamide provides excellent control of quackgrass.

TABLE 23-1. Herbicides Used on Pome and Stone Fruit Crops[a]

Pome and Stone Fruit Crops	Herbicides																						
	Dazomet	2,4-D	Dichlobenil	Diuron	Fluazifop	Glufosinate	Glyphosate	Isoxaben	MSMA	Metolachlor	Napropamide	Norflurazon	Oryzalin	Oxyfluorfen	Paraquat	Pendimethalin	Pronamide	Pelargonic Acid	Sethoxydim	Simazine	Sulfosate	Terbacil	Trifluralin
Apple	x	x	x	x	x	x	x	x	x	x	x	x	x	x	x	x	x	x	x	x	x	x	
Apricot	x				x	x	x	x	x	x	x	x	x	x	x	x		x	x		x		x
Cherry	x	x	x			x	x	x	x	x	x	x	x	x	x	x	x	x	x		x		
Nectarine	x	x	x	x		x	x	x		x	x	x	x	x	x	x	x	x			x		
Peach		x	x	x		x	x	x	x	x	x	x	x	x	x	x	x	x	x	x	x	x	x
Pear		x	x	x		x	x	x	x	x		x	x	x	x	x	x	x	x	x	x		
Plum	x	x	x			x	x	x	x	x	x	x	x	x	x	x	x	x	x		x		
Prune	x	x	x			x	x	x	x	x	x	x	x	x	x	x	x	x	x		x		

[a]Most of these herbicides control annual weeds; however, specific ones also control perennial weeds. Several have restrictions as to soil type, geographic location, application techniques, tree age, use on bearing and nonbearing crops, and minimum time between application and harvest. Always read and follow all instructions on the herbicide label.

TABLE 23-2. Herbicides Used on Nut Crops[a]

	Herbicides																						
Nut Crop	Atrazine	Dazomet	Dichlobenil	Diuron	EPTC	Fluazifop	Glufosinate	Glyphosate	Isoxaben	MSMA	Metolachlor	Napropomide	Norflurazon	Oryzalin	Oxyfluorfen	Paraquat	Pelargonic Acid	Pendimethalin	Sethoxydim	Simazine	Sulfosate	Terbacil	Trifluralin
Almond			x		x	x	x	x	x	x	x	x	x	x	x	x	x	x	x	x	x		x
Chestnut												x			x		x	x			x		
Filbert		x	x			x	x	x	x			x	x	x	x	x	x		x	x	x		x
Hazelnut												x									x		
Hickory Nut							x														x		
Macadamia	x			x		x	x	x	x		x			x	x	x	x	x	x	x	x		
Pecan			x	x		x	x	x	x		x	x	x	x	x	x	x		x	x	x	x	x
Pistachio						x		x	x			x		x	x	x	x	x	x		x		
Walnut, English		x	x	x	x	x	x	x	x	x	x	x	x	x	x	x	x	x	x	x	x		x
Walnut, Black		x		x		x	x		x			x		x		x	x	x	x	x	x		

[a]Most of these herbicides control annual weeds; however, specific ones also control perennial weeds. Several have restrictions as to soil type, geographic location, application techniques, tree age, use on bearing and nonbearing crops, and minimum time between application and harvest. Always read and follow all instructions on the herbicide label.

TABLE 23-3. Herbicides Used on Citrus Crops[a]

Citrus Crop	Herbicides																								
	Ametryn	Bromacil	DeVine	Dichlobenil	Diuron	DSMA	EPTC	Fluazifop	Glufosinate	Glyphosate	Isoxaben	Metolachlor	MSMA	Napropamide	Norflurazen	Oryzalin	Oxyfluorfen	Paraquat	Pelargonic Acid	Pendimethalin	Sethoxydim	Simazine	Sulfosate	Terbacil	Trifluralin
Grapefruit	x	x	x	x	x	x	x	x	x	x	x	x	x	x	x	x	x	x	x	x	x	x	x	x	x
Kumquat								x	x	x		x				x		x	x	x			x		
Lemon		x	x	x	x	x	x	x	x	x	x	x	x	x	x	x	x	x	x	x	x	x	x	x	x
Lime		x	x	x		x	x	x		x		x	x		x		x	x	x	x	x		x	x	
Orange	x	x	x	x	x	x	x	x	x	x	x	x	x	x	x	x	x	x	x	x	x	x	x	x	x
Tangelo					x			x		x		x					x	x	x	x	x		x		x
Tangerine					x	x	x	x		x		x	x	x	x		x	x	x	x	x		x		

[a]Most of these herbicides control annual weeds; however, specific ones also control perennial weeds. Several have restrictions as to soil type, geographic location, application techniques, tree age, use on bearing and nonbearing crops, and minimum time between application and harvest. Read labels carefully and follow directions.

TABLE 23-4. Herbicides Used on Other Orchard Fruits[a]

	Herbicides												
Crop	Diclobenil	Fluazifop	Glyphosate	Napropamide	MSMA	Norflurazon	Oryzalin	Oxyfluorfen	Paraquat	Pelargonic Acid	Sethoxydim	Simazine	Sulfosate
Avocado	x	x	x	x			x		x	x	x	x	x
Date		x						x		x	x		x
Fig	x	x		x			x	x	x	x	x		x
Mango			x							x			x
Olive		x	x	x	x		x	x	x	x	x	x	x
Papaya			x						x	x			x
Persimmon						x				x			x

[a]Most of these herbicides control annual weeds; however, specific ones also control perennial weeds. Several have restrictions as to soil type, geographic location, application techniques, tree age, use on bearing and nonbearing crops, and minimum time between application and harvest. Read labels carefully and follow directions.

The postemergence herbicides glyphosate and sulfosate provide excellent broad-spectrum control of many emerged annual and perennial weeds. When applying glyphosate and sulfosate, the operator must be careful to avoid contacting leaves or green bark with the spray, as severe tree injury can result. Fluazifop and sethoxydim control emerged grasses, and 2,4-D controls emerged broadleaf weeds. Pelargonic acid, glufosinate and paraquat provide top kill of most annual and perennial species, but do not provide total kill of perennials, as they do not translocate to the plant roots.

Resistant Weed Species

The key to successful chemical weed control in orchards is closely related to the weed species present. Once a herbicide program has been selected, it will usually have to be revised periodically to prevent an increase of resistant weed species; for example, Baron and Monaco (1986) reported that continual use of terbacil for weed control in blueberries resulted in a buildup of tolerant goldenrod. Skroch et al. (1974) observed a buildup of brambles and Virginia creeper with 5 years of continual terbacil usage in apple orchards. The first triazine-resistant weed was found in an orchard after repeated applications of simazine over many years. Herbicide rotation and the use of combinations of two or more herbicides with differing mechanisms of action will prevent or greatly slow the development of resistant weeds in tree fruit plantings (see Chapter 18).

SMALL FRUITS

Grapes

As in orchards, annual weeds in crop rows of grapes are often controlled with herbicides, and the interrow area is cultivated or mowed. Cultivation close to the grapevine often injures the plant. In some soils, repeated cultivations for weed control can hasten the development of a hardpan, which can impede water penetration and affect grape growth. A mowed cover crop or sod between rows with a herbicide-treated strip in the row is common practice in some areas.

Annual weeds in newly established vineyards can be controlled by preemergence soil application of metolachlor, napropamide, oryzalin, dichlobenil, isoxaben, or trifluralin (Table 23-5). After grapevines have become established, several additional herbicides can be applied, depending on type of grape, weeds to be controlled, age of planting, and geographic location. For example, prodiamine can be used preemergence on 1-year-old grapes, norflurazon can be used on 2-year-old grapes, and simazine and diuron can be used after grapevines have been established for 3 years. Oxyfluorfen can be used when grapes are dormant.

Postemergence herbicides include paraquat, for control of a wide array of annual weeds, and fluazifop or sethoxydim, for control of many annual grasses and certain perennials such as johnsongrass, quackgrass, and bermudagrass. Glyphosate, sulfosate, and glufosinate control many weeds, but as with the use of paraquat, contact with green tissue must be avoided to prevent crop injury. Thus, postemergence

TABLE 23-5. Herbicides for Small Fruits[a]

	Herbicides				
	Strawberries	Blackberries and Raspberries	Blueberries	Grapes	Cranberries
PREEMERGENCE					
Paraquat	X	X	X	X	
Pelargonic acid		X		X	
DCPA	X				
Napropamide	X	X	X	X	X
Terbacil	X	X	X		
Diclobenil		X	X	X	X
Norflurazon		X	X	> 2 yrs	X
Oryzalin		X	X	X	
Simazine		X	X	> 3 yrs	X
Hexazinone			X		
Pronamide			X	> 1 yr	
Diuron				> 3 yrs	
Trifluralin				X	
Oxyfluorfen				DORM[b]	
Pendimethalin				NB	
POSTEMERGENCE					
Paraquat	X				
2,4-D amine	X				X
Sethoxydim	X	X	X		X
Clethodim	X	NB[b]	NB		X
Fluazifop	NB	NB	NB		X
Glyphosate		DIR[b]	X		
Glufosinate					X
Sulfosate		NB	NB	X	NB

[a]Always read and follow all instructions on the herbicide label.
[b]NB = non bearing fruit, DORM = dormant treatment, DIR = directed spray application

applications of these herbicides must be shielded or directed to the base of the vine. When applied properly in regard to the stage of weed growth, glyphosate and sulfosate will control many troublesome perennial weeds such as field bindweed, johnsongrass, bermudagrass, artemisia, Canada thistle, quackgrass, and Russian knapweed. Repeat applications of fluazifop, glyphosate, sulfosate, and sethoxydim to weeds that regrow are required in order to obtain acceptable levels of perennial weed control.

Blueberries

Weed problems and general methods of weed control in blueberries depend on the species of blueberry. Lowbush blueberries (*Vaccinium angustifolium*) grown in the northeastern region of the United States are managed in the wild and are grown as solid stands. They have unique weed problems and rely heavily on herbicides because tillage is not possible; however, pruning (e.g., mowing and burning) is also used. In highbush (*V. corymbosum*) and rabbiteye (*V. ashei*) blueberry plantings, weeds are managed by cultivation, mowing, mulching and the use of herbicides. Highbush blueberries are grown in the Mid-Atlantic States through Florida and in other states, such as Michigan. Rabbiteye blueberries are grown primarily in the southeastern states.

Cultivation is sometimes used in highbush and rabbiteye blueberries; however, because blueberries are shallow rooted, cultivation has to be shallow to avoid injury to the plants. Some hand weeding is used in severe infestations. Mulch of sawdust or pine bark can be used under highbush and rabbiteye blueberries for annual weed control, which reduces the need for herbicides, although in mature plantings perennial weeds can become a problem. Many growers maintain a mowed sod strip between rows.

A wide variety of herbicides are registered for use in blueberries (Table 23-5). Once the crop is planted, many preemergence herbicides, including dichlobenil, diuron, hexazinone, isoxaben, napropamide, norflurazon, oryzalin, oxyfluorfen, pronamide, simazine, and terbacil, can be used. These herbicides primarily control annual weeds; however, dichlobenil, hexazinone, pronamide and terbacil also control certain perennial weeds (Figure 23-2). Postemergence applications of fluazifop, clethodim, and sethoxydim can be made to control annual grasses and certain perennial grasses. Carefully directed or shielded postemergence applications of paraquat, sulfosate, and glyphosate control all annual weeds, and glyphosate and sulfosate also control many perennial weeds. The selection of a herbicide program is dependent on type of blueberry being grown, weed(s) to be controlled, geographic location, and age of crop; labels must be consulted for specifics.

Strawberries

Weed control in strawberries may cost up to several hundred dollars per acre a year. By using the proper herbicides and other management practices, this cost can be reduced drastically. Lack of labor for hand weeding has stimulated the use of herbicides in this crop. Black polyethylene plastic or straw is also used to control weeds in strawberries, as well as to conserve moisture and keep the berries clean (Figure 23-3). Weeds that emerge through holes in the plastic are usually removed by hand. Many areas of the United States have turned to the use of single-row plantings grown on plastic for strawberry culture because of the limited number of herbicides available and the cost of hand weeding. In most instances, this is a raised plastic mulch production system in which plants are set on the plastic, usually in double rows, at high plant densities in the fall of the year. The plants are then harvested the next spring or during the growing season, depending on location (Figure 23-3). The planting is renewed every year, so the crop is treated as an annual. Plants are removed at the end of the growing season, the plastic is removed, and this is followed by methyl bromide fumigation under plastic mulch a few months later prior to establishing a new

Figure 23-2. Weed control with terbacil and paraquat in highbush blueberries.

planting. A crop grown with the use of this system is often referred to as "plasticulture strawberries."

Depending on location in the United States, matted row strawberry plantings are maintained for 1 to 3 years. The matted row plantings are usually removed after 3 years as a result of either disease or severe weed infestation and reduced vigor of the stand. Appropriate site preparation and use of herbicides can considerably extend the productive life of many plantings, especially if more effective disease controls become available.

In many areas, fields intended for strawberries are fumigated with a mixture of methyl bromide and chloropicrin before planting. This treatment not only kills many weed seeds but also kills soil-borne disease organisms (e.g., *Verticillium*) and nematodes. Methyl bromide is the primary weed killer of this mixture. Malva, burclover, filaree, field bindweed, and morning glory seeds are often not killed by this treatment.

Methyl bromide and chloropicrin are gases under normal temperatures and pressures. They are sold in pressurized containers as liquids. Both are injected into the soil about 6 inches deep through chisels spaced about 12 inches apart on a tool bar. Immediately after treatment, the treated area is sealed with a tarpaulin (polyethylene)

Figure 23-3. Plasticulture with strawberries. The field was fumigated with methyl bromide, and plastic was placed on the beds prior to planting the strawberries. Ryegrass was planted between the beds as a living mulch. The left two row middles of ryegrass were sprayed with sethoxydim to desiccate the grass. (D. Monks, North Carolina State University.)

for at least 48 hours. Injection of the fumigant and covering with the tarpaulin are completed in one operation, with all equipment mounted on a single tractor (Figure 23-4). Strawberries may be planted 3 days after removing the tarpaulin. If weeds resistant to this treatment emerge later, they have to be controlled by cultivation and hand weeding. New techniques include the use of plastic on beds, with a living mulch or a killed mulch used between rows to aid in weed management (Figure 23-3).

Weed management in many areas relies heavily on herbicide application, with cultivation limited to early season and at time of renovation after the harvest of matted rows. There has been some research to investigate the use of fall-seeded grain cover crops to assist in weed management; however, this practice is not widespread.

Preemergence herbicides that can be used in strawberries are DCPA and napropamide. Simazine is labeled in the states of Washington and Oregon. DCPA and napropamide can be applied to actively growing strawberries but must be applied before weeds emerge. Neither herbicide controls large-seeded broadleaves well, so cultivation, hand hoeing, and spot treatment with postemergence herbicides are required to supplement these herbicides. Terbacil is registered for application to dormant plantings in the winter or immediately following renovation after harvest and after strawberry leaves have been removed by mowing. There are considerable differences in varietal responses to terbacil of which growers must be aware, and

Figure 23-4. Methyl bromide fumigation equipment. The methyl bromide is injected into the soil on the plant beds and immediately covered with a plastic tarp. (D. Sanders, North Carolina State University.)

because the degree of dormancy differs in various locations, some injury can occur if plants are not totally dormant at the time of application.

The number of postemergence herbicides for weed control in strawberries is limited. Sethoxydim can be applied over the top of strawberries for control of emerged annual grasses and certain perennial grasses. However, sethoxydin cannot be applied within 10 days preceding harvest. Clethodim and fluazifop can be applied to nonbearing strawberries for grass control. Postemergence applications of 2,4-D when strawberries are winter dormant will control broadleaf weeds. Strawberry plants must be dormant to avoid injury, and only the amine formulation of 2,4-D is labeled for such use. After harvest and prior to renovation, 2,4-D can be applied to control emerged broadleaf weeds.

Paraquat can be applied as a directed spray between strawberry rows to control annual weeds. Shields are generally used to eliminate contact with the crop. Paraquat cannot be applied more than three times per year or within 21 days preceding harvest.

Cranberries

Cranberries are a unique crop grown in acid bogs, which present interesting problems for weed control. Cranberry is a low-growing, trailing, woody evergreen vine that forms a thick canopy over the soil surface. A good site for cranberries has a soil pH between 4.0 and 5.5, a large supply of water available, and either an organic or a sandy soil. Weed management begins with site selection and relies heavily on a vigorous stand of intertwining vines and the use of herbicides. Weeds can compete with the crop and reduce yield and quality, interfere with harvest, and result in damage to the berry skin.

Cultural practices include site preparation to remove problem weeds. As cranberry beds are prepared, the subsoil is excavated down to about 18 inches above the final water table and rectangular beds are formed for cranberry culture. The removal of the soil for bed preparation reduces later weed and disease problems. Maintaining good drainage in the beds helps to control certain problem weeds, such as wiregrass sedge and arrowhead. Increasing the soil moisture reduces ragweed and goldenrod numbers. The best weed control is obtained by maintaining good fertilization levels that allow the cranberry plants to form a solid stand that shades weeds and prevents weed seed germination and thus allows the cranberry plants to compete successfully against any weeds present. Weeds are also removed as needed by hand, and seed heads are mowed off to prevent seed formation.

Herbicides registered for use in cranberries are listed in Table 23-5. The most common practice in mature stands is to use preemergence applications of napropamide, diclobenil, and granular 2,4-D. These herbicides control many annual and some perennial weeds. Preemergence herbicides are applied in the early spring and can be applied again after harvest in the fall. Postemergence applications of sethoxydim are used for annual and perennial grass control, and wiper applications of 2,4-D or glyphosate are common for control of emerged annuals and perennials. Non bearing plantings can be treated with norflurazon, fluazifop, clethodim, and sulfosate.

Caneberries

Caneberries, or brambles, include mainly blackberries, boysenberries, currants, elderberries, gooseberries, huckleberries, loganberries, and raspberries, all of which are grown widely throughout the United States. Weed control in caneberries can involve some initial cultivation shortly after establishment. However, only light cultivation should be used, as these crops have shallow roots that can be easily injured by cultivation. During the first season, hand hoeing between plants in the row and mechanical cultivation along the rows is practiced, inasmuch as few preemergence herbicides can be safely used in new plantings. By the second year, plants have grown together, preventing further between-plant cultivation. A vigorous caneberry stand will provide shade that inhibits the growth of many weeds within the row and prevents the growth of many annual weeds. Growers often establish a mowed grass sod between caneberry rows which reduces problem weeds and helps stabilize the soil within the planting.

Herbicides registered for use in caneberries are listed in Table 23-5. Napropamide is the most commonly used preemergence herbicide in the first year, and sethoxydim and fluazifop are used to control emerged grass. After the first year, caneberries become more tolerant to a number of excellent preemergence herbicides that provide good control of annual weeds and certain perennials (diclobenil, norflurazon, and terbacil). Emerged annual and certain perennial grasses such as johnsongrass, quackgrass, and bermudagrass can be controlled by applications of fluazifop and sethoxydim. Broad-spectrum herbicides such as glyphosate, sulfosate and paraquat can be directed at the base of the canes, or shielded to prevent crop injury, for control of problem annual and perennial weeds.

LITERATURE CITED AND SUGGESTED READING

Baron, J. J., and T. J. Monaco. 1986. Uptake, translocation and metabolism of hexazinone in blueberry and goldenrod. *Weed Sci* **34**: 824–829.

Shribbs, J. M., and W. A. Skroch. 1986. Influence of 12 ground cover systems on young "Smoothee Golden Delicious" apple trees. II. Nutrition. *J. Am. Soc. Horti. Sci.* **111**: 529–533.

Skroch, W. A., T. J. Sheets, and T. J. Monaco. 1975. Weed populations and herbicide residues in apple orchards after 5 years. *Weed Sci.* **23**:52–56.

Smeda, R. J., and L. A. Weston. 1995. Weed management systems for horticulture crops. In *Handbook of Weed Management Systems*, ed. by A.E. Smith, Chapter 14. Marcel Dekker, New York.

WEB SITES

North Carolina State University: *North Carolina Agricultural Chemicals Manual*

http://ipmwww.ncsu.edu

Select "Pesticide Recommendations"

University of California: *Pest Management Guidelines*

http://www.ipm.ucdavis.edu

Select "How to Manage Pests," select "Pests of Agricultural Crops," select "Fruit or Nut Crop of Choice"

The Ohio State University: *Midwest Small Fruit Pest Management Handbook.* 1997. ed. by R. C. Funt, M. A. Ellis, C. Welty. Bulletin 861. Cooperative Extension Service, The Ohio State University

http://www.ag.ohio-state.edu

Select "Agriculture and Natural Resources," scroll to and select "*Midwest Small Fruit Pest* Management Handbook," scroll to "Weed Management"

University of Florida: *Weed Management in Fruit Crops*

http://edis.ifas.ufl.edu

Select "Pest Management Guides," select "Weed Management Guides," select "Weed Management in Fruit and Nut Crops," select "Fruit Crop of Choice"

AgChemical Label Information: Crop Data Management Systems, Inc. Marysville, CA

http://www.cdms.net/manuf/manuf.asp

For chemical use, see the manufacturer's or supplier's label and follow these directions. Also see the Preface.

24 Lawn, Turf, and Ornamentals

Weed management is important in commercial turf and ornamentals (in nurseries, landscapes, greenhouses, and outdoor cut flower production), recreational areas, and public and private lawns and gardens. Because there are millions of consumers, numerous turf species, and a multitude of ornamental plants, the job of educating users in this area is more complex than in other areas of weed control. This chapter describes weed management in various situations that involve a complexity of plants both desirable and undesirable.

LAWNS AND TURF

Weed control is the greatest agronomic concern in the lawn and turf industry. There are an estimated 93 million single dwellings and more than 46 million acres of turf in the United States. These turf areas include home lawns, commercial lawns, golf courses, athletic fields, parks, campuses, cemeteries, sod farms, recreational areas, and roadsides. Nearly every site has weed problems that must be managed to some degree in order to render the site useable and aesthetically pleasing.

Lawn and turf weed problems have been largely underestimated, and no other type of weed control directly affects so many people. Nearly 53 million households participate in do-it-yourself lawn care, and an additional 9.3 million lawns are maintained by professional lawn care operators (totally ~ 7 million acres). The home lawn care sector is by far the most intensively managed for weeds, and customer complaints regarding weeds are common in the professional lawn care business.

The importance of management practices that produce a strong and vigorous turf cannot be overemphasized. In general, these practices are well understood for each area of the United States and tend to be specific for each geographic area. They include choice of an adapted lawn grass, proper grading and seedbed preparation, fertilization, mowing, and watering, as well as control of insects, diseases, and weeds.

Turfgrass Species

Turfgrass species grown in the United States can be divided into two categories: cool season and warm season. Cool-season grasses are adapted to regions of the country with moderate summer temperatures and cold winter temperatures. These grasses provide the best turf quality in spring and fall when temperatures are cool, and endure hot summer temperatures while suffering a decline in performance. Cool-season grasses have a C-3 photosynthetic pathway.

Warm-season grasses are adapted to regions of the country with hot summer temperatures and mild winter temperatures. These grasses provide the best quality turf in summer and begin to go dormant in mid-fall, remain dormant during winter, and green-up in mid-spring. Warm-season grasses have a C-4 photosynthetic pathway.

Lawn and Turf Care

Weed control alone cannot guarantee a beautiful lawn. Other recommended practices must also be followed. For example, ridding a lawn of crabgrass may leave a bare area unless plans are made to encourage desirable turfgrasses to become established. With proper fertilization, mowing, and watering, lawn grasses such as Kentucky bluegrass or bermudagrass will soon cover the area and be better able to exclude weed encroachment.

New Lawns

Establishment After new homes or buildings are finished, only subsoil may remain for starting a lawn. Topsoil is often hauled in to cover the area 2 to 4 inches deep. Most topsoil contains weed seeds and weedy plants, which soon infest the area. The owner can (1) attempt to get topsoil from fields known to be free of seriously troubling perennial lawn weeds, (2) use a soil fumigant such as methyl bromide to rid the soil of weeds, (3) not use the topsoil before planting, or (4) develop a management plan to control most weeds after planting.

Proper fertilization, including thorough mixing into the upper 3 to 4 inches of soil, makes it possible to grow many turf plants on poor soils. Turf plants are favored by adding peat moss at the rate of 1 bale ($7 ft^3$) per 200 ft^2 of surface, plus liberal amounts of fertilizer, and mixing these 3 to 4 inches into the subsoil. Well-rotted sawdust is as good as peat; however, the fertilization program depends on the degree of sawdust decomposition. Well-decomposed sawdust will require less fertilizer than fresh sawdust. Alternately, well-rotted manure can be used in place of peat moss. Fresh manure will contain weed seeds. A surface mulch of peat moss, wheat straw, or other materials at seeding time will reduce soil erosion and improve turfgrass seedling establishment.

Before Planting Some weeds that cannot be controlled after turf is established should be eliminated before planting. The methods required may delay planting but will reduce the work necessary after the turf is planted.

Weed control before seeding may include shallow cultivation after emergence of most annual weeds, or the use of herbicides. To be effective, the herbicide needs to control all weedy growth and to have no residual toxicity to the turf to be seeded or sodded later. Herbicides especially useful for this purpose are glyphosate and glufosinate. Depending on the weed species present at the site, some phenoxy-type herbicides are often necessary to kill all the weeds present.

Glyphosate is a nonselective, broad-spectrum, foliar-applied herbicide that kills annual and most perennial plants. It is absorbed principally through foliage, requiring 7 to 10 days for translocation to the roots in some species. Application to foliage of desirable plants should be avoided. Glyphosate has no residual soil activity and does not control dormant seeds in the soil. This herbicide reacts with galvanized steel to form hydrogen gas, which is explosive. Stainless steel, aluminum, fiberglass, and plastic are suitable sprayer components.

Glufosinate is a nonselective contact herbicide that kills the tops of plants but seldom kills the roots of perennial plants; consequently, it is less effective as a site preparation treatment. At usual rates of application it has no residual herbicidal activity through the soil.

There are some important cultural considerations in seeding a site. Use only seed that has been certified to have a minimal percentage of weed seed contamination. Contamination with broadleaf and grassy weed seeds can lead to major weed problems during turf establishment and later. When establishing a lawn with sod, it is important that the sod be placed evenly on the surface, as any gaps in the sod can result in germination of crabgrass or other weed seeds in open areas of the sod (Figure 24-1). Following these preparation and planting procedures, the appropriate cultural practices and herbicide treatments may be applied to continue the weed control program.

At Planting Siduron can be applied at seeding of cool-season turf (Kentucky bluegrass, ryegrass, fescue, and some bentgrass) to control annual weedy grasses. Siduron selectively controls annual grasses such as crabgrass, foxtail, and barnyardgrass for about 1 month.

Figure 24-1. A poor job of laying sod leads to uneven turf and gaps where weed seed can germinate and reduce turf quality (Z. Reicher, Purdue University).

After Planting Postplanting control is often needed after turfgrass species have emerged, but before they become well established. This can be achieved by hand weeding, mowing, the use of herbicides, or a combination of these practices. Mowing of newly planted turf will control some erect broadleaf weeds but is not effective on grasses or prostrate broadleaf weeds. While the lawn is young, mowing height should be kept high (2 to 4 inches) so that a minimum of foliage of the turf species is removed, but not so high that certain weeds can become established (Figure 24-2).

Herbicides must be used with care on young turfgrass species to avoid injury. Uniform distribution and proper rate of application are essential. Many emerged broadleaf weeds can be controlled by low rates of bromoxynil, 2,4-D, mecoprop, or dicamba once the grass seedlings have reached the three- to four-leaf stage. At this stage, dosage rates should be reduced to about one-fourth to one-half of those recommended for established turf. As the grass becomes established, these rates can gradually be increased.

Established Turf

A turf is established when the grasses have developed an extensive root system and are well tillered or when the rhizome (runner) system is well developed. The first and most important step in controlling weeds in turf is to properly manage the turf to produce a dense, healthy, actively growing stand of grass. With good management practices, weeds in established turf can often be controlled by hand pulling or cutting the occasional weed out of a home lawn. Large populations of weeds in turf are a symptom of a management problem.

Figure 24–2. Yellow rocket or other similar weeds can become established in newly seeded lawns when improper mowing or no mowing is done after grass seed emergence. Proper mowing height easily eliminates these weeds (Z. Reicher, Purdue University).

Proper Management Common management mistakes in caring for a lawn include mowing the grass too short, mowing infrequently, using improper species for the region, low levels of fertility, soil pH extremes, poor irrigation practices, thatch accumulation, and soil compaction. The optimum mowing height is between 2 and 4 inches, depending on turfgrass species, and a healthy lawn should be mowed every 7 days (Figure 24-3). Cool-season lawns perform best if fertilized in the fall, as this stimulates root growth, resulting in stronger stands. Spring fertilization stimulates excessive shoot growth, depletes carbohydrate reserves in the roots, and necessitates more frequent mowing. Soil pH should be maintained near 6.0 to provide the best nutrient availability and grass growth. Excessive watering or infrequent watering also weakens the grass. Generally, cool-season grasses need 0.8 cm of water per day during periods of high water use, and warm-season grasses need 0.6 cm per day. Thatch buildup reduces air movement in the soil and water penetration, resulting in poor grass root function and poor shoot growth. High-traffic areas that are not maintained at the proper mowing height can, with fertilizer and water, become weedy (Figure 24-4). If weeds become a problem, there are many herbicides that may be used; established turf tolerates herbicides much better than a new planting.

Herbicide Use Herbicides used on established turf, including both soil-active and foliar-active, are listed in Table 24-1. Generally, soil-active, preemergence-type herbicides are applied about 2 to 3 weeks before germination of the weed seeds to be controlled. Some require immediate sprinkle irrigation to minimize foliar injury to desirable turf species, and all must be leached into the soil within a few days after

Figure 24-3. Wild garlic is a problem in lawns that are mowed too short. Proper mowing height of 2 to 4 inches allows the grass to compete against such weeds (Z. Reicher, Purdue University).

Figure 24-4. White clover is a problem in high-traffic turf areas that have been improperly mowed, watered, and fertilized (Z. Reicher, Purdue University).

application (by sprinkler irrigation or rainfall) to be effective. There are many formulations of turf fertilizers available that contain preemergence herbicides that are commonly used in the home and commercial turf market.

Many foliar-active (postemergence) herbicides are available for control of emerged weeds in turf. They tend to be most effective if applied during periods of active weed growth, rather than when weeds are stressed. Postemergence herbicides should not be applied if rain is expected or if irrigation is to be done shortly after application, as they must be thoroughly absorbed into the weed leaves to be effective.

Grass and Sedge Control with Herbicides Grasses can be grouped in three categories: winter annual, summer annual, and perennial. Annual bluegrass is a winter annual that germinates when the soil temperature falls below 70°F in the fall, and herbicides to control it must be applied prior to fall germination. Summer annuals (crabgrass, goosegrass, foxtails, barnyardgrass) germinate in the spring, and herbicides for their control are applied in the early spring. Perennial grasses (quackgrass, bermudagrass, creeping bentgrass, orchardgrass, nimblewill, tall fescue, and zoysiagrass) are present throughout the year and grow in the same site over many years; they can spread by seed and vegetative reproduction. Perennial grasses are difficult to control once present in a lawn. With heavy infestations, complete lawn renovation is necessary. In light infestations, selective postemergence herbicides can provide some control or nonselective herbicides can be used, followed by turf reestablishment. Sedges can be annual or perennial, but the perennial yellow and purple nutsedges are the most problematic. Purple nutsedge is

TABLE 24-1. Herbicides Registered for Use in Lawns and Turf [a]

Preemergence	Postemergence
Benefin	2,4-D + triclopyr
Benefin + trifluralin	2,4-D + clopyralid + dicamba
Bensulide	Bentazon + atrazine
Bensulide + oxadiazon	Metsulfuron
Dithiopyr	Chlorsulfuron
Metolachlor	Diquat
Napropamide	Glyphosate
Oryzalin	Fluazifop
Oxadiazon	Clethodim
Oxadiazon + benefin	Glufosinate
Oxadiazon + prodiamine	Fenoxaprop
Pendimethalin	Metribuzin
Prodiamine	Sethoxydim
Siduron	Quinclorac
Fenarimol	Diclofop
Ethofumesate	DSMA
Atrazine	MSMA
Simazine	MSMA + metribuzin
Isoxaben	MSMA + 2,4-D + dicamba
	Asulam
	Bentazon
	Imazaquin
	Halosulfuron
	Imazapic
	2,4-D amine
	MCPP
	Dicamba
	Clopyralid
	MCPA + MCPP + dicamba
	MCPA + MCPP + dichlorprop
	MCPA + triclopyr + dicamba
	thiclopyr + clopyralid
	2,4-D + dicamba
	2,4-D + dichlorprop
	2,4-D + MCPP
	2,4-D + MCPP + dicamba
	2,4-D + MCPP + dichlorprop

[a]Herbicides listed are registered for use, but labels must be referred to for turfgrass tolerance, application timing, type of planting, rates, formulations, and all other instructions and precautions.

found only in warm, subtropical regions of the United States. Nutsedges are controlled with selective postemergence hebicides applied in late spring and/or early summer.

Preemergence and postemergence herbicides commonly used to control annual grasses, some perennial grasses, and sedges in turf are listed in Table 24-2. The tolerances of established cool-season and warm-season turfgrasses to

TABLE 24-2. Herbicides for Control of Grassy Weeds in Turf [a]

Large Crabgrass, Foxtails, Goosegrass		Annual Bluegrass		Nutsedge and Annual Sedges
Preemergence	Postemergence	Preemergence	Postemergence	Postemergence
Benefin	Fenoxaprop	Benefin	Diquat	MSMA
Benefin + trifluralin	Metribuzin	Benefin + trifluralin	Glypohsate	Imazaquin
Bensulide	Sethoxydim	Bensulide	Metribuzin	Halosulfuron
Bensulide + oxadiazon	Quinclorac	Dithiopyr	Ethofumesate	Imazapic
Dithiopyr	Diclofop	Metolachlor	Atrazine	Bentazon
Metolachlor	DSMA	Napropamide	Simazine	
Napropamide	MSMA	Oryzalin		
Oryzalin	MSMA + metribuzin	Oxadiazon		
Oxadiazon	MSMA + 2,4-D	Pendimethalin		
Oxidiazon + benefin	MCPP + dicamba	Prodiamine		
Oxidiazon + prodiamine	Asulam	Ethofumesate		
Pendimethalin	Imazapic	Atrazine		
Prodiamine		Fenarimol		
Siduron				

[a]Always read and follow all instructions on the herbicide label.

preemergence herbicides used for annual grass are summarized in Table 24-3 and Table 24-4. The tolerances of both types of turfgrasses to postemergence herbicides for grass and/or broadleaf weed control are listed in Table 24-5. Information provided in this table indicates that only a few of the turfgrasses are tolerant to herbicides such as fenoxaprop, sethoxydim, DSMA, MSMA, and pronamide, which are used for control of grassy weeds. Fenarimol is a fungicide used in turf that can also exhibit herbicidal effects by inhibiting the growth of seedling annual bluegrass.

Broadleaf Weed Control with Herbicides Many types of broadleaf weeds may be found in turf, including winter annuals, summer annuals, biennials, and perennials, and there are many herbicides available for their control in turf (Table 24-1). The tolerance of cool- and warm-season turfgrass species to specific broadleaf weed herbicides is shown in Table 24-5 for non-growth-regulator types and in Table 24-6 for growth-regulator type herbicides. Table A-1 in the Appendix provides an extensive list of broadleaf weeds and their susceptibility to these various herbicides. Many commercial formulations of turf grass herbicides for broadleaf weed control contain two-way and three-way mixes (Table 24-6) which provide control of a wide spectrum of these weeds, and several commercially available lawn fertilizers are formulated with broadleaf and grass herbicides.

TABLE 24-3. Tolerance of Established Cool-Season Turfgrasses to Preemergence Herbicides for Control of Annual Weedy Grasses[a]

Herbicide	Kentucky Bluegrass	Tall Fescue	Fine Fescue	Perennial Ryegrass	Bentgrass Golf Greens
Benefin	T	T	M	T	NR
Benefin + oryzalin	NR	T	NR	NR	NR
Benefin + trifluralin	T	T	M	T	NR
Bensulide	T	T	T	T	T
Bensulide + oxadiazon	T	T	NR	T	T
DCPA	T	T	M	T	NR
Dithiopyr	T	T	T	T	T
Napropamide	NR	T	T	NR	NR
Oryzalin	NR	T	NR	NR	NR
Oxadiazon	T	T	NR	T	NR
Pendimenthalin	T	T	T	T	NR
Prodiamine	T	T	T	T	NR
Siduron	T	T	T	T	M

[a]Always read and follow all instructions on the herbicide label.
T = tolerant, when used properly according to the label; M = marginally tolerant, may cause injury or thinning of the turf; NR = not registered. Apply only to established grasses.
Only benefin, bensulide, DCPA, and oxadiazon may be applied in the spring to grasses seeded the previous fall. Siduron may be applied when seeding tolerant grasses. Do not use dithiopyr on Chewings fescue, colonial bentgrass, or unamended golf greens.

TABLE 24-4. Tolerance of Established Warm-Season Turfgrasses to Preemergence Herbicides for Control of Annual Weedy Grasses

Herbicide	Bahiagrass	Bermudagrass	Bermudagrass Putting Greens	Centipedegrass	St. Augustinegrass	Zoysiagrass
Benefin	T	T	NR	T	T	T
Benefin + oryzalin	T	T	NR	T	T	T
Benefin + trifluralin	T	T	NR	T	T	T
Bensulide	T	T	T	T	T	T
Bensulide + oxadiazon	NR	T	T	NR	NR	T
DCPA	T	T	NR	T	T	T
Dithiopyr	T	T	T	T	T	T
Metolachlor	T	T	NR	T	T	T
Napropamide	T	T	NR	T	T	NR
Oryzalin	T	T	NR	T	T	T
Oxadiazon	NR	T	NR	NR	T	T
Pendimenthalin	T	T	T	T	T	T
Prodiamine	NR	T	NR	T	T	T
Siduron	NR	NR	NR	NR	NR	T
Simazine and atrazine	NR	T	NR	T	T	T

T = tolerant, when used properly according to the label; NR = not registered for use on this turfgrass. Do not use dithiopyr on unamended golf greens.
From Bulletin #AG-348, Turfgrass Pest Management Manual: A Guide to Major Pests of Turfgrass by A. H. Bruneau, North Carolina Extension Service.

TABLE 24-5. Tolerance of Turfgrasses to Postemergence Herbicides for Grass and/or Broadleaf Weed Control[a]

Turfgrass	DSMA, MSMA, CMA	Asulam	Atrazine	Bentazon	Chlorsulfuron	Diclofop	Dithiopyr	Ethofumesate	Fenoxaprop	Glyphosate	Imazapic	Imazaquin	Metsulfuron	Metribuzin	Pronamide	Quinclorac	Sethoxydim
Cool Season																	
Bentgrass	I	S	S	T	T	S	T	I	T	S	S	S	S-I	S	S	I	S
Kentucky Bluegrass	I	S	S	T	T	S	T	T	T	S	S	S	I	S	S	T	S
Tall Fescue	I	S	S	T	S	S	T	T	T	S	S-I	S	S-I	S	S	T	S
Fine Fescue	I	S	S	T	T	S	T	S	T	S	S-I	S	I	S	S	I	T
Perennial Ryegrass	T	S	S	I	S	S	T	T	T	S	S	S	S	S	S	T	S
Warm Season																	
Bahiagrass	S	S	S	T	I	S	T	S	S	D	S-I	S	S	S	S	S	S
Bermudagrass	T	T[b]	T	T	T	T	T	D	S	D	I	T	T	T	T	T[c]	S
Centipedegrass	S	S	T	T	I	S	T	S	S	S	T	T	I	S	S	S	T
St. Augustinegrass	S	T	T	T	I	S	T	T	S	S	S	T	T	S	S	S	S
Zoysiagrass	I	S	T	T	I	S	T	S	T	S	I	T	T	S	S	T	S

[a]Always read and follow all instructions on the herbicide label.
[b]Use only on Tifway 419 bermudagrass.
[c]Hybrid bermudagrass is intermediately tolerant to quinclorac.
D = Apply only during dormant season; I = Intermediately tolerant; use with caution, use at reduced label rates, or minium label rates; S = sensitive–do not use this herbicide; T = tolerant.

TABLE 24-6. Tolerance of Turfgrasses to Postemergence Herbicides for Broadleaf Weed Control[a]

Turfgrass	2,4-D	Mecoprop	Dicamba	Bentazon + Atrazine	2,4-D + Dichlorprop	2,4-D + Mecoprop	2,4-D + Triclopyr	Triclopyr + Clopyralid	2,4-D + Clopyralid + Dicamba	2,4-D + Mecoprop + Dicamba	2,4-D + Mecoprop + Dichlorprop	MCPA + Mecoprop + Dichlorprop
Cool Season												
Bentgrass	S-I	T	S-I	S	S-I	I	S-I	S	T[b]	I	I	I
Kentucky Bluegrass	T	T	T	S	T	T	T	T	T	T	T	T
Tall Fescue	T	T	T	S	T	T	T	T	T	T	T	T
Fine Fescue	T	T	T	S	T	T	I	S	T	T	T	T
Perennial Ryegrass	T	T	T	S	T	T	T	T	T	T	T	T
Warm Season												
Bahiagrass	S-I	T	T	S	I	T	S	S	T	I-T	T	T
Bermudagrass	T	T	T	T	T	T	S	T	T	I-T	T	T
Centipedegrass	S-I	T	I	T	S-I	S-I	S	T	S	S-I	I	I
St. Augustinegrass	S-I	S-I	S-I	T	S-I	S-I	S	S	S	S-I	I	I
Zoysiagrass	T	T	T	T	S-I	S-I	S	T	T	T	T	T

[a]Always read and follow all instructions on the herbicide label.
[b]Do not use on golf course greens and trees.
I = intermediate tolerance; use with caution, use at reduced label rates or minimum label rates; S = sensitive—*do not* use this herbicide; T = tolerant; tolerance frequently depends on rate used and formulation selected; therefore, carefully check label.

GOLF COURSES

Weed control on golf courses is important; however, less herbicide is routinely applied in golf courses than in lawns. Golf course greens are seldom treated with herbicide unless goosegrass or annual bluegrass is a problem. Fairways may be treated with preemergence herbicides for annual grass control but are usually only spot treated for broadleaf weeds, and rough areas generally receive only spot treatments for broadleaf weed control as necessary.

NURSERY AND LANDSCAPE ORNAMENTALS

Weeds in ornamental nurseries and landscapes present serious problems for nursery operators, lawn care professionals, and home gardeners. In the United States, there is increasing emphasis on providing quality ornamentals for planting in well-designed and maintained landscapes for improved aesthetics and enjoyment of the public. This requires effective weed management. A balanced weed management program requires an integrated approach including *prevention, sanitation, hand weeding, mulching, mowing, cultivation*, and *the use of herbicides*. Selecting the most suitable methods of controlling weeds in ornamental nurseries is a difficult task, although herbicides do play a major role. In home landscapes that utilize a balanced approach to weed management, herbicides may have a lesser role.

Prevention and Sanitation

As with all crop sites, it is important to eliminate problem weeds from nurseries and home landscapes prior to planting. In nurseries this effort may include cropping the site with agronomic crops and maintaining a high level of weed control, following the site and using nonselective herbicides to eliminate problem weeds, cover cropping with cereal grains or legumes (green manures), or use of chemical fumigation prior to planting. Establishing plants in a weed-free site is critical to long-term success. A second form of prevention is to exclude weed introduction into the nursery. Weed seeds and other propagules can be introduced into a nursery by wind-blown seed, washed in by runoff, deposited by animals, or introduced in contaminated potting soil or in root balls of purchased plants. In nurseries and in home or public landscapes, it is important never to purchase trees, liners, or container plants that contain weeds. Many weeds, especially perennials, are introduced into nurseries and home landscapes from contaminated stock plants and, once introduced, can cause a continuing weed management problem. Other types of prevention include not allowing weeds to produce seeds, scouting to identify infestations of new weeds that can be targeted for control, removal of containers with perennial weeds, and disposal of the biomass of weeds pulled from containers, especially if they have seedheads or can reproduce vegetatively.

Cultivation

Cultivation involves removing weeds by manual or mechanical methods. Manual removal is tedious and expensive but is occasionally necessary in containers and home landscapes. Mechanical techniques for liners and other in-ground plantings in nurseries can include rototilling, disking, plowing, and hoeing, but care must be employed to avoid physical damage to valuable plants during weed removal.

Cover Crops

Cover crops are used in nurseries to improve weed management and reduce erosion. Common cover crops include seeded grain crops (rye, oats, wheat) and legumes that can be mowed and maintained as a living mulch or killed by a herbicide and used as a dried surface mulch. Some difficulties with establishment and stand vigor have limited these techniques in northern latitudes.

Mowing

Mowing is a form of weed control used in nurseries and home landscapes to reduce the reproductive potential of weeds. Mowing has to be timed prior to seed production and must be repeated when used against perennial weeds in order to weaken the underground rhizomes and roots. Mowing is not effective against low-growing weeds or weeds that flower below the mowing blade height. Furthermore, mowing does not reduce the competitiveness of weeds with field-grown nursery crops.

Mulching

Mulching is a very effective method of weed control for home and public landscape beds, as well as in certain nursery beds and container production. Mulching the soil surface reduces the light necessary for the germination of many weed seeds and presents a physical barrier to emergence. Other beneficial properties include moisture retention, soil stabilization, enhanced aesthetics, a reduction in the need for herbicides, and effectiveness of the herbicides used.

Mulches include natural organic, natural inorganic, and synthetic materials. Natural organic mulches can be composted yard wastes or animal wastes, various grain straws (weed free), peat moss, and pine straw, but the most common types are wood chips, hardwood bark, and softwood bark. Organic mulches, placed on the soil at 2- to 3-inch or greater depths, provide good weed control and have good stability over 1 to 2 years. Bark mulches are extremely popular in landscaping.

Inorganic mulches are not commonly used in commercial nurseries but are fairly common in home and public landscapes. Inorganic mulches can include sand, pebbles, stones, shale, and other rock types and are often available in a variety of colors. Inorganics are stable over a long time but require either a plastic mulch placed on the soil surface beneath them or the use of a herbicide, because weeds do germinate and emerge through these materials.

Synthetic mulches (also called geotextile mulches) are generally some type of polyethylene or woven synthetic fabric. These mulches are available as solid sheets, mats, or webbed material and come in a variety of colors, thicknesses, and service lives. Solid sheets are impermeable to water, but mats or webbed materials are permeable. The main purpose of synthetic materials is to prevent weeds from germinating and becoming a problem. These mulches are commonly used in landscapes, especially under inorganic mulches, and as soil covers in nurseries under containers to reduce weed problems between pots.

Use of Herbicides

Herbicides are still the most widely used means of weed control in nurseries and are safely used in a variety of other landscape situations as well. Weed control is essential in nurseries in order to provide healthy, weed-free plants to the consumer.

The primary herbicides used in nurseries and landscapes include both preemergence and postemergence compounds (Table 24-7). The best weed management program includes careful planning and implementation of an integrated approach that uses all available tactics. There are a wide variety of herbicide choices

TABLE 24-7. Ornamental Herbicides (See label for specific ornamentals)[a]

Preplant to All Ornamentals	Preemergence	Postemergence
Methyl bromide	Benefin+ oryzalin	Clethodim
Dazomet	Bensulide	Diclobenil
Diquat	Diclobenil	Fenoxaprop
Glufosinate	Imazaquin	Fluazifop
Glyphosate	Isoxaben	Glyphosate (no green bark or foliage)
Paraquat	Isoxaben + trifluralin	Diquat
Pelargonic acid	Metolachlor	Sethoxydim
	Napropamide	Asulam
	Napropamide + oxadiazon	Bentazon
	Norflurazon	Clopyralid
	Oryzalin	EPTC
	Oxadiazon	Imazaquin
	Oxyfluorfen	Oxyfluorfen
	Oxyfluorfen + oryzalin	Paraquat
	Oxyfluorfen + oxadiazon	Halosulfuron
	Oxyfluorfen + pendimethalin	
	Prodiamine	
	Pronamide	
	Simazine	
	Trifluralin	
	DCPA	
	Prodiamine + oxadiazon	
	Pendimethalin	
	Dithyopyr	

[a]Always read and follow all instructions on the herbicide label.

available; however, there are also a wide variety of landscape ornamental plants with significant differences in tolerance to herbicides. A manager should plan ahead by preparing the site properly and planting species having similar herbicide tolerances together in order to simplify herbicide applications and improve the effectiveness of the weed control program (Figure 24-5). The wide variety of plants in most landscapes results in many limitations relating to a simple herbicide-only approach to weed control; therefore, many home owners choose to use mulches, hand removal, and spot sprays with broad-spectrum herbicides. Nonselective sprays such as glyphosate, glufosinate, diquat, and pelargonic acid are especially effective for killing vegetation in brick walks, along landscape bed borders, and under woody ornamentals. Best results are obtained when weeds are 1 to 2 inches tall and actively growing, and care must be used to keep the spray off the foliage of valued ornamentals.

Preemergence herbicide use requires the appropriate choice of chemical for the weeds at the site (based on scouting and previous experience) and the crops planted at the site and application at the proper rate prior to any weed emergence. The selectivity of herbicides for ornamentals is often achieved by formulation. Many preemergence herbicides in woody ornamentals are formulated as granules, and foliar contact is avoided if the granules are applied when foliage is dry, because they tend to bounce off the leaves. Most preemergence herbicides work best when uniformly applied and irrigation follows immediately to wash the herbicide off the plant foliage. The irrigation moves the herbicide into the growth medium or soil, where it can inhibit

Figure 24-5. Use of banded preemergence herbicides within the row of nursery trees, combined with mowed grass strips between tree rows, results in effective and economic weed management.

weed seed germination. Many preemergence herbicides for nurseries provide acceptable weed control for most of the season in field production and for 8 to 12 weeks in containers. Containers are hand weeded prior to reapplication of preemergence herbicides.

Postemergence herbicides available for use in container production will not control all weeds present in the plantings, as very few postemergence herbicides are labeled for use in container production. Grasses can be controlled postemergence, but no herbicides are available for postemergence broadleaf weed control in containers. In field nurseries, more postemergence herbicide options are available; however, directed applications or spot treatments of nonselective herbicides such as glyphosate, glufosinate, paraquat, or pelargonic acid are commonly used to control emerged annual and perennial weeds. Selectivity is obtained by carefully directing or shielding the application to the base of woody ornamentals.

The key component in nursery weed management is to start weed free and maintain the site as weed free as possible. Once weeds become widespread in a nursery or landscape, it is very difficult to ever remove all the weeds. Use an integrated weed management approach for best results, and when using herbicides, obtain help in selecting the proper herbicides for weed control from local county extension agents, other public agencies, or local landscape management companies. There are many excellent weed management guides available from the university extension service in most states. One should always read the herbicide manufacturer's literature and labels to ensure proper use and obtain the most effective weed control with no injury to valued ornamentals.

GREENHOUSES

Weeds in greenhouses can diminish the quality of potted plants, harbor insects such as whitefly, mites, and thrips, and contribute to increased workloads for employees. Weeds do not have to be a major problem in greenhouses if operators follow proper management practices. The best weed control plan for greenhouses is based on prevention, sanitation, and removal of weeds by hand or with the use of selective postemergence herbicides.

Prevention means keeping weed problems out of the greenhouse. Do not let weed seeds or weed propagules enter the greenhouse. The use of screens on vents and open areas will keep weed seeds from blowing in. If the floor is not concrete, a weed block material can be used on the soil to prevent weed seed germination. A third method of prevention is to control weeds around the outside of the greenhouse (see the following discussion). *Sanitation* involves using sterile media for plant growth, planting weed-free planting stock and/or using certified seeds.

Weed control cannot rely on herbicides alone, as few are available for use in greenhouses. When weeds are present in pots, removal will by necessity involve hand pulling. If weeds are already established within the greenhouse, they can be pulled or the greenhouse can be emptied and the weeds allowed to dry. In extremely weedy situations, the greenhouse can be emptied and fumigated, although this is expensive.

A more common approach is to use postemergence herbicides between cropping cycles to remove troublesome weeds.

There are six postemergence herbicides registered for use in greenhouses, but no soil-applied residual herbicides. The six herbicides are glyphosate, glufosinate, pelargonic acid, diquat, clethodim, and fluazifop. Glyphosate, glufosinate, pelargonic acid, and diquat can be used only when the crop is not present because they are nonselective and if sprayed on the crop will cause damage. Glyphosate should be used if there are perennial weeds present; the other three are contact herbicides and are most effective against annual weeds. Clethodim and fluazifop are selective herbicides for annual and perennial grass control.

Weed control outside the greenhouse is important to remove a major source of weed seeds that may enter. Most operators maintain a 10- to 20-foot weed-free buffer zone around the perimeter of the greenhouse. This buffer is either mowed to prevent weed seed formation or maintained as a weed-free gravel zone. Most growers use a geotextile fabric cloth under gravel. The fabric prevents weed germination and does not require the use of a soil-applied herbicide.

Sometimes a buffer zone is maintained by placing an organic or inorganic mulch material on the soil and applying a preemergence herbicide, such as oryzalin, to reduce the emergence of weed seedlings. Caution should be used to avoid applying any herbicide that can volatilize and enter the greenhouse and damage the crop.

CUT FLOWER PRODUCTION

Field production of cut flowers requires good weed control in order to be successful. These filed-grown flowers include such common plants as alyssum, asters, chrysanthemum, campanula, gladiolus, geranium, daisy, and zinnia, to name a few. Most growers use a combination of cultural practices, cultivation, mulching, and herbicides to manage weed problems. The best approach is to start with a weed-free site. Because only a limited number of herbicides are available for use in cut flowers, a site with a heavy annual and/or perennial weed population will not be suitable for profitable cut flower production.

Cultural practices include the use of cover crops the year prior to flower production to reduce weed infestations at the site and to add organic matter to the soil, and planting transplants, vegetative propagules, or seeds early in the season so that the crop is established early and thereby has an advantage over the weeds. Growers generally use narrow between-row spacing to allow greater crop competition and soil shading to reduce the growth of weeds. Shallow cultivation (to avoid crop root damage) between rows is common; however, in-row weed removal requires hand hoeing or pulling. Some growers use geotextile fabric on the soil surface. The fabric is most effective with perennial flower production. Another common practice is to place some type of organic straw mulch on the soil surface for weed control. It is important that any organic mulch be free of weed seeds.

Herbicides can be used in association with all these practices in many types of cut flowers. Because there is a wide variety of flowers grown, it is important that the

TABLE 24-8. Herbicides for Use in Cut Flower Production[a]

Herbicide[b]	Weeds Controlled	Type of Application	Period of Weed Control Effectiveness
Glyphosate	Annual grasses, annual broadleaves, many perennials	Site preparation postemergence to weeds	Rapid kill of all foliage and perennial parts
Glufosinate	Annual emerged weeds	"	Rapid contact kill of emerged tissue, no soil residual
Diquat	"	"	"
Pelargonic acid	"	"	"
Sethoxydim	Emerged annual and some perennial grasses	Postemergence to broadleaved crop and grassy weeds	Kill of emerged grass, no soil residual
Clethodim	"	"	"
Fluazifop	"	"	"
Bensulide	Annual grasses, some small-seeded broadleaves	Apply after crop is emerged and established but preemergence to the weeds	3 to 4 months soil residual
Dithiopyr	"	"	2 to 3 months soil residual
Napropamide	"	"	3 to 4 months soil residual
Oryzalin	"	"	2 to 3 months soil residual
Trifluralin	"	"	1 to 3 months soil residual
Metolachlor	Annual grasses, some broadleaves and control of yellow nutsedge	"	"

[a]Always read and follow instructions on the herbicide label.
[b]Refer to product label for specific ornamental uses.

grower read each herbicide label to be certain it is registered for the target crop. Herbicides available for cut flower production are listed in Table 24-8. There are several nonselective site-preparation herbicides for control of annual and perennial weeds, including glyphosate, glufosinate, diquat, and pelargonic acid. Selective postemergence annual and perennial grass herbicides include sethoxydim, clethodim, and fluazifop. There are six soil-applied preemergence herbicides available for use in many cut flowers: bensulide, dithiopyr, napropamide, oryzalin, trifluralin, and metolachlor. To reduce the potential for flower damage from these herbicides, they should be applied only after the crop has emerged and become established, but prior to any weed emergence. These herbicides are most effective against small-seeded annual grasses and broadleaf weeds; therefore, it is important that appropriate cultural practices be a part of any effective weed management program. Weed control in cut flower production must include an integrated management approach, as outlined in Chapter 3.

SUGGESTED READING

Bingham, S. W., W. J. Chism, and P.C. Bhowmik. 1995. Weed management systems for turfgrass. In *Handbook of Weed Management Systems*, ed. by A.E. Smith, Chapter 15. Marcel Dekker, New York.

Bruneau, A.H. and D.C. Bowman. 1997. *Turfgrass Management Manual: A Guide to Major Pests of Turfgrass*. Bulletin AG 348. North Carolina Extension Service, North Carolina State University, Raleigh.

Neal, J.C., W. A. Skroch, J.F. Derr, and A.F. Senesac. 1999. *Weed Control Suggestions in Christmas Trees, Woody Ornamentals and Flowers*. North Carolina Cooperative Extension Publication. No. AG424.

Smeda, R.J., and L.A. Weston. 1995. Weed management systems for horticulture crops, In *Handbook of Weed Management Systems*, ed. by A.E. Smith, Chapter 14. Marcel Dekker, New York.

WEB SITES

Turf

University of Kentucky: *Weed Control Recommendations for Kentucky Bluegrass and Tall Fescue Lawns and Recreational Turf*, A.J. Powell, Jr., J.D. Green, and J.R. Martin. Bulletin # 78. University of Kentucky Cooperative Extension Service.

http://www.ca.uky.edu/agc/pubs/agr/agr78/agr78.htm

The Ohio State University: *Management of Turfgrass Pests: Weeds, Diseases and Insects.* 2001. B. Bloetscher, M.J. Boehm, J.W. Rimelspach, D.J. Shetlar, J.R. Street. Bulletin L-187-01. Cooperative Extension Service, The Ohio State University

http://www.ag.ohio-state.edu

Select "Agriculture and Natural Resources," scroll to and select "Management of Turfgrass Pests: Weeds, Diseases, and Insects."

University of Florida: *Weed Management for Florida's Turfgrass Professionals.* by J.B. Unruh and J.B. Brecke. 1997. Publication ENH-86, Cooperative Extension Service, Institute of Food and Agricultural Sciences.

http://edis.ifas.ufl.edu

Select "Pest Management Guides," then "Weed Management Guides;" select "Miscellaneous

Weed Management Topics" and select "Weed Management for Florida's Turfgrass Professionals."

Landscapes

University of California: *Weed Management in Landscapes.* 1997. C. Wilen and C.L. Elmore. Publication 7441. University of California Davis, Agriculture and Natural Resources Statewide Integrated Pest Management Project

http://www.ipm.ucdavis.edu

Select for "Publication 7441." scroll to "Weed Management in Landscapes."

University of Florida: *Weed Management in Ornamentals and Turf.* 1999. J.G. Norcini. Publication EHN-93. Cooperative Extension Service, Institute of Food and Agricultural Sciences.

http://edis.ifas.ufl.edu

Select "Pest Management Guides," select "Weed Management Guides," and select "Weed Management in Ornamentals and Turf."

University of Florida: *Weed Management in Enclosed Greenhouses.* 1999. J.G. Norcini. Publication EHN-96. Cooperative Extension Service, Institute of Food and Agricultural Sciences.

http://edis.ifas.ufl.edu

Select "Pest Management Guides," select "Weed Management Guides," select "Weed Management in Turf and Ornamentals," then select "Weed Management in Enclosed Greenhouses."

Cornell University: *Plan before You Plant—A 5-step Process for Developing a Landscape Weed Management Plan," by J.C. Neal*

http://www.hort.cornell.edu

Select "Flowers, Ornamentals and Turfgrass," select "Weed Facts," select "Landscape Weed Management: Plan before You Plant."

Utah State University: *Landscape and Garden Weed Control.* 1997. S. Dewey, D. Drost, L. Rupp and L. Sagers. Publication # HG-508. Utah State University Cooperative Extension Service

http://www.ext.usu.edu/

Select "Cooperative Extension," Search for "Yard and Garden," select "Yard and Garden Publications," scroll to "HG-508" and Select

The Ohio State University: *Controlling Weeds in Nursery and Landscape Plantings.* 1999. L.J. Kuhns, T. Harpster, S. Guiser, M.A. Rose. Bulletin 867-99. Cooperative Extension Service, The Ohio State University

http://www.ag.ohio-state.edu

Select "Agriculture and Natural Resources," scroll to "Controlling Weeds in Nursery and Landscape Plantings"

AgChemical Label Information: Crop Data Management Systems, Inc.
Marysville, CA
http://www.cdms.net/manuf/manuf.asp

For herbicide use, see the manufacturer's or supplier's label and follow these directions. Also see Preface.

25 Pastures and Rangelands

Hundreds of kinds of weeds infest pastures and ranges. These include trees, brush, broadleaf herbaceous weeds, poisonous plants, and undesirable grasses. The control of trees and brush in pastures and range lands are discussed in Chapter 26; this chapter is primarily limited to control of broadleaf weeds and grasses.

The herbaceous weeds of greatest concern in pastures and rangelands include winter weeds that grow and can produce seeds when many desirable forages are dormant, and summer annuals that tend to emerge and grow earlier than the desirable pasture plants. Winter weeds include mustards, burdocks, docks, buttercups, primroses, thistles, henbit, pepperweeds, prickly lettuce, wild garlic, and horseweed. The summer weeds include horsenettle, dogfennel, smartweeds, bitter sneezeweed, amaranths, wooly croton, jimsonweed, common milkweed, and hemp dogbane.

Almost half of the total land area of the United States is used for pasture and grazing. Nearly all of this forage land is infested with weeds, some of it seriously. Weeds interfere with grazing, lower the yield and quality of forage, increase the costs of managing and producing livestock, slow livestock gains, and reduce the quality of meat, milk, wool, and hides. Some weeds are poisonous to livestock (see Figure 1-4). The total cost of these losses is hard to estimate. Losses from undesirable woody plants are discussed in Chapter 26. Controlling heavy infestations of some woody weeds has increased forage yields two to eight times.

Grass yields increased 400% after the removal of sagebrush in Wyoming. Forage consumed by cattle increased 318% on native Nebraska pasture after perennial broadleaved weeds were controlled by improved agronomic practices and the use of 2,4-D. This 318% increase was a result of better pasture species, deferred and rotational grazing, and effective weed control. These results emphasize the usual need for both improved pasture management practices along with an effective weed control program.

Weed control programs for pastures and ranges often include a combination of good management practices, mechanical and chemical methods, and prescribed fires. Biological control has been effective on certain species (see Figure 3-6).

CHEMICAL COMPOSITION OF GRASSLAND WEEDS

The nutrient or chemical composition of grassland weeds is important to livestock farmers for two reasons. First, weeds contribute to the livestock ration. Second, weeds compete for nutrients and water needed by more palatable and more desirable species and thus reduce the yields of desirable forage.

Scientists collected forage and weed samples before mowing in an intensive agricultural area of the Connecticut River Valley in Massachusetts and compared their nutrient contents through chemical analysis (Table 25-1). These data showed that weeds do compete well with forages for several essential nutrients (see also Table 3-5). In addition to composition, one must evaluate freedom from toxins, palatability, yield, and persistence, in determining the worth of a species for forage.

MANAGEMENT

Pasture and rangeland weed control depends on an integrated weed management (IWM) approach using many tools to control unwanted plants. The IWM approach must be well planned and implemented to reduce the impacts of weeds.

TABLE 25-1. Chemical Composition of Grassland Weeds as Compared with Timothy and Red Clover (Sampling Dates June 5–10)

Plant	Growth Stage	Number of Samples	Mean Percentage Composition (Air-Dry Basis)				
			N	P	K	Ca	Mg
Timothy	Early heading	19	1.55	0.26	2.17	0.34	0.10
Red clover	In buds, before bloom	19	2.84	0.25	1.09	1.88	0.42
Tufted vetch	Early bloom	3	3.58	0.30	1.52	1.52	0.30
Yarrow	In buds, before bloom	8	1.56	0.31	2.35	0.82	0.18
Oxeye daisy	50% heads in bloom	7	1.63	0.34	2.48	0.94	0.21
Daisy fleabane	In buds, before bloom	11	1.47	0.38	2.12	1.12	0.20
Common dandelion	Mostly leaves	14	2.25	0.44	3.39	1.21	0.43
Yellow rocket	After bloom	7	1.44	0.24	1.55	1.23	0.17
Plantain	Mostly leaves	11	1.48	0.30	2.10	2.55	0.46
Narrowleaf plantain	Mostly leaves	5	1.85	0.37	1.90	1.90	0.33
Yellow dock	50% heads in bloom	13	1.84	0.30	2.29	1.11	0.42
Tall buttercup	In bloom	7	1.45	0.31	1.98	0.94	0.25
Wild carrot	Vegetative growth, leaves	5	2.52	0.54	2.37	1.92	0.44
Mouse-ear chickweed	In bloom	11	1.73	0.41	3.14	0.70	0.26
Cinquefoil	Early bud stage	2	1.49	0.28	1.31	2.08	0.33
Common milkweed	Vegetative growth	2	3.02	0.47	3.08	0.80	0.45
Sensitive fern	Vegetative growth	6	2.27	0.48	2.50	0.65	0.39
Quackgrass	Before heading	11	1.82	0.28	2.14	0.36	0.10

From Vengris et al., 1953.

Choosing the most desirable forage species for an area is the first step in pasture improvement. Proper management, as outlined in the following list, favors heavy growth of pasture species, which reduces weed presence. For example, broomsedge is favored by low pH soils. This weed disappears from pastures of the southeastern United States when the soils have a proper pH and are fertilized and seeded to high-yielding species such as ladino clover, orchard grass, or fescue. Some weeds are favored by the recommended agronomic programs. For example, broadleaf dock responds to high soil fertility and favorable moisture, often better than the desired pasture species.

The presence of weeds will vary, depending on the situation and previous management practices. In all pasture and rangeland situations, several management steps should be implemented to obtain satisfactory weed control:

1. Know the weed problem areas and species present. This baseline information can be obtained by field surveys, mapping, and geographic information systems and then used to implement a plan.
2. Implement a weed control plan based on problems identified and solutions available, then prioritize problem areas to be managed based on infestation severity and the costs involved.
3. Prevent movement of problem weeds into pastures and rangelands through good management practices. Use clean seed and propagation material, minimize equipment movement from heavily weedy areas to less weedy areas, carefully graze animals in weedy fields when weeds are not reproducing, minimize field soil disturbance, eradicate new noxious weeds before they spread, and maintain vigorous growth of pasture grass.

In some cases weedy pastures should be revegetated with desirable plants to minimize invasions of weeds and to maximize forage production. These steps involve killing especially difficult weeds with herbicides, followed by plowing, fertilization, and reseeding.

A final step is to employ proper management by alternating the seasons of grazing to achieve moderate forage use, and rotate stock to allow grass to recover before it is regrazed. These practices allow pastures and rangelands to remain healthy and vigorous for long-term use.

Mowing

Mowing was often recommended to control pasture weeds in the past. Today, mowing is seldom used for rangeland weed control, and it is steadily becoming less important on more intensively grazed areas.

Mowing will kill annual weeds if the growing point is eliminated. It is more effective on broadleaf than grass weeds and on annuals as compared with perennials. Mowing of perennials must be frequent enough to deplete their food reserves, but frequent mowing is not desirable for maximum forage health and productivity.

Results achieved by mowing are often disappointing; initially, the appearance of the area is improved, but few perennial weeds are actually killed. In North Carolina, 5 years of monthly mowing was required to control horsenettle. Weekly mowing for 3 years (about 18 times during each growing season) reduced wild garlic plants in bermudagrass turf by only 52%.

In Nebraska, mowing native grass pasture in either June or early July for 3 years had reduced the perennial broadleaved weeds by only 35% at the end of the experiment. After 20 years of mowing, 24 to 38% of the ironweed plants persisted.

Mowing may be used to good advantage in new grass-legume seedlings to lessen weed competition. Clipping the tops off broadleaved weeds may sufficiently reduce weed competition to permit survival of seedling grasses and legumes. However, mowing also clips the tops off the forage species, reducing their vigor to some extent.

Grazing

Grazing can be used to manage a pasture and encourage competitive grass. Overgrazing will weaken the pasture and can lead to weakened forage, bare spots where weeds can invade, and increased disease and insect problems. Proper grazing encourages rapid lateral branch proliferation by removal of apical dominance, which promotes a healthy, thick pasture.

Prescribed Fire

Prescribed fire, although not widely used, is a method employed to help maintain a quality rangeland. It is not used on pastures to the same extent as it is on rangelands. Fire is carefully employed at specific times of the year and under specific relative humidities, air temperatures, and wind speeds. Applied under specific conditions, prescribed fire is a safe and inexpensive management tool to manage rangeland weeds by eliminating unwanted species and standing dead forage. Fire releases nutrients from dormant standing forage (phosphorus and potassium) for a brief period of time, resulting in a somewhat increased nutritive value of subsequent forage. The forage on the burned surface generally emerges earlier than it does on nonburned areas, allowing earlier grazing.

If soil moisture is not adequate at the time of a fire or replaced soon after, areas subjected to the fire may produce less forage than unburned areas. Soil moisture is a critical aspect of the fire prescription and should be carefully considered in a fire plan (Redmon and Bidwell, 1999).

HERBICIDE CONTROL

Pasture and grazing areas are well suited to chemical weed control. It is often possible to control a weed with little or no injury to desirable forage species. These forage species then respond with increased ground cover and larger yields (Figures 25-1 and 25-2).

Figure 25-1. Herbicide use for weed control in pastures. *Left*: Before herbicide treatment. *Right*: After herbicide treatment. (Aventis CropScience.)

Herbicides should be used to control weeds not controlled by other pasture management practices. Herbicides are available to control most of the broadleaved weeds in pastures, although grass control is more difficult as few registered herbicides are available.

The proper choice of herbicide and the correct use rate are important for best results. Preemergence herbicides must be applied before the weeds emerge, and

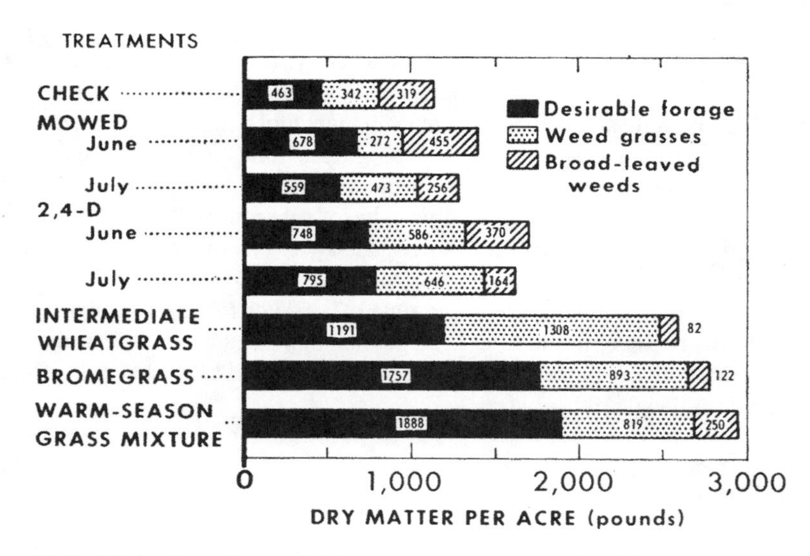

Figure 25-2. Effect of weed control treatments on average pounds of dry matter per acre eaten by cattle. Seedbed for wheatgrass, bromegrass, and warm grass mixture was initially plowed and sprayed annually with 2,4-D. (D.L. Klingman and M.K. McCarty, USDA and University of Nebraska.)

postemergence herbicides are most effective when applied to small, vigorously growing annual weeds. Control of perennials is best when systemic postemergence herbicides are applied prior to flowering or in the fall. This practice allows good spray coverage on the plant foliage, good absorption, and subsequent translocation to underground perennial plant organs. Herbicides can be applied broadcast over the entire pasture or as a spot spray to weedy areas.

Most broadleaf weeds can be controlled with phenoxy-type sprays. Control of undesirable grasses, especially annual grasses such as downy brome, has become an increasingly serious problem. Few preemergence herbicides are registered for use in pastures and rangelands, although use of a preemergence herbicide, such as diuron, before weed seed germination, may provide some control.

Table 25-2 lists the herbicides and methods of treatment for weed control, including woody plants, in permanent pastures and rangeland. Check product labels for specifics regarding use procedures and selectivity.

Livestock Poisoning and Abortion

Most herbicides used on grazing lands are relatively nonpoisonous. If recommendations on the label are followed, no poisoning should occur. Some herbicides, such as 2,4-D and similar compounds, are thought to temporarily increase the palatability of some poisonous plants. Livestock may eat these plants after treatment, whereas before spraying animals would have avoided them. Under such

TABLE 25-2. Herbicides for Weed Control of Woody and Herbaceous Plants, on Permanent Pastures and Rangeland by Various Application Methods[a,b]

Herbicide	Foliage Spray	Basal Bark Spray	Stump Spray	Frill, Notching	Dormant Stem Spray	Tree Injection	Soil Treatment
2,4-D	X	X	X	X	X	X	
Clopyralid	X						
Dicamba	X		X	X		X	X
Diuron	X						X
Fosamine	X						
Glyphosate	X		X	X		X	
Hexazinone	X						X
MCPA	X						
Metsulfuron	X						
Paraquat	X						
Picloram	X		X	X		X	X
Tebuthiuron							X
Triasulfuron	X						
Triclopyr	X	X	X		X		

[a]Commercial herbicide labels list different weeds controlled, give instructions for use, and state precautions. *Follow the label instructions.*
[b]See Chapter 26 for description of methods.

conditions livestock may be poisoned after plants have been sprayed, even though the herbicide itself is nonpoisonous.

Hydrocyanic acid (HCN), or prussic acid, may be produced from glucosides that are found in various species of the sorghum family and in wild cherry. If cattle, sheep, and other animals with ruminant stomachs eat these plants, HCN is produced and the animals may be poisoned. Studies of wild cherry indicated no increase in HCN content following treatment with 2,4-D.

"Tift" Sudan grass was treated when 8 inches tall with 1lb/acre of 2,4-D and MCPA. Chemical analysis showed that HCN increased after both chemical treatments (Swanson and Shaw, 1954). The use of phenoxy herbicides evidently does not change the usual precautions necessary to prevent HCN poisoning. Ruminant animals should not be permitted access to plants containing HCN, whether treated or not.

Livestock may also be poisoned by eating plants having a high nitrate content. Nitrate is reduced to nitrite by microorganisms in the animal's intestinal tract. Nitrite in the bloodstream interferes with effective transport and use of oxygen, and the animal dies of asphyxia (suffocation). The lethal dose of potassium nitrate is 25g/100 lb of animal weight.

In Mississippi, a number of pasture weeds were analyzed for chemical content. The scientists reported that redroot pigweed and horsenettle contained enough nitrate nitrogen to be toxic to livestock. Even though many weeds were found to have sufficient crude protein to meet animal needs, the authors concluded that most weeds should be considered detriments and should be removed. Factors cited as detrimental were bitterness, the presence of spines, toxic mineral levels, other toxic components of weeds, and the fact that weeds caused lower yields of forage (Carlisle et al., 1980).

A high nitrate content in plants has been associated with abortions in cattle in Wisconsin. In Portage County, 400 abortions in cattle were reported in 1954. Reproductive diseases and pathogens accounted for only a very small number of abortions. "Poisonous weeds" were considered a possible explanation. Further study revealed that the muck soils were high in nitrogen but lacking in phosphorus and potassium; this condition is conducive to nitrate storage in plants. Weed species were analyzed for nitrate nitrogen and classified according to nitrate content (Table 25-3). Pastures were treated with 2,4-D to eliminate weeds thought to contribute to the high abortion rate. Pasture areas were divided for experimental purposes. On one pasture, 2,4-D was applied in both 1956 and 1957. The area was weed free in 1957. Ten heifers that grazed on this area calved normally in 1956, but all 11 heifers that grazed on nontreated and weedy pastures aborted the same year.

A feeding trial was conducted to test the effectiveness of dosing pregnant cattle with nitrate to induce abortions. Three 700-pound heifers that were given 3.56 ounces of potassium nitrate each day aborted after 3 to 5 days. The aborted fetuses and placentas were similar to those aborted on the weedy pastures (Simon et al., 1958).

In summary, herbicides are generally nonpoisonous to livestock if used as directed on the labels. Poisoning may occur if palatability is increased following spraying so that livestock consume larger-than-usual quantities of poisonous weeds. Killing poisonous weeds with herbicides may reduce the poisoning hazard.

TABLE 25-3. Nitrate Nitrogen Content of Plants

High NO^3 Content (above 1000 ppm)	Medium NO^3 Content (300–1000 ppm)	Little or No NO^3 Content (below 300 ppm)
Elderberry	Goldenrod	Linaria
Canada thistle	Cinquefoil	Meadow rue
Stinging nettle	Boneset	Yarrow
Lamb's quarters	Mints	Vervain
Redroot pigweed	Foxtail	Dandelion
White cockle	Aster	Milkweed
Burdock	Groundcherry	Willow
Smartweed	Toadfax	Dogwood
		Spirea

From Sund and Wright, 1959.

Preventing Livestock Poisoning by Weeds

Livestock should be immediately isolated from poisonous plants to prevent poisoning. It may be necessary to fence the infested area or to remove livestock from the area. A small number of poisonous plant species can be cut and removed from a pasture or killed by herbicide treatment.

Hundreds of plants cause livestock poisoning. Among these are perilla mint (*Perilla frutescens*), common pokeweed (*Phytolacca americana*), dogbane (*Apocynum* sp.), milkweed (*Asclepias* spp.) nightshades (*Solanum* spp.), hemlocks (*Cicuta* spp.), jimsonweed (*Datura stramonium*), bitter sneezeweed (*Helenium amarum*), and boneset (*Eupatorium perfolistum*). Usually eradication, or at least very effective control, is needed.

Cutting and removal of plants or treatment with an effective herbicide may be most desirable. If the area is small, a soil sterilant may be the best answer, as discussed in Chapter 26. Small, isolated areas can probably be treated most effectively with hand equipment or with granular materials. Large areas may be treated best by broadcast-type equipment, through either ground or aerial application.

LITERATURE CITED AND SUGGESTED READING

Bovey, R. W. 1977. *Response of Selected Woody Plants in the United States to Herbicides*. Agriculture Handbook No. 443. USDA, ARS., p. 101.

Carlisle, R. J., V. H. Watson, and A. W. Cole. 1980. Canopy and chemistry of pasture weeds (Cover of bare soil, may serve as forage) *Weed Sci.* **28**(2):139–141.

Crop Protection Reference, 17th ed. Chemical and Pharmaceutical Press, New York.

Frank, P. A., and B. H. Grigsby. 1957. Effects of herbicidal sprays on nitrate accumulation in certain weed species. *Weeds* **5**(3):206–217.

Klingman, D. L. 1970. Brush and weed control on forage and grazing lands. *FAO International Conference on Weed Control*. WSSA, Lawrence, KS., p. 401–424.

Klingman, D. L., and M. K. McCarty. 1958. *Interrelations of Methods of Weed Control and Pasture Management at Lincoln, Nebraska 1949–55*. USDA Technical Bulletin 1180, p. 49.

Peters, E. J., and J. F. Strizke. 1971. *Effects of Weed Control and Fertilization on Botanical Composition and Forage Yields of Kentucky Bluegrass Pasture*. USDA Technical Bulletin 1430, p. 28.

Redmon, L. A., and T. G. Bidwell. 1999. *Management Strategies for Rangeland and Introduced Pastures*. Oklahoma State University Bull. F-2869.

Simon, J., J. M. Sund, M. J. Wright, A. Winter, and F. D. Douglas. 1958. Pathological changes associated with the lowland abortion syndrome in Wisconsin. *J. Am. Vet. Med. Assoc.* **132**:164–171.

Sund, J. M., and M. J. Wright. 1959. Control weeds to prevent lowland abortion in cattle. *Down to Earth* (Dow Chemical) **15**(1):10–13.

Swanson, C. R., and W. C. Shaw. 1954. The effect of 2,4-dichlorophenoxy-acetic acid on the hydrocyanic acid and nitrate content of Sudan grass. *Agron. J.* **46**(9):418–421.

Vengris, J., M. Drake, W. G. Colby, and J. Bart. 1953. Chemical Composition of weeds and accompanying crop plants. *Agron J.* **45**(5):213–218.

WEB SITES

Montana State University: Rangeland Weed Management by R.L. Sheley, 1997. Publication, MT9504. Cooperative Extension Service Montana State University

http://www.montana.edu

Select "Extension Service," then select "Publications," select "Publications "on-line"," select "Agriculture," select "Farm, Land and Wildlife Management," scroll to: "Rangeland Weed Management" and "select"

University of Florida: Weed Management in Pastures and Rangelands. 2001. J. A. Tredaway and B. J. Brecke. Publication SS-AGR-08. Cooperative Extension Service, Institute of Food and Agricultural Sciences

http://edis.ifas.ufl.edu

Select "Pest Management Guides," select "Weed Management Guides," and select "Weed Management in Crops and Pastures."

AgChemical Label Information, Crop Data Management Systems, Inc., Marysville, CA.

http://www.cdms.net/manuf/manuf.asp

For chemical use, see the manufacturer's or supplier's label and follow these directions. Also see the Preface.

26 Brush and Undesirable Tree Control

Control of woody-plant growth is a problem affecting most types of property. This includes grazing and recreational areas; telephone, highway, and railroad rights-of-way, and industrial plant and home sites.

There are about 1 billion acres of pasture, pasturelands, and grazing lands in the United States. On parts of nearly all of this area, woody plants present some problem. In range and pasture areas it is often desirable to eliminate all or most of the woody plants, leaving only grasses and legumes for livestock grazing.

On many western dryland ranges, native grasses increase rapidly where brush is controlled. In Wyoming, range forage yields doubled the first year after sagebrush was treated and increased fourfold during a 5-year control program. In Oklahoma, control of brush with chemicals plus proper management (principally keeping livestock off during the first summer) increased the growth of grass four to eight times during the first 2 years after treatment. Figure 26-1 shows the relationship between mesquite control and increase of perennial grass forage on southwestern rangeland. With 150 mesquite trees/acre, production of grass forage was reduced by about 85% and total production, including that produced by the mesquite, was reduced by nearly 60%.

In the southeastern United States, more than 100 million acres of land well suited to loblolly and short-leaf pine are being invaded by less-desirable hardwoods and heavy brush undergrowth. The same is true in northwestern United States and Canada, where dominant, but inferior, hardwoods may invade stands of Douglas fir, balsam fir, and spruce.

Some brush species act as alternate hosts for disease organisms affecting other plants. Common barberry harbors the organism causing stem rust in wheat and some other grasses. Other brush plants such as poison ivy and poison sumac are poisonous to humans. Wild cherry and locoweed are poisonous to ruminant animals.

Control methods for woody plants vary with the nature of the plant and size of the area infested. Some plants can be easily killed by one cutting. For example, many conifer trees have no adventitious buds on the lower parts from which new shoots or sprouts may develop. However, on large acreages even these species are controlled more efficiently by herbicidal sprays than by mechanical means. In Oregon and Washington, studies compared manual brush control with herbicide control in forests. Brush cut close to the ground by hand reached an average height of more than 4 feet in 6 months, and this brush overtopped more than 40% of young timber (crop) trees. Furthermore, this handwork damaged 22% of the small Douglas firs.

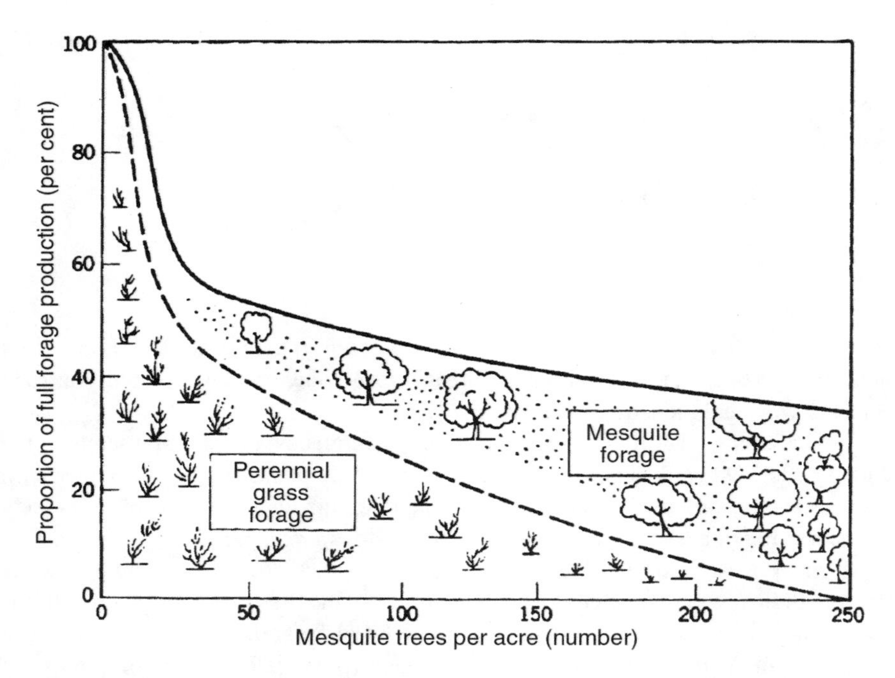

Figure 26-1. Effect of mesquite on forage production. (Adapted from USDA Leaflet No. 421, 1957.)

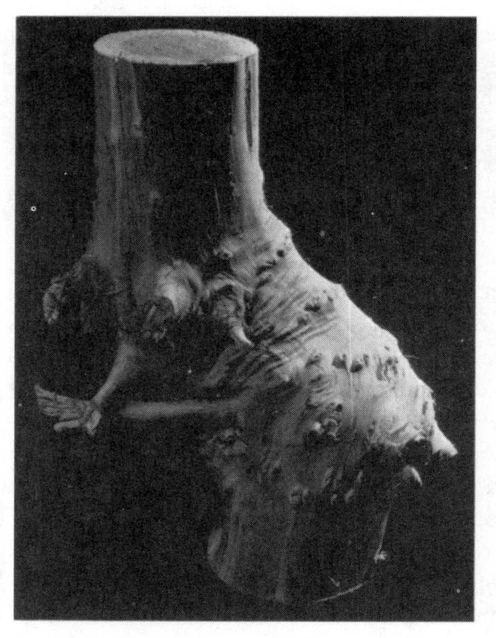

Figure 26-2. Section of mesquite stump. The wartlike structures are buds that produce sprouts. (Texas A&M University, Department of Forest Science.)

The plants hardest to control have underground buds from which new shoots or sprouts may develop. Some woody plants, such as velvet and honey mesquite, have these buds on the lower part of the trunk; other plants have buds on underground horizontal roots or stems (Figure 26-2). Most of the serious woody weeds are in one of these categories. As long as such plants have enough stored food and are not inhibited in some other way, new shoots will develop for as long as 3 years or more, even when all top growth is repeatedly removed. If photosynthetic plant organs, mainly the leaves, are repeatedly destroyed on woody weeds, the plant eventually dies of starvation.

Control methods for woody species are similar to those for other weeds, except that they must be adapted to woody and heavy type of growth. Methods include repeated cutting or defoliation, digging or grubbing, chaining and crushing, burning, girdling, and herbicide treatments.

REPEATED CUTTING AND DEFOLIATION

The object of repeated cutting or repeated defoliation is to reduce food reserves within the plant until it dies. Plants commonly die during winter months if reserve foods have been reduced enough to make them more susceptible to winter injury (Figure 3-3).

Most woody species bear their leaves well aboveground, making repeated cutting or defoliation effective. Most species must be defoliated several times each season if they are to be controlled. Plants may be cut by hand, by mowers, by special roller-type cutters, or by saws. They may be defoliated with special equipment or by chemicals. Species having palatable leaves may be sufficiently defoliated by livestock. Sheep and goats have eradicated undesirable shrubs and tree sprouts, and insects may also defoliate a plant, gaining control as a biological predator.

Plants have well-established annual patterns of food usage and food storage. A knowledge of root reserves, as they vary during the year, is valuable in reducing the number of treatments needed. With most deciduous plants, root reserves are highest during the fall. These reserves gradually diminish through the winter months as the plant uses this energy to survive. In the spring, the plant needs considerable energy to send up new stems and leaves. Thus, food reserves in roots are usually lowest just before full-leaf development. Many plants reach this stage at about spring flowering or slightly before; therefore, mowing or defoliation is best started at that time (Figure 3-3). Repeat treatments are ordinarily needed for one or more seasons. Treatments are usually repeated when leaves reach full development. For more precise timing, the root reserves of each plant species must be studied. As leaves become full grown, they start synthesizing more food than they need. As the excess accumulates, at least part of it is moved back to replenish the depleted root reserves.

Apical Dominance

The terminal bud of a twig or a shoot may hold back development of buds lower on the stem or roots. When the apical bud is cut off, lateral buds are no longer inhibited and may develop shoots. This mechanism is controlled by hormones within the plant. When apical dominance is destroyed, more stems may develop than were present originally. Following cutting, the stand may appear thickened rather than thinned. However, if cuttings are persistently continued so as to keep root reserves low, the stand will start to thin as the food supply is depleted. Under such conditions, the increase in number of shoots may actually speed the reduction of root reserves.

DIGGING OR GRUBBING

Digging or grubbing can effectively control plant species that sprout from an individual stem or root and remain as a "bunch" or "clump." Whether this method is practical depends on the density of the stand and the ease of removing the clump. Mesquite is an example of this type of plant. Young plants can easily be removed by grubbing.

Giant bulldozers have been equipped with heavy steel blades or fingers (root plows) that run under the mesquite, lifting out the bud-forming crown along with the rest of the tree. The operation is expensive and hard on desirable forage species. Other weeds may quickly invade the disturbed area. Reseeding the disturbed area with desirable grasses will help reduce invasion by weeds. Disadvantages of this method are that (1) remaining roots may resprout, (2) the soil surface is severely disturbed, (3) the method is difficult in rocky soils, and (4) dormant weed seeds are brought to the surface to germinate and invade the area.

CHAINING AND CRUSHING

Chaining is a technique of pulling out trees without digging. A heavy chain is dragged between two tractors. This action tends to uproot most of the trees in an area if the soil is moist. Then the chain may be pulled in the opposite direction to finish tearing trees loose from the soil. This method cannot be used on large, solidly rooted trees, and it is not effective on small, flexible brush or on any species that sprouts from underground roots or rhizomes.

Crushing mashes and breaks solid stands of brush and small trees and is accomplished by driving over them with a bulldozer and/or with rolling choppers. Rolling choppers have been developed with drums 3 to 8 feet in diameter and weighing at least 2000 lb/ft of length. Chopper blades, about 8 inches wide, are attached across the entire length of the drum. This treatment crushes and breaks most of the branches and main stems. The chopper is usually pulled by a large track-type tractor.

Crushing has disadvantages: It disturbs the soil, releasing buried dormant weed seeds to germinate and infest the area, and it requires expensive equipment. In experiments in Texas, roller chopping before, during, or after herbicide application did not increase herbicide effectiveness in control of live oak. Roller chopping was followed by no increase in livestock grazing per acre (Meyer and Bovey, 1980). Similar results have been observed with other species, including Texas white brush, Mcartney rose, and mesquite. The main advantage to chopping was a temporary increase in visibility.

BURNING

Prior to the advent of modern weed science (about 1950), many U.S. grasslands and woodlands were burned regularly. These fires were started by lightning, Native Americans, ranchers, and farmers. Some fires were accidental; others were intentional.

Effective chemical methods are available to control many brush species on rangeland, yet burning is still used in many places. There are two reasons: The cost is low, and certain species resistant to most herbicides can be controlled by burning (e.g., red cedar). In Kansas experiments, late-spring burning controlled brush best and injured desirable forage species least.

Many people consider forest fires only as destructive. This is usually true of large, uncontrolled fires that occasionally ravage forests. However, fire in forest areas can produce benefits. Since about 1950, "prescribed burning" or "controlled burning" has been used periodically as a method of control. When done properly, this method leaves desirable forest trees uninjured. Such burning removes some undesirable trees, brush, leaves, branches, and other debris and serves other useful purposes: (1) It results in less competition from weed species, (2) the seedbed is improved for desirable tree seedlings and made less favorable for others, (3) less fuel remains for wildfires, and (4) some disease organisms may be controlled.

Thickness of bark is important in determining susceptibility to fire. If the phloem and cambium of the trunk are killed by heat, the tree may show an effect similar to that of girdling.

Most pine seedlings develop best with less than 1.5 inches of debris on the forest floor, but oak seedlings need more debris. One reason for prescribed burning is to prepare seedbeds that are favorable to pine and unfavorable to oak. All tree seedlings, when small, are susceptible to fire. Most pine seedlings become resistant after the first year and will withstand considerable fire after the trunk has reached a 2-inch diameter at breast height (DBH). Oak and most other hardwoods of the same age are still susceptible; therefore, where pine forests are desired, controlled fire can effectively and selectively kill the deciduous hardwoods. The frequency of prescribed burning will depend on local conditions; usually every 2 to 3 years is best.

Local authorities must be consulted before planning a fire, as regulations may limit this approach. When prescribed burning is allowed, precautions such as establishing firebreaks, ensuring the availability of fire-fighting equipment, and consideration of environmental conditions are important. A large share of the cost of burning is for personnel and equipment on standby for emergency use.

GIRDLING

Girdling consists of completely removing a band of bark around the woody stem. This method is effective on most woody dicotyledonous plants, but is too laborious to be efficient on thick stands of smaller-stemmed species. In most species, sprouting near the base of the tree is eliminated, or at least considerably reduced, on large, mature trees. Young trees may sprout profusely following girdling.

Plant foods move down to the roots through the phloem, and water and nutrients move upward through the xylem (see Chapter 5). When bark is removed, phloem tissue is destroyed and plant foods can no longer move to the roots. Because roots depend on tops for food, the roots eventually die of starvation. This process usually takes from 1 to 3 years. The length of time depends on the rate of food usage and amount of food stored in the roots at the time of girdling. If girdling is properly done, the top of the tree shows little injury until the roots are near death. The top also succumbs as soon as the roots die.

The girdle must be wide enough to prevent healing and deep enough to prevent the cambium from developing new phloem. The girdled area should dry in a short time. Chemicals similar to those used for frill treatment or heat may help kill an active cambium. The xylem should not be injured. If the top is killed too rapidly through injury to the xylem (sapwood), root reserves may not be depleted enough to prevent shoot or sprout growth. If sprouts or lateral branches develop below the girdle, these will soon provide enough food for the roots to keep the tree alive. Thus, any sprouts that develop may be cut or treated with suitable chemical sprays to prevent food from returning to the roots.

HERBICIDES

The use of herbicides to control brush and undesirable trees has expanded rapidly since 1950. Proper use of chemicals has generally proven to be more effective and less costly than most other methods.

Herbicides registered for use to control brush are generally not labeled for use in the home landscape where desirable plants are present. The use of brush control herbicides requires the applicator to consider the environmental impact of the product and how to best apply them. The herbicides available for brush control vary in environmental stability, leachability, selectivity, and handling requirements, and if improperly used may damage surrounding vegetation, contaminate groundwater, and have a very long soil residual activity period.

Herbicides can be applied in many different ways to accomplish different purposes. These can be classified as foliage sprays, bark treatments, trunk injections, stump treatments, and soil treatments. The herbicides used for such treatments are given in Table 26-1 and Table 26-2.

The main steps in any brush control program involve evaluating the target species and the target site and determining the appropriate type of herbicide(s) and how best to apply them. Some brush species cannot be effectively controlled with herbicides

TABLE 26-1. Herbicides Used to Control Some Common Woody Plants[a,b]

	2,4-D	Hexazinone	Imazapyr	Dicamba[c]	Picloram[c,d]	Fosamine	Amitrole[d]	Glyphosate	Tebuthiuron[d]	Triclopyr[c]
Alder	F, B			C, S		F			S	
Ash		S	F, B, C	C, S	S	F	F	C, F	S	F, S
Blackberry		S		F, S		F	F	F	S	F, S
Buckberry	F, B								S	
Elderberry	F, B	S								F, S
Elm	C	S			F	F		C, F	S	F, S
Locust	F, B	S			S	F	F		S	F, S
Manzanita	F, B				F					
Maple		S	F, B, C	C, S	S	F	F	C, F		F, S
Mesquite				F, S	F				S	
Oak	F, C	S	F, B, C	C, S	S	F		C, F	S	
Poison ivy or oak	F, B	S		F, S	F		F		S	
Rabbitbrush	F, B								S	
Rose			F, B, C	S	S	F		F	S	
Sagebrush	F, B								S	
Saltcedar	F, B, C			S	S				S	
Sassafras			F, B, C			F				
Sumac	F, B	S	F, B, C	F, S		F	F		S	F, S
Willow	F, B	S	F, B, C	F, S		F		C, F	S	F, S

[a]F, foliar; B, stems; C, cut surfaces (frill, injection, or stump treatment); S, soil.
[b]Not all formulations of these herbicides are suitable for all the uses indicated. *Check manufacturer's label* for specific species controlled with various formulations and for use precautions. *Follow label instructions.*
[c]Dicamba, picloram, and triclopyr are often combined with 2,4-D.
[d]Noncropland only.

TABLE 26-2. Labeled Sites for Brush Control Herbicides

	Sites				
Brush Herbicides[a]	Grazing	Fence-rows	Noncrop	Ditch Banks	Haying
Picloram ester + triclopyr ester		X	X		
Glyphosate		X	X	X	
Metsulfuron (Ally)	X				X
Metsulfuron (Escort)		X	X		
Imazapyr (Arsenal)		X	X		
Dicamba (Banuel)	X	X	X	X	
Triclopyr ester + 2,4-D ester (Crossbow)		X	X	X	
Triclopyr amine (Garlon 3A)		X	X	X	
Triclopyr ester (Garlon 4)	X	X	X	X	X
Picloram amine + 2,4-D amine (Grazon P+D)	X				X
Bromacil (Hyvar)			X		
Fosamine (Krenite)			X	X	
Triclopyr ester (Pathfinder II)	X	X	X	X	X
2,4-D amine + picloram amine (Pathway)		X	X		
Triclopyr ester (Remedy)	X	X	X	X	X
Tebuthiuron (Spike)		X	X		
2,4-D	X	X	X	X	X
Hexazinone (Spike)			X		
2,4-D amine + dicamba (Weedmaster)	X	X	X	X	X

[a]Trade names of herbicides registered for these uses.
Adapted from J.W. Everest and M. Peterson, *Brush Control*, Bull. #ANR-1058 (Auburn, AL: Auburn University).

(e.g., persimmon, sassafras), and addition management techniques are necessary. Moreover, complete brush control is seldom achieved with a single herbicide application because of the diversity of species and their age at most sites. Generally, applications include a mixture of herbicides and follow-up treatments. The effectiveness of herbicides should not be evaluated too quickly, as a tree often does not die until the next growing season. In some cases, trees may die 2 to 3 years after herbicide treatment, after repeated development of new leaves, followed by defoliation.

Foliage Sprays

Foliage sprays require uniform coverage to be most effective. The need for thorough wetting depends on the chemical used and the species treated. In most cases, uniform coverage is more important than thorough wetting. The effectiveness of foliage sprays varies considerably among species and with the size of woody plants. Several chemicals kill aboveground parts but are less effective for killing roots, which results in sprouting in some species and the need to repeat the treatment as new sprouts develop.

Absorption by leaves is the first major problem of herbicide effectiveness. Many woody plants have coarse, thick leaves with a heavy cuticle. On such plants a nonpolar substance is absorbed more effectively than a polar substance. Thus, esters of 2,4-D, dichlorprop, and other oil-soluble herbicides are absorbed better than salt formulations. An effective wetting agent added to the spray solution often increases absorption. Once absorbed, water-soluble or polar forms (especially of 2,4-D) are translocated more readily through the phloem than the esters. Esters appear to be hydrolyzed to their respective acids on entering the plant. A second important problem is acute versus chronic toxicity. An excessively high application rate may kill or injure the phloem and inhibit symplastic translocation. The chemical will not be translocated to the roots, and roots will not be killed. Absorption is also discussed in Chapters 5 and 7.

For foliar application, this suggests three practical guidelines: (1) Low-acute-toxicity herbicides and wetting agents should be used for maximum absorption and translocation of growth substances through phloem tissues, (2) herbicides should be applied at rates that cause chronic toxicity, but not acute toxicity, and (3) low rates of treatment and repeated treatment should be used for some species, as they often give superior final results.

Foliage sprays of growth regulator type herbicides (such as 2,4-D and triclopyr) have usually given best root kills when large quantities of food are being translocated to the roots. With species that become somewhat dormant in summer, this peak translocation period usually occurs at about the time of full-leaf development in the spring; a second peak period may occur in the fall. Plants that continue rapid growth and translocation throughout the summer can be treated anytime during the summer.

Chemicals are applied by hand sprayers, mist nozzles, power-mist applicators, truck-, trailer-, or tractor-mounted power-spray equipment, and airplanes. Water is normally used as the carrier with ground equipment. Many plants develop a thickened, waxy cuticle. On such plants oil-water mixes are more effective. Airplane application may use water, oil, or oil-water emulsions, depending on the type of foliage to be treated and the relative humidity.

Herbicides commonly used for woody plant control include amitrole, picloram, dicamba, fosamine, glyphosate, and triclopyr. Amitrole is effective on poison ivy, poison oak, kudzu, ash, locust, blackberry, dewberry, and sumac. Picloram and dicamba are effective as a foliar or soil application for certain woody species, effective on species resistant to phenoxy-type herbicides, and, when combined with 2,4-D, control a broad range of woody species. Fosamine is effective when applied to foliage in late summer or early fall. Response is not normally observed until the next spring, and susceptible plants fail to refoliate and die. Glyphosate, a foliar-applied herbicide that translocates from the foliage to the roots, is effective for control of many species. It is applied when plants are actively growing and when most plants are at or beyond the full-bloom stage of growth; it is particularly effective against blackberries, honeysuckle, kudzu, maple, multiflora rose, trumpet creeper, and willow. Triclopyr, an analog of picloram, is effective on a number of woody species, with the ester formulation being more active than the amine formulation on woody plants. Triclopyr is often tank mixed with picloram for foliar applications for a broader spectrum of control.

Bark Treatment

In bark treatments the herbicide is applied to the stem near the ground and is absorbed through the bark. Two methods are used: broadcast bark treatment and basal bark treatment.

Broadcast Bark Treatment In broadcast bark treatment, the herbicide is applied to the entire stem area of the plant. This treatment is especially adapted to thick stands of small-stemmed woody plants. Spray is usually applied between fall leaf drop and midwinter. Phenoxy herbicides are effective on several species and are applied with diesel fuel oil or kerosene. The spray is applied so that the greatest amount of it will cover the stems evenly. For example, treating upright stems with a horizontally directed spray exposes the greatest stem surface to the spray.

Basal Bark Treatment The basal bark treatment method and most of the following methods have an advantage in that the operator can selectively treat only those trees he or she wishes to kill, without injury to other trees. With this method, the herbicide is applied so as to wet the stem base, 10 to 15 inches aboveground, until rundown drenches the stem at the ground line. The chemical can be applied from a sprayer or from a container that lets the chemical "trickle" onto the stem base. A small number of trees can easily be treated with a small container that pours the chemical onto the base of the tree trunk, leaving brush or larger trees standing. This method is especially effective on brush and trees less than 6 inches in diameter. It also controls some plants that sprout from horizontal underground roots and stems, especially if applied during early summer.

Herbicides used most often for basal bark treatment are 2,4-D, picloram, dicamba, triclopyr, imazapyr and various combinations of 2,4-D with dichlorprop, picloram, dicamba, or triclopyr. 2,4-D is mixed in enough diesel oil to make 100 gallons of mixture, which is applied at the rate of 1 gal/100 inches of tree diameter. This quantity can treat 50 trees 2 inches in diameter or 33 trees 3 inches in diameter.

Scientists do not fully understand how the basal bark treatment kills plants. It is possible that the treatment either kills the phloem or immobilizes it to the point that the tree is chemically girdled. In addition, the herbicide may inhibit bud formation and sprout development. If the tree is chemically girdled, the tree root dies from starvation, as it does with physical girdling.

Trunk Injections

Trunk injections often help the herbicide to penetrate through the bark. In some cases, injections may serve as a girdle, with the herbicide acting as a chemical girdle. The techniques used in this method are called *frilling* or *notching* and are done with an ax or a tree injector.

Frill Treatment A frill treatment consists of a single line of overlapping downward ax cuts around the base of the tree (Figure 26-3). The herbicide is then sprayed or

Figure 26-3. Frill application is used to apply various herbicides to trees too large for effective basal bark applications (DowAgro Sciences.)

squirted into the cut around the entire tree. The method is effective on trees too large in diameter for a basal bark treatment.

Phenoxy compounds such as 2,4-D have been the primary herbicides applied in this manner, but now a wide range of materials are being used. Frill treatment herbicides include dicamba, glyphosate, imzapyr, picloram, triclopyr, and various combinations of 2,4-D+picloram, and 2,4-D+triclopyr.

Notching A tree trunk may be notched with an ax, as in a frill treatment, except that one notch is cut for every 6 inches of trunk circumference. Notches are filled with one teaspoonful of ammonium sulfamate crystals, which will kill many kinds of broadleaved trees (Figure 26-4). This method is less effective than frill treatment, which uses a continuous notch around the tree.

A tree-injector tool speeds notch treatment, and when properly used does a satisfactory job (Figure 26-5). The oil-soluble amine or ester form of the phenoxy compounds plus diesel oil (2:9 ratio) can be used, with one cut being made for every 2 inches in trunk diameter. This tool is especially effective against elm, post oak, white oak, live oak, and willow. More resistant trees are treated by spacing injections closer together. Trees that are more resistant include ash, cedar, hackberry, hickory, blackjack oak, red oak, persimmon, and sycamore.

Figure 26-4. Application of ammonium sulfamate in notches. (F.A. Peevy, USDA-ARS.)

Figure 26-5. The tree injector speeds notch application of herbicides. (W.C. Elder, Oklahoma State University.)

Other herbicides that can be injected include dicamba, glyphosate, hexazinone, imazapyr, triclopyr, 2,4-D, and 2,4-D+picloram. Dicamba is injected as the undiluted commercial liquid concentrate added to water at a rate of 1:1 for resistant trees and 1:4 for susceptible trees, with 0.5 to 0.1 ml applied per injection cut.

Stump Treatment

Stumps of many species may quickly sprout after trees or brush are cut. Most trees can be prevented from sprouting by proper stump treatment (Figure 26-6). However, weedy trees that can develop sprouts from underground roots or stems are difficult to control by stump treatment alone. With most such species, stump treatment is effective if followed by a midsummer basal bark treatment of the sprouts when they develop.

Herbicides such as 2,4-D and 2,4-D+triclopyr are usually used in stump treatment in enough diesel oil to make 100 gal of spray solution. Enough spray is applied to "wet" the tops and sides of the stump so that rundown drenches the stem at the ground line.

Stump treatments are most effective if applied immediately after the tree is cut. Increasing the amount of chemical used can usually control sprouting from older cuts.

Figure 26-6. Hand spray application of a herbicide to prevent stumps from sprouting. (DowAgro Sciences.)

Soil Treatment

Some herbicides control woody plants when applied to the soil. They require rainfall to leach them into the soil as deep as the feeder roots. Therefore, they are usually applied just before or early in the rainy season. Herbicides used this way usually persist in the soil for more than 1 year for maximum efficacy. Effects can develop slowly and may not be apparent for 1 to 2 years after treatment (Figure 26-7).

Research has shown that bromacil, dicamba, hexazinone, picloram, and tebuthiuron can be used effectively with this method. The species to be controlled determines which of these herbicides is used, and sometimes combinations of these agents have been more effective than the use of any one alone.

Since the early 1970s, soil-applied herbicides have been used increasingly as a method to control undesirable brush and trees to improve grazing capacity on rangelands. Some of these herbicides are formulated as pellets or granules. They can be applied by hand, ground equipment, or airplane.

Many woody plants have extensive finely branched root systems, whereas many desirable grasses and forbs have root systems restricted to a small area. Some soil-applied herbicides used for woody plant control are also phytotoxic to desirable grasses and forbs. Therefore, studies have been conducted to determine the most efficient spacing of pellets for good woody plant control with minimal injury to desirable grasses and forbs. In Texas, tebuthiuron placed in rows from 4.5 to 18.0 feet apart was equally effective as a broadcast treatment for control of several woody

Figure 26-7. Brush and tree control with a soil treatment of tebuthiuron. *Right*: Treated. *Left*: Not treated. Photograph taken 2 years after application.

species. Researchers concluded that this spacing principle makes application of herbicides from airplanes in more or less grid patterns commercially possible (Meyer et al., 1978).

LITERATURE CITED AND SUGGESTED ADDITIONAL READING

Crop Protection Reference, 17th ed. 2001. Chemical and Pharmaceutical Press, New York.

Meyer, R. W., and R. W. Bovey. 1980. Control of live oak and understory vegetation with soil-applied herbicides. *Weed Sci.* **28**:51–58.

Meyer, R. W., R. W. Bovey, and J. R. Baur, 1978. Control of an oak complex with herbicide granules. *Weed Sci.* **26**:444–453.

WEB SITES

Texas A&M University: *Reference Guide for Texas Rangers.* Rangeland weed and brush control, by A. McGinly. Texas Agricultural Extension Service.

http://agpublications.tamu.edu

Select "Extension Publications," select "Rangelands," scroll to "Reference Guide for Texas Ranchers" (B6101).

Auburn University: *Brush Control* by J.W. Everest and M. Patterson. 1997. Bull. ANR-1058. Alabama Cooperative Extension, Auburn, AL.

http://www.aces.edu/

Search using "Brush Control."

AgChemical Label Information: Crop Data Management Systems, Inc., Marysville, CA.

http://www.cdms.net/manuf/manuf.asp

For chemical use, see the manufacturer's or supplier's label and follow these directions. Also see the Preface.

27 Aquatic Weed Control

Aquatic plants, as the term is used here, include those plants that normally start in water and complete at least part of their life cycle in water. They have both positive and negative aspects in regard to human welfare.

On the positive side, aquatic plants may reduce erosion along shorelines, and some plant species provide food and protection for aquatic invertebrates, fish, fowl, and game. Algae are the original source of food for nearly all fish and marine animals; and swamp smartweed, wild rice, wild millets, and bulrushes provide food and protection for waterfowl, especially ducks.

On the negative side, excessive growth of aquatic plants causes many serious problems for people who use ponds, lakes, streams, and irrigation and drainage systems. Weeds can (1) obstruct water flow in irrigation and drainage ditches and increase water loss through transpiration of water from leaf surfaces, (2) interfere with navigation, boating, fishing, swimming, and water skiing and pose safety hazards due to slippery surfaces, (3) destroy wildlife habitat, especially through decreases in species diversity caused by invasives, (4) cause undesirable odors and flavors and discoloration of water and fish (blue-green algae), (5) lower real estate values because of odors, unsightly vegetation, and problems in residential retention ponds, (6) create health hazards such as mosquito outbreaks, and blue-green algae toxicity to livestock, and harbor snails and leeches carrying disease organisms, and (7) speed up the rate of silting by increasing the accumulation of debris and sediment.

Controlling aquatic weeds with chemicals sometimes causes other problems. For example, the rapid killing of dense weedy growth may kill fish, which happens even though the chemical is nontoxic to the fish. During photosynthesis, living plants release oxygen, and fish depend on this oxygen for respiration. When the plants are killed, they produce no more oxygen. In addition, dead plants are decomposed by microorganisms that require oxygen for respiration. These two actions can reduce the oxygen content in the water, causing the fish to suffocate. The answer is to treat only a part of very heavily infested areas at one time; fish will move to the untreated part. Moreover, treating as early as feasible in the season will reduce the likelihood of a fish kill. The water is cooler at this time and there is generally less plant mass to decompose.

A recreation area suitable for both swimming and fishing presents management problems. For example, the right fertilization favors microscopic plants, which through a food chain are ultimately used as food by fish. A heavy growth of microscopic plants makes the water appear cloudy or dirty and may give it an undesirable odor; hence, it is less desirable for swimming. Therefore, it is difficult to manage a body of water so that it is optimal for both swimming and fishing.

AQUATIC PLANT GROUPS AND IDENTIFICATION

Aquatic plants are categorized in two main groupings, based on botanical relationships and growth habits. Those classified by botanical relationships (Table 27-1) include algae, mosses, ferns, and vascular flowering plants. *Algae* are simple plants that lack true roots, leaves, or flowers. They may be microscopic or visible to the naked eye, exist as single cells or in clusters or filaments, and be free floating (plankton) or attached to soil, rocks, or vegetation. Filamentous algae may be unbranched, slightly or highly branched, or netlike. Certain types (Macroalgae) may be large, very coarse, and resemble submersed vascular plants. Algae do not contain vascular tissues; consequently, all chemicals used for algae control have only contact activity. Algae reproduce by cell division, by fragmentation, or by spores. Algae may annoy bathers, causing a type of dermatitis and symptoms of hay fever. Blue-green algae (Cyanobacteria) have been known to cause poisoning of horses, cattle, sheep, dogs, and poultry. Odors and fishy tastes often result from decaying algae in water reservoirs. Extremely heavy algae growth may suffocate fish by depleting the supply of oxygen in the water at night. *Mosses* are plants that are visible to the eye and resemble delicate, leafy submersed vegetation. Mosses lack vascular tissue and roots and are usually attached to the soil. They reproduce by spore production. *Ferns* are visible to the eye and can be free-floating or rooted on the pond bottom and occasionally form loose, floating mats. Ferns have vascular tissues and reproduce by vegetative and sexual (spores) means. *Vascular flowering aquatic plants* have stems, roots in some cases, leaves, and flowers and can reproduce by seed, tubers, turions, or fragmentation of rhizomes, roots, or stolons. Flowering plants are generally placed in four groups: emersed (including marginal plants), woody, floating, and submersed aquatics. These plants have a vascular system that varies from rudimentary (duckweed) to complex (annual and perennial herbaceous and woody).

Classification based on growth habit includes five groupings: submersed, free-floating, floating-leaf, emergent, and woody (Table 27-2). *Submersed plants* are found below the water surface and may be rooted in bottom sediments or be free-floating with or without roots. Flowers are usually produced above the surface of the water and may occasionally be supported by specialized floatation structures. (Figure 27-1). Submersed plants have poorly developed vascular systems and very limited structural tissue and depend on the buoyancy of the water for support. Filamentous algae and macroalgae can be considered submersed plants. Examples are Eurasian water milfoil and hydrilla (Figure 27-2). *Free-floating plants* are found on the water surface and may lie flat on the water or have tissue well above the surface. These plants, with the exception of duckweeds, watermeal, and mosquito ferns, have well-developed vascular systems and strong support tissues and most form true roots. Flowers extend above the water surface. Examples are water hyacinth, waterlettuce, and duckweed. *Floating-leaf plants* are rooted into bottom sediments and have leaves attached to long, tough stems that extend to the surface from depths of 6 feet or more. Leaves float directly on the water surface, and mature leaves can be found well above the surface. Most of these plants have well-developed root, rhizome, stem, and

TABLE 27-1. Aquatic Weed Groups—Based on Botanical Relationships

Category	Examples
	Filamentous Algae
Algae	Blue-greens or *Cyanobacteria* Giant *Lyngbya* Green algae *Oedegonium* *Hydrodictoyon* (water net) *Sprirogyra* *Pithophora*
	Planktonic Algae
	Blue-greens or *Cyanobacteria* *Lyngbya* *Anabaena* *Oscillatoria* *Microcystis* Euglenoids (*Euglena*)
	Macroalgae
	Muskgrass (*Chara*) Stonewort (*Nitella*)
Mosses	*Fontinalis* *Sphagnum* (peat moss)
Ferns	Giant salvinia (*Salvinia molesta*) Mosquito fern (*Azolla* spp.) Water clover (*Marsilea quadrifolia*) Water spangles (*Salvinia minima*)
Vascular Flowering Plants	Bald cypress (*Taxodium distichum*) Bladderwort (*Utricularia* spp.) Bulrushes (*Scirpus* spp.) Cattail (*Typha* spp.) Duckweed (*Lemna* spp. and *Spirodela* spp.) Hydrilla (*Hydrilla verticillata*) Naiads (*Najas* spp.) Pondweeds (*Potamogeton* spp.) Rushes (*Juncus* spp.) Spikerushes (*Eleocharis* spp.) Water hyacinth (*Eichhornia crassipes*) Water milfoils (*Myriophyllum* spp.)

From: Aquatic Weed Control 2001. S. H. Kay. *North Carolina Agricultural Chemicals Manual*, p. 411. College of Agriculture and Life Sciences, North Carolina State University, Raleigh.

TABLE 27-2. Aquatic Weed Groups—Based on Growth Habit

Category	Examples
Submersed Plants	American elodea (*Elodea canadensis* and *E. nuttallii*) Bladderwort (*Utricularia* spp.) Brazilian elodea (*Egeria densa*) Brittle naiad (*Najas minor*) Coontail (*Ceratophyllum demersum*) Creeping rush (*Juncus repens*) Eurasian water milfoil (*Myriophyllum spicatum*) Fanwort (*Cabomba caroliniana*) Hydrilla (*Hydrilla verticillata*) Parrotfeather (*Myriophyllum aquaticum*) Pondweeds (*Potamogeton* spp.) Proliferating spikerush (*Eleocharis baldwinii*) Southern naiad (*Najas guadalupensis*) Variable-leaf milfoil (*Myriophyllum heterophyllum*) Widgeongrass (*Ruppia maritima*) Wild celery (*Vallisneria americana*)
Free-Floating Plants	Duckweeds (*Lemna* spp. and *Spirodela* spp.) Floating heart (*Nymphoides aquatica*) Frogbit (*Limnobium spongia*) Giant salvinia (*Salvinia molesta*) Mosquito fern (*Azolla caroliniana*) Water hyacinth (*Eichhornia crassipes*) Waterlettuce (*Pistia stratiotes*)
Floating-Leaf Plants	American lotus (*Nelumbo lutea*) Fragrant waterlily (*Nymphaea odorata*) Illinois pondweed (*Potamogeton illinoiensis*) Spatterdock (*Nuphar luteum*) Waterchestnut (*Trapa natans*) Water clover (*Marsilea quadrifolia*) Watershield (*Brasenia schreberi*)
	Broadleaf Species
Emergent Plants	Arrow arum (*Peltandra virginica*) Arrowhead (*Sagittaria* spp.) Asian spiderwort (*Murdannia keisak*) Frogbit (*Limnobium spongia*) Lizard's tail (*Saururus cernuus*) Pickerelweed (*Pondederia cordata*) Smartweeds (*Polygonum* spp.)
	Mat-Forming Broadleaf Species
	Alligatorweed (*Alternanthera philoxeroides*) Creeping water primrose (*Ludwigia hexapetala*) Water pennywort (*Hydrocotyle* spp.) Water willow (*Justicia americana*)

TABLE 27-2. Continued

Category	Examples
	Sedges, Rushes, Spikerushes, and Grasses
	Bulrushes (*Scirpus* spp.)
	Cattail (*Typha* spp.)
	Common reed (*Phragmites australis*)
	Flat sedge (*Carex* spp.)
	Foursquare (*Eleocharis quadrangulata*)
	Maidencane (*Panicum hemitomon*)
	Rushes (*Juncus* spp.)
	Sedge (*Cyperus* spp.)
	Soft rush (*Juncus effusus*)
	Softstem bulrush (*Scirpus validus*)
	Southern wild rice (*Zizaniopsis miliacea*)
	Spikerushes (*Eleocharis* spp.)
	Threesquare bulrush (*Scirpus americanus*)
	Torpedograss (*Panicum repens*)
	Water paspalum (*Paspalum repens*)
	Woolgrass (*Scirpus cyperinus*)
	Other Common Species
	Burreed (*Sparganium americanum*)
	Scouring rush (*Equisetum hymale*)
Woody Plants	Bald cypress (*Taxodium distichum*)
	Pond cypress (*Taxodium ascendens*)
	Tupelo (*Nyssa aquatica*)

From: Aquatic Weed Control 2001. S.H. Kay. *North Carolina Agricultural Chemicals Manual*, p. 411. College of Agriculture and Life Sciences, North Carolina State University, Raleigh.

Figure 27-1. Several submersed aquatic weeds. *Left to right*: Water milfoil, coontail, eelweed, pondweed, and large pondweed.

Figure 27-2. Hydrilla infestation at a North Carolina Piedmont Reservoir (S.H. Kay, North Carolina State University.)

vascular systems. *Emerged plants* are rooted in the bottom sediments at depths of 1 to 5 feet and have floating or erect leaves with showy and conspicuous flowers that extend well above the water surface (Figure 27-3). A few species may form floating mats. All have extensive root and rhizome, vascular and stem systems, and

Figure 27-3. Emerged and floating aquatic weeds. *Left to right*: Burreed, duckweed, white water lily, floating pondweed.

reproduction can be both vegetative and sexual. Examples are water lily, alligatorweed, American pondweed, cattails, and bulrushes. *Woody plants* include obligate aquatic species growing in totally flooded or saturated soils. Some form systems to provide aeration for their root systems ("knees"). Examples include bald cyprus and tupelo.

Identification of aquatic weeds is essential for selecting the appropriate control measures. Lists of common aquatic weeds and their scientific names are provided in Tables 27-1 and 27-2. Many excellent manuals for identification of aquatic weeds are available (Pieterse and Murphy, 1990; Westerdahl and Getsinger, 1988).

METHODS OF CONTROLLING AQUATIC WEEDS

Methods of controlling aquatic weeds include (1) prevention, (2) cultural control, (3) mechanical control, (4) biological control, and (5) chemical control.

Prevention

A good method of aquatic weed control is to not allow extremely noxious aquatic weeds to be introduced into water by boats, persons, or in runoff water in the first place. This is not always possible; however, movement of such plants can be restricted. In addition, it is possible to restrict the import of animal manure, septic tank field material, and nutrients from lawns or farm fields contained in runoff. Nutrients can enrich the water, resulting in algae blooms or more vigorous growth of flowering plants. Nutrient runoff can be reduced by the use of filter strips, restriction of animal movement, conservation tillage in adjacent fields, and construction of retention ponds near homesites, parking lots, and other potential high-traffic runoff sites near the body of water.

Proper Pond Construction Proper pond construction is very important in controlling weeds. Many rooted aquatic plants are not easily established in deep water. A pond should be built so that as much water as possible is at least 3 feet deep. A pond can have water 3 feet deep only 9 feet from the shoreline if all the edges of the pond have a slope of 3 to 1. Such a slope greatly reduces the area where cattails, rushes, and sedges first start growing. However, steep banks are hazardous for swimming. Gentle slopes should be provided for swimming areas.

Bank Management Establishing desirable vegetation around a pond or other body of water is economical and effective in controlling marginal aquatic grasses, weeds, and some weedy species. A good legume-grass pasture mixture, if properly managed, will give the banks and dam a lawnlike appearance. A good sod also protects the banks against erosion and helps to control undesirable species.

Cultural or Physical Control

Submerged weeds can be shaded by use of plastics or substrate liners and by nontoxic chemical dyes. The objective is to reduce the light necessary for photosynthesis and thus reduce plant growth. These techniques are most useful in small ponds with no outflow.

Fertilization is used in commercial catfish farming in the southeastern United States to shade plants. Ponds adequately fertilized develop millions of tiny plants and animals that give the water a cloudy appearance (bloom). If the water has a bloom and is at least 3 feet deep, submerged aquatic weeds have almost no chance to grow because of inadequate light. The benefits of fertilization include (1) increased growth of beneficial microscopic life, including phytoplankton and zooplankton, (2) increased food supply for fish from the food chain that develops from these planktons, and (3) effective weed control by shading. Plants that do reach the surface should be cut off; otherwise, they will be stimulated by the fertilizer. This technique is not recommended for most bodies of water because it requires constant monitoring and management. In most bodies of water the goal is to reduce nutrient levels to prevent unwanted plant growth and improve the overall quality of the water. Never fertilize a pond if weeds are already present.

Drawdown Drawdown, or water manipulation, is a simple way to control many submerged aquatics. If the water can be withdrawn from the pond or ditch, the leaf and stem growth of submerged weeds may be killed after 7 to 10 days of exposure to sun and air. Drying usually must be repeated to control regrowth from roots or propagules in the bottom mud or sand. In ditches this operation may be repeated several times per season. Especially in cold climates, if the water is drawn down in late fall and the lake not allowed to refill until early spring, many aquatic weeds will be killed. As the lake refills, reinfestation may occur from weed propagules from the deeper part of the lake.

Burning Burning may control ditch-bank weeds such as cottonwoods, willows, perennial grasses, and many annual weeds. Green plants are usually given a preliminary searing. After 10 to 14 days, the vegetation may be dry enough to burn from its own heat. Burning can also be combined with chemical or mechanical control programs. Burning the previous year's debris allows better spray coverage of regrowth. It may be desirable to burn the dead debris after chemical treatment. Burning the dried weeds after mowing may increase the effectiveness of the mowing; however, burning may be restricted in some areas.

Hand Removal In lightly infested areas, hand cleaning may be the most practical method of control. A few hours spent in pulling out an early infestation may prevent a weed from spreading. This method is particularly effective on new infestations of emergent weeds such as cattail, arrowhead, and willow.

Mechanical Methods

Many mechanical methods effectively control some ditch-bank weeds. Power equipment can be most easily used where the banks are relatively smooth and not too steep. Underwater power-driven weed saws and weed cutters are available. However, the effects usually last for only a short time. Mowing is generally required at rather frequent intervals, and disposal of the mowed weeds is often difficult.

Mechanical weed harvesters cut off underwater rooted vegetation 4 to 5 feet below the water surface and collect the plant material, which is an advantage as it reduces the effects of decomposition on fish and the collected biomass can be used as mulch or fertilizer in crop fields. Two or more cuttings per growing season are usually required.

Chaining Chaining aquatic weeds resembles chaining woody plants (Chapter 26). A heavy chain, attached between two tractors, is dragged in the ditch. The chain tears loose the rooted weeds from the bottom. This method is effective against both submerged and emergent aquatic plants, but is expensive and must be repeated every 3 to 4 years. Chaining should be started whenever new shoots of emersed weeds rise about 1 foot above the water or when submersed weeds reach the water surface. It should be repeated at regular intervals. Dragging the chain both ways may be effective in tearing loose most of the weeds. The method is limited primarily to ditches that are of uniform width, accessible from both sides with tractors, and free of trees and other obstructions. After chaining it is usually necessary to remove plant debris from the ditch to keep it from accumulating and stopping the flow of water.

Dredging Dredging is a common method of cleaning ditches that are accessible from at least one side. The dredge may be equipped with the usual bucket, or a special weed fork may be used. Dredging may solve two problems: removal of weeds and removal of silt and debris. Dredging has been tried in ponds from specially built pontoons, but in general the pontoon dredge has not proven practical. Dredging is an expensive operation, because of high equipment costs and the large amount of labor involved.

Biological Control

Aquatic weeds have been controlled by fish, snails, insects, microorganisms, and animals. Such biological control has appeal because of the continuing control potential and the nonuse of chemicals in the water. However, as with other biological control methods, care must be taken not to introduce a controlling organism that will have undesirable side effects—for example, a fish that reduces the population of game fish.

Some freshwater fish will eat aquatic vegetation. The white amur (Chinese grass carp), tilapia, and silver dollar fish are used to control aquatic weeds in certain areas of the world. The white amur has been used in the People's Republic of China, Czechoslovakia, Poland, and the Soviet Union. It is now being used in most of the continental United States for weed control in ponds. The grass carp is useful for control of most submersed plants, including macroalgae (*Chara* and *Nitella*). Most states require that grass carp be of the triploid, sterile variety; a permit may be required for stocking (see Figure 3-8). The relative effectiveness of the grass carp against specific weeds is provided in Table 27-3.

For alligatorweed control, a fleabeetle (*Agasicles hybrophila*) and a moth (*Vogtia malloi*) are used in most of the southeastern United States.

There is a tremendous amount of research investigating possible biocontrol agents for aquatic weeds. Although biological control holds great potential, its actual use has been limited. The federal and state governments now support considerable research on this method. Hydrilla and Eurasian water milfoil are two aquatic weeds receiving considerable biological control funding attention from federal and state agencies.

Chemical Control

Herbicides are effective for many aquatic and ditch-bank weeds. The following information is needed to use this method of control: (1) the name or names of the weed species, (2) use(s) of the water, (3) identification of the appropriate chemical, recommended rate, and time of treatment, and (4) the amount of water or size of area to be treated. The two most important factors to consider in selecting the appropriate herbicide are the correct weed identification and the uses of the body of water to be treated.

TABLE 27-3. Biological Control of Aquatic Weeds with Triploid Grass Carp

Weed	Relative Effectiveness
Algae	
Filamentous (green and blue-green) and planktonic	Poor
Algae	
Macro, chara, and nitella	Good to Excellent
Floating and Floating-Leaved Weeds	
Duckweeds, watermeal	Poor
Water ferns (Azolla and Salvinia)	Fair to Poor
Alligatorweed, water lilies, water primrose, lotus watershield, spadderdock, water hyacinth	Poor
Emergent and Marginal Weeds	
Cattails, rushes, common reed, bulrushes, pickerelweed, pennywort, arrowhead	Poor
Submersed Weeds	
	Good to Excellent

From: Aquatic Weed Control. 2001. S. H. Kay. *North Carolina Agricultural Chemicals Manual*, p. 412. College of Agriculture and Life Sciences, North Carolina State University, Raleigh.

Emerged weeds can be treated with handgun sprayers and boom applications. Submersed weeds require treating the water column. Water surface areas are usually measured in acres, like field areas. One acre is 43,560 ft^2, or an area 208.7 feet square (on each side). One acre-ft of water means 1 acre of water 1 foot deep, or 43,560 ft^3 of water (325,828 gallons, or 2,719,450 pounds). Thus, a chemical concentration of 1 ppm would require 2.7 pounds of the chemical (active ingredient) per acre-ft of water. One ppm to an average depth of 3 feet would require 8.11 pounds. A closely related technique involves treating the "bottom acre-foot." Two methods of application are used. Using formulations that are heavier than water, the chemical can be applied to the water surface as either a spray or a granule. The chemical sinks to the bottom. The second method involves drop hoses or weighted hoses dragged behind a boat, releasing the chemical at the lake bottom. Rates of application are usually based on a bottom acre (43,560 ft^2), much the same as with field-crop application. For conversion factors to other measurements, see Table A-5 in the Appendix.

Running water is measured by several methods. Rates are usually given as cubic feet per second; 1 ft^3/sec is equal to 450 gal/min. Usually the rate of water flow is determined by the use of a weir and a gauge.

CHEMICALS USED IN AQUATIC-WEED CONTROL

Application Techniques

The more commonly used herbicides in aquatic weed control are discussed in the following paragraphs. The general uses of specific chemicals for control of certain aquatic weeds are given in Table 27-4, with their relative effectiveness provided in Table 27-5. Use restrictions are provided in Table 27-6; however, always refer to the product label before using any herbicide in or near water. Restrictions on the use of water that has been treated with an aquatic herbicide are *extremely important*. Detailed instructions and restrictions are printed on the labels. *Those instructions must be followed.*

Acrolein

Acrolein use is restricted although it is sometimes used for treating weed-infested irrigation canals in the western United States. It controls most submersed water weeds and many snails. Acrolein is metered into irrigation canals from pressurized cylinders, and weed control with such treatments may extend 20 to 50 miles downstream. Acrolein cannot be used when fish are present, and other restrictions also apply (see label) which greatly limits its use.

Acrolein is dissipated within 24 to 48 hours, and when applied at low concentrations, treated water subsequently used for irrigation does not harm crops. Higher use concentrations may cause injury to susceptible crops, such as cotton. Acrolein is not effective in the control of water plantain.

TABLE 27-4. Chemical Control of Aquatic Plants

Weed	Herbicide
Algae	
Blue-green	Copper sulfate
Algae	
Filamentous and planktonic	Copper complex, various Copper sulfate Diquat
Algae	
Macro, *chara*, and *nitella*	Copper complex
Floating Weeds	
(except watermeal)	Diquat Fluridone
Floating Weeds	
(except duckweed, *azolla*, *salvinia*, and watermeal)	Glyphosate
Floating Weeds	
Watermeal	Fluridone
Emergent Marginal and Ditch-Bank Weeds	
	2,4-D granular Diquat Diuron Glyphosate 2,4-D amine Imazapyr
Submersed Weeds[a]	
	Diquat Endothall Fluridone 2,4-D granular

[a]Grass carp give the most cost-effective control of the majority of the weeds in this group and should be given consideration before using herbicides. A permit is required to purchase more than 150 grass carp or for stocking in impoundments larger than 10 acres in most states. Grass carp usually are not effective on filamentous algae, duckweed, watermeal, or any of the plants in the emersed and marginal group.

Copper Sulfate and Complexes

Copper sulfate is very effective for control of algae and is formulated as copper sulfate pentahydrate. The activity of copper sulfate is reduced when used in hard water (high concentrations of carbonates), as the copper can be quickly precipitated. In extremely hard water, higher use rates may be necessary. Although copper sulfate can be toxic to certain fish species, kills are relatively rare because of the precipitation of copper that occurs in hard water, which is common in most U.S. waters. Fish kills are possible in areas where heavy infestations of algae are controlled, as a result of oxygen depletion.

Copper chelates are formulations of copper that do not precipitate in hard water. These are effective against algae and, depending on the chelate formulation, on submersed plants such as hydrilla, naiad, and elodea (Table 27-5). Chelates are less toxic to fish, but more expensive than copper sulfate, and are formulated as liquids and granules.

Both copper sulfate and copper chelates are contact herbicides and are effective at low concentrations. There are no use restrictions for the treated water, as the compounds are fairly short-lived in their active form because of their rapid adsorption to soil and organic materials and their quick movement into sediments (Table 27-6).

2,4-D

Many aquatic plants are susceptible to 2,4-D dissolved in water (Table 27-5). Water-soluble liquid and granular formulations of 2,4-D may be used. 2,4-D has greater efficacy on emerged than underwater plants; however, granular formulations work well on water milfoil and coontail.

2,4-D should be exposed in water to susceptible weeds for at least 10 hours. In nonmoving water, 2,4-D in the water-soluble form will tend to distribute itself equally over several days for distances of up to 400 feet. In contrast, the granular form falls to the pond bottom, where a relatively high concentration may develop at the soil-water interface.

Ester formulations of 2,4-D are 50 to 200 times more toxic to fish than amine formulations, but toxic effects have rarely been experienced under field conditions. Esters are oil-like and oil-soluble materials, and oils are known to be toxic to most fish. The esters may be acting like oils; or the solvents, emulsifying agents, or other additives in the formulation may be killing the fish. When pure 2,4-D ester is used in granular form, without surface-active agents, there is considerably less hazard to fish. There are restrictions for the use of 2,4-D, but labels vary as to specifics (Table 27-6[b]). 2,4-D should not be used in water intended for irrigation, human drinking, or watering of livestock.

Diquat

Diquat controls many submersed and free-floating aquatic weeds and algae in static water. Bladderwort, coontail, elodea, naiad, pondweed, water milfoil, spirogyra, and pithophora are among the species controlled (Table 27-5).

TABLE 27-5. Effectiveness of Herbicides Recommended for Control of Weeds in Aquatic Situations

	Aquathol	Hydrothol	Diquat	2,4-D	Copper Compounds	Fluridone	Glyphosate
Filamentous Algae	NR	E	E	NR	G	NR	NR
			Floating Plants				
Duckweed	NR	NR	G	P	P	E	NR
Watermeal	NR	NR	P	NR	NR	G	NR
Alligatorweed	NR	NR	NR	P	NR	F	G
Creeping water primrose	NR	NR	NR	E	NR	ID	E
Water hyacinth	NR	NR	G	E	NR	F	G
			Emersed Plants				
American lotus	NR	NR	NR	G	NR	G	E
Cattail	NR	NR	F	F	NR	G	E
Common reed	NR	NR	NR	NR	NR	NR	G
Fragrant water lily	NR	NR	NR	G	NR	G	E
Rush	NR	NR	NR	P	NR	NR	G
Spatterdock	NR	NR	NR	P	NR	G	E
Water pennywort	NR	NR	F	G	NR	G	E
Most grasses	NR	NR	F	P	NR	F	E

TABLE 27-5. Continued

	Endothall Aquathol	Hydrothol	Diquat	2,4-D	Copper Compounds	Fluridone	Glyphosate
			Submersed Plants				
Bladderwort	P	P	G	P	NR	E	NR
Brazilian elodea	P	P	E	NR	F	E	NR
Coontail	E	E	E	G	NR	E	NR
Hydrilla	E	E	E	NR	F	E	NR
Parrotsfeather	E	E	G	E	NR	E	NR
Pondweed	E	E	E	NR	NR	E	NR
Proliferating spikerush	NR	NR	NR	NR	NR	E	NR
Brittle naiad	E	E	E	NR	NR	E	NR
Southern naiad	P	P	P	NR	NR	G	NR
Variable leaf milfoil	E	E	E	E	NR	E	NR

Key: NR = Not recommended; ID = Insufficient data; P = Poor; F = Fair; G = Good; E = Excellent

TABLE 27-6. Use Restrictions for Aquatic Herbicides[a]

	Uses of Water				
Herbicide	Drinking	Fishing	Livestock watering	Swimming	Irrigation
Copper	0	0	0	0	0
Diquat	14	0	14	NR	3–5
Endothall	7–25[b]	3	7–25[b]	NR	7–25[b]
2,4-D	*[b]	*	*	*	*
Fluridone	Do not apply within $\frac{1}{4}$ mile of intake.	0	0	0	7–30[b]
Glyphosate	Do not apply within $\frac{1}{2}$ mile of intake.	0	0	0	0

[a]Days after application or distance from application; always check label and follow all instructions before using an aquatic herbicide.
[b]Label restrictions vary, depending on formulation and manufacturer.
NR = Not recommended.

Diquat is a contact herbicide, so good distribution in the water and on the plant foliage is required. Quick burndown of filamentous algae and emerged plants can be obtained, but it is temporary as no diquat is translocated to the plant roots. Because diquat is a contact spray, repeated applications may be necessary to give season-long control. Diquat is often mixed with copper chelates to provide broader-spectrum weed control. Diquat should not be used in muddy water because it is rendered ineffective by its tight adsorption on soil particles.

Diquat is not harmful to most fish at the recommended rate. In heavy weed infestations only one-third to one-half of the area should be treated so as to avoid fish kill through oxygen depletion, and treatments should be spaced 10 to 14 days apart. There are restrictions on water use (i.e., for animal and human consumption, irrigation, swimming, and domestic purposes) following application of diquat; these are continually being revised, thus the most recent label must be consulted. This is true for all aquatic herbicides.

Endothall

Endothall is available in two salt forms (dimethylamine (Hydrothol) and dipotassium (Aquathol)) and in several liquid and granular formulations. The dimethylamine salt controls filamentous algae and *Chara* and submersed flowering plants. The dipotassium salt controls submersed flowering plants but not algae. Elodea is not controlled by either form. The dimethylamine salt can be toxic to fish (especially in liquid formulations), which restricts its use. Endothall is applied at a broad range of rates, with lower rates for control of algae and higher rates for coontail, water milfoil, pondweeds, naiad, bassweed, and burreeds (Table 27-5). Treated water should not be used for irrigation, agricultural sprays, animal consumption, or domestic purposes, as specified on the most recent label.

Fluridone

Fluridone, available as an AS formulation or as a slow-release granule, and controls at least 14 submersed and emerged plants (Table 27-5). Weed control for up to 1 year after initial application can often be obtained. Fluridone is applied during early stages of rapid plant growth. After application, chlorosis and a pinkish color appears on foliage in 7 to 10 days, but full herbicidal effects may take up to 90 days. About 4 weeks after treatment, susceptible weeds begin to sink to the bottom. This slow herbicidal response reduces the potential of fish kill resulting from oxygen depletion. Desirable plants in the water or along the shoreline can sometimes absorb fluridone from the soil and show injury, but they usually recover.

Fluridone is adsorbed by the hydrosoil (lake-bottom soil), reaching a maximum adsorption by 1 to 4 weeks after treatment. These hydrosoil residues decline to a nondetectable level after 16 to 52 weeks. Restrictions on use include applications near water source intakes and the use of treated water for irrigation; consult the most current label.

Glyphosate

Glyphosate is primarily effective for control of emerged aquatic weeds such as alligatorweed, cattail, maidencane, paragrass, spatterdock, water hyacinth, giant cutgrass, and torpedograss (Table 27-5). Glyphosate must be applied to plant foliage to be effective, and use of a surfactant is required for optimum results. Glyphosate must be retained on the foliage of actively growing plants for approximately 6 hours for best activity and is not effective for control of submersed or mostly submersed vegetation. Good kill of the entire plant can be obtained because glyphosate translocates throughout the plant. The only restriction on glyphosate use is that no applications can be made within $\frac{1}{2}$ mile of a drinking water intake site (Table 27-6).

LITERATURE CITED AND SUGGESTED READING

Kay, S. H. 1991. *Using Grass Carp for Aquatic Weed Management*. North Carolina Cooperative Extension Service Pub. # AG-456.

Kay, S. H. 1992. *Hydrilla, a Rapidly Spreading Aquatic Weed in North Carolina*. North Carolina Cooperative Extension Service Pub. # AG-449.

Kay, S. H. 1997. *Weed Control in Irrigation Water Supplies*. North Carolina Cooperative Extension Service Pub. # AG-438.

Pieterse, A. H., and K. J. Murphy. 1990. *Aquatic Weeds: The Ecology and Management of Nuisance Aquatic Vegetation*. Oxford University Press, New York.

Sculthorpe, C. D. 1971. *The Biology of Aquatic Vascular Plants*. Edward Arnold, London.

Westerdahl, H. E., and K. D. Getsinger. 1988. *Aquatic Plant Identification and Herbicide Use Guide*. Vol. 11. *Aquatic Plants and Susceptibility to Herbicides*. U.S. Army Corps of Engineers, Tech. Rep. A-88-9.

WEB SITES

Grass Carp Information

North Carolina State University: Using Grass Carp for Aquatic Weed Management in North Carolina. S. H. Kay and J. A. Rice. North Carolina State University Cooperative Extension http://www.ces.ncsu.edu/

Search "Grass Carp," Select "Grass Carp."

Weed Management

University of Missouri: *Aquatic Weed Control in Missouri*. 1996. A. Kendig and M. S. DeFelice. Publication G4856. Cooperative Extension Service, University of Missouri-Columbia.

http://muextension.missouri.edu/

Go to "Search Extension," then to Publication Search, Enter "Aquaculture Weed Control." Select "Missouri Aquaculture Publications," Select "Aquatic Weed Control in Missouri (M99)."

Auburn University: *Chemical Weed Control for Lakes and Ponds*. 1999. J. W. Everest. ANR-48. Alabama Cooperative Extension Service, Auburn University.

http://www.aces.edu

Search for "Aquatic Weed Control," Select "ANR-48, Chemical Weed Control for Lakes and Ponds."

North Carolina State University: North Carolina Agricultural Chemicals Manual. North Carolina State University Cooperative Extension.

http://ipmwww.ncsu.edu

Go to "Pesticide Recommendations," then to "Chemical Weed Control."

University of Florida: Aquatic Weed Management Guide. 2000. V. V. Vandiver, Jr., and D. H. Teem. Fact Sheet AGR 70. Cooperative Extension Service, Institute of Food and Agricultural Sciences.

http://aquat1.ifas.ufl.edu/

Select "Pest Management Guides-Aquatic Weed Management Guide."

AgChemical Label Information: Crop Data Management Systems, Inc., Marysville, CA.

http://www.cdms.net/manuf/manuf.asp

For chemical use, see the manufacturer's or supplier's label and follow these directions. Also see the Preface.

28 Industrial Vegetation Management

TOTAL VEGETATION CONTROL

Industrial vegetation management covers many nonagronomic and nonaquatic situations. These sites include forests (also Christmas trees), railway roadbeds, utility rights-of-way, pipelines, highway brims, fencerows, irrigation and drainage banks, and industrial plants and storage sites. In many of these sites, total vegetation control is desired, although weed management is often more realistic.

Total control of vegetation is the removal of all higher green plants in an area and maintenance of the area vegetation free. Complete absence of vegetation is desirable on many sites for a variety of reasons (Figure 28-1).

Vegetation can be controlled totally by mechanical or chemical methods, and sometimes both methods are used. Brush and trees may be removed by mechanical means at first clearance, although this method adds considerable cost, and following initial removal of vegetation, such areas are then maintained weed free with herbicides. Herbaceous species can also be eliminated by either mechanical or chemical methods. Disking or other mechanical means may have to be repeated several times a season to keep the area weed free, and certain residual herbicides can be applied at high rates on an annual basis or as needed to control the weeds.

Figure 28-1. Total vegetation control on a railroad with tebuthiuron, also used on industrial areas and noncrop sites. (DowAgro Sciences.)

Foliar Herbicides in Total Vegetation Management

To obtain total vegetation management, certain nonselective foliar-applied herbicides are used together with persistent, nonselective, soil-applied herbicides. The foliar herbicides include contact herbicides such as paraquat or diquat and translocated materials such as 2,4-D, picloram, dicamba, triclopyr, and glyphosate (Table 28-1). Combination products are available for broad-spectrum control with foliar herbicides and persistent soil-applied herbicides.

Soil Herbicides in Total Vegetation Management

Persistent nonselective herbicides used for vegetation management are listed in Table 28-1. The chemical and physical properties of these compounds, as well as various other uses, have been discussed in previous chapters. Although some of these herbicides are also used as selective herbicides, the rates of application for nonselective uses are much higher to allow weed control for a greater length of time.

There are a number of herbicides used for industrial weed control (Table 28-4). Because food crop tolerances do not need to be determined, a herbicides's use for industrial areas has sometimes been the first registered use of a herbicide so that sales can be generated while crop residue data are developed for food use crop registration.

TABLE 28-1. Herbicides Commonly Used for Total Vegetation Control of Weeds on Noncropland[a]

Soil-Applied Herbicides[b]	Foliar-Applied Herbicides[b]
Bromacil	Ametryn
Dicamba	Amitrole
Diuron	Asulam
Imazapyr	Bromoxynil
Metsulfuron	Dicamba
Picloram	2,4-D
Prometon	Dichlorprop
Simazine	Diquat
Sodium chlorate	Fosamine
Sulfometuron	Glyphosate
Tebuthiuron	Hexazinone
	Linuron
	MSMA
	Paraquat
	Triclopyr

[a]Various combinations of these herbicides are often used to provide greater persistence and/or a broader spectrum of weeds controlled. Trade names are given in Table A-3 of the Appendix.

[b]Some herbicides have considerable foliar activity as well as soil activity. See the manufacturer's label for rates, method of usage, species susceptibility, and precautions. Follow the label—regardless of any statements in this book.

The high rate of use for total vegetation management often results in these soil-applied herbicides remaining toxic to plants for more than 1 year. Factors affecting the length of time that a herbicide remains toxic in the soil are discussed in Chapter 6. In general, dry weather with little or no leaching, cool or cold temperatures, and heavy soils tend to lengthen the time a herbicide will remain toxic. Under any given conditions, the length of time a herbicide will remain toxic can be predicted with reasonable accuracy (Table 6-2). Annual applications of most of these herbicides at rates that persist somewhat longer than 1 year are usually more economical than massive rates that will persist for 2 years or more.

To be effective, these persistent, nonselective soil-applied herbicides must be leached into the rooting of the seed-germination zone of the weeds. Therefore, they are usually applied just before or during the rainy season. Persistent soil-applied herbicides gradually lose phytotoxicity, and as this occurs, weeds begin to reinfest the area. Typically, the weedy plants are stunted and grow slowly, and the use of a broad-spectrum foliar herbicide usually kills these weeds and extends the period of total vegetation control.

INDUSTRIAL MANAGEMENT SITES

Forests

The 400$^+$ million acres of forestland in the United States (Table 28-2) are not subjected to total vegetation management. Conifer forests are managed more intensively than hardwood forests, and vegetation control is more likely to be part of the management plan. These forests are located in the following regions: (1) Southern United States (Virginia to Texas), (2) Pacific Northwest (northern California to Washington), (3) Lake States (northern Minnesota to northern Michigan), and (4) Northeast (northern New York to Maine). The conifer species of the Great Lake States and the Northeast also extend northward across Canada (some 250 million acres). Management of these forests, whose ownership is largely by private industry, can be intensive, and herbicide use is widespread. Slow growth and federal ownership

TABLE 28-2. Estimated Acres of Selected Forest Regions

Forest Location	Acres (millions)
Conifers	
Southern	180.0
Pacific Northwest	56.5
Lake States	24.0
Northeast	26.0
Hardwoods—Central	124.0
Total	410.5

From Walstad and Kuch (1987) and U.S. Department of Agriculture (1979).

TABLE 28-3. Estimated Right-of-Way Acres

Type	Acres (thousands)
Railroad	2,417
Highway (vegetation area)	8,255
Electric transmission	4,954
Pipelines	2,188
Total	17,814

From U.S. Department of Agriculture, 1979.

characterize the coniferous forests of the Rocky Mountain region of the United States, where herbicide use is more limited.

Hardwood forests dominate in the broad area between the Great Lake States and the Southern conifers, where small nonindustrial ownership is common and herbicide use is much lower than in coniferous regions.

Christmas tree production occurs nationwide. However, because the crop species are conifers, major production tends to be located in the Pacific Northwest, across the northern United States, and through the Appalachian mountains and foothills.

Rights-of-Way

Major rights-of-way include those of railroads, highways, electric transmission lines, and pipelines (Table 28-3). Most right-of-way acres are located in the eastern United States.

Industrial Sites

Industrial sites include industrial plant sites, storage yards of all types, electric transmission substations, pumping stations, and parking areas.

HERBICIDE PROGRAMS

Forests

Herbicide use in forestry can be categorized as (1) site preparation, (2) plantation establishment, (3) conifer release, and (4) timber stand improvement. Despite the large forest acreage, an individual site may actually be treated with herbicides only once or twice throughout a 30- to 60^{+}-year rotation, although Christmas tree production is an exception.

Site Preparation Site preparation is particularly important in coniferous forests. Most conifers are harvested in blocks. After the removal of all salable trees, the remaining vegetation, including noncommercial stems and undesired species, are controlled mechanically, chemically, with fire, or by some combination of these before the new crop of trees is planted. Herbicide treatments effective on hardwood species include nonselective or selective chemicals, and application can be made by aerial or ground equipment. After herbicide application, burning may be done to remove obstructing plant material. Herbicides used in such situations include glyphosate, hexazinone, imazapyr, picloram, and triclopyr.

Plantation Establishment Weed control at the time of planting, or very soon after, is important for optimizing the survival and growth of conifers and hardwoods. The planted sites are already covered with weeds or immediately become occupied with invading pioneer weed species. Weed management is achieved with selective herbicides applied as broadcast, banded, or spot treatments. Common herbicides used in tree establishment are atrazine, glyphosate, hexazinone, imazapyr, simazine, and sulfometuron methyl.

Conifer Release Because conifers tend to be subclimax species, invasion by climax hardwood species is the natural progression of plant succession. A herbicide application may be used to release (provide an establishment advantage) the conifers if the hardwood component of the stand is excessive. The herbicides used must be selective for the conifer species involved. Conifer release can be done at any time, but is usually done only once or twice during the life of the plantation. Good site preparation with herbicides such as 2,4-D, glyphosate, hexazinone, imazapyr, and triclopyr tend to reduce the need for a release treatment.

Timber Stand Improvement Timber stand improvement is a procedure by which individual trees are selected for culling to improve stand quality. It is practiced throughout the forested regions and is of particular importance in hardwood management. Historically, this has been done by individual tree injection or by applying herbicides to frills or girdles. However, low-volume basal and foliage treatments are becoming more common. Herbicides used in these treatments include 2,4-D, dicamba, glyphosate, imazapyr, picloram, and triclopyr.

Christmas Trees

A general consideration in Christmas tree production weed control is the selection and preparation of a weed-free site to minimize weed problems in a young planting. After planting, growers often use organic mulches and carefully choose herbicides for weed control, (Table 28-4) as trees less than 4 years old tend to be more vulnerable to certain labeled herbicides than older trees (see herbicide table for use restrictions based on tree age). The general approach to herbicide use is to apply soil-active herbicides such as simazine, oxyfluorfen, napropamide, or oryzalin directly over the transplants or to use directed sprays with herbicides like glyphosate or triclopyr. Weed control the first year after transplanting is essential to obtain vigorous trees. Banded applications within tree rows with soil-applied herbicides are common each spring, but as trees grow, herbicide use is gradually reduced after the third year. However, annual treatments are continued (to some degree) for herbaceous weed control over a 5- to

TABLE 28-4. Examples of Herbicides Used for Industrial Weed Control

	Forest	Rights-of-Way	Industrial	Christmas Trees
Asulam				X
Atrazine				X
Basagran				X
Bromacil			X	
Clethodim				X
2,4-D	X	X	Turf	
Dicamba		Highway	Turf	X
Diquat			X	
Diuron		Railroad	X	
Fosamine		X		
Fluazifop				X
Glyphosate	X	X	X	X
Hexazinone	X			
Imazapyr	X	X	X	
Isoxaben				X
Metsulfuron methyl		X		
MSMA		Highway		
Napropamide				X
Norflurazon		X	X	
Oryzalin		Highway		X
Oxyfluorfen	Forest nursery			X
Paraquat	X	X	X	X
Pendimethalin		X		X
Picloram	X	X		
Prodiamine		X	X	X
Prometon			X	
Sethoxydim				X
Simazine				X
Sulfometuron methyl	X	X	X	
Tebuthiuron		X	X	
Triclopyr	X	X	Turf	X

12-year period. Postemergence herbicides are used as needed to control escaped and troublesome perennials throughout the life of the planting, as weed control is important to allow a full, uniform crown development from the treetop to the ground.

Rights-of-Way

Railroad Vegetation control on railroad rights-of-way is done for employee safety, fire hazard reduction, increased tie life, public safety, improved visibility, and to allow easy inspection of the facility. Railroad vegetation programs are concerned with (1) yards, (2) bridges, (3) line-of-road (ballast), (4) road crossings, and (5) brush. Long-term total vegetation control is desired in railroad yards and under bridges, especially wooden bridges, and herbicide treatments are applied early in the year as broadcast preemergent applications. A line-of-road is treated for short-term total vegetation control, and the ballast area is spot treated annually to control emerged weeds. A second treatment in mid to late summer is commonly applied to control encroaching vines and emerged weeds. Road crossings tend to be treated at the same time as the line-of-road, with treatment providing short-term total vegetation control or release of low-growing desirable vegetation. Brush control treatments are typically applied in mid to late summer and use both selective and nonselective herbicides, which are applied infrequently, only as needed. Herbicides used in these various applications include bromacil, 2,4-D, diuron, glyphosate, imazapyr, sulfometuron methyl, and triclopyr.

Highway Highway vegetation control is especially concerned with public safety. Increased visibility and improved road maintenance result from controlling vegetation. Although aesthetics are of constant concern because of public visibility, aesthetic acceptability differs from state to state. Total vegetation control is generally important around guardrails and barriers, and may be applied to the road shoulder area. Most highway acres are managed with selective herbicides to control broadleaf weeds and woody plants, along with selective control of grassy weeds and grass seedhead suppression. Common herbicides used include 2,4-D, dicamba, glyphosate, oryzalin, pendimethalin, sulfometuron methyl, and triclopyr.

Electric Uninterrupted electrical service is a requirement of electric transmission rights-of-way. Consequently, trees capable of reaching the electrical conductors must be controlled. Herbicide treatments may be selective for grassy vegetation or nonselective for broad-spectrum weed control. Treatment cycles tend to be 4 to 8 years, and low-volume backpack sprayers are rapidly replacing high-volume hydraulic sprayers for application.

Pipelines Woody vegetation in proximity to pipelines is controlled so that the rights-of-way can be inspected and maintenance operations completed. Weed control programs for pipelines are similar to those for rights-of-way of electric transmission equipment. Herbicides used in electric and pipeline areas include 2,4-D, glyphosate, imazapyr, metsulfuron, picloram, and triclopyr.

Industrial Sites Industrial sites include turf management areas, as well as sites requiring total vegetation control. Aesthetics, reduced fire hazard, rodent control, and employee safety are reasons for industrial weed control. Total vegetation control treatments tend to be applied annually and herbicide combinations are used with bromacil, 2,4-D, dicamba, diquat, diuron, glyphosate, imazapyr, sulfometuron methyl, and triclopyr.

Herbicide Resistant Weeds Resistant biotypes of kochia and Russian thistle are problems on western sites managed for total vegetation control, and marestail is becoming a concern in the eastern United States. With residual soil activity so important to industrial weed control, herbicide resistance will be a continuing problem.

LITERATURE CITED AND SUGGESTED READING

Cantrell, R. L. 1985. *A Guide to Silvicultural Herbicide Use in the Southern United States*. School of Forestry, Auburn University, Auburn, AL. 482 pp.

Glover, G. R., H. E. Quicke, and D. K. Lauer. 1994. Economics of early vegetation control in southern pine forests. *Southern Weed Science Proc.* **47**:83–84.

Lauer, D. K., and G. R. Glover. 1995. Pine and competing vegetation response to herbaceous weed control and levels of shrub control in the flatwoods. In *Proceedings of the Eighth Biennial Southern Silvicultural Research Conference*, November 1–3, 1994, Auburn, AL, Comp. By M. Boyd Edwards.

Quicke, H. E., D. K. Lauer, and G. R. Glover. 1996. Growth responses following herbicide release of loblolly pine from competing hardwoods in the Virginia Piedmont. *Southern Journal of Applied Forestry* **20**(4):177–181.

U.S. Department of Agriculture. 1979. *The Biological and Economic Assessment of 2,4,5-T*. U.S. Department of Agriculture. Agriculture Technical Bulletin 1671. 445 pp.

Walstad, J. D., and P. J. Kuch. 1987. *Forest Vegetation Management for Conifer Production*. John Wiley & Sons, Inc., New York. 523 pp.

WEB SITES

Michigan State University: *Some Fundamentals for Successful Weed Control in Forest Crops*. 2001. D. O. Langtagne. Michigan State University Extension Forestry Bulletin FTE 22180

http://www.msue.msu.edu

Search "Forest Crops Weed Control," scroll to "FTE 22180"

Indiana Department of Natural Resources: *Forestry Best Management Practices, Tree Planting and Weed Control*. 2000

http://www.state.in.us/dnr/forestry/bmp/nonplant.htm

University of Florida: *Vegetation Management in Florida's Private Non-Industrial Forests*. 1998. P. Campbell and A. Long. Publication SS-FOR-10. Cooperative Extension Service, Institute of Food and Agricultural Sciences

http://edis.ifas.ufl.edu/

Select "Publications by Department," then select "Forest Resources and Conservation," select "Regeneration, Fertilization, Vegetation and Pest Management" and select "SS-FOR-10"

Weed Management in Fence Rows and Non-Cropped Areas. 2000. J. A. Tredaway. Publication SS-AGR-111. Cooperative Extension Service, Institute of Food and Agricultural Sciences

http://edis.ifas,ufl.edu/

Select "Pest management Guides," then select "Weed Management Guide," then select "Miscellaneous Weed Management Topics," select "Weed Management in Fence Rows and Non-Cropped Areas"

The Ohio State University: *Ohio Christmas Tree Producers Manual.* 1991. J. H. Brown, W. F. Cowen, Jr., R. B.Heiligmann. Bulletin 670, Cooperative Extension Service, The Ohio State University

http://www.ag.ohio-state.edu

Select "Agriculture and Natural Resources," scroll to "Ohio Christmas Tree Producers Manual"

North Carolina State University: *Weed Management in Conifer Seedbeds and Transplant Beds.* 1999. J. C. Neal. Bulletin HIL-449. Cooperative Extension Service, North Carolina State University

http://www.ces.ncsu.edu/

Search for "Conifer Beds," select "HIL-449"

AgChemical Label Information: Crop Data Management Systems, Inc. Marysville, CA

http://www.cdms.net/manuf/manuf.asp

For chemical use, see the manufacturer's or supplier's label and follow these directions. Also see the Preface.

29 Diagnosis of Herbicide Injury

There are many opportunities for agronomists, horticulturalists, and other plant professionals to advise farmers, home owners, and the general public about how various biotic (insects, animals, nematodes, plant pathogens) and abiotic (pesticides, air pollution, nutritional disorders, environmental factors, mechanical factors) stresses can influence the growth of desirable plants. There are many situations in which desirable plants are damaged in some way and a professional is called to make a determination about whether some agricultural practice is the cause of the reduced plant growth observed. Often, the first suggested cause of plant injury is herbicidal, and to avoid jumping to conclusions and making the wrong diagnosis, it is imperative that agricultural professions use a systematic process for accurate determination of the cause(s) of the plant injury observed. This chapter describes a systematic approach to diagnosing plant injury and determining whether a herbicide is the cause. Suggestions as to the appropriate actions to employ to alleviate a herbicide-related problem are also offered.

One challenge facing any practitioner using herbicides is to correctly determine whether damage to plants adjacent to herbicide-treated areas was caused by chemical application or by some other factor. Plants respond to a wide variety of environmental stresses (both biotic and abiotic) in similar ways. Damage can involve leaf, stem, and root symptoms, which need to be carefully observed. Herbicides can cause plant injury, but so can nutrient imbalances, air pollution, mechanical stresses, diseases, insects, viruses, nematodes, compacted soil, other pesticides, extremes of environmental conditions (temperature and light), water stress, and many other factors.

When dealing with a diagnostic problem, it is always important to conduct yourself in a professional manner, which includes proper appearance and demeanor. Never become involved in a plant injury claim with preconceived ideas and opinions. Gather all the information regarding the plant injury event and perform a thorough analysis prior to making a recommendation. Remember, *never* jump to conclusions and *never* provide a recommendation without having as much information as possible. The person whose plants are affected often has a hunch as to the cause, and because herbicides are so widely used, they often are initially blamed for all types of problems. Herbicides are an easy target, and it is the job of the diagnostician to determine whether they are in fact the cause.

Diagnosticians need a good background of knowledge about how plants grow and respond to their environment. In the case of insect- or pathogen (fungal, bacterial, viral, and nematodes)-caused injury, the causal agent can often be identified by an expert. If no causal agent is identified, then other factors must be investigated. With herbicides and nutrients, determination of the specific causal agent may be more difficult. A tissue analysis can be done, but such an analysis is expensive and time-consuming; therefore, a good knowledge of plant response symptomology can be valuable in identifying the causal agent. Herbicide injury usually results from the drift of spray to adjacent areas, herbicide runoff, herbicide carryover, misapplication, contamination, or the uptake of a herbicide by the roots of trees. In cases of spray drift, plant death is not common and, usually, after initial symptoms, plants regrow with the new growth being normal. All these scenarios and their outcomes require a good knowledge base and some level of experience to aid in the appropriate diagnosis.

DIAGNOSTIC PROCEDURE

Diagnosis involves a systematic approach to a probable cause. Most situations involve analysis of samples sent to a diagnostic laboratory or a field injury that requires a site visit. In both circumstances, the best approach is to gain as much knowledge as possible. This is more difficult with samples submitted to laboratories, as the information provided is generally limited, which makes a definitive diagnosis of herbicide injury difficult. A field visit allows the use of a more systematic approach, which in most cases is critical to the proper diagnosis and recommendation for remedial action.

The Four-Step Approach

A four-step approach is recommended for determining the causal agent of a plant injury. Herbicides are chemicals that are designed to kill plants and as such can, under certain circumstances, cause damage to unintended targets such as adjacent crops and landscapes. Many factors can lead to off-site movement of herbicides and plant injury (see Chapter 7), and an awareness of these factors is important in any approach to diagnosis.

A diagnosis of injury should not be based on visual symptoms alone. There are five major groups of causal agents that must be considered as possible causes of injury to plants: (1) infectious pathogens, (2) environmental factors, (3) chemical factors, (4) mechanical factors, (5) insects, mites, and other animals. In approaching any situation, a diagnostician should have a knife, a shovel, sample bags, an ice chest, a camera, and a notebook and should come armed with questions. Remember, although many abiotic and biotic factors cause similar plant responses, the key to a proper diagnosis is to be observant, look at the appearance and pattern of the injury at the site, gain as much information as possible, and never jump to conclusions before all the information is gathered and analyzed. The appropriate diagnosis is especially important in cases of herbicide injury, as plant loss due to herbicides is often the result of human error and fault determination can result in litigation.

When approaching a diagnosis of plant injury, it is important to make a series of observations in a systematic manner. The four-step approach described here is the recommended procedure to follow in determining the causal agent of plant injury whether herbicidal or not.

Step 1 *Determine what is normal in regard to plant growth.* Determine the plant genus, species, and cultivar, then gather the affected plants and make a careful comparison with a normal, unaffected plant to determine what symptoms are associated with the injury. This initial step allows a determination that a problem actually does exist (Figure 29-1).

Step 2 *Check the pattern of injury observed and determine the time frame of the injury occurrence and development.* This is important for gaining insight into the nature of the causal agent. Gather as much information as possible about the history of the site, including when the crop was planted, what was done to the crop, when did the injury first occur, and has the injury changed since it was first noticed. This information should include specific questions regarding cultural practices, as listed in Table 29-1.

In addition to questions regarding cultural practices, there are a series of observations that should be made. These observations are summarized in Table 29-2

Figure 29-1. What is normal? This is zebra grass, an ornamental grass, which is supposed to have variegated spots on its leaves. Some people could confuse this with paraquat drift injury. (G. Ruhl, Purdue University.)

and constitute a vital component in determining the nature of the problem in regard to causal agents.

Answers to the questions and observations suggested in Tables 29-1 and 29-2 can provide important insight into whether the causal agents of the injury observed are biotic or abiotic in nature. Biotic agents (insects, animals, nematodes, plant pathogens) cause injury in a *random* pattern (Figure 29-2 and Figure 29-3). There is no defined pattern, and symptoms develop slowly over time. Abiotic agents (pesticides, air pollution, nutritional disorders, environmental factors, mechanical factors) result in a more defined or *nonrandom* pattern of injury (Figure 29-2). Infectious plant diseases are usually host specific. If the pattern of injury is random and it is seen in many different species (or plant types), it is most likely not caused by a biotic agent. Abiotic injury tends to appear at one time and does not spread to other plants. There is often a clear definition of injured versus normal tissue on a plant. Moreover, an injury may be specific to certain plants or, depending on the agent, may be a general injury to all plants. The lack of a biotic agent, as shown by a laboratory analysis, indicates that

TABLE 29-1. Questions to Ask Regarding Cultural Practices Employed at Site Where Injury Has Occurred

1. When was the injury first observed, and has it changed, gotten worse, or improved?
2. How was the field site prepared prior to planting, and what cultural practices have been employed during the cropping cycle?
3. What herbicides were used this year and in past years?
4. How much herbicide was added to the spray tank, and what was the tank volume?
5. Were other herbicides or pesticides applied, and if so, when and how? (These questions are important whether the agents were custom applied or not.)
6. What was the sprayer used for immediately before the application, and was it cleaned and flushed prior to use?
7. When and how was the sprayer calibrated?
8. What are the specifics regarding application equipment (sprayer output in gallons per acre, nozzle types and spacing, speed of application, pressure used, and boom length and height)?
9. When and how (broadcast or banded) was the application made in relation to crop planting and emergence?
10. What were the weather conditions prior to, during, and after application (temperature, wind, soil moisture, and any unusual environmental factors)?
11. What variety (cultivar) of crop was planted, and how (planter used, depth of seeding or transplanting, and any special treatments to the seed or transplants)?
12. How old was the seed, or how were the transplants handled prior to planting?
13. Is there soil analysis information available for the field?
14. What is the history of field cropping, and have there been disease or nematode problems in the field?
15. What fertilizer (type) was applied and how?
16. What crops are in adjacent fields, and what cultural and pesticide application practices were used?

Adopted from Monaco et al., 1986.

TABLE 29-2. Observations to Make Regarding Injury Patterns

1. Does the injury appear in a recognizable pattern, such as in bands or strips in the field (for example, the width of a spray boom), and does the injury vary with soil type and organic matter?
2. Is the injury more severe at the ends of the field, and does it occur on the sides of fields adjacent to other crops, or on ditch banks, or in rights-of-way?
3. Can injury symptoms be observed on weeds in and adjacent to the field and on crops in adjacent fields?
4. Is there a slope to the field that could result in downhill washing or leaching of a herbicide?
5. Does injury occur on low, wet spots rather than on high, dry spots in the field?
6. Is there obvious damage to the crop such as from insects, diseases, hail, cold weather, root pruning, or cultivation?
7. Is the equipment used in the field (especially spray equipment) properly maintained and operational for the uses described?

Adapted from Monaco et al., 1986.

some abiotic agent is involved, and the questions posed in the diagnosis scheme can provide important clues as to the cause.

Step 3 *Examine injured plants for specific symptomology.* This step involves observing the plant's leaves, stem, roots, vascular system, and flowers or fruits for the type of injury expressed. Leaf injury can involve necrosis, off-color (yellowing, light or dark green, water-soaked), small or irregular or enlarging spots, spots with fungal fruiting bodies, abnormal size and shape (including irregular edges and venation pattern), twisting, or tissue loss. Stem injury may involve cankers or lesions, breakage, discoloration, twisting (epinasty), brittleness or swelling, especially at the soil-air interface, or distortion of leaf or meristematic tissue (Figure 29-4). Root injury can

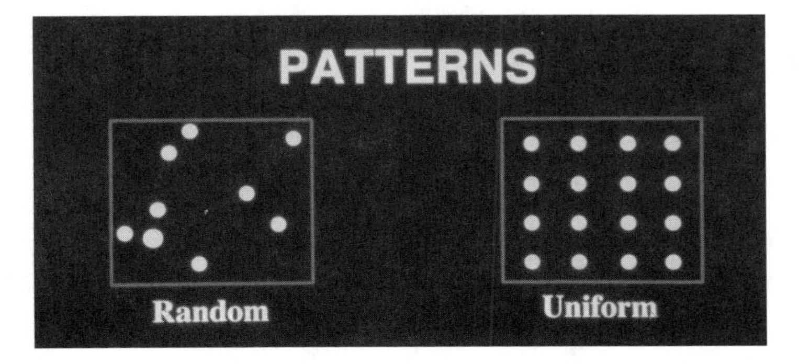

Figure 29-2. Patterns of plant injury can be random or uniform. A random pattern usually indicates a biotic factor, whereas a uniform pattern usually indicates an abiotic factor. (P. Sellers, Purdue University.)

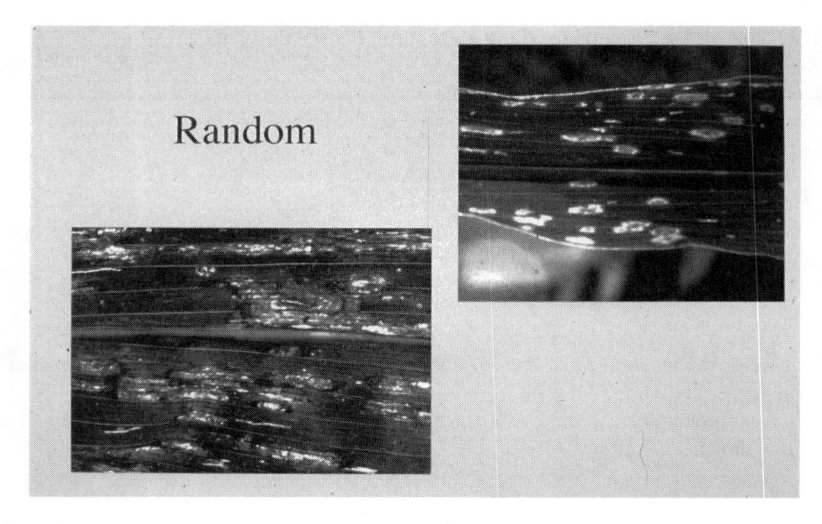

Figure 29-3. Random infectious disease (gray leaf spot) is typical of a random biotic agent. (P. Sellers, Purdue University.)

involve stunting, lack of lateral roots (Figure 29-5), clubbed tips (Figure 29-6), abnormal growth with lesions or enlarged regions, or rotting tissue with an irregular texture. Vascular tissue may be irregularly colored or the vascular fluids may be sticky or stringy. Flowers and fruit may be misshapen or discolored or have visible injury such as tissue loss, lesions, and the like. The injury observed can provide important

Figure 29-4. Leaf epinasty, or distortion, caused by a growth-regulator-type herbicide. Grapes are very sensitive to spray drift from herbicides of this type. (G. Ruhl, Purdue University.)

Figure 29-5. Root injury and stunting caused by branch-chain amino acid inhibitor herbicides. Figure shows injury from carryover of imazaquin onto corn.

Figure 29-6. Enlarged root tips and stem swelling injury, which is typical of a herbicide like trifluralin. (T.N. Jordan, Purdue University.)

clues as to the cause. A uniform versus random pattern, the position of the injury on older versus younger tissue, and the type of injury can often be associated with particular herbicides, nutrients, water stress, weather, compacted soils, nematodes, insects or disease factors. Always collect samples and have the appropriate laboratory examination performed to determine whether biotic agents are involved, because these causal agents can be specifically determined.

Step 4. *Determination of causal agents.* Once the information from Steps 1 through 3 has been gathered and analyzed, the investigator can begin to piece together the causes of the injury to make a diagnosis. This analysis uses the pattern of injury observed in the field, symptoms observed on the plant, the time frame of injury occurrence, the subsequent plant response, management practices employed in or around the site, environmental conditions before, during, and after the injury was observed, soil factors, questions asked and answers received, and laboratory results. The problem analysis usually results in identification of the cause(s) and allows the determination of an appropriate resolution of the problem. In some situations, the solution is not always obvious, as there may be missing or partial information. Even in these cases, use of the systematic approach described here will provide possible answers and allow some recommendations toward a solution. Multifaceted problems of an abiotic nature are, without a doubt, the most difficult to diagnose. However, by piecing together all the information, a good diagnostician can narrow the alternatives and work toward a determination of cause.

Herbicide Look-Alike Symptoms

There are many factors in the environment that can cause injury to plants that looks like herbicide injury. These factors include diseases, nematodes, viruses, mites, insects, air pollution, nutrients, adverse weather conditions (e.g., wind, frost, sand blasting, lightning, hail), soil compaction, drought, root stress, soil pH, and a variety of others caused by interactions between factors. It is relatively easy to determine whether the biotic factors in this group are involved, because a causal agent can, in most cases, be identified. The other factors can be evaluated specifically as to their involvement on the basis of answers to questions, observations, and checking farm and weather bureau records. For example, frost usually occurs in low areas of a field or on darker soils early in the season. Lightning may cover a circular area, and plants near the strike are often killed and wilted, with damage diminishing toward the edge of the area. Wind can cause growth-regulator-type injury on the meristematic regions of certain crops. Once these factors have been eliminated, then the steps outlined for herbicide diagnosis should be followed to determine cause.

CAUSES AND PREVENTION OF HERBICIDE INJURY

As mentioned, plant injury can be caused by many factors. Herbicide injury most often results from equipment problems, drift or vaporization, carryover, interactions and incompatibilities between chemicals, and miscellaneous mistakes. These causes can be avoided when the applicator pays attention to detail by reading and following all herbicide label recommendations, using appropriate application techniques, and maintaining equipment in proper operating condition. Improper application or misuse causes most herbicide injury. The applicator, whether a farmer or a commercial firm, is responsible for any damage that occurs.

Equipment must be carefully and frequently calibrated and maintained in order to avoid possible injury to plants from herbicides. A sprayer that is not properly calibrated will apply the wrong amount of herbicide, which can result in unanticipated results—for example, an underdose or an overdose. The sprayer itself must be properly maintained by having appropriate and well-maintained nozzles, a boom adjusted to the right height, and sufficient tank agitation. Application must be done at the right pressure and speed to ensure uniform herbicide application, and the sprayer must be cleaned thoroughly after each use to eliminate the chemical from the tank boom and nozzles (Figure 29-7). If equipment is maintained and operated properly, there is a much-reduced chance for a herbicide to injure plants. Application equipment should be used only for herbicides and not used for other pesticides, as problems of contamination can occur.

Figure 29-7. An example of injury due to poor sprayer maintenance. Glyphosate residual was not cleaned from the spray boom and nozzles after a previous application. The residual glyphosate in the boom killed the corn at the beginning of the subsequent herbicide application.

Herbicide drift is the movement of a herbicide away from the place of application, which can result in injury to adjacent plants (Figure 29-8). Herbicide movement can be caused by the drifting of spray particles by wind and by movement of the herbicide off the soil by vaporization. Drift usually occurs under windy conditions, with the greatest injury occurring in fairly close proximity to the area of application.

Drift is minimized when herbicides are applied under low to no wind conditions and when droplet size is greater than 200μM (see Chapter 7). Vapor drift is most likely to occur with herbicides that are volatile (have a vapor pressure of $<10^{-5}$ mm Hg). Movement of herbicides by vaporization generally occurs over a larger range than by drift. Herbicide movement off-site can be minimized by applying herbicides only when there is no wind, using nonvolatile formulations, applying them at low pressures to ensure larger droplets, choosing the appropriate nozzle type, and maintaining the spray boom at an appropriate height.

Herbicide residues, high levels of chemicals remaining at the site into the following growing season, are referred to as carryover. Such residues can result in injury to subsequent crops planted on the site that are not tolerant of the herbicide. Carryover can occur over an entire field or, more likely, in spots throughout the field and can result in stunted or malformed plants or lack of seedling emergence. Growers should be aware of herbicide soil-residual properties, crop tolerance to particular herbicides, and the proper application procedures in designing their herbicide and crop rotation programs to avoid carryover problems.

Most residue problems can be associated with the use of long-lived soil-residual herbicides. Factors contributing to carryover involve misapplications (overdosing, overlapping of applications in the field or at the ends of fields), adverse environmental

Figure 29-8. An example of injury due to paraquat drift onto a variety of ornamental plants. The key to diagnosis was that the injury was not host specific. (G. Ruhl, Purdue University.)

conditions that reduce herbicide breakdown in the soil (cold temperatures, moisture variations), and ignoring replant restrictions for sensitive crops.

Interactions and incompatibilities can also cause plant damage when various chemicals are mixed together and applied. This can occur when herbicides are mixed with other herbicides, insecticides, fungicides, or fertilizers. Always check the product label to determine whether a herbicide can be mixed and applied with other chemicals. Prepackaged mixtures are available, but most other combinations are either not labeled or have use restrictions that must be carefully followed. Problems, such as unexpected injury or lack of performance, can occur when nonrecommended adjuvants are used with herbicides. Applicators should always test combinations or sequential applications on a small section of a field before using them on the entire field so as to avoid unexpected plant damage.

Contamination can occur if herbicides are stored adjacent to other pesticides, seeds, or fertilizers, or if water sources used for application and cleaning of equipment contain residues. Maintaining separate storage areas for herbicides can avoid contamination of other pesticides, seeds, and fertilizer due to leaky herbicide containers or bags. Avoid the use of contaminated water sources as carriers for applying herbicides and for cleaning equipment. Applicators must avoid applying herbicides into irrigation ditches, ponds, or lakes at rates that can cause contamination and/or death of desirable organisms.

Other factors that must be considered when assessing herbicide injury and determining cause include knowledge of cultivar variation in response to certain herbicides, the effects of soil components such as clay and organic matter on herbicide activity, and how weather can influence herbicide activity. The best example of weather effects is the greater susceptibility of tomatoes to metribuzin injury if the herbicide is applied after a period of cool, wet, cloudy weather. Injury can be minimized or eliminated if metribuzin is not applied unless the tomato plants have been growing under warm, clear, and sunny conditions prior to application. Always check to make sure that the proper herbicide is actually applied. Sometimes applicators may put the wrong herbicide in the spray tank and thus apply the wrong herbicide to the crop, causing injury. Remember, if the labeled herbicide is applied in accordance with the label and under good environmental conditions, the number of herbicide injury cases will be minimal.

HERBICIDE INJURY SYMPTOMS BY PRIMARY MECHANISM OF ACTION

Photosynthesis Inhibitors

Atrazine, simazine, cyanazine, prometon, ametryn, prometryn, hexazinone, metribuzin, diuron, fluometuron, linuron, tebuthiuron, bromacil, terbacil, bentazon, bromoxymil, desmedipham, phenmedipham, pyrazon, pyridate, and propanil are photosynthesis-inhibiting herbicides. When they are soil applied, weed seeds germinate, their roots absorb the herbicide and translocate it in the xylem to the leaves, and the plant slowly dies as photosynthesis is inhibited. When these herbicides are applied postemergence, the action is by contact, requiring complete wetting of the foliage for complete kill. Susceptible plants turn yellow, then die from the bottom to the top. Leaves yellow between the veins and then turn brown from the base and outer leaf edges toward the center, eventually falling off the plant and leaving only a stem with an apical bud (Figure 29-9).

Pigment Inhibitors

Amitrole, norflurazon, isoxaflutole, clomazone, fluridone, and mesotrione are pigment-inhibiting herbicides. Injury caused by these herbicides is a bleached white to translucent appearance of the leaves (Figure 29-10). Sometimes the bleaching is not complete on the entire leaf but is interveinal with pink or red highlights along the margins.

Cell Membrane Disrupters and Inhibitors

Dilute sulfuric acid, monocarbamide dihydrogen sulfate, herbicidal oil, diquat, paraquat, acifluorfen, fomesafen, lactofen, oxyfluorfen, oxadiazon, carfentrazone, flumiclorac, sulfentrazone, azafenidin, fluthiacet, flumioxazin and glufosinate are the cell membrane disrupters and inhibitors. Complete coverage of the leaf is important for best activity by foliar applications, and the rate of plant death is more rapid under

Figure 29-9. Metribuzin injury on soybean. *Left*: Untreated; *Right*: Treated. The injury appears slowly after root uptake and is exhibited by initial leaf chlorosis, followed by leaf necrosis, and plant death. (University of Illinois.)

Figure 29-10. Injury from a pigment inhibitor herbicide. The white tissue is typical of spray drift from these compounds.

high light and warm environmental conditions. Injury symptoms include an initial water-soaked-tissue appearance followed by desiccation of leaf tissue (see Figure 29-8).

Cell Growth Disrupters and Inhibitors

Mitoic disrupters (benefin, ethalfluralin, oryzalin, pendimethalin, prodiamine, trifluralin, dithiopyr, thiazopyr, pronamide, and DCPA) inhibit shoot elongation when effective, and susceptible weeds never see the light of day. Root inhibition is observed as root pruning, and roots can be swollen and expanded at the tip (clubbed shaped) (see Figure 29-6). The underground portion of the stem can be thickened and shortened (Figure 12-6), and stems often have callus growth thickenings at the soil surface and become brittle (Figure 12-4). Inhibitors of roots only (napropamide, siduron, bensulide) or roots and shoots of seedlings (acetochlor, dimethenamid, alachlor, metolachlor, propachlor, butachlor, flufenacet) result in root pruning and growth inhibition, but no root swelling. The inhibition of shoots by the carbamothioates (EPTC, butylate, cycloate, molinate, pebulate, thiobencarb, triallate) and the chloroacetamides and oxyacetamides results in a lack of seedling shoot emergence. If shoots do emerge, they tend to be twisted and leaves are tightly rolled, with stems sometimes rupturing and new growth protruding from the ruptured tissue (Figure 29-11 and Figure 12-7).

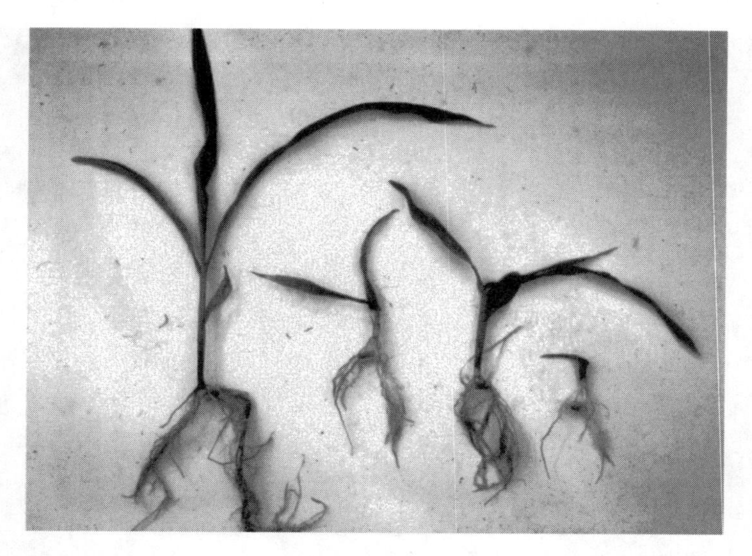

Figure 29-11. Shoot injury on corn from a cell growth inhibitor herbicide (chloroacetamide). Note shoot twisting, tightly rolled leaves and horizontal leaf orientation. *Left*: Normal plant; *Right*: Injured plants.

Cellulose Biosynthesis Inhibitors

The cellulose biosynthesis inhibitors, dichlobenil, isoxaben, and quinclorac are, in most cases, used for preemergence control and result in the inability of weed seedlings to grow. Symptoms include stunted growth and root swelling. Dichlobenil and isoxaben are used preemergence and are most effective against dicots, whereas quinclorac is used both preemergence and postemergence. Quinclorac as a cellulose biosynthesis inhibitor is most active against monocots, although it has a proposed second mechanism against dicots as a growth regulator.

Growth Regulators

The growth-regulator herbicides, 2,4-D, MCPA, dichlorprop (2,4-DP), 2,4-DB, mecoprop (MCPP), MCPB, dicamba, picloram, triclopyr, clopyralid, and quinclorac can be absorbed from the soil by plant roots; however, most of these compounds are applied as postemergence treatments. Translocation can be in both the xylem and the phloem to active growth regions, but their action tends to be localized on the shoot system. They selectively kill broadleaf weeds but can injure grass crops if applied at the wrong time. In the case of perennial weeds, many of these herbicides translocate to below-ground portions of the plant for systemic kill. Initial symptomology is quickly apparent on newly developing leaves and shoot regions as a twisting and epinasty of the shoot (Figure 14-3), cupping and crinkling of leaves, elongated leaf strapping (sometimes called "buggy whip") with parallel veins, stem swelling, and a disruption of phloem transport (see Figure 29-4). Secondary effects can be a fusion of brace roots, such as observed with corn (Figure 14-2). Root injury is expressed as a proliferation or clustering of secondary roots and overall inhibition of root growth.

Lipid Biosynthesis Inhibitors (Grass-Specific Herbicides)

Diclofop-methyl, fenoxaprop-ethyl, fenoxaprop-p-ethyl, fluazifop-p-butyl, quizalofop-p, clethodim, sethoxydim, and tralkoxydim are lipid biosynthesis inhibitors. These herbicides have specific activity against grass species only. Dicots and nongrass monocots are tolerant. Some of these herbicides have shown minimal soil activity; however, the main activity occurs after postemergence application to emerged grass. Death of the grass is slow, requiring a week or more for complete kill. Symptoms include rapid cessation of shoot and root growth, pigment changes (purpling or reddening) on the leaves occurring within 2 to 4 days, followed by a progressive necrosis beginning at meristematic regions and spreading over the entire plant (Figure 29-12).

Amino Acid Synthesis Inhibitors

The amino acid synthesis inhibitor herbicides are potent inhibitors of plant growth and are effective on both dicots and monocots. Glyphosate and sulfosate have only foliar activity (no soil activity), and the ALS inhibitors (bensulfuron, chlorimuron, chlorsulfuron, ethametsulfuron, halosulfuron, metsulfuron, nicosulfuron, oxasulfuron, primisulfuron, prosulfuron, rimsulfuron, sulfometuron, sulfosulfuron, thifensulfuron, tribenuron, triflusulferon, imazamethabenz, imazamox, imazapic, imazapyr, imazaquin, imazethapyr, cloransulam, flumetsulam, dicolusulam, pyrithiobac, and

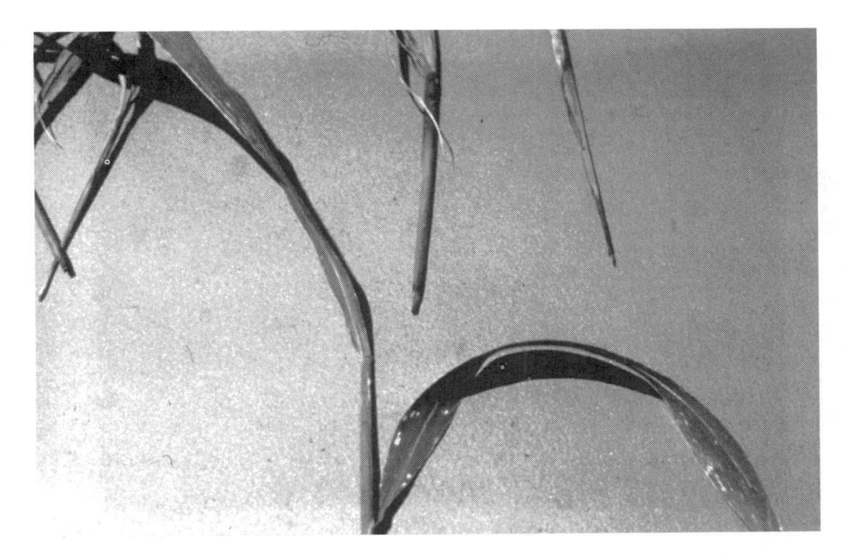

Figure 29-12. Injury on susceptible grasses from a lipid-biosynthesis-inhibiting herbicide. The plant initially stops growing, followed by the appearance of red pigmentation and then gradual leaf necrosis. The meristematic region of the grass leaf becomes necrotic within a couple of days.

flucarbazone) have members with foliar, soil, or both foliar and soil activity. Treated plants stop growing almost immediately after application. In the case of (5-enolpyruvylshikimate-3-phosphate synthase (EPSPS) inhibitors, plants may show a small amount of bleaching around new growth areas leaf distortion at shoot tip and they die slowly (1 to 2 weeks), and turn a uniform harvest brown color (Figure 29-13). With acetolactate synthase (ALS) inhibitors, 2 to 4 days after treatment the growing point (apical meristem) becomes chlorotic and later necrotic. Plants may also have shortened internodes, reduced root growth ("bottle brushing," and pigment changes, including yellowing, purpling, or reddening. Plant death begins in the growing point and gradually spreads to the entire plant, with death occurring within 7 to 10 days.

Auxin Transport Inhibitors

Naptalam is soil applied, whereas diflufenzopyr is foliar applied. A common symptom of plants treated with these auxin transport inhibitor herbicides, in addition to reduced plant growth, is the upward turning of the root tip.

Modes of Action Not Clear

DSMA, MSMA, asulam, difenzoquat, fosamine, TCA, pelargonic acid, endothall, and ethofumesate are herbicides whose modes of action are not clear. A variety of symptoms

Figure 29-13. Injury on maple caused by a soil application of glyphosate near the tree roots and stem. Absorption through exposed root tips, green bark, or foliage can result in growth distortion, as observed here. (G. Ruhl, Purdue University.)

are possible with these compounds. DSMA and MSMA can cause rapid yellowing and necrosis of plant tissue.

ACTION TO ADDRESS HERBICIDE INJURY

Once herbicide damage to a plant has occurred, there is often little that can be done to eliminate the problem. In severe cases of injury, the crop will be lost. Frequently, there is localized damage from carryover or misapplication in particular areas of a field. Sometimes the farmer will choose to do nothing and allow the crop to recover. In other cases, some remedial action will be taken to address the problem immediately or to reduce the effect in future years.

Techniques often employed for herbicide problems include *tillage, cultural practices, use of antidotes*, and *replanting*. These techniques are useful in the case of drift or vapor injury, misapplication, or carryover from a previous application.

Drift damage to plants is usually most severe in areas closest to the source of drift (along an edge of a field). Plants farther into the field may show some damage but often do not die and usually recover over time. Recovery is especially likely in the case of drift to ornamental plantings or perennial fruiting plants. In the case of immediate damage or death, a determination of loss or cost of replacement must be made. The most immediate solution to a drift problem in an annual crop (agronomic or horticulture) is to replant a nonsensitive crop in the affected parts of the field.

Herbicide carryover is a potential problem when a soil-applied residual herbicide is used. Carryover potential is often determined by collecting a uniform soil sample in the field well in advance of the growing season. The soil is then analyzed through laboratory tests and soil bioassays to determine what effect the residues may have on plants grown in the soil. If minor injury is observed on the indicator species in the bioassay or a known concentration of herbicide is identified in the laboratory, remedial actions can be employed in the field prior to planting the crop. Deep *tillage* is often used to mix the herbicide-contaminated soil with a larger volume of uncontaminated soil so as to dilute the residue and minimize potential injury to the crop. In some cases, if the injury is widespread over the field early in the season, *replanting* of another crop tolerant to the herbicide is possible.

Cultural practices can be used to minimize a reduction in crop growth caused by herbicide injury. If a crop is injured but remains in the field, the farmer must reduce crop stress and allow the crop to quickly recover vigor. Such practices include use of irrigation, optimal fertilization, and good pest control.

Activated carbon is often used as the universal *antidote* to herbicide contamination. Activated carbon is very effective in inactivating residues of nonionic chemicals with low water solubility. Most herbicides fall into this category (Weber, 1980). Activated carbon is used when the crop is of high value, or the site is of high value, and replanting another crop is not an option (e.g., gardens, landscapes, golf courses, and other high-value sites). Activated carbon is thoroughly mixed into the soil by disking or rototilling to deactivate the herbicide residues. Activated carbon is used at a rate of approximately 100 pounds per acre for each 1 pound of herbicide residue in the soil.

In some situations, activated carbon slurries (1 pound of activated carbon per 1 gallon of water) is used to protect transplant roots or seeds from a toxic herbicide (e.g., root dips for tomatoes to protect against trifluralin or root dips of strawberry transplant roots to protect against simazine). Activated carbon slurries have been used to coat seeds or have been sprayed as a band over seeded rows for protection against diuron.

In summary, knowledge of the herbicide used and the factors related to a plant injury are important in determining what, if any, remedial action to take. Following the recommendations presented in this chapter will allow a determination of the cause of plant injury and the specific actions to implement to reduce immediate and long-term economic loss.

LITERATURE CITED AND SUGGESTED READING

Crop Injury Diagnostic Guides with "Look-alike" Symptoms. 1992. (Corn, cotton, small grains, soybeans.) Agri-Growth Research, Inc., Rt. 1, Box 33, Hollandale, MN 56045.

Green, J. L., O. Maloy, and J. Capizzi. 1990. A systematic approach to diagnosing plant damage. *Cooperative Extension Service, Oregon State University, North West Ornamental Newsletter* **13**:1–32.

Monaco, T. J., A. R. Bonanno, and J. J. Baron. 1986. Herbicide injury: Diagnosis, causes, prevention and remedial action. In *Research Methods in Weed Science*, 3rd ed., ed. by N. D. Camper, Chapter 21. Southern Weed Science Society (SWSS), Champaign, IL.

Plant Injury Picture Sheets (Insect, disease, and herbicide damage.) 1980–1989. University of Illinois College of Agriculture, Cooperative Extension Service and Vocational Agriculture Service, Urbana-Champaign.

Poling, E. B., and T. J. Monaco. 1985. Use of activated carbon to safen herbicide applications in newly set strawberries. *HortSci.* **20**:251–252.

Shurtleff, N. C., and C. W. Averre III. 1996. *The Plant Disease Clinic and Field Diagnosis of Abiotic Diseases*. ASP Press, St. Paul, MN.

Walker, S. E., and T. S. Schubert. 1997. *Assessing Plant Problems in Cropping Systems: A Systematic Approach.* Pathology Circular. No. 381. Florida Department of Agriculture and Consumers Service, Division of Plant Industry.

Weber, J. B. 1980. Use of activated carbon to clean up herbicides in greenhouses. *North Carolina Flower Growers Bull.* **24**:9–10.

WEB SITES

University of Missouri: *Herbicide Injury Websites*

http://www.psu.missouri.edu

Select "Agronomy," select "Extension," select "Weeds and Weed Control," Scroll to and select "Herbicide Injury"

University of Minnesota: *Herbicide Mode of Action and Injury Symptoms Interactive CD Homepage*. 1998. J. L. Gunsolus and W. S. Curran. North Central Regional Publication 377.

http://www.agro.agri.umn.edu

Select "Extension," select "Weed Control," scroll to "Herbicide Mode of Action and Injury Symptoms"

Iowa State University: *Herbicide Mode of Action Reference Table*

http://www.weeds.iastate.edu

Select "Mode of Action Reference"

Purdue University: *Herbicide Injury Symptomolgy*

http://www.btny.purdue.edu

Select "Extension," scroll to "Weed Science" and select "Herbicide Injury Symptoms on Corn and Soybean"

AgChemical Label Information: Crop Data Management Systems, Inc. Marysville, CA

http://www.cdms.net/manuf/manuf.asp

For chemical use, see the manufacturer's or supplier's label and follow these directions. Also see the Preface.

30 Weed Science in the Future

The various aspects of weed science discussed in the first 29 chapters of this book cover the current status of our knowledge relating to weeds and the methods used for their management. Although our knowledge regarding plant biology, physiology, and biochemistry is increasing, there is still much to learn, especially concerning weeds.

Weeds are ever present in our agriculture endeavors. Weed removal has progressed from a system totally based on the physical efforts of humans through the use of animals, mechanical implements, chemicals, and, to some extent, biological methods. Although this book has addressed most of these efforts and the need to integrate the various methods, much of the discussion has centered on the use of herbicides for managing weeds. Herbicides, although widely used, have not stopped weeds from being major pests in agriculture. However, they have reduced the amount of direct human effort needed for weed removal. As Warren (1998) pointed out, a good deal of increased agricultural productivity has resulted from improved crop breeding, nutrition, and pest management, of which weed management has been a major factor. Herbicides have allowed people once wedded to the farm to pursue other employment activities. The reduced number of people needed for direct involvement in production agriculture has allowed the expansion of other sectors of developing countries.

Yet even with all our advances in technologies for weed removal, there are many challenges that face weed scientists in developing new weed management tactics. The first seven chapters of this book covered our current knowledge relating to weeds and the methods used for their management. Although we have improved our removal techniques, the challenges that lie ahead are directly related to the development of knowledge-based management systems rather than simply the easiest way to kill weeds. The greatest challenge is to develop control techniques that are in concert with sound ecological and environmental principles. These methods must also be environmentally and economically sustainable, must allow the production of a safe food supply, and must be acceptable by society. Society will dictate what agricultural practices are acceptable, and the weed science community should strive to develop better practices. These challenges will necessitate an improved scientific knowledge of weeds, their interaction with the environment, and their direct response to our imposed control tactics. A control approach based solely on herbicides is not sustainable; instead, herbicides, when used, must be a part of a fully integrated and effective control strategy as outlined in Chapter 3. To accomplish this end, an emphasis on research that increases our knowledge of weed ecology and biology is necessary. The management of weeds based on sound scientific knowledge integrated with affordable technology will allow the development of weed control programs that are in concert with the environment. Such an approach will allow many research and outreach opportunities that not only advance our scientific knowledge base for weeds but will also result in practical and acceptable weed management systems that society will accept.

There are many opinions on what the future holds for weed science. Several excellent reviews on this subject are available and are recommended reading for students (Hall et al., 2000; Gressel, 2000; Buhler et al., 1997; and Shaner, 2001). Each of these discusses many areas of research that will allow us to attain our goals for weed science. In the following sections we offer a brief overview of future research areas that are important for the discipline of weed science.

RESEARCH AREAS

The basic aim of research within the discipline of weed science is to understand the fundamental biology, ecology, biochemistry, physiology, and genetics of weeds and to use this knowledge to develop effective weed management systems. How will this be accomplished? Regardless of our best efforts, weeds are able to eventually respond to any tactic that is used repeatedly, and after repeated use of the same herbicide, this tactic becomes ineffective. The development of herbicide-resistant weeds is a perfect example. To avoid the rapid loss of effective tactics, an integration of many control methods is necessary. How do we integrate a variety of tactics into a holistic approach to weed science? The basis of such an approach is a better understanding of weed growth and interaction in relation to plant biology and ecology. Weed science research should never lose sight of this basic principle, as all further research will evolve from this knowledge.

The many tools of molecular biology will be an integral part of all weed research programs, well beyond the development of transgenic herbicide-resistant plants. Molecular techniques will allow us to investigate more thoroughly the basic biology and genetics of weedy plants. This will lead to a better understanding of the genes involved in weed growth and the biology of weed populations. Molecular tools are currently being used to taxonomically classify weeds and to differentiate between species. Current advances have been reported in understanding the taxonomy of the *Amaranthaceae* in relation to herbicide resistance among species, and the genetic variation in *Euphorbia esula* by use of DNA markers, to name just two examples. The use of modern molecular analysis will be able to clarify taxonomic uncertainty in weedy species and will lead to better understanding of the evolution and development of weedy species.

Weed Biology

The design of improved management practices relies on a better understanding of the basic areas of weed biology: weed genetics, reproductive biology, introgression potential between weeds and crops, factors controlling seed germination and longevity in the soil, internal and external controls of seed and bud dormancy, and the nature of perennial weeds. In addition to this basic information on growth and survival, physiological studies related to factors involved in weed growth can contribute to improved management practices. These studies can include weed growth requirements for light, nutrients, water, and many other factors that, if understood, can be manipulated for improved crop growth and establishing production systems that are detrimental to weed growth.

Weed Ecology

Ecological studies can be conducted to determine what factors contribute to invasiveness, a weed's ability to compete, its adaptation to cultural practices, and its response to imposed selection pressures. Improved understanding of weed biology coupled with weeds' responses to the agroecosystem will provide a much more scientific basis for the development of management systems. Management models can use weed thresholds and can be developed to concentrate on the detrimental weeds. Additional ecological research can study how the soil environment (or soil health) affects the growth, survival, and reproductive ability of weeds. There is a need to determine the effect of cropping sequences, intercropping, cover crops, composts, and other management techniques on weed seed production and survival. These studies will not only allow a better understanding of the agroecosystem but will also provide the basic knowledge required for the development of crop/weed interactive models.

Many researchers are already investigating the problem of weeds developing resistance to existing herbicides. This research, as described in Chapter 18, will continue to be an important aspect of weed control programs and basic ecological studies relating to weed adaptation to management practices.

KNOWLEDGE-BASED SYSTEMS APPROACH TO WEED MANAGEMENT

Improved biological and ecological knowledge is vital to the development of useful weed management models. These models must be able to effectively predict the degree of weed infestation and whether there is a real need for weed management. As described in Chapter 3, there are many computer-based decision aids in use that in the future will be further refined to allow modeling of weed-crop interactions in various production systems. Currently, there are bioeconomic models that are assisting growers in determining weed control programs. These will become even more useful and widely accepted in the future as they are shown to be consistently reliable in weed management. Models are now available to predict what conditions (field and climate) govern weed seed germination and subsequent growth. Models will be developed incorporating multiple components related to specific cultural practices and their influences on weed competition. Models of the same type will eventually be developed in regard to seed and bud dormancy, seed longevity in the soil, and the effects of management practices on perennial weed growth. Such an approach to agriculture will truly be based on sound science and ecological principles.

The practices that farmers are willing to use to manage weeds will affect how appropriate (in a practical sense) a particular model is for any given situation. There is interest in developing new and improved approaches to weed management that are not wholly dependent on herbicides. Research is needed relating to improved cultivation tactics (see Chapter 22), use of cover crops, use of living mulches, development of more competitive crops, potential for allelopathy, site-specific weed management, the development of more effective and environmentally benign herbicides, the potential for natural-product herbicides, and the examination of potential biological control organisms.

What will development of these control tactics involve? There will have to be a greater emphasis on understanding how these various cultural practices, targeting the cause rather than the result of weed competition, can be used to manage weeds more efficiently. The use of knowledge-based decision aids will allow farmers and other weed managers to efficiently remove weeds if and when they are a problem.

Another tool that has great potential is the use of geographic information systems (GIS) for weed management. Mapping of fields to identify specific locations of problem weeds, coupled with imaging equipment and smart sprayers that differentiate weeds from crops, will allow precise applications of herbicides or other treatments to the problem. Targeting weedy areas in a field rather than treating the entire field will cut expenses and reduce the environmental load of herbicides. Such capabilities can eliminate the need for broadcast applications of herbicides and result in overall greater production efficiency.

INVASIVE WEEDS

The presence of invasive weeds in both agriculture lands and noncrop areas is a major area of concern throughout the world. A great deal of research deals with plant invasions. A better understanding of how native and exotic species interact in invasions is an area of rapidly advancing research. Research in molecular biology and gene function will play an important role in providing a better understanding of the processes involved.

GENES FOR CROP IMPROVEMENT

The identification of genes conferring traits of competitive advantage to weeds will allow the use of such genes in crop improvement programs. Although this area of research seems to have been neglected in most crop improvement programs, there is potential for identification of useful genes in weeds that control germination timing, rapid growth rate, hardiness, and other traits. Genes influencing the dispersal and persistence of annual and perennial species constitute fertile ground for research. The identification of genes related to the perennial phenotype could be extremely important in crop improvement programs. Competitive genes identified and isolated from weeds could be useful in transformation of crop plants. Conversely, the identification of genes that would weaken a weed's competitive advantage could be introduced into a community of outcrossing weeds and, over time, through sexual recombination result in a less-fit weed type. The concept of loss of competitive advantage in weeds must be linked with a specific selective pressure, such as a chemical introduced into the environment. This approach, discussed in Gressel (2000), may involve the use of certain antigrowth plant genes induced by chemical application (chemically assisted suicide genes).

Biological Control

Biological manipulation and engineering of insects and pathogens for improved utility in programs to control weeds will be possible and will assist in reducing farmers' dependence on chemical weed control. At present there are isolated instances of biocontrol agents that have some utility against weeds; however, most have serious limitations in achieving complete control. Genomics research will allow the isolation of genes that may be useful in increasing the virulence and host specificity of fungal and bacterial agents and insects. Another approach to biological control will involve study of the genetic diversity of weedy species and the application of such knowledge to select biocontrol agents. The study of genetic diversity of a weedy invasive population within a given country and in its country of origin is playing an important role in the search for natural control agents and potentially useful new biocontrol agents.

Allelopathy

Isolated genes that are involved in the competitive ability of weeds will not only be useful in improving the growth habit and resource utilization of crops, but will allow the development of crops that produce allelochemical compounds for attaining natural weed control. Much interest has been generated for allelopathy in weed control, although few examples of its actual application in agriculture are available beyond the use of certain cover crops. Engineering plant allelopathic pathways to produce natural herbicides has great potential; however, knowledge about the specifics of allelochemical biosynthetic pathways is lacking, which may slow progress in this potentially exciting area of research.

Seed and Bud Dormancy

An improved understanding of the genes involved in weed seed and bud dormancy, perennial plant growth habit, and factors controlling their regulation can be used to manipulate the system, either culturally or through biotechnology, to attain more effective weed management. For example, recent molecular examinations of bud dormancy-associated gene expression in *E. esula* show how a clearer understanding of the genes involved in bud dormancy control and their regulation will be important in the design of improved control programs for management.

Parasitic Weeds

Recent interest in the molecular mechanisms involved in parasitic weed/host plant interactions will lead to a clearer understanding of the genes involved in the interaction between parasitic weeds and their hosts. The expression of defense-related genes in host plants parasitized by *Orobanche* spp., the cytochemical aspects of cellulose distribution in resistant and susceptible host tissues in cell surface interactions between sorghum roots and *Striga hermonthica*, and the development of infection structures and the mechanism of penetration of *Striga gesneriodides* into cowpea are examples of this type of research.

Herbicide-Related Issues

The development of low-use-rate, environmentally benign, yet effective herbicides is a major objective of all chemical companies in their herbicide discovery programs. Sophisticated techniques have been developed to screen chemicals quickly and efficiently to determine their biological activity. The search for unique sites of action specific to various plants is also a priority. Many of the newer techniques involve the use of molecular biology tools to screen and determine the effectiveness of candidate herbicides.

Genomics will allow the isolation of new herbicide-resistance genes (including site of action, metabolism, sequestration, and exclusion genes) that can be used in the study of weed adaptation to herbicide use and allow better predictions of the evolution of weed resistance to all classes of herbicides. Resistance genes may also be useful in engineering herbicide-resistant crop plants or, more important, the identification of specific sites of action for certain herbicide groups, such as the growth regulators and some pigment inhibitors. The identification of such gene functions will ultimately be useful in the design of novel chemical and nonchemical approaches to the control of weeds. These techniques are thoroughly described by Shaner (2001) and will play a major role in future weed science programs.

CONCLUSIONS

The ideas discussed in this chapter are only a few of the many exciting possibilities available to weed scientists for using the power of research to study weed/crop interactions and develop improved management practices. In any research program we must be mindful that our pest management practices impact the environment and our ability to provide an abundant and safe food supply. We must design our research programs so they can have a positive influence on society and contribute to the betterment of humankind. We never want our research to have a negative effect. There are many exciting weed-related research opportunities. The only real limitation to improved agricultural pest management (specifically weed management) is a limited imagination of the researchers themselves. The big challenge is how to feed an increasing world population. We believe that future generations are ready to meet this challenge and that the discipline of weed science will play an important role in the success of efficient agricultural production systems.

LITERATURE CITED AND SUGGESTED READING

Buhler, D. D., R. G. Hartzler, and F. Forcella, 1997. Implications of weed seedbank dynamics to weed management. *Weed Sci.* **45**:329–336.

Gressel, J. 2000. Molecular biology of weed control. *Transgenic Res.* **9**:355–382.

Hall, C. J., L. L. van Eerd, S. D. Miller, M. D. K. Owen, T. S. Prather, D. L. Shaner, M. Singh, K. C. Vaughn, and S. C. Weller. 2000. Future research directions for weed science. *Weed Technol.* **14**:647–658.

Shaner, D. L., ed. 2001. Symposium on the effect of biotechnology and genomics on weed science. *Weed Sci.* **49**:248–289.

Warren, G. F. 1998. Spectacular increases in crop yields in the United States in the twentieth century. *Weed Technol.* **12**:752–760.

WEB SITES

Weed Science Society of America:

http://wssa.net

Contains Numerous Links to Other Sites Related to Weed Science

Refer to Chapter 18 for Biotechnology Related Web Sites

For chemical use, see the manufacturer's or supplier's label and follow these directions. Also see the Preface.

Appendix

TABLE A-1. Herbicides That Control Common Broadleaf Weeds as Listed on the Manufacturer's Label

	Weed																									
	Beggarweed	Bindweed, field	Carpetweed	Chickweed, common	Chickweed, mouse-ear	Cocklebur, common	Dandelion	Dock, curly	Dog fennel	Galinsoga	Groundcherry, cutleaf	Groundsel spp.	Henbit	Jimsonweed	Knotweed, prostrate	Kochia	Lamb's-quarters, common	Lettuce, prickly	Mallow, little	Morning glory, ivyleaf	Mustard, wild	Nightshade, black	Nightshade, hairy	Pennycress, field	Pigweed, redroot	Puncture vine
1. Acetachlor	X		X			X				X			X									X	X		X	
2. Acifluorfen		X	X			X					X			X			X			X	X	X			X	
3. Alachlor			X								X			X		X	X					X	X			
4. Ametryn				X									X								X					
5. Amitrole			X									X		X		X										
6. Asulam																										
7. Atrazine						X			X				X	X		X	X				X	X		X		X
8. Benefin			X	X											X		X								X	
9. Bensulide			X														X								X	
10. Bentazon		X								X		X		X		X	X			X	X	X	X		X	
11. Bromacil							X		X				X			X										X
12. Bromoxynil						X								X	X		X			X				X	X	
13. Butylate																										
14. Carfentrazone						X					X			X		X	X			X		X			X	
15. Chlorimuron	X													X						X						
16. Chlorsulfuron				X	X		X	X				X	X		X	X	X	X			X			X	X	X
17. Clethodim																										
18. Clomazone			X											X				X			X	X			X	
19. Clopyralid							X	X				X		X				X				X	X			
20. Cloransulam						X								X			X		X	X	X				X	
21. Cyanazine			X	X		X		X	X			X	X	X	X	X	X	X			X	X			X	X
22. Cycloate													X				X					X	X		X	
23. 2,4-D		X	X	X		X	X	X		X			X	X		X	X	X	X	X	X			X		X
24. 2,4-DB						X		X				X		X		X	X	X			X	X	X	X		
25. DCPA			X	X													X								X	
26. Desmedipham				X																	X	X			X	
27. Dicamba	X	X	X	X	X	X	X	X	X				X	X		X	X	X		X	X	X		X	X	X
28. Dichlobenil			X					X	X			X	X				X				X				X	
29. Diclofop																										
30. Difenzoquat																										
31. Difluenzopyr	X	X	X			X	X	X						X		X	X		X	X		X	X		X	
32. Dimethenamid			X							X			X				X			X		X	X		X	
33. Endothall													X			X									X	
34. EPTC			X	X									X	X		X	X					X	X		X	X
35. Ethalfluralin			X	X									X	X		X	X				X	X	X		X	

TABLE A-1. Continued

	Weed																									
	Beggarweed	Bindweed, field	Carpetweed	Chickweed, common	Chickweed, mouse-ear	Cocklebur, common	Dandelion	Dock, curly	Dog fennel	Galinsoga	Groundcherry, cutleaf	Groundsel spp.	Henbit	Jimsonweed	Knotweed, prostrate	Kochia	Lamb's-quarters, common	Lettuce, prickly	Mallow, little	Morning glory, ivyleaf	Mustard, wild	Nightshade, black	Nightshade, hairy	Pennycress, field	Pigweed, redroot	Puncture vine
36. Ethofumesate				X												X	X					X			X	
37. Fenoxaprop																										
38. Fluazifop-P																										
39. Flumiclorac						X								X			X								X	
40. Flumioxazin	X		X				X							X		X	X		X	X		X			X	X
41. Fluometuron														X												
42. Fomesafen			X			X								X			X			X	X	X			X	
43. Glufosinate	X	X	X	X	X	X	X	X		X	X			X		X	X		X	X	X	X		X	X	X
44. Glyphosate	X	X		X	X	X		X	X	X		X	X			X	X	X						X	X	
45. Halosulfuron						X										X	X		X						X	
46. Hexazinone				X		X	X	X	X			X	X					X						X	X	
47. Imazamethabenz	X																				X			X		
48. Imazamox			X	X		X								X		X	X				X	X	X		X	
49. Imazapyr	X	X	X	X		X	X		X							X	X	X	X							
50. Imazethapyr				X	X	X					X	X	X	X	X	X	X	X	X	X	X	X	X	X	X	
51. Imazaquin						X								X			X			X					X	X
52. Isoxaflutol			X	X		X	X			X			X	X		X	X		X	X	X	X	X	X	X	
53. Lactofen	X		X			X				X	X		X								X					X
54. Linuron	X								X	X		X								X		X				
55. Mesotrione			X	X		X				X				X	X	X	X			X		X	X		X	
56. Metolachlor			X							X												X	X		X	
57. Metribuzin	X	X	X	X	X	X		X	X	X			X	X	X	X	X	X	X	X	X	X		X	X	
58. Metsulfuron		X		X				X	X			X	X		X	X	X	X	X		X			X	X	
59. MSMA				X		X																X		X	X	
60. Napropamide			X	X								X	X		X		X	X	X				X		X	
61. Naptalam		X	X	X		X				X	X						X				X	X	X		X	
62. Nicosulfuron														X						X					X	
63. Norflurazon			X			X			X			X	X			X	X		X							X
64. Oryzalin		X	X	X					X			X	X	X	X	X		X	X		X	X	X		X	X
65. Oxyfluorfen			X			X					X	X	X	X	X		X		X	X	X	X			X	
66. Pebulate													X										X		X	

TABLE A-1. Continued

	Weed																									
	Beggarweed	Bindweed, field	Carpetweed	Chickweed, common	Chickweed, mouse-ear	Cocklebur, common	Dandelion	Dock, curly	Dog fennel	Galinsoga	Groundcherry, cutleaf	Groundsel spp.	Henbit	Jimsonweed	Knotweed, prostrate	Kochia	Lamb's-quarters, common	Lettuce, prickly	Mallow, little	Morning glory, ivyleaf	Mustard, wild	Nightshade, black	Nightshade, hairy	Pennycress, field	Pigweed, redroot	Puncture vine
67. Pendimethalin			X										X			X	X									X
68. Phenmedipham				X							X						X				X					
69. Primsulfuron	X					X								X		X	X			X		X	X		X	
70. Prometryn																						X				
71. Pronamide			X	X	X								X				X		X		X	X	X			
72. Propachlor			X											X		X	X					X	X			
73. Propanil												X				X	X				X				X	
74. Prosulfuron	X	X		X		X							X	X		X	X	X	X	X	X			X	X	X
75. Pryazon													X											X		
76. Pyridate			X	X		X				X			X	X		X	X					X			X	
77. Quizalofop																										
78. Rimsulfuron						X							X			X	X				X		X		X	
79. Sethoxydim			X								X			X			X				X	X			X	
80. Siduron																										
81. Simazine			X	X								X	X				X	X			X	X	X			
82. Sulfentrazone	X		X			X					X			X		X	X			X		X	X		X	
83. Sulfometuron				X			X	X	X							X		X	X							
84. Sulfosate			X	X	X	X	X	X				X		X	X	X	X	X		X	X	X	X	X	X	
85. Tebuthiuron			X						X			X	X			X		X								X
86. Terbacil				X					X			X	X	X				X								
87. Thifensulfuron						X								X			X				X				X	
88. Thiobencarb																										
89. Triallate																										
90. Triclopyr		X	X					X					X						X							
91. Trifluralin		X	X						X				X	X		X					X	X	X		X	X

TABLE A-1. Continued

	Weed																		
	Purslane	Pulsey, Florida	Radish, wild	Ragweed, common	Ragweed, giant	Rocket, London	Shepherd's purse	Sicklepod	Sida, prickly	Smartweed, Penn.	Sowthistle, spp.	Speedwell, corn	Spurge, spp.	Sunflower, spp.,	Thistle, bull	Thistle, Canada	Thistle, Russian	Velvetleaf	Wild buckwheat
1. Acetachlor	X	X		X															
2. Acifluorfen	X	X		X						X			X			X		X	X
3. Alachlor		X							X									X	
4. Ametryn	X	X								X	X			X				X	
5. Amitrole							X				X		X			X			X
6. Asulam																			
7. Atrazine	X			X				X		X	X	X	X	X	X	X	X	X	X
8. Benefin	X	X																	
9. Bensulide	X						X											X	
10. Bentazon	X			X			X		X	X				X		X		X	X
11. Bromacil		X																	
12. Bromoxynil			X	X		X	X			X				X		X	X	X	X
13. Butylate																			
14. Carfentrazone				X									X					X	
15. Chlorimuron				X				X						X					
16. Chlorsulfuron	X		X			X	X			X	X			X		X	X		X
17. Clethodim																			
18. Clomazone	X	X		X					X	X			X					X	X
19. Clopyralid				X	X			X		X		X		X	X	X	X		
20. Cloransulam				X	X			X		X				X					
21. Cyanazine	X	X	X	X		X	X		X	X	X		X	X			X	X	X
22. Cycloate	X						X												
23. 2,4-D			X	X			X	X		X	X		X	X	X	X	X	X	
24. 2,4-DB				X			X		X								X	X	
25. DCPA	X	X											X						
26. Desmedipham	X			X		X	X				X								X
27. Dicamba	X	X	X	X			X	X	X	X	X		X	X	X	X	X	X	X
28. Dichlobenil		X	X	X			X		X	X			X		X	X	X		
29. Diclofop																			
30. Difenzoquat																			
31. Difluenzopyr	X			X	X			X	X	X	X		X	X		X	X	X	
32. Dimethenamid	X	X		X															
33. Endothall																			X
34. EPTC	X	X					X		X									X	
35. Ethalfluralin	X	X		X			X		X	X							X	X	X

TABLE A-1. Continued

	Weed																		
	Purslane	Pulsey, Florida	Radish, wild	Ragweed, common	Ragweed, giant	Rocket, London	Shepherd's purse	Sicklepod	Sida, prickly	Smartweed, Penn.	Sowthistle, spp.	Speedwell, corn	Spurge, spp.	Sunflower, spp.,	Thistle, bull	Thistle, Canada	Thistle, Russian	Velvetleaf	Wild buckwheat
36. Ethofumesate	X						X			X	X						X		X
37. Fenoxaprop																			
38. Fluazifop-P																			
39. Flumiclorac				X					X				X					X	
40. Flumioxazin	X	X		X	X		X		X	X			X				X	X	
41. Fluometuron		X						X	X	X									
42. Fomesafen	X	X		X					X	X								X	
43. Glufosinate	X			X	X		X	X	X	X	X			X		X	X	X	X
44. Glyphosate	X	X		X		X	X	X	X	X	X	X		X		X	X	X	
45. Halosulfuron				X	X					X				X				X	
46. Hexazinone				X		X	X			X			X			X			
47. Imazamethabenz																			X
48. Imazamox	X	X	X	X	X				X	X	X			X		X	X	X	
49. Imazapyr	X			X				X			X			X	X	X			X
50. Imazethapyr			X			X	X			X			X	X				X	
51. Imazaquin		X		X					X	X			X	X				X	
52. Isoxaflutol	X		X	X	X		X						X	X				X	
53. Lactofen	X	X		X					X	X			X			X		X	
54. Linuron	X	X	X	X				X	X									X	X
55. Mesotrione		X		X	X				X	X				X				X	
56. Metolachlor	X	X																	
57. Metribuzin	X	X		X		X	X	X	X	X	X		X	X			X	X	X
58. Metsulfuron	X						X				X			X	X	X	X		X
59. MSMA							X												
60. Napropamide	X	X		X		X	X				X								
61. Naptalam	X	X		X			X							X				X	
62. Nicosulfuron										X									
63. Norflurazon	X	X		X		X	X		X	X	X		X				X	X	X
64. Oryzalin	X	X	X	X		X	X		X	X	X	X	X		X		X	X	X
65. Oxyfluorfen	X			X		X	X		X	X	X		X		X		X	X	X
66. Pebulate	X	X					X												

TABLE A-1. Continued

	Weed																		
	Purslane	Pulsey, Florida	Radish, wild	Ragweed, common	Ragweed, giant	Rocket, London	Shepherd's purse	Sicklepod	Sida, prickly	Smartweed, Penn.	Sowthistle, spp.	Speedwell, corn	Spurge, spp.	Sunflower, spp.,	Thistle, bull	Thistle, Canada	Thistle, Russian	Velvetleaf	Wild buckwheat
67. Pendimethalin	X	X											X					X	
68. Phenmedipham	X			X		X	X				X								
69. Primsulfuron			X	X	X			X	X	X				X		X	X	X	
70. Prometryn		X							X	X									
71. Pronamide	X		X			X	X												
72. Propachlor				X			X		X	X								X	X
73. Propanil		X																	
74. Prosulfuron	X	X	X	X	X		X	X	X	X				X		X	X	X	
75. Pryazon							X											X	
76. Pyridate	X					X	X							X					
77. Quizalofop																			
78. Rimsulfuron	X			X															
79. Sethoxydim	X	X		X			X		X				X			X		X	X
80. Siduron																			
81. Simazine	X	X					X			X							X		
82. Sulfentrazone	X	X		X					X				X					X	
83. Sulfometuron											X		X	X		X	X		
84. Sulfosate	X	X		X	X	X	X	X	X	X	X	X	X	X			X	X	
85. Tebuthiuron	X			X			X		X	X	X		X	X		X			
86. Terbacil		X					X												
87. Thifensulfuron										X				X				X	
88. Thiobencarb																			
89. Triallate																			
90. Triclopyr											X	X	X		X	X			
91. Trifluralin		X		X			X		X	X			X				X	X	

TABLE A-2. Herbicides That Control Twenty-six Common Grasses and Sedges as Listed on the Manufacturer's Label

	Weed																									
	Barnyardgrass	Bermudagrass	Bluegrass, annual	Brome, downy	Cheatgrass	Crabgrass; large, smooth	Crowfootgrass	Cupgrass	Dallisgrass	Foxtail, yellow	Goosegrass	Guineagrass	Johnsongrass	Millet, proso	Oats, wild	Panicum, fall	Red rice	Ryegrass, Italian	Quackgrass	Sandbur	Shattercane	Signalgrass	Sprangletop	Witchgrass	Nutsedge, yellow	Nutsedge, purple
1. Acetachlor	X				X		X	X		X	X		X	X	X		X					X	X	X		
2. Acifluorfen							X			X			X			X					X					
3. Alachlor	X					X	X	X		X	X			X		X	X			X	X	X	X	X	X	
4. Ametryn		X				X			X	X	X		X			X					X					
5. Amitrole		X		X		X					X				X				X							
6. Asulam	X					X					X		X			X										
7. Atrazine	X		X	X	X	X				X					X				X					X	X	
8. Benefin	X	X	X			X	X			X	X					X		X		X					X	
9. Bensulide	X		X			X					X					X							X			
10. Bentazon	X	X				X				X	X		X		X	X			X		X				X	
11. Bromacil	X	X	X	X	X	X	X		X		X		X		X				X				X			
12. Bromoxynil																										
13. Butylate	X	X				X	X			X	X		X			X				X	X				X	X
14. Carfentrazone																										
15. Chlorimuron																									X	
16. Chlorsulfuron		X								X																
17. Clethodim	X	X		X		X	X	X		X	X		X	X	X	X	X	X	X	X	X		X	X		
18. Clomazone	X	X		X		X		X		X	X					X	X				X					
19. Clopyralid																										
20. Cloransulam																										
21. Cyanazine	X		X	X	X	X				X	X				X	X		X	X		X			X	X	X
22. Cycloate	X		X			X				X					X						X				X	X
23. 2,4-D																				X						
24. 2,4-DB																										
25. DCPA	X		X			X				X						X				X						
26. Desmedipham											X															
27. Dicamba																										
28. Dichlobenil			X			X													X							
29. Diclofop	X			X		X				X	X				X	X		X				X		X		
30. Difenzoquat															X											
31. Difluenzopyr																										
32. Dimethenamid	X					X		X		X	X		X	X	X	X	X			X	X	X	X	X	X	
33. Endothall	X		X																							
34. EPTC	X	X	X			X				X	X				X	X		X	X	X	X	X			X	X
35. Ethalfluralin	X		X			X				X	X		X		X	X		X			X	X			X	X

TABLE A-2. Continued

	Weed																									
	Barnyardgrass	Bermudagrass	Bluegrass, annual	Brome, downy	Cheatgrass	Crabgrass; large, smooth	Crowfootgrass	Cupgrass	Dallisgrass	Foxtail, yellow	Goosegrass	Guineagrass	Johnsongrass	Millet, proso	Oats, wild	Panicum, fall	Red rice	Ryegrass, Italian	Quackgrass	Sandbur	Shattercane	Signalgrass	Sprangletop	Witchgrass	Nutsedge, yellow	Nutsedge, purple
36. Ethofumesate	X		X							X					X										X	X
37. Fenoxaprop	X					X				X	X		X		X	X							X			
38. Fluazifop-P	X	X				X				X	X	X	X		X	X		X	X	X	X					
39. Flumiclorac																										
40. Fluometuron	X					X	X				X					X										
41. Fomesafen	X					X				X	X					X									X	
42. Glufosinate	X		X			X		X		X			X	X	X	X	X		X	X	X	X	X	X		
43. Glyphosate	X	X	X	X	X	X	X	X	X		X	X	X	X	X	X	X	X	X	X	X	X	X	X	X	X
44. Halosulfuron																									X	X
45. Hexazinone	X	X	X	X	X	X						X			X				X							
46. Imazamethabenz															X											
47. Imazamox	X					X	X	X		X	X		X		X	X				X	X	X		X		
48. Imazapyr		X	X	X	X	X		X	X		X	X	X		X	X		X	X			X	X	X		
49. Imazethapyr	X					X				X			X	X			X				X	X				
50. Imazaquin	X									X	X										X	X			X	
Isoxaflutol	X					X		X		X	X		X	X		X				X	X	X		X		
51. Lactofen																										
52. Linuron						X					X					X						X				
53. Metolachlor	X						X	X		X	X			X		X	X			X	X	X			X	
54. Metribuzin	X		X	X	X	X	X			X	X	X	X		X	X		X	X	X	X				X	
55. Metsulfuron																										
56. MSMA	X					X			X		X		X		X					X					X	
57. Napropamide	X	X	X	X		X				X	X	X	X		X	X		X		X					X	X
58. Naptalam	X					X				X	X					X									X	
59. Nicosulfuron	X							X		X			X		X	X		X	X	X	X	X				
60. Norflurazon	X	X	X	X	X	X	X				X	X	X			X			X			X		X		
61. Oryzalin	X	X	X	X		X	X	X		X	X	X	X		X	X		X	X	X			X	X		X
62. Oxyfluorfen	X		X			X					X				X	X			X	X						
63. Pebulate		X				X				X	X				X										X	X
64. Pendimethalin						X	X			X	X					X				X	X	X		X		
65. Phenmedipham																										

TABLE A-2. Continued

	Weed																									
	Barnyardgrass	Bermudagrass	Bluegrass, annual	Brome, downy	Cheatgrass	Crabgrass; large, smooth	Crowfootgrass	Cupgrass	Dallisgrass	Foxtail, yellow	Goosegrass	Guineagrass	Johnsongrass	Millet, proso	Oats, wild	Panicum, fall	Red rice	Ryegrass, Italian	Quackgrass	Sandbur	Shattercane	Signalgrass	Sprangletop	Witchgrass	Nutsedge, yellow	Nutsedge, purple
67. Primsulfuron										X			X			X		X	X	X	X				X	
68. Prometryn	X					X		X			X				X	X						X				
69. Pronamide	X		X	X		X				X	X		X		X	X		X	X							
70. Propachlor	X					X				X	X					X				X						
71. Propanil	X					X				X	X											X				
72. Prosulfuron																										
73. Pryazon																										
74. Pyridate																										
75. Quizalofop	X							X		X	X			X	X	X	X		X		X		X	X		
76. Rimsulfuron	X					X				X					X											
77. Sethoxydim	X	X				X		X		X	X			X	X	X	X		X	X	X	X	X	X	X	
78. Siduron	X					X																				
79. Simazine	X		X	X		X					X				X	X			X			X		X		
80. Sulfentrazone	X									X						X									X	
81. Sulfometuron	X		X	X		X							X					X							X	
82. Sulfosate	X	X	X	X	X	X	X			X	X	X	X	X	X	X	X	X		X	X	X	X	X		
83. Tebuthiuron	X			X	X	X			X						X	X		X		X						
84. Terbacil	X	X		X		X	X					X							X			X			X	X
85. Thifensulfuron																										
86. Thiobencarb	X																						X			
87. Triallate															X											
88. Triclopyr																										
89. Trifluralin	X	X	X	X	X	X		X		X	X	X	X		X	X	X	X		X	X	X	X		X	X

TABLE A-3. Common Name Index of Major Manufacturers and Suppliers of Herbicides[a]

Common Name	Manufacturer	Trade Name
2,4-D	Helena	2,4-D Amine
2,4-D	Helena	Butyl Ester 6
2,4-D	PBI Gordon	Orchardmaster, CA
2,4-D	Agriliance	2,4-D Amine 4
2,4-D	Agriliance	2,4-D LV4, LV Ester 6, LV6, LV Ester 4
2,4-D	Agriliance	SWB, 2,4-D LV4
2,4-D	Albaugh, Inc.	2,4-D Amine 4, Amine 6
2,4-D	Albaugh, Inc.	2,4-D LV4, LV6
2,4-D	Albaugh, Inc.	Five Star
2,4-D	Albaugh, Inc.	Solve 2,4-D
2,4-D	BASF	Oasis
2,4-D	Cenex-Land-o-Lakes	40A Phenoxy
2,4-D	Cenex-Land-o-Lakes	LV4 Phenoxy
2,4-D	Cenex-Land-o-Lakes	LV6 Phenoxy
2,4-D	Cenex-Land-o-Lakes	Yardmaster 4LG 2,4-D
2,4-D	Cenex-Land-o-Lakes	Yardmaster Garden Weeder
2,4-D	Cenex-Land-o-Lakes	Yardmaster Lawn & Garden Weeder
2,4-D	Cerexagri	Aqua-Kleen
2,4-D	DowAgro Canada only	Attain B
2,4-D	Helena	Barrage, HF
2,4-D	Helena	Weed Rhap A 4-D
2,4-D	Helena	Weed Rhap LV6D
2,4-D	PBI Gordon	Dymec
2,4-D	PBI Gordon	Hi-Dep, 1VM
2,4-D	Riverdale	2,4-D granules
2,4-D	Riverdale	AM-40
2,4-D	Riverdale	Dri-clean
2,4-D	Riverdale	Solution water soluble
2,4-D	UAP/PLATT	Amine 4
2,4-D	UAP/PLATT	Low Vol 4 Ester, 6 Ester
2,4-D	UAP/PLATT	Salvo
2,4-D	UAP/PLATT	Savage, Dry Soluble
2,4-D	United Hort Supply	Amine 4 2,4-D
2,4-D + 2,4-DB	Agriliance	Phenoxy 088
2,4-D + 2,4-DB	Albaugh, Inc.	D-638
2,4-D + 2,4-DP	Riverdale	Turf Weed & Brush
2,4-D + 2,4-DP	Scotts	Fluid Broadleaf Weed Control
2,4-D + dicamba + MCPP	United Hort Supply	Mec Amine BG
2,4-D + dichlorprop	Riverdale	Tri-Ester, II
2,4-D + dichlorprop	UAP/PLATT	DPD Ester Brush Killer
2,4-D + dichlorprop + dicamba	PBI Gordon	Brushmaster, 875
2,4-D + dichlorprop + dicamba	PBI Gordon	Super Brush Killer

TABLE A-3. Continued

Common Name	Manufacturer	Trade Name
2,4-D + dichlorprop + dicamba	PBI Gordon	Super Trimec
2,4-D + MCPA + dichloprop	Lesco	Granular Broadleaf Herbicide
2,4-D + MCPP + dicamba	Agriliance	Strike 3
2,4-D + MCPP + dicamba	Lesco	Three-Way, Selective
2,4-D + MCPP + dicamba	PBI Gordon	Trimec 899, Bentgrass, Classic, DSC, SI, Southern, Turf Ester, 992
2,4-D + MCPP + dicamba	PBI Gordon	Trimec LAF 637
2,4-D + MCPP + dicamba	Scotts	Fertilizer Plus Dicot Weed Control III/32-3-2
2,4-D + MCPP + dicamba	UAP/PLATT	Mec-amine-D, Plus
2,4-D + MCPP + dicamba	United Hort Supply	Trimec-992, -Bent, -Classic, -Plus, -Super, -Southern, Trimec Plus
2,4-D + mecoprop + dicamba	Riverdale	Triplet, WSP
2,4-D + mecoprop + dicamba	United Hort Supply	Mec Amine - D
2,4-D + mecoprop + dichlorprop	Riverdale	Dissolve
2,4-D + mecoprop + dichlorprop	Riverdale	Triamine Jet-Spray, G
2,4-D + prometon	PBI Gordon	Vegemec
2,4-D + triclopyr	DowAgro	Crossbow
2,4-D + triclopyr	United Hort Supply	Chaser
2,4-D + triclopyr + clopyralid	Lesco	Momentum
2,4-DB	Agriliance	2,4-DB 1.75, 200
2,4-DB	Albaugh, Inc.	Butyrac 175, 200
2,4-DB	Cedar Chemical	Butoxone 200, 7500
Acetochlor	Monsanto	Harness, 20G
Acetochlor	Monsanto	Degree
Acetochlor + atrazine	Monsanto	Degree Xtra
Acetochlor + atrazine	Monsanto	Harness Xtra, Xtra 5.6
Acetochlor + atrazine	Syngenta	Fultime
Acetochlor + EPTC + safener	Syngenta	Doubleplay
Acifluorfen	BASF	Blazer, Ultra
Acifluorfen	BASF	Conclude B
Acifluorfen + bentazon	BASF	Galaxy
Acifluorfen + bentazon + sethoxydim	BASF	Conclude Ultra
Alachlor	Monsanto	Lasso, II
Alachlor	Monsanto	Micro-Tech
Alachlor	Monsanto	Partner WDG
Alachlor + atrazine	Monsanto	Bullet
Alachlor + atrazine	Monsanto	Lariat
Alachlor + trifluralin	Monsanto	Freedom
Ametryn	Syngenta	Evik, DF
Asulam	Aventis	Asulox
Atrazine	Agriliance	Atrazine 4L, 90DF

TABLE A-3. Continued

Common Name	Manufacturer	Trade Name
Atrazine	Albaugh, Inc.	Atrazine 4F, 40DF
Atrazine	Cenex-Land-o-Lakes	Aatrex Nine-D
Atrazine	Drexel Chemical	Atra-5
Atrazine	Drexel Chemical	Atrazine 4L, 80, 90-DF
Atrazine	Helena	Aatrex 4L
Atrazine	Lesco	St. Augustine Weed & Feed
Atrazine	Syngenta	Aatrex, Nine-O, 4L
Atrazine	UAP/PLATT	Atrazine 4L, 90 WDG
Atrazine	UAP/PLATT	Conifer 90
Atrazine	UAP/PLATT	Stubble
Atrazine	United Hort Supply	Atrazine 4L
Atrazine + 2,4-D	UAP/PLATT	Shotgun
Atrazine + acetamide	Syngenta	Bicep II Magnum FC
Atrazine + acetamide	Syngenta	Bicep Lite II Magnum
Atrazine + bentazon	United Hort Supply	Promp
Atrazine + metalachlor + benoxacor	Syngenta	Bicep II
Benefin	UAP/PLATT	Balan DF, 2.5G
Benefin	United Hort Supply	Balan 2.5G
Benefin + oryzalin	Setre	XL 2G
Benefin + trifluralin	Lesco	Team
Benefin + trifluralin	United Hort Supply	Team 2G
Bensulfuron	DuPont	Londax
Bensulide	Gowen	Prefar
Bensulide	PBI Gordon	Bensumec
Bensulide	PBI Gordon	Pre-San 7G, 12.5G
Bensulide	Scotts	Weedgrass Preventer
Bensulide	UAP/PLATT	Betasan 12.5 G, 4E, 7-G, 36
Bensulide	United Hort Supply	Betasan 4EC, 7G, 12.5G
Bensulide + oxadiazon	Scotts	Goosegrass/Crabgrass Control
Bentazon	BASF	Basagran, SG, TO
Bentazon	BASF	Fortune
Bentazon	BASF Canada only	Basagran Forte
Bentazon	Lesco	Lescogran
Bentazon	United Hort Supply	BasagranT/O
Bentazon + acifluorfen	BASF	Storm
Bentazon + acifluorfen + clethodim	BASF	Conclude Xact
Bentazon + aciflurafen	BASF	Maifest B
Bentazon + sethoxydim	BASF	Rezult B, G
Bromacil	DuPont	Hyvar X, X-L
Bromacil	United Hort Supply	Hyvar XL
Bromacil + diuron	DuPont	Krovar I DF
Bromacil + diuron	UAP/PLATT	Weed Blast
Bromacil + diuron	United Hort Supply	Krovar I DF
Bromoxynil	Agriliance	Moxy 2E
Bromoxynil	Aventis	Buctril, 4 cereals, 4EC
Bromoxynil	UAP/PLATT	Broclean

TABLE A-3. Continued

Common Name	Manufacturer	Trade Name
Bromoxynil	Aventis	Connect 20 WSP
Bromoxynil + atrazine	Agriliance	Moxy + Atrazine
Bromoxynil + atrazine	Albaugh, Inc.	BROX-AT, M, 2EC
Bromoxynil + atrazine	Aventis	Buctril + Atrazine
Bromoxynil + atrazine	UAP/PLATT	Brozine
Bromoxynil + MCPA	Aventis	Bronate, Bronate Pro
Bromoxynil + MCPA	UAP/PLATT	Bromac
Butylate	Cedar	Sutan +
Caporal + MSMA	Agriliance	Prometryn + MSMA
Carfentrazone-ethyl	FMC	Aim
Chelated copper	Lesco	Lescocide Plus
Chloransulam-methyl	DowAgro	Firstrate
Chlorimuron	DuPont	Classic
Chlorimuron + metribuzin	DuPont	Canopy
Chlorimuron ethyl + sulfentrazone	DuPont	Canopy XL
Chlorimuron ethyl + thifensulfuron	DuPont	Synchrony STS
Chloropropham	UAP/PLATT	CIPC 7, 700
Chlorsulfuron	DuPont	Glean
Chlorsulfuron	DuPont	Telar DF
Chlorsulfuron + metasulfuron	DuPont	Finesse
Clethodim	United Hort Supply	Envoy
Clethodim	Valent	Envoy
Clethodim	Valent	Prism, Envoy, Select
Clodinafop - propargyl	Syngenta	Discover
Clomazone	FMC	Command 3ME, 4EC
Clopyralid	DowAgro	Reclaim
Clopyralid	DowAgro	Lontrel T&O
Clopyralid	DowAgro	Stinger
Clopyralid	DowAgro	Transline
Clopyralid	DowAgro Canada only	Curtail M
Clopyralid	DowAgro Canada only	Lontrel, 360
Clopyralid	United Hort Supply	Stinger
Clopyralid + 2,4-D	DowAgro	Curtail, M
Clopyralid + flumetsulam + nicosulfuron + rimsulfuron	DuPont	Accent Gold
Clopyralid + glyphosate	DowAgro Canada only	Eclipse
Clopyralid + MCPA	DowAgro Canada only	Prestige B
Clopyralid + MCPA	DowAgro Canada only	Prevail B
Copper chelate	Griffin	Komeen
Copper ethylenediamine complex + copper sulfate pentahydrate	PBI Gordon	Aquacure
Cyanazine	DuPont	Bladex, 4L, 90DF
Cyanazine	Griffin	Cy-Pro 4L
Cyanazine + atrazine	DuPont	Extrazine II 4L, II DF
Cyanazine + MSMA	Setre	Bladex—MSMA

TABLE A-3. Continued

Common Name	Manufacturer	Trade Name
Cycloate	Cedar Chemical	Ro-Neet
Cyclosulfamuron	BASF	Invest
Dazomet	BASF	Basamid
Dazomet	United Hort Supply	Basamid
DCPA	Amvac	Dacthal W-75
DCPA	UAP/PLATT	Dacthal W75
Desmedipham	Aventis	Betanex
Desmedipham + phenmedipham	Aventis	Betamix
Dicamba	Agriliance	Sterling
Dicamba	Albaugh, Inc.	Dicamba DMA, 6G
Dicamba	BASF	Banvel SGF
Dicamba	BASF Canada only	Banvel—corn
Dicamba	BASF Canada only	Banvel II—cereal
Dicamba	Scotts	K-O-G Weed Control
Dicamba	Syngenta	Rave
Dicamba + 2,4-D	Albaugh, Inc.	Range Star
Dicamba + 2,4-D	BASF	Weedmaster
Dicamba + 2,4-D + MCPP	Lesco	Bentgrass Selective
Dicamba + 2,4-D + mecoprop	BASF Canada only	DyVel DS
Dicamba + atrazine	BASF	Marksman
Dicamba + dimethenamid	BASF	OpTill TM
Dicamba + MCPA	BASF Canada only	DyVel
Dicamba + MCPA	United Hort Supply	Four Power Plus
Dicamba diglycolamine	BASF	Clarity
Dichlobenil	PBI Gordon	Barrier
Dichlobenil	Uniroyal	Casoron
Dichlobenil	United Hort Supply	Casaron 4G, 50W
Dichlobenil	United Hort Supply	Dyclomec 4G
Diclofop-methyl	Aventis	Hoelon 3EC
Diclofop-methyl	Aventis Canada only	Hoe Grass 284
Diclofop-methyl	Aventis Canada only	Hoe Grass II
Diclofop-methyl	Aventis	Illoxam
Diclosulam	DowAgro	Strongarm
Difenzoquat	BASF	Avenge
Diflufenzopyr	BASF	Distinct
Dimethenamid	BASF	Frontier 6.0
Dimethenamid + atrazine	BASF	Guardsman
Dimethenamid + atrazine	DuPont	Leadoff
Dimethenamid-p	BASF	Outlook
Dimethenamid-p + atrazine	BASF	Guardsman Max
Dimethipin	Uniroyal	Harvade 5F
Diquat	United Hort Supply	Reward
Dithiopyr	Lesco	Dimension
Dithiopyr	Dow Agro	Dimension, Ultra WSP
Dithiopyr	United Hort Supply	Dimension

TABLE A-3. Continued

Common Name	Manufacturer	Trade Name
Dithiopyr	Scotts	Nitrogen Potassium Fertilizer with Dimension/14-0-14
Diuron	Agriliance	Diuron 4L, 80DF
Diuron	DowAgro	Diuron 4L IVM, 80DF IVM, 4L, 80DF
Diuron	Drexel Chemical	Diuron, 4L, 80, DP
Diuron	Griffin	Direx 4L, 80DF
Diuron	Griffin	Karmex DF
Diuron	Helena	Diuron 80W
Diuron	UAP/PLATT	Diuron 80 WDG, 80WP
Diuron	United Hort Supply	Karmex 80
DSMA	Drexel Chemical	DSMA liquid, liquid 4
DSMA	Helena	DSMA, 4, liquid
DSMA	Setre	DSMA 4, Liquid DSMA
DSMA	UAP/PLATT	DSMA Plus
DSMA	United Hort Supply	Methar 30
Endothall	Cerexagri	Accelerate
Endothall	Cerexagri	Aquathol K, Super K, Granular
Endothall	Cerexagri	Desicate II, Des-i-cate
Endothall	Cerexagri	Herbicide 273
Endothall	Cerexagri	Hydrothol 191, 191G
EPTC	Syngenta	Eptam 7-E, 20G
Ethalfluralin	DowAgro	Sonalan 10G, HFP
Ethalfluralin	DowAgro Canada only	Edge
Ethalfluralin	UAP/PLATT	Curbit
Ethametsulfuron-methyl	DuPont Canada only	Muster
Ethametsulfuron-methyl + quizalofop-p-ethyl	DuPont Canada only	Muster Gold II
Ethofumesate	Aventis	Nortron SC
Ethofumesate	United Hort Supply	Prograss 1.5EC
Ethofumesate + phenmedipham + desmedipham	Aventis	Progress, Prograss EC
Fenoxaprop	Aventis	Bugle
Fenoxaprop + MCPA + 2,4-D + thifensulfuron methyl	DuPont Canada only	Champion Plus, Extra
Fenoxaprop-ethyl	Aventis	Horizon 1EC
Fenoxaprop-ethyl	United Hort Supply	Acclain Extra
Fenoxaprop-p-ethyl	Aventis	Acclaim
Fenoxaprop-p-ethyl	Aventis	Silverado
Fenoxaprop-p-ethyl	Aventis	Whip 360
Fenoxaprop-p-ethyl	Aventis Canada only	Puma
Fenoxaprop-p-ethyl	Aventis Canada only	Puma 120 Super
Fenoxaprop-p-ethyl	Aventis Canada only	Puma One Pass
Fenoxaprop-p-ethyl	Aventis Canada only	Puma Super
Fenoxaprop-p-ethyl + 2,4-D + MCPA	Aventis	Tiller

TABLE A-3. Continued

Common Name	Manufacturer	Trade Name
Fenoxaprop-p-ethyl + MCPA + thifensulfuron + tribenuron	Aventis	Cheyenne FM, Cheyenne X-tra
Fenoxaprop-p-ethyl + safener	Aventis	Puma 1E
Fenoxaprop-p-ethyl+fluazifop-p-butyl	Syngenta	Fusion
Fenoxaprop-p-ethyl+MCPA	Aventis	Dakota
Fluazifop	United Hort Supply	Fusilade II T/O
Fluazifop-p-butyl	PBI Gordon	Ornamec, 170
Fluazifop-p-butyl	Syngenta	Fusilade DX
Fluazifop-p-butyl + fomesafen	Syngenta	Typhoon
Flucarbazone	Bayer	Everest
Flufenacet	Aventis	Define
Flufenacet	Bayer	Domain
Flufenacet + isoxaflutole	Bayer	Epic DF
Flufenacet + metribuzin	Bayer	Axiom DF
Flufenacet + metribuzin + atrazine	Bayer	Axiom AT
Flumetsulam	DowAgro	Broadstrike
Flumetsulam	DowAgro	Python WDG
Flumetsulam + chloransulam-methyl	DowAgro	Frontrow
Flumetsulam + clopyralid	DowAgro	Broadstrike Plus
Flumetsulam + clopyralid	DowAgro	Hornet, WDG
Flumetsulam + clopyralid	DowAgro Canada only	Fieldstar, DG, WSP
Flumetsulam + clopyralid + 2,4-D	Dow Agro	Broadstrike Post
Flumetsulam + clopyralid + 2,4-D	DowAgro Canada only	Striker
Flumetsulam + metolachlor	DowAgro Canada only	Broadstrike + Dual
Flumetsulam + trifluralin	DowAgro	Broadstrike + Treflan
Flumiclorac-pentyl	Valent	Resource
Flumioxazin	Valent	Valor
Fluometuron	Agriliance	Fluometuron 4L, 80DF
Fluometuron	Griffin	Meturon 4L
Fluroxypyr	DowAgro Canada only	Prestige A
Fluroxypyr	DowAgro	Starane
Fluroxypyr	DowAgro	Vista
Fluroxypyr	DowAgro Canada only	Attain A
Fluroxypyr + 2,4-D	DowAgro	Starane + Esteron
Fluroxypyr + 2,4-D	DowAgro	Starane + Salvo
Fluroxypyr + MCPA	DowAgro	Starane + MCPA
Fluroxypyr + MCPA	DowAgro	Starane + Sword
Fluxofenim	Syngenta	Concept III
Fomesafen	Syngenta	Reflex
Fomesafen + isolink technology	Syngenta	Flexstar
Fosamine	DuPont	Krenite S

TABLE A-3. Continued

Common Name	Manufacturer	Trade Name
Glufosinate	Aventis	Finale VM
Glufosinate	Aventis	Liberty, ATZ
Glufosinate	Aventis	Rely
Glufosinate	Aventis	Remove
Glufosinate	United Hort Supply	Finale
Glyphosate	Cerexagri	Aqua-Neat
Glyphosate	DowAgro	Glyphomax, Plus
Glyphosate	DowAgro	Glypro
Glyphosate	DowAgro Canada only	Vantage, Plus
Glyphosate	DuPont	Glyphosate
Glyphosate	Griffin	Glyphosate Original
Glyphosate	Helena	Rattler
Glyphosate	Monsanto	Accord, Site Prep, SP, Concentrate
Glyphosate	Monsanto	Aqua Master
Glyphosate	Monsanto	Campaign
Glyphosate	Monsanto	Honcho
Glyphosate	Monsanto	Polado L
Glyphosate	Monsanto	Protocol
Glyphosate	Monsanto	Rodeo
Glyphosate	Monsanto	Roundup as: Custom, D-Pak, Dry Pak, Original, Original RT, Pro, Pro Dry, Ultra, Ultra RT, Ultradry, Ultramax
Glyphosate	Monsanto Canada only	Roundup Transorb, Fast Forward
Glyphosate	Monsanto Canada only	Vision
Glyphosate	Scotts	Roundup
Glyphosate	UAP/PLATT	Dead-n-Gone
Glyphosate	United Hort Supply	Roundup DRY PACK, Pro
Glyphosate + 2,4-D	Monsanto	Landmaster BW
Glyphosate + acetochlor + atrazine	Monsanto	Field Master
Glyphosate + atrazine	Monsanto	Ready Master
Glyphosate + dicamba	Monsanto	Fallow Master, Broadspectrum
Glyphosate + imazethapyr	BASF	Extreme
Halosulfuron	Monsanto	Manage Turf
Halosulfuron	Monsanto	Permit
Halosulfuron	United Hort Supply	Manage WSB
Hexazinone	DuPont	Velpar, DF, L, ULW
Hexazinone	United Hort Supply	Velpar
Hexazinone + sulfometuron/methyl	DuPont	Oustar
Holosulfuron	Monsanto	Sempra
Imazameth	BASF	Cadre DG
Imazamethabenz - methyl	BASF	Assert
Imazamox	BASF	Raptor
Imazapyr	BASF	Arsenal, AC, Railroad
Imazapyr	BASF	Chopper

TABLE A-3. Continued

Common Name	Manufacturer	Trade Name
Imazapyr	BASF	Habitat Release, 75 SG
Imazapyr	United Hort Supply	Arsonal
Imazaquin	BASF	Image 70 DG, 1.5 LC
Imazaquin	BASF	Scepter 70DG
Imazaquin + dimethenamid	BASF	Detail
Imazaquin + glyphosate	BASF	Backdraft
Imazethapyr	BASF	Newpath
Imazethapyr	BASF	Plateau, DG
Imazethapyr	BASF	Pursuit, DG, W, WDG
Imazethapyr	BASF	Stalker
Imazethapyr + diuron	BASF	Sahara
Imazethapyr + imazapyr	BASF	Lightning
Imazethapyr + pendimethalin	BASF	Pursuit Plus EC
Imidazolinone	United Hort Supply	Image
Isoxaben	DowAgro	Gallery 75DF, T&V
Isoxaben	United Hort Supply	Gallery 75DF
Isoxaben + trifluralin	DowAgro	Snapshot 2.5TG
Isoxaflutole	Aventis	Balance Pro, WDG
Lactofen	Valent	Cobra
Lactofen + flumiclerac-pentyl	Valent	Stellar, 2EC
Linuron	Griffin	Lorox DF
Linuron + monolinuron	Aventis Canada only	Afalon S
MCPA	Agriliance	MCPA—Amine
MCPA	Agriliance	SWB MCPA Ester
MCPA	Albaugh, Inc.	MCPA Amine 4, Ester 4, Sodium Salt
MCPA	Albaugh, Inc.	Solve MCPA
MCPA	Aventis	Chiptox
MCPA	Aventis	Rhomene
MCPA	Aventis	Rhonox
MCPA	Cenex-Land-o-Lakes	MCPA Phenoxy
MCPA	Riverdale	Dagger
MCPA	Riverdale	MCPA-4
MCPA	UAP/PLATT	MCP 4 Amine
MCPA	UAP/PLATT	MCP 2 Sodium
MCPA	UAP/PLATT	MCP 4 Ester
MCPA	UAP/PLATT	Sword
MCPA + MCPP + dicamba	PBI Gordon	Trimec Encore
MCPA + MCPP + dicamba	Syngenta Canada only	Target
MCPA + MCPP + MSMA + dicamba	United Hort Supply	Trimec Encore
MCPA + mecoprop + dicamba	Riverdale	Tri-Power Selective, Dry
MCPA + mecoprop + dichlorprop	Riverdale	Triamine II, G
MCPA + triclopyr + dicamba	Lesco	Eliminate
MCPA + triclopyr + dicamba	Lesco	Three-Way Ester II
MCPB	Aventis	Thistrol MCPB

TABLE A-3. Continued

Common Name	Manufacturer	Trade Name
MCPB	Cenex-Land-o-Lakes	MCPB Phenoxy
MCPP	Riverdale	MCPP-4
MCPP	UAP/PLATT	MCPP 4K
MCPP	United Hort Supply	MCPP 4K
MCPP	United Hort Supply	Mecomec
MCPP + 2,4-D	United Hort Supply	Two + Two
MCPP + MCPA + dicamba	United Hort Supply	Hat Trick
Mecoprop	PBI Gordon	Mecomec 2.5
Mefluidide	PBI Gordon	Embark 2S 1VM
Mesotrione	Syngenta	Callisto
Metham	Amvac	Vapam HL
Metham	UAP/PLATT	Metam Sodium
Metobromuron	Syngenta Canada only	Patoran
Metolachlor	Syngenta	Dual II G Magnum
Metolachlor	Syngenta	Pennant
Metolachlor	United Hort Supply	Pennant
Metolachlor + atrazine + safener	Syngenta Canada only	Paimextra II
Metribuzin	Bayer	Sencor 4, DF, SoluPac
Metribuzin	DuPont	Lexone DF
Metribuzin	United Hort Supply	Sencor 75WP
Metsulfuron	DuPont	Ally
Metsulfuron	DuPont	Escort
Molinate	Syngenta	Ordram 8E, 15-G, 15-GM
MSMA	Agriliance	120 Herbicide
MSMA	Albaugh, Inc.	Weed Hoe 108,120
MSMA	Drexel Chemical	MSMA-6 Plus, 6.6
MSMA	Helena	MSMA, Plus, Plus H.C.
MSMA	Lesco	MSMA Soluble Granules
MSMA	Setre	MSMA, Plus
MSMA	UAP/PLATT	MSMA 6 Plus, 6.6, 600 Plus
MSMA	United Hort Supply	MSMA 6.6 Turf, Turf
Napropamide	United Hort Supply	Devrinol 50DF
Napropamide + oxadiazon	United Hort Supply	Pre Pair
Naptalam	Uniroyal	Alanap
Nicosulfuron	BASF	Celebrity Herbicide, Plus
Nicosulfuron	DuPont	Accent
Nicosulfuron	DuPont	Steadfast
Nicosulfuron + rimsulfuron + atrazine	DuPont	Basis Gold
Norflurazon	Setre	Zorial Rapid 80
Norflurazon	Syngenta	Evital
Norflurazon	Syngenta	Solicam DF
Norflurazon	Syngenta	Zorial
Norflurazon	Syngenta	Zorial Rapid 80, 5G
Norflurazon	United Hort Supply	Predict
Oryzalin	DowAgro	Surflan AS, AS Specialty
Oryzalin	United Hort Supply	Surflan AS
Oryzalin	United Hort Supply	XL 2G

TABLE A-3. Continued

Common Name	Manufacturer	Trade Name
Oxadiazon	Scotts	Fertilizer Plus 1% Ronstar 15-5-10
Oxadiazon	Scotts	Fertilizer Plus 1% Ronstar 16-0-16
Oxadiazon	Scotts	Fertilizer Plus 1% Ronstar 5-5-20
Oxadiazon	Scotts	Fertilizer Plus 1.5% Ronstar 21-0-20
Oxadiazon	Scotts	Fertilizer Plus 1.5% Ronstar 5-5-20
Oxadiazon	Lesco	Ronstar
Oxadiazon	United Hort Supply	Ronstar 2G, 50WP
Oxadiazon	Aventis	Ronstar 50 WSP, G
Oxadiazon + pendimethalin	Scotts	Kansel + 28-0-0
Oxyfluorfen	Dow Agro	Goal, 1.6, 2XL
Oxyfluorfen	United Hort Supply	Goal T/O
Paraquat	Syngenta	Cyclone Max
Paraquat	Syngenta	Gramoxone Extra, MAX
Paraquat + diuron	UAP/PLATT	Surefire
Pebulate	Cedar Chemical	Tillam 6E
Pebulate	Syngenta	Tillam 6E
Pelargonic acid	United Hort Supply	Scythe
Pelargonic acid	DowAgro	Scythe
Pendimethalin	BASF	Pendulum, 2G, 3.3EC, WDG
Pendimethalin	BASF	Pentagon
Pendimethalin	BASF	Prowl
Pendimethalin	DowAgro	Pendimax 3.3
Pendimethalin	Lesco	Pre-M, 3.3EC, 60DG
Pendimethalin	Scotts	High K Turf Fertilizer Plus Preemergent Weed Control/0-0-13
Pendimethalin	Scotts	NK Fertilizer Plus Turf Weed Control/14-0-14
Pendimethalin	Scotts	NPK Fertilizer Plus Preemergence Weed Control/11-3-11
Pendimethalin	Scotts	NPK Fertilizer Plus Preemergence Weed Control/13-3-7
Pendimethalin	Scotts	Turf Fertilizer Plus PreemergenceWeed Control/22-0-11
Pendimethalin	Scotts	Turf Fertilizer Plus PreemergenceWeed Control/22-0-6
Pendimethalin	Scotts	Turf Weedgrass Control
Pendimethalin	United Hort Supply	Pendimethalin + Fertilizer

TABLE A-3. Continued

Common Name	Manufacturer	Trade Name
Pendimethalin	United Hort Supply	Pendulum 60 WDG, 60 WSB, 2G, 3.3EC
Pendimethalin + imazaquin	BASF	Squadron
Pendimethalin + imazaquin + imazethapyr	BASF	Steel
Phenmedipham	Aventis	Betanol, AM
Phenmedipham	Aventis	Spin-aid
Picloram	DowAgro	Tordon 22K, K
Picloram	United Hort Supply	Tordon RTU
Picloram + 2,4-D	DowAgro	Grazon P + D
Picloram + 2,4-D	DowAgro	Pathway
Picloram + 2,4-D	DowAgro	Tordon RTU, 101 Mixture
Primisulfuron	Syngenta	Beacon
Primisulfuron + dicamba	Syngenta	North Star
Prodiamine	Scotts	21-0-20 + Barricade
Prodiamine	United Hort Supply	Barricade WDG
Prodiamine	United Hort Supply	Factor
Prometon	Cenex-Land-o-Lakes	Pramital 25
Prometon	UAP/PLATT	Pramitol 25E, 5PS
Prometon	Agriliance	Pramitol 25E
Prometon	United Hort Supply	Pramitol 25EC, 5PS
Prometryn	Agriliance	Prometryn 4L
Prometryn	Griffin	Cotoran 4L, DF
Prometryn	Griffin	Cotton-Pro
Prometryn	Syngenta	Caparol 4L
Pronamide	Dow Agro	Kerb WSP, 50W
Pronamide	United Hort Supply	Kerb WSP
Propachlor	Monsanto	Ramrod FL, 20G
Propanil	Agriliance	Propanil 4E, 80DF
Propanil	Dow Agro	Stam M-4, 4E, 80EDF, Pro
Propanil	Dow Agro	Stampede
Prosulfuron	Syngenta	Peak
Prosulfuron	Syngenta Canada only	Peak Plus
Prosulfuron + primisulfuron	Syngenta	Exceed
Prosulfuron + primisulfuron	Syngenta	Spirit
Pyrazon	BASF	Pyramin DF, SC
Pyridate	Syngenta	Tough
Pyrithiobac	DuPont	Staple, Plus
Quinclorac	BASF	Drive
Quinclorac	BASF	Facet 75DF, GR
Quinclorac	BASF	Paramount
Quinclorac	BASF Canada only	Accord
Quinclorac	United Hort Supply	Drive
Quizalofop-p-ethyl	DuPont	Assure II
Quizalofop-p-ethyl + thifensulfuron + bentazen	DuPont Canada only	Hat Trick
Rimsulfuron	DuPont	Matrix
Rimsulfuron	DuPont	Shadeout

TABLE A-3. Continued

Common Name	Manufacturer	Trade Name
Rimsulfuron	DuPont Canada only	Elim EP
Rimsulfuron	DuPont Canada only	Prism
Rimsulfuron + nicosulfuron	DuPont Canada only	Ultim
Rimsulfuron + nicosulfuron + striker	DuPont Canada only	Ultimax
Rimsulfuron + thifensulfuron	DuPont	Basis
Sethoxydim	BASF	Conclude G
Sethoxydim	BASF	Maifest G
Sethoxydim	BASF	Poast, HC, Plus
Sethoxydim	BASF	Vantage
Sethoxydim	BASF Canada only	Poast Ultra
Sethoxydim	United Hort Supply	Vantage
Sethoxydim + clopyralid + MCPA	BASF Canada only	Flax Max Ultra
Siduron	Gowen	Tupersan
Siduron	PBI Gordon	Tupersan
Siduron	Scotts	Starter Fertilizer with Preemergent Weed Control/16-21-4
Siduron	United Hort Supply	Tupersan
Simazine	Agriliance	Simazine 4L, 90DF
Simazine	Drexel Chemical	Simazine 4L, 90DF
Simazine	Syngenta	Princep 4L, Calibar 90
Simazine	UAP/PLATT	Simazine 4L, 90 WDG, 80W
Simazine	United Hort Supply	Princep 4LT & D
s-Metolachlor	Syngenta	Dual II Magnum, Magnum SI, Dual Magnum
s-Metolachlor + metribuzin	Syngenta	Boundry
Sodium chlorate	Helena	Chlorate
Sulfentrazone	DuPont	Authority
Sulfentrazone	FMC	Spartan
Sulfentrazone + chloransulam	FMC	Gauntlet
Sulfometuron	DuPont	Oust
Sulfometuron	United Hort Supply	Oust
Sulfosate	Syngenta	Touchdown, 5
Sulfosulfuron	Monsanto	Maveric
Sulfosulfuron	Monsanto	Outrider
Sulfosulturon	Monsanto Canada only	Sundance
Tebuthiuron	DowAgro	Spike 20P, 80W
Terbacil	DuPont	Sinbar
Thiazopyr	Dow Agro	Visor
Thidiazuron	Aventis	Dropp 50, Ultra
Thidiazuron + diuron	Aventis	Ginstar EC
Thifensulfuron	DuPont	Pinnacle
Thifensulfuron	DuPont Canada only	Refine Extra
Thifensulfuron + tribenuron	DuPont	Harmony Extra, GT
Thifensulfuron + tribenuron	DuPont Canada only	Harmony Total

TABLE A-3. Continued

Common Name	Manufacturer	Trade Name
Thifensulfuron + tribenuron + metsulfuron	DuPont	Canvas
Thiobencarb	Valent	Bolero, 8EC, 10G
Tralkoxydim	DowAgro Canada only	Prevail A
Tralkoxydim	Syngenta	Achieve, 40DG, 80DG
Triallate	Monsanto	Far-go, G
Triallate	Monsanto Canada only	Avadex
Triallate + trifluralin	Monsanto	Buckle
Triallate + trifluralin	Monsanto Canada only	Fortress
Triasulfuron	Syngenta	Amber
Tribenuron	DuPont	Express
Tribenuron + 2,4-D	DuPont Canada only	Express Pack
Tribufos	Bayer	Def 6
Triclopyr	DowAgro	Forestry Garlon 4
Triclopyr	DowAgro	Garlon 3A, 4
Triclopyr	DowAgro	Grandstand R, CA
Triclopyr	DowAgro	Pathfinder II
Triclopyr	DowAgro	Remedy
Triclopyr	DowAgro Canada only	Release
Triclopyr + 2,4-D	United Hort Supply	Crossbow
Triclopyr + 2,4-D	United Hort Supply	Turflon Ester
Triclopyr + clopyralid	DowAgro	Confront
Triclopyr + clopyralid	Scotts	Fertilizer plus Confront Broadleaf Weed Control/30-5-5
Triclopyr + clopyralid	United Hort Supply	Confront
Triclopyr ester	DowAgro	Turflon Ester
Trifluralin	Albaugh, Inc.	Trifluralin 10G, 4ED
Trifluralin	Aventis Canada only	Rival 10 G Superflow
Trifluralin	Aventis Canada only	Rival EC
Trifluralin	DowAgro	Treflan HPF, TR-10
Trifluralin	DowAgro Canada only	Advance 10G
Trifluralin	DowAgro Canada only	Heritage
Trifluralin	DowAgro Canada only	Treflan QR5
Trifluralin	Gowen	Trifluralin 5, 4, 10G
Trifluralin	Griffin	Trilin, 5, 10G
Trifluralin	Lesco	Treflan 5G
Trifluralin	Setre	Trifluralin 4EC
Trifluralin	UAP/PLATT	Legacy
Trifluralin	UAP/PLATT	Trifluralin HF, 10G, 4EC
Trifluralin	United Hort Supply	Treflan 5G
Trifluralin + clomazone	FMC	Commence
Trifluralin + isoxaben	United Hort Supply	Snapshot 2.5TG
Triflusulfuron	DuPont	Upbeet

[a]Does not include soil fumigants or most herbicides sold only for the home market.

TABLE A-4. Trade Name Index of Major Manufacturers and Suppliers of Herbicides[a]

Trade Name	Manufacturer	Common Name
120 Herbicide	Agriliance	MSMA
2,4-D Amine	Helena	2,4-D
2,4-D Amine 4	Agriliance	2,4-D
2,4-D Amine 4, Amine 6	Albaugh, Inc.	2,4-D
2,4-D granules	Riverdale	2,4-D
2,4-D LV4, LV Ester 6, LV6, LV Ester 4	Agriliance	2,4-D
2,4-D LV4, LV6	Albaugh, Inc.	2,4-D
2,4-DB 1.75, 200	Agriliance	2,4-DB
21-0-20 + Barricade	Scotts	Prodiamine
40A Phenoxy	Cenex-Land-o-Lakes	2,4-D
Aatrex 4L	Helena	Atrazine
Aatrex Nine-D	Cenex-Land-o-Lakes	Atrazine
Aatrex, Nine-O, 4L	Syngenta	Atrazine
Accelerate	Cerexagri	Endothall
Accent	DuPont	Nicosulfuron
Accent Gold	DuPont	Clopyralid + flumetsulam + nicosulfuron + rimsulfuron
Acclaim	Aventis	Fenoxaprop-p-ethyl
Acclain Extra	United Hort Supply	Fenoxaprop-ethyl
Accord	BASF Canada only	Quinclorac
Accord, Site Prep, SP, Concentrate	Monsanto	Glyphosate
Achieve, 40DG, 80DG	Syngenta	Tralkoxydim
Advance 10G	DowAgro Canada only	Trifluralin
Afalon S	Aventis Canada only	Linuron + monolinuron
Aim	FMC	Carfentrazone-ethyl
Alanap	Uniroyal	Naptalam
Ally	DuPont	Metsulfuron
AM-40	Riverdale	2,4-D
Amber	Syngenta	Triasulfuron
Amine 4	UAP/PLATT	2,4-D
Amine 4 2,4-D	United Hort Supply	2,4-D
Aqua Master	Monsanto	Glyphosate
Aquacure	PBI Gordon	Copper ethylenediamine complex + copper sulfate pentahydrate
Aqua-Kleen	Cerexagri	2,4-D
Aqua-Neat	Cerexagri	Glyphosate
Aquathol K, Super K, Granular	Cerexagri	Endothall
Arsenal	United Hort Supply	Imazapyr
Arsenal, AC, Railroad	BASF	Imazapyr
Assert	BASF	Imazamethabenz-methyl
Assure II	DuPont	Quizalofop-p-ethyl
Asulox	Aventis	Asulam
Atra-5	Drexel Chemical	Atrazine

TABLE A-4. Continued

Trade Name	Manufacturer	Common Name
Atrazine 4F, 40DF	Albaugh, Inc.	Atrazine
Atrazine 4L	United Hort Supply	Atrazine
Atrazine 4L, 80, 90-DF	Drexel Chemical	Atrazine
Atrazine 4L, 90 WDG	UAP/PLATT	Atrazine
Atrazine 4L, 90DF	Agriliance	Atrazine
Attain A	DowAgro Canada only	Fluroxypyr
Attain B	DowAgro Canada only	2,4-D
Authority	DuPont	Sulfentrazone
Avadex	Monsanto Canada only	Triallate
Avenge	BASF	Difenzoquat
Axiom AT	Bayer	Flufenacet + metribuzin + atrazine
Axiom DF	Bayer	Flufenacet + metribuzin
Backdraft	BASF	Imazaquin + glyphosate
Balan 2.5G	United Hort Supply	Benefin
Balan DF, 2.5G	UAP/PLATT	Benefin
Balance Pro, WDG	Aventis	Isoxaflutole
Banvel—corn	BASF Canada only	Dicamba
Banvel II—cereal	BASF Canada only	Dicamba
Banvel SGF	BASF	Dicamba
Barrage, HF	Helena	2,4-D
Barricade WDG	United Hort Supply	Prodiamine
Barrier	PBI Gordon	Dichlobenil
Basagran Forte	BASF Canada only	Bentazon
Basagran, SG, TO	BASF	Bentazon
BasagranT/O	United Hort Supply	Bentazon
Basamid	BASF	Dazomet
Basamid	United Hort Supply	Dazomet
Basis	DuPont	Rimsulfuron + thifensulfuron
Basis Gold	DuPont	Nicosulfuron + rimsulfuron + atrazine
Beacon	Syngenta	Primisulfuron
Bensumec	PBI Gordon	Bensulide
Bentgrass Selective	Lesco	Dicamba + 2,4-D + MCPP
Betamix	Aventis	Desmedipham + phenmedipham
Betanex	Aventis	Desmedipham
Betanol, AM	Aventis	Phenmedipham
Betasan 12.5 G, 4E, 7-G, 36	UAP/PLATT	Bensulide
Betasan 4EC, 7G, 12.5G	United Hort Supply	Bensulide
Bicep II	Syngenta	Atrazine + metalachlor + benoxacor
Bicep II Magnum FC	Syngenta	Atrazine + acetamide
Bicep Lite II Magnum	Syngenta	Atrazine + acetamide
Bladex—MSMA	Setre	Cyanazine + MSMA
Bladex, 4L, 90DF	DuPont	Cyanazine
Blazer, Ultra	BASF	Acifluorfen
Bolero, 8EC, 10G	Valent	Thiobencarb

TABLE A-4. Continued

Trade Name	Manufacturer	Common Name
Boundry	Syngenta	s-Metolachlor + metribuzin
Broadstrike	DowAgro	Flumetsulam
Broadstrike + Dual	DowAgro Canada only	Flumetsulam + metolachlor
Broadstrike + Treflan	DowAgro	Flumetsulam + trifluralin
Broadstrike Plus	DowAgro	Flumetsulam + clopyralid
Broadstrike Post	Dow Agro	Flumetsulam + clopyralid + 2,4-D
Broclean	UAP/PLATT	Bromoxynil
Bromac	UAP/PLATT	Bromoxynil + MCPA
Bronate, Bronate Pro	Aventis	Bromoxynil + MCPA
BROX-AT, M, 2EC	Albaugh, Inc.	Bromoxynil + atrazine
Brozine	UAP/PLATT	Bromoxynil + atrazine
Brushmaster, 875	PBI Gordon	2,4-D + dichlorprop + dicamba
Buckle	Monsanto	Triallate + trifluralin
Buctril + Atrazine	Aventis	Bromoxynil + atrazine
Buctril, 4 cereals, 4EC	Aventis	Bromoxynil
Bugle	Aventis	Fenoxaprop
Bullet	Monsanto	Alachlor + atrazine
Butoxone 200, 7500	Cedar Chemical	2,4-DB
Butyl Ester 6	Helena	2,4-D
Butyrac 175, 200	Albaugh, Inc.	2,4-DB
Cadre DG	BASF	Imazameth
Callisto	Syngenta	Mesotrione
Campaign	Monsanto	Glyphosate
Canopy	DuPont	Chlorimuron + metribuzin
Canopy XL	DuPont	Chlorimuronethyl + sulfentrazone
Canvas	DuPont	Thifensulfuron + tribenuron + metsulfuron
Caparol + MSMA	UAP/PLATT	Caparol + MSMA
Caparol 4L	Syngenta	Prometryn
Casaron4G, 50W	United Hort Supply	Dichlobenil
Casoron	Uniroyal	Dichlobenil
Celebrity Herbicide, Plus	BASF	Nicosulfuron
Champion Plus, Extra	DuPont Canada only	Fenoxaprop + MCPA + 2,4-D + thifensulfuron methyl
Chaser	United Hort Supply	2,4-D + triclopyr
Cheyenne FM, Cheyenne X-tra	Aventis	Fenoxaprop-p-ethyl + MCPA + thifensulfuron + tribenuron
Chiptox	Aventis	MCPA
Chlorate	Helena	Sodium chlorate
Chopper	BASF	Imazapyr
CIPC 7, 700	UAP/PLATT	Chloropropham
Clarity	BASF	Dicamba diglycolamine
Classic	DuPont	Chlorimuron
Cobra	Valent	Lactofen
Command 3ME, 4EC	FMC	Clomazone

TABLE A-4. Continued

Trade Name	Manufacturer	Common Name
Commence	FMC	Trifluralin + clomazone
Concept III	Syngenta	Fluxofenim
Conclude Xact	BASF	Bentazon + acifluorfen + clethodim
Conclude B	BASF	Acifluorfen
Conclude G	BASF	Sethoxydim
Conclude Ultra	BASF	Acifluorfen + bentazon + sethoxydim
Confront	DowAgro	Triclopyr + clopyralid
Confront	United Hort Supply	Triclopyr + clopyralid
Conifer 90	UAP/PLATT	Atrazine
Connect 20 WSP	Aventis	Bromoxynil
Cotoran 4L, DF	Griffin	Prometryn
Cotton-Pro	Griffin	Prometryn
Crossbow	DowAgro	2,4-D + triclopyr
Crossbow	United Hort Supply	Triclopyr + 2,4-D
Curbit	UAP/PLATT	Ethalfluralin
Curtail M	DowAgro Canada only	Clopyralid
Curtail, M	DowAgro	Clopyralid + 2,4-D
Cyclone Max	Syngenta	Paraquat
Cy-Pro 4L	Griffin	Cyanazine
D-638	Albaugh, Inc.	2,4-D + 2,4-DB
Dacthal W75	UAP/PLATT	DCPA
Dacthal W-75	Amvac	DCPA
Dagger	Riverdale	MCPA
Dakota	Aventis	Fenoxaprop-p-ethyl + MCPA
Dead-n-Gone	UAP/PLATT	Glyphosate
Def 6	Bayer	Tribufos
Define	Aventis	Flufenacet
Degree	Monsanto	Acetochlor
Degree Xtra	Monsanto	Acetochlor + atrazine
Desicate II, Des-i-cate	Cerexagri	Endothall
Detail	BASF	Imazaquin + dimethenamid
Devrinol 50DF	United Hort Supply	Napropamide
Dicamba DMA, 6G	Albaugh, Inc.	Dicamba
Dimension	Lesco	Dithiopyr
Dimension	United Hort Supply	Dithiopyr
Dimension, Ultra WSP	Dow Agro	Dithiopyr
Direx 4L, 80DF	Griffin	Diuron
Discover	Syngenta	Clodinafop-propargyl
Dissolve	Riverdale	2,4-D + mecoprop + dichlorprop
Distinct	BASF	Diflufenzopyr
Diuron 4L IVM, 80DF IVM, 4L, 80DF	DowAgro	Diuron
Diuron 4L, 80DF	Agriliance	Diuron
Diuron 80 WDG, 80WP	UAP/PLATT	Diuron
Diuron 80W	Helena	Diuron
Diuron, 4L, 80, DP	Drexel Chemical	Diuron

TABLE A-4. Continued

Trade Name	Manufacturer	Common Name
Domain	Bayer	Flufenacet
Doubleplay	Syngenta	Acetochlor + EPTC + safener
DPD Ester Brush Killer	UAP/PLATT	2,4-D + dichlorprop
Dri-clean	Riverdale	2,4-D
Drive	BASF	Quinclorac
Drive	United Hort Supply	Quinclorac
Dropp 50, Ultra	Aventis	Thidiazuron
DSMA 4, Liquid DSMA	Setre	DSMA
DSMA liquid, liquid 4	Drexel Chemical	DSMA
DSMA Plus	UAP/PLATT	DSMA
DSMA, 4, liquid	Helena	DSMA
Dual II G Magnum	Syngenta	Metolachlor
Dual II Magnum, Magnum SI, Dual Magnum	Syngenta	s-Metolachlor
Dyclomec 4G	United Hort Supply	Dichlobenil
Dymec	PBI Gordon	2,4-D
DyVel	BASF Canada only	Dicamba + MCPA
DyVel DS	BASF Canada only	Dicamba + 2,4-D + mecoprop
Eclipse	DowAgro Canada only	Clopyralid + glyphosate
Edge G	DowAgro Canada only	Ethalfluralin
Elim EP	DuPont Canada only	Rimsulfuron
Eliminate	Lesco	MCPA + triclopyr + dicamba
Embark 2S 1VM	PBI Gordon	Mefluidide
Envoy	United Hort Supply	Clethodim
Envoy	Valent	Clethodim
Epic DF	Bayer	Flufenacet + isoxaflutole
Eptam 7-E, 20G	Syngenta	EPTC
Escort	DuPont	Metsulfuron
Everest	Bayer	Flucarbazone
Evik, DF	Syngenta	Ametryn
Evital	Syngenta	Norflurazon
Exceed	Syngenta	Prosulfuron + primisulfuron
Express	DuPont	Tribenuron
Express Pack	DuPont Canada only	Tribenuron + 2,4-D
Extrazine II 4L, II DF	DuPont	Cyanazine + atrazine
Extreme	BASF	Glyphosate + imazethapyr
Facet 75DF, GR	BASF	Quinclorac
Factor	United Hort Supply	Prodiamine
Fallow Master, Broadspectrum	Monsanto	Glyphosate + dicamba
Far-go, G	Monsanto	Triallate
Fertilizer Plus 1% Ronstar 15-5-10	Scotts	Oxadiazon
Fertilizer Plus 1% Ronstar 16-0-16	Scotts	Oxadiazon
Fertilizer Plus 1% Ronstar 5-5-20	Scotts	Oxadiazon

TABLE A-4. Continued

Trade Name	Manufacturer	Common Name
Fertilizer Plus 1.5% Ronstar 21-0-20	Scotts	Oxadiazon
Fertilizer Plus 1.5% Ronstar 5-5-20	Scotts	Oxadiazon
Fertilizer Plus Confront Broadleaf Weed Control/30-5-5	Scotts	Triclopyr + clopyralid
Fertilizer Plus Dicot Weed Control III/32-3-2	Scotts	2,4-D + MCPP + dicamba
Field Master	Monsanto	Glyphosate + acetochlor + atrazine
Fieldstar, DG, WSP	DowAgro Canada only	Flumetsulam + clopyralid
Finale	United Hort Supply	Glufosinate
Finale VM	Aventis	Glufosinate
Finesse	DuPont	Chlorsulfuron + metasulfuron
Firstrate	DowAgro	Chloransulam-methyl
Five Star	Albaugh, Inc.	2,4-D
Flax Max Ultra	BASF Canada only	Sethoxydim + clopyralid + MCPA
Flexstar	Syngenta	Fomesafen + isolink technology
Fluid Broadleaf Weed Control	Scotts	2,4-D + 2,4-DP
Fluometuron 4L, 80DF	Agriliance	Fluometuron
Forestry Garlon 4	DowAgro	Triclopyr
Fortress	Monsanto Canada only	Triallate + trifluralin
Fortune	BASF	Bentazon
Four Power Plus	United Hort Supply	Dicamba + MCPA
Freedom	Monsanto	Alachlor + trifluralin
Frontier 6.0	BASF	Dimethenamid
Frontrow	DowAgro	Flumetsulam + chloransulam-methyl
Fultime	Syngenta	Acetochlor + atrazine
Fusilade DX	Syngenta	Fluazifop-p-butyl
Fusilade II T/O	United Hort Supply	Fluazifop
Fusion	Syngenta	Fenoxaprop-p-ethyl + fluazifop-p-butyl
Galaxy	BASF	Acifluorfen + bentazon
Gallery 75DF	United Hort Supply	Isoxaben
Gallery 75DF, T&V	DowAgro	Isoxaben
Garlon 3A, 4	DowAgro	Triclopyr
Gauntlet	FMC	Sulfentrazone + chloransulam
Ginstar EC	Aventis	Thidiazuron + diuron
Glean	DuPont	Chlorsulfuron
Glyphomax, Plus	DowAgro	Glyphosate
Glyphosate	DuPont	Glyphosate
Glyphosate Original	Griffin	Glyphosate
Glypro	DowAgro	Glyphosate
Goal T/O	United Hort Supply	Oxyfluorfen

TABLE A-4. Continued

Trade Name	Manufacturer	Common Name
Goal, 1.6, 2XL	Dow Agro	Oxyfluorfen
Goosegrass/Crabgrass Control	Scotts	Bensulide + oxadiazon
Gramoxone Extra, MAX	Syngenta	Paraquat
Grandstand R, CA	DowAgro	Triclopyr
Granular Broadleaf Herbicide	Lesco	2,4-D + MCPA + dichloprop
Grazon P + D	DowAgro	Picloram + 2,4-D
Guardsman	BASF	Dimethenamid + atrazine
Guardsman Max	BASF	Dimethenamid-p + atrazine
Habitat Release, 75 SG	BASF	Imazapyr
Harmony Extra, GT	DuPont	Thifensulfuron + tribenuron
Harmony Total	DuPont Canada only	Thifensulfuron + tribenuron
Harness Xtra, Xtra 5.6	Monsanto	Acetochlor + atrazine
Harness, 20G	Monsanto	Acetochlor
Harvade 5F	Uniroyal	Dimethipin
Hat Trick	DuPont Canada only	Quizalofop-p-ethyl + thifensulfuron + bentazen
Hat Trick	United Hort Supply	MCPP + MCPA + dicamba
Herbicide 273	Cerexagri	Endothall
Heritage	DowAgro Canada only	Trifluralin
Hi-Dep, 1VM	PBI Gordon	2,4-D
High K Turf Fertilizer Plus Preemergent Weed Control/0-0-13	Scotts	Pendimethalin
Hoe Grass 284	Aventis Canada only	Diclofop-methyl
Hoe Grass II	Aventis Canada only	Diclofop-methyl
Hoelon 3EC	Aventis	Diclofop-methyl
Honcho	Monsanto	Glyphosate
Horizon 1EC	Aventis	Fenoxaprop-ethyl
Hornet, WDG	DowAgro	Flumetsulam + clopyralid
Hydrothol 191, 191G	Cerexagri	Endothall
Hyvar X, X-L	DuPont	Bromacil
Hyvar XL	United Hort Supply	Bromacil
Illoxam	Aventis	Diclofop-methyl
Image	United Hort Supply	Imidazolinone
Image 70 DG, 1.5 LC	BASF	Imazaquin
Invest	BASF	Cyclosulfamuron
Kansel + 28-0-0	Scotts	Oxidiazon + pendimethalin
Karmex 80	United Hort Supply	Diuron
Karmex DF	Griffin	Diuron
Kerb WSP	United Hort Supply	Pronamide
Kerb WSP, 50W	Dow Agro	Pronamide
K-O-G Weed Control	Scotts	Dicamba
Komeen	Griffin	Copper chelate
Krenite S	DuPont	Fosamine
Krovar I DF	DuPont	Bromacil + diuron
Krovar I DF	United Hort Supply	Bromacil + diuron
Landmaster BW	Monsanto	Glyphosate + 2,4-D

TABLE A-4. Continued

Trade Name	Manufacturer	Common Name
Lariat	Monsanto	Alachlor + atrazine
Lasso, II	Monsanto	Alachlor
Leadoff	DuPont	Dimethenamid + atrazine
Legacy	UAP/PLATT	Trifluralin
Lescocide Plus	Lesco	Chelated copper
Lescogran	Lesco	Bentazon
Lexone DF	DuPont	Metribuzin
Liberty, ATZ	Aventis	Glufosinate
Lightning	BASF	Imazethapyr + imazapyr
Londax	DuPont	Bensulfuron
Lontrel T&O	DowAgro	Clopyralid
Lontrel, 360	DowAgro Canada only	Clopyralid
Lorox DF	Griffin	Linuron
Low Vol 4 Ester, 6 Ester	UAP/PLATT	2,4-D
LV4 Phenoxy	Cenex-Land-o-Lakes	2,4-D
LV6 Phenoxy	Cenex-Land-o-Lakes	2,4-D
Maifest B	BASF	Bentazon + acifluorfen
Maifest G	BASF	Sethoxydim
Manage Turf	Monsanto	Halosulfuron
Manage WSB	United Hort Supply	Halosulfuron
Marksman	BASF	Dicamba + atrazine
Matrix	DuPont	Rimsulfuron
Maveric	Monsanto	Sulfosulfuron
MCP 4 Amine	UAP/PLATT	MCPA
MCP 2 Sodium	UAP/PLATT	MCPA
MCP 4 Ester	UAP/PLATT	MCPA
MCPA—Amine	Agriliance	MCPA
MCPA Amine 4, Ester 4, Sodium Salt	Albaugh, Inc.	MCPA
MCPA Phenoxy	Cenex-Land-o-Lakes	MCPA
MCPA-4	Riverdale	MCPA
MCPB Phenoxy	Cenex-Land-o-Lakes	MCPB
MCPP 4K	UAP/PLATT	MCPP
MCPP 4K	United Hort Supply	MCPP
MCPP-4	Riverdale	MCPP
Mec Amine-D	United Hort Supply	2,4-D + mecoprop + dicamba
Mec Amine BG	United Hort Supply	2,4-D + dicamba + MCPP
Mec-amine-D, Plus	UAP/PLATT	2,4-D + MCPP + dicamba
Mecomec	United Hort Supply	MCPP
Mecomec 2.5	PBI Gordon	Mecoprop
Metam Sodium	UAP/PLATT	Metham
Methar 30	United Hort Supply	DSMA
Meturon 4L	Griffin	Fluometuron
Micro-Tech	Monsanto	Alachlor
Momentum	Lesco	2,4-D + triclopyr + clopyralid
Moxy + atrazine	Agriliance	Bromoxynil + atrazine
Moxy 2E	Agriliance	Bromoxynil
MSMA 6 Plus, 6.6, 600 Plus	UAP/PLATT	MSMA

TABLE A-4. Continued

Trade Name	Manufacturer	Common Name
MSMA 6.6 Turf, Turf	United Hort Supply	MSMA
MSMA Soluble Granules	Lesco	MSMA
MSMA, Plus	Setre	MSMA
MSMA, Plus, Plus H.C.	Helena	MSMA
MSMA-6 Plus, 6.6	Drexel Chemical	MSMA
Muster	DuPont Canada only	Ethametsulfuron-methyl
Muster Gold II	DuPont Canada only	Ethametsulfuron-methyl + quizalofop-p-ethyl
Newpath	BASF	Imazethapyr
Nitrogen Potassium Fertilizer with Dimension/14-0-14	Scotts	Dithiopyr
NK Fertilizer Plus Turf Weed Control/14-0-14	Scotts	Pendimethalin
North Star	Syngenta	Primisulfuron + dicamba
Nortron SC	Aventis	Ethofumesate
NPK Fertilizer Plus Preemergence Weed Control/11-3-11	Scotts	Pendimethalin
NPK Fertilizer Plus Preemergence Weed Control/13-3-7	Scotts	Pendimethalin
Oasis	BASF	2,4-D
OpTill TM	BASF	Dicamba + dimethenamid
Orchardmaster, CA	PBI Gordon	2,4-D
Ordram 8E, 15-G, 15-GM	Syngenta	Molinate
Ornamec, 170	PBI Gordon	Fluazifop-p-butyl
Oust	DuPont	Sulfometuron
Oust	United Hort Supply	Sulfometuron
Oustar	DuPont	Hexazinone + sulfometuron/methyl
Outlook	BASF	Dimethenamid-p
Outrider	Monsanto	Sulfosulfuron
Paimextra II	Syngenta Canada only	Metolachlor + atrazine + safener
Paramount	BASF	Quinclorac
Partner WDG	Monsanto	Alachlor
Pathfinder II	DowAgro	Triclopyr
Pathway	DowAgro	Picloram + 2,4-D
Patoran	Syngenta Canada only	Metobromuron
Peak	Syngenta	Prosulfuron
Peak Plus	Syngenta Canada only	Prosulfuron
Pendimax 3.3	DowAgro	Pendimethalin
Pendimethalin + Fertilizer	United Hort Supply	Pendimethalin
Pendulum 60 WDG, 60 WSB, 2G, 3.3EC	United Hort Supply	Pendimethalin
Pendulum, 2G, 3.3EC, WDG	BASF	Pendimethalin
Pennant	Syngenta	Metolachlor
Pennant	United Hort Supply	Metolachlor

TABLE A-4. Continued

Trade Name	Manufacturer	Common Name
Pentagon	BASF	Pendimethalin
Permit	Monsanto	Halosulfuron
Phenoxy 088	Agriliance	2,4-D + 2,4-DB
Pinnacle	DuPont	Thifensulfuron
Plateau, DG	BASF	Imazethapyr
Poast Ultra	BASF Canada only	Sethoxydim
Poast, HC, Plus	BASF	Sethoxydim
Polado L	Monsanto	Glyphosate
Pramital 25	Cenex-Land-o-Lakes	Prometon
Pramitol 25E	Agriliance	Prometon
Pramitol 25E, 5PS	UAP/PLATT	Prometon
Pramitol 25EC, 5PS	United Hort Supply	Prometon
Pre Pair	United Hort Supply	Napropamide + oxadiazon
Predict	United Hort Supply	Norflurazon
Prefar	Gowen	Bensulide
Pre-M, 3.3EC, 60DG	Lesco	Pendimethalin
Pre-San 7G, 12.5G	PBI Gordon	Bensulide
Prestige A	DowAgro Canada only	Fluroxypyr
Prestige B	DowAgro Canada only	Clopyralid + MCPA
Prevail A	DowAgro Canada only	Tralkoxydim
Prevail B	DowAgro Canada only	Clopyralid + MCPA
Princep 4L, Calibar 90	Syngenta	Simazine
Princep 4LT & D	United Hort Supply	Simazine
Prism	DuPont Canada only	Rimsulfuron
Prism, Envoy, Select	Valent	Clethodim
Prometryn + MSMA	Agriliance	Caporal + MSMA
Prometryn 4L	Agriliance	Caporal
Prograss 1.5EC	United Hort Supply	Ethofumesate
Progress, Prograss EC	Aventis	Ethofumesate + phenmedipham + desmedipham
Promp	United Hort Supply	Atrazine + bentazon
Propanil 4E, 80DF	Agriliance	Propanil
Protocol	Monsanto	Glyphosate
Prowl	BASF	Pendimethalin
Puma	Aventis Canada only	Fenoxaprop-p-ethyl
Puma 120 Super	Aventis Canada only	Fenoxaprop-p-ethyl
Puma 1E	Aventis	Fenoxaprop-p-ethyl + safener
Puma One Pass	Aventis Canada only	Fenoxaprop-p-ethyl
Puma Super	Aventis Canada only	Fenoxaprop-p-ethyl
Pursuit Plus EC	BASF	Imazethapyr + pendimethalin
Pursuit, DG, W, WDG	BASF	Imazethapyr
Pyramin DF, SC	BASF	Pyrazon
Python WDG	DowAgro	Flumetsulam
Ramrod FL, 20G	Monsanto	Propachlor
Range Star	Albaugh, Inc.	Dicamba + 2,4-D
Raptor	BASF	Imazamox
Rattler	Helena	Glyphosate

TABLE A-4. Continued

Trade Name	Manufacturer	Common Name
Rave	Syngenta	Dicamba
Ready Master	Monsanto	Glyphosate + atrazine
Reclaim	DowAgro	Clopyralid
Refine Extra	DuPont Canada only	Thifensulfuron
Reflex	Syngenta	Fomesafen
Release	DowAgro Canada only	Triclopyr
Rely	Aventis	Glufosinate
Remedy	DowAgro	Triclopyr
Remove	Aventis	Glufosinate
Resource	Valent	Flumiclorac-pentyl
Reward	United Hort Supply	Diquat
Rezult B, G	BASF	Bentazon + sethoxydim
Rhomene	Aventis	MCPA
Rhonox	Aventis	MCPA
Rival 10 G Superflow	Aventis Canada only	Trifluralin
Rival EC	Aventis Canada only	Trifluralin
Rodeo	Monsanto	Glyphosate
Ro-Neet	Cedar Chemical	Cycloate
Ronstar	Lesco	Oxadiazon
Ronstar 2G, 50WP	United Hort Supply	Oxadiazon
Ronstar 50 WSP, G	Aventis	Oxadiazon
Roundup	Scotts	Glyphosate
Roundup as: Custom, D-Pak, Dry Pak, Original, Original RT, Pro, Pro dry, Ultra, Ultra RT, Ultradry, Ultramax	Monsanto	Glyphosate
Roundup DRY PACK, Pro	United Hort Supply	Glyphosate
Roundup Transorb, Fast Forward	Monsanto Canada only	Glyphosate
Sahara	BASF	Imazethapyr + diuron
Salvo	UAP/PLATT	2,4-D
Savage, Dry Soluble	UAP/PLATT	2,4-D
Scepter 70DG	BASF	Imazaquin
Scythe	DowAgro	Pelargonic acid
Scythe	United Hort Supply	Pelargonic acid
Sempra	Monsanto	Haosulfuron
Sencor 4, DF, SoluPac	Bayer	Metribuzin
Sencor 75WP	United Hort Supply	Metribuzin
Shadeout	DuPont	Rimsulfuron
Shotgun	UAP/PLATT	Atrazine + 2,4-D
Silverado	Aventis	Fenoxaprop-p-ethyl
Simazine 4L, 90 WDG, 80W	UAP/PLATT	Simazine
Simazine 4L, 90DF	Agriliance	Simazine
Simazine 4L, 90DF	Drexel Chemical	Simazine
Sinbar	DuPont	Terbacil
Snapshot 2.5TG	DowAgro	Isoxaben + trifluralin
Snapshot 2.5TG	United Hort Supply	Trifluralin + isoxaben
Solicam DF	Syngenta	Norflurazon

TABLE A-4. Continued

Trade Name	Manufacturer	Common Name
Solution water soluble	Riverdale	2,4-D
Solve 2,4-D	Albaugh, Inc.	2,4-D
Solve MCPA	Albaugh, Inc.	MCPA
Sonalan 10G, HFP	DowAgro	Ethalfluralin
Spartan	FMC	Sulfentrazone
Spike 20P, 80W	DowAgro	Tebuthiuron
Spin-aid	Aventis	Phenmedipham
Spirit	Syngenta	Prosulfuron + primisulfuron
Squadron	BASF	Pendimethalin + imazaquin
St. Augustine Weed & Feed	Lesco	Atrazine
Stalker	BASF	Imazethapyr
Stam M-4, 4E, 80EDF, Pro	Dow Agro	Propanil
Stampede	Dow Agro	Propanil
Staple, Plus	DuPont	Pyrithiobac
Starane	DowAgro	Fluroxypyr
Starane + Esteron	DowAgro	Fluroxypyr + 2,4-D
Starane + MCPA	DowAgro	Fluroxypyr + MCPA
Starane + Salvo	DowAgro	Fluroxypyr + 2,4-D
Starane + Sword	DowAgro	Fluroxypyr + MCPA
Starter Fertilizer with Preemergent Weed Control/16-21-4	Scotts	Siduron
Steadfast	DuPont	Nicosulfuron
Steel	BASF	Pendimethalin + imazaquin + imazethapyr
Stellar, 2EC	Valent	Lactofen + flumiclerac-pentyl
Sterling	Agriliance	Dicamba
Stinger	DowAgro	Clopyralid
Stinger	United Hort Supply	Clopyralid
Storm	BASF	Bentazon + acifluorfen
Strike 3	Agriliance	2,4-D + MCPP + dicamba
Striker	DowAgro Canada only	Flumetsulam + clopyralid + 2,4-D
Strongarm	DowAgro	Diclosulam
Stubble	UAP/PLATT	Atrazine
Sundance	Monsanto Canada only	Sulfosulturon
Super Brush Killer	PBI Gordon	2,4-D + dichlorprop + dicamba
Super Trimec	PBI Gordon	2,4-D + dichlorprop + dicamba
Surefire	UAP/PLATT	Paraquat + diuron
Surflan AS	United Hort Supply	Oryzalin
Surflan AS, AS Specialty	DowAgro	Oryzalin
Sutan +	Cedar Chemical	Butylate
SWB MCPA Ester	Agriliance	MCPA
SWB, 2,4-D LV4	Agriliance	2,4-D
Sword	UAP/PLATT	MCPA

TABLE A-4. Continued

Trade Name	Manufacturer	Common Name
Synchrony STS	DuPont	Chlorimuron ethyl + thifensulfuron
Target	Syngenta Canada only	MCPA + MCPP + dicamba
Team	Lesco	Benefin + trifluralin
Team 2G	United Hort Supply	Benefin + trifluralin
Telar DF	DuPont	Chlorsulfuron
Thistrol MCPB	Aventis	MCPB
Three-Way Ester II	Lesco	MCPA + triclopyr + dicamba
Three-Way, Selective	Lesco	2,4-D + MCPP + dicamba
Tillam 6E	Cedar Chemical	Pebulate
Tillam 6E	Syngenta	Pebulate
Tiller	Aventis	Fenoxaprop-p-ethyl + 2,4-D + MCPA
Tordon 22K, K	DowAgro	Picloram
Tordon RTU	United Hort Supply	Picloram
Tordon RTU, 101 Mixture	DowAgro	Picloram + 2,4-D
Touchdown, 5	Syngenta	Sulfosate
Tough	Syngenta	Pyridate
Transline	DowAgro	Clopyralid
Treflan 5G	Lesco	Trifluralin
Treflan 5G	United Hort Supply	Trifluralin
Treflan HPF, TR-10	DowAgro	Trifluralin
Treflan QR5	DowAgro Canada only	Trifluralin
Triamine II, G	Riverdale	MCPA + mecoprop + dichlorprop
Triamine Jet-Spray, G	Riverdale	2,4-D + mecoprop + dichlorprop
Tri-Ester, II	Riverdale	2,4-D + dichlorprop
Trifluralin 10G, 4ED	Albaugh, Inc.	Trifluralin
Trifluralin 4EC	Setre	Trifluralin
Trifluralin 5, 4, 10G	Gowen	Trifluralin
Trifluralin HF, 10G, 4EC	UAP/PLATT	Trifluralin
Trilin, 5, 10G	Griffin	Trifluralin
Trimec 899, Bentgrass, Classic, DSC, SI, Southern, Turf Ester, 992	PBI Gordon	2,4-D + MCPP + dicamba
Trimec -992, -Bent, -Classic, -Plus, -Super, -Southern	United Hort Supply	2,4-D + MCPP + dicamba
Trimec Encore	PBI Gordon	MCPA + MCPP + dicamba
Trimec Encore	United Hort Supply	MCPA + MCPP + MSMA + dicamba
Trimec LAF 637	PBI Gordon	2,4-D + MCPP + dicamba
Trimec Plus	PBI Gordon	2,4-D + mecoprop + dicamba
Triplet, WSP	Riverdale	2,4-D + mecoprop + dicamba
Tri-Power Selective, Dry	Riverdale	MCPA + mecoprop + dicamba
Tupersan	Gowen	Siduron

TABLE A-4. Continued

Trade Name	Manufacturer	Common Name
Tupersan	PBI Gordon	Siduron
Tupersan	United Hort Supply	Siduron
Turf Fertilizer Plus PreemergenceWeed Control/22-0-11	Scotts	Pendimethalin
Turf Fertilizer Plus PreemergenceWeed Control/22-0-6	Scotts	Pendimethalin
Turf Weed & Brush	Riverdale	2,4-D + 2,4-DP
Turf Weedgrass Control	Scotts	Pendimethalin
Turflon Ester	DowAgro	Triclopyr ester
Turflon Ester	United Hort Supply	Triclopyr + 2,4-D
Two + Two	United Hort Supply	MCPP + 2,4-D
Typhoon	Syngenta	Fluazifop-p-butyl + fomesafen
Ultim	DuPont Canada only	Rimsulfuron + nicosulfuron
Ultimax	DuPont Canada only	Rimsulfuron + nicosulfuron + striker
Upbeet	DuPont	Triflusulfuron
Valor	Valent	Flumioxazin
Vantage	BASF	Sethoxydim
Vantage	United Hort Supply	Sethoxydim
Vantage, Plus	DowAgro Canada only	Glyphosate
Vapam HL	Amvac	Metham
Vegemec	PBI Gordon	2,4-D + prometon
Velpar	United Hort Supply	Hexazinone
Velpar, DF, L, ULW	DuPont	Hexazinone
Vision	Monsanto Canada only	Glyphosate
Visor	Dow Agro	Thiazopyr
Vista	DowAgro	Fluroxypyr
Weed Blast	UAP/PLATT	Bromacil + diuron
Weed Hoe 108, 120	Albaugh, Inc.	MSMA
Weed Rhap A 4-D	Helena	2,4-D
Weed Rhap LV6D	Helena	2,4-D
Weedgrass Preventer	Scotts	Bensulide
Weedmaster	BASF	Dicamba + 2,4-D
Whip 360	Aventis	Fenoxaprop-p-ethyl
XL 2G	Setre	Benefin + oryzalin
XL 2G	United Hort Supply	Oryzalin
Yardmaster 4LG 2,4-D	Cenex-Land-o-Lakes	2,4-D
Yardmaster Garden Weeder	Cenex-Land-o-Lakes	2,4-D
Yardmaster Lawn & Garden Weeder	Cenex-Land-o-Lakes	2,4-D
Zorial	Syngenta	Norflurazon
Zorial Rapid 80	Setre	Norflurazon
Zorial Rapid 80, 5G	Syngenta	Norflurazon

[a]Does not include soil fumigants or most herbicides sold only for the home market.

TABLE A-5. Conversion Factors

Liquid Measure

1 gallon (U.S.) = 3785.4 milliliter (ml); 256 tablespoons; 231 cubic inches;
128 fluid ounces; 16 cups; 8 pints; 4 quarts; 0.8333 imperial gallon; 0.1337 cubic foot;
8.337 pounds of water
1 liter = 1000 milliliters; 1.0567 liquid quarts (U.S.)
1 gill = 118.29 milliliters
1 fluid ounce = 29.57 milliliters; 2 tablespoons
3 teaspoons = 1 tablespoon; 14.79 milliliters; 0.5 fluid ounce
1 cubic foot of water = 62.43 pounds; 7.48 gallons

Weight

1 gamma = 0.001 milligram (mg)
1 grain (gr) = 64.799 milligrams
1 gram (g) = 1000 milligrams; 15.432 grains; 0.0353 ounce
1 pound = 16 ounces; 7000 grains; 453.59 grams; 0.45359 kilogram
1 short ton = 2000 pounds; 0.097 metric ton
1 long ton = 2240 pounds; 1.12 short ton
1 kilogram = 2.2046 pounds

Linear Measure

12 inches = 1 foot; 30.48 centimeters
36 inches = 3 feet; 1 yard; 0.914 meter
1 rod = 16.5 feet; 5.029 meters
1 mile = 5280 feet; 1760 yards; 160 rods; 80 chains; 1.6094 kilometers (km)
1 chain = 66 feet; 22 yards; 4 rods; 100 links
1 inch = 2.54 centimeters (cm)
1 meter = 39.37 inches; 10 decimeters (dm); 3.28 feet
1 micron = (μm) = 1/1000 millimeter (mm)
1 kilometer = 0.621 statue miles; 0.5396 nautical miles

Area

1 township = 36 sections; 23,040 acres
1 square mile = 1 section; 640 miles
1 acre = 43,560 square feet; 160 square rods; 4840 square yards
1 hectare = 2.471 acres

Capacity (Dry Measure)

1 bushel (U.S.) = 4 pecks; 32 quarts; 35.24 liters; 1.244 cubic feet; 2150.42 cubic inches

Pressure

1 foot lift of water = 0.433 pound pressure per square inch (psi)
1 pound pressure per square inch will lift water 2.31 feet
1 atmosphere = 760 millimeters of mercury; 14.7 pounds; 33.9 feet of water

TABLE A-5. Continued

Geometric Factors

($\pi = 3.1416$; r = radius; d = diameter; h = height)
Circumference of a circle = $2\ \pi r$ or πd
Diameter of a circle = $2r$
Area of a circle = πr^2 or $1/4\pi d2$ or $0.7854d^2$
Volume of a cylinder = $\pi r^2 h$
Volume of a sphere = $1/6\pi d^3$

Other Conversions

Multiply	by	To Obtain
Gallons per minute	2.228×10^{-3}	Cubic feet per second
Gallons per acre	9.354	Liters per hectare
Kilograms per hectare	0.892	Pounds per acre
Liters	1.05	U.S. quarts
Liters	0.2642	U.S. gallons
Liters per hectare	0.107	Gallons per acre
Miles per hour	88.0	Feet per minute
Miles per hour	1.61	Kilometers per hour
Pounds per gallon	0.12	Kilograms per liter
Pounds per square inch	0.068046	Atmosphere (Atm)
Pounds per 1000 square feet	0.489	Kilograms per acre
Pounds per square inch	6.89476	kPa
Pounds per acre	1.12	Kilograms per hectare
Square inch	6.452	Square centimeter
Parts per million	2.719	Pounds acid equivalent per acre foot of water

Temperature Degrees

F° = C° + 17.78 × 1.8
C° = F° – 32.00 × 5/9

°C	°F	°C	°F
100	212	30	86
90	194	20	68
80	176	10	50
70	158	0	32
60	140	–10	14
50	122	–20	–4
40	104	–30	–22

TABLE A-6. Herbicide Concentration Calculations

Conversion Factors

1 lb = 454 g	1 qt = 946 ml
1 kg = 1000 g	1 L = 1000 ml
1000 mg = 1 g	1000 μl = 1 ml
1 mg = 0.001 g	1 μl = 0.001 ml

Calculations

I. 1 molar = the molecular weight of a pure substance (in grams) dissolved in enough water to make 1 L (1000 ml). Molar is abbreviated M.

Example: If a herbicide's molecular weight is 250, then 250 g dissolved in enough water to equal 1 L (1000 ml) is 1 M.

1 mM = 10^{-3} M
1 μM = 10^{-6} M

II. 1 mole = the molecular weight of a pure substance in grams. There is no abbreviation for mole.

Example: If a herbicide's molecular weight is 300, then 300 g = 1 mole. Volume is not a consideration.

III. ppm = part per million

1 ppm = 1 lb/1,000,000 lb
1 ppm = 1 mg/l of water (1g = 1ml)
1 ppm = 1mg/kg

Example: A 25 ppm solution of herbicide is 25 mg/l (1000 ml) or 2.5 mg/100 ml. The molecular weight is not involved in the calculation.

IV. Conversion from ppm to M.

75 ppm herbicide solution
75 mg/l = 0.075 g/l
Herbicide molecular weight = 250

$$\frac{0.075 \text{ g/l} = x \text{ M}}{250 \text{ g/l} = 1 \text{ M}}$$

(0.075) (1) = (250) (x)
3×10^{-4} M = x

or 0.3 mM

V. Conversion from M to ppm

10 μM herbicide solution
Herbicide molecular weight = 250

$$\frac{250 \text{ g/l}}{x \text{ g/l}} = \frac{1 \text{ M}}{10^{-5} \text{ M}}$$

TABLE A-6. Continued

$(250)(10^{-5}) = (x)(1)$
$0.0025 \text{ g/l} = x$
$0.0025 \text{ g/l} = 2.5 \text{ mg/l}$
$= 2.5 \text{ ppm}$

VI. Comparing ppm versus M for herbicide concentration

10 ppm solution of oxyfluorfen: molecular weight = 361.7
10 ppm solution of bensulide : molecular weight = 397.5

Oxyfluorfen	Bensulide
$\underline{361.7 \text{ g/l} = 1 \text{ M}}$	$\underline{397.5 \text{ g/l} = 1 \text{ M}}$
$0.01 \text{ g/l} = x \text{ M}$	$0.01 \text{ g/l} = x \text{ M}$
$x = 2.8 \times 10^{-5} \text{ M}$	$x = 2.8 \times 10^{-5} \text{ M}$

VII. Approximate soil concentration between lb/A, ppm, and M.

The approximate weight of soil is 2,000,000 lb/acre furrow slice (6-inch slice of soil).

1 lb/A in the upper 3 inches of soil is equivalent to 1 ppm.

Problem 1: What is the ppm concentration of 1/2 lb/A in 1/4 inch of soil?

1/2 lb/A in 1/4 inch of soil is equivalent to 2 lb per 1 inch of soil or 6 lbs per 3 inches of soil. One lb per 3 inches is 1 ppm, so 6 lb per 3 inches is 6 ppm.

Problem 2: What is the approximate molarity of herbicides (molecular weight = 250) applied at 1/4 lb/A and present in the upper 1/2 inch of soil?

1/4 lb/A in 1/2 inch is equivalent to 1/2 lb/A per inch or 1.5 lb/A per 3 inches, which is 1.5 ppm (1.5 mg/kg).

$\underline{0.0015 \text{ g/kg} = x \text{ M}}$
$250 \text{ g/kg} = 1 \text{ M}$

$(250)(x) = (0.0015)(1)$

$= 6 \times 10^{-6} \text{ M}$
or $= 6\ \mu\text{M}$

Example Problem:

What is the approximate field concentration (incorporated 1.5 inch) of a herbicide at a concentration of 5 μM; molecular weight = 250?

$\underline{5 \times 10^{-6} \text{ M} = x \text{ g/kg}}$
$1 \text{ M} = 250 \text{ g/kg}$

$(x)(1) = (250)(5 \times 10^{-6})$
$= 0.00125 \text{ g/kg}$
$= 1.25 \text{ mg/kg}$
$= 1.25 \text{ ppm}$

1.25 ppm = 1.25 lb/1,000,000 lb
1.5 inches = 500,000 lb soil

TABLE A-6. Continued

0.625 lb/1.5 inches soil

The soil concentration calculations are correct based on the soil weight. However, two important points must be remembered. (1) These calculations do not consider herbicide binding to soil. Because of soil binding, not all of the applied herbicide is available for plant uptake. (2) The herbicide taken into the plant is usually dissolved in the soil water. These calculations do not consider the herbicide concentration in the available soil water.

TABLE A-7. Weight of Dry Soil

Type	Pounds per Cubic Foot	Pounds per Acre, 7 Inches Deep
Sand	100	2,500,000
Loam	80–95	2,000,000
Clay or silt	65–80	1,500,000
Muck	40	1,000,000
Peat	20	500,000

TABLE A-8. Length of Row Required for One Acre

Row Spacing (inches)	Length or Distance
24	7260 yards = 21,780 ft
30	5808 yards = 17,424 ft
36	4840 yards = 14,520 ft
42	4149 yards = 12,445 ft
48	3630 yards = 10,890 ft

TABLE A-9. Available Commercial Materials in Pounds Active Ingredient per Gallon Necessary to Make Various Percentage Concentration Solutions[a]

Pounds of Active Ingredient in 1 gal of Commercial Product	Pounds of Active Ingredient/Pint[a]	Liquid Ounces of Commercial Product to Make One Gallon of				
		1/2%	1%	2%	5%	10%
2.00	0.25	2.68	5.36	10.72	26.80	53.60
2.64	0.33	2.02	4.05	8.10	20.25	40.50
3.00	0.375	1.78	3.56	7.12	17.80	35.60
3.34	0.72	1.59	3.18	6.36	15.90	31.80
4.00	0.50	1.34	2.68	5.36	13.40	26.80
6.00	0.75	0.89	1.78	3.56	8.90	17.80

[a]Based on 8.4 lb/gal (weight of water) and 128 liquid oz = 1 gal, 16 liquid oz = 1 pint.

TABLE A-10. Equivalent Quantities of Liquid Materials When Mixed by Parts

Water	1–400	1–800[a]	1–1600
100 gal	1 qt	1 pt	1 cup
50 gal	1 pt	1 cup	1/2 cup
5 gal	3 tbs	5 tsp[a]	2 1/2 tsp
1 gal	2 tsp	1 tsp	1/2 tsp

TABLE A-11. Equivalent Quantities of Dry Materials (Wettable Powders) for Various Quantities of Water

Water	Quantity of Materials					
100 gal[a]	1 lb	2 lb	3 lb	4 lb[a]	5 lb	6 lb
50 gal	8 oz	1 lb	24 oz	2 lb	2.5 lb	3 lb
5 gal[a]	3 tbs[b]	1.5 oz	2.5 oz	3.25 oz[a]	4 oz	5 oz
1 gal	2 tsp[b]	3 tsp	1.5 tbs	2 tbs	3 tbs	3 tbs

[a]*Example*: If a recommendation calls for a mixture of 4 lb of a wettable powder to 100 gal of water, it would take 3.25 oz (approximately 12 tsp) to 5 gal of water to give 5 gal of spray mixture of the same strength.
bWettable materials vary considerably in density. Therefore, the teaspoonful (tsp) and tablespoonful (tbs) measurements in this table are not exact dosages by weight but are within the bounds of safety and efficiency for mixing small amounts of spray.

TABLE A-12. Common Cropland Weeds in North America

Northern Corn Belt (Minnesota, Wisconsin, Michigan, New York, Pennsylvania, Ontario)

Common Name	Scientific Name	Family Name
Giant foxtail	*Setaria faberi*	Poaceae
Green foxtail	*Setaria viridis*	
Yellow foxtail	*Setaria glauca*	
Crabgrass	*Digitaria* spp.	
Barnyardgrass	*Echinochloa crus-galli*	
Fall panicum	*Panicum dichotomiflorum*	
Quackgrass	*Agropyron repens*	
Wild proso millet	*Panicum miliaceum*	
Volunteer corn		
Yellow nutsedge	*Cyperus esculentus*	Cyperaceae
Pigweeds	*Amaranthus* spp.	Amaranthaceae
Common lambsquarter	*Chenopodum album*	Chenopodiaceae
Common ragweed	*Ambrosia artemisiifolia*	Asteraceae
Canada thistle	*Cirsium arvense*	
Wild mustard	*Sinapis arvensis*	Brassicaceae
Velvetleaf	*Abutilon theophrasti*	Malvaceae
Wild buckwheat	*Polygonum convolvulus*	Polygonaceae
Annual smartweeds	*Polygonum* spp.	
Common purslane	*Portulaca oleracea*	Portulacaceae
Black nightshade	*Solanum nigrum*	Solanaceae
Annual nightshades	*Solanum* spp.	

Central Corn Belt (Ohio, Indiana, Illinois, Iowa, Eastern Nebraska, Missouri, Kentucky, Maryland, Delaware, Virginia)

Common Name	Scientific Name	Family Name
Giant foxtail	*Setaria faberi*	Poaceae
Fall panicum	*Panicum dichotomiflorum*	
Crabgrass	*Digitaria* spp.	
Johnsongrass	*Sorghum halepense*	
Shattercane	*Sorghum bicolor*	
Volunteer corn		
Wild garlic	*Allium vineale*	Liliaceae
Pigweeds	*Amaranthus* spp.	Amaranthaceae
Common lambsquarter	*Chenopodum album*	Chenopodiaceae
Common ragweed	*Ambrosia artemisiifolia*	Asteraceae
Giant ragweed	*Ambrosia trifida*	
Common cocklebur	*Xanthium strumarium*	
Canada thistle	*Cirsium arvense*	
Morning glories	*Ipomea* spp.	Convolvulaceae
Velvetleaf	*Abutilon theophrasti*	Malvaceae
Annual smartweeds	*Polygonum* spp.	Polygonaceae
Jimsonweed	*Datura strumonium*	Solanaceae
Black nightshade	*Solanum nigrum*	
Eastern black nightshade	*Solanum ptycanthum*	

TABLE A-12. Continued

Southern States (Eastern Texas, Arkansas, Louisiana, Mississippi, Alabama, Georgia, Tennessee, Florida, North Carolina, South Carolina)

Common Name	Scientific Name	Family Name
Crabgrass	*Digitaria* spp.	Poaceae
Goosegrass	*Eleusine indica*	
Fall panicum	*Panicum dichotomiflorum*	
Texas panicum	*Panicum texanum*	
Broadleaf signalgrass	*Brachiaria platyphylla*	
Barnyardgrass	*Sorghum halepense*	
Bermudagrass	*Cynodon dactylon*	
Red rice	*Oryza sativa*	
Yellow nutsedge	*Cyperus esculentus*	Cyperaceae
Purple nutsedge	*Cyperus rotundus*	
Pigweeds	*Amaranthus* spp.	Amaranthaceae
Common ragweed	*Ambrosia artemisiifolia*	Asteraceae
Common cocklebur	*Xanthium strumarium*	
Morning glories	*Ipomea* spp.	Convolvulaceae
Crotons	*Croton* spp.	Euphorbiaceae
Spurges	*Euphorbia* spp.	
Sicklepod	*Cassia obtusifolia*	Leguminosae
Hemp sesbania	*Sesbania exaltata*	
Florida beggarweed	*Desmodium tortuosum*	
Prickly sida	*Sida spinosa*	Malvaceae

Northern Great Plains of the United States and Canada (North Dakota, South Dakota, Montana, Wyoming, Manitoba, Saskatchewan, Alberta)

Common Name	Scientific Name	Family Name
Wild oats	*Avena fatua*	Poaceae
Green foxtail	*Setaria Viridis*	
Yellow foxtail	*Setaria glauca*	
Downy brome	*Bromus tectorum*	
Quackgrass	*Agropyron repens*	
Redroot pigweed	*Amaranthus retroflexus*	Amaranthaceae
Cowcockle	*Vaccaria pyramidata*	Caryophyllaceae
Common lambsquarters	*Chenopodium album*	Chenopodiaceae
Kochia	*Kochia scoparia*	
Russian thistle	*Salsola iberica*	
Canada thistle	*Cirsium arvense*	Asteraceae
Perennial sowthistle	*Sonchus arvensis*	
Wild mustard	*Sinapis arvensis*	Brassicaceae
Tansy mustard	*Descurainia pinnata*	
Flixweed	*Descuraninia sophia*	
Shepherd's purse	*Capsella bursa-pastoris*	
Field pennycress (stinkweed)	*Thlaspi arvense*	
Wild buckwheat	*Polygonum convolvulus*	Polygonaceae

TABLE A-12. Continued

Central and Southern Great Plains (Kansas, Western Nebraska, Colorado, Oklahoma, Western Texas, New Mexico)

Common Name	Scientific Name	Family Name
Foxtails	*Setaria* spp.	Poaceae
Crabgrass	*Digitaria* spp.	
Panicums	*Panicum* spp.	
Downy brome	*Bromus tectorum*	
Cheat	*Bromus secalinus*	
Sandbar	*Cenchurs* spp.	
Jointed goatgrass	*Aegilops cylindrica*	
Wild oats	*Avena fatua*	
Barnyardgrass	*Echinochloa crus-galli*	
Johnsongrass	*Sorghum halepense*	
Shattercane	*Sorghum bicolor*	
Volunteer grain		
Yellow nutsedge	*Cyperus esculentus*	Cyperacceae
Pigweeds	*Amaranthus* spp.	Amaranthaceae
Kochia	*Kochia scoparia*	Chenopodiaceae
Russian thistle	*Salsola iberica*	
Canada thistle	*Cirsium arvense*	Asteraceae
Woolyleaf bursage	*Ambrosia grayi*	
Texas blueweed	*Helianthus ciliaris*	
Field bindweed	*Convolvulus arvensis*	Convolvulaceae
Mustards	*Descurainia* spp.	Brassicaceae
Wild buckwheat	*Polygonum convolvalus*	Polygonaceae
Silverleaf nightshade	*Solanum elaeagnifolium*	Solanaceae
Buffalobur	*Solanum nostratum*	
Annual nightshades	*Solanum* spp.	

Pacific Northwest (Washington, Oregon, Idaho)

Common Name	Scientific Name	Family Name
Downy brome	*Bromus tectorum*	Poaceae
Ripgut brome	*Bromus diandrus*	
Wild oats	*Avena fatua*	
Green foxtail	*Setaria viridis*	
Barnyardgrass	*Echinochloa crus-galli*	
Quackgrass	*Agropyron repens*	
Italian ryegrass	*Lolium multiflorum*	
Volunteer wheat		
Pigweeds	*Amaranthus* spp.	Amaranthaceae
Kochia	*Kochia scoparia*	Chenopodiaceae
Common lambsquarters	*Chenopodium album*	
Canada thistle	*Cirsium arvense*	Asteraceae
Prickly lettuce	*Lactuca serriola*	
Field bindweed	*Convolvulus arvensis*	Convolvulaceae
Dodder	*Cuscuta* spp.	
Tumble mustard	*Sisymbrium altissimum*	Brassicaceae
Birdsrape	*Brassica rapa*	
Wild radish	*Raphanus raphanistrum*	
Shepherd's purse	*Capsella bursa-pastoris*	
Annual nightshades	*Solanum* spp.	Solanaceae

TABLE A-12. Continued

Southwest (California, Arizona, Nevada, Utah)		
Common Name	Scientific Name	Family Name
Barnyardgrass	*Echinochloa crus-galli*	Poaceae
Yellow foxtail	*Setaria glauca*	
Green foxtail	*Setaria viridis*	
Johnsongrass	*Sorghum halepense*	
Bermudagrass	*Cynodon dactylon*	
Canarygrass	*Phalaris* spp.	
Wild oats	*Avena fatua*	
Volunteer cereals		
Yellow nutsedge	*Cyperus esculentus*	Cyperaceae
Purple nutsedge	*Cyperus rotundus*	
Pigweeds	*Amaranthus* spp.	Amaranthaceae
Common lambsquarters	*Chenopodium album*	Chenopodiaceae
Kochia	*Kochia scoparia*	
Sowthistles	*Sonchus* spp.	Asteraceae
Prickly lettuce	*Lactuca serriola*	
Common groundsel	*Senecio vulgaris*	
Field bindweed	*Convolvulus arvensis*	Convolvulaceae
London rocket	*Sisymbrium irio*	Brassicaceae
Shepherd's purse	*Capsella bursa-pastoris*	
Mallows	*Malva* spp.	Malvaceae
Common purslane	*Portulaca oleraceae*	Portulacaceae
Black nightshade	*Solanum nigrum*	Solanaceae
Hairy nightshade	*Solanum sarrachoides*	

Weedy Families	
Most Common	Common in One or Two Regions
Poaceae	Caryophyllaceae
Cyperaceae	Euphorbiaceae
Amaranthaceae	Leguminaceae (Fabaceae)
Chenopodiaceae	Liliaceae
Asteraceae	Portulacaceae
Convolvulaceae	
Brassicaceae	
Malvaceae	
Polygonaceae	

INDEX

Major coverage in **bold type**, herbicides and corps in use tables in *italics*, and major use table for crops or herbicides in ***bold-italics***.